Chemistry *of* Modern Papermaking

Chemistry *of* Modern Papermaking

Cornel Hagiopol

James W. Johnston

CRC Press
Taylor & Francis Group
Boca Raton London New York

CRC Press is an imprint of the
Taylor & Francis Group, an **informa** business

CRC Press
Taylor & Francis Group
6000 Broken Sound Parkway NW, Suite 300
Boca Raton, FL 33487-2742

First issued in paperback 2021

Version Date: 20110815

ISBN 13: 978-1-03-209926-2 (pbk)
ISBN 13: 978-1-4398-5644-4 (hbk)

Visit the Taylor & Francis Web site at
http://www.taylorandfrancis.com

and the CRC Press Web site at
http://www.crcpress.com

Contents

Foreword xi
Authors... xiii
Abbreviations xv

Chapter 1 Introduction 1

Acknowledgments 3
References 3

Chapter 2 From Wood to Paper: A General View of the Papermaking Process 5

2.1 From the *Papyrus* Era to Modern Times: A Brief History of Making Paper 5
2.2 Pulp: The Support for Paper Chemicals 7
    2.2.1 Sulfite Pulping 9
    2.2.2 Sulfate Pulping (KRAFT).. 11
    2.2.3 The Bleaching Process 12
    2.2.4 Wet End: Sheet Formation and White Water 12
    2.2.5 Paper Drying and Finishing (Dry End)..................................... 13
2.3 Paper Structure and Composition.. 14
2.4 The Chemistry of Poly-Carbohydrates.. 17
    2.4.1 Chemical Reactions That Keep the Molecular Weight Unchanged 18
    2.4.2 Chemical Reactions That May Alter the Molecular Weight............. 22
2.5 Synthetic Polymers: Everywhere in Papermaking Process........................... 28
    2.5.1 Polymer Synthesis 28
    2.5.2 Chemical and Physical Properties of Polymers 32
2.6 Paper Testing: A Difficult Task 35
References 39

Chapter 3 The Fate of Paper Chemicals at the Wet End.. 51

3.1 Friends and Foes at the Wet End 51
3.2 Polymers in Heterogeneous Systems... 53
    3.2.1 Polyelectrolyte Interactions in a Continuous Phase 55
    3.2.2 Polyelectrolyte Adsorption at an Interface............................... 56
    3.2.3 The Polymer Retention Mechanism 58
    3.2.4 Polymer Particles Retained on Cellulose Fibers 61
    3.2.5 Colloidal Titration 62
3.3 Retention Aids 63
    3.3.1 Electrophoretic Mobility 64
    3.3.2 Fiber Flocculation Mechanisms 65
    3.3.3 Paper Chemicals as Retention Aids 67
        3.3.3.1 Nonionic Flocculants.. 68
        3.3.3.2 Aluminum Compounds as Retention Aids.................. 70
        3.3.3.3 Anionic Retention Aids 71

            3.3.3.4   Cationic Polymers as Retention Aids 72

            3.3.3.5   Amphoteric Retention Aids 78

   References 85

Chapter 4   Temporary Wet-Strength Resins 97

   4.1   A Look at the Paper Wet-Strength Concept 97

   4.2   The Synthesis of Temporary Wet-Strength Resins: General Chemistry 100

       4.2.1   Strong Bonds and Weak Bonds in Organic Chemistry 100

       4.2.2   The Backbone Structure for Carriers of Aldehyde Groups 104

            4.2.2.1   Backbone with Aldehyde Functionality Bonded

                    through Strong Bonds 105

            4.2.2.2   Carriers of Aldehyde Group through a Weaker Bond

                    (Hemiacetal or Amidol) 111

   4.3   The Synthesis of Polyacrylamide 113

       4.3.1   Cationic Polyacrylamide through Free Radical

            Copolymerization 113

       4.3.2   Acrylamide Copolymers with a "Diluter" 118

       4.3.3   Polyacrylamide Molecular Weight 120

       4.3.4   Polymer Blends as TWSR 120

   4.4   Polyaldehyde Copolymers from Polyacrylamide 121

       4.4.1   Glyoxalation of Polyacrylamide 121

       4.4.2   The Glyoxalated Polyacrylamide Stability 125

   4.5   Paper Wet Strength and Its Decay 127

   References 131

Chapter 5   Wet-Strength Resins 137

   5.1   Prepolymer Synthesis 139

       5.1.1   Prepolymers with a Hetero-Atom in the Backbone 139

            5.1.1.1   Urea–Formaldehyde Resins 139

            5.1.1.2   Melamine–Formaldehyde Resins 141

            5.1.1.3   Polyamines and Polyethylene Imines 142

            5.1.1.4   Polyamidoamine 146

            5.1.1.5   Polyamidoamine Esters 151

            5.1.1.6   Polysaccharides 152

            5.1.1.7   Polyisocyanates 155

            5.1.1.8   Polycarboxylic Acids 161

            5.1.1.9   Polyethers 163

       5.1.2   Backbone with Carbon–Carbon Bonds Only 164

            5.1.2.1   Homopolymers as Wet-Strength Resins 164

            5.1.2.2   Copolymers as Wet-Strength Resins 168

       5.1.3   Polymer-Analogous Reactions 171

       5.1.4   Polymer Latexes 176

   5.2   Ionic Charge Addition 178

       5.2.1   Anionic Groups 179

       5.2.2   Cationic Groups 179

       5.2.3   PAE Resins Synthesis: The Epichlorohydrin Ability to Add

            Cationic Charges 182

       5.2.4   The Synthesis of PAE-Type Resin without Epichlorohydrin as

            Raw Material 184

5.3   Polyamidoamine Epichlorohydrin Polymers as Wet-Strength Resins 185
        5.3.1   Chemical Structure of PAE Resins 186
        5.3.2   Molecular Weight 187
        5.3.3   Resin Stability and Shelf Life 189
        5.3.4   By-Products (DCP and CPD) and How to Lower Their
                 Concentration 193
                 5.3.4.1   WSR with Low AOX by Adjusting the Synthesis
                            Parameters 194
                 5.3.4.2   The Reduction of the Concentration of DCP and CPD
                            by Their Hydrolysis 195
                 5.3.4.3   Producing WSR with Reduced AOX via Physical
                            Processes 196
                 5.3.4.4   Epichlorohydrin-Free Resins as Paper
                            Wet-Strengthening Agents 196
5.4   WSR Made from Blends 197
        5.4.1   Blends of Resins with Similar Chemistry and No Synergetic
                 Effect 197
        5.4.2   Synergetic Effects Provided by Blends of Resins with
                 Different Chemistries 198
5.5   Paper Wet-Strengthening Mechanisms 201
        5.5.1   The Strength of Wet and Dry Paper 201
        5.5.2   WSR Retention Mechanism 203
        5.5.3   Diverging Views on the Wet-Strength Mechanism 209
                 5.5.3.1   Are Cellulose Fibers Involved in New Covalent Bond
                            Formation? 210
                 5.5.3.2   To What Extent Does Hydrogen Bonding Explain the
                            Paper Wet Strength? 214
                 5.5.3.3   What Would a Protective Mechanism Look Like? 216
5.6   Paper Repulpability 219
        5.6.1   Fighting the Chemicals that Yield Permanent Wet Strength 220
        5.6.2   The Repulping Mechanism 221
        5.6.3   Repulpable Paper 222
        5.6.4   Improved Recycled Fibers 224
References 224

Chapter 6   Dry-Strength Resins 241

6.1   Involvement of Chemicals in the Dry Strength Mechanism of Paper 241
6.2   Anionic Dry-Strength Additives 245
6.3   Cationic Dry-Strength Additives 248
        6.3.1   Cationic Starch as a Dry-Strength Additive 248
        6.3.2   Cationic Polyvinyl Alcohol as Dry-Strength Additive 250
        6.3.3   Cationic Polyacrylamide as a Dry-Strength Resin 251
        6.3.4   Blends of Cationic Resins as Dry-Strength Additives 252
        6.3.5   Polyamine 253
        6.3.6   Cationic Latexes as Dry-Strength Additives 254
6.4   Amphoteric Dry-Strength Resins 255
6.5   Blends of Anionic and Cationic Resins 256
References 260

Chapter 7   Internal Sizing Agents 267

    7.1   The Chemistry of Alum in the Papermaking Processes 268
    7.2   Rosin is Back on the Cellulose Fibers 271
        7.2.1   Exploring the Organic Chemistry of Rosin: Rosin Derivatives
              as Sizing Agents 272
              7.2.1.1   Reactions at Double Bonds 272
              7.2.1.2   Reactions at Carboxyl Group 273
              7.2.1.3   Rosin Neutralization 274
        7.2.2   Anionic Rosin Size 274
        7.2.3   Cationic Rosin Dispersions and Amphoteric Stabilizers 276
        7.2.4   Rosin Sizing Mechanism 276
        7.2.5   Technological Consequences of the Rosin Sizing Mechanism 279
        7.2.6   Other Carboxylic Acids as Sizing Agents 280
    7.3   Reactive Internal Size (1): Alkyl Ketene Dimer 282
        7.3.1   AKD Synthesis 282
        7.3.2   The Emulsification of AKD 283
              7.3.2.1   Stabilizers for AKD Emulsion 283
              7.3.2.2   AKD Dispersion with Higher Solids Content 285
              7.3.2.3   AKD Emulsion Stability 286
        7.3.3   AKD Retention 287
        7.3.4   AKD Sizing Mechanism 289
              7.3.4.1   Investigating the Formation of Covalent Bond
                     between AKD and Cellulose 290
              7.3.4.2   Alternative Suggestions for an AKD Sizing
                     Mechanism 294
    7.4   Reactive Internal Size (2): Akenyl Succinic Anhydride 297
        7.4.1   The Synthesis of ASA-Type Compounds 298
        7.4.2   ASA Emulsification 300
        7.4.3   Effects of ASA Hydrolysis on Its Application 302
        7.4.4   ASA Sizing Mechanism 304
    7.5   Other Chemical Compounds Able to Fit the General Concept for an
        Internal Sizing Agent 307
        7.5.1   Other "Potentially Reactive" Compounds as Internal Sizing
              Additives 307
        7.5.2   Other Nonreactive Compounds as Internal Sizing Agents 311
    References 316

Chapter 8   Creping Adhesives and Softeners 327

    8.1   Creping Adhesives 328
    8.2   Composition of Creping Adhesives 329
        8.2.1   Adhesives for the Yankee Dryer 330
              8.2.1.1   Nonreactive Creping Adhesives 330
              8.2.1.2   Reactive Self-Cross-Linkable Creping Adhesives 332
              8.2.1.3   Creping Adhesives with a Cross-Linker 333
              8.2.1.4   How to Control the Cross-Linking Reaction on the
                     Yankee Dryer 336
        8.2.2   Modifiers 337
        8.2.3   Release Aids 339

8.3   Debonders/Softeners 340
      8.3.1 Softeners Retention and Softening Mechanism 340
      8.3.2 Paper Softness Evaluation 342
      8.3.3 Chemical Structure of Softeners/Debonders 342
References 346

Chapter 9   Chemicals for the Treatment of Paper Surface 351

9.1   Surface Sizing Agents 352
      9.1.1 Starches for Size-Press Solutions 354
      9.1.2 Nonreactive Surface Sizing Agents 356
            9.1.2.1 Emulsions of Nonreactive Small Molecules
                     as Sizing Materials 358
            9.1.2.2 Surface Size Obtained through Emulsion
                     Polymerization 359
            9.1.2.3 Surface Treatment for Oil-Resistant Paper 367
      9.1.3 "Reactive" Surface Sizing Agents 368
            9.1.3.1 Anionic Water-Soluble Polymers 369
            9.1.3.2 Dispersions of Nonreactive Sizing Agents Stabilized
                     with Reactive Sizing Agents 374
            9.1.3.3 Internal Sizing Agents for Surface Treatment 378
            9.1.3.4 Surface Sizing Mechanism 381
      9.1.4 Effect of the Defoamer 385
9.2   Surface Strength Agents 386
9.3   Porosity Builders 388
9.4   Polymers in Paper Coatings 390
      9.4.1 Natural and Synthetic Binders 390
      9.4.2 Binder Migration 393
      9.4.3 Hydrophobic and Cross-Linked Binders 394
      9.4.4 Coating Hydrophobicity and Its Repulpability 396
      9.4.5 Coating Surface Properties 396
References 398

Index 411

Foreword

Readers of this book are in for a joyful experience! The authors of *Chemistry of Modern Papermaking* clearly have a profound enthusiasm for their subject matter, of which they care about deeply and want to share with their audience.

Though there have been other textbooks dealing with the chemistry of papermaking, this book achieves an important new milestone in bringing together a wealth of insight concerning the chemical strategies that can have practical use in a state-of-the-art papermaking facility. Never before has a textbook compiled, carefully digested, and lucidly explained such a deep collection of details from both the patent and scientific literature. This synthesis is achieved not only through diligent work, but also reflects the years of industrial experience of the authors. Readers will also quickly come to respect Hagiopol and Johnston's gifts for teaching—especially the teaching of chemistry.

As the authors themselves state more effectively in their book, one of the important principles of the chemistry of papermaking is that of "leveraging." With typically only about 3% of the mass of a paper product invested in water-soluble chemicals, the papermaker can achieve dramatic effects. On the one hand, he or she can greatly increase the efficiency of the process—including the production rate. For instance, by the use of retention aids, the efficiency of retaining fine particles in the paper can be improved. Not only does this help to minimize wasted materials, but it also helps paper-makers to avoid significant discharges of waterborne substances as liquid effluent. An optimized wet-end chemistry program can also achieve higher rates of water removal, which often allows papermakers to speed up the process. On the other hand, papermakers are able to differentiate a paper product in terms of its appearance, resistance to fluids, strength properties, and myriad other attributes that are needed by specific customers. The latter changes are brought about by various "functional additives," which can include dyes, sizing agents, starch, and various wet-strength additives. This book does a particularly good job at describing the options that papermakers have with respect to strength-contributing additives.

The subject of "wet strength," which is treated in detail by the authors in Chapters 4 and 5, is a delightful paradox. As the authors themselves note throughout the book, paper can be defined as a hydrogen-bonded material. As such, one of its great positive features is that ordinarily it can be made to come apart just by making it thoroughly wet. The practice of paper recycling depends on this attribute. But there are specific applications in which paper's inherent nature has to be turned on its head. Rather than just relying on the hydrogen bonds, such bonds need to be supplemented. How often, in your experience, has one or more bills or your own paper money passed through your own washing machine? The fact that it still was undamaged after such treatment was no accident. Similarly, wet-strength chemical treatments are also needed in order to achieve the kind of properties needed in tissue papers. While some wet-strength products (such as dollar bills) need to retain their strength indefinitely, those products that are designed to be flushed, especially when flushed into a septic system, need to be designed to fall apart a few minutes after they have become wet. In addition to such mundane concerns, there have been many recent initiatives to avoid certain toxic monomers associated with some wet-strength treatments. This book is especially authoritative and complete in its treatment of subjects related to chemical details underlying these important issues.

Readers of this textbook will appreciate the authors' concern regarding sustainability. Here the authors help to illuminate a key fact: Whereas many sectors of industry have only recently begun converting their thinking toward the incorporation of more sustainable practices, the paper chemicals industry can trace its involvement in such practices at least back to the first major industrial wet-end additive—in a sense the first "modern" papermaking chemical—back in 1807. That was the year in which the rosin-alum system for internal sizing of paper was patented. Rosin is a by-product of the production of paper pulp. Instead of getting rid of rosin as a waste product, the industry instead transformed it into a useful additive—helping the paper to resist water, inks, and other fluids. To this day there are many important additives in the papermaking process that are based on renewable, photosynthetic source materials.

Open a page of this book at random and you are more likely than not to encounter carefully selected and redrawn chemical formulae and reaction paths, illuminating many of the most promising strategies for the use of chemicals in a papermaking process. Too many authors have shied away from such a graphic and explicit approach to explaining important concepts underlying chemical technology. Though the detailed chemistry of various copolymers may not be every reader's "cup of tea," one needs to bear in mind that a book without such effective chemical notations and reaction schemes would require a great many more words—and probably achieve less clarity and utility. And this is a book that is clearly intended to be useful. The extensive literature references can also serve as a starting point for those readers who wish to pursue related research in new and interesting directions.

Martin Hubbe
Department of Forest Biomaterials
North Carolina State University, Raleigh, North Carolina

Authors

Cornel Hagiopol received his PhD in macromolecular chemistry from Polytechnic University, Bucharest, Romania, in 1983. His expertise lies in polymer chemistry. He joined Lehigh University in 1998 and came to Georgia-Pacific Chemicals LLC (paper chemicals group) in 2000 to work on the synthesis of copolymers for surface sizing agents and wet/dry strength resins. He authored the book *Copolymerization* (Plenum/Kluwer, New York, 1999) and was a contributor to *Encyclopedia of Condensed Matter Physics* (Elsevier, Oxford, 2005). He is the coauthor of more than 30 publications and the coinventor of more than 20 patents.

James (Jim) W. Johnston is currently a research and development manager for Georgia-Pacific's paper chemicals business in Decatur, Georgia. He is responsible for the development of intellectual property, project management, and technology development within the paper chemicals market. Jim's expertise lies in chemistry applications, chemical engineering, and paper properties. He has held various technical, operations, marketing, and R&D positions over the years within Georgia-Pacific, Hercules, Inc., and International Paper. He is a past lecturer for TAPPI's Wet End Chemistry short course and is the coauthor of several patents. Jim is a graduate of Syracuse University and SUNY ESF Chemical Engineering and Paper Science and Engineering Program.

Abbreviations

AKD	Alkenyl ketene dimer
AOX	Absorbable organic halogen
ASA	Alkenyl succinic anhydride
AZC	Ammonium zirconium carbonate
CAS	Cationic aldehyde starch
CCD	Chemical composition distribution of the copolymers
CMC	Carboxymethyl cellulose
CPD	3-Chloropropane-1,2-diol
CTA	Chain transfer agent
DCP	1,3-Dichloro-2-propanol
DETA	Diethylene triamine
DSR	Dry strength resin
EAA	Ethylene acrylic acid copolymer
EVA	Ethylene-vinyl acetate copolymer
FPR	First pass retention
GC	Gas chromatographie
G-PAm	Glyoxalated polyacrylamide
GPC	Gel permeation chromatography
HLB	Hydrophilic–hydrophobic (lipophilic) balance
HPLC	High-performance liquid chromatography
HST	Hercules sizing test
MF	Melamine formaldehyde resins
MFFT	Minimum film formation temperature
MMA	Methyl methacrylate
NMR	Nuclear magnetic resonance
PAA	Polyamidoamine
PAAm	Polyacrylamide
PAC	Polyaluminum chloride
PAE	Polyaminoamine epichlorohydrin polymer
PEG	Polyethylene glycol
PEI	Polyethylene imine
PEO	Polyethylene oxide
PPG	Polypropylene glycol
PVA	Polyvinyl alcohol
PVSK	Polyvinyl sulfonate, potassium salt
RBA	Relative bonded area
SAE	Styrene acrylic ester
SBR	Styrene-butadiene rubber
SEC	Size exclusion chromatography
SMA	Styrene maleic anhydride copolymer
TAPPI	Technical Association for Pulp and Paper Industries
TEMPO	2,2,6,6-Tetramethylpiperidine-1-oxyl
THF	Tetrahydrofuran

TWSR	Temporary wet-strength resins
UF	Urea formaldehyde resins
WL	Waterleaf
WSR	Wet strength resin

1 Introduction

We can thank Cai Lun in the year 105 for the earliest description of papermaking and we can thank a myriad of inventors through the millennia for moving this invention from being produced one piece at a time to tens of thousands of tons per hour globally. As we embark on the reuse of fiber and the design enhancements of the paper, the novelty of innovating chemistry needed to attain the properties required is developed at an increasing rate.

This first sheet of paper was not intended for mass consumption but gave rise to a process that would produce billions of tons of paper using billions of tons of wood, which touches another global issue: wood is one of the most important materials of our time. It is a raw material for many other industries and a balancing factor for climate.

How we manage the papermaking process is actually translated in how the wood is protected. The general concept of wood protection relates to other hot topics of our time: sustainability, renewable materials, green chemistry, recycling processes, environmental protection, the marriage of synthetic and natural products, etc.

Paper is a "hydrogen-bond-dominated solid" [1] (Figure 1.1), a random network of natural cellulosic fibers [2,3], a complex composite [4], with an average composition (all paper grades) of 89% fiber, 8% filler, and 3% paper chemicals (performance and process chemicals) [5–7].

Performance paper chemicals impart various physical attributes and make the difference between paper expected to last for centuries and paper discarded within a matter of days [8]. The chemical compounds can provide high wet strength (paper towel) or easy disintegration in water (toilet paper), high graphic or poor printing quality, degree of repulpability, easy penetration by water-based liquids, or perfect containers for milk.

The main trends in the papermaking industry are [9]: balance of expensive virgin fibers with lower secondary cost (de-inked) fibers, more filler and/or pigments, lower grammage (basis weight), lower density [10], and higher speed. These trends will lead to increased paper chemical consumption [11]: in the United States, the forecast is to advance with 1.0% annually to 19.8 million tons in 2011 [12] with specialty additives showing the most rapid growth [13].

The structure of paper chemicals must accommodate not only the diversity of demands for paper quality but also the papermaking process. Paper chemicals must not react unintentionally with water and must be adsorbed on the cellulose fiber in the presence of water. The interactions between chemicals and cellulose (solubilization, adsorption, chemical reactions) must take place in the temperature ranges from 20°C to about 105°C and moisture from 99.5% to 5%. Thus, it is obvious that there are no two papermaking systems alike and there is no paper chemical that serves all functions and performs under every set of conditions [14].

There are specific types of paper chemicals for each particular paper property that needs to be improved. Since the average content of chemicals (excluding fillers) in paper is less than 3% by mass (including starch [5]), each chemical with a specialized job is used in a much lower concentration. Organic chemistry (especially polymer science) is able to offer the diversity of chemical compositions to accommodate the diversity of paper grades. Paper chemistry is developed by keeping in mind not only the paper performances and costs, but also the wood preservation and the environment protection. Thus, paper additives have recently become more significant [15].

The organic chemistry of paper chemicals must be seen as parts of the papermaking process where chemical engineering, colloid and surface science, and materials science must also be considered [8]. The manipulation of the paper chemical composition, within the papermaking process, using organic chemistry is the background of this book.

FIGURE 1.1   Paper structure by electronic microscopy (magnification ×35 and ×80 in red frame).

To review the literature published in this area is a challenging task, because the volume and diversity of the published data are describing the infancy of paper chemistry. Due to the lack of information, several theories have been put forward, for which the balance pro–against is still undecided. Moreover, it is a great disparity between those who work in applications and who publish patents and those working in academia who publish scientific articles and reviews [16,17]. Seldom is a patent referenced in these articles. There appear to be two different worlds with different tools, different targets, and a different sense of urgency. Sometimes these worlds are disconnected.

The authors believe that the addition of the patent literature can consolidate and enrich the picture about the papermaking process. Moreover, patent information is incorporated in the structure information provided by the academic environment. The target is to make an articulate and comprehensive inventory of ideas concerning the papermaking process. The general review on the information about making paper was enlarged and assembled by using the challenging scheme and the creative distraction tool [18]. The graphs are converted to the SI units according to the conversion factors approved by the Steering committee of the Process Product and Product Quality Division—TAPPI [19].

This book is focused on the structures of paper chemicals and their significance in the papermaking process: the interactions with cellulose, the type of organization on the cellulose surface or in the inter-fiber area. It is an attempt to connect the local microstructure to the macro-properties of paper.

We focused on the chemistry behind each application, on what has been done, and on what can be done. If we can control the chemistry, the application step will look much easier. The chemist's point of view is moderated by the application demands, whereby a spectacular opening from the chemistry side should be balanced by the reality of the application.

The miraculous effectiveness of chemicals at very low concentrations is a fascinating topic in paper chemistry. A magical paper chemical, able to improve each and every property, does not exist! However, there is a world of innovation in the papermaking process. Now, for the first time, the inventions concerning paper chemistry are put together in this book. Thus, a new foundation is laid for a fresh, new start.

ACKNOWLEDGMENTS

This book went through an accumulation of the publications and a revision of the organized data. We would like to thank Kathy Dalburg for her constant and gracious support, Rita White and Carrie Doyle (NTC), Professor A. Isogai (The University of Tokyo), and our colleagues Dr. Clay Ringold, Dr. Gregg Reed, Karla Favors, Fred Potter, Dr. Ramji Srinivasan, Gene Lou, and Don Jenkins for helping us to get all these 3400 references incorporated in this book. Special thanks are due to Dr. David Townsend for his review of this text and his balanced comments and to Dr. Ellen Nagy for her assistance in understanding the paper structure. The authors would also like to thank to Peter Williams, Dr. Larry Gollob, and Dr. Kurt Gabrielson for their enthusiastic encouragements, Francis Giknis, Diane Brower, Ellen Whitefield, Mike Kerns, Perry McGuire, and Christine Cason, because without their effort and guidance the publishing of this book would not be possible. And last, but not least, we are grateful to Georgia-Pacific for their on-going support.

REFERENCES

1. A.H. Nissan, *Macromolecules*, **9(5)**, 840–850 (1976).
2. D.H. Page, *Tappi*, **52(4)**, 674–681 (1969).
3. N.L. Salmen, Mechanical properties of wood fibers and paper, in *Cellulose Chemistry and its Applications*, T.P. Nevell and S.H. Zeronian (Eds.), Ellis Horwood Ltd., Chichester, England, 1985, p. 505.
4. F. Tiberg, J. Daicic, and J. Froberg, Surface chemistry of paper, in *Applied Surface and Colloid Chemistry*, K. Holmberg (Ed.), John Wiley, New York, 2002.
5. P.H. Brouwer, J. Beugeling, and B.C.A. Ter Veer, *PTS-Symposium*, Dresden, Germany, 1999, pp. 1–15/28.
6. M.A. Hubbe, Pulp and paper chemical additives, in *Encyclopedia of Forest Sciences*, J. Burley, J. Evans, and J. Youngquist (Eds.), Elsevier, Oxford, U.K., 2005.
7. M.A. Dulany, G.L. Batten, M.C. Peck, and C.E. Farley, *Kirk-Othmer Encyclopedia of Chemical Technology*, Vol. 18, 4th edn., John Wiley & Sons, New York, 1996, pp. 35–60.
8. M.A. Hubbe and O.J. Rojas, *Chem. Eng. Educ.*, **39(2)**, 146–155 (2005).
9. P.H. Brouwer, T. Wielema, B. Ter Veer, and J. Beugeling, *PTS-Symposium*, Munich, Germany, 2001, pp. 14/1–14/214.
10. T. Tadokoro, Y. Ikeda, T. Nishimori, and H. Takahashi, US 7,122,098 (2006).
11. M. Voith, *Chemical & Engineering News*, April 19, 2010, pp. 13–17.
12. The Freedonia group, *Pulp & Paper*, April 2008, p. 21.
13. P. Pruszynski, Recent developments in papermaking chemicals, in *Conference Presented at Wood, Pulp and Paper Conference*, Bratislava, Slovakia, 2003.
14. E. Strazdins, *Pulp & Paper*, March 1984, pp. 73–77.
15. T. Kitaoka and H. Tanaka, *J. Wood Sci.*, **47**, 322–324 (2001).
16. O.J. Rojas and M.A. Hubbe, *J. Dispers. Sci. Technol.*, **25(6)**, 713 (2004).
17. G.J.C. Potter and J. Weiner, *Adv. Chem. Ser.*, **78**, 296–348 (1968).
18. C. Koch, *The Science of Success*, John Wiley, New York, 2007.
19. TAPPI Steering Committee of the Process Product and Product Quality Division, Units of measurement and conversion factors, in *TAPPI Papermakers Conference and Trade Fair*, Vancouver, BC, Canada, 2000, pp. 823–838.

2 From Wood to Paper
A General View of the Papermaking Process

Wood, the material paper is made of, is a heterogeneous material in both the chemical and physical sense of the term: it incorporates different chemical compounds distributed in separate phases [1].

The papermaking process deals with the heterogeneous wood structure and with water as a continuous phase of a heterogeneous system. Water, the "active" component, is further involved in many chemical and/or physical processes during papermaking, and it is retained as moisture in the final paper.

In this heterogeneous system, chemical reactions and physical processes take place at the same time: covalent bond (ester, imid, hemiacetal, etc.), hydrogen bond, and ionic bond formations occur simultaneously with adsorption, solubilization, precipitation, filtration, diffusion, evaporation, and changes of the surface tension.

Macromolecular compounds are always present, and their behavior in solution, or at the interface, their interactions with the small molecule, and their chemical reactions are critical for the papermaking process.

2.1 FROM THE *PAPYRUS* ERA TO MODERN TIMES: A BRIEF HISTORY OF MAKING PAPER

The word *paper* itself comes from the name of the *papyrus* plant, which in ancient times was prevalent along the Nile River in Egypt. Papyruses, made from thin slices of this plant, are pressed, dried, and then used as a substrate for the written word, in a process that preserves the original distribution of cellulose fibers, and which came to be around two and a half millennia prior to the discovery of paper.

The invention of paper is undoubtedly the most important invention of mankind during the first millennium. Since paper production spread all over the world, the contribution to novelty has crossed all borders.

Nearly 2000 years ago, the official imperial court records made note of the invention presented to the Chinese emperor by Cai Lun (AD 105). It concerned a pulping mulberry bark and created a fiber suspension that, after filtration, became the very first piece of paper ever made. None of the new randomly distributed cellulose fibers retained the same neighbor from the original material [2–5].

Although, the earliest sheets of paper were not intended for mass consumption, it gave rise to a process that would produce billions of tons of paper over the course of the next two millennia. Papermaking made its way to Korea, Japan, and the Middle East between AD 600 and 800. The Chinese exported this technology to Baghdad in 793. These early papers were untreated with any chemistry and remained so until the eighth century when the Chinese began utilizing a tub-sizing agent derived from flour starch.

From the Middle East, the papermaking process spread to Egypt (around AD 900), then to Morocco, and in 1150 reached Spain. In Europe, the papermaking process was spread toward North and East: France (1157), Italy (1276), Germany (1390), Poland (1491), Great Britain (1495), and Russia (1576). In the United States, the first paper mill was erected in 1690 (Germanstown, PA).

Only from the mid-nineteenth century on was wood considered as a raw material for papermaking. Pulping, the separation of cellulose from wood, has a short but vigorous history: countless new technologies were developed in a very short period of time. Soda pulping was invented by Watt and Burgess in 1853 (Brit 1942) [6]. The first patent on sulfite pulping was granted to Benjamin Tilghman (US 70,485/1867) [7]. In 1889, the German chemist Carl F. Dahl invented the sulfate pulping process (Kraft) by using the sodium sulfate cake available as a by-product from the manufacture of hydrochloric acid [8].

The use of paper has touched different areas of everyday life. To address the diversity of applications, paper needs more ingredients along with cellulose fibers.

The common perception is that the development of paper technology was targeting a material for the written records. Papermaking technologies continued to evolve: a more effective gelatin sizing technology was used in Italy in 1337. It was a finishing technique that produced paper with a tough, opaque surface, well suited to the use of the quill pen, but which deteriorated quickly in hot weather. Around the sixteenth century, alum was added to control bacterial growth and mold.

However, there are many other uses for that new material: balloons, mats, toilet paper, clothing, fireworks, lanterns, etc. [9]. Paper sacks were first referenced in 1630, but the use of the grocery sack produced from paper became popular in the eighteenth century. The paper bag as we know it today was designed and produced by Margaret Knight who founded the Paper Bag Company in 1870. Luther Crowell also patented a machine to produce a square bottom bag in 1872. The cardboard box was invented in 1870 by the American Robert Gair. This type of invention took flight as patents were filed by Albert James for a corrugated paper used to ship glass goods and shortly after Oliver Long patented a lined corrugated cardboard [10].

The first patent was granted to S. Hooper (British Patent nr. 1632 in late eighteenth century) in the modern era for using fillers in the papermaking process (cited by Beazley) [11].

Toward the end of the eighteenth century, casein sizing was used in Germany, where in 1807 Moritz Friedrich Illig developed a practical method of adding rosin size and alum at the wet end. This new technology opened the era of what we now call the "wet end chemistry"; with a particularly interesting idea: chemicals separated from cellulose during pulping are now returning next to the fibers. This new concept was taken over and other chemicals, obtained during wood processing, were introduced in the papermaking process: modified rosin, lignin [12], and hemicellulose.

As sciences developed, and cultural exchanges involved a greater need for communication and record keeping, the demands on paper as a substrate of information became greater as became the need for different paper types and properties. In 1838 Charles Fenerty of Halifax made the first newsprint, using ground wood as a fiber source, after he was unable to supply a local paper mill with enough rags for paper production. He failed to patent this invention, which was later patented by others [13].

Since the hydrophilic nature of the paper should be protected against water, in 1886 the articles made of paper were water-proof after they were coated with a paraffin wax layer [14].

As the world of single-use products was evolving, the invention of the paper plate came about in 1904. Disposable paper cups were first introduced in 1908 by Hugh Moore who owned a paper cup factory that happened to be located next to the Dixie Doll Company. Permission was granted by the Dixie Doll Company for the marketing of the cup. Prior these cups were marketed as health cups around water fountains and for hospital use.

Regardless of the important progress made over the centuries and the development of modern papermaking technologies, physical processes were its dominant part: the major properties of paper were obtained in a beater [15]. Only in the twentieth century did paper chemicals and chemists became key factors in the papermaking process.

Technological developments in the papermaking industry were accompanied by the parallel evolution of organic and colloidal chemistry: chemically modified starches, newly developed small molecules (i.e., hydrophobic compounds), new (co)polymers (anionic, cationic, low and high molecular weights), water-soluble compounds or dispersions. The importance of the hydrogen bonding for paper strength and for the entire papermaking domain waiting for new paper chemicals was suggested in 1949 [16]. Some authors have seen paper as a "hydrogen-bonded solid" [17].

A strong interest in improving the paper wet strength by adding a resin "to glue" the point of contact between cellulose fibers [18–21] is noticed from the 1930s. The first synthetic material used as a wet-strength resin was a urea-formaldehyde compound. Melamine-formaldehyde resins (MFR) (cationic) are introduced only in 1943 [22].

Rosin dispersions with casein (Wieger, 1932) are more substantive to cellulose fibers and the alum usage in paper mills started declining. The reactive sizing has developed since 1935 (Nathansohn, US 1,996,707) [23] when a fatty acid anhydride (a precursor of ASA) was patented as sizing agent.

Carothers synthesized (1939) [24] the water-soluble polyamidoamine that will make a strong impact on the papermaking process around 1960, when the new wet-strength resins were developed (alkaline currying, without formaldehyde).

A "dual system" as wet-strength resin is suggested by Maxwell in 1946 [25] as a blend of anionic urea-formaldehyde resin (UFR) and cationic MFR.

AKD as sizing agent for paper (Keim, Thompson and separately Downey, all with Hercules) is mentioned for the first time in two patent applications filled the same day of October 6, 1949. In 1957, Hercules offered AKD emulsion "Aquapel" [26].

Polyamidoamine–epichlorohydrin wet-strength resins are developed by Keim with patent application filed in 1957 and 1959, respectively (Hercules, US 2,926,154 and US 2,926,116 both granted in 1960). In 1957, the first wet-strength resin based on the polyaminoamine epichlorohydrin (PAE) polymer was introduced on the market under the trade name KYMENE [27]. The first patent on glyoxalated polyacrylamide as a temporary wet-strength resin was granted in 1971 to Coscia and Williams [28].

The existence of the wet-strength resin and sizing agents effective in neutral or alkaline pH open a new era in papermaking. Since the early 1970s, there has been a rapid process to convert papermaking from operating in acid pH (and alum chemistry) to neutral or even slightly alkaline pH [29], involving a tremendous effort for paper chemical usage [30].

Knowledge about the papermaking process was consolidated over the years by publications as the monumental work of Casey [31], or critical additions on paper chemistry by Roberts, Eklund, Lindstrom, Au, Thorn, Scott, or Neimo [29,32–35]. Last but not least, the noteworthy effort made by the TAPPI organization to publish monographies on different topics related to paper chemistry and the papermaking process [36–40] should be mentioned.

2.2 PULP: THE SUPPORT FOR PAPER CHEMICALS

Cellulose fibers are recovered from wood through a process called pulping. During wood extraction with water, only <5% of soluble material goes into the water phase (as in thermo–mechanical pulp). The wood cell wall has a heterogeneous structure and consists primarily of three polymeric materials: cellulose (42%–45%), hemicellulose (27%–30%), and lignin (20%–28%) [41]. Fibers within wood are glued together with a natural phenolic resin, lignin [3,42].

Cellulose is composed of linear and parallel macromolecules (β-D-glucopyranosyl units with (1-4)-β-D-linkage), which are very tightly packed (crystallite) due to inter- and intramolecular hydrogen bonds [43,44]. Cellulose is not soluble in water and the intermolecular hydrogen bonding reduces water access to its functional groups [45].

Cellulose fibers do not lose their integrity in water. Because cellulose is highly crystalline, only certain specialized solvents can put it into solution without chemically degrading it [46]. Polyamines make a complex with cellulose [47–51] but they cannot bring it into solution. In order to dissolve cellulose, unusual solvents (aqueous solvents: ethylene diamine in water with cadmium hydroxide [52], lithium chloride/*N*,*N*-dimethyl acetamide [46,53], cuprammonium hydroxide, iron-sodium tartrate [44]), organic solvents [54–56], or ionic liquids [57] are required.

Cellulose fibers need to be separated from the wood structure by selective extraction, which is the core of the pulping process. The pulping process is based on the difference in reactivity of the wood component toward the chemical compounds. In order to preserve the valuable component—the cellulose fibers—the pulping process is an extraction of lignin derivatives and soluble polysaccharide.

Selective extraction with a solvent ("organosolv pulping" [58] or "ester pulping" [59,60]) can dissolve lignin and hemicellulose. The most common organic solvent is ethanol [61–63] and the pulping process is faster in the presence of an acid [64]. Only 0.34% residual lignin remains after extraction with monoethanol amine at 180°C [65–67] or ethylenediamine–sodium sulfide [68]. The extraction can be performed simultaneously with a grafting reaction in the presence of vinyl monomers (styrene, acrylonitrile) [69]. However, the volatile, flammable, and harmful organic solvents are still considered dangerous for the papermaking process.

Lignin has a phenolic structure with weaker covalent bonds, while hemicellulose, due to its amorphous structure, shows a higher reactivity toward water-soluble compounds. For instance, hemicelluloses are preferentially etherified or esterified in the presence of cellulose [70]. On the other hand, lignin reacts with formaldehyde and amines [71–74] or polyamine [75–77] (Mannich reaction), epoxides, ethyleneimine, thimethylolmelamine [77–79], diepoxides [73,75,80–81]. Most of those compounds are water soluble because more hydrophilic groups are attached to lignin (quaternary ammonium salt [82], hydroxyl, amine, etc.). Those reactions support the concept of the selective reaction of lignin and hemicellulose and their migration into water, which separates them from cellulose.

In order to extract the undesired material, the pulping process involves chemical reactions able to make those macromolecular compounds soluble in water. A large molecule becomes water soluble when the molecular weight is reduced (fragmentation) simultaneously with the addition of new hydrophilic groups and the transformation of the acid groups into their corresponding salts. Due to chemical reactions and extractions [83], pulping can be seen as a simplifying process: pulp ("water-swollen gel" [84]) is a material less complex than wood.

Wood, which is subjected to the pulping process, is different from one area to the other, from one season to the other: the southern pine fibers have poor performance characteristics [85] and therefore requires a special treatment with paper chemicals. A large variability in extractions and chemical interactions is also recorded during the pulping process.

Three are four different kinds of wood pulp: mechanical, sulfite, sulfate (or Kraft), and soda pulp. Mechanical pulp contains substantially all of the wood except the bark and what is lost during storage and transportation. That is the reason for a different interaction between the mechanical pulp and paper chemicals [86] and for a higher cationic demand [87].

Chemical pulps are, more or less, pure cellulose: noncellulosic components of wood are dissolved away by the treatment [42]. However, the pulping process is performed "at equilibrium," and, therefore, the reaction yield (far from 100%) and the extraction yield (the pulping yield) depend on the phase ratio (wood/water), time, temperature, etc., thus the pulp will contain residual lignin and hemicellulose (both with a partially modified structure). About 3% lignin remains in the bleached softwood Kraft pulp and about 38% in the thermo–mechanical pulp [88]. A pulping process in two steps with a pretreatment seems to be more flexible and more effective [89].

The pulping process involves high temperatures, high pressure, and chemical compounds. Water as a plasticizer (to reduce the T_g), high temperatures of 120°C–130°C, and corresponding pressure are the regular conditions helping to free cellulose fibers. The chemical compounds are sodium hydroxide, sodium sulfite or sodium sulfide (S^{2-}), or bleaching agents like sodium hypochlorite.

Wood pulping must prevent not only the degradation of valuable material (cellulose fibers), but also reduce the environmental issues. During wood pulping, an "unintentional chemical modification" of the cellulose may take place (mainly hydrolysis and oxidation) [90–93]. Selectivity is not always good: oxidants (such as peracetic acid and chlorine dioxide [94]) may oxidize indiscriminately all the wood components. Anthraquinone [92,95,96], methylanthraquinone

[97] anthraquinone monosulfonic acid [99], and hydroquinone compounds [98] are stabilizers for cellulosic compound and increase the pulp yield.

The changes in the composition and chemical structure of cellulose fibers should be taken into account when interaction with paper chemicals is considered. The fiber porosity decreases with increasing the pulping yield [100] and the fiber swelling wall increases with the number of carboxylic groups [101,102].

The characteristics of the fiber surface, the ratio between crystalline and amorphous cellulose, the number of carboxylic groups [103], and the amount of residual lignin or hemicellulose are set during the pulping process and will have an important impact on the paper chemicals adsorption at the wet end. That is why the interaction between pulp and paper chemicals is very different from one type of pulp to the other [104].

Pulp properties can be altered by adding recycled paper [105]. In order to clean recycled fibers, old paper chemicals should be processed [106] (see Section 5.6).

2.2.1 SULFITE PULPING

The sulfite pulping process [107] is performed on a wide pH range: from about pH=2 to over pH=10. Most procedures are used in acidic pH. The morphology of fiber and lignin distribution (on fines and fibers) depend on the pulping yield [108].

Lignin has a very complex macromolecular structure with different substituted phenol units and many branches (M_w about 20,000). It is easy to see the potential hydrophilic part of the molecule (II)1: that there are many hydroxyl groups (in blue), which can be involved in hydrogen bonding [109].

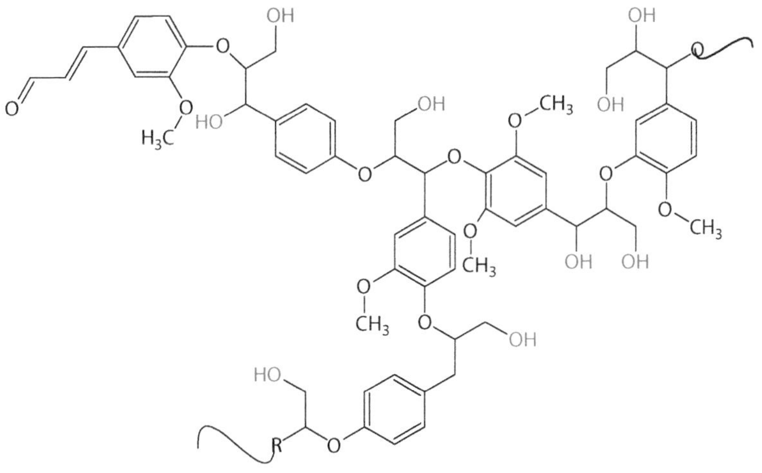

The number of free phenolic hydroxyl (in red, (II)2) is about 25 at 100 phenolic units.

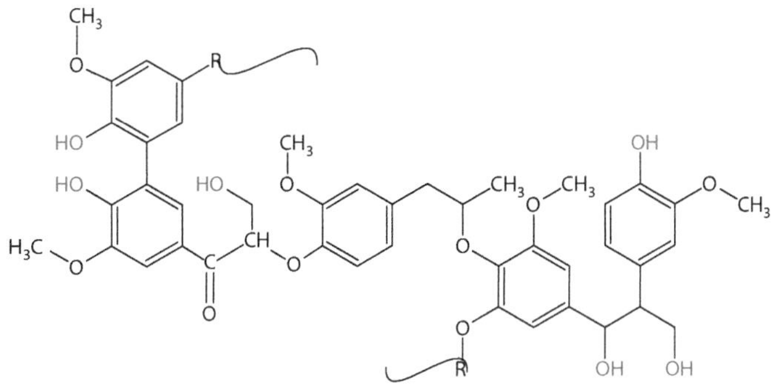

and few carboxylic groups (in green, (II)3) [110,111].

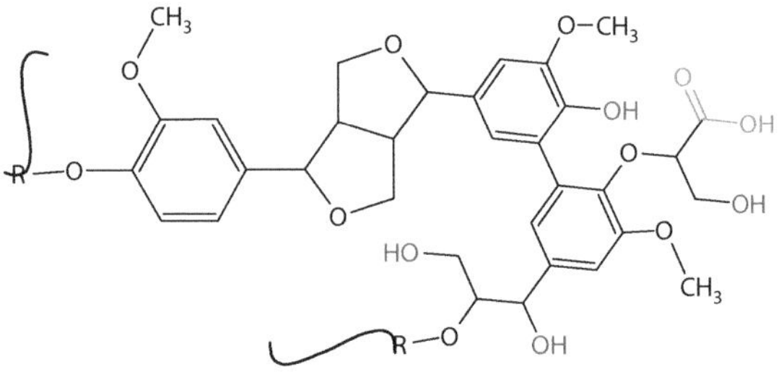

(II)3

The weak bond in the lignin structure is the ether linkage (especially that connected to a benzylic carbon). As a result, the sulfonic group is attached to in the side chain and not to the aromatic ring [112].

Phenolic and carboxyl groups are acids and can be neutralized with caustic (or any other base). The sodium salt of lignin is supposed to be more soluble in water and the pulping process is facilitated based on solubility. During the sulfite pulping process [113], macromolecule fragmentation takes place along with the addition of hydrophilic groups (II)4:

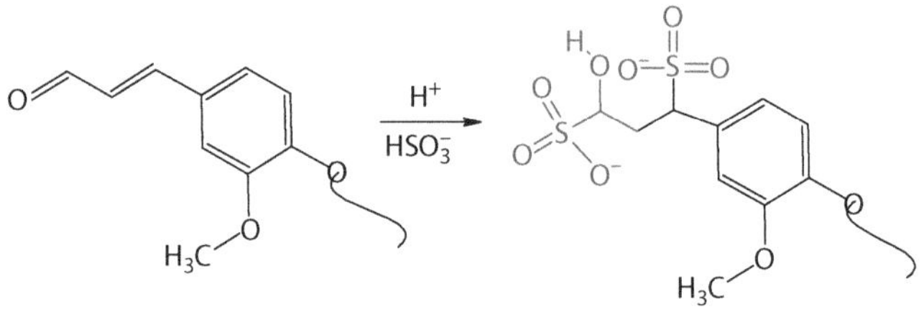

(II)4

and a reaction of breaking the macromolecular chain (II)5:

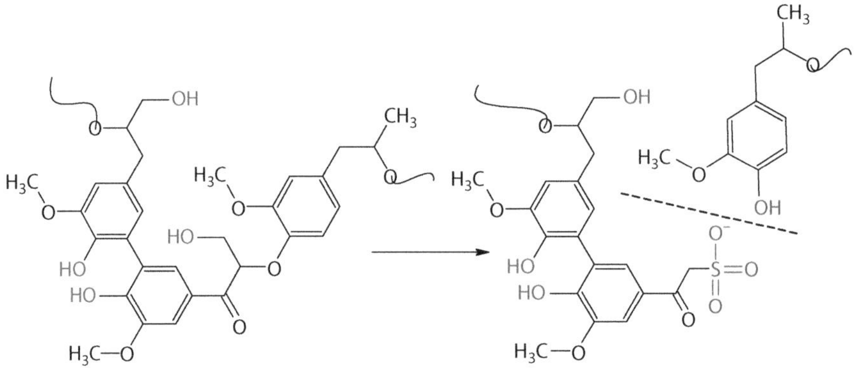

(II)5

After a cooking time of 2 h at 135°C, the lignosulfonate material contains a large number of methoxy groups and the number of sulfonic groups exceeds the number of phenolic hydroxyl groups [114]. In the presence of formaldehyde, the methylol group can be added to the phenolic structure [115].

All these reactions are performed at high temperatures [7] where their selectivity is lower. The partly converted lignin remains in pulp and will be carried to the paper formation stage. The sulfonic groups were identified in the sulfite pulp and estimated to be in higher concentration than the carboxylic groups [116].

The glycosidic bond of hemicellulose is easily cleaved by acids and de-polymerization cannot be avoided. Degraded hemicellulose dissolves in cooking liquor and its hydrolysis is accelerated. Generally, no cellulose is lost in the acid sulfite process.

2.2.2 Sulfate Pulping (KRAFT)

Acid sulfite pulping was replaced by alkaline pulping (sodium hydroxide and sodium sulfide or polysulfide [117]). The dissolution of lignin from wood under KRAFT cooking includes the fragmentation of lignin macromolecules, the solubilization of the fragments, and the restricted migration of those fragments through the fiber wall [118]. Basically, the same concept of macromolecule fragmentation and hydrophilic group addition [113] is also valid for this procedure. The polysaccharides depolymerization (peeling) increases the dissolved carbohydrate in alkaline pH [119]. The oxidation process of the end groups is accelerated by the presence of oxygen [92]. Lignin undergoes several different reactions. The β-ether bond is broken (II)6:

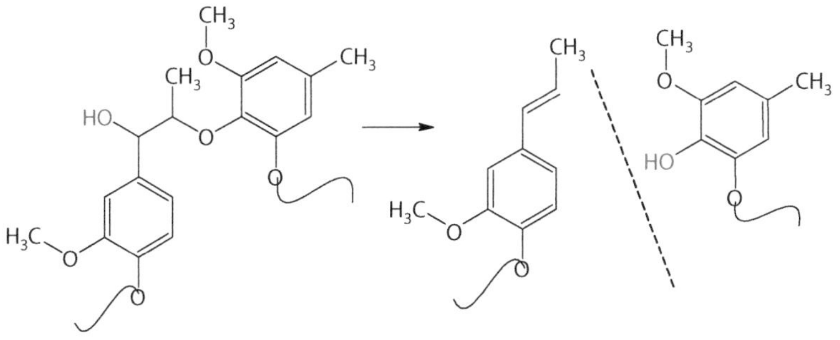

and a condensation is also possible (II)7:

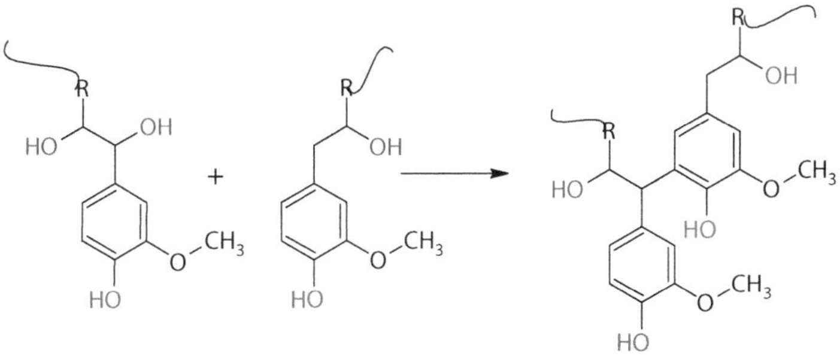

Kraft lignin contains a more hydrophilic, soluble fraction, and a more hydrophobic dispersion of microgel particles swollen in water [120]. The presence of an organic solvent (ethanol or ethylene glycol) [121–123] helps the extraction of the modified lignin. The maximum extraction is a function of pH.

The carbon–carbon double bond located in the α-position of the aromatic cycle and in the para-position related to the phenolic functionality opens the possibility of making structures with extended conjugation, which are responsible for the dark color of the black liquor [112]. The stability of lignin fragments depends on pH, electrolyte concentration, and temperature.

Pulping, defibration, and refining of coarse pulps are wet processes: water is always present and cellulose fibers are swollen in water. Lignin fragments can precipitate, be adsorbed on the cellulose fibers [118], and form complexes with transitional metals like iron [124]. Lignin and hemicellulose fragments help the swelling process of cellulose fibers. The swollen fiber will interact with bleaching chemicals in the next step of pulp preparation.

2.2.3   THE BLEACHING PROCESS

Due to residual lignin, the bleaching process is intended to enhance the delignification and decolorize pulp [125]. Hypochlorite (chlorine in alkaline pH) can make substitution reactions in the phenol ring, addition reactions at the aliphatic double bonds, and oxidation reactions.

During bleaching, residual lignin undergoes a new fragmentation and new carboxylic groups are attached to those smaller molecules. Colored fragments lose their color because the extended conjugation systems are destroyed during the oxidation (II)8.

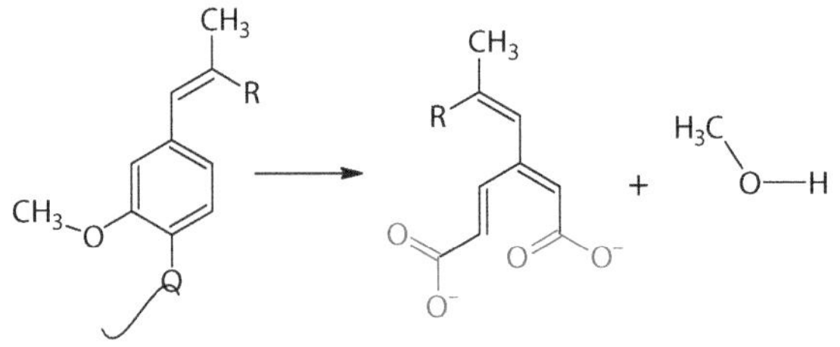

Cellulose is quite a reactive compound [126]. As expected, it reacts with the bleaching agents, which results in some oligomeric sugars containing 2–7 glucose moieties, several oligomeric aldonic acids, and other monocarboxylic acids (glyceric, glycolic, and formic) [127]. It is easy to understand that modified cellulose contains carboxylic groups as well [128] (see polysaccharide oxidation).

Bleaching agents increase the content of carboxylic groups through direct oxidation. Carboxylic groups help fiber swelling in water, increase fiber plasticity and specific strength [129]. The topochemistry of the acidic group is also important for their interaction with the cationic compounds at the wet end. More carboxylic groups can be added by specific reactions [129] (see Section 2.4). In order to provide more anionic functional groups, the fiber surface can be enriched with sulfonyl groups [130–133] by a reaction with sulfonyl groups carriers (II)9:

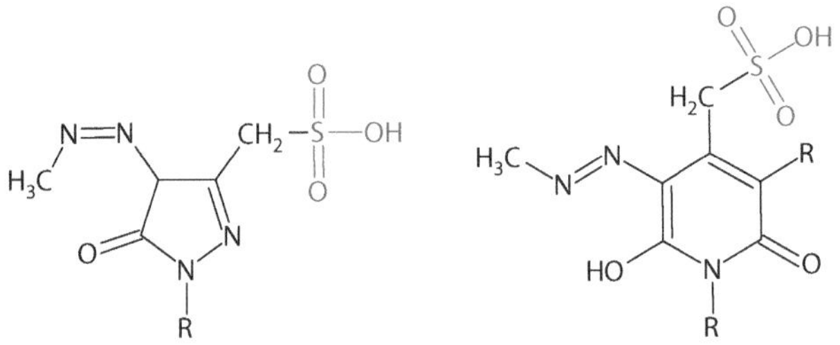

Extracted, refined, bleached fibers are still swollen in water and ready to interact with paper chemicals. Pulp dispersion may undergo a beating step, which increases the fines concentration, lowers the freeness (pulp holds more water), and increases the sheet density and strength [15,134,135].

2.2.4   WET END: SHEET FORMATION AND WHITE WATER

At the end of the pulping and bleaching processes, the cellulose fiber is different: new functional groups were added [136–138], the amorphous regions were enlarged [139] and some lignin and hemicellulose patches are still present [140]. Lignin fragments can be present in water as well; an unintended result of such changes can include strongly anionic "pitch" particles [141].

At the wet end, pulp is dispersed in water and the interaction between fibers and other chemicals must take place within a temperature range of about 25–55°C, and very low concentrations. Preselected fibers are used for better paper strength [142]. The dispersion of cellulose fibers is diluted in order to prevent flocculation (interactions between fibers occur at concentrations of over 0.05% only [143]).

Now an environment has been created for paper chemicals: acids or urea sulfate (or urea hydro-chloride) [144–148] are used to adjust the pH and inorganic compounds or cellulose powder [149] are added as fillers.

At the wet end, both process and performance chemicals are added. Process chemicals include pitch control chemicals (polysiloxane to reduce the wet press picking [150]), foam control chemicals, retention aid, etc. Performance chemicals enhance paper characteristics (strength, softness, hydro-phobicity, etc.) [151]. The interactions between cellulose fibers and paper chemicals or between different paper chemicals are critical at this point [152]. Those interactions can be manipulated by using intermediate chemicals, such as pre-adsorbed carboxymethyl cellulose to enhance the adsorp-tion of other paper chemicals [153].

If the pulping process simplified pulp composition, starting at the wet and finishing at the dry end, paper chemicals are used in order to make cellulose fibers more functional and more complex. These efforts are not targeted at recomposing the wood structure, but at making a novel product: paper.

The paper sheet (wet web) is formed on a conventional Fourdrinier wire, after the well-dispersed fiber is deposited upon the fine mesh, woven forming wire. Water passes readily through the wire because there is relatively little resistance to the flow [154]. Material that was not caught on the wire forms the "white water." The wet web contains random cellulose fibers. The uniformity of the web structure is identified through the light transmission method [155].

Capillary forces (surface tension) cause the fibers to get closer to each other and wet web strength increases during drying [15,156]. Surface tension can be manipulated by adding a surfactant [157]. At that point, the fibers are swollen in water and free water is still present. Paper chemicals are adsorbed on the fiber surface. Along with the solubilized cellulosic chains and hemicellulose mol-ecules form a gel-like layer [158]. In the drying section, gel will collapse and the final sheet proper-ties are set based on hydrogen bonding and polymer compatibilities.

The material in white water can be roughly divided into fibers, fines, hemicellulose, lignin deriv-atives [118], and un-retained paper chemicals (paper chemicals are rarely 100% retained: the reten-tion of rosin [159], or AKD [160,161] is <70%). Soluble species and colloidal materials are not able to diffuse through the dialysis membrane [162]. Anionic compounds, such as lignin derivatives (se above), fatty acids, and hemicellulose fragments are constituents of the "detrimental substances" [163]. Some are nonionic and are also able to interfere with paper chemicals. In other words, the so-called anionic trash has a wider meaning.

The reuse of water multiple times within a papermaking operation is called "system closure". The closing of the white water system has a positive effect on the production costs but a strong nega-tive effect on retention, sizing, paper properties, and corrosion [164]. Unless white water is purified by treating with a cationic chemical ("anionic trash catchers") [165–167], the un-retained chemicals are exposed to water and high shear rate for a longer time, which may result in hydrolysis (see the by-product resulting from AKD, ASA, etc., Chapter 7).

2.2.5 Paper Drying and Finishing (Dry End)

The removal of water from the fiber suspension is the main purpose of the papermaking process. Until the higher consistence technology (up to 15% fibers in water [168]) is implemented, the current papermaking implies very low consistence: <1%. The cost of water removal increases exponentially as the web moves down to the dryer: more than 75% of the cost of water removal is in the drying sec-tion [169]. Moreover, the paper web reaches higher temperatures only in the drying section where several chemical reactions may take place.

During the drying step, the loss of water (<1% of water is left [134]) and the development of interfiber bonding [170,171] are accompanied by fiber shrinkage: small in the longitudinal direction and large in the transverse direction. The average pore size and the specific surface are reduced with about 40% after drying [172], the amount of water swollen in fibers is lower [173,174], and, therefore, hydroxyl groups are less accessible [45]. Fiber bonding and web shrinkage are quantitatively related [175]. After the drying step (when the pore size and distribution are dramatically changed [176–178]), the surface is supposed to get smaller and the adsorbed chemicals "relocate" accordingly. On the dried fiber, desorption is unlikely and the interactions of paper chemicals with the cellulose support is critical for their effectiveness in the final sheet.

Broadly speaking, when cellulose and another compound (additive) are blended together, it is possible to produce two types of products: one in which the continuous phase is cellulosic, and another in which the continuous phase is the new additive [179]. From a papermaker's standpoint, in order to classify a product as "paper," the cellulosic part should be the continuous phase.

Paper sheet displays a large surface/volume ratio. The paper surface has special functionalities and for that reason, sometimes, paper needs additional surface treatment: sizing and coating.

2.3 PAPER STRUCTURE AND COMPOSITION

Paper structure depends on the pulping, web formation, and drying processes. Thus, wood cells (pore size and pore size distribution [176] and lignocellulosic gel [133]) undergo major changes during pulping and drying [84,133]. After the fiber selection [142], the speed and shake of the wire, the pressure applied to the calendars, the drying temperature, and the rate of drying have important effects on the paper structure [180]. The fibers tend to collapse and become ribbon shaped on drying of the sheet, thus largely eliminating the lumen voids. On pulping, the average un-collapsed fiber width is about 30μ and the average collapsed fiber thickness is about $2.5\mu m$.

Paper incorporates cellulose fibers with a broad size distribution and those fibers are randomly distributed in a three-dimensional (3D) network (each fiber crosses about 20–40 other fibers [181]). That geometric arrangement of cellulose fibers shows interfiber spaces or pores [182], which is why paper is considered as a continuous solid phase containing air voids [183]: the specific area is of 0.5–$10\,m^2/g$ and the voids represent 25%–70% of the paper volume [135,184,185]. The wood pulp contains about 65% crystalline material [134] with a density of about $1.6\,g/cm^3$. Due to those voids and despite the high density of crystalline cellulose, the apparent density of paper is $<0.8\,g/cm^3$ [186,187].

Fibers lie in virtually parallel lamina in the direction of the thickness of the paper, but are almost randomly oriented within the plane of the sheet. The voids within the paper structure consist primarily of volume generated by the crossing of three or four fibers to form triangular or rectangular voids. The mean radius of those pores is about $30–50\mu m$ [188]. The breaking stress for the machine direction is higher than that for the cross-machine direction [16] and therefore the paper sheet is anisotropic.

The internal surface of the cellulosic material is about $2\times10^3\,cm^2/g$. Paper structure is sensitive to water [126], for instance, the internal surface of sulfite pulp increases by 1000 times ($2.7\times10^6\,cm^2/g$) when the cell walls are swollen in water [189] and the specific surface drops by 2%–11% when paper is subjected to the wetting–drying cycle [185]. This is important because the adsorption of paper chemicals at the wet end is performed on the swollen cellulosic fibers.

When the paper properties are evaluated, its geometric structure should be associated with the chemical structure of cellulose fibers. Cellulose fibers contain linear chains of polysaccharide [43] with a large number of inter- and intramolecular hydrogen bonds. Hemicellulose comprises about 50% of the cell wall in wood [190] but much of it becomes solubilized during pulping and bleaching.

Paracrystalline regions within native cellulose are partially transformed into amorphous regions during pulping (organic amines help this process [49,50]). Kraft pulp has a greater proportion of amorphous cellulose than sulfite pulp [139].

The cellulose fiber strength is governed by the following important factors [187]: the cellulose degree of polymerization [191], the size of crystalline and amorphous regions, the orientation of macromolecules, and the presence of "foreign" substance (water, hemicellulose [192]), the intramolecular and intermolecular hydrogen bonds, and the interchain van der Waals bonds.

Several other factors would need to be added to that list: the geometric dimensions of the fiber (the average length and width, the average fiber diameter in cross section), the average hydrogen bonds, the degree of fiber "softening" (water retention due to the chemical modification of lignin and hemicellulose left on fibers [193]), and the relative bonded area [135]. The presence of inorganic fillers should also be considered [194].

Paper strength is a result of cellulose fiber strength [195], fiber orientation, fiber wall structure and interfiber interaction [196]. Fiber–fiber bonds are the weak link in paper: the calculated sheet strength based on fiber strength exceeds by far the value measured experimentally on the sheet [197].

Despite the fact that any paper sheet is a 3D network [198], it is often viewed as an essentially two-dimensional structure: its surface is much larger than its thickness. For a basis weight of $60\,g/m^2$, the fibrous material (wood fibers of 1–4 mm length and 5–10 microns thick) consists of about 10 layers of fibers [199]. Paper tensile strength is expressed as the force required to rupture a strip of unit width: kg/cm. The breaking length is merely the tensile strength per unit width, divided by the sheet basis weight.

The overwhelming number of parameters involved, make it next to impossible to find a phenomenological relationship between paper structure and its properties. Therefore, efforts were made to define empirical equations tested for fitting the experimental data.

The relationship between paper structure and its mechanical properties can be studied probabilistically [180,182,200]. The simplest model that can be used is a random network of infinite straight lines lying in a single plane. If the mean fiber length is ranging from 0.1 to 4.0 mm and the mean fiber width is ranging from 0.01 to 0.04 mm, then the estimated average number of fibers per gram of paper is about 10 millions. These data are processed in terms of the average number of fibers within a chosen area, the average number of crossings per fiber, and the size of the bonded area of the fiber-to-fiber contacts. The number of crossings per fiber ranges from about 20 to about 60 depending on the pulp type and is higher for the sheet obtained at higher pressure [201]. The chemical structure of the fiber is not considered in this model.

The strength of paper (T) derives from both the tensile strength of individual fibers (F) and the strength of the forces that hold them to other fibers (B) [200]:

$$\frac{1}{T} = \frac{1}{F} + \frac{1}{B} \qquad (2.1)$$

The fiber strength is even more important (of the order higher than 2.5) for the paper tear strength [202]. The extended form (Page) of the Equation 2.1 is [200,203]

$$\frac{1}{T} = \frac{9}{8Z} + \frac{12A\rho g}{bPL(\text{RBA})} \qquad (2.2)$$

where
  T is the paper tensile strength (N m/g)
  Z is the zero span tensile strength (N m/g)
  b is the share strength of the fiber–fiber bond (N/m^2)
  P is the fiber perimeter (m) of the average fiber cross section
  L is the average fiber length
  A is the average fiber cross section
  ρ is the density of the fibrous material
  g is acceleration due to gravity
  RBA is the relative bonded area in the sheet (the fraction of the bonded external surface area of
    the fibers [204,205])

The RBA was defined as the ratio of the specific conductivity of the sheet to the conductivity of the "infinite beating time" [175] or as the ratio of the optical scattering coefficient of the sheet and the optical scattering coefficient of the pulp fibers in the un-bonded state [206] (<1% of the optical contact area is really bonded [197]).

In spite of the difficulty to measure each variable (in particular, the bond strength b, Equation 2.2 is validated by experimental data [207]. However, the probabilistic approach does not take into account the chemical structure of cellulose. On the other hand, based on the effect of acetylated pulp, the involvement of hydrogen bonds in paper strength seems to be important [187,208]. Page pointed out that the variation in acetylated cellulose density should also be taken into account [209].

Cellulose (II)10 hydroxyl groups (C2, C3, and C6) are differently engaged in inter- or intramolecular hydrogen bonding [210]. The parallel chain structure in crystalline cellulose contains two intrachain hydrogen bonds [211–213] (O3-H....O5 and O6.....H-O2 with bond lengths 2.7 and 2.8 Å) along both sides of each cellulose chain, and one interchain hydrogen bond with a length of 2.79 Å (one for each glucose unit). For each macromolecule of cellulose with a molecular weight of one million, over 5500 interchain hydrogen bonds are present, which explains the very tight crystalline structure. This structure is less sensitive to water.

(II)10

Cellulose fibers include both crystalline and amorphous regions. Amorphous regions show a less ordered structure (lower number of intermolecular hydrogen bonds), with lower densities than crystalline regions [214]. The amorphous structure is much more sensitive to water and, during the pulping process, the proportion of amorphous cellulose increases [139]. Some cellulose macromolecular chains belong to both crystalline and amorphous regions. Semi-ordered regions are also present at the periphery of the crystallites in the cellulose fibers [215].

Due to the hydrogen bond length (about 2.7 Å) and the bond angles (102°–127° [212], a very tight contact between two fibers is needed to develop interfiber hydrogen bonds. Paper with lower residual moisture, processed at higher drying temperatures and higher pressures, will result in a larger number of hydrogen bonds and higher dry strength [216].

The "molecular approach" is the theory that takes into account hydrogen bonds to explain paper strength [217–222]. In order to reach a quantitative equation, several concepts were developed about hydrogen bonding in paper [187]: the average load per apparent area of fiber-to-fiber contact is about 5.5×10^4 g/cm^2, the effective interfiber bonds are <10% from the total hydrogen bonds [223], the number of effective hydrogen bonds depends (among other parameters of the papermaking process) on the apparent paper density [186] and on residual moisture.

For a total of randomly oriented hydrogen bonds (N_T), the effective number of bonds (N), which may be considered to resist uniaxial strains in the body, is given by $N = N_T/3$. The Young's modulus (E, measured in dyn/cm^2) was shown to be related to the number of effective hydrogen bonds per unit volume (cm^3) by the following equation [17,224,225]:

$$E = kN^{1/3}$$

(2.3)

where k is a constant (7.8×10^3 dyn/cm) for bonds occurring in cellulosic-based materials. That constant incorporates the characteristics of hydrogen bonds: a dissociation energy of 4.5 kcal/mol.

The molecular theory better fits the discussion of the moisture effect [226,227] and therefore the wet-strength mechanism. However, the two theories deal with the same process, and, eventually, a link between them has been identified [226–228].

The interfiber bonds are very strong when paper is dried and weak after soaking in water. That aggregate of cellulose fibers includes covalent bonds (50–100 kcal/mol), ionic bonds (10–30 kcal/mol), and hydrogen bonds (4–6 kcal/mol) [229]. The attractive forces between polar molecules are developed only at very short distance [230] and they decrease with the inverse sixth power of the distance. That is why the pressing force, the moisture, and the temperature are important in developing paper strength by bringing the fibers into intimate molecular contact.

Both the geometric aspect and the chemical structure of paper must be taken into account for the paper chemical adsorption and, eventually, the reactions of paper chemicals. The adsorption process may be altered by residual lignin or hemicellulose. The geometric structure has an important contribution through the surface area available for adsorption and the surface tension in capillary tubes and pores. The chemical structure of cellulose is involved in the adsorption of paper chemicals through hydrogen and ionic bonds.

The addition of chemicals to paper is supposed to change fiber surface (in terms of surface tension), the cellulose swollen ability, the interfiber bonds, the pore size, and the pore size distribution.

RBA is about 4%–34% of the total fiber surface [204]. Paper chemicals designed as dry-strength, wet-strength, or temporary wet-strength resin will be retained on the entire fiber surface (at the wet end), but their effect is limited to just the bonded area [231].

For a fiber of 0.02 mm in diameter and 2 mm in length, the surface area is 1.2×10^{-3} cm^2. For 40 crossings with other identical fibers, only <15% of the fiber surface area will be in contact with other fibers. If the amorphous region is <10% of the cellulose and we assume that the amorphous region is evenly distributed, it is highly unlikely that each fiber's amorphous region will cross the amorphous regions of another fiber (lower than 2%).

The incorporation of fines from 0% to 30% in paper can double the internal bond strength [232]. The apparent paper density is also increased. Moreover, fines possess an enormous surface area and are capable of adsorbing a major proportion of wet end additives.

Virtually all papers (except the absorbent types) must have a finely ground filler added to them, the purpose of which is to occupy the space between fibers. Thus, the paper surface becomes smoother, and its printability and opacity are improved.

Are paper chemicals able to change the fundamentals of paper structure? That is a rather challenging question. For instance, are paper chemicals able to extend the bonded area, to reduce the pore size and narrow their distribution, to change the surface tension, and, therefore, the water absorbency, etc.? The goal of the present review is to give an answer to these questions.

2.4 THE CHEMISTRY OF POLY-CARBOHYDRATES

Over 95% of the final paper weight consists of poly-carbohydrates: cellulose and starch. Cellulose fibers are the largest raw materials for paper manufacturing and the starch is the largest paper chemical.

Starch and cellulose have a very similar chemical structure and there is a good compatibility between them. There are many chemical reactions performed on polysaccharides and most of them are used to improve paper processing and quality. The unintentional chemical modification during the pulping process is followed by "intentional chemical modification" of the cellulose fibers [93] or starch.

The chemistry of starch and cellulose (or chitosan [233]) allows us to better understand the papermaking process: what kinds of chemicals are compatible with this particular environment, what type of functionalities are needed for paper chemicals' adsorption and effectiveness, what reactions can take place, and what new chemicals compounds can be obtained by polysaccharide processing [234].

Starch, with its extended amorphous structure, can be seen as a carrier for paper chemicals and as "glue" between them and cellulose. Different grades of paper consume different amounts of starch, for instance newsprint requires an average starch content of 0.1%, while packaging paper has about 5% starch content [235].

Cellulose, chitosan, and starch can undergo chemical reactions, which get them ready for a better interaction with paper chemicals. The chemical structure of cellulose (and/or starch) can be altered to the extent of making it a "paper chemical": for instance carboxymethyl cellulose.

Most chemical reactions on cellulose or starch are performed in water. Because starch and cellulose [236] are less soluble in water, their chemical reactions are developed mainly at the interface, and the reaction product shows a heterogeneous composition: a large amount of cellulose (or starch) remains unchanged. However, the esters and ethers of cellulose can be obtained by performing the reaction in homogeneous solution [237].

Carbohydrates contain two different functionalities: acetal (glycosidic linkages) and hydroxyl (chitosan has an additional amino functionality in each repeated unit). There are chemical reactions that do not change significantly the polymer molecular weight and the chemical reaction able to reduce or increase the molecular weight of carbohydrates. The acetal functionality is incorporated in the main chain and any reaction of those bonds will result in a reduction in the polymer molecular weight. Hydroxyl groups are involved in both types of chemical reactions. Polysaccharides may also be involved in successive reactions, the sequence of which is also important [238,239].

2.4.1 CHEMICAL REACTIONS THAT KEEP THE MOLECULAR WEIGHT UNCHANGED

When pendant hydroxyl groups are involved in the chemical reaction with mono-functional compounds (or complex with iodine [240]), the polymer molecular weight remains unchanged.

The etherification of carbohydrates changes the hydroxyl group into ether bonds. Starch reacts with ethylene oxide [241] in alkaline pH to make ethylated starch. Hydrophilicity is modified and the new functional group attached through the ether bond may bring new properties to that poly-saccharide. Hydrophobic starch [242], cationic starches [243–248], polygalactomannans [249], amphoteric starch [250], or cationized pulp [251] is obtained through an etherification reaction (II)11:

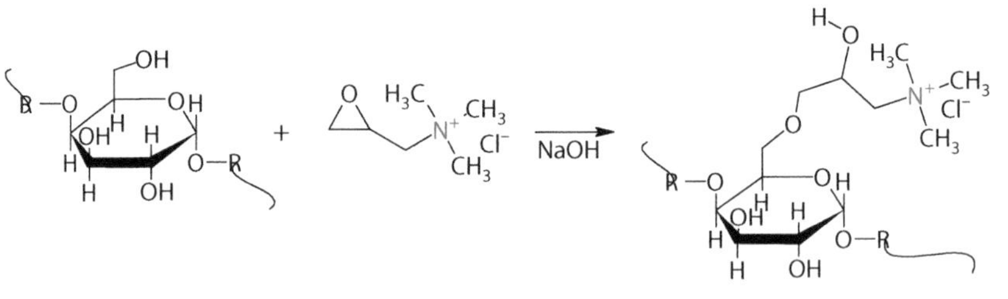

(II)11

Starch (as a slurry) reacts with diethyl-aminoethyl chloride hydrochloride (or 3-chloro-2-hydroxypropyl trimethyl ammonium chloride [252,253], or epichlorohydrin-ammonia condensate [254,255]), in the presence of sodium hydroxide [256] (or sodium methoxide [257]) to form cationic starch [258] (II)12:

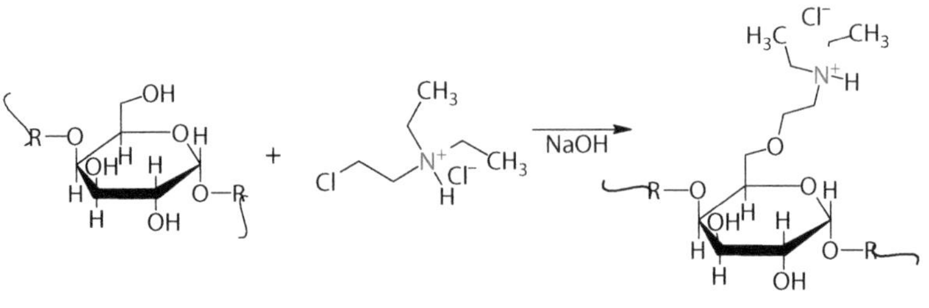

(II)12

N-(4-chloro, 2-butenyl) aziridine reacts with starch in the same way to obtain a cationic starch with reactive aziridine groups [259].

Due to the system heterogeneity (starch slurry), the degree of substitution is typically <0.05. To increase the degree of substitution over 0.3, the starch paste should be used [260]. The alkyl group connected to the nitrogen atom can be replaced by a polysiloxane moiety [261,262] (II)13:

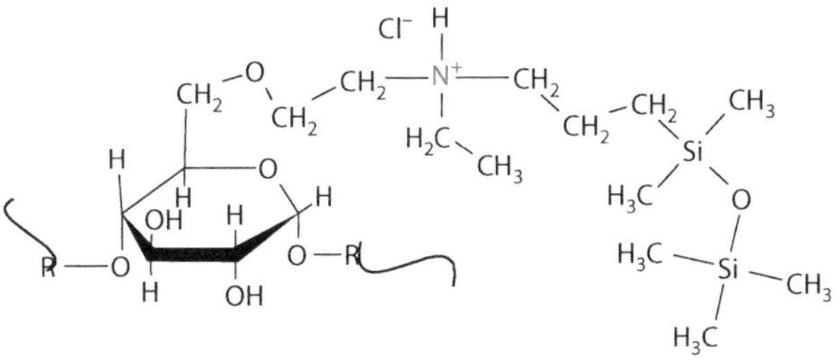

(II)13

Cationic starch ethers are obtained by a similar method (in the presence of sodium hydroxide) with 2-chloroethyl-methyl-ethyl sulfonium iodide [263] or 2-chloroethyl-tributil phosphonium chloride [264] (II)14:

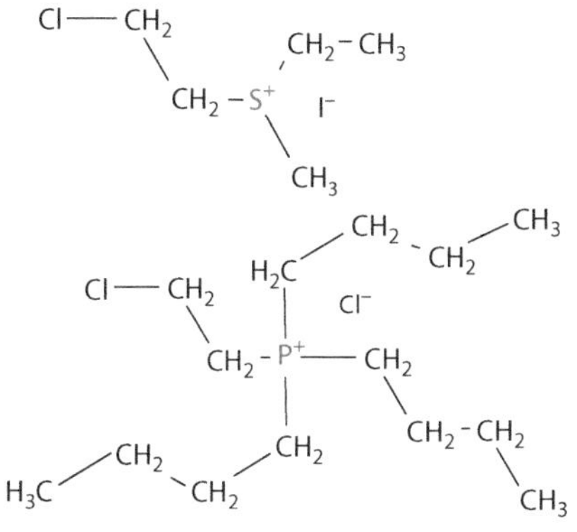

(II)14

The etherification reaction can also be performed by a dry state process (powder of starch and powder of sodium hydroxide) with aqueous solution of 3-chloro-2-hydroxypropyl-trimethyl ammonium chloride [265]. The etherification may also add anionic groups (carboxyl, sulfonic) and in a successive reaction a cationic group and the resulting starch is amphoteric [266,267].

Starch and modified polysaccharides (when soluble in water) make their hydroxyl functions available within an (almost) homogeneous system. Cellulose is not water soluble and it swells to a limited degree in water. Despite those limitations, chemical reactions are developed at the cellulose surface: epichlorohydrin [268], β-chloroethyl diethylamine [269], butadiene di-epoxide [270] or epichlorohydrin–dimethyl amine condensed resins react with the cellulose pulp, resulting in a cationic cellulose with a nitrogen content of about 0.05% [271]. This content is only apparently very low, because if we take into account the fact that all the cationic groups are on the surface, that means an important amount of new cationic charges were in fact attached to cellulose fibers.

Because the high concentration of caustic is a drawback of that process, cationic charges are added through an ionic interaction between the anionic charges of oxidized starch and the cationic charges of condensed resin (EPI-DMA-NH$_3$) [272].

Additional carboxylic groups were introduced in the cellulose structure through etherification with the halo-acetic acid [273] (or 2,3-epoxysuccinic acid [274]) to form carboxymethyl cellulose. For a degree of substitution of 0.6 or higher, carboxymethyl cellulose becomes water soluble [275].

In order to add different anionic groups, starch is modified with 3-chloro-2-sulfopropionic acid [276], or with sodium cyanide and hydrolysis [277]. When cyanide is C^{14} labeled, the resulting polysaccharide is labeled as well.

Hydroxyl groups from polysaccharides have different reactivities in the etherification reaction. In the case of a reaction with monochloro-acetic acid in the presence of sodium hydroxide [278], the reactivity of the hydroxyl groups is in the following order $C_2 > C_6 > C_3$ [279].

If the fiber cell wall is in a dry collapsed state, substitution occurs primarily at the fiber surface. If the fiber is in a more expanded state, then the substitution occurs uniformly across the cell wall [129]. More carboxylic groups are added to pulp with chloroacetic acid [129] (II)15:

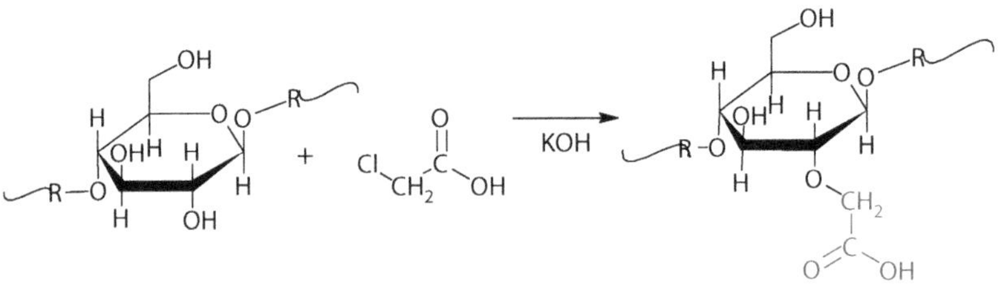

(II)15

The etherification reaction may involve a diversity of halogen derivatives. Blocked aldehyde groups can be attached to the carbohydrates macromolecules (reaction with N-(2,2-dimethoxyethyl)-N-methyl-2-chloracetamide) and the aldehyde functionality is released by hydrolysis [280] (II)16:

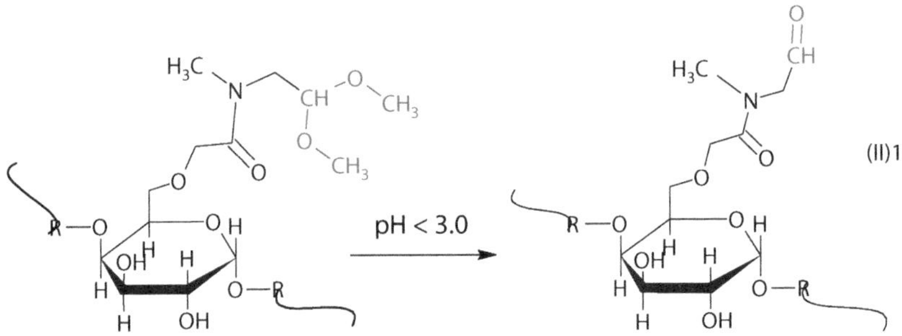

(II)16

Pectin is a highly branched polysaccharide that consists of D-galacturonic acid residues linked together through α-1,4-glycosidic bonds [281]. The methyl ester of the polygalacturonic acid reacts with amines and amino-acids (L-lysine) in the presence of an enzyme (protease) as a catalyst, at room temperature (II)17:

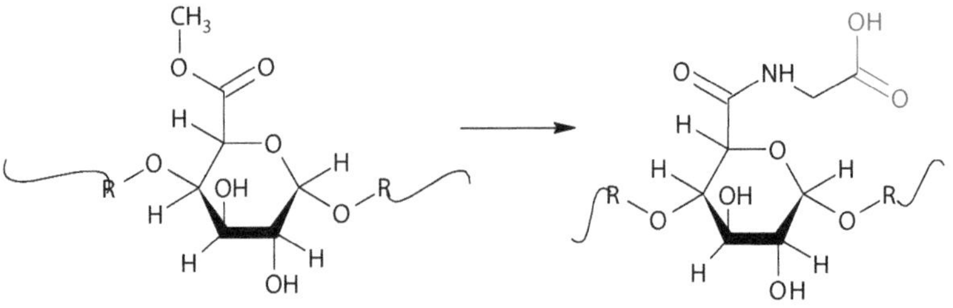

(II)17

Ethers are also obtained through Michael addition, which involves acrylic double bonds. Acrylamide (and/or N,N'-methylene bis-acrylamide, acrylonitrile [282,283]) reacts with starch [241,246], cellulose [284–286], or hydroxyethyl cellulose [70,287,288] in the presence of sodium hydroxide. If the reaction is performed at room temperature, the amide group can be preserved. At higher temperatures (70°C), the amide hydrolysis takes place and the carboxy-ethylated compounds are obtained (II)18.

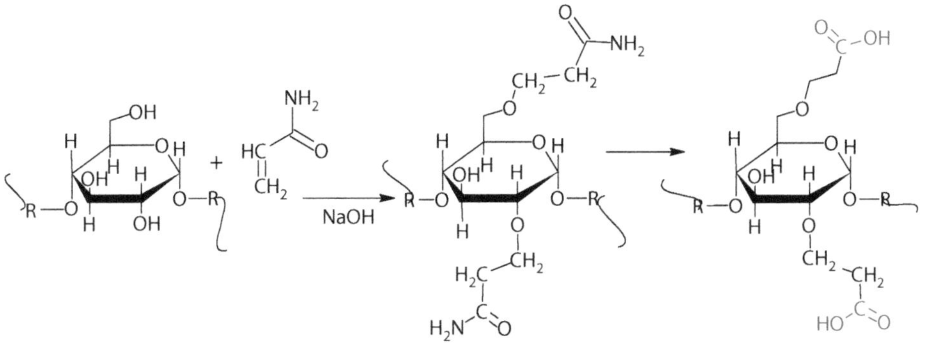

Starch with amide functionalities makes it possible to have a stronger hydrogen bonding from both hydroxyl and amide groups.

The esterification reaction, including nitration, involves only the hydroxyl groups of the poly-carbohydrates and no change in the cellulose molecular weight is recorded [289]. Cellulose esterification with acid chlorides is performed in dimethylformamide [290]. Starch or pulp esterification using acetic [253,291,292], succinic [293], alkenyl succinic [294–296], maleic [297], or phthalic anhydride [298,299] (II)19 are performed in the presence of pyridine or sodium hydroxide:

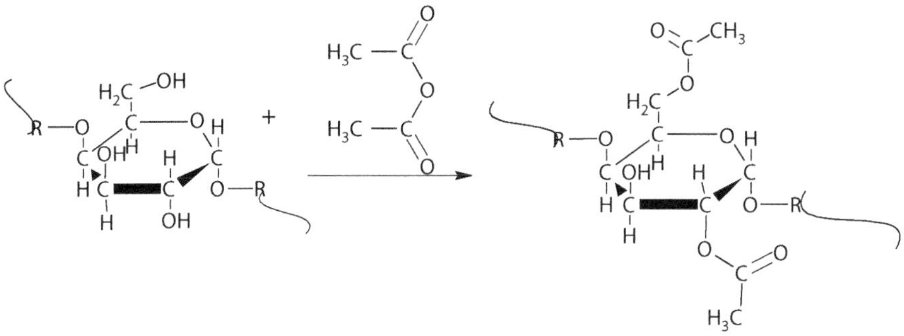

When cyclic anhydride is used, the resulting polymer contains more carboxylic groups. The presence of water reduces the reaction yield down to 10% (for isatoic anhydride) [300].

The esterification reaction with a carboxylic acid should be performed at high temperatures (120°C–180°C) in the presence of a catalyst (sodium hypophosphite) [301]. Phosphoric salts (such as monosodium phosphate or sodium methaphosphate) react with starch (30 min at 160°C) to make starch esters with anionic charges [302–304]. Esterification with sulfamic acid (II)20 takes place in the presence of urea [305]:

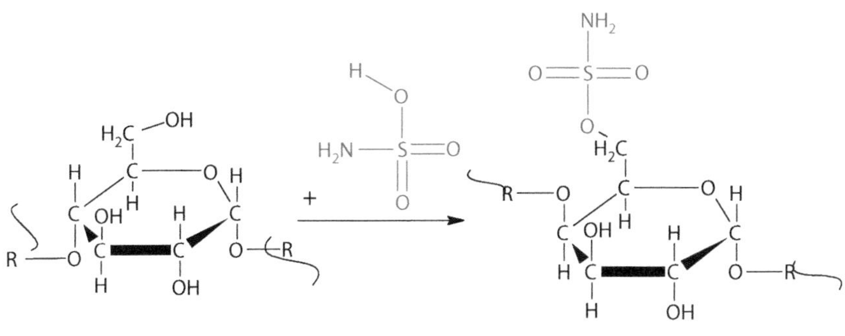

The amino group (from chitosan) reacts first with an anhydride (phtalic anhydride) to form an amide (II)21 and the hydroxyl groups remain unchanged [233]:

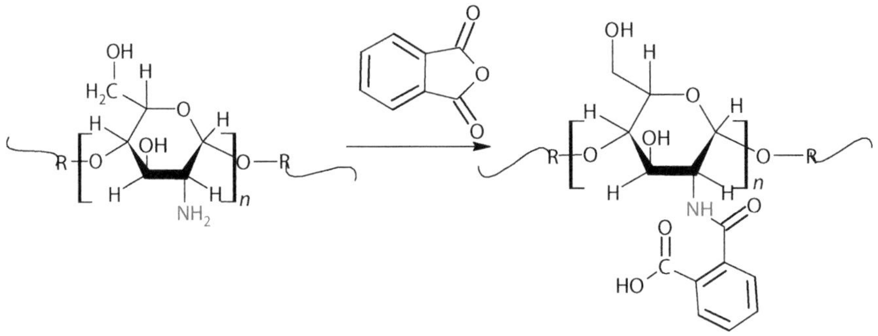

Cellulose esterification is a different case because it is not soluble in water. Due to the reaction at interface (with acetic anhydride in the presence $ZnCl_2$ or pyridine [282] at 100°C–120°C [306]), partial esterification is expected. Maleic anhydride (about 1%) is reacted with pulp at 65°C–115°C [307] and moisture content <5%. Modified pulp shows higher surface anionic charges. The water content in the nitric acid has a strong effect on the degree of substitution of nitrated cellulose [308].

Esterification with organic or inorganic acids is retarded by the water formation, and, therefore, it must be removed from the system to force the reaction to completion [110]. The transesterification reaction with a methyl ester is an easier reaction [309] because methanol has a lower boiling point.

Cellulose esterification with unsaturated fatty acids (such as sorbic acid) is performed in the presence of trifluoroacetic anhydride [310] and that with butyric anhydride is performed in the presence of butyric acid and sulfuric acid [311]. As expected, the molecular weight of the cellulose ester is smaller due to the acidic hydrolysis of the cellulose backbone (see Section 2.4.2).

Despite the similar reactivity and identical organic compounds used for esterification (acids and anhydrides [70]), in the absence of a solvent for cellulose, esterification takes place on the surface only. The final product shows a very broad compositional distribution.

Cellulose esterification performed in solution is more effective. Cellulose acetate is synthesized [312] in solvent/non-solvent mixture (methylene chloride–methanol–xylene), or lithium chloride-N,N-dimethylacetamide blend is used as a solvent for cellulose esterification with a fatty acid to a high degree of substitution [53].

2.4.2 CHEMICAL REACTIONS THAT MAY ALTER THE MOLECULAR WEIGHT

Chemical reactions that may change the molecular weight of the macromolecular compound are of two types: one involving the acetal bonds from the backbone (i.e., the hydrolysis of α-D-glycosidic bonds results in lower molecular weight: "thinned starch") and the other involving pendant hydroxyl groups in cross-linking or grafting reactions. Grafted or cross-linked polysaccharides show a higher molecular weight.

The hydrolysis of glycosidic linkage is easily performed with acids [233,313–316] or enzymes as catalysts [317–319]. Dextrin is a starch decomposition product and consists of glucose chains, is formed by incomplete hydrolysis of starch with diluted acids, or by the action of heat [320].

Polysaccharides (such as cellulose and hemicellulose) are degraded in alkaline pH by an endwise mechanism, known as the "peeling reaction" [110]. The peeling process results in a loss of about 50 monosaccharide units from a single molecule, a reaction terminated by converting glucose groups into carboxylic acid end groups that render cellulose stable in alkaline pH [321,322]. The presence of oxygen or hydrogen peroxide improves the stability of polysaccharide by helping the carboxyl group formation [90].

Starch hydrolysis [323,324] can also be performed in the presence of 1% concentration of enzyme solution (α-amylase A). Enzymatic starch degradation is controlled by adding 7.5 g of glacial acetic acid. Actually, the glycosidic ether has split to two hydroxyls. As a consequence, the molecular weight decreases.

The hydroxyl groups of poly-carbohydrates can be oxydized to aldehyde and eventually to carboxyl groups. Oxidized polysaccharides show a different moisture absorption [325], different interactions with water and with neutral or ionic polymers. Oxidation is also associated with a reduction in the molecular weight. Reaction selectivity is an important problem posed by this process.

Starch oxidation is performed at room temperature with sodium hypochlorite (pH 8.5–9.5) [326] or with hypochlorous acid (pH = 1–5) [327]. The oxidation reaction with sodium periodate is run at 0°C [328]. In order to balance the molecular weight reduction, starch can be cross-linked (with epichlorohydrin [329]) before oxidation with sodium hypochlorite. Although starch particles are unevenly oxidized, oxidized starch shows a better dispersability in water [330].

Pulp oxidation (the addition of aldehyde groups) is performed in order to improve paper properties [331–334]. During the oxidation of cellulose (chromic acid an oxalic acid), the fiber morphology and the accessibility of the cellulose chemical bonds not only affect the extent of oxidation but also influence the aldehyde/carboxyl group ratio [335]. The amorphous zone becomes more accessible and the oxidation reaction enlarges the noncrystalline part of the fiber.

More carboxylic groups obtained through pulp oxidation increase fiber anionicity [336–338]. The oxidation method (with TEMPO as a primary oxidant) is targeting the C_6 carbon [339,340], which may result in a little change in the molecular weight of cellulose. In a more flexible process, oxidation is performed in two steps: TEMPO and NaOCl in the first, and sodium periodate in the second [341].

The oxidation at C_6 carbon [342–345] (with chlorine dioxide or hypochlorite [346] and TEMPO as catalyst) results in aldehyde as an intermediate and eventually carboxyl functionality (II)22 (of the uronic acid type [327]). The reaction selectivity can be improved by lowering the temperature to 5°C [347].

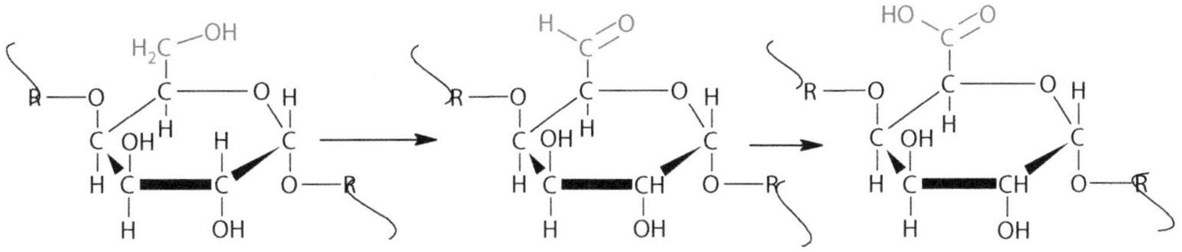

Starch oxidation with periodic acid results in di-aldehyde starch [240,271,348–351]. Aldehyde groups have been introduced into starch without a significant lowering of the molecular weight. Cellulose oxidation with sodium periodate [331,352] results in a di-aldehyde compound (II)23 as well.

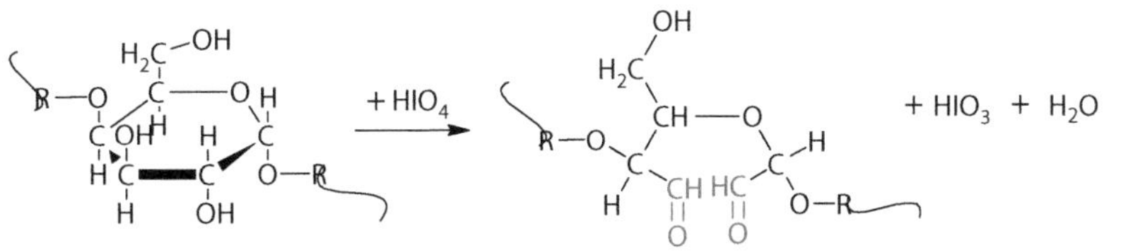

Di-aldehyde starches present a major disadvantage: they undergo rapid depolymerization in alkaline solution, yielding acidic products. Actually, depolymerization starts with the opening of the glucopyranose cycle during oxidation. In this case, the carboxylic group is located at the end of the macromolecule. In order to avoid depolymerization, the aldehyde groups are attached *via* an etherification reaction [353] (II)24:

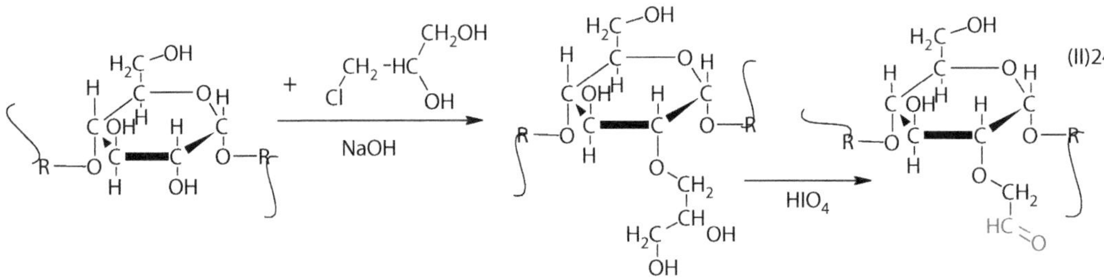

It is difficult to prevent the oxidation of aldehyde groups because they are easily oxidized to a carboxylic acid (dicarboxy cellulose or tricarboxy cellulose [352]) with bromine or nitrogen tetroxide.

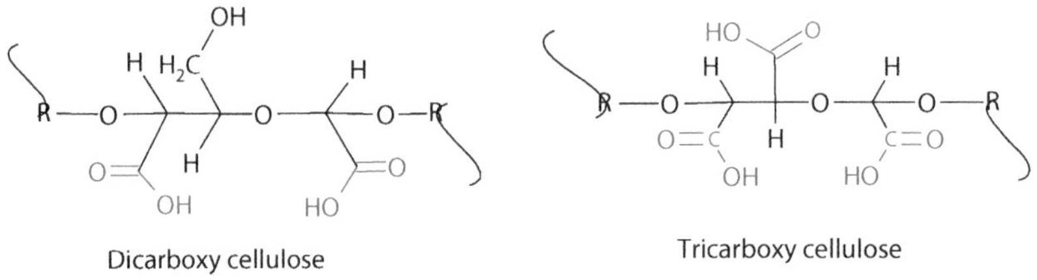

Dicarboxy cellulose   Tricarboxy cellulose

Oxidized polysaccharides free of organic halogen compounds are obtained by oxidation with hydrogen peroxide in the presence of copper sulfate in alkaline pH [241,354].

The anionic carboxylic groups of the cellulose pulp can be converted to nonionic functionalities by reaction with water-soluble carbodiimide at pH=4.75 [355,356] (II)25. That reaction was performed in order to obtain neutral pulp.

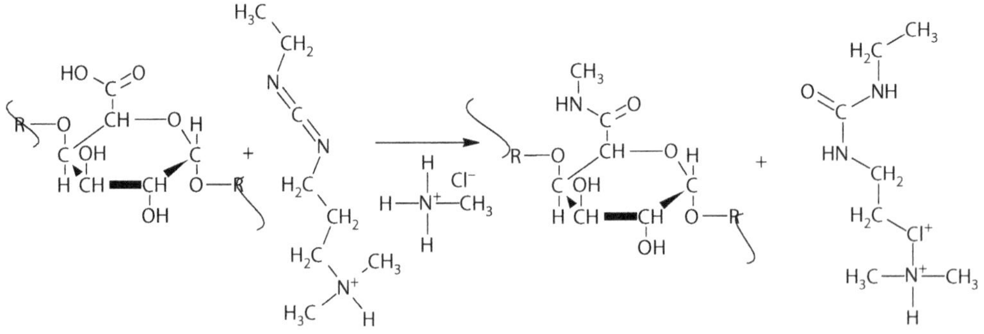

The increase in the molecular weight of starch can be achieved by reacting regular starch with difunctional compounds, or by grafting. The reaction presented in the previous chapter (Section 2.4.1. etherification, esterification, etc.) involves polysaccharides (as a macromolecular compound) and a small molecule. When the small molecule is a multifunctional compound the polysaccharide is cross-linked.

The cross-linking process with difunctional compounds results in larger molecules with basically the same chemical structure (polysaccharide). Examples of multifunctional reactive compounds able to cross-link starch (or cellulose) are polyepoxides [357,358] (such as 1,4 butandiol diglycidyl ether), polyfunctional silanes [359], polyanhydride of polyacids, polyacids (adipic acid [354], citric

acid [360,361], vinyl acetate-maleic anhydride copolymer [362], acrylic acid-maleic acid copolymers [363,364]), formaldehyde [282,365–367], copolymers of methylol acrylamide [368], polyaldehyde, epichlorohydrin [369–374], UFR, polyisocyanates, phosphorous oxychloride, glyoxalated
polyacrylamide [375–377]. Starch can be cross-linked after fictionalization [378,379] or the cross-
linking process can be performed simultaneously with the functionalization step. In that case, the
cationic charges are added by the cross-linking compound [380] (II)26. The balance between fictionalization and cross-linking is kept by adding cationizing monofunctional compounds (see the
etherification reaction).

When nonreactive cationic compounds (such as softeners) are associated with a cross-linker
[381,382], interfiber reactions are partially prevented: the cross-linking process is developed mainly
within the cellulose fiber.

The starch (or cellulose) grafting reaction results in a higher molecular weight product, but with
the polysaccharide structure in lower amount: the branches have a different chemical structure. It is
commonly understood that the grafted copolymer [383] retains the basic properties of the existing
polymer to which it adds the properties of the new polymer (such as polystyrene grafted on cellulose
by irradiation [384]).

The grafting reaction involves a polymer (in this case, starch or cellulose) and a monomer able to
polymerize [261,385] in a free radical mechanism or in anionic polymerization [386,387]. The final
product is supposed to have a higher molecular weight.

Due to its lower selectivity, the grafting reaction will result in a blend of un-reacted starch,
un-grafted linear polymer and grafted copolymer [388,389]. The polymerization of methyl methacrylate in the presence of cellulosic pulp shows a grafting effectiveness of maximum 68% [390].
The extraction of grafted starch with methyl acrylate (42% starch) indicated that about 34% of the
poly(methyl acrylate) was in the form of a free homopolymer [391].

The monomers used in a free radical grafting polymerization are nonionic (acrylamide
[246,392–394], acrylonitrile [395], vinyl acetate, styrene, ethyl acrylate, alkyl methacrylates [396–
398]), anionic (acrylic and methacrylic acid [399]) or cationic (DADMAC [388], cationic methacrylates [400,401], 2-hydroxy-3-methacryloyloxypropyl trimethylammonium chloride [402]) or
potential cationic after the hydrolysis of grafted copolymers of *N*-vinyl formamide [403].

The grafting process can be performed in homogeneous systems: water as a solvent [396]. Due
to the viscosity of the starch solution, the reaction system must be diluted to a concentration lower
than 15%. The diluted systems reduce the grafting yield.

Heterogeneous systems allow for a higher grafting yield. Styrene grafted polymerization on
pulp is improved by the presence of acrylonitrile [69]. Starch and monomers are dispersed [402] in
a water-immiscible solvent (xylene or C_8 isoparaffins [404]) and then the oil phase is dispersed in
water. In this case, the starch concentration can be much higher (50%) and the grafting efficiency
increases.

There are several approaches to improve the grafting efficiency. The simplest method to obtain
a grafted polymer is to put all three components: the monomers, the grafting support (a macromolecular compound as starch or cellulose), and the initiator, in water. Initiators, such as persulfate or hydrogen peroxide, are water-soluble compounds. Free radicals are generated in the water

phase and the grafting effectiveness depends on statistical parameters: total concentration and relative concentrations. Two processes are simultaneously occurring: the homopolymerization of the monomer and the grafting copolymerization. The grafted copolymer is formed by chain transfer to the substrate (II)27.

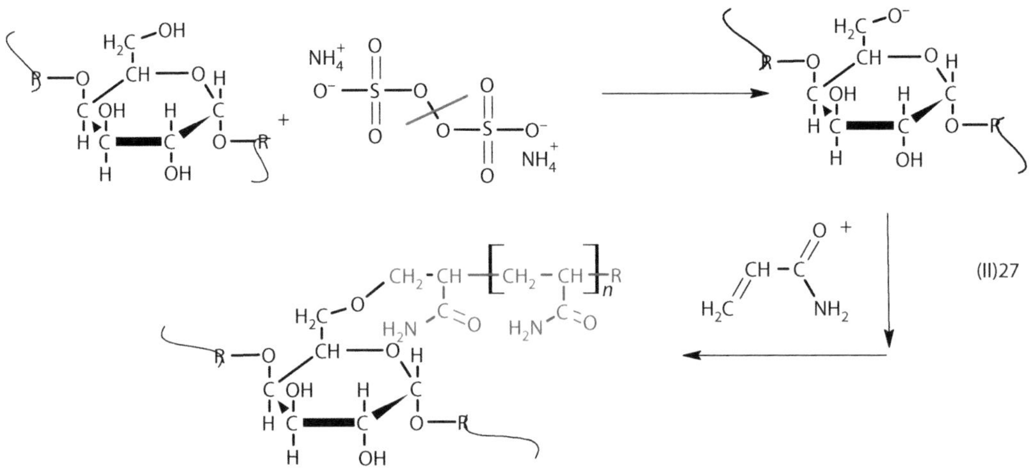

(II)27

Gamma rays are also used [386,400,405] to initiate the free radical reaction. There is a good chance to homolitically break the carbon–carbon or carbon–oxygen or oxygen–hydrogen bonds of the poly-carbohydrates. At a high level of irradiation (up to 4 Mrad), the cationic starch molecular weight is drastically reduced [406]. Microwave irradiation has a strong impact on the grafting process as well [407].

Under moderate irradiation performed in the presence of oxygen, a high concentration of hydroperoxide is obtained [408]. Hydroperoxide groups are located on the starch molecule and thus the substrate is involved in the initiation step. Free radicals are trapped into the polysaccharide structure and, in the presence of water, the concentration of free radicals decays. More hydrophobic substrates preserve free radicals for a longer time [409]. The involvement of the substrate in the initiation step reduces the homopolymerization and increases the grafting effectiveness [410].

Starch is also involved in the initiation step of the ring opening polymerization of caprolactone performed in the presence of Tin(II)—2-ethylhexanoate as catalyst; and a grafted starch—polycaprolactone is obtained [411] (II)28:

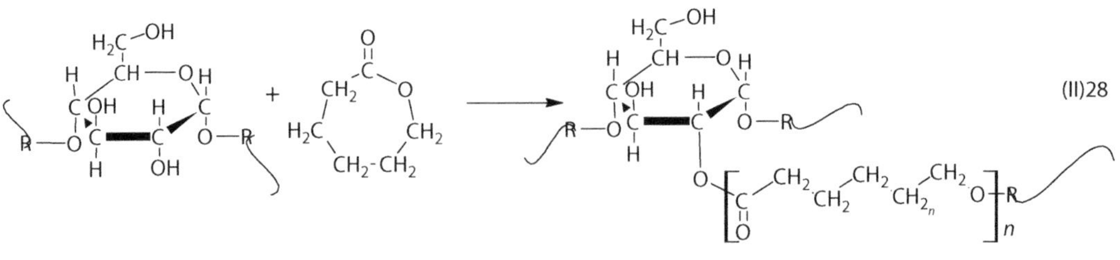

(II)28

However, in order to improve the grafting process, another approach has been taken: a condensation type reaction between starch and di-isocyanate terminated polycaprolactone [412].

In order to generate free radicals on the substrate (starch or cellulose [413,414]), researchers used the same concept of reacting the macromolecular compound and the catalyst first (ceric ammonium nitrate [415]) (II)29. Reactive species are formed on starch only [395,416–420] and over 90% of the polymer formed during the reaction was grafted [292] (II)30. The free, un-grafted, polymer is formed exclusively by chain transfer to the un-reacted monomer.

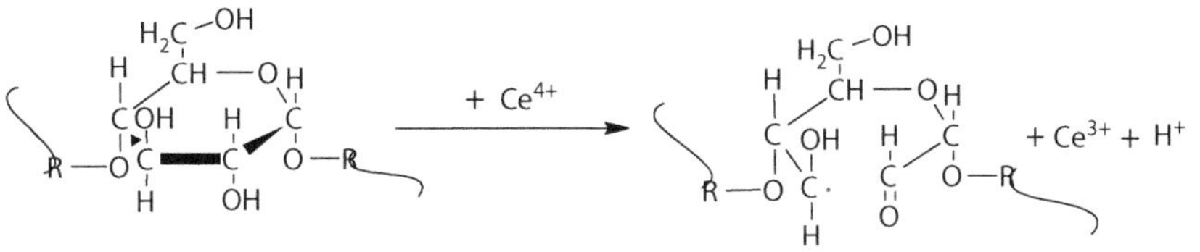

(II)29

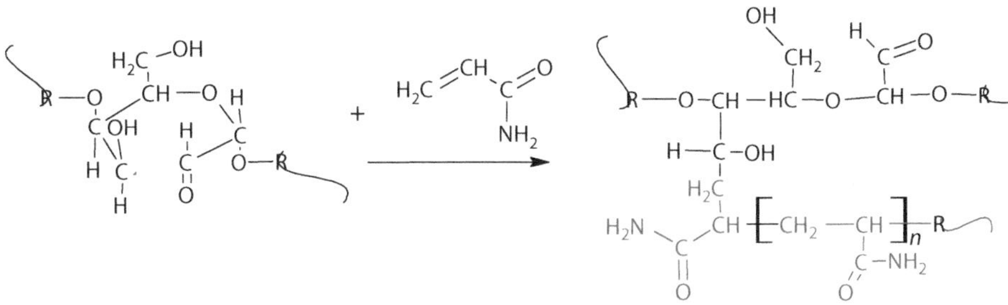

(II)30

In the previous examples, poly-carbohydrates were involved in the initiation step. Macromolecular compounds (such as starch and cellulose) can also be derivatized with AGE [421] or methylol acrylamide [422] (II)31 to become macro monomers.

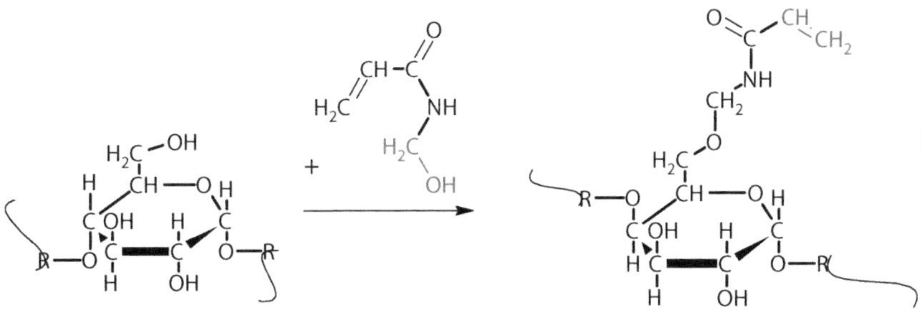

(II)31

Derivatized compounds are grafted through copolymerization or chain transfer [423]. A more effective transfer of free radicals to cellulose is obtained by derivatization with ethylene sulfide [386].

The chemical composition of the grafted copolymer can be different from that obtained in the absence of polysaccharides. The branched formation changes the solution composition and the local concentration of the comonomers. That will result in a different chemical composition of copolymers obtained at low and, respectively, high conversion [398].

Starch solutions have a tendency toward retrogradation during storage. That process develops within hours. Retrogradation means a return of well-dissolved starch molecules to a more orderly state. Retrogradation is mainly due to the presence of amylose molecules, which are linear and therefore easier to be organized based on hydrogen bonding. Higher starch concentrations can speed up the retrogradation process [323].

All parameters able to increase the agitation of macromolecules (higher temperature, higher shear rate), the presence of another compound that can be involved in hydrogen bonding, grafted branches, or pendant groups will slow retrogradation down.

2.5   SYNTHETIC POLYMERS: EVERYWHERE IN PAPERMAKING PROCESS

The synthesis of derivatives from a natural product (starch and cellulose) is limited in terms of functional groups and their distribution, as well as in terms of its capabilities to manage the molecular weight and the molecular weight distribution. Synthetic polymers are obtained in much more flexible processes in which their molecular weight and structure can be more easily managed. The molecular weights, the sequence of the monomer units, the linear structure, and/or the number of branches are designed during their synthesis [424].

Synthetic polymers used as paper chemicals must perform under particular conditions: various pH values, the presence of electrolytes and polyelectrolytes, and in a suspension with various surfaces (including cellulose fibers). There should be a right ratio between their hydrophobic and hydrophilic parts, between the ionic and the nonionic pendant groups, with well-defined molecular weights and chemical composition distributions.

Polymer synthesis directly on paper (polymerization [425] or polycondensation [426]) has not been used on a large scale. It is usually developed in a chemical plant and shipped to a paper mill as a powder, water solution, or dispersion. The synthesis of synthetic polymers and their properties may provide a better insight into the accessibility and usage of paper chemicals.

2.5.1   POLYMER SYNTHESIS

Although polyethylene is used in coating formulas, most of the macromolecular compounds used as paper chemicals have a backbone and many pendant groups. The chemical structure of both the backbone and its moieties are crucial for the synthesis protocol and for its final application as a paper chemical.

Molecular modeling can be used to find the right structure for the right application in papermaking process [427]. There are two general concepts used in synthesizing a macromolecular compound: starting from a small molecule and building up a macromolecular backbone (polymerization, copolymerization, or polycondensation) or using a polymer as starting material and synthesizing a new polymer by changing the chemical structure of the pendant groups (polymer analogous reactions).

Polymerization processes are developed through chain reaction (free radical, ionic, or coordinative polymerization) with three steps: initiation, propagation, and termination. Most of the monomer is consumed in the propagation step. No small molecule is released during the free radical propagation. In this case, functional groups preexist in the monomer structure.

Most polymers resulting from the polymerization process of a carbon–carbon double bond (like vinyl or acryl monomers) and the resulting polymers have only carbon atoms in their backbone. The carbon–carbon backbone is hydrophobic, and, therefore, not soluble in water (polyethylene, for instance). Hydrophobic polymers are delivered as dispersions (see "Emulsion polymerization"). In order to make it water soluble, polar–hydrophilic groups must be attached to the carbon–carbon double bond as in the case of polyacrylamide:

$$R \left[CH_2 - CH \right]_n R$$
$$\underset{H_2N}{\overset{C}{\diagdown}} \diagup O$$

The number of units (n) included in the backbone is usually larger than 100, but for special applications in the papermaking process (as flocculants or dry-strength resins), n may reach 100,000.

Water-soluble monomers (acrylamide, acrylic acid, NVP, etc.) are easily polymerized in water as solvent (solution polymerization). The viscosity of the polymer solution will be higher at higher concentrations and/or higher molecular weights.

In water solution or at the wet end, that type of polymer can undergo some chemical reactions on the pendant groups (like hydrolysis) but molecular weight will not change because carbon–carbon bonds cannot be broken under papermaking conditions.

In solution, polymers behave as individual macromolecules; their properties are related to the macromolecule properties: the molecular weight and its distribution.

Copolymerization is a polymerization process when at least two monomers are involved in building the macromolecular structure. For instance, the copolymerization between acrylamide and the acrylic acid (II)32, results in the formation of an anionic copolymer with two different functionalities. In most cases, the comonomer molar ratio is different from one ($n \neq m$).

$$R-CH_2-CH-CH_2-CH-R \qquad (II)32$$

Copolymers are characterized by the distribution of both their molecular weight and chemical composition [428]. In other words, they are a complex mixture of macromolecules having different molecular weights and chemical compositions.

The molar ratio of the comonomer (n/m) and its distribution both within a macromolecule and between different macromolecules strongly influence their water solubility, chemical reactivity, and physical properties. By copolymerization with hydrophilic comonomers (like the acrylic acid, sodium ethylenesulfonate [429] or the maleic acid), hydrophobic monomers (like ethylene or styrene) can be converted to water-soluble copolymers. Styrene-acrylic acid and styrene–maleic anhydride (SMA) copolymers are widely used as paper chemicals.

Some comonomers are effective even at low concentrations. Paper chemicals (such as retention aids) have a very low concentration of one comonomer (lower than 5% of cationic comonomer), but that amount is critical for the copolymer performances. It is not easy to synthesize such copolymers with low concentrations of comonomer, which should be evenly distributed to all macromolecules. It is also challenging to estimate the small concentration of the comonomer by elemental or spectral analysis [430].

Hydrophobic type (co)polymers are also of interest as paper chemicals: for sizing, as wet-strength resins or as binder in coatings. Hydrophobic monomers are not soluble in water, and, therefore, the "solution" polymerization procedure is no longer usable. The un-soluble comonomers are dispersed in water in the presence of anemulsifier (emulsions) and a water-soluble initiator is used to start their emulsion copolymerization [431,432]. The final product is a dispersion of copolymer in water (latex).

Polymer latexes bring some other properties along with the intrinsic properties of the polymer itself: particle size and distribution, specific area, functional groups on the particle surface, amount and nature of the stabilizer, and the free emulsifier in water.

The functional groups located on the particle surface are hydrophilic and can be anionic (COO^-), cationic (quaternary ammonium salt), or nonionic (poly-ether). In those cases, the stabilizer should be chosen accordingly (anionic, cationic, or nonionic).

The effectiveness of the dispersed (co)polymer depends less on molecular characteristics (molecular weight) and more on particle sizes (smaller particles result in larger number of particles tending to reach the molecular level for homogeneous systems), and the surface functional groups.

Polymer latexes are also made by emulsifying, in water, the preexisting polymer brought into a solution in an organic solvent (such as toluene), in the presence of emulsifier followed by the distillation of the solvent [433].

Most macromolecular compounds, resulting from **polycondensation**, have both carbon and het-ero-atoms (nitrogen, oxygen, etc.) in their backbone structure. The condensation between a diacid and a diamine involves the release of water within an equilibrium reaction (II)33:

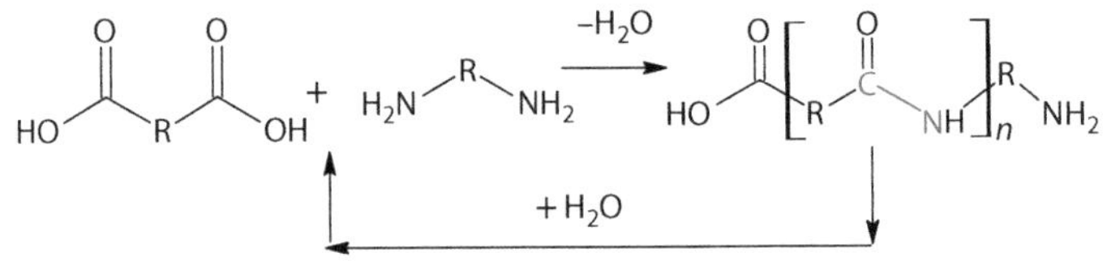

Therefore, the new amide bond (in red in the poly-amide structure) can be hydrolyzed in the presence of water and a catalyst, at an appropriate temperature (reversible reaction). Hydrolysis results in a lower molecular weight compound because the breakable bond is located in the back-bone. The reduction in the polymer molecular weight (and in its performances) can take place, at lower rates, even at room temperature and may affect the shelf life of the product.

Equilibrium reactions also involve bonds successively formed and broken during a condensation reaction (such as amide bond). Thus, the interchange reactions (polyamide-polyamide or polyester-polyester or even polyamide-polyester) [434] are easily performed in the presence of a catalyst. In other words, if two polyamides with different structures (DETA-adipic acid and DETA-glutaric acid) are blended at high temperatures and in the presence of an acid catalyst, a new polyamide is obtained: a ternary compound DETA-adipic-glutaric (co-polyamide). After a reasonable reaction time, the molecular weight of the new polyamide is the average value of the molecular weights of the two initial polyamides. The component distribution into the new inter-polymers is derived by their reactivity [435] and the reaction time: a longer reaction time will result in a more random distribu-tion of the units [436].

An interchange reaction can be performed between a high molecular weight polymer (obtained through polycondensation) and a reactive small molecule. Polyesters react with alcohols (alcoholy-sis) or with acids (acidolysis). Polyamides (or polyesters) react with amine (aminolysis). Interchange reactions can occur between two polyamides, between a polyamide and a polyester; and between two or more polyesters [437–439].

One amide may have a low molecular weight (such as urea). The interchange reaction (*trans-amidation*) [440] is performed with the polyamide prepared in the previous step. A mixed structure is obtained for the new polyamide (II)34.

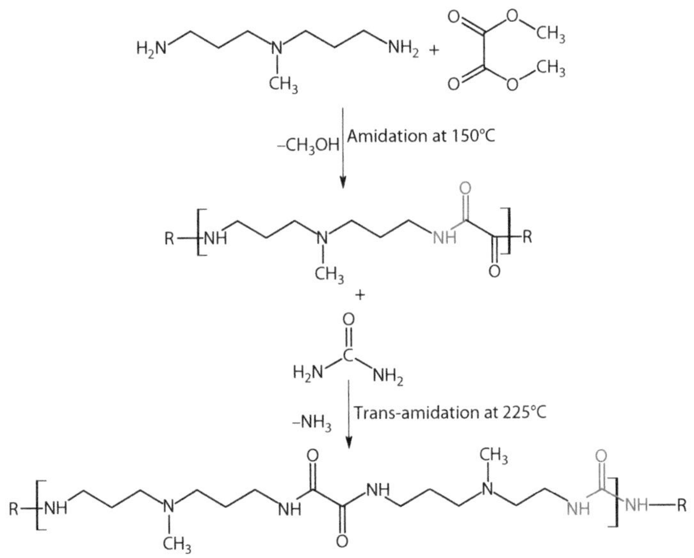

During the polymerization, copolymerization, and polycondensation reactions, a new backbone is built up starting from small molecules. The future pendant groups of the backbone are carried by those small molecules (monomers). New macromolecular compounds can also be obtained by adding new moieties to a macromolecular compound (polymer analogous reactions).

The organic chemistry of high polymers [441,442] deals with changes in the pendant group structure of a polymer, while keeping the molecular weight at the same value. For instance, the hydrolysis of polyacrylamide to polyacrylic acid [443–449] or the Hofmann degradation of polyacrylamide [444,450,451] to make a cationic polyacrylamide (II)35:

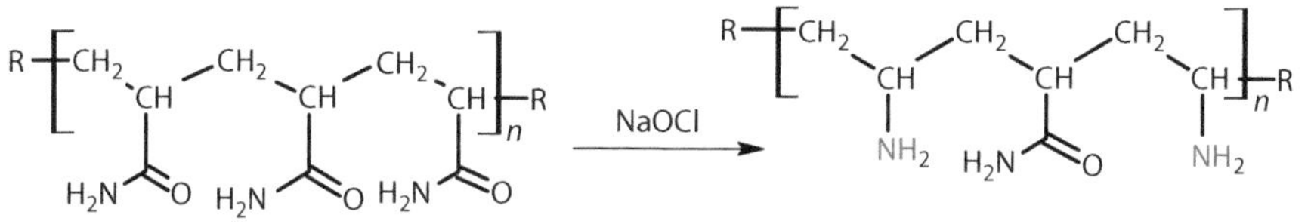

(II)35

At a lower conversion, a copolymer is synthesized (acrylamide-vinyl amine copolymer, which is also a "poly-amido-amine").

The addition of a new functionality through a reaction with formaldehyde [444] may change the reactivity of the new polymer: it becomes self-cross-linkable (II)36.

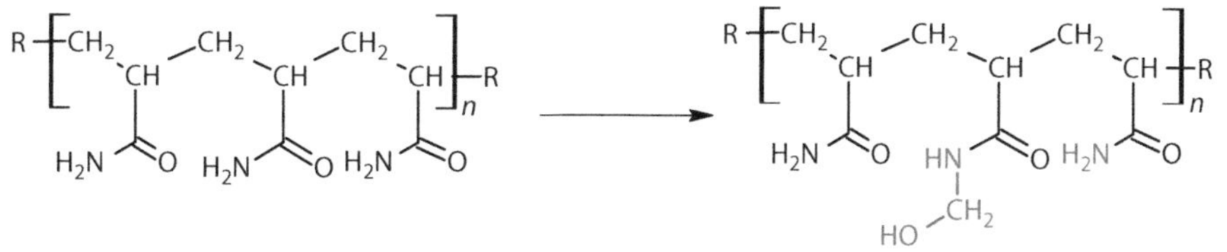

(II)36

Paper chemicals must be retained on cellulose fibers, which is why they need cationic charges. Most industrial polymers are nonionic, such as polyacrylamide, polyethylene oxide, polyvinyl alcohol (PVA), etc. To synthesize an anionic polyelectrolyte, polyvinyl alcohol reacts with acrylic acid (Michael addition) [452]. To obtain a cationic polymer, PVA reacts with 2,3-epoxypropyl trimethylammonium chloride, at 25°C–75°C, in the presence of sodium hydroxide [453,454] (see also cationic starch synthesis) (II)37:

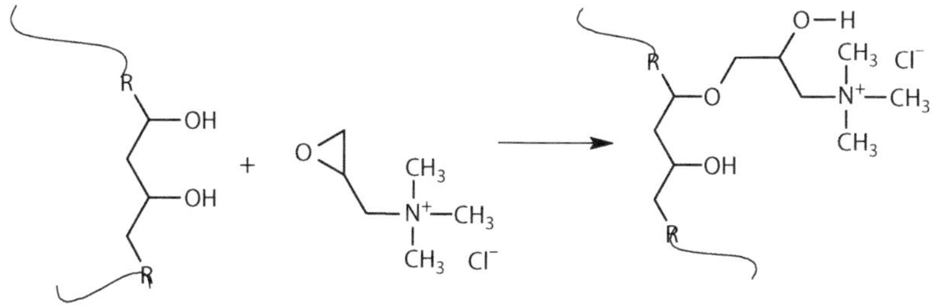

(II)37

Polymer-analogous reactions can change all functionalities; the resulting compound is a homopolymer. However, that reaction type passes—at lower conversion—through a "copolymer step" when the old functionality coexists with the new one (the hydrolysis of polyacrylamide to acrylamide–acrylic acid copolymer or partially nitrated cellulose [308]). It is difficult to convert all functionalities because of the macromolecular coil shape that shifts the accessibility of its functionalities during the reaction. For instance, the amide groups in polyacrylamide cannot be converted 100% to carboxylic groups due to electrostatic repulsions (hydrolysis is performed through OH⁻, which has no access to the macromolecule due to the presence of the carboxyl group—COO⁻).

As conversion progresses, the new copolymer shows a different distribution of its segments in solution. The solvent and the reaction temperature are also important for the final function group distribution Thus it is possible to have an uneven distribution of functionalities. The distribution of newly formed amine groups may not be the same during the Hofmann degradation of two poly-acrylamides with two different molecular weights (10^4 and 10^6) [451]. Lower molecular weight poly-acrylamide reacts much faster than higher molecular polymers [443]. However, Tanaka et al. [450] report that cationic charges created through a Hofmann reaction of a polyacrylamide, and a fluorescent labeling through a polymer analogous reaction, result—after multiple re-precipitations—in a homogeneous distribution of the functionalities [450].

Polymers with acid or base moieties are sensitive to the pH and/or salt concentration [455]. When the dissociation equilibrium takes place, both undissociated and dissociated forms coexist. Ethyl acrylate–acrylic acid dissolved in the presence of caustic [456] is actually a ternary copolymer: ethyl acrylate–acrylic acid–sodium acrylate.

If a type of chemical reaction is useful for the papermaking process, it can be partially performed before the paper chemical is added to the cellulose fibers. A preliminary reaction reduces the reaction conversion needed to be developed on paper. Thus, polymers with reactive functional groups may react with another (co)polymer with reactive groups in water solution. The resulting inter-polymer may combine the properties of the initial reactants [457,458] and the final properties are reached during the paper drying step.

2.5.2 Chemical and Physical Properties of Polymers

Why are polymers preferred as paper chemicals? What makes them so special? In an attempt to answer those questions, let us start from the fact that the cellulose is also a macromolecular compound. From that perspective, a chemical compound able to improve the papermaking process and the paper quality should be a compound with high molecular weight and multiple functional groups. The effectiveness of a paper chemical is related to its molecular weight, reactivity, and compatibility with cellulose fibers.

The improvement in paper strength by paper chemicals comes from their macromolecular size, and their functionalities involved in ionic, hydrogen, and covalent bonds. However, macromolecular compounds with higher molecular weight show higher viscosity for their water solution. For a reasonable viscosity, their solutions should be diluted to lower solids content. Regardless of the difficulties generated by handling solutions with high viscosity or low concentration, the higher molecular weight for polymers as paper chemicals is a desirable feature.

The molecular weight of the polymer is an average value and should be associated with the distribution of the molecular weights. Polymers with a broad molecular weight distribution are "blends" of macromolecules with very different molecular weights. Polymer adsorption on cellulose fibers is a function of the molecular weight and charge density of paper chemicals. At the wet end, macromolecules with different dimensions and different charge densities compete for the same spot on the surface of the cellulose fiber.

Paper chemicals are used in complicated blends both at the wet and dry ends of the papermaking process. Water solutions are often a mixture of two or more polymers. The behavior of the macromolecular coil in the presence of another polymer should always be a concern [459]: the polymer–polymer interactions [460] are noticed by changes in viscosity and precipitation.

Most paper chemicals are copolymers. Their properties depend largely on the comonomer ratio. Glass transition temperature for binary copolymers is calculated with the following Fox equation [461]:

$$\frac{1}{T_g} = \frac{W_1}{T_{g1}} + \frac{W_2}{T_{g2}} \tag{2.4}$$

Where T_{g1} and T_{g2} are the glass transition temperature for the corresponding homopolymers and W_i is the concentration of the i comonomer in the copolymer. That equation considers a copolymer as a blend of homopolymers, which is a questionable approach. For the SMA copolymers, a simpler equation was published for the T_g as a function of maleic anhydride content ([MA]) [462]:

$$T_g(^\circ C) = 100 + 3.367\,[MA] \tag{2.5}$$

It has been shown [463–465] that the T_g of a copolymer does not always follow the Fox equation. The T_g of a copolymer is better described by an equation that takes into account the sequence distribution of diads (AA, BB, AB, and BA linkages in the backbone). Possible chemical reactions developed during the T_g measurements should also be considered [466].

A broad copolymer composition may create phase separation due to the lack of miscibility between different copolymer fractions having different compositions. For instance, a difference of only 2.5% in the maleic content makes two SMA copolymers immiscible [462].

At the wet end, we have a complex mixture of polymers (WSR, DSR, TWSR, starch, CMC, etc.) dissolved or dispersed in water along with other organic or inorganic small molecules. All these paper chemicals are facing dispersed cellulose fibers. There are also interactions between the polymers used in the papermaking process. Interactions are also developed in organic solvents (hemicellulose acetate reduces the solubility of the cellulose acetate in organic solvents [70]).

Most macromolecular paper chemicals are polyelectrolytes. The shape of the polyelectrolyte coil depends on the presence and the concentration of the other electrolytes or polyelectrolytes [429,467–469]. The retention of the melamine–formaldehyde resins depends on the electrolyte concentration, and on the pH [468]. The change in shape of the macromolecular coil can be very drastic: cationic polyacrylamide precipitates due to the presence of potassium carbonate, sodium or ammonium sulfate [470,471].

Viscosity and phase separation are crucially important in papermaking. Higher polymer concentrations in solution or dispersion with low viscosities are a major target for polymer application in this industry. Viscosity depends on the polymer concentration and on the interaction between the polymer and the solvent.

Polyacrylamide is well known in the papermaking industry. Figure 2.1 shows the intrinsic viscosity for the polymer measured in water and in ethylene glycol [472], as a function of the molecular weight. For the same polymer molecular weight, the intrinsic viscosity is several times higher in water. That is an indication of the interactions between polyacrylamide and water molecules.

The interactions between polyelectrolytes and the water solution environment can be manipulated by inorganic electrolytes, organic polyelectrolytes, or nonionic polymer (hydrogen bonding)

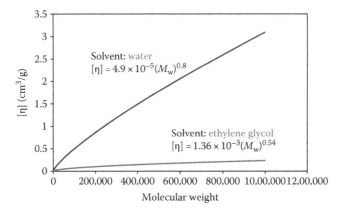

FIGURE 2.1   The intrinsic viscosity of polyacrylamide. (Klein, J. and Konrad, K.D.: *Makromol. Chem.*, 179, 1635–1638, 1978. Copyright Wiley-VCH Verlag GmbH & Co. KGaA. With permission.)

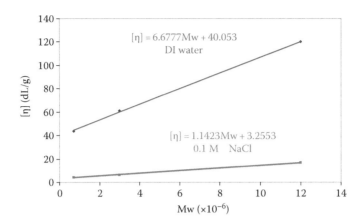

FIGURE 2.2   The electrolyte effect on the intrinsic viscosity of the cationic polyacrylamide.

[473,474]. In the case of polyamines [475], the viscosity of their water solutions depend on the pH as well.

The estimation of the intrinsic viscosity of the polyelectrolytes should be carefully performed because the reduced viscosity for that type of polymer (maleic acid–vinyl acetate copolymer [476,477]) increases with the decreasing polyelectrolyte concentration by dilution with water. The presence of an electrolyte (NaCl) changes the shape of the viscosity curve.

The presence of the electrolyte (such as NaCl) changes the relationship between the molecular weight of the polyelectrolyte and the measured intrinsic viscosity [472,478,479]. Electrolytes have an effect even on the viscosity of nonionic polymers solution (such as polyacrylamide [472,480]). In order to have a straight line for $\eta_{sp}/c = f(c)$, where c stands for the polymer concentration, the salt solution at a given concentration should be used as "solvent" [478].

The intrinsic viscosity for the same cationic polyacrylamide was measured in water and in NaCl solution [481] (Figure 2.2): for the same molecular weight, the molecular coil is about 10 times smaller in the presence of an electrolyte. That is why the intrinsic viscosity is measured in 1.0 M sodium nitrate solution as a standard solvent [482]. The same trend is noted for polyamine polymers in the presence of inorganic electrolytes [483]. As expected, amphoteric compounds show a different behavior [484]: they precipitate at their isoelectric point but, at high concentrations of NaCl, the amphoteric starch does not precipitate at any pH.

The polyacrylamide degree of hydrolysis (or the concentration of sodium acrylate units) is important for the molecular weight estimation [485]. The electrolyte effect depends on the copolymer composition. For acrylamide–sodium acrylate copolymers [486], the intrinsic viscosity in 1.0 M NaCl is always smaller than that in 0.5 M NaCl (Figure 2.3).

The effect of the electrolyte concentration is negligible for the homopolymer of polyacrylamide and for the copolymer with low concentrations of sodium acrylate (10%). However, it is not recommended to extend the use of the Mark–Houwink–Sakurada equation established for nonionic polyacrylamide ($[\eta] = 3.73 \times 10^{-4} M^{0.66}$ for 0.05 M sodium sulfate as a solvent at 30°C) to another anionic copolymer of acrylamide [487].

In order to suppress the ionic effects, size-exclusion chromatography method uses an aqueous eluent containing 0.3 M acetic acid and 0.3 M sodium acetate [100].

The composition and the amount of white water depend on the wood quality, the type of pulping process, and the impurities brought in by paper chemicals. These make every paper mill a unique operating system, and the biodegradability of water-soluble compounds should be a concern [488].

At the wet end, polyelectrolytes as paper chemicals are facing the so-called anionic trash. The shape of the cationic polyelectrolyte molecular coil (cationic starch) is changing in the presence of

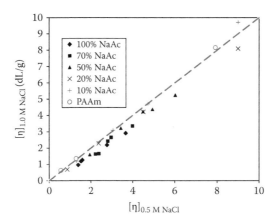

FIGURE 2.3 The effect of the copolymer composition on the intrinsic viscosity of anionic polyacrylamide.

anionic compounds (SLS) [489]. The combined interactions (such as those between ionic polymers and ionic small molecules [470]) are very common in papermaking. The molecular coil is shrinking (cationic charges are neutralized) or expanding (micelles formation) as a function of the ratio between the anionic environment and the cationic charges. Cationic and anionic macromolecular compounds interact, and precipitation is often noticed [490–492].

Nonionic polymers can also interact in solution through hydrogen bonding or hydrophobic type interactions [458,493,494]. PVA forms a supramolecular aggregate with the polyacrylic acid [495]. But, those interactions change [496] in the presence of a polar surface such as that of cellulose fibers. The anionic cellulose surface is an "active" partner in any type of interactions.

2.6 PAPER TESTING: A DIFFICULT TASK

Paper is a heterogeneous material: different components are unevenly distributed in a web-like structure. There are three different broad distributions, overlapping within the paper structure: the fiber size distribution, the filler size distribution, and the paper chemical distribution. As a result of that heterogeneity, strong sections are distributed alongside weak sections. Mechanical stress and water penetration (to consider only two paper characteristics) will be affected by that heterogeneity.

Web formation depends on temperature, shear rate [497,498], and consistency. The ideal pulp concentration should be <0.01% to prevent partial flocculation [180]. The real consistency is 0.5% or higher, which results in partial flocculation, and translates into a variation of the fiber number and length distribution in the rupture zone [499]. The machine speed and nip pressure are aggravating factors for variability in fiber distribution, fiber crossing number, pore size, void portion, and RBA.

Adsorption and flocculation [500] differ on different cellulose fiber fractions, and particularly during the mixing of paper stock with paper chemicals at the wet end, which enhances the heterogenic character. For instance: different dry-strength resins show different capabilities not only in strengthening the interfiber bonds but also in sheet formation [152].

Some paper chemicals bring along their own polydispersity: molecular weight, composition, and charge density distributions (for (co)polymers), alkyl length distribution (AKD, ASA), particle size, and particle size distribution (for latexes).

The wet end is a continuous process operating with very low concentrations (stock is below 1% and all chemicals taken together are below 5% based on pulp), which makes it hard to control the consistency of those concentrations over time; the consistency of the white water quality is also a factor [501].

The concentrations of paper chemicals in the final paper sheet are so low (most of them are below 1% based on cellulose) that the chemical and/or physical analytical methods [502] can hardly measure their amount and distribution. Moreover, the distribution of paper chemicals added during the surface treatment (dry end) is obviously uneven in the Z direction.

When the effect of an additive at the wet end is evaluated, care must be taken to determine to what extent variables such as sheet formation, fine retention, and wet pressing can be controlled. Moreover, each group of investigators used different types of fibers [335]. Thus, it became very difficult to compare the effects of wet end additives based on experiments performed in different laboratories and under different sheet-forming conditions [503].

Therefore, nonuniformity is an intrinsic property of randomly formed fiber networks. In order to acknowledge this heterogeneity, it is essential to understand the behavior of paper during testing [180]. Testing methods are hard to standardize: due to the behavior of interfiber bonds, paper properties are time-dependent [214,504]. In other words, mechanical properties are rate-dependent (for instance, the response to one particular load versus time is different from one type of paper to another). A different error structure may be present for different paper types.

All the above statements would lead to the idea that paper testing is a difficult task and special precautions must be taken when an experiment is designed. Closer attention should be paid to standard deviation values and the number of replicates for each experimental point.

TAPPI did remarkable work in trying to standardize testing methods. However, standardized procedures are valid mostly for a stable papermaking process (steady state) with repeated measurements on basically the same stock, raw materials, and process parameters. In other words, those methods are valid for "standardized" expectations. When standard equipment is not available (for the evaluation of paper softness, for instance) and an "experienced panelist" is used, the paper characterization [505] is even less accurate: the experimental errors are very large ($0.4 < R^2 < 0.7$) [506].

Due to experimental errors, for lab-scale papermaking systems, the above-mentioned parameters make it very difficult to get consistent replicates for an experiment. When a new raw material or a new procedure are used (in original papers or patents), it is clear that the "standard" number of replicates or the "standard" number of experimental points are valid for the preliminary evaluation only. More prudent authors [507] are aware of the fact that unexpected results can be due to the experimental errors.

For a higher confidence in new data, any experimental design must include a larger number of replicates of which the average is calculated. The experimenter is in an uncomfortable position: working with a complicated multistep process, and compelled to run many replicates associated with a high number of experimental points.

Unfortunately, many of the articles and patents published fail to follow an accuracy-oriented experimental work. Some authors go even deeper in building up the confusion: they are stubbornly looking for conclusions. Sometimes their conclusions are reached without any information about experimental errors, are based on graphs that have only three [497] or two points [501], etc. However, even when the data concerning the experimental errors are available, and the authors are aware about their implications, that does not prevent them from jumping to conclusions such as "...opacity decreases, albeit the opacities are in the same range within the error quoted" [508,509].

An experiment should be designed based on the intrinsic performances of the analytical methods and on the handling of that particular method. In most cases, an arbitrary approach is taken: "Each point (in a graph) represents the average of two specimens from each of the five sheets" [510].

Experimental errors are generated not only by the analytical method but also by the experimental procedures used, such as the temperature effect on the sizing test [511]. The fact that PAE resins are adsorbed by glassware [512] adds more errors when a retention study is performed.

In what is to follow, some examples of cursory interpretations of experimental data and their effect on the final conclusions published in technical literature are presented.

The water contact angle is considered as a measure of the hydrophobic character of the sized sheet. The pH of the testing water is very important [513]. This piece of information is always

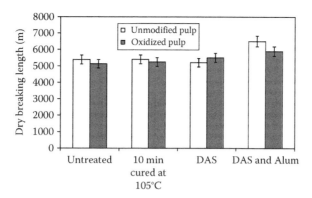

FIGURE 2.4   Experimental errors in paper dry-strength measurements.

missing, which may result in confusing conclusions. In rare cases, when the standard deviation (SD) was published for the contact angle [514], a value of more than 10% should be a sign that special attention needs to be paid to the statistical interpretation of data.

Rarely, when the SD was calculated (5% based on an arbitrary number: 20 strips per condition!) [332] the author simply ignored it, and jumped to conclusions such as "the oxidized groups decrease interfiber hydrogen bonding." Figure 2.4 shows that, within the experimental errors, oxidized pulp does not make any difference in the dry strength (the error bars are overlapping) regardless of what dry-strength additive is used (dialdehyde starch (DAS) or DAS and alum).

A relationship between the roughness of the print surface (measured objectively) and the visual (subjective) ranking is sought [515]. If the data are recorded in a graph, a very poor correlation for a straight line ($R^2 = 0.65$) is obtained (Figure 2.5).

The trend is still obvious: as the print surface roughness is higher, a higher visual ranking will result. However, the poor correlation coefficient invalidates any attempt at translating the visual ranking into objective measurements: for a print surface roughness of 2.5, there are two visual rankings of 12 and 25. The experimental design is also un-balanced: there are about 12 points around the 2.5 value for the print surface roughness and only two points between 3.5 and 4.5. That relationship is valid only for the print surface roughness between 2 and 4.5: any extrapolation is hazardous (see the intercept in the straight line equation).

The effect of cellulose acetylation on paper strength was studied for bleached Kraft pulp [306]. The degree of acetylation was estimated indirectly from the reduction in swelling in water. Experimental errors are involved in both the swelling reduction and the dry (or wet) tensile

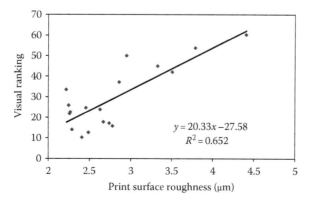

FIGURE 2.5   The correlation between print surface roughness and the visual ranking.

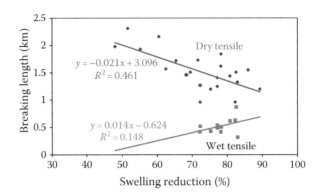

FIGURE 2.6   The effect of acetylation of bleached Kraft paper on its strength and swelling index.

measurements. Figure 2.6 shows that as the degree of acetylation increases (higher reduction in the swelling ratio), the dry strength decreases (reduced number of hydrogen bonds) and the paper wet strength is higher (less access of water). That would make sense, but the correlation coefficients (R^2) are unacceptably low, and, therefore, the experimental data should be taken only as a preliminary value.

Experimental errors are rather large for the measurements of bending stiffness [516] both for machine (MD) and cross direction (CD). The experimental points are almost identically distributed as a function of the starch concentration. Surprisingly, the authors chose a linear model for MD and a curve with a maximum for CD.

The factorial experimental design is an excellent strategy. Unfortunately, that type of experimental design has several limitations, the most important of which is the number the experimental points run in the center. When only one experimental point is run for the center [517], no experimental error analysis can be performed and the related data are rather unreliable.

The effect of the internal size concentration on paper strength is studied based on three experimental points only and conclusions are reached based on a difference of only 8% between maximum and minimum strength values [518]. Due to the much larger experimental errors (standard deviation), it appears that the internal size does not have any impact on paper strength (within experimental errors), which could be a hasty conclusion.

Mixed measurements like dry and wet tensile strength and their ratio (wet/dry) have a powerful significance. On the other hand, wet and dry strength are measured on separate strips and different error profiles are involved. Generally, wet tensile strengths were found to be less scattered than the dry-strength values [519]. Due to the different error distribution, the interpretation of the ratio value is rather difficult.

The large number of potential errors makes the experiment more tedious: in every set of experiment, a control sample should be added. It is already very common to notice differentiations between data collected for the same (control) samples placed in different sets of experimental points [520].

The rosin–alum sizing efficiency is supposed to be a function of pH [521]. However, the usual errors of about 30% mentioned in the Tappi protocols are painting a very different picture. Figure 2.7 shows that it is hard to separate between samples even if only 20% errors are considered (differences of 33%—200 and 300s for HST are considered "as consistent with normal run-to-run variation" [522]). The samples are equivalent within those experimental errors. That is also valid for the hydrocarbon resin dispersions used as surface size [523].

On the other hand, this type of experimental data is very common in paper testing and the scientist must accept it as such for a while. Eventually, the authors' honesty is always the preferable approach: as in the case of Espy [524]: "The plot of the paper wet strength vs. resin content was scattered: values could not be estimated."

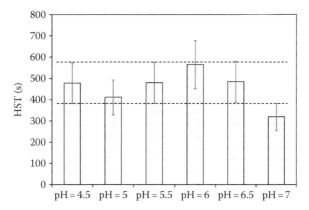

FIGURE 2.7   The effectiveness of rosin as sizing agent as a function of pH.

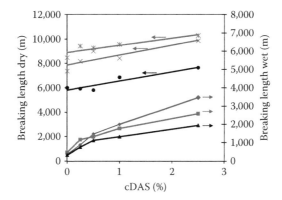

FIGURE 2.8   The effect of pulp type on the properties of paper strengthened with cationic di-aldehyde starch.

The effect of pulp on the properties of paper is overwhelming [525]: Northern bleached Kraft (in blue, Figure 2.8), Northern unbleached Kraft (in red), and Southern unbleached Kraft (in green) show different behaviors when treated with cationic di-aldehyde starch. Moreover, there is a change between unbleached and bleached Northern Kraft: bleached pulp shows a higher dry strength, but a lower wet strength than the unbleached material. An aggravating factor is that the switch takes place only at higher concentrations of strengthening agent.

Quite frequently, published data involve mixtures of different types of pulps. Figure 2.8 shows how difficult it would be to align different data published by different teams in one comprehensive and coherent collection of data.

The observations presented in this chapter have covered most paper testing methods and pointed to the many hindrances in obtaining reliable experimental data. They are also intended as a warning on the difficulties related to the correlation between paper properties and the papermaking parameters.

REFERENCES

1. H. Tarkow and A.J. Stamm, *Colloid Chemistry*, **7**, 619–640 (1950).
2. D. Hunter, *Papermaking: The History and Technique of an Ancient Craft*, Dover Publications, Inc., New York, 1978, pp. 463–584.
3. O.J. Rojas and M.A. Hubbe, *J. Disp. Sci. Technol.*, **25(6)**, 713–732 (2004).

4. F.L. Moore, *J. Ind. Eng. Chem.*, **7(4)**, 292–293 (1915).
5. Rod Ewins, The origins of paper, Lecture presented at the University of the Third Age in Hobart, Tasmania, May 15, 2001.
6. C. Watt and H. Burgess, Brit. 1942 (1853).
7. B.C. Tilghman, US 70,485 (1867).
8. F.G. Sawyer, W.F. Holzer, and L.D. McGlothlin, *Modern Chemical Processes*, Vol. 2, Reinhold Publishing, New York, 1952, pp. 255–266.
9. M.A. Hubbe and C. Bowden, *BioResources*, **4(4)**, 1736–1792 (2009).
10. M. Bells, History of papermaking, The invention of paper and the history of papermaking machinery. http://www.about.com
11. K.M. Beazley, *Paper Technol. Ind.*, **26(6)**, 266, 268 (1985).
12. K.G. Forss, A.G.M. Fuhrmann, and M. Toroi, US 5,110,414 (1992).
13. C.B. Ferguson, *Charles Fenerty: The Life and Achievement of a Native of Sackville, Halifax County, N.S.*, William Macnab, Halifax, 1955, p. 12.
14. F.C. Robinson and W.H. Cothren, US 347,200 (1886).
15. W.B. Campbell, *Paper Trade J.*, **95(8)**, 29–33 (1932).
16. J.A. Van Den Akker, *Tappi*, **53(3)**, 388–400 (1970).
17. A.H. Nissan, *Fibre-Water Interact. Paper-Making, Trans. Symp. 1977*, Vol. 2, Oxford, U.K., 1978, pp. 609–640.
18. G.A. Richter, M.O. Schur, and R.H. Rasch, US 1,745,557 (1930).
19. G.A. Richter and M.O. Schur, US 1,971,274 (1934).
20. M.O. Schur, CA 393,326 (1940).
21. Scott Paper Company, GB 502,724 (1939).
22. K.W. Britt, *Tech. Assoc. Paper*, **31**, 594–596 (1948).
23. A. Nathansohn, US 1,996,707 (1935).
24. W.H. Carothers, US 2,149,273 (1939).
25. C.S. Maxwell, US 2,407,376 (1946).
26. M.O. Schur, US 3,006,806 (1961).
27. R. Lacher, *Wochenbl. Papierfabr.*, **134(5)**, 192–195 (2006).
28. A.T. Coscia and L.L. Williams, US 3,556,932 (1971).
29. J.C. Roberts, Applications of paper chemistry, in *Paper Chemistry*, 2nd edn., J.C. Roberts (Ed.), Blackie, London, U.K., 1996, pp. 1–8.
30. P. Pruszynski, Recent developments in papermaking chemicals, in *Wood, Pulp and Paper Conference*, Bratislava, 2003.
31. J.P. Casey, *Pulp and Paper Chemistry and Chemical Technology*, 3rd edn., Wiley, New York, 1980.
32. C.O. Au and I. Thorn, *Applications of Wet-End Paper Chemistry*, Blackie Academic & Professional, London, U.K., 1995, *Applications of Wet-End Paper Chemistry*, 2nd edn., I. Thorn and C.O. Au (Eds.), Springer, Dordrecht, the Netherlands, 2009.
33. W.E. Scott, *Principles of Wet-End Chemistry*, Tappi Press, Atlanta, GA, 1996.
34. D. Eklund and T. Lindstrom, *Paper Chemistry, An Introduction*, DT Paper Science Publications, Grankulla, Finland, 1991.
35. L. Neimo (Ed.), *Papermaking Chemistry*, Fapet Oy, Helsinki, Finland, 1999.
36. W.F. Reynolds (Ed.), *The Sizing of Paper*, Tappi Press, Atlanta, GA, 1989.
37. W.F. Reynolds (Ed.), *Dry Strength Additives*, Tappi Press, Atlanta, GA, 1980.
38. J.P. Weidner (Ed.), *Wet Strength in Paper and Paperboard*, Tappi Press, Atlanta, GA, 1965.
39. L.L. Chan (Ed.), *Wet Strength Resins and Their Application*, Tappi Press, Atlanta, GA, 1994.
40. J.M. Gess and J.M. Rodriquez (Eds.), *The Sizing of Paper*, Tappi Press, Atlanta, GA, 2005.
41. R.J. Thomas, Wood: Structure and chemical composition, in *Wood Technology: Chemical Aspects, ACS Symposium*, Series 43, pp. 1–23 (1977).
42. R.H. Pelton, Polymer-colloid interactions in pulp and paper manufacture, in *Colloid–Polymer Interactions*, R.S. Farinato and P.L. Dubin (Eds.), Wiley, New York, 1999, pp. 51–82.
43. K.H. Gardner and J. Blackwell, *Biopolymers*, **13**, 1975–2001 (1974).
44. M.R. Kasaai, *J. Appl. Polym. Sci.*, **86**, 2189–2193 (2002).
45. Y. Sumi, R.D. Hale, and B.G. Ranby, *Tappi*, **46(2)**, 126–130 (1963).
46. C.L. McCormick, P.A. Callais, and B.H. Hutchinson, *Macromolecules*, **18**, 2394–2401 (1985).
47. L. Segal and J.J. Creely, *J. Polym. Sci.*, **50**, 451–465 (1961).
48. L. Segal, *J. Polym. Sci. B*, **1(5)**, 241–244 (1963).
49. L. Segal, L. Loeb, and C.M. Conrad, US 2,955,014 (1960).

50. K. Ward, C.M. Conrad, and L. Segal, US 2,580,491 (1952).
51. J. Blackwell, D.M. Lee, D. Kurz, and M.Y. Su, Structure of cellulose-solvent complex, in *Cellulose: Structure, Modification and Hydrolysis*, R.A. Young and R.M. Rowell (Eds.), Wiley, New York, 1986, pp. 51–66.
52. W.J. Brown, *Tappi*, **49(8)**, 367–373 (1966).
53. C. Vaca-Garcia, S. Thiebaud, M.E. Berredon, and G. Gozzelino, *J. Am. Oil Chem. Soc.*, **75(2)**, 315–319 (1998).
54. B. Philipp, *J. Macromol. Sci. Pure Appl. Chem.*, **A30(9–10)**, 703–714 (1993).
55. B. Philipp, *Polym. News*, **15**, 170–175 (1990).
56. W. Berger, M. Keck, and B. Philipp, *Cellul. Chem. Technol.*, **22**, 387–397 (1988).
57. K. Massonne, V. Stegmann, G.D' Andola, W. Mormann, M. Wezstein, and W. Leng, US 2009/0281303.
58. S. Aziz and K. Sarkanen, *Tappi J.*, **72(3)**, 169–175 (1989.
59. S. Aziz and T.J. McDonough, *Tappi J.*, **70(3)**, 137–138 (1987).
60. R.A. Young, *Tappi J.*, **72(4)**, 195–200 (1989).
61. T. Kleinert and K. Tayenthal, US 1,856,567 (1932).
62. T.N. Kleinert, US 3,585,104 (1971).
63. V.B. Diebold, W.F. Cowan, and J.K. Walsh, US 4,100,016 (1978).
64. P.-C. Chang and L. Paszner, US 4,594,130 (1986).
65. F.C. Paterson and L.E. Wise, US 2,192,202 (1940).
66. F.C. Paterson and L.E. Wise, US 2,218,479 (1940).
67. J. Gordy, US 4,548,675 (1985).
68. R.G. Nayak and J.L. Wolfhagen, *J. Appl. Polym. Sci. Appl. Polym. Symp.*, **37**, 955–965 (1983).
69. R.A. Young and S. Achmadi, *J. Wood Chem. Technol.*, **12(4)**, 485–510 (1992).
70. K.B. Gibney, J. Howard, and R.S. Evans, US 3,870,703 (1975).
71. E.G. Wiest and W.J. Balon, US 2,709,696 (1955).
72. J.C. Ball, US 2,863,780 (1958).
73. P. Schilling and P.E. Brown, US 4,781,840 (1988).
74. C.H. Ludwig, US 4,017,475 (1977).
75. P. Schilling and P.E. Brown, US 4,789,523 (1988).
76. H.W. Hoftiezer, D.J. Watts, and A. Takahashi, US 4,455,257 (1984).
77. P. Schilling and P.E. Brown, US 4,787,960 (1988).
78. G.G. Allan, US 3,600,308 (1971).
79. G.G. Allan, US 3,470,148 (1969).
80. W.S. Briggs, US 3,935,101 (1976).
81. W.S. Briggs, US 3,857,830 (1974).
82. G.A. Cavagna, US 3,407,188 (1968).
83. Y. Sumi, Y. Sawaguchi, and M. Nagata, *Tappi*, **57(3)**, 146–149 (1974).
84. J.E. Stone and A.M. Scallan, *Cellulose Chem. Technol.*, **2**, 343–358 (1968).
85. R. Ellis and A.W. Rudie, *21st Southern Forest Tree Improvement Conference*, Knoxville, TN, 1991, pp. 295–302.
86. A. Gibbs, R. Pelton, and R. Cong, *Tappi Proceedings—Papermaking Conference*, 2000, pp. 445–457.
87. M. Laleg and I.I. Pikulik, *Nord. Pulp Paper Res. J.*, **8(1)**, 41–47 (1993).
88. R. Subramanian, A. Kononov, T. Kang, J. Paltakari, and H. Paulapuro, *BioResources*, **3(1)**, 192–203 (2008).
89. R. Alen, K. Niemela, and E. Sjostrom, *J. Wood Chem. Technol.*, **5(3)**, 335–345 (1985).
90. T. Vuorinen and E. Sjostrom, *J. Wood Chem. Technol.*, **2(2)**, 129–145 (1982).
91. S.I. Andersson, O. Samuelson, M. Ishihara, and K. Shimizu, *Carbohydr. Res.*, **111**, 283–288 (1983).
92. B. Bihani and O. Samuleson, *Polym. Bull.*, **3**, 425–430 (1980).
93. A. Meller, W.E. Cohen, H.G. Higgins, and C.M. Stewart, *Appita*, **13(6)**, 19–27 (1960).
94. P.N. Yiannos and K.W. Britt, US 3,458,394 (1969).
95. F.L.A. Arbin, L.B. Schroede, N.S. Thompson, and E.W. Malcolm, *Tappi*, **63(4)**, 152–153 (1980).
96. O.W. Gordon, E. Plattner, and F. Doppenberg, US 5,595,628 (1997).
97. R.S. Roberts, J.D. Muzzy, and G.S. Faass, US 4,746,401 (1988).
98. S. Kenig, US 3,888,727 (1975).
99. N.N. Vanderhoek, P.F. Nelson, and A. Farrington, US 4,213,821 (1980).
100. A. Swerin and L. Wagberg, *Nord. Pulp Paper Res. J.*, **9(1)**, 18–25 (1994).
101. S. Katz, N. Liebergott, and A.M. Scallan, *Tappi*, **64(7)**, 97–100 (1981).
102. S. Katz and A.M. Scallan, *Tappi J.*, **66(1)**, 85–87 (1983).

103. J.A. Loyd and C.W. Horne, *Nord. Pulp Paper Res. J.*, **8(1)**, 48–52 (1993).
104. G. Carlsson, T. Lindstrom, and C. Soremark, *Sven. Papperstidn.*, **80(6)**, 173–177 (1977).
105. N. Wistara and R.A. Young, *Cellulose*, **6(4)**, 291–324 (1999).
106. S.A. Fischer, *TAPPI Proceedings—Papermakers Conference*, Atlanta, GA, 1996, pp. 403–409.
107. J.R.G. Bryce, Sulfite pulping, in *Pulp and Paper Chemistry and Chemical Technology*, Vol. 1, 3rd edn., J.P. Casey (Ed.), Wiley, New York, 1980, p. 291.
108. T. Iwamida and Y. Sumi, *Cellul. Chem. Technol.*, **14(2)**, 253–268 (1980).
109. L. Jurasek, *J. Pulp Paper Sci.*, **24(7)**, 209–212 (1998).
110. E. Sjostrom, *Wood Chemistry*, 1st edn., Academic Press, New York, 1981.
111. E. Sjostrom, *Wood Chemistry*, 2nd edn., Academic Press, New York, 1993.
112. J. Gierer, *Sven. Papperstidn.*, **73(18)**, 571–596 (1970).
113. A. Swerin, G. Glad-Nordmark, and L. Odberg, *J. Pulp Paper Sci.*, **23(8)**, 389–393 (1997).
114. T. Lindstrom and C. Soremark, *J. Colloid Interface Sci.* (*Proc. Int. Conf. 50th*) **5**, 217–230 (1976).
115. J. Nakano, Y. Sumi, and M. Nagata, US 3,711,366 (1973).
116. S. Katz, R.P. Beatson, and A.M. Scallan, *Sven. Papperstidn.*, **87(6)**, R48–R53 (1984).
117. J. Jiang, US 2009/0126883 (2009).
118. M. Norgren, H. Edlund, L. Wagberg, and G. Annergren, *Nord. Pulp Paper Res. J.*, **17(4)**, 370–373 (2002).
119. Y.Z. Lai, A.R. Czerkies, and I.L. Shiau, *J. Appl. Polym. Sci. Appl. Polym. Symp.*, **37**, 943–953 (1983).
120. R.E. Lappan, R. Pelton, I. McLennan, J. Patry, and A.N. Htymak, *Ind. Eng. Chem. Res.*, **36**, 1171–1175 (1997).
121. L.D. Starr and R.L. Casebier, Brit. 1,217,573 (1970).
122. L.D. Starr and R.L. Casebier, US 3,513,068 (1970).
123. K.V. Sarkanen, US 4,329,200 (1982).
124. J. Zhuang and C.J. Biermann, *Tappi J.*, **76(12)**, 141–147 (1993).
125. R.P. Singh, Bleaching-Chlorination Stage, in *Pulp and Paper Technology*, K.W. Britt (Ed.), Van Nostrand Reinhold, London, U.K., 1970, p. 249.
126. N. Gurnagul, R.C. Howard, X. Zou, T. Uesaka, and D.H. Page, *J. Pulp Paper Sci.*, **19(4)**, J160–J166 (1993).
127. S.I. Andersson and O. Samulson, *Cellul. Chem. Technol.*, **10(2)**, 209–218 (1976).
128. S.I. Andersson and O. Samuelson, *Tappi J.*, **66(7)**, 87–89 (1983).
129. D. Barzyk, D.H. Page, and A. Ragauskas, *J. Pulp Paper Sci.*, **23(2)**, J59–J61 (1997).
130. T. Sun and J.D. Lindsay, US 5,935,383 (1999).
131. G. Hegar and G. Back, US 4,141,890 (1979).
132. G. Hegar, US 4,092,308 (1978).
133. G.G. Allan and W.M. Reif, *Sven. Papperstidn.*, **74 (18)**, 563–570 (1971).
134. A.J. Stamm, *Tappi*, **33(9)**, 435–439 (1950).
135. J.W. Swanson and A.J. Steber, *Tappi*, **42(12)**, 986–994 (1959).
136. D.W. Clayton, The chemistry of alkaline pulping, in *Pulp and Paper Manufacture*, Vol. 1, R.G. Macdonald (Ed.), McGrow-Hill, New York, 1969, p. 347.
137. F. Kraft, Bleaching of wood pulps, in *Pulp and Paper Manufacture*, Vol. 1, R.G. Macdonald (Ed.), McGrow-Hill, New York, 1969, p. 628.
138. N.L. Salmen, Mechanical properties of wood fibers and paper, in *Cellulose Chemistry and its Applications*, T.P. Nevell and S.H. Zeronian (Eds.), Ellis Horwood Ltd., Chichester, U.K., 1985, p. 505.
139. D.H. Page, *J. Pulp Paper Sci.*, **9(1)**, 15–20 (1983).
140. R.A. Young, *Cellulose*, **1(2)**, 107–130 (1994).
141. R. Pelton, *Nord. Pulp Paper Res. J.*, **8(1)**, 113–119 (1993).
142. K.D. Vison and J.P. Erspamer, US 5,228,954 (1993).
143. O.L. Forgacs, A.A. Robertson, and S.G. Mason, *Pulp Paper Mag. Can.*, **59(5)**, 117–128 (1958).
144. M.S. Williams and R.R. Sargent, US 7,029,553 (2006).
145. R.R. Sargent and J.R. Alender, US 5,919,375 (1999).
146. R.R. Sargent and J.R. Alender, US 5,7339,463 (1998).
147. R.R. Sargent, J.R. Alender, and T.H. Moss, US 5,672,279 (1997).
148. R.R. Sargent and J.R. Alender, US 5,234,466 (1993).
149. G.A. Richter, US 1,880,045 (1932).
150. E.D. Mazzarella, US 4,028,172 (1977).
151. R. Bates, G.J. Broekhuisen, E.R. Hensema, and M.J. Welch, US 6,165,320 (2000).
152. R. Pelton, *Appita J.*, **57(3)**, 181–190 (2004).
153. M. Watanabe, T. Gondo, and O. Kitao, *Tappi J.*, **3(5)**, 15–19 (2004).

154. T.N. Kershaw, Sheet formation and drying, in *Pulp and Paper: Chemistry and Chemical Technology*, Vol. 2, J.P. Casey, (Ed.), Wiley, New York, 1980, p. 965.
155. H. Tomimasu, D. Kim, M. Suk, and P. Luner, *Tappi J.*, **74(7)**, 165–176 (1991).
156. J.W. Swanson, *Tappi*, **39(5)**, 257–269 (1956).
157. G. Broughton and J.P. Wang, *Tappi*, **38(7)**, 412–415 (1955).
158. B. Alince, A. Vanerek, M.H. de Oliveira, and T.G.M. van de Ven, *Nord. Pulp Paper Res. J.*, **21(5)**, 653–658 (2006).
159. E. Strazdins, *Tappi*, **64(1)**, 31–34 (1981).
160. J.E. Maher, *TAPPI Seminar Notes—Alkaline Papermaking*, Atlanta, GA, 1985, pp. 89–92.
161. U.D. Akpabio and J.C. Roberts, *Tappi J.*, **72(7)**, 141–145 (1989).
162. R.H. Pelton, L.H. Allen, and H.M. Nugent, *Sven. Papperstidn.*, **83(9)**, 251–258 (1980).
163. F. Linhart, W.J. Auhorn, H.J. Degen, and R. Lorz, *Tappi J.*, **70(10)**, 79–85 (1987).
164. T. Lindstrom, C. Soremark, and L. Westman, *Sven. Papperstidn.*, **80(11)**, 341–345 (1977).
165. B. Ayukawa, US 3,997,439 (1976).
166. J. Konig, J. Kopp, U.-W. Hendricks, J. Reiners, and P. Nowak, US 5,688,371 (1997).
167. S. Bencheva, D. Todorova, and K. Todorova, *J. Univ. Chem. Technol. Metal.*, **43(2)**, 223–226 (2008).
168. T. Cichoracki, J. Gullichsen, and H. Paulapuro, *Tappi J.*, **84(3)**, 61–68 (2001).
169. C.D. Smith and M. Callejo, *Pulp & Paper*, August 2008, pp. 31–35.
170. J.A. Van Den Akker, *Tappi*, **42(12)**, 940–947 (1959).
171. A.K. Vainio and H. Paulapuro, *BioResources*, **2(3)**, 442–458 (2007).
172. L. Gardlund, J. Forsstrom, and L. Wagberg, *Nord. Pulp Paper Res. J.*, **20(1)**, 36–42 (2005).
173. R. Grundelius and O. Samuelson, *Sven. Papperstidn.*, **65**, 310–312 (1962).
174. A.A. Robertson, *Sven. Papperstidn.*, **66**, 477–497 (1963).
175. J.V. Robinson, Fiber bonding, in *Palp and Paper: Chemistry and Chemical Technology*, Vol. 2, J.P. Casey (Ed.), Wiley, New York, 1980, p. 915.
176. B. Alince, *Tappi J.*, **74(11)**, 200–202 (1991).
177. G.G. Allan, Y.C. Ko, and P. Ritzenthaler, *Tappi J.*, **74(11)**, 202–203 (1991).
178. G.G. Allan, Y.C. Ko, and P. Ritzenthaler, *Tappi J.*, **74(3)**, 205–212 (1991).
179. R.H. Walsh, H.H. Abernathy, W.W. Pockman, J.R. Galloway, and E.P. Hartsfield, *Tappi*, **33(5)**, 232–237 (1950).
180. O.J. Kallmes, *Tappi*, **44(7)**, 516–519 (1961).
181. M.A. Hubbe and O.J. Rojas, *Chem. Eng. Educ.*, **39(2)**, 146–155 (2005).
182. O. Kallmes and H. Corte, *Tappi*, **43(9)**, 737–752 (1960).
183. B. Alince, J. Porubska, and T.G.M. van de Ven, *J. Pulp Paper Sci.*, **28(3)**, 93–98 (2002).
184. J.-F. Lafaye, *Peintures Pigments Vernis*, **43(6)**, 375–384 (1967).
185. O. Polat, R.H. Crotogino, A.R.P. van Heiningen, and W.J.M. Douglas, *J. Pulp Paper Sci.*, **19(4)**, J137–J142 (1993).
186. S. Dasgupta, *Tappi J.*, **77(6)**, 158–166(1994).
187. A.H. Nissan and S.S. Sternstein, *Tappi*, **47(1)**, 1–6 (1964).
188. A.J. Stamm, *Wood Sci. Technol.*, **13**, 41–47 (1979).
189. A.J. Stamm and M.A. Millett, *J. Phys. Chem.*, **45**, 43–54 (1941).
190. S. Asunmaa and P.W. Lange, *Sven. Papperstidn.*, **57**, 501–516 (1954).
191. R. Evans and A.F.A. Wallis, *J. Appl. Polym. Sci.*, **37**, 2331–2340 (1989).
192. M.N. Fineman, *Tappi*, **35(7)**, 320–324 (1952).
193. T. Iwamida, Y. Sumi, and J. Nakano, *Cellul. Chem. Technol.*, **14(4)**, 497–506 (1980).
194. Y. Xu, X. Chen, and R. Pelton, *Tappi J.*, **4(11)**, 8–12 (2005).
195. J.A. van den Akker, A.L. Lathrop, M.H. Voelker, and L.R. Dearth, *Tappi*, **41(8)**, 416–425 (1958).
196. R.W. Perkins and R.E. Mark, *J. Appl. Polym. Sci. Appl. Polym. Symp.*, **37**, 865–880 (1983).
197. R.W. Davison, *Tappi*, **55(4)**, 567–573 (1972).
198. S. Onogi and K. Sasaguri, *Tappi*, **44(12)**, 874–880 (1961).
199. R.W. Davison, Theory of dry strength development, in *Dry Strength Additives*, W.F. Reynolds (Ed.), Tappi, Atlanta, GA, 1980, pp. 1–31.
200. D.H. Page, *Tappi*, **52(4)**, 674–681 (1969).
201. O. Kallmes, H. Corte, and G. Bernier, *Tappi*, **44(7)**, 519–528 (1961).
202. D.H. Page and J.M. MacLeod, *Tappi J.*, **75(1)**, 172–174 (1992).
203. D.H. Page, *J. Pulp Paper Sci.*, **19(4)**, J175–J176 (1993).
204. O. Kallmes and C. Eckert, *Tappi*, **47(9)**, 540548 (1964).
205. O. Kallmes, H. Corte, and G. Bernier, *Tappi*, **46(8)**, 493–502 (1963).

206. D.H. Page and R.S. Seth, *Tappi*, **63(6)**, 113–116 (1980).
207. D.H. Page, *Tappi J.*, **71(10)**, 182–183 (1988).
208. H.G. Higgins, A.W. McKenzie, and K.J. Harrington, *Tappi*, **41(5)**, 193–204 (1958).
209. D.H. Page, *Tappi*, **46(12)**, 750–756 (1963).
210. T. Kondo, C. Sawatari, R.St.J. Manley, and D.G. Gray, *Macromolecules*, **27**, 210–215 (1994).
211. K.H. Gardner and J. Blackwell, *Biochim. Biophys. Acta*, **343**, 232–237 (1974).
212. F.J. Kolpak and J. Blackwell, *Macromolecules*, **9(2)**, 273–278 (1976).
213. J. Blackwell, F.J. Kolpak, and K.H. Gardner, *Tappi*, **61(1)**, 71–72 (1978).
214. H.W. Haslach, *Mech. Time-Depend. Mater.*, **4(3)**, 169–210 (2000).
215. P.K. Chidambareswaran, N.B. Patil, and V. Sundaram, *J. Appl. Polym. Sci.*, **20**, 2297–2298 (1976).
216. H. Corte, H. Schaschek, and O. Broens, *Tappi*, **40(6)**, 441–447 (1957).
217. A.H. Nissan, *J. Chem. Soc. Faraday Trans.*, **53(5)**, 700–709 (1957).
218. A.H. Nissan, V.L. Byrd, and G.L. Batten, *Tappi J.*, **68(9)**, 118–124 (1985).
219. A.H. Nissan and G.L. Batten, *Tappi J.*, **73(2)**, 159–164 (1990).
220. A.H. Nissan, *J. Chem. Soc. Faraday Trans.*, **53**, 710–721 (1957).
221. A.H. Nissan, *Nature*, **175**, 424 (1955).
222. A.H. Nissan, *Surf. Coat. Relat. Paper Wood Symp.*, 221–265 (1967).
223. A.H. Nissan, *Tappi*, **42(12)**, 928–933 (1959).
224. A.H. Nissan, *Macromolecules*, **9(5)**, 840–850 (1976).
225. A.H. Nissan, *Macromolecules*, **10(3)**, 660–662 (1977).
226. A.H. Nissan and G.L. Batten, *Tappi J.*, **80(4)**, 153–158 (1997).
227. G.L. Batten and A.H. Nissan, *Tappi J.*, **70(9)**, 119–123 (1987).
228. A.H. Nissan and G.L. Batten, *Tappi J.*, **70(10)**, 128–131 (1987).
229. E. Delgado, F.A. Lopez-Dellamary, G.G. Allan, A. Andrade, H. Contreras, H. Regla, and T.C Resson, *J. Pulp Paper Sci.*, **30(5)**, 141–144 (2004).
230. J.W. Swanson, *Tappi*, **44(1)**, 142–181 (1961).
231. B. Alince, *Tappi J.*, **74(8)**, 221–223 (1991).
232. E. Strazdins, *TAPPI Proceedings—Papermakers Conference*, Atlanta, GA, 1982, pp. 145–152.
233. C. Cravotto, S. Tagliapietra, B. Robaldo, and M. Trotta, *Ultrason. Sonochem*, **12**, 95–98 (2005).
234. H.W. Moeller, *Paper Technol.*, **6(4)**, 279–284 (1965).
235. P.H. Brouwer, J. Beugeling, and B.C.A. Ter Veer, *PTS-Symposium*, Munich, Germany, 1999, pp. 1–15/28.
236. A. Meller, *Tappi*, **36(6)**, 264–268 (1953).
237. J.K. Smith, US 4,495,226 (1985).
238. A. Meller, *Sven. Papperstidn.*, **60(17)**, 611–615 (1957).
239. S. Kamel, N. Ali, K. Jahangir, S.M. Shah, and A.A. El-Gendy, *eXPRESS Polym. Lett.*, **2(11)**, 758–778 (2008).
240. T.J. Schoch, *Wallerstein Lab. Commun.*, **32(109)**, 149–171 (1969).
241. H.W. Durand, US 3,655,644 (1972).
242. K.M. Gaver, D.V. Tieszen, and E.P. Lasure, US 2,671,781 (1954).
243. L.L. Williams and A.T. Coscia, US 3,597,313 (1971).
244. J.C. Roberts, C.O. Au, and G.A. Clay, *Tappi J.*, **69(10)**, 88–93 (1986).
245. E.F. Paschall, US 2,876,217 (1959).
246. G.L. Deets and W.G. Tamalis, US 4,684,708 (1987).
247. H. Meisel, US 3,017,294 (1962).
248. H. Ikeda, F. Suzuki, Y. Watanabe, M. Matsumura, Y. Takahashi, H. Murakami, and K. Maeda, US 4,840,705 (1989).
249. R.N. DeMartino and A.B. Conciatori, US 4,031,307 (1977).
250. M.M. Tessler, US 4,017,460 (1977).
251. J. Marton and T. Marton, *Tappi*, **59(12)**, 121–124 (1976).
252. J. Tsai, W. Maliczyszyn, T. Capitani, and C. Kulp, US 5,723,023 (1998).
253. J. Tsai and W. Maliczyszyn, US 5,595,631 (1997).
254. W. Jarowenko and M.W. Rutenberg, US 3,737,370 (1973).
255. W. Jarowenko, US 3,770,472 (1973).
256. C.G. Caldwell and O.B. Wurzburg, US 2,813,093 (1957).
257. R.J. Berni, R.R. Benerito, J.B. McKelvey, T.L. Ward, and D.M. Soignet, US 3,644,082 (1972).
258. E.J. Barber, R.H. Earle, and G.C. Harris, US 3,219,519 (1965).
259. R.E. Gramera and D.H. le Roy, US 3,464,974 (1969).
260. R.D. Harvey and R.E. McPherson, US 4,579,944 (1986).
261. W.Z. Schroeder, D.A. Clarahan, M.T. Goulet, and T.G. Shannon, US 6,398,911 (2002).

262. T.G. Shannon, D.A. Clarahan, M.T. Goulet, and W.Z. Schroeder, US 6,596,126 (2003).
263. M.W. Rutenberg and J.L. Volpe, US 2,989,520 (1961).
264. A. Aszalos, US 3,077,469 (1963).
265. J.C. Rankin and B.S. Phillips, US 4,127,563 (1978).
266. D.L. Elliott, R.J. Falcione, and W.E. Hunter, US 5,501,772 (1996).
267. J.W. Adams, US 3,180,787 (1965).
268. J.B. McKelvey, R.R. Benerito, R.J. Berni, and C.A. Hattox, *Textile Res. J.*, **34(9)**, 759–767 (1964).
269. D.M. Soignet, R.R. Benerito, and J.B. McKelvey, US 3,526,475 (1970).
270. R.J. Berni, J.B. McKelvey, and R.R. Benerito, US 3,150,920 (1964).
271. M.J. Harding, R.C. Gaines, and J.M. Gess, US 4,505,775 (1985).
272. P.D. Buikema, US 4,029,885 (1977).
273. L. Beghello and T. Lindstrom, *Nord. Pulp Paper Res. J.*, **13(4)**, 269–273 (1998).
274. S.L. Greene, R.M. Kaylor, and K.R. Smith, US 5,770,711 (1998).
275. T. Sun, US 6,361,651 (2002).
276. M.M. Tessler, US 4,119,487 (1978).
277. H.S. Isbell, US 2,635,931 (1953).
278. M. Luo, A.N. Neogi, S.A. Weerawarna, and A.J. Dodd, US 7,541,396 (2009).
279. F.A. Buytenhuys, and R. Bonn, *Papier*, **31(12)**, 525–527 (1977).
280. D.B. Solarek, P.G. Jobe, and M.M. Tessler, US 4,675,394 (1987).
281. H.N. Cheng, Q.M. Gu, and R.G. Nickol, US 6,159,721 (2000).
282. A.J. Stamm and W.E. Cohen, *Austr. Pulp Paper Ind. Tech. Assoc. Proc.*, **10**, 366–393 (1956).
283. Y. Matsunaga, T. Sugiyama, and E. Maekawa, US 4,632,984 (1986).
284. R.A. Jewell, US 5,667,637 (1997).
285. R.F. Schwenker and E. Pacsu, US 3,312,642 (1967).
286. J.E. Jayne, US 2,994,634 (1961).
287. C.L.P. Vaughan, US 2,618,633 (1952).
288. C.L.P. Vaughan, US 2,618,635 (1952).
289. E.H. Immergut, B.G. Ranby, and H.F. Mark, *Ind. Eng. Chem.*, **45(11)**, 2483–2490 (1953).
290. J.B. McKelvey, R.R. Benerito, and R.J. Berni, US 3,432,252 (1969).
291. D.G. Barkalow, R.M. Rowell, and R.A. Young, *J. Appl. Polym. Sci.*, **37(4)**, 1009–1118 (1989).
292. D.K. Ray-Chaudhuri, US 3,332,897 (1967).
293. F.C. McIntire, US 2,505,561 (1950).
294. C.G. Caldwell and O.B. Wurzburg, US 2,661,349 (1953).
295. P. Fuertes, A. Lambin, and J.L. Dreux, US 5,647,898 (1997).
296. P.T. Trzasko, M.M. Tessler, R. Trksak, and W. Jarowenko, US 4,721,655 (1988).
297. W. Jarowenko and H.R. Hernandez, USA 4,029,544 (1977).
298. F. Schulze, US 2,069,974 (1937).
299. J.J. Tsai and W. Maliczyszyn, US 5,658,378 (1997).
300. E.L. Speakmen, US 3,511,830 (1970).
301. D.J. Smith and J.E. Ruth, US 5,698,688 (1997).
302. H. Neukom, US 2,884,412 (1959).
303. R.W. Kerr and F.C. Cleveland, US 2,884,413 (1959).
304. R.W. Kerr and F.C. Cleveland, US 2,961,440 (1960).
305. J.C. Thomas, US 2,511,229 (1950).
306. A.J. Stamm and J.N. Beasley, *Tappi*, **44(4)**, 271–275 (1961).
307. M.A. Hubbe, D.G. Wagle, and E.R. Ruckel, US 5,958,180 (1999).
308. L. Segal, J.D. Timpa, and J.I. Wadsworth, *J. Polym. Sci.*, **A-1(8)**, 25–35 (1970).
309. A. Laine and H. Lattikedde, US 2009/0270606.
310. R.J. Berni, R.R. Benerito, and J.B. McKelvey, US 3,617,201 (1971).
311. M.C. Shelton, J.D. Posey-Dowty, L.G.R. Perdomo, D.W. Dixon, P.L. Lucas, A.K. Wilson, K.R. Walker, J.E. Lawniczak, R.G. Foulk, H.D. Phan, and C.C. Freeman, US 2009/0286095.
312. D.G. Barkalow, R.M. Rowell, and R.A. Young, *Polym. Mater. Sci. Eng.*, **57**, 52–56 (1987).
313. K.W. Kirby, *TAPPI Proceedings—Papermaking Conference*, Atlanta, GA, 1985, pp. 59–64.
314. J.F. Walsh and W.L. Morgan, US 2,170,272 (1939).
315. J.F. Walsh and W.L. Morgan, US 2,170,271 (1939).
316. M. Wayman, Comparative effectiveness of various acids for hydrolysis of cellulosics, in *Cellulose: Structure, Modification and Hydrolysis*, R.A. Young and R.M. Rowell (Eds.), John Wiley, New York, 1986, pp. 265–279.

317. P.H. Brouwer, M.A. Johnson, and R.H. Olsen, Starch and retention, in *Retention of Fines and Fillers during Papermaking*, J.M. Gess (Ed.), Tappi Press, Atlanta, GA, 1998, pp. 199–242.
318. B. Lindberg, J. Lonngren, and S. Svensson, *Adv. Carbohydr. Chem. Biochem.*, **31**, 185–240 (1975).
319. K. Niemela and E. Sjostrom, *Carbohydr. Res.*, **180(1)**, 43–52 (1988).
320. G. Rinck, K. Moller, S. Fullert, F. Krause, and H. Koch, US 5,147,907 (1992).
321. M.H. Johansson and O. Samuelson, *J. Appl. Polym. Sci.*, **19**, 3007–3013 (1975).
322. M.H. Johansson and O. Samuelson, *Cabohydr. Res.*, **34**, 232–237 (1974).
323. P.H. Brouwer, *PTS-Manuskript*, 12/1–12/37 (2003).
324. H.J. Degen, F. Reichel, U. Riebeling, and L. Hoehr, US 4,835,212 (1989).
325. L. Kuniak, B. Alince, V. Masura, and J. Alfoldi, *Sven. Papperstidn.*, **72(7)**, 205–208 (1969).
326. V. Prey and S.K. Fischer, *Starke*, **27(6)**, 192–196 (1975).
327. M.E. McKillican and C.B. Purves, *Can. J. Chem.*, **32**, 312–321 (1954).
328. R.L. Mellies, C.L. Mehltretter, and I.A. Wolff, *Ind. Eng. Chem.*, **50(9)**, 1311–1314 (1958).
329. F.R. Senti, R.L. Mellies, and C.L. Mehltretter, US 2,989,521 (1961).
330. R.L. Mellies, C.L. Mehltretter, and F.R. Senti, *J. Chem. Eng. Data*, **5(2)**, 169–171 (1960).
331. B. Alince, *Sven. Papperstidn.*, **78(7)**, 253–257 (1975).
332. R.A. Young, *Wood and Fiber*, **10(2)**, 112–119 (1978).
333. A. Meller, *Tappi*, **41(11)**, 679–683 (1958).
334. M.O. Schur and R.M. Levy, *Paper Trade J.*, **124(20)**, 43–46 (1947).
335. P. Luner, K.P. Vemuri, and B. Leopold, *Tappi*, **50(3)**, 117–120 (1967).
336. R.A. Jewell, J.L. Komen, Y. Li, and B. Su, US 6,379,494 (2002).
337. A.A. Weerawarna, J.L. Komen, and R.A. Jewell, US 7,135,557 (2006).
338. M. Medoff, US 2009/0283229.
339. H. Hamada, T. Enomae, I. Shibata, A. Isokai, and F. Onabe, *PITA Coating Conference*, Edinburg, Great Britain, 2003, pp. 41–45.
340. T. Jaschinski, US 6,409,881 (2002).
341. T. Jaschinski, S. Gunnars, A.C. Besemer, and P. Bragd, US 6,635,755 (2003).
342. J.L. Komen, A.A. Weerawarna, and R.A. Jewell, US 7,109,325 (2006).
343. J.P. Van Der Lugt, J.M. Jetten, C. Besemer, and H.A. Van Doren, US 6,518,419 (2003).
344. T. Jaschinski, S. Gunnars, A.C. Besemer, P. Bragd, J.M. Jetten, R. Van Den Dool, and W. Van Hartingsveldt, US 6,824,645 (2004).
345. J.M. Jetten, R. Van Den Dool, W. Van Hartingsveldt, and A.C. Besemer, US 6,716,976 (2004).
346. A.C. Besemer and A.E.J. De Nooy, WO 95/07303 (1995).
347. A.L. Cimedoglu and J.S. Thomaides, US 7,247,722 (2007).
348. R.A. Jeffreys and B.E. Tabor, US 3,057,723 (1962).
349. K.J. Pickard and A. Smith, US 3,993,640 (1976).
350. P.J. Borchert and J. Mirza, *Tappi*, **47(9)**, 525–528 (1964).
351. C.L. Mehltretter, US 2,713,553 (1955).
352. K. Tajima, *J. Appl. Polym. Sci. Appl. Polym. Symp.*, **37**, 709–721 (1983).
353. H.J. Roberts, US 3,329,672 (1967).
354. H. Ketola and P. Hagberg, US 6,670,470 (2003).
355. A. Isogai, Fundamentals of papermaking materials, in *Transactions of the Fundamental Research Symposium 11th*, Cambridge U.K., September 1997, pp. 1047–1071.
356. A. Isogai, *Proceedings of 52nd APPITA Annual General Conference*, Vol. 1, Appita, Brisbane, Australia, 1998, pp. 71–76.
357. F.E. Condo, E. Cerrito, and C.W. Schroeder, US 2,752,269 (1956).
358. J.B. McKelvey, R.J. Berni, and R.R. Benerito, US 3,382,030 (1968).
359. K.R. Anderson, US 5,122,231 (1992).
360. C.M. Herron and D.J. Cooper, US 5,183,707 (1993).
361. C.M. Herron, D.J. Cooper, T.R. Hanser, and B.S. Hersko, US 5,190,563 (1993).
362. C.M. Xiao, J. Tan, and G.N. Xue, *eXPRESS Polym. Lett.*, **4(1)**, 9–16 (2010).
363. C.M. Herron and W.L. Dean, US 5,549,791 (1996).
364. J.A. Westland, R.A. Jewell, and A.N. Neogi, US 5,998,511 (1999).
365. C.H. Lesas and M. Pierre, US 4,113,936 (1978).
366. C.H. Lesas and M. Pierre, US 4,204,054 (1980).
367. F.H. Steiger, US 3,658,613 (1972).
368. N. AIbrahim and N.H. Abo-Shosha, *J. Appl. Polym. Sci.*, **49**, 291–298 (1993).
369. J.B. McKelvey, R.J. Berni, and R.R. Benerito, US 3,382,029 (1968).

370. P. Hakansson, US 6,585,859 (2003).
371. D.B. Solarek, L.R. Peek, M.J. Henley, R.M. Trksak, and M.T. Philbin, US 5,523,339 (1996).
372. D.B. Solarek, L.R. Peek, M.J. Henley, R.M. Trksak, and M.T. Philbin, US 5,368,690 (1994).
373. J.A. Harpham and H.W. Turner, US 3,069,311 (1962).
374. I.L.-A. Croon and N.V. Blomqvist, US 3,700,549 (1972).
375. D.L. Johnson, US 4,810,785 (1989).
376. D.L. Shaw and E.A. Wodka, US 3,819,470 (1974).
377. R. Chung, US 3,440,135 (1969).
378. K.R. Anderson and D.E. Garlie, US 6,524,440 (2003).
379. K.R. Anderson and D.E. Garlie, US 6, 451,170 (2002).
380. H. Ketola, T. Andersson, A. Karppi, Sauli Laakso, and A. Likitalo, US 6,187,144 (2001).
381. P.A. Graef and F.R. Hunter, US 5,225,047 (1993).
382. P.A. Graef and F.R. Hunter, US 5,399,240 (1995).
383. A. Bhattacharya, J.W. Rawlins, and P. Ray (Eds.), *Polymer Grafting and Crosslinking*, John Wiley, New York, 2009.
384. E.M.L. Ehrnrooth, P. Kolseth, and A. de Ruvo, *Fibre-Water Interact. Paper-Making Trans. Symp.*, Vol. 2, Oxford, U.K., September 1977, 1978, pp. 715–740.
385. E. Schwab, V. Stannett, and J.J. Hermans, *Tappi*, **44(4)**, 251–256 (1961).
386. A. Zilkha, B.-A. Feit, and A. Bar-Nun, US 3,341,483 (1967).
387. G. Ezra and A. Zilkha, *J. Appl. Polym. Sci.*, **13**, 1493 (1969).
388. G.L. Deets and W.G. Tamalis, US 4,744,864 (1988).
389. D.R. Patil and G.F. Fanta, *Starch/Starke*, **47(3)**, 110–115 (1995).
390. B. Ranby, New methods for graft copolymerization onto cellulose and starch, in *Modified Cellulosics*, R.M. Rowell and R.A. Young (Eds.), Academic Press, New York, 1978, pp. 171–196.
391. G.F. Fanta and F.H. Otey, US 4,839,450 (1989).
392. C.E. Farley, G. Anderson, and K.D. Favors, 6,787,574 (2004).
393. G.F. Touzinsky and H.W. Maurer, US 3,640,925 (1972).
394. A.R. Reid, US 4,028,290 (1977).
395. Y. Sugahara, T. Ohta, *J. Appl. Polym. Sci.*, **82**, 1437–1443 (2001).
396. C.E. Brockway, R.R. Estes, and D.R. Smith, US 3,095,391 (1963).
397. C.C. Nguyen, V.J. Martin, and E.P. Pauley, US 5,003,022 (1991).
398. I. Goni, M. Gurruchaga, B. Vazquez, M. Valero, G.M. Guzman, and J. San Roman, *Polymer*, **35(7)**, 1535–1541 (1994).
399. F. Ide, T. Kodama, and Y. Kotake, US 3,785,921 (1974).
400. G.F. Fanta and R.C. Burr, US 3,976,552 (1976).
401. J.C. Rankin, US 4,330,443 (1982).
402. G.F. Fanta and R.C. Burr, US 3,809,664 (1974).
403. H. Hartmann, W. Denzinger, M. Kroener, C. Nilz, F. Linhart, and A. Stange, US 5,334,287 (1994).
404. C.P. Iovine and D.K. Ray-Chaudhuri, US 4,131,576 (1978).
405. C.G. Beddows, M.H. Gil and J.T. Guthrie, *Polym. Photochem.*, **7**, 213–230 (1986).
406. B.T. Hofreiter, H.D. Heath, M.I. Schulte, and B.S. Phillips, *Starch/Staerke*, **33(1)**, 26–30 (1981).
407. V. Singh, A. Tiwari, S. Pandey, and S.K. Singh, *eXPRESS Polym. Lett.*, **1(1)**, 51–58 (2007).
408. A. Restaino, L. Township, and W.N. Reed, US 3,635,857 (1972).
409. C. Hagiopol, T. Deleanu, and T. Memetea, *J. Appl. Polym. Sci.*, **37**, 947–959 (1989).
410. R.L. Walrath, Z. Reyes, and C.R. Russell, *Ad. Chem. Ser.*, **34**, 87–95 (1962).
411. E.J. Choi, C.H. Kim, and J.K. Park, *Macromolecules*, **32**, 7402–7408 (1999).
412. D.-K. Kweon, D.-S. Cha, H.-J. Park, and S.-T. Lim, *J. Appl. Polym. Sci.*, **78**, 986–993 (2000).
413. S.B. Vitta, E.P. Stahel, and V.T. Stannett, *J. Macromol. Sci. Chem.*, **A22(5–7)**, 579–590 (1985).
414. M.M. Cruz, US 3,372,132 (1968).
415. V.D. Athawale and V. Lele, *Carbohydr. Polym.*, **41**, 407–416 (2000).
416. L. Mr, S. Silong, W.M.Z.W. Yunus, M.Z.A. Rahman, M. Ahmad, and M.J. Haron, *J. Appl. Polym. Sci.*, **77**, 784–791 (2000).
417. L. Rahman, S. Silong, W.M. Zin, M.Z.A. Rahman, M. Ahmad, and J. Haron, *J. Appl. Polym. Sci.*, **76**, 516–523 (2000).
418. M.R. Lutfor, S. Sidik, W.M.Z. Wan Yulus, M.Z.A. Rahman, A. Mansor, and M.J. Haron, *J. Appl. Polym. Sci.*, **79**, 1256–1264 (2001).
419. A. Fakhru-Razi, I.Y.M. Qudsieh, W.M.Z. Wan Yulus, M.B. Ahmad, and M.Z.A. Rahman, *J. Appl. Polym. Sci.*, **82**, 1375–1381 (2001).

420. G.B. Butler, T.E. Hogen-Esch, J.J. Meister, and H. Pledger, US 4,400,496 (1983).
421. A.H. Young and F. Verbanac, US 4,079,025 (1978).
422. F. Verbanac, US 4,060,506 (1977).
423. J.J. Tsai and E.A. Meier, US 4,973,641 (1990).
424. G. Odian, *Principles of Polymerization*, Wiley-Interscience, New York, 2004.
425. W. Charlton, S.G. Jarrett, and E.E. Walker, US 2,406,454 (1946).
426. W.A. Schenck, US 2,540,352 (1951).
427. G. Rodden, *Pulp Paper*, **80(4)**, 26–27 (2006).
428. C. Hagiopol, *Copolymerization: Towards a Systematic Approach*, Kluwer-Academic Press, New York, 1999.
429. D.S. Breslow and A. Kutner, *J. Polym. Sci.*, **27**, 295–312 (1958).
430. C.L. McCormick, G.-S. Chen, and B.H. Hutchinson, *J. Appl. Polym. Sci.*, **27**, 3103–3120 (1982).
431. D.I. Lee, Latex, in *Papermaking Science and Technology*, Vol. 11, Fapet Oy, Helsinki, 2000, pp. 196–217.
432. D.I. Lee, Y. Chonde, Ionic and ionogenic polymer colloids, in *Microspheres, Microcapsules and Liposomes*, Vol. 4, Citus Books, London, U.K., 2002, pp. 137–170.
433. T.C. Bissot, US 3,347,811 (1967).
434. H.R. Kricheldorf and Z. Denchev, Interchange reactions in condensation polymers and their analysis by NMR spectroscopy, in *Transreactions in Condensation Polymers*, S. Fakirov (Ed.), Wiley-VCH Verlag, New York, 1999, pp. 1–78.
435. L.F. Beste and R.C. Houtz, *J. Polym. Sci.*, **8(4)**, 395–407 (1952).
436. L.F. Beste, *J. Polym. Sci.*, **34**, 313–323 (1959).
437. A.M. Kotliar, *J. Polym. Sci. Macromol. Rev.*, **16**, 367–395 (1981).
438. M.J. Han, H.C. Kang, and K.B. Choi, *Polymer*, **11(4)**, 349–355 (1987).
439. I.K. Miller, *J. Polym. Sci. Polym. Chem. Ed.*, **14**, 1403–1417 (1976).
440. G.I. Keim, US 4,537,657 (1985).
441. R.W. Lenz, *Organic Chemistry of Synthetic High Polymers*, Wiley, New York, 1967, pp. 687–770.
442. Y. Gnanou and M. Fontanille, *Organic and Physical Chemistry of Polymers*, Wiley-Interscience, New York, 2008, pp. 357–376.
443. H.-K. Lee and W.-L. Jong, *J. Polym. Res.*, **4(2)**, 119–128 (1997).
444. A.M. Swift, *Tappi*, **40(9)**, 224–227 (1957).
445. K. Yasuda, K. Okajima, and K. Kamide, *Polymer J.*, **20(12)**, 1101–1107 (1988).
446. H. Kheradmand, J. Francois, and V. Plazanet, *Polymer*, **29**, 860–870 (1988).
447. G. Muller, J.C. Fenyo, and E. Selegny, *J. Appl. Polym. Sci.*, **25**, 627–633 (1980).
448. S. Sawant and H. Morawetz, *Macromolecules*, **17**, 2427–2431 (1984).
449. W.R. Cabaness, T.Y.-C. Lin, and C. Parkanyi, *J. Polym. Sci.*, **A-1(9)**, 2155–2170 (1971).
450. H. Tanaka, A. Swerin, L. Odberg, and M. Tanaka, *J. Appl. Polym. Sci.*, **86(3)**, 672–675 (2002).
451. H. Tanaka, A. Swerin, L. Odberg, and S.B. Park, *J. Pulp Paper Sci.*, **23(8)**, J359–J365 (1997).
452. P. Lagally and J.W. Brook, US 3,348,997 (1967).
453. A. Beresniewicz and T. Hassall, US 4,775,715 (1988).
454. R. Stober, E. Kohn, and D. Bischoff, US 4,822,851 (1989).
455. E. Stone and B. Wasyliw, US 5,180,782 (1993).
456. A.D. Armstrong, Bleaching mechanical pulp", in *Pulp and Paper Manufacture*, Vol. 1, R.G. Macdonald (Ed.), McGrow-Hill, New York, 1969, pp. 191–225.
457. W.C. Floyd and L.R. Dragner, US 5,147,908 (1992).
458. R.L. Adelman, US 4,461,858 (1984).
459. C. Hagiopol, M. Georgescu, T. Deleanu, and V. Dimonie, *Colloid Polym. Sci.*, **257(11)**, 1196–202 (1979).
460. B. Philipp and N. Lang, *Tappi*, **52(6)**, 1179–1183 (1969).
461. T.G. Fox, *Bull. Am. Phys. Soc.*, **1**, 123 (1956).
462. J. Pionteck, V. Reid, and W.J. MacKnight, *Acta Polym.*, **46**, 156–162 (1995).
463. N.W. Johnston, *J. Macromol. Sci. Rev. Macromol. Chem. Phys.*, **C14(2)**, 215–250 (1976).
464. N.W. Johnston, *Macromolecules*, **6(3)**, 453–456 (1973).
465. J. Guillot, A. Guyot, and C. Pichot, *Macromol. Symp.*, **92**, 354–367 (1995).
466. H. Stutz, K.-H. Illers, and J. Mertens, *J. Polym. Sci. Part B Polym. Phys.*, **28**, 1483–1498 (1990).
467. W. Rodriguez, US 5,543,446 (1996).
468. B. Steenberg, *Sven. Papperstidn.*, **49(14)**, 311–323 (1945).
469. H. Lauer, A. Stark, H. Hoffmann, and R. Donges, *J. Surfactants Deterg.*, **2(2)**, 181–191 (1999).
470. K. Fujimura and K. Tanaka, US 3,790,529 (1974).

471. E.E. Maury, R. Buelte, and C.C. Johnson, US 6,171,505 (2001).
472. J. Klein and K.D. Conrad, *Makromol. Chem.*, **179**, 1635–1638 (1978).
473. M.S. Ghafoor, M. Skinner, and I.M. Johnson, US 6,031,037 (2000).
474. M.S. Ghafoor, M. Skinner and I.M. Johnson, US 6,001,920 (1999).
475. Y. Zhong, J. Jachowitcz, P. Wolf, and R. McMullen, *Polym. Prepr.*, **41(2)**, 1136–1137 (2000).
476. G. Nemtoi, C. Beldie, C. Tircolea, I. Popa, I. Cretescu, I. Humelnicu, and D. Humelnicu, *Eur. Polym. J.*, **37**, 729–735 (2001).
477. S.A. Rice and M. Nagasawa, *Polyelectrolyte Solutions*, Academic Press, New York, 1961.
478. H. Terayama and F.T. Wall, *J. Polym. Sci.*, **16**, 357–365 (1955).
479. H. Terayama, *J. Polym. Sci.*, **15**, 575–590 (1955).
480. J. Klein and K.D. Conrad, *Makromol. Chem.*, **181**, 227–240 (1980).
481. T. Lindstrom and C. Soremark, *J. Colloid Interface Sci.*, **55(2)**, 305–312 (1976).
482. J.B. Wong Shing, R.T. Gray, A.S. Zelenev, and J. Chen, US 6,592,718 (2003).
483. K.J. McCarthy, C.W. Burkhardt, and D.P. Parazak, *J. Appl. Polym. Sci.*, **34**, 1311–1323 (1987).
484. M.E. Carr, B.T. Hofreiter, and C.R. Russell, *J. Appl. Polym. Sci.*, **22**, 897–904 (1978).
485. X.Y. Wu, D. Hunkeler, A.E. Hamielec, R.H. Pelton, and D.R. Woods, *J. Appl. Polym. Sci.*, **42**, 2081–2093 (1991).
486. K.J. McCarthy, C.W. Burkhardt, and D.P. Parazak, *J. Appl. Polym. Sci.*, **33**, 1699–1714 (1987).
487. X. Chen, R. Huang, and R. Pelton, *Ind. Eng. Chem. Res.*, **44**, 2078–2085 (2005).
488. G. Swift, *Polym. Degrad. Stab.*, **45**, 215–231 (1994).
489. J. Merta and P. Stenius, *Colloid Surf.*, **A122**, 243–255 (1997).
490. G.C. Maher, US 3,436,305 (1969).
491. M.E. Carr, W.M. Doane, G.E. Hamerstrand, and B.T. Hofreiter, *J. Appl. Polym. Sci.*, **17**, 721–735 (1973).
492. M.E. Carr, B.T. Hofreiter, and C.R. Russell, *J. Polym. Sci. Polym. Chem. Ed.*, **13**, 1441–1456 (1975).
493. A. Rudin, H.L.W. Hoegy, and H.K. Johnston, *J. Appl. Polym. Sci.*, **16**, 1281–1293 (1972).
494. S.C. Ng and K.K. Chee, *Eur. Polym. J.*, **33(5)**, 749–752 (1997).
495. L. Daniliuc, C. De Kesel, and C. David, *Eur. Polym. J.*, **28(11)**, 1365–1371 (1992).
496. R. Cherrabi, A. Saout-Elhak, and M. Benhamou, *Eur. Phys. J.*, **E2**, 91–101 (2000).
497. R.W. Davison, *J. Pulp Paper Sci.*, **14(6)**, J151–J159 (1988).
498. J. Marton and F.L. Kurrle, *TAPPI Proceedings Papermakers Conference*, Atlanta, GA, 1985, pp. 197–207.
499. R.H. Moffett, US 5,584,966 (1996).
500. M.A. Hubbe, *Paper Technol.*, **44(8)**, 20–34 (2003).
501. H.H. Espy, *Tappi J.*, **78(4)**, 90–99 (1995).
502. Y. Ozaki and A. Sawatari, *Nord. Pulp Paper Res. J.*, 12(4), 260–266 (1997).
503. T. Lindstrom, L. Wagberg, and T. Larsson, On the nature of joint strength in paper, in *Proceedings of 13th Fundamental Research Symposium*, Cambridge, U.K., 2005, pp. 457–562.
504. H.W. Haslach, *Mech. Time-Depend. Mater.*, **13(1)**, 11–35 (2009).
505. J.P.O' Brien, E.T. Reaville, and F.B. Erickson, US 3,296,065 (1967).
506. J. Liu and J. Hsieh, *Tappi J.*, **3(4)**, 3–8 (2004).
507. J. Russell, O.J. Kallmes, and C.H. Mayhood, *Tappi*, **47(1)**, 22–25 (1964).
508. Z.R. Zhang, R.W. Wygant, A.V. Lyons, and F.A. Adamsky, *Proceeding of TAPPI Coating Conference*, Toronto, Ontario, Canada, 1999, pp. 275–285.
509. Z.R. Zhang, R.W. Wygant, and A.V. Lyons, *Tappi J.*, 84(3), 48–71 (2001).
510. A.J. Stamm, *Tappi*, **42(1)**, 44–50 (1959).
511. R.W. Kumler and J.M. Gess, Testing paper and board for sizing, in *The Sizing of Paper*, 2nd edn., W.F. Reynolds (Ed.), Tappi Press, Atlanta, GA, 1989, pp. 103–132.
512. N.A. Bates, *Tappi*, **52(6)**, 1157–1161 (1969).
513. P.D. Garrett and K.I. Lee, *Tappi J.*, 81(4), 198–203 (1998).
514. J.J. Krueger and K.T. Hodgson, *Tappi J.*, 78(2), 154–161 (1995).
515. J. Aspler, B. Jordan, and M.O' Neill, *Pulp Paper Can.*, **101(2)**, T40–T45 (2000).
516. J. Lipponen, J. Gron, S.E. Bruun, and T. Laine, *J. Pulp Paper Sci.*, **30(3)**, 82–90 (2004).
517. L.J. Barker, R.J. Proverb, W. Brevard, I.J. Vazquez, O.S. dePierne, and R.B. Wasser, *TAPPI Proceedings Papermakers Conference*, Atlanta, GA, 1994, pp. 393–397.
518. S. Iwasa, *Paper Technol.*, **43(3)**, 31–37 (2002).
519. A. Jurecic, C.M. Hou, K. Sarkanen, C.P. Donofrio, and V. Stannett, *Tappi*, **43**, 861–865 (1960).
520. D.L. Taylor, *Tappi*, **51(9)**, 410–413 (1968).

521. J.M. Gess, Rosin, in *The Sizing of Paper*, J.M. Gess and J.M. Rodriquez (Eds.), Tappi Press, Atlanta, GA, 2005, pp. 57–73.
522. B. Brungardt, *Pulp Paper Can.*, **98(12)**, 152–155 (1997).
523. P.H. Aldrich, US 4,017,431 (1977).
524. H.H. Espy, *Tappi J.*, **70(7)**, 129–133 (1987).
525. P.J. Borchert, W.L. Kaser, and J. Mirza, US 3,269,852 (1966).

3 The Fate of Paper Chemicals at the Wet End

The papermaking process is essentially a filtration process [1] involving the dynamic separation of the paper web from the white water. If chemicals are to influence the web and paper properties, they must be retained on the cellulose fibers [2]. Poor retention results in more water removal from the wet web [3], low paper quality and a high concentration of chemicals in white water, and the lack of attainment of the desired functional properties.

Wet end chemistry deals with the adsorption of small and/or large molecules on cellulose fibers [4]. The heterogeneous system at the "wet end" can be defined as "the miraculous loci": the cellulose fibers are dispersed (less than 1% based on water) in a huge volume of water and they must interact with paper chemicals, which are in very low concentration (less than 5% based on dry cellulose). If the paper chemical is also a dispersion, the wet end process turns into a hetero-coagulation.

Cellulose fibers (and other dispersed components) will generate a huge surface area. All processes developed in this step will take place at the interface: adsorption, flocculation, anion–cation interactions, hydrophobic interactions, etc.

3.1 FRIENDS AND FOES AT THE WET END

The dispersion components at the wet end are either ionic or nonionic. Many of them show antagonistic properties: hydrophilic–hydrophobic, ionic–nonionic, cationic–anionic, small molecule–macromolecular compounds, soluble compounds, and dispersions. In the presence of cellulose pulp, the wet end additives will compete with each other in terms of mobility, ionic and hydrophobic interactions, and eventually the adsorption capabilities. They may also interact with each other and with some synergetic effects.

At the wet end, there is a continuous water phase; a dispersed phase of cellulose fibers. Cellulose fibers, which are non-soluble, but swollen in water, have a broad particle size distribution, a porous structure, and reactive groups on the surface. The continuous phase contains soluble compounds left from the pulping process and paper chemicals additives.

The most common interactions at the wet end are developed between cellulose fibers and cationic macromolecules (or small molecules), cellulose fibers and nonionic polymers, cationic polymers and anionic trash, etc. Even the nonionic small molecules, such as organic halogen compounds, are about 50% retained on cellulose [5].

In order to prevent uncontrolled flocculation, the cellulose fiber concentration in water is very low. Fibers are isolated from one another in water only at concentrations lower than 0.01% [6]. However, that concentration is not commercially attractive, and therefore the papermaking process deals with potentially uneven flocculation.

The dispersed phase of cellulose fibers is defined by the fiber size and the size distribution and by the concentration of anionic functional groups. The structural elements of cellulose fibers are the fibrils and microfibrils having diameters ranging from 3 to 30 nm. At the wet end, those fibers are swollen in water, which pushes the fibrils apart and generates a large internal surface area (pore area). The average diameter for newly formed pores is about 100 nm [7]. The swelling and beating processes make fibers more accessible. As expected, after beating, the adsorption of cationic polymer increases [8], and pigments are more evenly distributed [9]. After the drying of the initially beaten pulp, pore sizes change and the cationic polymer adsorption decreases to the level of unbeaten pulp.

The geometric structure of the fiber and its capability to adsorb water are evaluated by the "fiber saturation point" [10]. The average cellulose pore size is about 13 nm in diameter with 80% of the pore volume within ±1.2 nm and only 20% of the pore volume is due to the pores larger than 10 nm [11]. It is easy to understand the huge specific surface area developed by that porous material: this represents over 200 m^2/g [12] (50 times larger than the outer surface).

Adsorption on cellulose fibers deals with a wide variety of molecules present in the continuous phase. These molecules are differentiated by their molecular size, charge density, and/or polarity. Cellulose fibers have an external surface area (available for any kind of molecule but especially for high molecular weight polymers [13]), an internal surface area (due to the micropores and macropores [14]) accessible for moderately sized macromolecules, and the surface area displayed by small pores and available only to small molecules [12].

The ratio between the pore size and the size of a polymer coil is essential for this wet end chemistry to be effective. Polymer retention within cellulose pores is a function of the polymer molecular weight and the cellulose pore size [12]. It was experimentally proven that larger molecules have less access to a given pore structure [13]. Due to the fact that all polymers show a molecular weight distribution, a selection of the macromolecule is expected during the adsorption process. Failure to take into account the accessibility of the internal surface area to the paper chemicals may lead to erroneous conclusions concerning the adsorption process [15].

The retention of any cationic polymer is directly proportional to the carboxyl content of cellulose [16,17] and their degree of dissociation. About 90% of the cellulose carboxyl groups are dissociated at pH 5.0 [18]. At a higher pH, the degree of dissociation is even higher, but the dissociation degree as a function of pH does not follow the theoretical curve: the real pKa is higher than 3.5 at pH = 4.0 and higher than 4.0 at pH = 5.0. At pH = 2, cellulose fibers become neutral [19]. Therefore, the cationic resin retention depends on the pH [20].

The number of anionic (carboxyl) groups on cellulose fiber can be increased by oxidation [20,21], by etherification [22,23], or by enriching the fiber surface with CMC bonded with aluminum or zirconium salts [24,25]. Both oxidation and etherification are performed in the presence of water (swollen fibers) with small molecules (such as sodium chloroacetate), which may result in the formation of carboxylic groups on both the fiber surface and inside the pores. The fact that carboxylic groups are immobilized on the surface or trapped in the cellulose structure is important for their dissociation and supposed interaction with cationic species. The anionic groups trapped into the fiber structure are accessible only to low molecular cationic polymers (such as PEI) and only after a longer period of time needed for their diffusion (24 h) [26].

The presence of anionic groups on cellulose fibers explains their electrophoretic mobility. The presence of electrolytes changes the magnitude of electrophoretic mobility [19,27,28]: a higher concentration of the electrolyte results in a lower mobility of the cellulose fibers. Pulp readily exchanges an ion even in very diluted solutions and preferentially picks up the higher valence cations [29]. The adsorption of paper chemicals and, eventually, the paper properties depend on the quality of water. The water/cellulose fiber interface is far from continuous or smooth.

The fiber is swollen in water and the amorphous cellulose part has a much larger swollen index. The amorphous zone has segments of cellulose molecules (loops and tails) released into the water phase showing higher capability to interact with water soluble polymers.

The pulp includes lignin fragments, hemicelluloses, fatty acids, and rosin acids [30] to which chemicals from recycled white water are added ("anionic trash"). Primary fines (from unbeaten pulp) are richer in lignin than secondary fines, which were created during beating [31]. Bleached sulfate pulp releases organic substances during slashing and refining [32]. Hardwood pulp releases more anionic trash with a higher molecular weight than softwood pulp. This release process depends on the electrolyte concentration. Anionic trash behaves as a polyelectrolyte. This topic is revisited under the sections on polyelectrolyte adsorption–desorption and on interactions of anionic and cationic species to make a polyelectrolyte complex. Cationic demand is directly related to the amount of anionic contaminants in paper furnish [30,33–35].

Anionic small molecules (such as Direct Red 2 dye) are retained on cellulose fibers almost 100% by aluminum salts [17]. This experimental data proves that the anionic compounds are involved in the retention process by reducing the cationic retention aids efficiency and by bringing on paper all kinds of undesired molecules. The anionic trash will compete with anionic cellulose fibers for cationic paper chemicals [30].

Wet end chemistry involves water-soluble polymers and dispersions of hydrophobic compounds (polymers and/or small molecules) [36]. Examples of soluble paper chemicals used at the wet end are retention aids, temporary wet-strength resins, wet-strength resins, dry-strength resins, and rosin salts as internal sizing agents. Dispersed paper chemicals are internal sizing agents, polymer latexes, and fillers.

Most paper chemicals are polyelectrolytes whose adsorption on a solid surface depends on their chemical structure, molecular weight, molecular weight distribution, cationic (or anionic) charge density and distribution, and on the presence of inorganic electrolytes. Their molecular weight and charge density distributions will generate a selective adsorption on cellulose fibers [37].

At the wet end, in a very diluted system, a heterogeneous mixture is characterized by a strong circulation. The shear rate has an important effect on diffusion, adsorption, and precipitation of alum [38]. In an ideal picture (no chemical reaction is 100% retention), the final moment (before the Fourdrinier wire) would be a drastic reduction in the number of components: all chemicals should be adsorbed on cellulose fibers and the ideal white water would be a clean liquid with no residual solids.

However, the real wet end systems involve, unfortunately, only a partial adsorption of chemicals. An important amount of un-retained chemicals remain in white water. The art of wet end chemistry is to place the real system as close as possible to the ideal picture.

3.2 POLYMERS IN HETEROGENEOUS SYSTEMS

Paper chemicals added at the wet end are adsorbed on the surface of cellulose fibers, interact with each other in the continuous phase, or remain unchanged. Due to the diversity of species at the wet end, a competition takes place between different interactions and processes (dissociation, precipitation, adsorption, coagulation, and hetero-coagulation) [39]. The "art" of "wet end chemistry" relies on the ability to control their retention in order to obtain the desired effect on the wet web and the final paper.

The adsorption of paper chemicals on cellulose fibers is the ultimate goal of wet end chemistry. Adsorption is driven by ionic interactions, and hydrogen bonding; hydrophobic interactions also appear to be very strong [40–42]. The ionic bonds (cationic surfactant and anionic cellulose carboxyl) will force hydrophobic tails to interact with the hydrophobic part of the cellulose fiber. Increasing the number of CH_2 units in the hydrophobic tail (alkyl chain) increases the contribution of the hydrophobic interaction.

This is a random adsorption due to the fact that the adsorbed molecules do not interfere with each other. The specific area occupied by a hydrophobic tail depends on the surface polarity: the hydrophobic segment is longer on a polar surface [43]. By adding more cationic surfactant, the small molecules start to organize (redistribution) in hemi-micelle. They are able to cover the entire surface, while the hydrophobic tails are interacting with each other:

The hydrophobic pack of tails will accept more hydrophobic tails. New molecules are oriented with the cationic charges toward water (admicelle):

During the adsorption of the cationic surfactant, the cellulose fiber surface changes its charge density and its hydrophobicity. This surfactant bilayer overcomes the anionicity of cellulose and a positive ξ-potential value is recorded [40].

The "hydrophobic" character of the cellulose fiber is a fascinating topic. Regardless of the large number of hydroxyl groups in the cellulose structure, nonionic but hydrophobic polymers (such as polyvinyl acetate) are adsorbed onto cellulose fibers from a benzene solution [44,45] even when cellulose had swollen in water.

Cellulose fibers are considered as a "rigid" polyelectrolyte with anionic functional groups and a large hydrophobic/polar surface. Macromolecular compounds such as paper chemicals are dissolved in the water phase, and adsorbed onto cellulose fibers. Before the interaction model is decided ("ladder," "scrambled eggs," "entrapment," "surface excess," [46,47] etc.), we have to look at the ionic interactions first.

Ion exchange adsorption is an equilibrium process between bonded and un-bonded small molecules on a macromolecular compound [48] or on cellulose fibers. Those interactions are reversible. When cationic charges are attached to a polymer, which interacts with an anionic charge located on the cellulose fiber, the reversibility of those interactions is put in a new light. The macromolecular character of the cationic polyelectrolyte adds two new variables: a larger number of ionic charges per macromolecule and their distribution within that macromolecule. In this case, the ratio between dissociated and undissociated bonds is different from that in which a small molecule is involved. Apparently, only one undissociated ionic bond is sufficient to generate a "permanent" connection between two polyelectrolytes with different charges. To make the interaction reversible, all ionic bonds should be broken simultaneously, which is very unlikely. The process can be considered "irreversible."

Individual ionic bonds are sensitive to dilution, to pH change, and to the presence of other electrolytes. This means that the individual ionic bonds are still reversible, which allows for macromolecular re-conformation. During re-conformation, several ionic bonds are broken while some other are formed. For a polyanion/polycation complex, it is assumed that the number of ionic bonds remain relatively constant for a given concentration, pH, salt concentrate, and temperature. This constant number of ionic bonds provides the "irreversible" character of the link between two polyelectrolytes. Because the individual ionic bonds are reversible, but there always exist a number of ionic bonds between two opposite polyelectrolytes, the general interaction can be defined as "pseudo-irreversible" (quasi-irreversible [49]).

This macromolecular re-conformation, which may result in an increased number of ionic charges available for interactions, is a time-controlled process. In other words, the adsorption process and the amount of adsorbed polymer are expected to be time dependent. The molecular weight of the polyelectrolyte is also important: a polymer with a higher molecular weight able to displace a molecule with a lower molecular weight [50]. The higher number of ionic bonds for a macromolecule makes it more difficult to be desorbed.

Although we assume that the number of ionic bonds is constant, the availability of ionic charges changes during the dynamic equilibrium of individual ionic bond dissociation. This dynamic process is even more complicated when two or more ionic or poly-ionic species are competing for the same anionic charge on cellulose.

Regardless of the time required for desorption, the adsorbed polymer is able to leave the cellulose fiber surface and migrate into water. The "pseudo-irreversible" adsorption involves, at the other end, a desorption process [51,52] that confirms the reversible character of the ionic bonds.

3.2.1 POLYELECTROLYTE INTERACTIONS IN A CONTINUOUS PHASE

Nonionic polymers take a random coil structure (I) in solution. At low polymer concentrations, the coil diameter depends only on the solvent–polymer interactions. Polyelectrolytes are carrying electric charges and, due to repulsive forces, the molecular coil is much longer (II), with large loops.

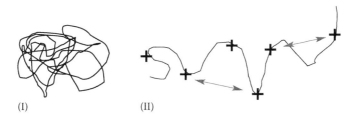

(I)   (II)

The presence of a hydrophobic part in the polyelectrolyte molecule [53] changes the ionic charges distribution in the macromolecular coil [54].

Water-soluble anionic polyelectrolytes (including anionic trash) may interact with water-soluble cationic polyelectrolytes used as paper chemicals. Polyanion–polycation complex is the so-called SYMPLEX [55], "complex coacervates" [47], "polysalt complex" or "polyelectrolyte complex" [56].

Different states of such systems include soluble complexes, stable dispersions, and precipitates [57]. One such example is the anionic polyacrylic acid, which causes partial precipitation of polyethyleneimine (PEI) from its aqueous solution in the pH range from 4 to 8. As expected, no precipitation occurred at higher alkalinities [58].

The complex solubility and its residual charge density depend on the molecular weight of the polyelectrolyte [59], type of ionic site, the charge distribution (see the copolymer structure) and accessibility, total concentration, molar ratio (anion/cation), pH [60], ionic strength of the continuous phase [57,61,62], temperature, order of addition, etc. The presence of a branched structure will result in more large loops and a poorer compensation of the opposite charges [63].

Polyelectrolytes interact with small molecules as well. Anionic polymers interact with cationic emulsifiers (cetyltrimethyl ammonium bromide) [64] or with aluminum sulfate [65]. These interactions are so strong that the anionic insoluble copolymer (ethyl acrylate (46%), vinyl acetate (50%), acrylic acid (4%) copolymer) is brought into solution in the presence of lauryl-trimethylammonium chloride [66].

When a cationic polymer is added to an anionic polymer solution, the anionic polymer is in excess and the polyelectrolyte complex coexists with un-bonded polyanion [67]. The free polyanion disappeared when the ratio between cationic to anionic polymers was about 0.9. The 0.9 ratio means that there are uncompensated charges on both anionic and cationic polymer. The experimental values of the charge titration of the complex are higher by about 40% than the theoretically calculated numbers [67].

The polyelectrolyte ratio at the neutralization point is also crucial for the wet end chemistry because it will set the amount of paper chemicals needed, the maximum retention, and the amount of material lost in water.

The stoichiometric reaction between a polyanion and a polycation depends on that macromolecular coil conformation [68] and the dissociation of the ionic functionalities. Both depend on pH and salt concentration [69]. The stoichiometric reaction occurs only when the polyelectrolytes have a sufficiently open, extended conformation. The pairing of segments from many different entangled macromolecules would make it impossible to achieve high degrees of reaction. Primary polyelectrolyte complex particles involve only a small number of molecules. Any other interactions restrict the motion of ionic species (self-restricted structure).

The electrolyte ($NaCl$, $CaCl_2$) changes the macromolecular coil shape [70]: salt effect and the diffusion of counter-ions will slow down the re-conformation process. The presence of a polyvalent cation (Ca^{2+}) results in a deviation from stoichiometry and a higher $NaCl$ content helps the dissolution of the polyelectrolyte complex [57].

During papermaking, the interactions in the water phase compete with the possible interactions with cellulose fibers as a "rigid polyelectrolyte," which displays an interface able to adsorb paper chemicals.

3.2.2 POLYELECTROLYTE ADSORPTION AT AN INTERFACE

Polyelectrolytes (such as retention aids, cationic starches, WSR, TWSR, or DSR) interact with all other components at the wet end: cellulose fibers (including fines), internal sizes, latex particles, aluminum compounds (precipitated form), fillers, etc. The properties of the wet web and the dry paper depend on the amount and quality of retained paper chemicals. The pulp slurry drainage rate, the filler (and/or other chemicals) retention, and the white water clarification, all depend to a large extent on the manipulation of the paper chemicals adsorption on the cellulose fibers.

The amount of polymer adsorbed onto cellulose fibers depends on the type of pulp [71,72] (the fines concentration [73,74], the cellulose pore size [12,75,76]), water phase (salt concentration [77–79], pH, the amount of dissolved wood-based compounds pine xylan, lignosulfonic acid [80]), temperature [48], the concentration, chemical structure, charge density and molecular weight of the polymer [13,71,81–86], as well as the size and type of filler [87].

For a given system, the amount of polyelectrolyte adsorbed on cellulose fibers is limited by the fiber charge density, the charge density of the polyelectrolyte and the ionic strength [82,88]. At a certain level of polyelectrolyte addition, the surface is saturated and further addition of additive remains in the white water. Some molecules may desorb as a result of short-range repulsion [89].

The adsorption of a polyelectrolyte on a smooth surface (polystyrene particles) is developed in the first seconds and remains constant for about 4000 min afterward. Cellulose fibers display a rough surface, with patches of crystalline alternating with amorphous structures, with spots of lignin and hemicellulose, and with pores having different sizes. Thus, the adsorption of a polyelectrolyte onto cellulose is a function of time; the polymer concentration at the interface reaches a plateau only after about 100 min. The adsorbed cationic polyacrylamide (PAAm) reached the equilibrium after 1 week [90]. The presence of fines further complicates the picture: cationic resins are preferentially attached to the fines [91].

The amount of cPAAm with a higher molecular weight adsorbed on cellulose is less than that of polymers with lower molecular weight. Higher molecular weight polymers can only reach the outer surface and the larger pores of the fiber [90,92]. Low molecular weight polymers can penetrate the pores. Therefore, cellulose fibers behave as molecular sieves: to penetrate into the fibers' pores, the diameter of polymer coil should be 3–5 times smaller than the diameter of the pore [51].

The shape and the adsorption of the macromolecular coil depends on the composition of the water phase and on the pH value [93,94]. Cellulose has negative charges that can be protonated in the acid pH. Retention is much stronger in alkaline pH [95] when the carboxylic groups are more dissociated [96]. Changing the pH of aqueous buffer solutions in contact with the pulp fibers also alters the wettability of the pulp fibers [97].

Water is present in the continuous phase and in the swollen fiber. Any adsorbed molecule must replace the existing water molecule in the fiber vicinity. The adsorbed molecule must have a stronger affinity for the solvent (water) and a weaker interaction with water. In the presence of salt, when ionic species are less dissociated, the polyelectrolytes interaction with water changes. At higher salt concentrations, the polyelectrolyte can precipitate.

Salt concentration regulates both the amount of adsorbed polymer [98] and its conformation [99]. Low salt concentrations improve the polyelectrolyte adsorption [51]. The amount of poly-DADMAC adsorbed on cellulosic fibers is about six times higher in 10^{-2} mol/L $CaCl_2$ than in deionized water [79].

However, the polymer–cellulose interaction is driven by the ion-exchange process; higher conductivity and higher hardness lead to lower adsorption [100], therefore the adsorption of a cationic polymer [101] vs. the salt concentration shows a maximum. At higher salt concentrations, the adsorbed polymer will have larger loops and tails, the adsorbed layer becomes thicker and the desorption process starts. For cationic starch, a higher concentration of alum results in a lower starch adsorption [102,103]. Ionic small molecules, such as anionic emulsifier [104] or anionic trash, can moderate the interactions between the cationic (co)polymers and the cellulose fibers.

Due to the adsorption process, the polymer concentration on the fiber surface is higher than the concentration in solution. The adsorbed layer thickness [105] increases with the increasing polymer concentration in the water phase. Based on the segment density profile at the interface [106], the polymer concentration falls monotonously with increasing distance from the solid surface until it reaches the concentration in the continuous phase.

The cellulose fiber can be seen as an anionic polyelectrolyte. Cationic polyelectrolytes adsorb to such extent as to neutralize charges on the surface of the cellulose fiber. Therefore, the adsorbed amount of polymer increases with the surface charge density but decreases with the increasing of the polymer charge density [77,107,108].

The addition of 0.5% cationic starch (degree of substitution 0.047 mol/mol) can change to zero the zeta potential for cellulose fibers. However, starch retention continues to be 100% up to 1.5% starch based on fibers [100]. It seems that, after the neutralization of anionic charges, cationic starch continues to adsorb on cellulose fibers. Much more cationic polymers (like *p*-DADMAC) are adsorbed on a slightly anionic support (silica) after the isoelectric point was reached [109].

Anionic charges are randomly distributed on the cellulose fibers surface. The adsorbed cationic polyelectrolyte also has a random distribution of its charges. Obviously, the randomness of the polyelectrolyte charge distribution cannot match the randomness of the anionic charge distribution of the cellulose fiber. An aggravating factor is the rigidity of the cellulose structure and its complicated "solid" shape (pores with different sizes and uneven distribution).

Regardless of the very close match between the charge on the fibers and charge on the polyelectrolyte, the adsorption stoichiometry was found to be about 90% [75]. This will give us an idea about the quantitative titration and the "absolute" value of the charge density. Below the equivalence point, cationic starch with a higher charge density is better adsorbed on a given pulp [110]. Obviously, at the equivalence point, the amount of cationic starch adsorbed decreases with its increasing cationic charge density [73,82,90,98]. Polymers with a higher charge density will develop repulsive forces, which further extend molecular coil, exposing cationic charges for interactions with the anionic fibers.

Branched polymers show a different shape for their macromolecular coil. Cationic starch is a blend of cationic amylose and cationic amylopectin. The linear, cationic amylose adsorbs preferentially on cellulose fibers more so than the branched amylopectin [82,102]. For a branched PEI [76], due to the steric hindrance, the cationic charges cannot reach that short distance to anionic charges and set an ionic bond.

Anionic polymers (such as CMC or oxidized starch [86]) are difficult to adsorb onto pulp because of the electric repulsion with the negatively charged groups on pulp. The CMC adsorption increases at lower pH in the presence of alum [111] or calcium chloride [112]. At lower pH [113] and at high temperatures, the CMC adsorption onto the cellulose fibers (in the presence of 0.05 M CaCl$_2$) is almost quantitative [114]. Thus, the charge density of cellulose fibers can be controlled by the pre-adsorption of CMC [111].

There is a tendency for better adsorption of higher molecular weight polymers (probably on the outer surface) based on a higher number of potential ionic bonds. At the same time, smaller molecules are better adsorbed into the pore structure [15]. Therefore, the polymer molecular weight and molecular weight distribution [50] is important for their diffusion through water and adsorption. For those polymers with broad molecular weight distribution, the adsorption process could have a fractionation effect; longer chains can displace shorter ones [78,83].

3.2.3 The Polymer Retention Mechanism

The adsorption of polyelectrolyte onto cellulose is a function of time. Due to the fast continuous papermaking process, the adsorption of polymers onto cellulose fibers during short time intervals is of great practical interest [115,116]. Moreover, wet end chemistry is very complex: cellulose fibers and fillers [117] are competing to adsorb polyelectrolytes; the presence of an electrolyte (NaCl) changes that process [118].

For a system containing pulp of 0.2% consistency and 0.1% polymer (based on cellulose fibers) with a molecular weight of 10 million, the collision frequency is 8×10^{17} collisions/m$^3 \cdot$s. Polymer molecules have a high probability of being adsorbed [115,119] in this system. Thus, an important amount of cPAAm was adsorbed after 0.5 s [120]. However, the adsorption equilibrium is reached after a very long period of time [121]. For smaller molecules (such as polyionene with $M_w = 5900$), the adsorption equilibrium has been reached in about 5 days [75]. Cationic polymers with low molecular weights ($M_w = 6,000$), such as 3,6-polyionene, are adsorbed at a rate about three times higher than a poly-DADMAC with the same charge density but with a much higher molecular weight ($M_w = 150,000$) [108].

For higher molecular weight, the adsorption time did not reach a real plateau even after 10 days [51,90,122]. Therefore, there is a "short" and a "long time" adsorption. This is translated into an evolution of the adsorbed layer over time: the amount of adsorbed polymer and the conformation of the molecular coil are changing during the papermaking process.

During the first adsorption stage (seconds), only a small amount of polymer is adsorbed [120] and the randomly adsorbed coils initially have the same size as in solution.

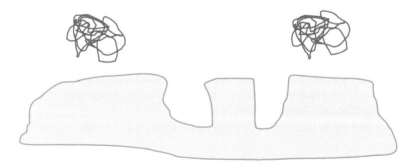

The radius of gyration in solution for cationic PAAm is approximately 125 nm [90]. The area covered by 400 µg of random coils is approximately 3 m^2 for one gram of cellulose fibers. This corresponds well to the measured value of the specific outer hydrodynamic surface area of cellulosic fibers [115].

Polyelectrolyte adsorption is driven by the presence of an anionic charge onto cellulose fibers [19]: the more anionic charges (carboxylic groups) on a cellulose fiber, the more cationic polymer is adsorbed [51,73,75,123]. Ionic bonds developed at the peripheral of the macromolecular coil will start to change the coil shape. The random distribution of the anionic charge on the fiber surface does not match the charge density distribution of a homopolymer (such as pDADMAC). Adsorbed polyelectrolytes form loops and tails so that many polymer segments are still in solution [78]. The tail distribution is very wide and the fraction of the tail segments is over half of the chain length [124].

The presence of the macromolecular tail helps the re-conformation of the coil when new ionic bonds are formed. The re-conformation of the molecular coil is the rate determining step [120]: high molecular weight polymers need a longer time to reach the final position. Therefore, a shorter contact time at the wet end may result in a poorer retention [125]. Coil re-conformation allows for more polymers to be adsorbed.

After re-conformation, the macromolecular coil is often flatter than the random coil configuration of the same polymer in solution. The adsorbed polyelectrolyte layer is more compact than the adsorbed layer of a nonionic polymer, which is thicker [106].

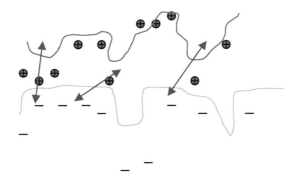

In the case of adsorbed cationic polymers, in order to reach a flat position on the surface and a 1:1 charge ratio, the distance between cationic charges must match the distance between anionic charges on the cellulose surface, which is very unlikely. The distance between charges of cPAAm shows an average value of 0.89 nm [90]. cPAAm is a random copolymer, and behind the "average" number there is the distribution of that distance. The cationic PAAm adsorption must fit a larger distance between the anionic charges onto the polystyrene latex (1.71 or 1.44 nm). Even for many ionic bonds, the re-conformed molecules have large loops and long tails.

In the case of a homopolymer (pDADMAC), with a more rigid structure (poly-cycles), the distance between charges is set to 0.4 nm [55] or 0.5 nm [56] and a slower re-conformation is expected.

Reduced free rotation, bond angles, hydroxyl–carboxyl interaction [126] and hydrophobic–hydrophobic interaction will restrain the free re-conformation of the adsorbed molecule. After the first step (adsorption) and before any re-conformation, we have to consider the system as self-restricted because the initial few ionic bonds are able to reduce the mobility of the molecule.

For a highly charged polymer of low molecular weight (10^4), the re-conformation was completed during the shortest times (5–10 s). The time for re-conformation and penetration into the fiber pores is longer for higher molecular weight polymers with higher charge densities [90].

The outer surface of cellulosic fibers is of about 3–4 m^2/g, while the total surface area of cellulosic fibers (including the pores area) is over 200 m^2/g. If we assume that the charge density is evenly distributed on the total surface, the number of charges located in pores is about 50 times higher than the number of charges located on the outer surface. In order to reach anionic charges trapped in the pore structure, the macromolecular paper chemical must penetrate into the pores. Pore penetration by polymers is a time-dependent process as well [48,75]: the diffusion occurs within several hours for low charge density cationic dextran, and within months for high charge density cPAAm [118].

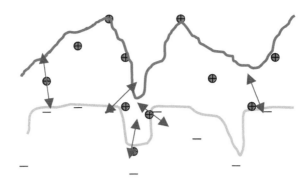

Re-conformation may involve the formation of a larger number of ionic bonds for one macro-molecule. However, each ionic bond is reversible and therefore the total number of ionic bonds increases but individual bonds may change their position (see the pseudo-irreversible process). During re-conformation, some anionic charges (on cellulose the fiber) may become available and more polymers can be adsorbed.

It might be conceptually difficult to distinguish between re-conformation and penetration on a surface as complicated as that of cellulose fibers. Re-conformation is a process in which rotations around chemical bonds are involved. A new ionic bond will freeze the new loop. The remaining tail will try to make another loop following the formation of a new ionic bond. The more flexible the macromolecular structure, the faster is the re-conformation process. Macromolecular flexibility depends on the nature of chemical bonds existing in the backbone. PAE seems to be more flexible than PEI [16].

Polymer adsorption is a dynamic process, which includes the polymer desorption, polymer dif-fusion, and re-adsorption. The polyelectrolyte dissolved in water can replace the polyelectrolyte already adsorbed [119]. The non-adsorbed polyelectrolyte can be removed by centrifugation and washing [127], and during the papermaking process, the un-retained polymer will remain in the white water, where its biodegradability should be a concern [128].

Some adsorbed polymer (PEI on cellulose) can also be desorbed by washing with water but the major amount of adsorbed PEI is irreversibly bonded on the cellulose surface [121]. A longer time for re-conformation [115] results in a stronger bonding and the inter-changing process becomes less effective [117]. The re-conformation process also results in shorter tails and less effectiveness in fines retention [125]. Pseudo-irreversible adsorption is reached when a certain number of ionic bonds are developed for a given polymer molecular weight: the desorption rate decreases with the polymers' molecular weight [129].

During the dynamic adsorption process, a new solid support (such as anionic polystyrene latex) may transfer adsorbed polyelectrolytes from cellulose to the particles of polystyrene latex. This is a slow process: it takes more than 3 weeks [130]. The transfer rate depends on the charge density and the molecular weight of the cationic polymer and the latex particle size [131]. As expected, the transferred polymer is that with a lower molecular weight [115,119]. The cleavage of the covalent bonds within a macromolecule [117,131] is difficult to explain.

The desorption process is more difficult for cationic starch: retained starch could not be removed from the fiber by washing or even prolonged boiling in water [103]. Cationic starch has a very large molecule and after adsorption, the similarity of the chemical structure might develop some other hydrogen bonding along with the ionic interactions.

Inorganic salts have a strong effect on the shape of the molecular coil of polyelectrolytes [132], on their adsorption–desorption process [101], on the amount of water incorporated in the cellulose fibers, and on the dissociation of ions inside the fiber [133–135].

The neutralization of the anionic charges with cationic polyelectrolytes at the wet end has impor-tant consequences on the papermaking process: reduction in the electrokinetic charge of fibers, and increase in the drainage rate and paper strength [16,28]. Since the papermaking process is a dynamic system with short contact time between paper chemicals and cellulose fibers, the kinetics of adsorption as well as changes in the particle charge must be of decisive importance [136]. They should be quantitatively measured (see Sections 3.2.4 and 3.3.1).

At any moment of the re-conformation of the polyelectrolyte molecular coil, there are uncompensated anionic charges on the cellulose fibers and uncompensated cationic charges on polyelectrolyte. A new anionic polyelectrolyte can be adsorbed on top of the cationic, making a multilayer structure [137,138]. The polyelectrolyte complex (PAE/CMC) with a slight excess of anionic compound is retained on the pretreated cellulose fibers with a cationic polymer (PAE) [139].

Due to the geometry of the cellulose surface and the macromolecular character of the paper chemicals, the neutralization process is a kinetically controlled process. The polymer is believed to interact first with easily accessible outer surfaces. The adsorption and re-confirmation of cationic polyelectrolytes result in charge decay, which is a function of time. The charge density decay depends on the structure of the paper chemical, the polymer molecular weight [140], the polymer charge density, the pulp carboxyl content, the agitation rate [141], the polymer dosage level, pulp refining, fines content, and pH [142]. The charge decay vs. time can be an indirect evaluation of the polymer re-conformation.

PEI (0.05%) added to a bleached sulfite pulp will move the electrophoretic mobility into the cationic range within the first 5 s. During the following 300 s, the electrophoretic mobility decays [16] back to the anionic range. The anionic surface will adsorb more cationic polymers [121]. An increase in charge decay rate is noted with an increase in temperature, consistency, stock agitation time [143], and with a decrease in the electrolyte concentration and the polymer molecular weight [144].

3.2.4 POLYMER PARTICLES RETAINED ON CELLULOSE FIBERS

There are dispersed paper chemicals like starch, internal sizing agents (cationic rosin, AKD and ASA emulsions), fillers, etc., utilized in the papermaking system. The retention mechanism of those dispersed particles is different and less complicated than water-soluble polymers. Starches used at the wet end are retained through mechanical entanglement [110].

There is no interaction between anionic latexes and cellulose fibers in the absence of an added retention aid [145]. Anionic latexes are deposited (heterocoagulation [146]) on negatively charged cellulose fibers by using electrolytes or coupling agents [147]. Styrene–glycidyl methacrylates copolymer latex [148,149] has an anionic charge due to the presence of an emulsifier (SLS) and of sulfate groups from a persulfate initiator. Due to its anionicity, this latex needs a retention aid like poly-DADMAC or a cationic small molecule (alkyl benzyl dimethyl ammonium chloride) to reduce the latex stability [150].

The retention process of the cationic latexes (such as styrene-butadiene-2(diethylamino)ethyl methacrylate copolymers [147,151]) depends on the latex concentration, pH [145,152–154], time [155–157], the latex particle size, electrolyte concentration [93], and on the amount of stabilizer (large amount of nonionic stabilizer prevents the heterocoagulation) [158], as well as the presence of lingo-sulfonate [159–161]. However, latex retention, as that of other polyelectrolytes, is always less than 100%.

Cationic latex retention is lower at higher pH [147], and it is also noticed in the presence of sodium chloride latex desorption (detachment [93,146,162]). Experimental data confirm the ionic bond formation: any change in the ionic pair dissociation affects the retention process. This is valid for polymer particles with high T_g values (such as polystyrene [163]) because their interactions with cellulose are limited to the ionic bonds.

Residual fragments of initiator (either anionic or cationic) and molecules of emulsifier (either anionic or cationic) from the water phase of the latexes interfere with the adsorption of polymeric particles [161,164]. Using the (self-stabilized) latex obtained in the absence of emulsifier, is a better option [93,147,162]. A small amount of polyelectrolyte may collect the water-soluble ionic species [164], or the technique of latex purification by serum replacement, should be considered [145].

The interaction between the cellulose fiber and latex particles is more complicated. The new type of interaction depends on the polymer T_g, the drying temperature, and the pressure applied to the paper web. The polystyrene particle ($T_g = 105°C$) retains its geometry during the drying step, the number of ionic bonds with the cellulose fiber is limited, and the particles desorb especially in the presence of salt [93].

When the polymer has a lower T_g [165,166], after its deposition on the cellulose fiber, the particle may lose its spherical shape and become flatter. This transformation will change the nature of the particle-cellulose interaction, which is no longer ionic only: hydrophobic interactions make the particle adsorption irreversible. Even polystyrene particles are spread on cellulose fibers at a higher drying temperature (140°C).

Hydrophobic polymer particles present in the sheet affect its interaction with water and make it water repellent [161]. Hydrophobic polymers in the form of latex are a multi-functional additives acting as both bonding and sizing agents [157].

3.2.5 Colloidal Titration

Polyelectrolyte interactions in homogeneous or heterogeneous systems are applied to measure the charge density and furnish cationic demand [167,168]. The standard method uses pDADMAC for the estimation of the cationic demand, and PVSK (anionic polyelectrolyte) for the titration of cationic paper chemicals (or for back titration [169]). Both titrations involve polyelectrolyte interactions and precipitations. The cellulose cationic demand is a polyelectrolyte adsorption on the surface of cellulose fibers. For the charge density estimate of any paper chemicals, polyelectrolytes interact in homogeneous phase.

Polyelectrolyte interaction with an ionic, small molecule is fundamentally important in establishing the standards for colloidal titration (metachromasy with a dye such as toluidine blue) [168]. This colloidal titration measures the changes in ionic interactions and the response is a change in color [168]. A streaming current detector can also be used [170–172]. However, the ionic interactions are perturbed by other factors such as the macromolecular coil shape, which depends on the ionic strength of the solution, and hydrophobic associations (micelle formation).

That titration poses all the problems mentioned in the previous chapters: overall concentration, paper chemical molecular weight and charge density [173], molecular weight distribution for PVSK (or pDADMAC), presence of electrolyte [38,174,175], pH [176] (see the constant of the carboxyl group dissociation or amine protonation [177] as a function of pH [167]), and, lastly, the time reserved for titration. For accurate measurements, the solution pH should be set before titration [34] and the electrical conductivity should be adjusted to 1000 μS/cm with sodium sulfate [178].

The largest number of cellulose charges is located in pores. The polyelectrolyte penetration (diffusion [75]) becomes the driving force in the titration method. The penetration is time dependent and so should be the titration (for a sample of 50 mL, the recommended titrant dosage rate is 1–2 mL/h [174]).

The surface of cellulose fibers includes some amorphous zones and some residual hemicelluloses. On those areas, the polysaccharide structure is swollen in water (like a gel) and the charge titration of a microgel structure [174] shows a larger deviation from the 1:1 stoichiometry at the endpoint.

The anionic–cationic ratio is often different from 1:1 [179]: the cationic pDADMAC–anionic polyacrylic acid ratio is 0.9 at the isoelectric point [56]. The stoichiometry of the polyelectrolyte complex formation also depends on pH, the ionic strength [59,172,180], the order of addition [46], and the alum concentration. The complexation between colloidal aluminum and polyelectrolytes of opposite charge may strongly deviate from the stoichiometric ratio [38,171].

In other words, the deviation from 1:1 stoichiometry in the polyelectrolyte complex formation is the rule rather than the exception [55,174,181]. Cationic—anionic charge interactions seem to be stoichiometrically possible only by accident [48]. The amount of adsorbed cationic polymer is not likely to rigorously reflect the amount of charge present on the fiber surface. Low molecular weight polymers are adsorbed close to a 1:1 stoichiometry. However, the adsorption onto the cellulose fiber of high molecular weight polymers is lower and it is not stoichiometrically proportional to the charge on fibers [108]. The charge density of Kraft lignin measured with a low molecular weight cationic polymer ($M_w = 7800$) is twice that obtained by titration with a high molecular weight poly-electrolyte ($M_w = 10^5$) [59].

That is why a low molecular weight polymer (such as 3,6-polyionene with $M_w = 6000$) can assess the total fibers' charge and a relatively high molecular weight polymer (poly-DADMAC with $M_w = 400,000$) can be used to estimate the surface charge [180].

Colloidal titration has an important benefit: the concentration and the cationic demand of anionic fibers can be estimated [48,181–183] by a special experimental procedure (which involves a filtration and a back titration). The above-mentioned downsides notwithstanding, colloidal titration can be a useful tool for a process operated in the steady state, when titration serves to identify variations within a standard formula (relative values). Papermaking can be optimized by combining a charge detector with the retention aid monitoring system [34].

3.3   RETENTION AIDS

The retention process has two major targets: to bring together most of the component at the wet end (quantitative target for flocculation) and to organize the floc structure in order to retain less water (qualitative target). The flocculation must include the largest amount possible of fibers, fines, and fillers. The floc structure prefigures the dewatering step, the production rate of the paper machine [184], the drying processes, and the paper strength.

The papermaking process starts with a dispersed system with more than 99% water at the wet end and the final dried paper has less than 6% moisture The handling of that large amount of water is handled (flocculation, filtration, and evaporation) is the key factor for both the economic and the technical aspects of the papermaking process.

If polymer adsorption at the wet end can be seen as a micro-selection of the chemicals, the flocculation [185], and filtration processes [186] (sheet formation) on the Fourdrinier wire are macro-selections. The art of wet end chemistry would be a "co-flocculation" of filler with cellulose fibers [187]: homo-flocculation of the filler particles results in poorer optical properties, while the homo-flocculation of the fibers results in their uneven distribution in the sheet and lower mechanical properties.

Wet end chemistry involves a diversity of compounds and processes in which the adsorption paper of chemicals onto the cellulose fibers is only a part. The paper chemicals can reduce the variability in moisture content [188] only if they are retained on the cellulose fibers. The filler distribution in the *z*-direction is much more uniform at higher first pass retention (FPR) [189,190]. Polymeric retention aids increase the web consistency after filtration but at higher retention they often cause poor sheet formation due to over-flocculation [191].

Retention is always less than 100%: the average retention varies between 30% and 80% [192]. Cationic cellulose fibers (treated with PEI to overcome the isoelectric point) retained less than 90% of anionic clay [193]. If paper is manufactured from 100% broke, the native starch is retained at only 30% while the retention of cationic starch is 90% [194,195].

Since mills differ greatly in terms of fiber type, fillers, and water quality, it is of extreme importance to select the proper retention aid for each mill situation [196]. In the absence of any coagulant, anionic charges located onto the cellulose fiber surface will generate electrostatic repulsive forces between particles [197]. Pulp flocculation is performed based on the mechanical entanglement of cellulose fibers. More anionic charges onto cellulose fibers result in a smaller floc size [198].

Without any retention aids, the components of the dispersion are not separated on the wire; most fillers and some fines pass through and remain in the white water [199]. In order to optimize the FPR, the zeta potential should be close to zero, preferably anionic. This is possible by adding a retention aid, (seen as a "control additive" able to improve the papermaking process [200]). However, no polymer can function as a retention aid in all cases [201]. For instance, mechanical pulp contains almost all the hemicellulose and lignin from the original wood and usually carries a much greater negative charge [202]. A better flocculation is ensured by changing the particle surface (such as kaolin) by adding successive layers (inorganic and organic compounds) [203].

The desired function of retention chemicals is to retain the colloidal particles at the cellulose surfaces without causing inter-fiber aggregation and thus bad formation [36]. The effectiveness of retention aids depends on the stock consistency, carboxyl content, concentration of anionic trash, electrolyte type, pH, polymer concentration, polymer charge density, polymer molecular weight, contact time, and shear rate [33,204,205]. A higher shear rate [206] decreases the FPR of fines and filler [190].

The retention of fibers and fillers involves the segment of the papermaking process located before the head box. There are pumps and mixing tanks (machine chest) and a circuit for a pulp slurry and recycled white water. The flow rates and the flow rate ratio will set the contact time. According to the usual protocol, fillers are added to the pulp slurry and the blend is diluted with white water to form the thin stock. The retention aid is then added to the thin stock upstream of the head box. Due to the high concentration of the anionic trash, a portion of retention aid may be added directly to the thick stock. For a more uniform flocculation [207] and depending on the type of pulp and filler, the streams can be divided and blended with different ratios of cellulose/filler/retention aid.

Fillers are very diverse and their particle shape, specific area, polarity, and charge density are different from those of cellulose fibers. A different adsorption of the retention aid on different solid surfaces should be expected. In order to have an even distribution of the filler in the paper web, the filler is recommended to be previously flocculated, coated with starch or treated with a phenolic resin enhancer [208–210]. Particle agglomeration [99] will improve the retention tremendously.

Most retention aids are chemicals useful to improve some other paper properties, such as dry strength and wet strength. Cationic starch [211], acrylamide copolymers, PEI or PAE resins are dry-strength or wet-strength resins as well. Besides, retention aids are important for deposition of internal sizing agents and sometimes are also used in the size press solution along with surface sizing agents [212–214].

In order to obtain effective flocculation, retention aids must be effective at very low concentrations, must have the right molecular weight, polarity and/or electric charges and to react in a very short time, in an adverse environment. They should have a high molecular weight and a low charge density [215].

Most retention aids are copolymers and thus the retention process involves a selection of the more effective fraction. The properties of copolymers should not be limited to their overall composition and average molecular weight [80]. The distribution of the chemical composition (fractions with higher or lower concentrations of ionic comonomers), and the molecular weight distribution should be considered as well.

The diversity of retention aids, in terms of chemical structures, molecular weight, and charge density, is trying solve various problems generated by using several pulp types in different conditions at the paper mills. The chemical structure of retention aids is accommodated in different protocols able to improve filler retention by using a special sequence: cellulose fibers and fillers are treated separately [216–218], or the filler is pre-flocculated [219].

3.3.1 Electrophoretic Mobility

The negative charge of cellulose generates repulsive forces between fibers dispersed in water, which makes the flocculation process extremely difficult. The heterogeneous surface is also generated by fillers. In acid pH (5–7), the aluminum compounds have a slightly positive surface while kaolin clay, titanium dioxide, and silica are slightly anionic. Calcium carbonate, which is stable only in alkaline pH, has a moderate cationic surface [26].

During the flocculation with ionic retention aids, a neutralization of the electric charges onto cellulose fibers takes place. That neutralization is recorded by electrophoretic mobility measurements. The electrophoretic mobility depends on the ion pair dissociation and on the ionic strength of the environment. Calcium carbonate is positively charged (+0.7) in distilled water and negatively charged (−1.7) in tap or white water [220–222]. The amount of PEI added to the fines in order to reach the isoelectric point is 3 mg/g fines in deionized water and 9 mg/g fines in tap water [223].

Electrokinetic measurements (the effects of chemical additives on the zeta potential [215,224,225]) are useful in estimating the cationic demand and therefore the amount of paper chemicals (i.e., the retention aid) needed for a good formation [26,33,215,224,226]. The pulp slurry drainage rate, the retention yield, and the white water clarification depend on the manipulation of the electric charges at the wet end [26].

The amount of resin needed to reach the isoelectric point depends on the cellulose surface, (obtained after the beating process), the number of anionic charges on the cellulose fibers, pH, the amount of anionic trash [33], and the presence of fillers.

The basic rules for optimizing the FPR are zeta potential close to zero [205], cationic (or anionic) demand close to zero, and a retention aid with high molecular weight and low charge density [215]. For a system using cationic resin (PAE) as a retention aid, and anionic PAAm as a dry-strength resin, the maximum value of the sizing properties and the wet-strength properties reach a maximum when the system approaches a charge of zero [227].

Alum is used to adjust the pH and to change the electrophoretic mobility of the components at the wet end. As expected, cellulose fibers and the filler particles show a different behavior: to be converted to positive charges, TiO_2 particles need less alum than cellulose fibers [228].

The papermaking process implies only slight, but delicate and very important adjustments. The amount of cationic macromolecular used as retention aid is very low (less than 0.1%), which makes it very sensitive to anionic trash [207]: due to diffusion, most of the cationic retention aid would interact with the dissolved anionic material before its adsorption on the cellulose fiber surface [35].

By adding too much cationic polymer at the wet end, the cellulose fiber charge is reversed to strongly positive [229]. The effect of the cationic polymer can be moderated by adding another anionic polymer, such as carboxymethyl cellulose.

Because polyelectrolytes are directly involved in the process, the cationic demand determination is better suited by the colloidal titration method, than by the zeta-potential method [179]. On the other hand, it is hard to establish a correlation between those two methods [48,182], mainly because the pH and the ionic strength have a different effect on each [33,230]. However, the zeta potential and colloidal titration methods should be viewed as complementary techniques [183,230].

3.3.2   FIBER FLOCCULATION MECHANISMS

Retention can be performed by filtration, mechanical attachment, and flocculation. Retention aids are mainly involved in flocculation. The retention mechanism belongs to a more general topic concerning flocculation by polymeric materials [185,231]. Flocculation is a destabilization process of a stable colloidal dispersion by the addition of a chemical compound.

Cellulose fibers and most paper chemicals and retention aids (including aluminum compound at pH > 4.5) are polyelectrolytes. The efficiency of the flocculation process depends on the properties of the cellulose fibers, pH, ionic strength, shear rate [11], and the presence of a retention aid [79,232–235]. At the wet end, the retention aid may remain in solution, adsorb onto the solids surface (fibers and filler), or interact with anionic trash [236].

The retention mechanism is related to the chemical structure and molecular weight of the retention aid. All the factors that influence the dissociation of the ionic pairs and the shape of the macromolecular coil (such as charge density, ionic strength, molecular weight, and pH) are important for a good flocculation [26].

In the absence of any chemicals at the wet end, the DLVO theory may apply [237] to the papermaking process: sheet formation is poor due to the charge repulsions between anionic cellulose fibers. One of the mechanisms to destabilize cellulose fibers dispersion is charge neutralization [49,100,221,243]. Thus, the anionic repulsive forces are shut down, which enables the weak attractive forces to initiate flocculation. This occurs when cationic small molecules (alum in acid pH, cationic surface active agent [238], or low molecular weight PEI [129,239]) neutralize the opposite sign of the anionically charged cellulose fibers. At the isoelectric point [16], the white water turbidity reaches a minimum [80].

Fiber flocculation is caused by weak forces, which is why resistance to shear forces is low. It is obvious that higher molecular weight retention aids will bring a higher efficiency (see the polyelectrolyte "pseudo-irreversible" adsorption). The neutralization and flocculation capabilities depend on the polymer molecular weight and the ratio between the cationic charge of the polymer and number of anionic charges of the cellulose fibers [114,126].

A polymeric retention aid has multiple cationic charges even for macromolecules with low molecular weight [100]. During the melamine–formaldehyde resin adsorption on cellulose fibers, essentially all the acid, used to convert to a cationic version, is released due to ionic exchange [240,241]. There is a charge reversal for slightly anionic particles (clay, titanium dioxide, calcium carbonate, etc.) after a cationic resin addition [242,243]. Due to electrostatic attraction, neutralized cellulose fibers can be coagulated with still other anionic fibers.

At the wet end, there is a competition between different components. Thus, while pulp is still negatively charged [193,244] (Figure 3.1), the PEI adsorbs much faster on clay particles, after which their charge switches to positive. Clay particles may act as bridging agents for still anionic fines [223] (co-flocculation). The positively charged clay is retained at low concentrations of PEI and lower pH [163,245,246]. The amount of retained material on cellulose fibers as a function of the amount of retention aid shows a maximum [247].

At higher concentrations of cationic retention aid, more cationic polymer is adsorbed, the cellulose fibers become cationic and their stability is improved again. The amount of retention aid is critical. Poly-aluminum compounds and pDADMAC [52] can also be included in this type of flocculation mechanism.

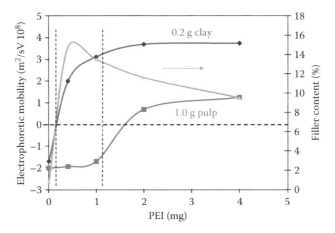

FIGURE 3.1   The retention of clay on cellulose fiber with PEI resin.

The competition for the retention aid may change when a different filler is involved at the wet end (cellulose fibers and anionic TiO_2 filler in the presence of cPAAm [248]). Cellulose fibers adsorb more cPAAm than filler and their electrophoretic mobility has a positive value, while the filler particles still preserve their anionic character. Electrostatic attraction results in co-flocculation.

The flocculation of furnishes with extremely high cationic demands may not be cost-effective when the charge neutralization approach is used [249]. Nonionic retention aids are very effective as flocculants in that case and therefore, a bridging flocculation mechanism should be considered.

Bridging flocculants [100,185,186] are ionic or nonionic polymers with high molecular weight. A bridge formation between two particles depends on both the polymer (flocculant) and the fiber properties. Differentiation should be expected between large fibers and fines (for instance, fines retain the largest amount of rosin [250]).

The charge density and the macromolecular structure of the retention aid are equally important. Low molecular weight polyelectrolytes, with a linear structure, are more likely to interact with contaminants first [236]. After adsorption on the cellulose fiber, the conformation of the polyelectrolytes shows large loops and long tails. High molecular weight cPAAm [239] flocculates by making bridges between surfaces that are highly resistant to hydraulic shear [233,235,251].

A linear polymer with a long molecule seems to be the most effective retention aid. If the molecular weight is large enough, the loops and tails penetrate beyond the electric double layer and are available for adsorption on another particle. However, the long molecule is seen in terms of the molecular coil: the most expanded molecule is more effective. Polyelectrolytes are very sensitive to the ionic strength of the solution and at the wet end, there are many ions able to change the shape of the molecular coil of the polyelectrolyte. Due to the repulsive forces, higher charge density on the polymer backbone results in a more extended coil, but higher ion concentrations in water shrink the macromolecular coil, and thus reduce its efficiency as a retention aid.

The re-conformation of the macromolecular coil takes a longer time [189] and there is a high probability for other particles to be captured by the loops and tails [124]. The tail extension [247] depends on the polymer/fiber ratio (an excess of polymer leads to the re-dispersion of the flocs [252]), on pH, the presence of electrolyte, temperature, etc.

Due to their insensitivity to the presence of ions in solution, nonionic, water-soluble (co)polymers represent a very interesting group of retention aids. For nonionic retention aids (polyethylene oxide), the adsorption mechanism on cellulose involves hydrogen bonding [1]. A nonionic polymer flocculates (through an asymmetric polymer bridging [35,253]). Both the cellulose fibers and the pigments (such as calcium carbonate) are usually anionic and, consequently, repel each other [254].

Regardless of the type of bonds between flocculants and fibers, (ionic or hydrogen bonding), polymer adsorption is a pseudo-reversible process. Retention obviously implies an adsorption step and therefore, it is also a pseudo-reversible process in which the shear rate is able to break down the flocs [52]. As far as filler retention is concerned, a deposition and also a detachment are noted [187,255]. At higher shear rate, the amount of PCC retained on cellulose fibers decreases in the presence of the same amount of cPAAm [141]. In order to increase the resistance to particle detachment, retention aids with high molecular weight are required [256].

However, for high molecular weight cationic polyelectrolytes, it is reasonable to consider both mechanisms. Low molecular weight PEI resin may help cellulose fiber aggregation based on charge neutralization, while high molecular weight PEI is also involved in a bridging mechanism [129].

3.3.3 PAPER CHEMICALS AS RETENTION AIDS

The components at the wet end of the papermaking process (such as fillers, fibers, and fines) have different specific surface areas, different chemical structures on their surface, different charge densities, and respond differently at the change in pH [251]. Enzymes can make the process even more complex [257]. Any particular flocculant will interact differently with each participant at the wet end; that diversity may result in a broad floc distribution. Beside the time-dependent adsorption of the polymers, the soluble chemicals may interfere with that adsorption process [258].

The chemical structure and the molecular weight of the retention aid must fit the complexity of wet end chemistry. There are limitations for the polymers' molecular weight, branched structure, or charge density [259]. The solids content, the solution viscosity, the water solubility, and the costs, are restraints capable of reducing the number of valuable candidates as retention aids. By using blends of two or more components (organic or inorganic compounds), it is possible to overcome those limitations.

In the following sections, retention aids are classified as nonionic, anionic, and cationic. Amphoteric compounds and blends of different ionic polyelectrolytes with specific interactions are presented in a separate section.

3.3.3.1   Nonionic Flocculants

Unlike a cationic polymer, which is consumed by anionic trash, and is thus less effective [244], the performance of a nonionic polymer should not be affected by the anionic molecule [254,260]. In order to be effective, nonionic polymers should have a very high molecular weight. The formation of intramolecular hydrogen bonding, which may reduce the size of the macromolecular coil, should be avoided [72]. There is a downside to having very high molecular weight polymers: they dissolve slowly in water and their solutions have very high viscosities even at lower concentrations.

The chemical structure of the nonionic retention aid must include hydrophilic groups such as polyether, polyamide (amide in the backbone or as a pendant group), etc. Common nonionic retention aids are polyethylene oxide (PEO) with a molecular weight higher than 1 million [35,261]), PAAm, polyamidoamine, polyvinyl pyrrolidone, and their grafted copolymers.

Nonionic PAAm with a high molecular weight ($500,000 < M_w < 7,000,000$) is a well-known retention aid [1]. Nonionic PAAm is much more effective (lowest drainage time in Figure 3.2), as compared with ionic acrylamide copolymers (anionic or cationic comonomer) with the same molecular weight (intrinsic viscosity of about 6 dL/g) [262].

Phenol-formaldehyde (PF) resin (as an enhancer) increases filler retention with about 10%, with a nonionic PAAm [208,209]. However, it is hard to make sure that the acrylamide polymer is a real nonionic homopolymer: the hydrolysis of PAAm can take place, and even at very low conversions the resulting material is an anionic polymer. That can explain the effectiveness of the acrylamide homopolymer (0.01%–0.3% based on dry pulp) in the presence of bentonite [263].

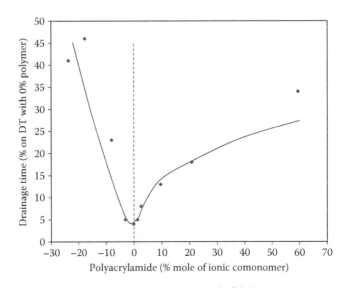

FIGURE 3.2   The drainage time vs. ionic comonomer content in PAAm.

A typical nonionic retention aid is PEO with molecular weights of over 1 million, which dramatically improves the retention and dewatering processes [264–267]. A small amount of polymer (0.05%) can reduce the drainage time from 145 s (without any additive) to 68 s. Nonionic PAAm seems to be less useful within the same experiment: only 109 s.

It was reported that PEO shows a synergetic effect in the presence of dissolved and colloidal materials (anionic trash) due to their association, which results in an aggregate with very high molecular weight [35]. Starting from that observation, the next step was to find an "enhancer" able to make PEO even more effective [260]. It was found out that PEO shows a synergistic effect as a retention aid in the presence of compounds with phenolic structure, such as Kraft lignin derivatives [254,268–270], poly(p-vinylphenol) [271], water-soluble PF resins [272–275], or modified PF resins with chloroacetic acid or sulfonic groups [276,277]. The recommended amount of phenolic resin to PEO is about 10:1. Poly(p-vinyl phenol) also has the benefit that no formaldehyde is involved in its synthesis [271,278].

Other enhancers are tannic acids, polystyrene sulfonated [279], tall oil rosin (sodium salt and a salt of polyvalent metal) [280], or poly(sodium naphthalene sulfonated) [275] (III)1 (or blends of all of the above [281]):

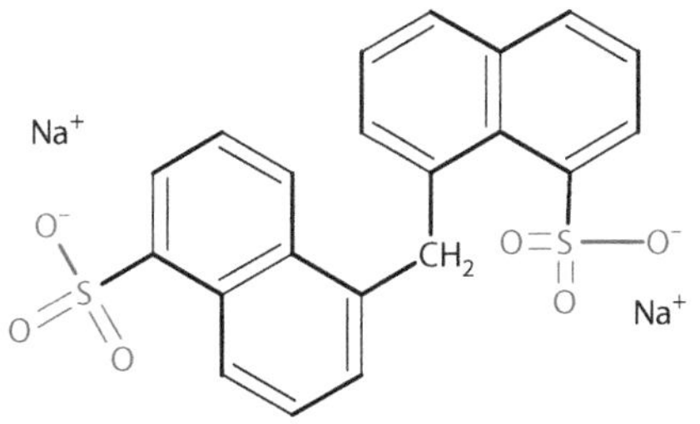

(III)1

Enhancers are polyfunctional compounds; the complex formation with PEO is based on multiple hydrogen bonding interactions [271]. The dimension of the aggregate is expected to be much larger and the PEO molecular coil shape to be different.

Clay retention is lower in the absence of any polymer, and the addition of PEO has a strong effect [272]:

$$\text{Clay retention } (\%) = 31.8 + 72.4 \cdot [\text{PEO}] \tag{3.1}$$

Retention of cellulose fibers reaches about 50% in the absence of PEO. PEO has a weaker effect on fiber retention, but an enhancer helps the process:

$$\text{Fiber retention } (\%) = 50.8 + 40.2 \cdot [\text{PEO}] + 3.38 \cdot [\text{PEO}] \cdot [\text{Enhancer}] \tag{3.2}$$

where [**PEO**] and [**Enhancer**] are the concentration of PEO and enhancer, respectively (in lb/ton of dry cellulose). The validity of that type of equation has been confirmed by other authors for disentangled PEO in the presence of a different enhancer [282].

The enhancer has a significant effect (about 20% more fiber retention [249]) only at high enhancer concentrations. However, the effectiveness of the enhancer seems to be related to the presence of a water-soluble inorganic salt [279] and their concentration should be considered for a new equation.

This relationship is valid only for reasonable PEO concentrations (below 0.05%) and the linear relationship with the PEO/enhancer ratio is also limited to certain concentrations. For instance, the PEO/enhancer shows an optimum [270]: for a constant concentration of PEO (0.023% based on dry pulp) [278] the efficiency of the system (PEO + poly(p-vinyl phenol)) has a maximum at about 0.05% enhancer (Figure 3.3).

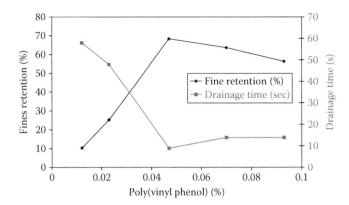

FIGURE 3.3 The effect of the enhancer on PEO effectiveness.

The adsorbed nonionic polymers interfere with the electric double layer of the clay suspension. If the electric double layer is thicker than the adsorbed polymer layer, electrostatic repulsions dominate, and no coagulation is recorded [253,260]. In the presence of an electrolyte, the double layer is compressed and the nonionic polymer is much more effective in the flocculation process [260].

The clay retention mechanism seems to be an "association-induced polymer bridging flocculation" [283]. Effective flocculation depends on the re-conformation of the polymer on the particle surface. The re-conformation (the flattening) process is time dependent [509], and retention also depends on the molecular size of the retention aid. In the case of PEO, the complexity of its dissolution is also a factor [268]. The PEO molecule can be entangled and behave as a larger macromolecular aggregate.

For more convenient handling, PEO is dispersed in water (20% solids, probably swollen in water) at a high concentration of salt (sodium formate, 20% in water) and a stabilizer (xanthan gum) [265]. Swollen PEO dispersed in water is easy to use as a retention aid for fines. The efficiency of high molecular weight PEO declined steadily over an increasing storage time. In order to prevent that process, short segments of PEO are attached to high molecular weight PAAm ($m \gg n$ and $p > 10$) [284] (III)2:

$$R \left[CH_2 - CH \right]_m \left[CH_2 - C(CH_3) \right]_n \left[CH_2 \quad O \quad CH_2 \right]_p R \qquad (III)2$$

In order to make a bridge between two particles (flocculation), the PEO molecular weight [1] must be large enough to extend beyond the thickness of the electrical double layer. There are at least two options for increasing the size of the molecule or the aggregate: to perform a grafting copolymerization (acrylamide grafted on poly-vinylpyrrolidone) [285] and to associate a nonionic polymer with other compounds. PEO forms association complexes with phenolic resins [286], polyureas, lignin derivatives [268], polymeric salts, and mercury salts [265]. Thus, hydrogen bonding between the enhancers and the ether groups of the PEO may result in larger aggregates [181,270,287,288], which can explain the cofactor efficiency. Larger aggregates as retention aids are also obtained from polysaccharides and boric acid or zirconium salts [289].

3.3.3.2 Aluminum Compounds as Retention Aids

Flocculation with a nonionic polymer does not involve the anionic charges of cellulose fibers. In order to take advantage of the ionic character of the cellulose fiber, ionic macromolecular compounds are used. In the past, papermaker's alum was the retention aid of choice because of its ability to reduce electrostatic repulsive forces [290]. The aluminum compound effectiveness depends on pH [18]. Alum can help the retention of filler (TiO_2) and fines up to 98.5% at pH 4.6 [234]. The coagulating ability of the aluminum sulfate solution decreases with aging [38] by increasing hydrodynamic forces.

Aluminum compounds are precipitated at the pH of the neutral to alkaline papermaking process ($6.0 < pH < 7.5$, see also Section 7.1). Precipitated flocs are cationic polyelectrolytes. The size and the shape of the precipitated aluminum hydroxide depend on the pH and on the presence of other ionic species in the water phase [291,292]: small amount of electrolyte influence the flocculation process [293].

Aluminum compounds must flocculate fillers, fines, and the sizing agent at the neutral or slightly alkaline pH [196,294]. Poly-aluminum chloride (PAC) is an useful coagulant at pH (6.5–7.5) [1] and it is effective in neutralizing anionic trash [215].

The sensitivity to pH of the aluminum compound, the ionic strength, and the shear rate made necessary their association with other macromolecular compounds. The combination between alum and a high molecular weight cationic polymer seems to be the best choice in systems with high anionic trash content [201].

Dispersions of aluminum silicates consist of three mineral layers containing a central layer of alumina sandwiched between two layers of silica, with an overall anionic lattice charge. Aluminum silicate (bentonite) [295] helps the flocculation process by mechanical entrapment. Their effect in flocculation is improved by the addition of a polymer [296–300] such as anionic PAAm [301].

A star-like hybrid (organic polymer–inorganic compound) is obtained by attaching PAAm molecules to aluminum hydroxide particles [302]. The links between inorganic particles having cationic charges and the PAAm molecules, can be made either through the residual anionic sulfate group located at the end of the molecule or through the few anionic carboxylic groups obtained by amide hydrolysis. The hybrid has cationic charges, a much higher molecular weight of the extended "PAAm," and a branched structure. It provides a better flocculation than straight PAAm.

3.3.3.3 Anionic Retention Aids

Anionic polyelectrolytes are generally used to flocculate the dispersion of particles with cationic charges [303]. Cellulose fibers are carrying anionic charges. Repulsive forces are developed between particles and an anionic coagulant. However, anionic macromolecular compounds (anionic PAAm [304], anionic polyacrolein [292], or alginate [216]) promote the coagulation of mineral dispersions when divalent metal ions are present or protect calcium carbonate in acid pH. Sodium alginate (III)3 flocculates the cellulosic dispersion only in the presence of calcium and barium cations [305].

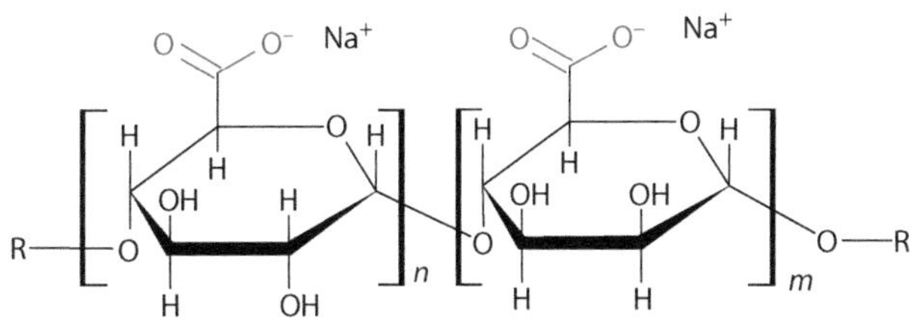

(III)3

Anionic starch (carboxy-alkyl ether or 2-sulfo-2-carboxyethyl ether) is effective as a retention aid in the presence of alum [306,307]. Starch is maleated with maleic anhydride and the added double bond reacts with sodium bisulfite to obtain a sulfosuccinated starch [308,309] (III)4:

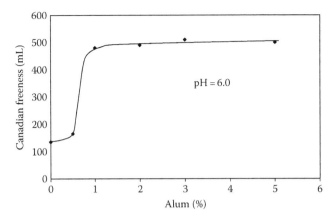

FIGURE 3.4   Canadian freeness for anionic PAAm (0.025%)—alum system.

Anionic PAAm is a synthetic polymer synthesized by partial hydrolysis of neutral PAAm; by copolymerization with acrylic acid [310], or with maleic (or fumaric) half esters [311]. Hydrolyzed PAAm must have an optimum carboxylic content [312], and its molecular weight should be over three million [313].

In the absence of alum, anionic PAAm does not have any impact on freeness. An amount of at least 1% alum (based on the dry fibers) [310] strongly impacts the flocculation of the system (Figure 3.4).

Alum concentrations over 1.5% do not have any effect on flocculation. For 2% alum (at the same value for pH = 6.0), the effect of the amount of anionic PAAm needed for a good flocculation [310] is about 100 times lower than that for alum (Figure 3.5).

PAAm with less than 3% acrylic acid [262,314] is more useful in the presence of bentonite [264]. A copolymer of methacrylic acid with allyl ether of ethoxylated fatty alcohol [315] is also used along with bentonite. Polymers with anionic groups can also inhibit the swelling of clay. That changes the slurry viscosity [316,317] and may improve its retention. Bentonite clay [318] (treated with polyacrylic acid), in the presence of high molecular weight PAAm, increases the fines retention from 34% to 65% [319].

3.3.3.4   Cationic Polymers as Retention Aids

Cationic macromolecular compounds are more effective in flocculation than the cationic inorganic compounds [188]. Due to their cationic charges, cationic retention aids can be adsorbed onto both inorganic filler particles and cellulose fibers. There is a competition between participants at the wet end to adsorb cationic species, which, given the time-dependent adsorption process (see short contact time in the case of high-speed paper machine) may result in a poor flocculation.

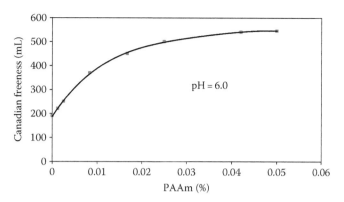

FIGURE 3.5   Canadian freeness for alum (2%)—anionic PAAm system.

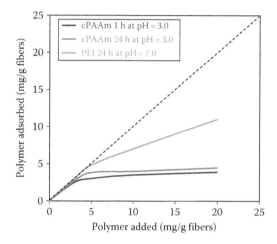

FIGURE 3.6   Cationic polymer adsorption on cellulose fibers.

The effectiveness of the retention aids depends on their adsorption on the cellulose fibers. The adsorption of cPAAm and PEI [320] was studied at different polymer/cellulose ratios, and different adsorption time (Figure 3.6). The amount of adsorbed polymer reaches a plateau, which depends on the polymer type, pH, and time.

Adsorption is practically 100% at very low polymer concentrations. At higher polymer concentrations, the adsorbed polymer is in equilibrium with the dissolved polymer and the saturation point has not been reached as yet. At a dosage of about 1100 g PAAm/ton fiber, only about 400 g cPAAm/tone fibers were adsorbed after 0.5 s [115]. After a longer period of time needed for re-conformation, the amount of cPAAm adsorbed on cellulosic fibers is much higher.

The adsorption of the retention aid on particles distributed at the wet end is the first step of flocculation. In the second step, the tails and loops of the adsorbed polymer must be connected with other cellulose fibers, and thus the dispersed system includes a lower number of dispersed particles.

There are different tools to manage the polymer retention and particles flocculation with cationic polyelectrolytes: the addition of an anionic small molecule (to moderate the charge density) [321], the electrolyte concentration, and the presence of another dispersed phase [322].

3.3.3.4.1   Cationic Starch as a Retention Aid

Due to their similar chemical structure, starch and cellulose have a strong affinity to each other, and thus starch is expected to be a good retention aid. In order to make it more effective, cationic charges are added to the starch molecule. Filler retention has been shown to increase from 53% to 82% when 0.7% of cationic starch was used in the example [323].

The retention value of cationic starch decreases to less than 20% when the cellulose carboxyl groups are blocked [195]. Therefore, cationic starches are retained primarily by forming ionic bonds. Cationic starches [324–326] with degrees of substitution of 0.017 and 0.047 mol/mol, 1.5% based on fibers [100,327], are 100% retained. At a 2% addition level, this retention level is only 63%–76%.

Starch is chemically modified by polymer analogous reactions (see Section 2.5.1). The degree of substitution is limited by the reaction conditions and the costs of the process including the purification of resulting cationic starch. A degree of substitution of about 5% (5 from 100 glucose units have a cationic charge) is a common number. This low degree of substitution explains the cationic starch sensitivity to the high concentration of anionic trash. Trash catchers, such as low molecular weight PEI resin or poly-DADMAC, improve the cationic starch activity [328].

The improvement of cationic starch can be done by increasing the cationic charge density. Cationic corn starch is cooked at 80°C–130°C with other cationic polymers [329–332] in the presence of an oxidizing agent (ammonium persulfate) in alkaline pH. Dry blends can also be made after cooking [333–335]. Regular starch is reacted with ethyleneimine [336] or cooked in the presence of PAC, aluminum sulfate, or ferric chloride [337] to make a cationic retention aid.

The molecular weight of the retention aid is important factor for its efficiency. Cationic starch with higher molecular weight is obtained by grafting with propylene oxide [338], by cross-linking [339–345], or by supramolecular aggregate formation.

A branched starch molecule (amylopectin) is cationized through a reaction with epichlorohydrin and then with ammonia in a heterogeneous system. The heterogeneous system suggests the idea of adding amine groups directly to the mechanical pulp, which will make the dispersion self-flocculated [192].

Large aggregates are obtained through interactions between an anionic compound and cationic starch (see Section 3.3.3.5.2). Due to its large surface area, silica will first retain the cationic starch (silica particles of about $500\,m^2/g$ and polysilicic acid over $1000\,m^2/g$ [346]) and enhance cationic starch activity as a retention aid [79,347–349]. Arguably, the silica–starch aggregate is still cationic because the starch–silica ratio is about six (alum can also be involved [349]); it also has many cationic tails able to grab the cellulose fibers.

3.3.3.4.2 Cationic Synthetic Polymers as Retention Aids

Synthetic polymers such as retention aids are obtained by polycondensation, polymerization, or copolymerization. The benefit of a cationic synthetic polymer obtained by condensation (urea-formaldehyde or melamine-formaldehyde resins) [350] was noted as retention aid for pigments or acid dyes since the 1950 [351].

Polyamine and polyamidoamine are also synthesized by condensation and are largely used as retention aids [352,353]. Polyamine can be obtained by amine (or polyamine) reactions with epichlorohydrin and/or di-chloro compounds [354–357] (or poly-epychlorydrin [358]), which lead to branched structures [359] (III)5.

(III)5

Mono-methyl amine is a trifunctional compound; its degree of branching (n/m ratio) depends on the amine-epi ratio, temperature, reaction time, and the presence of sodium hydroxide. The degree of branching is controlled by the order of addition [352,353]. The cross-linking process is stopped by adding an excess of mono-methyl amine. Linear polyamines are obtained through N-substituted ethyleneimine polymerization [360].

Modified polyamines (such as PAE thermosetting) can flocculate the cellulose fibers [1]. However, due to their reactive functional groups (azetidinium cycle), the broke is no longer repulpable. To reduce that risk, the polyamine prepolymer (see Section 5.1.1.4) is only partially cross-linked with epichlorohydrin (0.2 mol per 1 mol secondary amine) [361,362], or blends of PAE and glyoxalated PAAm were used as retention aids [363]. The azetidinium cycle from the reactive PAE resins can also be "neutralized" with ammonia (or amine) [364–366] in order to make it a non-thermosetting polyamine (III)6.

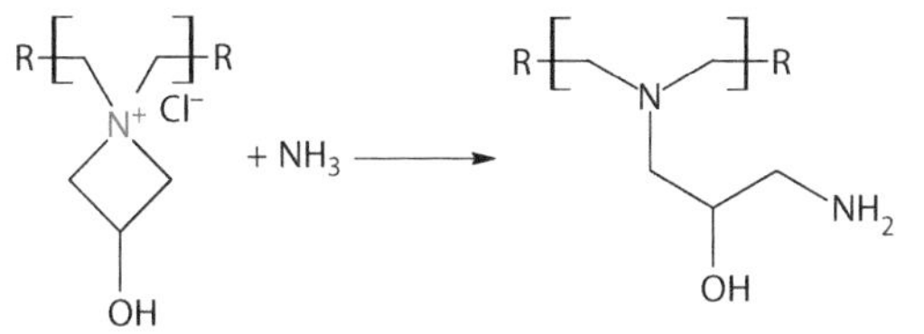

(III)6

A branched polyamine is obtained by grafting ethyleneimine on a polyamidoamine [367–369] or by partial cross-linking with a di-chloro polyether cross-linker [370,371]. The grafting and the cross-linking process generate compounds with higher molecular weight and higher tertiary amine content.

That type of condensation is difficult to control and branches are randomly distributed. An attempt to build up a symmetrical structure has been made by dendrimeric polymer synthesis [372–374]. Dendrimers are synthesized in a stepwise manner, which provides a unique control over the chemical structure. Acrylonitrile reacts with polyamine (di-amino butane, in red) through a Michael addition, and the nitrile is hydrogenated to a new polyamine. The Michael addition and the hydrogenation are repeated several times to get a dendrimeric structure (III)7:

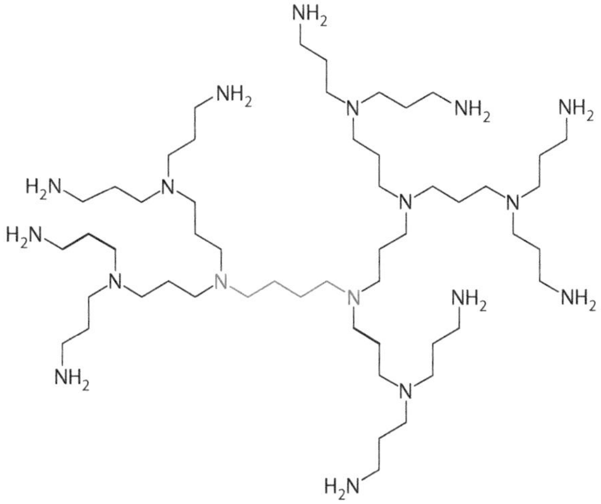

(III)7

The cationic character of these polyamines is preserved only at acid and neutral pH. Quaternary ammonium salts (such as poly-ionene [375]) are less sensitive to pH. The quaternary ammonium groups [376] are obtained by reacting of polyamine [377] or polyamidoamine [378,379] with glycidyl tri-methyl ammonium chloride (III)8.

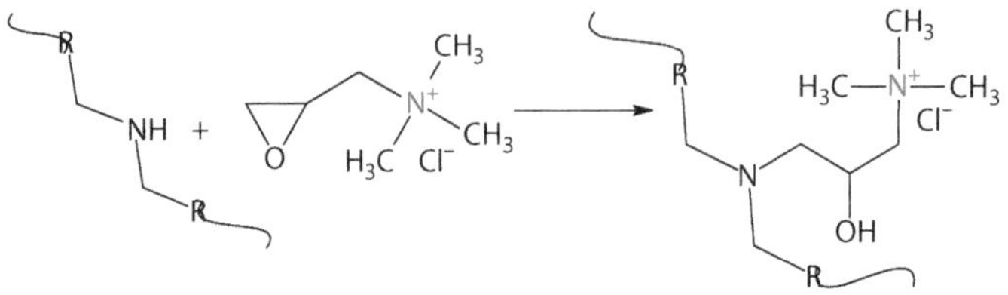

(III)8

The free radical copolymerization of acrylamide with a cationic comonomer is an easy process [380,381] and the resulted copolymers (3.1%–40% mol of cationic acrylate [80,255]) show a good flocculation performance [122,221,382,383]. Higher molecular weight cPAAm (2–12 millions) is preferred [122,384–387].

CPAAm (10% cationic comonomer) with high molecular weight ($M_w = 8 \times 10^6$) is more effective as a flocculant in the presence of low molecular weight cationic polyacrylates ($M_w = 10^4$) [388–390] or polyDADMAC [391]. When the cationic copolymer is obtained in the presence of another modified cationic copolymer, a synergetic effect as retention aid is noted [392].

High molecular weight polymers have a high viscosity in solution even at lower concentrations. It is a challenge to synthesize those polymers in solution or to dissolve high molecular weight polymers at high concentration in water. These retention aids are polyelectrolytes, which explain why they are sensitive to the presence of electrolytes. By changing the electrolyte concentration, the polyelectrolytes can be dissolved or dispersed. At higher dilution, a precipitated polyelectrolyte will go back into solution. The emulsion polymerization of acrylamide and dimethylaminoethyl acrylate benzyl chloride has been performed in the presence of a large amount of electrolytes (ammonium sulfate or sodium chloride) [392,393]. The cationic polyelectrolyte is obtained as a dispersion (15% solids and viscosity of about 2000 cPs), which turns into solution upon dilution down to 0.5% polymer [390].

Cationic polyacrylates are obtained in two steps: a polymerization and a polymer analogous reaction. Acrylates are copolymerized with N-vinyl formamide while the nonionic copolymer is hydrolyzed in a second step. The copolymers of N-vinyl formamide with acrylamide [394], acrylates [395], and acrylonitrile [396] are hydrolyzed to amines in acid conditions to protect the comonomer functionality. Cationic acrylic copolymers [397,398] and copolymers of N-vinyl formamide [395,399–403] are recommended as retention aids (0.1%) even when the pulp is contaminated with fragments of lignin [404].

Copolymers with a pendant amino functionality are modified with epoxides [405] or with aldehydes [406] (polymer analogous reactions) to make good retention aids. However, cationicity is provided by a primary or secondary amine, by protonation, which strongly depends on the pH value [407]. In an alkaline pH, primary amines are less protonated and the resin efficiency is much lower. When N-vinyl formamide is copolymerized with cationic comonomers having functional groups of the quaternarium ammonium salt, the cationic copolymer performs even at higher pH [408].

Poly-acrylonitrile is reacted with ethylene diamine to make poly(2-vinyl imidazoline) [409], which is used as a retention aid [410]. Cationic retention aids can be obtained by trans-amidation with dimethylaminomethyl amine [411] or by a Mannich reaction [412–414] (see Section 5.1.3) of a polyamide. The backbone can be a copolymer (III)9 of acrylamide and vinyl-lactams:

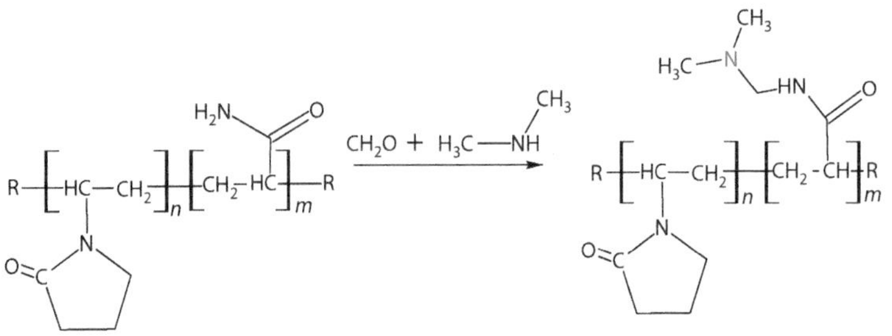

(III)9

The final step is a quaternization (III)10 with methyl chloride:

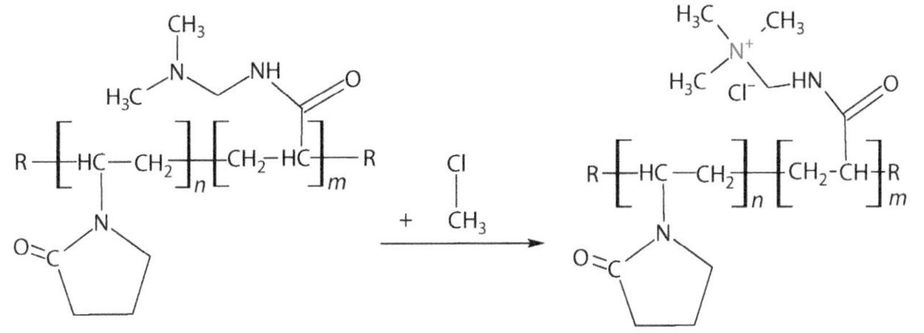

(III)10

3.3.3.4.3 Hybrid Retention Aids

An excess of anionic substances (anionic trash) decreases the flocculating ability of cationic retention aids (PAAm [222] or PEI [244]). Other potential changes in the polymer structure and the molecular coil shape in solution were also explored.

Higher molecular weight cationic polymers are obtained as a branched structure by a partial cross-linking process [259,415] (with 0.0005% methylene-bis-acrylamide) [416] or by copolymerization with a macro-monomer (allyloxy polyethyleneoxide monomethyl ether [417]). In the latter case, the branches are nonionic (hybrid macromolecular compounds) (III)11. Hybrid copolymers are also obtained by grafting a cationic comonomer onto a nonionic homopolymer [285].

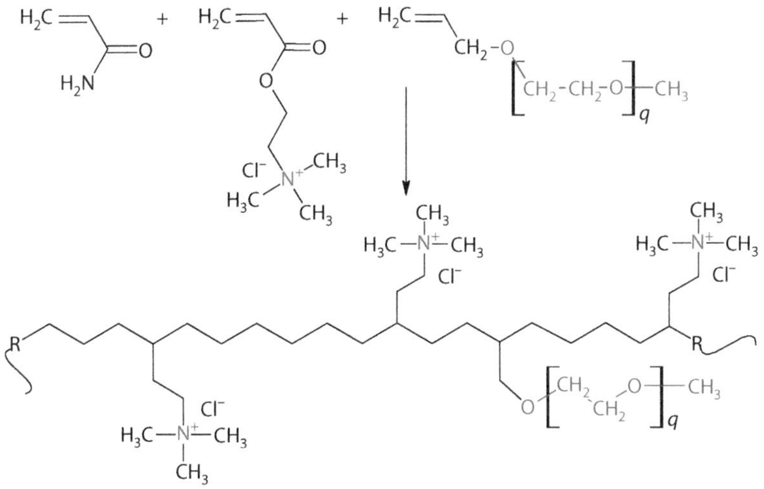

(III)11

This type of molecule seems to perform based on different flocculation mechanisms. The retention mechanism, in this case, appears to be more complex: the grafted copolymer is adsorbed on fibers by the interaction of its positive charge with the negative charge of the cellulose. Grafted PEO tails remain in water and are available for a bridging flocculation with another cellulose fiber [418].

The nonionic portion is provided by the segment from PEO or polyvinyl alcohol (PVA). The PEO grafted on polyamidoamine is synthesized by reacting of the secondary amine with a PEO having an epoxy group at one end [419] (III)12:

(III)12

Cationic PVA as retention aid is the obtained by the copolymerization of vinyl acetate with *N*-vinyl formamide [420], or with *N*-(3-dimethylaminopropyl) acrylamide [421], followed by copolymer hydrolysis.

Hybrid cationic inter-polymers are prepared from nonionic PAAm and a cationic polymer obtained by polycondensation (dicyandiamide, formaldehyde, urea with or without polyamine, and/ or epichlorohydrin) [422,423]. The cationic inter-polymer is not linear; it has a very large molecular weight as it combines the cationic condensed polymer with linear PAAm. Those inter-polymers are recommended for retention of pigments (TiO_2) or small particles of phosphate rock.

The fiber dispersion can be flocculated by polymer latex or with latexes and cPAAm [424] (hetero-coagulation [191]). Paper made with cationic latexes as retention aids for clay (up to 40% inorganic filler) [425] shows good tensile strength and is water repellant.

In order to improve the retention process, the diversity of the components at the wet end needs special attention with regard to the sequence of addition [184]. The constituents of the wet end composition are treated separately with cationic starches [426,427] or other synthetic polymers (such as PAE resins [242] or cationic copolymers [243,428]). It is recognized that one cationic polymer cannot meet the needs of all papermaking systems. Some other chemical compounds are involved in flocculation process along with cationic macromolecular compounds. Cationic polymers (such as PAAm, PEI [391,429,430] or ternary copolymer vinyl amine–*N*-vinyl formamide–DADMAC [431]) are mixed with inorganic anionic particles (bentonite, hectorite, or smectite) for a better retention [381,432,433]. Blends of cationic and anionic compounds suggest the need for amphoteric retention aids.

3.3.3.5   Amphoteric Retention Aids

As shown in the previous chapters, the balance between anionic and cationic charges is very important for retention purposes. It would be very useful to bring both types of ionic charges in the same molecule (amphoteric compound) or aggregate. That it is not an easy task. However, proteins as models of amphoteric compounds were successfully used as retention aids [434]. The presence of both cationic an anionic charges in the same molecule (zwitterionic compound) makes the retention aid effective on a broader range of pH values [435] and prevents it from collapsing at high concentrations of salt solution [436]. Amphoteric materials are obtained through a polymer analogous reaction (starch derivatization), by copolymerization, or by mixing an anionic polyelectrolyte with a cationic polyelectrolyte (dual system).

3.3.3.5.1   Amphoteric Macromolecular Compounds

Amphoteric compounds include anionic and cationic species bonded covalently. The chemical modification of the pendant groups of a preexisting polymer is performed through polymer analogous reactions. The preexisting polymer can be cationic or anionic; the added functionality will be anionic or cationic, respectively. This type of reaction can occur on both natural macromolecules and synthetic polymers.

Anionic native potato starch is amphoteric after cationization [437] (for starch etherification and esterification, see Section 2.3.1). Another route is also possible: amphoteric starch synthesis [211,435,438–440] starts from cationic starch obtained by etherification (with diethylamino ethyl chloride [441,442]) followed by esterification for anionic charges with sodium phosphate [443–445] or CS_2 to make anionic xanthate [446] (III)13 in successive reactions.

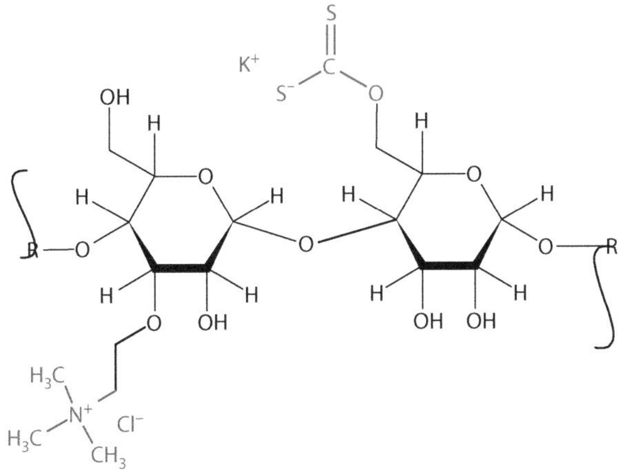

(III)13

The pendant group can also carry both anionic and cationic charges [447]. The reaction between starch and N-(2-chloroethyl)iminobis(methylene)-diphosphonic acid at a pH of 11–13 and a temperature of 20°C–95°C results in an amphoteric starch (III)14, in one step. Amphoteric starch is a very effective retention aid for pigments during the papermaking process.

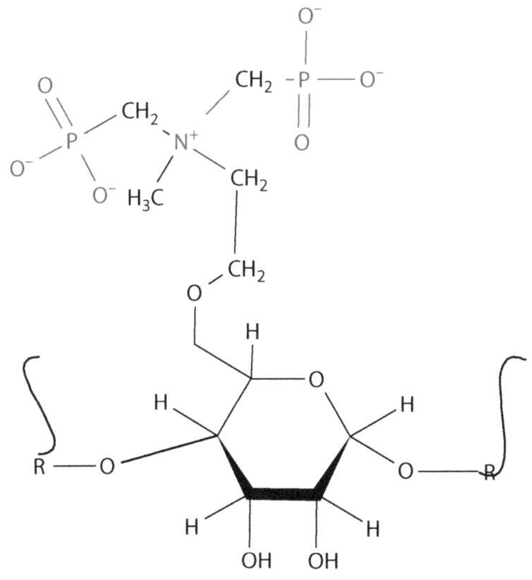

(III)14

The ratio between anionic and cationic charges is a critical parameter for amphoteric retention aids. Amphoteric starches contain a larger number of cationic charges than anionic charges: up to 0.5 mol of anionic groups per 1 mol of cationic group [439]. When a functionalized amino acid (2-chloroethylamino di-propionic acid or sulfosuccinate group) is used to modify the starch structure, the anionic groups are in excess and alum should be involved in the retention process [448–451].

It is interesting that lignin (or lignin sulfonate), a byproduct of wood processing, can return to paper as a retention aid after the addition of a cationic group through a Mannich reaction with formaldehyde and amines [452] (III)15.

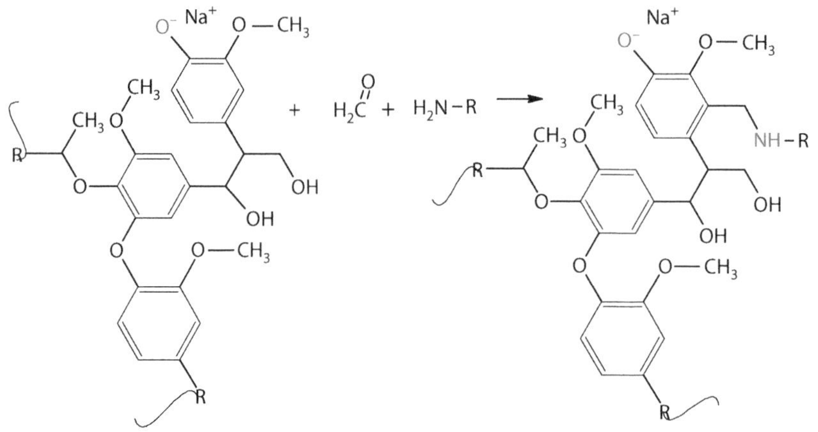

(III)15

The same concept is applied to synthetic polymers such as polyamidoamine, which reacts first with sodium chloroacetate and then with epichlorohydrin [453,454] (III)16.

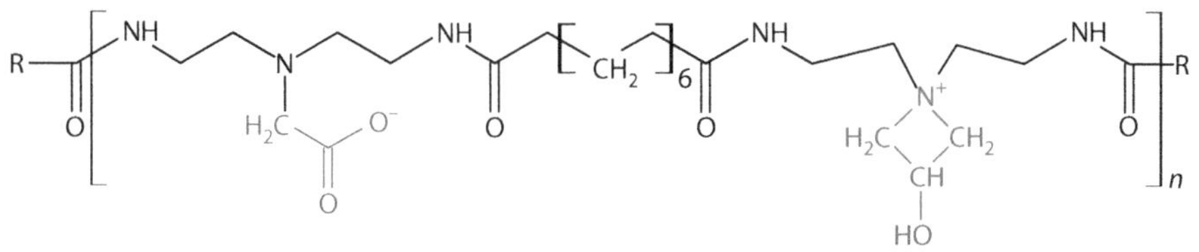

(III)16

In order to synthesize an amphoteric copolymer (polyampholyte [455]) from small molecules, a binary free radical copolymerization is used (acrylic acid–DADMAC [456], or acrylic acid dimethyaminoethyl methacrylate [457]). Monomers with lower tendency to homopolymerize (such as maleic acid–allyl amines, $r_{MA}=0.051$ and $r_{AA}=0.033$) [458,459] generate alternant amphoteric copolymers. Those copolymers show a reduced viscosity with a minimum at a pH of 3, when the amine is protonated and the carboxylic group is undissociated, resulting in a less extended molecular coil. This copolymerization reaction can be followed by a polymer analogous reaction: copolymers of vinyl-amine and acrylic acid are hydrolyzed in acidic pH [460], or the grafted PAAm on a cationic substrate (polyamine) is partially hydrolyzed to obtain an anionic species [353].

The binary copolymerization of ionic comonomers results in copolymers with a very high charge density. In order to make macromolecular compounds with a lower and adjustable charge density, a nonionic monomer (as a diluter) is added to make a ternary copolymer. Thus, nonionic acrylamide is copolymerized with an anionic (acrylic acid or itaconic acid) and cationic monomers (DADMAC, dimethyl aminopropyl acrylamide or acryloyloxyethyl trimethyl ammonium chloride) [461–465].

When the nonionic comonomer is not soluble in water (such as butyl acrylate) and is in excess of 20%, the amphoteric compound is dispersion (latex) [466].

It is difficult to manipulate a ternary copolymerization in order to obtain a desired ratio and distribution of ionic charges. For better control of the charge concentration and distribution, the copolymerization process was simplified to a binary system in which the neutral comonomer is copolymerized with an ionic comonomer having both cationic and anionic charges (amphoteric monomers). N,N-Dimethyl-N-acryloyloxyethyl-N-(3-sulfopropyl)-ammonium betaine (obtained through a reaction between N,N-dimethylaminoethyl acrylate and propane sultone (III)17) [436] and *N*,*N*-dimethyl-*N*-acrylamidopropyl-*N*-(2-carboxymethyl)-ammonium betaine (III)18 have been used for this purpose.

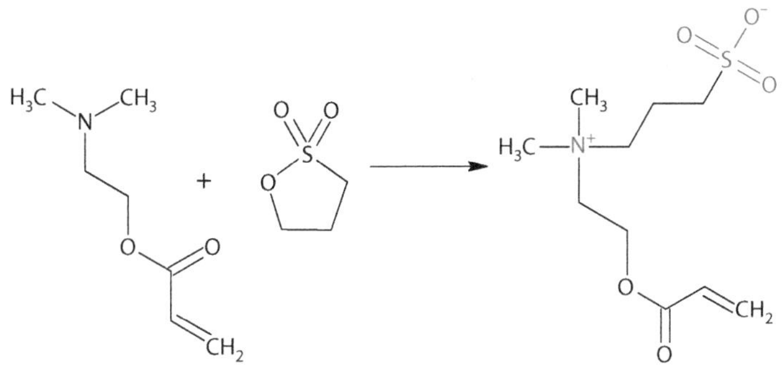

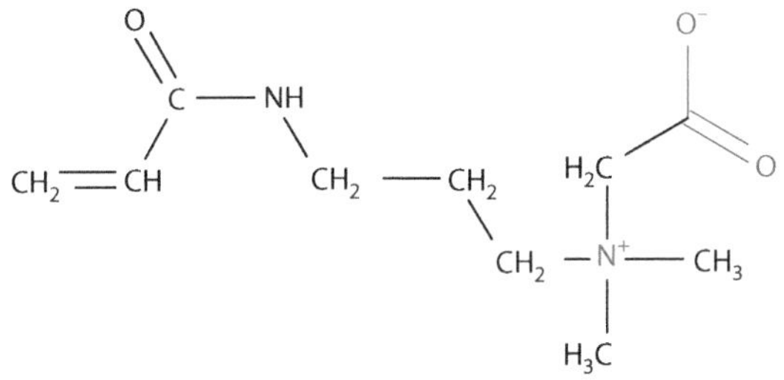

Unfortunately, the binary copolymerization of an amphoteric monomer results in a copolymer with a perfectly equal number of anionic and cationic charges. The copolymerization reaction must overcome some challenges related to comonomer interaction before polymerization, to the association of macromolecule during reaction, and to changes in the local composition of comonomers resulting in a variation of the copolymer composition as a function of conversion. That is why, an easier procedure is needed: blends of two ionic polymers (dual system).

3.3.3.5.2 Dual Systems as Retention Aids

The interaction between anionic and cationic water-soluble polymers [55] takes place at the wet end of the papermaking system. These interactions result in amphoteric aggregates (dual systems) [196] in which anionic and cationic charges are connected through both covalent and ionic bonds. The new aggregate has a larger size and a controlled ratio between ionic charges. For instance, low molecular weight cationic resins (*N*,*N*,*N*′,*N*′-tetramethyl ethylenediamine condensation with dichloroethyl ether) are less efficient as retention aids, but their efficiency increases when their molecular weight was built up by mixing with a poly-anion (such as polyacrylic acid) [467].

The cationic polymer (promoter or fixative [468]) is blended with an anionic organic (or inorganic) compound, such as PAAm [469–472], sulfonated lignin [473,474], or anionic (oxidized) starch [475,476]. The ratio between anionic and cationic charges is critical: a low charge cationic starch (DS=0.06) asks for a low charge anionic starch (DS=0.05) [477].

In other words, the amphoteric aggregates are built up *in situ*, in the presence of cellulose fibers, which can also be seen as anionic polyelectrolytes. The dual aggregate shows a much larger molecular weight than individual components do. In heterogeneous systems (wet end), the interaction between two different polyelectrolytes can occur in the continuous phase or at the fiber–water interface. The order of addition dictates where the interactions take place. If the anionic polyelectrolyte (anionic PAAm) is added first, it will remain in the water phase. The two polyanionic components will compete for the cationic polyelectrolyte (polyamine) [478]: some molecules adsorb at the water-cellulose interface and some will make an amphoteric aggregate with the anionic PAAm in water. The charge balance can be adjusted with the third polyelectrolyte (such as cationic starch).

The benefits of using blends of anionic (sodium salt of maleic anhydride—diisobutylene copolymer) and cationic polyacrylates in filler retention were identified in 1958 [398]. The addition sequence was also set based on common sense: first a cationic resin [79] (cationic MF resin [350], cationic starch [479–484], or cPAAm [485]), followed by an anionic compound (acrylamide copolymer with acrylic acid, blends of copolymers with different molecular weights [486]. Polyaluminum chloride and anionic silica [484,485,487]) were added for higher fines and filler retention [479,488]. Flocculation occurs very rapidly: the equilibrium is reached within less than 2 s after the second polymer is added [79].

The balance between ionic charges should be taken into account. Therefore, the "dual system" [33] is defined based on the charge ratio of the two polyelectrolytes. During the first step, a cationic compound such as poly-DADMAC [230], cationic colloidal silica sol [489], cationic starch [490], or a cationic acrylamide copolymer [491,492] is added to furnish [493]. The cellulose charge is reversed to positive if enough cationic compound is added. Afterward, an anionic condensed polymer [456], an anionic PAAm [230,490,493,494,495], anionic silica sol, or anionic montmorillonite clay [496] may be used to flocculate the dispersion [470].

When these steps are repeated under a controlled charge ratio, a multilayer structure is obtained on the fiber surface. This process results in a much higher concentration of polymers at the interface. The cationic compound is polyallylamine or polyDADMAC, which is combined with polyacrylic acid or polystyrene sulfonate [137]. Other multilayer systems can be formed with cationic and anionic starch [497,498], cationic starch and CMC [499,500], or PAE and CMC [138].

Both the amphoteric complex and the retained cationic pDADMAC (actually a complex with anionic cellulose fibers) show residual opposite charges. That is why, the polyelectrolyte complex is retained on the cellulose fiber even when the support has been saturated with cationic pDADMAC [178].

Retention using a "dual system" is a flexible process: it can accommodate any furnish within a broader pH range. Residual charges are balanced between homopolymers (cPAAm and anionic polyacrylic acid, or polystyrene sulfonate) and amphoteric copolymers (acrylic acid-DADMAC) [501] or sulfonated Kraft lignin [502]).

The solubility and the efficiency of the amphoteric aggregate depend on its molecular shape. This aggregate shape depends on the ratio between components, the charge density and molecular weight of each polyelectrolyte, the presence of a branched structure (anionic [503], or cPAAm partially cross-linked with MBA) [259,504], and the presence of electrolytes.

A polyelectrolyte complex ("polysalt coacervate") is soluble when a relatively large macromolecule (the "host") interacts with a smaller molecule of opposite charge (the "guest") [505] or when an ionization suppressor is present [506–509]. The solubility of the polyelectrolyte may be reduced up to a point where the polyelectrolyte precipitates: the salting out of the polymer [494,510]. The balance between cationic and anionic charges in that aggregate, and its solubility, can be controlled by adding an ionic (anionic or cationic) surfactant [511]. Even a nonionic polymer (such as polyethylene glycol) may reduce the solubility of PAAm in water [512].

The precipitation of the amphoteric aggregate and its effectiveness as a flocculant ("network flocculation" [355]) seems to be related to the size of the new, dispersed phase. On the other hand, each component of the amphoteric dual system can also be in a heterophase. A linear or branched [513], soluble cationic copolymer (cPAAm [205,514], copolymers of vinyl amine [515], or cationic starch [516,517]) is added first, and the dispersed anionic component is added second [518] (bentonite [259,504,519], polysilicate microgel [516], borosilicate [520,521], colloidal silica [483,522], polystyrene latex [523], or MF resins [524]). The cationic resin can also be in a dispersed phase (latex [525,526]), when the second component is anionic PAAm.

Anionic polyelectrolytes (polysilicate microgels [527]) are effective in the presence of cationic macromolecular compounds such as cationic guar gum or cationic starch [349]. Bentonite (0.25% based on cPAAm) is an enhancer due to its capability to make large aggregates with cationic polymers. At low bentonite concentrations, these aggregates can be seen as "cPAAm with a higher molecular weight" A giant aggregate is able to flocculate through a "bridging mechanism" [141].

For the adsorption of the anionic polymer to be effective, the adsorbed cationic polymer (the retention aid) must have a "loopy" configuration. If the cationic polymer is lying flat on the surface after re-conformation, the absorbed anionic polymer will be less. The usefulness of the blends of retention aids is a function of the blend ratio and depends especially on the order of addition. Sequential addition involves the time factor: the time between the two additions and the contact time [193,286]. In order to buy some time for re-conformation, a pre-flocculation (starch, cPAAm, and some cellulose fibers, which are also rigid polyelectrolytes [74]), is performed before the contact with the rest of the furnish [528]. Fibers and the fillers may also be treated separately with a cationic polymer (such as PAE). After mixing, the blend is flocculated with anionic starch [529].

The sequence addition is difficult to manage within the continuous papermaking process. That is why, other options have been explored [530]. The polyelectrolyte complex is formed in the water phase, in the presence of cellulose fibers, or is prepared in a separate reactor. If the *in situ* formation of the aggregate, the anionic carboxymethyl cellulose (or anionic PAAm) is added first and no coagulation or adsorption on cellulose fibers is recorded, then the cationic polymer (PAE or cPAAm) is added for a good flocculation [229]. However, in the presence of cellulose fibers, the efficiency of the amphoteric aggregate is diminished because the cationic polyelectrolyte can interact directly with the anionic fibers before the complex formation in water.

The formation of the "dual system" in a separate step seems to be a more reasonable approach. The largest improvement in cationic starch retention is obtained through the addition of a preformed complex of cationic starch and CMC [531]. Cationic starch is cooked in the presence of anionic PAAm and the blend is used at the wet end [532]. Depending on the ratio between the anionic (poly(sodium styrene sulfonate)) and the cationic (poly(vinyl benzyl trimethyl ammonium chloride)), the complex may have anionic in blue or cationic charges in red [533] (see the drawing). The two methods can also be combined: the pre-addition of the cationic resin (PAE) followed by the addition of preformed polyelectrolyte complex (CMC in excess over PAE) [534].

As a general example, anionic starch interacts with cationic polyamines through ionic bonds [535]. The cationic polyelectrolyte (in blue) and the anionic polyelectrolyte (in red) are interacting in solution [505]. Some charges are intermolecularly neutralized: cationic polyallylamine (charge density 9 meq/g) and polyacrylic acid make a polyelectrolyte complex with a charge density of 1.8 meq/g [88].

A number of both anionic and cationic charges may remain uncompensated and provide the amphoteric character of the aggregate. The aggregate looks like a tri-bloc-copolymer: a sequence of cationic branch, a neutral part, and a sequence of anionic branch. Thus, the dual system PEI-anionic polymer is equally efficient in the retention of both anionic and cationic molecules (such as Direct dye and Basic dye) [17].

Inorganic dispersions (such as colloidal silica, borosilicate [536], or polysilicic acid [347]) may change the shape of the aggregate. The cationic polyelectrolyte (cationic starch [346], cPAAm, or polyamine [537,538]) will form long tails on the surface of the inorganic particles.

Anionic starch is blended with a slight excess of cationic starch [211,539,540]. The efficiency of that dual system depends on the pH and the type of filler [541] and it is a proof of the synergetic effect (Figures 3.7 and 3.8): the single component (0% anionic or 0% cationic) is much less efficient than the dual system. The dual system can increase the clay (or CaCO$_3$) retention by up to 100% as compared with the single retention component. Figure 3.7 shows the maximum efficiency of clay retention close to 1:1 ratio cationic starch/anionic starch, at pH = 6.0, when the aggregate is slightly cationic.

The dual system is less effective when used with calcium carbonate (at pH = 7.8, the maximum retention is only 33%) and the aggregate shows, at the maximum efficiency, a zeta potential close to zero (Figure 3.8).

A more complicated system includes three components: a cationic polyelectrolyte (PEI), bentonite, and a high molecular weight anionic (or cationic) PAAm [542], or anionic starch, cationic polymer, and anionic inorganic particles (silica, bentonite, clay, etc.) [496,530,543].

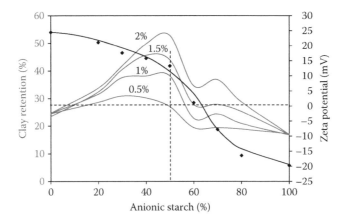

FIGURE 3.7   Dual system as a retention aid for clay (pH = 6.0).

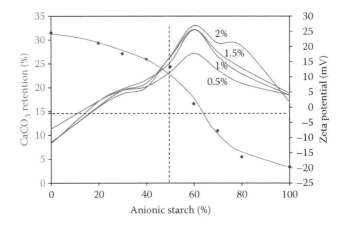

FIGURE 3.8   Dual system as a retention aid for CaCO$_3$ (pH = 7.8).

As far as retention aids are concerned, there is another perspective: their fate after web formation and their impact on the paper properties. The effectiveness of retention aids depends upon their adsorption and their concentration in the final paper, which is expected to be high. The concentration is in the same range (less than 1%) as that for any other paper chemical, such as wet- or dry-strength resins. Concerns have been raised about their effect on the effectiveness of other paper chemicals because retention aids are ionic compounds that can interfere with the retention of other paper chemicals. Most retention aids are nonreactive compounds and an improvement in paper strength is not expected. Moreover, they are very hydrophilic and their presence may increase water adsorption and may decrease the sizing effect of the internal sizing agents.

These concerns lead to an interesting topic: is any paper chemical able to perform as both a dry (or wet) strength resin and as a retention aid? Or, in other words, is there any retention aid with good performance as a dry (or wet) strength resin? Now we have the picture of an ideal paper chemical: it should act as a good retention aid, able to provide excellent wet strength (or dry strength [544]), while still allowing for the repulpability of the final paper (including the broke).

REFERENCES

1. B.E. Doiron, Retention Aid Systems in *Retention of Fines and Fillers during Papermaking*, J.M. Gess (Ed.), TAPPI Press, Atlanta, GA, 1998, pp. 159–176.
2. J.C. Roberts, *Paper Technology*, **34(3)**, 27–30 (1993).
3. K.W. Britt, J.E. Unbehend, and R. Sheridharan, *Tappi J.*, **69(7)**, 76–79 (1986).
4. L. Webb, *Pulp & Paper Industry*, January 1998, p. 33.
5. D.I. Devore, N.S. Clungeon, and S.A. Fischer, *Tappi J.*, **74**, 135 (1991).
6. O.L. Forgacs, A.A. Robertson, and S.G. Mason, *Pulp Paper Mag. Can.*, **59(5)**, 117–128 (1958).
7. B. Alince, *Nord. Pulp Paper Res. J.*, **17(1)**, 71–73 (2002).
8. H.L. Lee and S.B. Joo, *Nord. Pulp Paper Res. J.*, **15(5)**, 446–451 (2000).
9. B. Alince, *Tappi J.*, **70(10)**, 114–117 (1987).
10. J.C. Day, B. Alince, and A.A. Robertson, *Cellul. Chem. Technol.*, **13(3)**, 317–326 (1979).
11. H. Tanaka, A. Swerin, L. Odberg, and S.-B. Park, *J. Pulp Paper Sci.*, **25(8)**, 283–288 (1999).
12. B. Alince and T.G.M. van de Ven, Fundamentals of Papermaking Materials, Baker, C. F (Ed.), in *Transactions of the Fundamental Research Symposium, 11th*, Cambridge, U.K., Vol. 2, 1997, pp. 771–788.
13. T.G.M. van de Ven, *Nord. Pulp Paper Res. J.*, **15(5)**, 494–501 (2000).
14. T.C. Maloney and H. Paulapuro, *J. Pulp Paper Sci.*, **25(12)**, 430–436 (1999).
15. B. Alince, *J. Appl. Polym. Sci.*, **39(2)**, 355–362 (1990).
16. E. Strazdins, *Tappi*, **57(12)**, 76–80 (1974).
17. K.W. Britt, A.G. Dillon, and L.A. Evans, *TAPPI Proceedings—Papermakers Conference*, Atlanta, GA, 1977, pp. 39–42.
18. T.R. Arnson and R.A. Stratton, *Tappi J.*, **66(12)**, 72–75 (1983).
19. E. Strazdins, *Tappi*, **55(12)**, 1691 (1972).
20. M. Chene, J.P. Quiles, and J.F. Lafaye, US 3,657,066 (1972).
21. M. Medoff, US 2009/0283229.
22. T. Sun, US 6,361,651 (2002).
23. S.L. Greene, R.M. Kaylor, and K.R. Smith, US 5,770,711 (1998).
24. M. Luo, A. Weerawarna, J. Qin, and J.H. Wiley, US 7,604,714 (2009).
25. M. Luo, A. Weerawarna, J. Qin, and J.H. Wiley, US 7,608,167 (2009).
26. R.W. Davison, *Tappi*, **57(12)**, 85–89 (1974).
27. E. Strazdins, *Pulp & Paper*, March 1984, pp. 73–77.
28. E. Strazdins, *Tappi J.*, **78(8)**, 115–119 (1995).
29. A.M. Scallan and J. Grignon, *Sven. Papperstidn.*, **80(2)**, 40–47 (1979).
30. L. Wagberg and I. Asell, *Colloids Surf. A*, **104**, 169–184 (1995).
31. T. Lindstrom and G.G. Nordmark, *Sven. Papperstidn.*, **81(15)**, 489–492 (1978).
32. L. Sjostrom, J. Laine, and A. Blademo, *Nord. Pulp Paper Res. J.*, **15(5)**, 469–475 (2000).
33. E. Strazdins, in *Wet-Strength Resins Their Application*, L.L Chan (Ed.), Tappi Press, Atlanta, GA, 1994, pp. 63–83.
34. L. Bley, *Pulp Paper Can.*, **99(5)**, 51–55 (1998).
35. R.H. Pelton, L.H. Allen, and H.M. Nugent, *Sven. Papperstidn.*, **83(9)**, 251–258 (1980).

36. F. Tiberg, J. Daicic, and J. Froberg Surface chemistry of paper, in *Applied Surface and Colloid Chemistry*, K. Holmberg (Ed.), John Wiley, New York, 2002.
37. H. Tanaka, A. Swerin, L. Odberg, and S.B. Park, *J. Pulp Paper Sci.*, **23(8)**, J359–J365 (1997).
38. J. Chen, M.A. Hubbe, J.A. Heitmann, D.S. Argyropoulos, and O.J. Rojas, *Colloids Surf. A*, **246(1–3)**, 71–79 (2004).
39. D.E. Boardman, *Tappi J.*, **76(12)**, 148–152 (1993).
40. S. Alila, S. Boufi, M.N. Belgacem, and D. Beneventi, *Langmuir*, **21**, 8106–8113 (2005).
41. A.M. Crutzen, *J. Am. Oil Chem. Soc.*, **72(1)**, 137–143 (1995).
42. S. Friberg, H.H. Bruun, and R. Enquvist, *Sven. Papperstidn.*, **14**, 517–522(1975).
43. J.M. Stubbs, Y.G. Durant, and D.C. Sundberg, *Langmuir*, **15**, 3250–3255 (1999).
44. B. Alince and A.A. Robertson, *J. Appl. Polym. Sci.*, **14**, 2581–2593 (1970).
45. B. Alince, L. Kuniak, and A.A. Robertson, *J. Appl. Polym. Sci.*, **14**, 1577–1590 (1970).
46. J. Chen, J.A. Heitmann, and M.A. Hubbe, *Colloid Surf.*, **A-223**, 215–230 (2003).
47. A.S. Michaels, *J. Ind. Eng. Chem.*, **57(10)**, 32–40 (1965).
48. M.R.St. John, and T.M. Gallagher, *Tappi Proceedings—Papermakers Conference*, Atlanta, GA, 1992, pp. 479–502.
49. T. Kitaoka and H. Tanaka, *J. Wood Sci.*, **47**, 322–324 (2001).
50. C. Geffroy, J. Persello, A. Foissy, P. Lixon, F. Tournilhac, and B. Cabane, *Colloids Surf.*, **A-162**, 107–121 (2000).
51. L. Wagberg, *Nord. Pulp Paper Res. J.*, **15(5)**, 586–597 (2000).
52. M.A. Hubbe, *Nord. Pulp Paper Res. J.*, **15(5)**, 545–553 (2000).
53. L. Ghimici, I. Dranca, S. Dragan, T. Lupascu, and A. Maftuleac, *Eur. Polym. J.*, **37**, 227–231 (2001).
54. H. Lauer, A. Stark, H. Hoffmann, and R. Donges, *J. Surfact. Deterg.*, **2(2)**, 181–191 (1999).
55. J. Kotz, *Nord. Pulp Paper Res. J.*, **8(1)**, 11–14 (1993).
56. M. Mende, G. Petzold, and H.M. Buchhammer, *Colloid Polym. Sci.*, **280**, 342–351 (2002).
57. R. Gernandt, L. Wagberg, L. Gardlund, and H. Dautzenberg, *Colloids Surf. A Physicochem. Eng. Aspects*, **213**, 15–25 (2003).
58. K.V. Sarkanen, F. Dinkler, and V. Stannett, *Tappi*, **49(1)**, 4–9 (1966).
59. R.E. Lappan, R. Pelton, I. McLennan, J. Patry, and A.N. Htymak, *Ind. Eng. Chem. Res.*, **36**, 1171–1175 (1997).
60. H.-K. Lee and W.-L. Jong, *J. Polym. Res.*, **4(2)**, 119–128 (1997).
61. S. Dragan and M. Cristea, *Eur. Polym. J.*, **37**, 1571–1575 (2001).
62. S. Dragan and M. Cristea, *Polymer*, **43**, 55–62 (2002).
63. S. Dragan, M. Cristea, C. Luca, and B.C. Simionescu, *J. Polym. Sci. Part A Polym. Chem.*, **34**, 3485–3494 (1996).
64. X.P. Wang, J.A. Tang, and L. Jiang, *Chin. Chem. Lett.*, **7(9)**, 875–878 (1996).
65. T.G. Majewicz, US 4,309,535 (1982).
66. W. Shimokawa, US 3,759,861 (1973).
67. L. Gardlund, L. Wagberg, and R. Gernandt, *Colloids Surf. A Physicochem. Eng. Aspects*, **218**, 137–149 (2003).
68. A.S. Michaels, L. Mir, and N.S. Schneider, *J. Phys. Chem.*, **69(5)**, 1447–1455 (1965).
69. C.W. Neal, E. Aprahamian, and J.A. Cain, US 7,683,126 (2010).
70. K.J. McCarthy, C.W. Burkhardt, and D.P. Parazak, *J. Appl. Polym. Sci.*, **33**, 1699–1714 (1987).
71. M. Laleg and I.I. Pikulik, *Nord. Pulp Paper Res. J.*, **8(1)**, 41–47 (1993).
72. Y. Ishimaru and T. Lindstrom, *J. Appl. Polym. Sci.*, **29**, 1675–1691 (1984).
73. L. Wagberg and M. Bjorklund, *Nord. Pulp Paper Res. J.*, **8(4)**, 399–404 (1993).
74. M.A. Hubbe, *Proc. Sci. Tech. Advan. Wet End Chemistry*, PIRA, Barcelona, Spain, 2002.
75. L. Wagberg, L. Winter, L. Odberg, and T. Lindstrom, *Colloids Surf.*, **27**, 163–173 (1987).
76. G.G. Allan and W.M. Reif, *Sven. Papperstidn.*, **74(18)**, 563–570 (1971).
77. M.A. Cohen Stuart, *Nord. Pulp Paper Res. J.*, **8(1)**, 10 (1993).
78. M.A. Cohen Stuart, T. Cosgrove, and B. Vincent, *Adv. Colloid Interface Sci.*, **24**, 143–239 (1986).
79. L. Wagberg and T. Lindstrom, *Nord. Pulp Paper Res. J.*, **2(2)**, 49–55 (1987).
80. T. Lindstrom, C. Soremark, C. Heinegard, and S. Martin-Lof, *Tappi*, **57(12)**, 94–96 (1974).
81. J.C. Day, B. Alince, and A.A. Robertson, *Can. J. Chem.*, **56**, 2951–2958 (1978).
82. H.G.M. van de Steeg, A. de Keizer, M.A. Cohen Stuart, and B.H. Bjsterbosch, *Colloids Surf.*, **A-70**, 91–103 (1993).
83. A. Swerin and L. Wagberg, *Nord. Pulp Paper Res. J.*, **9(1)**, 18–25 (1994).
84. E.E. Morse, US 2,757,086 (1956).

85. J. Laine, T. Lindstrom, G.G. Nordmark, and G. Risinger, *Nord. Pulp Paper Res. J.*, **17(1)**, 50–56 (2002).
86. H.W. Moeller, *Paper Technol.*, **6(4)**, 279–284 (1965).
87. L. McLain and R. Wygant, *Pulp Paper*, **80(3)**, 46–51 (2006).
88. L. Gardlund, J. Forsstrom, and L. Wagberg, *Nord. Pulp Paper Res. J.*, **20(1)**, 36–42 (2005).
89. I. Borukhov, D. Andelman, and H. Orland, *Europhys. Lett.*, **32(6)**, 499–504 (1995).
90. H. Tanaka, L. Odberg, L. Wagberg, and T. Lindstrom, *J. Colloid Interface Sci.*, **134(1)**, 219–228 (1990).
91. M.J. Smith, J.M. Kaun, and V.R. Gentile, US 5,993,602 (1999).
92. R. Pelton, *Appita J.*, **57(3)**, 181–190 (2004).
93. B. Alince, *J. Colloid Interface Sci.*, **69(3)**, 367–374 (1979).
94. B. Bianchim, G. Gervason, P. Vallette, and G. Sauret, *Fibre–Water Interact. Paper-Making, Trans. Symp.*, Vol. 1, Oxford, U.K., 1977, pp. 151–161 (1978).
95. J.C. Roberts, C.O. Au, and G.A. Clay, *Tappi J.*, **69(10)**, 88–93 (1986).
96. J. Blaakmeer, M.R. Bohmer, M.A. Cohen Stuart, and G.J. Fleer, *Macromolecules*, **23**, 2301–2309 (1990).
97. P.N. Jacob and J.C. Berg, *Tappi J.*, **76(5)**, 133 (1993).
98. P.H. Brouwer, M.A. Johnson, and R.H. Olsen, Starch and retention, in *Retention of Fines and Fillers during Papermaking*, J.M. Gess (Ed.), TAPPI Press, Atlanta, GA, 1998, pp. 199–242.
99. P. Lepoutre and D. Lord, *J. Colloid Interface Sci.*, **134(1)**, 66–73 (1990).
100. P.H. Brouwer, *PTS-Manuskript*, 2/1–2/11 (2003).
101. H.G.M. van de Steeg, A. de Keizer, M.A. Cohen Stuart, and B.H. Bijsterbosch, *Nord. Pulp Paper Res. J.*, **8(1)**, 34–40 (1993).
102. H.G.M. van de Steeg, A. de Keizer, and B.H. Bijsterbosch, *Nord. Pulp Paper Res. J.*, **4(2)**, 173–178 (1989).
103. H.W. Moeller, *Tappi*, **49(5)**, 211–214 (1966).
104. F. Poschmann, US 3,256,140 (1966).
105. G.J. Fleer, J.M.H.M. Scheutjens, and M.A. Cohen Stuart, *Colloid Surf.*, **31**, 1–29 (1988).
106. M.A. Cohen Stuart, *J. Phys. France*, **49**, 1001–1008 (1988).
107. J. Zhang, R. Pelton, L. Wagberg, and M. Rundlof, *Nord. Pulp Paper Res. J.*, **15(5)**, 400–405 (2000).
108. L. Winter, L. Wagberg, L. Odberg, and T. Lindstrom, *J. Colloid. Interface Sci.*, **111(2)**, 537–543 (1986).
109. D.P. Parazak, C.W. Burkhardt, K.J. McCarthy, and M.P. Stehlin, *J. Colloid Interface Sci.*, **123**, 59–72 (1988).
110. J. Baas, P.H. Brouwer, and T.A. Wielema, *PTS-Symposium*, 38/1–38/30 (2002).
111. M. Watanabe, T. Gondo, and O. Kitao, *Tappi J.*, **3(5)**, 15–19 (2004).
112. J. Laine, T. Lindstrom, G.G. Nordmark, and G. Risinger, *Nord. Pulp Paper Res. J.*, **17(1)**, 57–60 (2002).
113. J. Laine, T. Lindstrom, G.G. Nordmark, and G. Risinger, *Nord. Pulp Paper Res. J.*, **15(5)**, 520–526 (2000).
114. J. Laine and T. Lindstrom, *IPW*, **1**, 40–45 (2001).
115. L. Odberg, H. Tanaka, and A. Swerin, *Nord. Pulp Paper Res. J.*, **8(1)**, 6–9 (1993).
116. B. Alince, *Tappi J.*, **79(3)**, 291–294 (1996).
117. L. Odberg, H. Tanaka, G. Glad-Nordmark, and A. Swerin, *Colloids Surf.*, **A-86**, 201–207 (1994).
118. A.T. Horvath, A.E. Horvath, T. Lindstrom, and L. Wagberg, *Langmuir*, **24**, 10797–10806 (2008).
119. A. Swerin and L. Odberg, Fundamentals of papermaking materials, in *Transaction of the Fundamental Research Symposium, 11th*, C.F. Baker (Ed.), Vol. 1, Cambridge, U.K., September 1997, pp. 265–350.
120. M. Falk, L. Oderg, L. Wagberg, and G. Risinger, *Colloids Surf.*, **40**, 115–124 (1989).
121. M.P. Nedelcheva and G.V. Stoilkov, *J. Appl. Polym. Sci.*, **20**, 2131–2141 (1976).
122. L. Wagberg and T. Lindstrom, *Nord. Pulp Paper Res. J.*, **2(4)**, 152–160 (1987).
123. J. Marton and T. Marton, *Tappi*, **59(12)**, 121–124 (1976).
124. J.M.H.M. Scheutjens, G.J. Fleer, and M.A. Cohen Stuart, *Colloid Surf.*, **21**, 285–306 (1986).
125. D. Abson and D.F. Brooks, *Tappi J.*, **68(1)**, 76–78 (1985).
126. D. Barzyk, D.H. Page, and A. Ragauskas, *J. Pulp Paper Sci.*, **23(2)**, J59–J61 (1997).
127. E. Donath, G.B. Sukhorukov, F. Caruso, S.A. Devis, and H. Mohwald, *Angew. Chem. Int. Ed.*, **37(16)**, 2202–2205 (1998).
128. G. Swift, *Polym. Degrad. Stab.*, **45**, 215–231 (1994).
129. B. Alince and A.A. Robertson, *Colloid Polym. Sci.*, **252(11)**, 920–927 (1974).
130. H. Tanaka and L. Odberg, *J. Colloid Interface Sci.*, **149(1)**, 40–48 (1992).
131. H. Tanaka, A. Swerin, and L. Odberg, *Nord. Pulp Paper Res. J.*, **10(4)**, 261–268 (1995).
132. J. Klein and K.D. Conrad, *Makromol. Chem.*, **179**, 1635–1638 (1978).
133. R. Grundelius and O. Samuleson, *Sven. Papperstidn.*, **65**, 273–281 (1962).
134. R. Grundelius and O. Samuelson, *Sven. Papperstidn.*, **65**, 310–312 (1962).

135. O. Samuelson and A. Viale, *Sven. Papperstidn.*, **66(18)**, 701–702 (1963).
136. T. Lindstrom and C. Soremark, *J. Colloid Interface Sci.*, **55(2)**, 305–312 (1976).
137. R. Lingstrom and L. Wagberg, *J. Colloid Interface Sci.*, **328**, 233–242 (2008).
138. L.E. Enarsson and L. Wagberg, *Nord. Pulp Paper Res. J.*, **22(2)**, 258–266 (2007).
139. L. Gardlund, M. Norgren, L. Wagberg, and A. Marklund, *Nord. Pulp Paper Res. J.*, **22(2)**, 210–216 (2007).
140. C.E. Farley, *Tappi Proceedings—Papermakers Conference*, Atlanta, GA, 1996, pp. 83–90.
141. T. Asselman, B. Alince, G. Garnier, and T.G.M. van de Ven, *Nord. Pulp Paper Res. J.*, **15(5)**, 515–519 (2000).
142. J.L. Koethe and W.E. Scott, *Tappi J.*, **76(12)**, 123–133 (1993).
143. E. Strazdins, *TAPPI Proceedings—Papermakers Conference*, Atlanta, GA, 1974, pp. 59–65.
144. C.E. Farley, *Tappi J.*, **80(10)**, 177–183 (1997).
145. B.W. Green, *Ind. Eng. Chem. Prod. Res. Dev.*, **21**, 592–598 (1982).
146. T.G.M. van de Ven, *Colloids Surf.*, **A138**, 207–216 (1998).
147. B. Alince, M. Inoue, and A.A. Robertson, *J. Appl. Polym. Sci.*, **20**, 2209–2219 (1976).
148. X. Zhang and H. Tanaka, *J. Appl. Polym. Sci.*, **81**, 2791–2797 (2001).
149. X. Zhang and H. Tanaka, *J. Appl. Polym. Sci.*, **80**, 334–339 (2001).
150. R.D. Patel, M.P. Breton, M.A. Hopper, G.E. Kmiecik-Lawrynowicz, and B.S. Ong, US 5,977,210 (1999).
151. B. Alince, *Tappi*, **60(12)**, 133–136 (1977).
152. B. Alince, *Paper Trade J.*, **163(5)**, 52–57 (1979).
153. B. Alince and P. Lepoutre, *Tappi J.*, **66(2)**, 101–102 (1983).
154. J.A. Maroto and F.J. de las Nieves, *Colloids Surf.*, **A96**, 121–133 (1995).
155. B. Alince, P. Arnoldova, and R. Frolik, *J. Appl. Polym. Sci.*, **76(11)**, 1677–1682 (2000).
156. B. Alince, A.A. Robertson, and M. Inoue, *J. Colloid Interface Sci.*, **65(1)**, 98–107 (1978).
157. B. Alince, *Paper Technology*, **46(6)**, 14–17 (2005).
158. M. Okubo, Y. He, and K. Ichikawa, *Colloid Polym. Sci.*, **269**, 125–130 (1991).
159. B. Alince, *Tappi J.*, **82(3)**, 175–187 (1999).
160. B. Alince, J. Kinkal, F. Bednar, and T.G.M. van de Ven, *ACS Symp. Ser.*, **801** (Polymer Colloids), 52–70 (2002).
161. B. Alince, *Paperi ja Puu*, **3**, 118–120 (1985).
162. B. Alince and A.A. Robertson, *Tappi*, **61(11)**, 111–114 (1978).
163. B. Alince, *Colloids Surf.*, **39(1–3)**, 39–51 (1989).
164. B. Alince and R. Tino, *Colloids Surf. A Physicochem. Eng.*, **218**, 1–9 (2003).
165. B. Alince, M. Inoue, and A.A. Robertson, *J. Appl. Polym. Sci.*, **23**, 539–548 (1979).
166. B. Alince, *J. Appl. Polym. Sci.*, **23**, 549–560 (1979).
167. K.H. Wassmer, U. Schroeder, and D. Horn, *Makromol. Chem.*, **192**, 553–565 (1991).
168. H. Terayama, *J. Polym. Sci.*, **7(2)**, 243–253 (1952).
169. J.A. Loyd and C.W. Horne, *Nord. Pulp Paper Res. J.*, **8(1)**, 48–52 (1993).
170. D.L. Kenaga, W.A. Kindler, and F.J. Meyer, *Tappi*, **50(7)**, 381–387 (1967).
171. M.A. Hubbe and J. Chen, *Paper Technol.*, **45(8)**, 17–23 (2004).
172. J. Chen, M.A. Hubbe, and J.A. Heitmann, *Proc. TAPPI Papermakers Conf.*, Atlanta, GA, 2001, pp. 1–14.
173. H. Tanaka and Y. Sakamoto, *J. Polym. Sci. Part A Polym. Chem.*, **31**, 2693–2696 (1993).
174. J. Kotz, B. Paulke, B. Philipp, P. Denkinger, and W. Burchard, *Acta Polym.*, **43**, 193–198 (1992).
175. H. Tanaka, A. Swerin, L. Odberg, and M. Tanaka, *J. Appl. Polym. Sci.*, **86(3)**, 672–675 (2002).
176. H. Tanaka and Y. Sakamoto, *J. Polym. Sci. Part A Polym. Chem.*, **31**, 2687–2691 (1993).
177. D.I. Devore, N.S. Clungeon, and S.A. Fischer, *ACS Symp. Ser.*, **548**, 394–405 (1994).
178. M.C. Lofton, S.M. Moore, M.A. Hubbe, and S.Y. Lee, *Tappi J.*, **4(9)**, 3–8 (2005).
179. M.A. Hubbe, *Tappi*, **62(8)**, 120–121 (1979).
180. L. Wagberg, L. Odberg, and G. Glad-Nordmark, *Nord. Pulp Paper Res. J.*, **4(2)**, 71–76 (1989).
181. R.H. Pelton, Polymer–colloid interactions in pulp and paper manufacture, in *Colloid–Polymer Interactions*, R.S. Farinato and P.L. Dubin (Eds.), Wiley, New York, 1999, pp. 51–82.
182. A. Kirbawy, *TAPPI proceedings—Papermaking Conference*, Atlanta, GA, 1992, pp. 473–477.
183. N.D. Sanders and J.H. Schaefer, *TAPPI Proceedings—Papermaking Conference*, Atlanta, GA, 1992, pp. 463–472.
184. M.A. Hubbe, H. Nanko, and M.R. McNeal, *BioResources*, **4(2)**, 850–906 (2009).
185. M.B. Hocking, K.A. Klimchuk, and S. Lowen, *J. Macromol. Sci. Rev. Macromol. Chem. Phys.*, **C39(2)**, 177–203 (1999).

186. D. Horn and F. Linhart, Retention aids, in *Paper Chemistry*, J.C. Roberts (Ed.), Blackie Academic, London, U.K., 1996, pp. 64–82.

187. S.R. Middleton and A.M. Scallan, *J. Pulp Paper Sci.*, **17(4)**, J127–J133 (1994).

188. R.E. Scalfarotto and R.F. Tarvin, *Tappi Proceedings—Papermakers Conference*, Atlanta, GA, 1984, pp. 65–75.

189. M.A. Hubbe and F. Wang, *Tappi J.*, **1(1)**, 28–33 (2002).

190. P. Luner, *Tappi Proceedings—Papermakers Conference*, Atlanta, GA, 1984, pp. 95–106.

191. B. Alince and P. Lepoutre, *J. Polym. Sci. Poly. Letters Ed.*, **20**,615–620 (1982).

192. M. Antal, M. Laleg, and I.I. Pikulik, US 6,455,661 (2002).

193. B.Alince, J. Petlicki, and T.G.M. van de Ven, *Colloids Surf.*, **59**, 265–277 (1991).

194. D. Glittenberg and A. Becker, *Paperi ja Puu*, **79(4)**, 240–243 (1997).

195. J. Yoshizawa, A. Isogai, and F. Onabe, *J. Pulp Paper Sci.*, **24(7)**, 213–218 (1998).

196. J.E. Maher, *TAPPI Seminar Notes—Alkaline Papermaking*, Atlanta, GA, 1985, pp. 89–92.

197. R. Evans and P. Luner, *J. Colloid Interface Sci.*, **128(2)**, 464–476 (1989).

198. L. Beghello and T. Lindstrom, *Nord. Pulp Paper Res. J.*, **13(4)**, 269–273 (1998).

199. T. Lindstrom, C. Soremark, and L. Westman, *Sven. Papperstidn.*, **80(11)**, 341–345 (1977).

200. P. Marimuthu, A. Gopalan, and G. Venkoba Rao, *Ippta J.*, **16(3)**, 65–74 (2004).

201. A. Springer, W. McMahon, and J. Boylan, *Tappi Proceedings—Papermakers Conference*, Atlanta, GA, 1984, pp. 13–22.

202. A. Gibbs, R. Pelton, and R. Cong, *Tappi Proceedings—Papermaking Conference.*, Atlanta, GA, 2000, pp. 445–457.

203. O. Sunden, US 4,294,885 (1981).

204. R.H. Pelton, *TAPPI Proceedings—Papermakers Conference*, Atlanta, GA, 1984, pp. 1–5.

205. E. Gruber and P. Muller, *Tappi J.*, **3(12)**, 17–22 (2004).

206. A.A. Robertson and S.G. Mason, *Pulp Paper Mag. Can.*, **55(3)**, 263–269 (1954).

207. R.D. Carter and B. Cicerchi, US 5,779,859 (1998).

208. S.W. Rosencrance, P. Pruszynski, S.M. Shawki, R. Jakubowski, and J.F. Lin, US 6,048,438 (2000).

209. P. Pruszynski, R. Jakubowski, J.R. Armstrong, and S.W. Rosencrance, US 6,033,524 (2000).

210. D. Gomez, US 4,943,349 (1990).

211. D. Gilttenberg, R.J. Tippett, and P. Leonhardt, *Paper Technol.*, **45(6)**, 27–33 (2004).

212. M. Bernheim, D. Strasilla, B. De Sousa, and P. Rohringer, US 4,956,049 (1990).

213. M. Bernheim, H. Meindl, P. Rohringer,H. Wegmuller, and D. Werthemann US 4,623,428 (1986).

214. R. Topfl, M. Bernheim, H. Meindl, H. Wegmuller, P. Rohringer, and D. Werthemann, US 5,201,998 (1993).

215. P.H. Brouwer, *Tappi J.*, **74(1)**, 170–179 (1991).

216. R.R. Davidson, US 4,115,187 (1978).

217. F. Zhang, W.S. Carey, E.A.S. Doherty, and J.T. Sarraf, US 2009/0229775.

218. M.Laleg, US 7,625,962 (2009).

219. B.M. Clitherow, T.W.R. Dean, J.A. Gascoigne, and B.E. Van Issum, US 5,017,268 (1991).

220. S. Suty, B. Alince, and T.G.M. van de Ven, *J. Pulp Paper Sci.*, 22(9), J321–J326 (1996).

221. A. Vanerek, B. Alince, and T.G.M. van de Ven, *J. Pulp Paper Sci.*, **26(9)**, 317–322 (2000).

222. A. Vanerek, B. Alince, and T.G.M. van de Ven, *J. Pulp Paper Sci.*, **26(4)**, 135–139 (2000).

223. J. Porubska, B. Alince, and T.G.M. van de Ven, *Colloids Surf. A Physicochem. Eng. Aspects*, **210(2–3)**, 223–230 (2002).

224. E. Strazdins, *TAPPI Proceedings—Papermakers Conference*, Atlanta, GA, 1976, pp. 49–55.

225. N.D. Sanders and J.H. Schaefer, *TAPPI Proceedings—Papermaking Conference*, Atlanta, GA, 1989, pp. 69–74.

226. M.A. Hubbe, US 5,936,151 (1999).

227. E. Strazdins, *Tappi*, **60(7)**, 113–114 (1977).

228. S.R. Middleton and A.M. Scallan, *Colloids Surf.*, **16**, 309–322 (1985).

229. E. Strasdins, *TAPPI Proceedings—Papermaking Conference*, Atlanta, GA, 1992, pp. 503–505.

230. M.A. Hubbe, *Nord. Pulp Paper Res. J.*, **16(4)**, 369–375 (2001).

231. M.A. Hubbe, *BioResources*, **2(2)**, 296–331 (2007).

232. T. Tripattharanan, M.A. Hubbe, R.A. Venditti, and J.A. Heitmann, *Appita J.*, **57(5)**, 404–410 (2004).

233. M.A. Hubbe, T.L. Jackson, and M. Zhang, *Tappi J.*, **2(11)**, 7–12 (2003).

234. K.W. Britt, *Tappi*, **56(3)**, 83–86 (1973).

235. K.W. Britt, *Tappi*, **56(10)**, 46–50 (1973).

236. R.I.S. Gill, *Nord. Pulp Paper Res. J.*, **8(1)**, 208–210 (1993).

237. R. Pelton, *Nord. Pulp Paper Res. J.*, **8(1)**, 113–119 (1993).
238. J.S. Michelman, US 6,255,375 (2001).
239. D. Horn and J. Melzer, *Fibre-Water Interact. Paper-Making, Trans. Symp.*, Vol., Oxford, U.K., 1977, pp. 135–150 (1978).
240. C.S. Maxwell and C.G. Landes, US 2,559,220 (1951).
241. C.S. Maxwell, US 2,559,221 (1951).
242. R.V. Lauzon, US 5,169,441 (1992).
243. C.A. Savino, US 4,874,466 (1989).
244. B. Alince, *Colloids Surf.*, **33(3–4)**, 279–288 (1988).
245. B. Alince and T.G.M. van de Ven, *J. Colloid Interface Sci.*, **155**, 465–470 (1993).
246. B. Alince, A. Vanerek, and T.G.M. van de Ven, *J. Pulp Paper Sci.*, **28(9)**, 315–321 (2002).
247. R.A. Stratton and C.E. Miller, *Nord. Pulp Paper Res. J.*, **8(1)**, 15–20 (1993).
248. B. Alince, *Colloids Surf.*, **23**, 199–210 (1987).
249. S.C.H. Tay, *Tappi J.*, **80(9)**, 149–156 (1997).
250. J. Marton and F.L. Kurrle, *TAPPI Proceedings—Papermakers Conference*, Atlanta, GA, 1985, pp. 197–207.
251. M.A. Hubbe, *Colloid Surf.*, **25**, 325–339 (1987).
252. O.J. Rojas and M. AHubbe, *J. Disp. Sci. Technol.*, **25(6)**, 713–732 (2004).
253. T.G.M. van de Ven and B. Alince, *J. Colloid Interface Sci.*, **181**, 73–78 (1996).
254. M. Cechova, B. Alince, and T.G.M. van de Ven, *Colloid Surf. A Physicochem. Eng. Aspects*, **141(1)**, 153–160 (1998).
255. B. Alince and F. Bednar, *J. Appl. Polym. Sci.*, **88(10)**, 2409–2415 (2003).
256. M.A. Hubbe, *Tappi Proceedings—Papermakers Conference*, Atlanta, GA, 1984, pp. 23–30.
257. B.A. Saville and R. Pelton, US 2005/0279469.
258. T.G.M. van de Ven, *Nord. Pulp Paper Res. J.*, **8(1)**, 130–134 (1993).
259. J.-H. Shin, S.H. Han, C. Sohn, S. Kyoun, and S. Mah, *Tappi J.*, **80(10)**, 185–189 (1997).
260. B. Alince and T. Van De Ven, *Tappi J.*, **80(8)**, 181–186 (1997).
261. J.A. Manley, US 3,141,816 (1964).
262. T.A. Cauley, J.G. Langley, and A. Nixon, US 5,514,249 (1996).
263. D.J. Pye, US 3,052,595 (1962).
264. J.G. Langley and E. Litchfield, US 4,305,781 (1981).
265. C.L. Burdick and E. Echt, US 5,578,168 (1996).
266. J.A. Manley, US 3,141,815 (1964).
267. T.S.C. Lindstrom and L.H. Westman, US 4,362,600 (1982).
268. D. Kratochvil, B. Alince, and T.G.M. van de Ven, *J. Pulp Paper Sci.*, **25(9)**, 331–335 (1999).
269. R.H. Pelton, L.H. Allen, and H.M. Nugent, US 4,313,790 (1982).
270. H. Xiao, R. Pelton, and A. Hamielec, *J. Pulp Paper Sci.*, **22(12)**, J475–J485 (1996).
271. R. Pelton, U. Xiao, M.A. Brook, and A. Hamielec, *Langmuir*, **12**, 5756–5762 (1996).
272. D.B. Braun and D.A. Ehms, *Tappi Proceedings—Papermakers Conference*, Atlanta, GA, 1984, pp. 57–63.
273. J.-P. Carrard and H. Pummer, US 4,070,236 (1978).
274. M. Owens, US 5,670,021 (1997).
275. A. Carignan, G. Garnier, and T.G.M. van de Ven, *J. Pulp Paper Sci.*, **24(3)**, 94–99 (1998).
276. G.V. Laivins and M. Polverari, US 6,372,088 (2002).
277. B.F. Satterfield and J.O. Stockwell, US 5,538,596 (1996).
278. E. Echt, US 5,472,570 (1995).
279. R. Gaudreault, T.G.M. van de Ven, and M.A. Whitehead, *Colloids Surf. A Physicochem. Eng. Aspects*, **268**, 131–146 (2005).
280. R.A. Fetters and H.F. Arledter, US 3,281,312 (1966).
281. G. Radu and P. Langevin, US 5,554,260 (1996).
282. A. Hermann, F. Bednar, M. Perrier, J. Paradis, and J. Paris, *Can. J. Chem. Eng.*, **84**, 597–603 (2006).
283. T.G.M. van de Ven and B. Alince, *J. Pulp Paper Sci.*, **22(7)**, J257–J263 (1996).
284. R. Pelton, A.E. Hamielec, and H. Xiao, WO 94/17243.
285. L.J. Guilbault and J.R. Zakrzewski, US 3,926,718 (1975).
286. M. Polverari, J. Vu, and D. Aston, US 2006/0000568.
287. T. Lindstrom, *Colloid Polym. Sci.*, **257**, 277–285 (1979).
288. H. Xiao, R. Pelton, and A. Hamielec, *J. Polym. Sci. Part A Polym. Chem.*, **33**, 2605–2612 (1995).
289. C. Bonnet-Gonnet, J.C. Castaing, and P. Le Cornec, WO 98/55694.
290. H. Tanaka, A. Swerin, and L. Odberg, *Tappi J.*, **76(5)**, 157 (1993).

291. P. Lagally and H. Lagally, US 3,074,843 (1963).
292. S.P. Malchick, W.G. James, and J.S. Munday, US 3,325,347 (1967).
293. C. Bjellfors, K.-E. Eriksson, and F. Johanson, *Sven. Papperstidn.*, **68(24)**, 865–869 (1965).
294. H.L. Jones, US 3,421,976 (1969).
295. L.M. Booth, US 2,368,635 (1945).
296. P.O. Eriksson, I. Eriksson, B. Hjalmarson, and J.G. Langley, US 6,063,240 (2000).
297. W. Hatton, US 5,234,548 (1993).
298. A.P. Derrick, US 5,015,334 (1991).
299. J.D. Rushmere, US 4,927,498 (1990).
300. S.C. Sofia, K.A. Johnson, M.S. Crill, M.J. Roop, S.R. Gotberg, A.S. Nigrelli, and L.S. Hutchinson, US 4,795,531 (1989).
301. E.K. Stilbert, M.W. Zembal, and L.H. Silvernail, US 2,972,560 (1961).
302. W.Y. Yang, J.W. Qian, and Z.Q. Shen, *J. Colloid Interface Sci.*, **273**, 400–405 (2004).
303. H.W. Walker and S.B. Grant, *Colloids Surf.*, **A119**, 229–239 (1996).
304. H. Li, J. Long, Z. Xu, and J.H. Masliyah, *Energy Fuels*, **19**, 936–943 (2005).
305. T. Lindstrom and C. Soremark, *J. Colloid Interface Sci.*, **55(1)**, 69–72 (1976).
306. P.R. Shildneck and M.L. Cushing, US 3,151,019 (1964).
307. M.M. Tessler, US 4,119,487 (1978).
308. M.M. Tessler, O.B. Wurzburg, and T.A. Dirscherl, US 4,379,919 (1983).
309. M.M. Tessler, O.B. Wurzburg, and T.A. Dirscherl, US 4,387,221 (1983).
310. W.F. Reynolds and N.T. Woodberry, US 3,019,157 (1962).
311. F.S. Munjat, US 3,201,304 (1965).
312. T.C. Spence and E.W. Malcolm, US 3,391,057 (1968).
313. F. Suyama, T. Yamamoto, K. Shimono, H. Kuwahara, M. Ishii, and I. Watanabe, US 3,725,195 (1973).
314. B. Walchuk, F. Zhang, J.C. Harrington, W.S. Carey, and R.L. Brady, US 7,250,448 (2007).
315. J. Cluyse, P. Ford, J.G. Langley, and P. Lowry, US 5,223,098 (1993).
316. W. Alexander, US 4,624,982 (1986).
317. W. Alexander, US 4,613,542 (1986).
318. J.B. Wong Shing, C. Maltesh, and J.R. Hurlock, US 6,331,229 (2001).
319. A.P. Derrick and W. Hatton, US 5,032,227 (1991).
320. G. Szeiffova and B. Alince, *Proceedings of the International Conference Chemical Technology of Wood, Pulp and Paper*, Bratislava, Slovakia, September. 17–19, 2003, pp. 225–228.
321. B. Danner, US 5,130,358 (1992).
322. B. Danner and J.R. Ruynon, US 4,696,962 (1987).
323. P.H. Brouwer, J. Beugeling, and B.C.A. Ter Veer, *PTS-Symposium*, 1–15/28 (1999).
324. W. Jarowenko, US 3,721,575 (1973).
325. W. Jarowenko and M.W. Rutenberg, US 3,737,370 (1973).
326. W. Jarowenko, US 3,770,472 (1973).
327. H. Ikeda, F. Suzuki, Y. Watanabe, M. Matsumura, Y. Takahashi, H. Murakami, and K. Maeda, US 4,840,705 (1989).
328. D. Gilttenberg, J.L. Hemmes, and N.O. Bergh, *Paper Technol.*, **35(7)**, 18–27 (1994).
329. T. Aitken and W.D. Pote, US 4,097,427 (1978).
330. T. Altken, D.R. Anderson, and M.J. Jursich, US 3,674,725 (1972).
331. T. Altken, US 3,854,970 (1974).
332. H.J. Degen, S. Pfohl, V. Weberndoerfer, G. Rehmer, M. Kroener, and A. Stange, US 4,818,341 (1989).
333. R.H. Moffett, US 6,048,929 (2000).
334. R.H. Moffett, US 5,928,474 (1999).
335. R.H. Moffett, US 5,859,128 (1999).
336. R. Nordgren, US 3,303,184 (1967).
337. F. Marzolini, US 5,976,322 (1999).
338. J. Tsai, W. Maliczyszyn, T. Capitani, and C. Kulp, US 5,723,023 (1998).
339. K.R. Anderson, US 5,122,231 (1992).
340. H. Ketola, T. Andersson, A. Karppi, Sauli Laakso, and A. Likitalo, US 6,187,144 (2001).
341. K.R. Anderson and D.E. Garlie, US 6,524,440 (2003).
342. K.R .Anderson and D.E. Garlie, US 6,451,170 (2002).
343. D.B. Solarek, L.R. Peek, M.J. Henley, R.M. Trksak, and M.T. Philbin, US 5,523,339 (1996).
344. D.B. Solarek, L.R. Peek, M.J. Henley, R.M. Trksak, and M.T. Philbin, US 5,368,690 (1994).
345. D.L. Johnson, US 4,810,785 (1989).

346. H.E. Johansson, US 5,127,994 (1992).

347. P.J. Svending, P.G. Batelson, H.E. Johansson, and H.M. Larsson, US 4,385,961 (1983).

348. O. Sunden, P.G. Batelson, H.E. Johansson, H.M. Larsson, and P.J. Svending, US 4,388,150 (1983).

349. G. Greenwood, US 6,238,520 (2001).

350. H.P. Wohnsidler and W.M. Thomas, US 2,345,543 (1944).

351. M.C. Hutchins, US 2,730,446 (1956).

352. D.E. Nagy, US 3,567,659 (1971).

353. D.E. Nagy, US 3,697,370 (1972).

354. J. Green, US 2,969,302 (1961).

355. J.H. Daniel, L.H. Wilson, R. Hastings, and C.G. Landes, US 2,601,597 (1952).

356. M. Matter and R. Oberholzer, US 3,632,559 (1972).

357. L. Hoppe, B. Boehmer, and R. Behn, US 4,093,605 (1978).

358. R.H. Symm and B.D. Sheffield, US 4,056,510 (1977).

359. A.T. Coscia, US 3,493,502 (1970).

360. K.G. Phillips, US 3,468,818 (1969).

361. S.P. Dasgupta and H.H. Espy, US 5,656,699 (1997).

362. L. Hoppe and R. Behn, US 4,052,259 (1977).

363. G.G. Spence, D.R. Palmateer, and J.R. Yarnell, US 6,077,394 (2000).

364. H.H. Espy and S.T. Putnam, US 3,951,921 (1976).

365. H.H. Espy and S.T. Putnam, US 3,966,694 (1976).

366. H.H. Espy and S.T. Putnam, US 3,966,684 (1976).

367. U. Steuerle, H. Meixner, R. Dyllick-Brenzinger, W. Reuther, H. Kanter, A. Hettche, J. Weiser, and G. Scherr, US 6,056,967 (2000).

368. H. Enders, R. Fikentscher, W. Maurer, E. Scharf, and U. Soemksen, US 3,642,572 (1972).

369. W. Auhorn, D. Monch, R. Dyllick-Brenzinger, R. Scholz, R. Blum, and H. Meixner, US 6,083,348 (2000).

370. G. Scherr, W. Reuther, P. Lorencak, D. Moench, F. Linhart, and J. Weiser, US 5,536,370 (1996).

371. J. Decker, N. Mahr, A. Esser, H. Meixner, R. Dyllick-Brenzinger, M. Kahmen, and M. Gercke, US 6,716,311 (2004).

372. L.H. Allen and M.S. Polverari, US 6,468,396 (2002).

373. L. Allen and M. Polverari, *Nord. Pulp Paper Res. J.*, **15**(**5**), 407–414 (2000).

374. X.C. Peng, X.H. Peng, S.M. Liu, and J.Q. Zhao, *eXPRESS Polym. Lett.*, **3**(**8**), 510–517 (2009).

375. R.J. Schaper, US 4,166,894 (1979).

376. I.A. Pudney, B.M. Stubbs, and M.J. Welch, US 5,393,338 (1995).

377. A.T. Coscia, US 3,248,353 (1966).

378. L. Hoppe and R. Behn, US 4,036,821 (1977).

379. I.A. Pudney, B.M. Stubbs, and M.J. Welch, US 5,912,306 (1999).

380. I. Sobolev, US 3,428,617 (1969).

381. M.B. Heard and G.C.I. Chen, US 6,406,593 (2002).

382. J.B.W. Shing, J.R. Hurlock, C. Maltesh, and R. Nagarajan, US 6,071,379 (2000).

383. K.S. Dell and J.R. Armstrong, CA 2,102,742 (1994).

384. W.A. Foster, J.E. Stout, US 3,323,979 (1967).

385. T.C. Fallon, US 5,571,380 (1996).

386. K.D. Vison, J.P. Erspamer, C.W. Neal, and J.P. Halter, US 5,611,890 (1997).

387. P. Lowry and D. Farrar, US 5,254,221 (1993).

388. M. Persson, J. Carlen, H. Johansson, and C. Cordoba, US 5,858,174 (1999).

389. M. Persson, J. Carlen, H. Johansson, and C. Cordoba, US 6,100,322 (2000).

390. E.E. Maury, R. Buelte, and C.C. Johnson, US 6,171,505 (2001).

391. A. Swerin, G. Glad-Nordmark, and L. Odberg, *J. Pulp Paper Sci.*, **23**(**8**), 389–393 (1997).

392. J.B. Wong Shing, R.T. Gray, A.S. Zelenev, and J. Chen, US 6,592,718 (2003).

393. H. Takeda and M. Kawano, US 5,006,590 (1991).

394. T. Itagaki, M. Shiraga, S. Sawayama, and K. Satoh, US 4,808,683 (1989).

395. E. Freudenberg, F. Linhart, R. Tresch, H. Hartmann, W. Dezinger, M. Kroener, and N. Sendhoff, US 5,098,521 (1992).

396. T. Itagaki, M. Shiraga, S. Sawayama, and K. Satoh, US 5,064,909 (1991).

397. D.T. Nzudie and C. Collette, US 6,225,395 (2001).

398. I.J. Gruntfest and D.B. Fordyce, US 2,838,397 (1958).

399. F. Linhart and W. Auhorn, Papier, 46(10A), 38–45 (1992).

400. F. Brunnmueller, R. Schneider, M. Kroener, H. Mueller, and F. Linhart, US 4,421,602 (1983).
401. W. Auhorn, F. Linhart, P. Lorencak, M. Kroener, N. Sendhoff, W. Denzinger, and H. Hartmann, US 5,145,559 (1992).
402. R. Hund and C. Jehn-Rendu, US 6,797,785 (2004).
403. D. Moench, H. Hartmann, E. Freudenberg, and A. Stange, US 5,262,008 (1993).
404. F. Linhart, H.-J. Degen, W. Auhorn, M. Kroener, H. Hartmann, and W. Heide, US 4,772,359 (1988).
405. J. Utecht, M. Niessner, D. Monch, and M. Rubenacker, US 6,057,404 (2000).
406. J.G. Smigo, R.K. Pinschmidt, A.F. Nordquist, and T.L. Pickering, US 5,232,553 (1993).
407. F. Wang, T. Kitaoka, and H. Tanaka, Tappi J., 2(12), 21–25 (2003).
408. L.L. Kuo, R.Y. Leung, and K.S. Williams, US 5,473,033 (1995).
409. M.J. Hurwitz and H. Aschkenasy, US 3,406,139 (1968).
410. M.J. Hurwitz, H. Aschkenasy, and L.E. Kelley, US 3,527,719 (1970).
411. J.C. McClendon, US 3,478,003 (1969).
412. K. Fujimura and K. Tanaka, US 3,790,529 (1974).
413. B. Danner, H. Gerber, and H. Pummer, US 4,251,410 (1981).
414. E.V. Hort and E.P. Williams, US 4,057,533 (1977).
415. R. Hund and C. Jehn-Rendu, US 6,579,417 (2003).
416. R. Hund and C. Jehn-Rendu, US 2004/0040683.
417. Y. Mori and K. Takeda, US 2005/0230319.
418. X. Peng, J. Shen, and H. Xiao, J. Appl. Polym. Sci., 101, 359–363 (2006).
419. E. Scharf, R. Fikentscher, W. Auhorn, and W. Streit, US 4,144,123 (1979).
420. W.L. Renz, R. Ravichandran, A.J. Naisby, J. Suhadolnik, M.G. Wood, and R. Xiong, US 7,572,843 (2009).
421. T. Moritani, J. Yamauchi, and M. Shiraishi, US 4,311,805 (1982).
422. S.A. Lipowski and J.C. Queen, US 3,703,563 (1972).
423. F.E. Woodward, US 3,509,021 (1970).
424. R.L. Post and R.G. Fort, US 4,445,970 (1984).
425. B. Alince, Tappi J., 3(1), 16–18 (2004).
426. K.D. Vison, J.P. Erspamer, C.W. Neal, J.A. Ficke, and J.P. Halter, US 5,672,249 (1997).
427. R.A. Gill and N.D. Sanders, US 4,892,590 (1990).
428. R.A. Gill, US 5,147,507 (1992).
429. P.J. Svending, US 4,946,557 (1990).
430. H.J. Bixler and S. Peats, US 5,178,730 (1993).
431. L.L. Kuo and R.Y. Leung, US 5,720,888 (1998).
432. G.V. Lopez, US 7,244,339 (2007).
433. H.J. Bixler and S. Peats, US 5,071,512 (1991).
434. S. Roy, M. Desrochers, and L. Jurasek, J. Wood Chem. Technol., 9(3), 407–420 (1989).
435. C.G. Caldwell, W. Jarowenko, and I.D. Hodgkin, US 3,459,632 (1969).
436. P.W. Carter, P.G. Murray, L.E. Brammer, and A.J. Dunham, US 6,313,246 (2001).
437. D. Glittenberg, Tappi J., 76(11), 215–219 (1993).
438. W. Bindzus and P.A. Altieri, US 6,365,002 (2002).
439. D.B. Solarek, T.A. Dirscherl, H.R. Hernandez, and W. Jarrowenko, US 4,876,336 (1989).
440. K.B. Moser and F. Verbanac, US 3,562,103 (1971).
441. E.F. Paschall, US 2,876,217 (1959).
442. C.G. Caldwell and O.B. Wurzburg, US 2,813,093 (1957).
443. H. Neukom, US 2,884,412 (1959).
444. R.W. Kerr and F.C. Cleveland, US 2,884,413 (1959).
445. R.W. Kerr and F.C. Cleveland, US 2,961,440 (1960).
446. M.E. Carr, US 4,093,510 (1978).
447. M.M. Tessler, US 4,243,479 (1981).
448. K.A. Bernard, J. Tsai, R.L. Billmers, and R.W. Sweger, US 5,455,340 (1995).
449. K.A. Bernard, J. Tsai, R.L. Billmers, and R.W. Sweger, US 5,500,087 (1996).
450. W. Jarowenko and H.R. Hernandez, USA 4,029,544 (1977).
451. B. Carre and U. Carlson, US 5,496,440 (1996).
452. P. Schilling and P.E. Brown, US 4,775,744 (1988).
453. W. Lehmann, G. Troemel, K. Ley, and F. Muller, US 3,932,363 (1976).
454. W. Lehmann, G. Troemel, K. Ley, and F. Muller, US 4,166,002 (1976).
455. M.A. Hubbe, O.J. Rojas, N. Sulic, and T. Sezaki, Appita J., 60(2), 106–111 (2007).

456. A.J. Begala, US 5,098,520 (1992).
457. K. Chujo, K. Tanaka, and K. Ohata, US 3,634,366 (1972).
458. M. Hahn, W. Jaeger, R. Schmolke, and J. Behnisch, *Acta Polym.*, **41(2)**, 107–112 (1990).
459. M.B. Jackson, *J. Macromol. Sci. Chem.*, **A10(5)**, 959–980 (1976).
460. T. Itagaki, M. Shiraga, S. Sawayama, and K. Satoh, US 4,957,977 (1990).
461. A.J. Allen, E. Echt, W.W. Maslanka, and J.C. Peters, US 6,294,645 (2001).
462. W.H. Schuller, W.M. Thomas, S.T. Moore, and R.R. House, US 2,884,058 (1959).
463. E. Echt and R.P. Geer, WO 98/06898.
464. Y. Mori, K. Adachi, K. Takeda, and T. Tsuzuki, US 2008/0004405.
465. M. Hollomon, B.L. Walchuk, and F.J. Sutman, US 7,396,874 (2008).
466. C.P. Iovine and D. Ray-Chaudhuri, US 4,305,860 (1981).
467. S.J. Buckman, J.D. Pera, F.W. Raths, and G.D. Mercer, US 3,784,649 (1974).
468. J. Terpstra and J. Hendrinks, WO 99/64677.
469. J.G. Langley and D. Holroyd, US 4,913,775 (1990).
470. M.T. Wilharm, *Tappi Proceedings—Papermakers Conference*, Atlanta, GA, 1984, pp. 7–11.
471. D.R. Cosper and P.R. Stoll, US 4,268,352 (1981).
472. W. Cheng and R.T. Gray, US 2009/0267258.
473. D.L. Elliott, W.E. Hunter, and R.J. Falcione, US 5,647,956 (1997).
474. D.L. Elliott, W.E. Hunter, and R.J. Falcione, US 5,567,277 (1996).
475. P.D. Buikema and T. Aitken, US 4,146,515 (1979).
476. D. Owen, WO 96/05373.
477. S. Gosset, P. Lefer, G. Fleche, and J. Schneider, EP 0 282 415 (1988).
478. D.K. Chung, US 5,126,014 (1992).
479. T. Lindstrom, L. Wagberg, and H. Hallgren, *Industria della Carta*, **25(5)**, 227–232 (1987).
480. K.A. Johnson, US 4,643,801 (1987).
481. L.E.R. Wagberg and T.S.C. Lindstrom, US 4,824,523 (1989).
482. F.J. Sutman and R.A. Hobirk, US 6,168,686 (2001).
483. K.A. Johnson, US 4,750,974 (1988).
484. J.H. Smith, US 5,221,435 (1993).
485. H. Johansson, US 4,964,954 (1990).
486. W.E. Hunter and R.J. Falcione, US 4,908,100 (1990).
487. L. Bourson, US 5,501,771 (1996).
488. W.H. Stauffenberg, US 3,021,257 (1962).
489. E.P. Moore and G. Vurlicer, US 3,956,171 (1976).
490. J.C. Harrington and M.G. Hollomon, US 7,615,135 (2009).
491. R.W. Novak and T.C. Fallon, US 5,266,164 (1993).
492. S. Frolich, F. Solhage, E. Lindgren, and H.E. Johansson-Vestin, US 6,918,995 (2005).
493. A.J. Begala, US 5,595,629 (1997).
494. R. Nagarajan and J.B.W. Shing, US 6,007,679 (1999).
495. J.D. Rushmere, US 4,798,653 (1989).
496. A. Swerin and L. Odberg, *Nord. Pulp Paper Res. J.*, **8(1)**, 141–147 (1993).
497. M. Eriksson, G. Pettersson, and L. Wagberg, *Nord. Pulp Paper Res. J.*, **20(3)**, 270–276 (2005).
498. S. Gosset, P. Lefer, G. Fleche, and J. Schneider, US 5,129,989 (1992).
499. G. Pettersson, H. Hoglund, and L. Wagberg, *Nord. Pulp Paper Res. J.*, **21(1)**, 115–121 (2006).
500. G. Pettersson, H. Hoglund, and L. Wagberg, *Nord. Pulp Paper Res. J.*, **21(1)**, 122–128 (2006).
501. A.J. Begala, US 5,185,062 (1993).
502. D.L. Elliott, R.J. Falcione, W.E. Hunter, US 5,501,772 (1996).
503. M. Heard, G. Chen, and J.O. Stockwell, US 5,958,188 (1999).
504. R. Hund and E. Philibert, US 5,393,381 (1995).
505. M.A. Hubbe, S.M. Moore, and S.Y. Lee, *Ind. Eng. Chem. Res.*, **44**, 3068–3074 (2005).
506. P. Economou, US 3,677,888 (1972).
507. P. Economou, US 3,660,338 (1972).
508. E. Strazdins, US 4,002,588 (1977).
509. P. Economou, US 3,790,514 (1974).
510. H. Takeda and M. Kawano, US 4,929,655 (1990).
511. K.P. Kehrer, S.M. Atlas, V.A. Kabanov, A. Zenin, and V. Rogachova, US 6,755,938 (2004).
512. Y. Hosoda, S. Ishihara, and S. Kobayashi, US 4,380,600 (1983).
513. R. Hund and C. Jehn-Rendu, US 2009/0283232.

514. J.G. Langley and D. Holroyd, US 4,753,710 (1988).
515. L.L. Kuo, R.Y. Leung, S.R. Prescott, and T. Hassler, US 6,273,998 (2001).
516. R.H. Moffett, US 5,584,966 (1996).
517. T. Lindström and H. Hallgren, US 4,911,790 (1990).
518. D.S. Honig and E. Harris, US 5,274,055 (1993).
519. B. Alince, F. Bednar, and T.G.M. van de Ven, *Colloid Surf. A Physicochem. Eng. Aspects*, **190**, 71–80 (2001).
520. B.A. Keiser and J.E. Whitten, US 6,361,652 (2002).
521. B.A. Keiser and J.E. Whitten, US 6,358,364 (2002).
522. R.H. Moffett, US 5,595,630 (1997).
523. D.S. Honig and E. Harris, US 5,167,766 (1992).
524. C.R. Hunter and C.W. Vaughan, US 6,719,881 (2004).
525. W.A. Foster, D.M. Pckelman, and R.A. Wessling, US 4,189,345 (1980).
526. D.M. Pickeman, W.A. Foster, and R.A. Wessling, US 4,187,142 (1980).
527. J.D. Rushmere, US 4,954,220 (1990).
528. P. Pruszynski, US 5,942,087 (1999).
529. P. Sinclair and A.J. Hayes, US 4,925,530 (1990).
530. A. Swerin, U. Sjodin, and L. Odberg, *Nord. Pulp Paper Res. J.*, **8(4)**, 389–398 (1993).
531. T.E. Taggart, M.A. Schuster, and A.J. Schellhamer, US 5,061,346 (1991).
532. J.E. Voigt and H. Pender, US 4,066,495 (1978).
533. A.S. Michaels, M.J. Lysaght, and S.A. Splitz, US 3,558,744 (1971).
534. A. Marklund, C. Garcia-Lindgren, S. Wannstrom, L. Wagberg, and L. Gardlund, EP 1,918,455 (2008).
535. P.H. Aldrich, US 4,263,182 (1981).
536. B.A. Keiser, and J.E. Whitten, US 6,270,627 (2001).
537. H. Hallstrom, R. Sikkar, and O. Struck, US 7,306,700 (2007).
538. K. Kettunen, A. Kjellander, and M. Norell, US 6,113,741 (2000).
539. D. Glittenberg and P. Leonhardt, US 7,022,174 (2006).
540. D. Glittenberg and P. Leonhardt, US 7,147,753 (2006).
541. W. Maliczyszyn, W. Bindzus, and P.A. Altieri, US 6,413,372 (2002).
542. R. Lorz, F. Linhart, W. Auhorn, and M. Matz, US 4,749,444 (1988).
543. H. Johnston and L. Collett, US 6,475,341 (2002).
544. M. Singh and R. Cockcroft, US 2009/0120601.

4 Temporary Wet-Strength Resins

4.1 A LOOK AT THE PAPER WET-STRENGTH CONCEPT

As a wet web of paper is being dried, the paper's strength progressively increases with the decrease in the moisture content [1,2]. A lower water content result in larger areas of interfiber contact and thus creates potential bonding sites. The strength of dried paper, which is then soaked in water (called "wet strength"), takes the reverse route: more water, less strength.

Water (in wet web and in the wet paper after soaking in water) wets the cellulose fibers that swell and reduces the paper strength [3,4]. The fiber–fiber intermolecular bonds (van der Waals, hydrogen bonding) are broken leaving the paper strength somewhere between 1% [5] and 10% of the original dry strength [6–10] (see Section 5.5.1).

The extent of this loss is such that ordinary paper is useless in the wet state. Until relatively recently, paper was a material considered to be useful only in its dry condition. However, paper wet strength is a desirable attribute of many disposable products (napkins, paper towels, tissues, etc.). A chemical compound needs to be added to improve the paper wet strength. In order to figure out what kind of paper chemical is needed, the water effect of the paper structure must be scrutinized.

The loss in strength on paper rewetting is not related to a decrease in the cellulose molecular weight [11]; it is rather a result of a reduction in the number of bonds between cellulose fibers, which alters their adhesion. The strongest bonds between fibers are the hydrogen bonds. Compounds with a hydroxyl group (water, alcohols) will interfere with the paper structure by competing for the interaction with cellulose hydroxyl groups. Therefore, paper loses its strength when soaked not only in water but also in ethyl alcohol, glycerol, formamide, or ethylene glycol [7,12,13]. In the case of alcohols, as the molecular weight of the mono-hydroxyl alcohol increases, the lost strength of paper soaked in alcohol is smaller (Figure 4.1). As the alcohol molecular weight is higher, the hydroxyl concentration (by weight) decreases and the molecule gets more hydrophobic. Toluene (a very hydrophobic solvent) caused virtually no change in the mechanical properties of paper [10].

Water absorption in paper [14] is different in blends of water and propanol [12]. The loss of paper strength after soaking in those blends does not follow the addition rule based on the average of the separate effect of water and propanol (Figure 4.2). Water is more effective in damaging paper strength because cellulose may absorb water preferentially.

The effect of the water content on the paper properties shows two regimes separated by "the critical moisture content" (the amount needed for the adsorbed water to make a monomolecular layer on the cellulose fiber). The critical moisture content varies for different papers and may have values between 3.5% and 6.5% [15].

At about 20% water content in paper, the fiber reaches the saturation point (no free water is present). At that point, the fiber increases 25%–30% in circumference, 10%–12% in diameter, and 1%–2% in length [10]. An aggravating factor is the presence of residual hemicellulose [16], which is responsible for dry strength in paper, but it is readily re-swellable in water.

Two mechanisms have been put forward in order to explain the wet-strength development in paper in the presence of a wet-strength resin: the protection of the existing bonds and the addition of new ionic or covalent bonds [17,18]. Therefore, the wet-strength resin must be reactive and/or able to irreversibly change its conformation following the interaction with cellulose fibers.

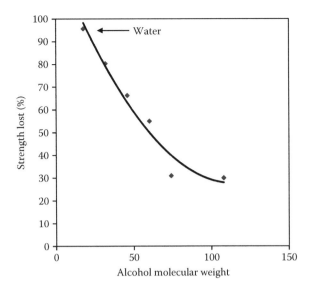

FIGURE 4.1 The paper strength loss after soaking in mono-hydroxyl alcohol with different molecular weights.

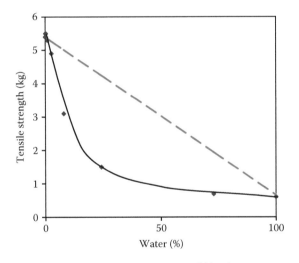

FIGURE 4.2 Paper strength after immersing in water–propanol blend.

Regardless of the mechanism (protection and/or reinforcement), if the criterion of the bond sensitivity to water is taken into account, paper strengthening may be classified as being either "temporary" or "permanent." The formation of strong covalent bonds between paper chemicals and cellulose results in permanent paper strengthening.

"Permanent strength" refers to the very long time needed to reduce the paper wet strength [19] after soaking in water. The wet-strength decay for paper treated with the urea-formaldehyde (UF) resin (Figure 4.3) and soaked in distilled water at 35°C is very slow: wet strength reaches about 50% decay in more than 1 year and even after 2 years, its wet strength is much higher than that of the untreated paper [20].

The pH value is important for the decay rate of the wet strength in water, for paper strengthened with cationic UF resin [21] (Figure 4.4). In spite of the more aggressive acid pH, it appears that the wet strength is still very high after 60 days.

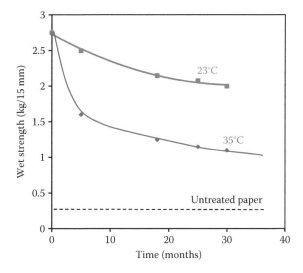

FIGURE 4.3 The loss of paper wet strength vs. soaking time (5.3% UF resin). (From Jurecic, A. et al., *Tappi*, 43, 861, 1960. With permission.)

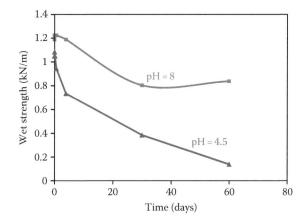

FIGURE 4.4 The effect of pH on paper wet-strength decay.

Polyamidoamine-epichlorohydrin resin imparts permanent wet strength, which makes it unsuitable for use in the preparation of recyclable paper [22–25]. The concept of "permanent wet strength" got a conventional definition [26] taking into account the practical needs of the sewer system: no significant wet strength is lost during 24 h of water soaking at ambient temperature.

Permanent wet strength is often an unnecessary and undesirable property. In commercial and industrial towel and tissue markets, there is a need for wet-strength resins having good initial wet strength and a large and fast decay rate (i.e., temporary wet strength).

The resin sensitivity to water of paper showing temporary wet strength is important in order to recycle the broke and get repulpable paper. Only for special applications a really permanent wet strengthening of the paper is needed. Paper chemicals that provide temporary wet strength to paper are called temporary wet-strength resins (TWSR).

These resins are soluble in water and/or react with cellulose through hydrolytically unstable or shear-sensitive bonds [27]. The temporary wet strength of paper means that it disintegrates over time in the presence of water and prevents the plugging of sewer system and septic tanks.

The concept of how "temporary" the strength should be has its own history. In 1973 [28], paper making standards indicated that the half of the wet strength should disappear after soaking in water at room temperature for 24 h. Today, there are different expectations for toilet paper made with a TWSR: 40% decay after 5 min of soaking in water at neutral pH, 60% decay, after 30 min, and about 85% after 24 h is a good enough decay rate for the actual sewer system.

A TWSR can be added to the wet end, can be sprayed on the paper surface [29], or applied by impregnation [30]. Chemical compounds binding the dried cellulose fibers disintegrate in cold water. In order to be retained on the pulp, the polymers added to the wet end should be cationic, while the polymers used for the immersion of the web can be either anionic or nonionic.

Cold water-soluble adhesives substantive to cellulose (polyvinyl alcohol, methyl cellulose, etc.) have been tried as sanitary pad wrappers and found to be readily disintegrable in excesses of water [31]. When the concentration of film-forming glue is high (up to 30% based on the cellulose), the paper is called "water-soluble paper" [32,33]. The interfiber bonding was replaced by fiber-adhesive hydrogen bonding, which is not sufficient to retain the wet-web integrity in water.

Anionic polyelectrolytes, such as acrylic copolymers added to the preformed web, are swollen and dispersible in water [34], or show inverse solubility in aqueous media (methacrylic acid copolymers) [35,36], or are ion-sensitive, water-dispersible polymers [37–39].

4.2 THE SYNTHESIS OF TEMPORARY WET-STRENGTH RESINS: GENERAL CHEMISTRY

The chemical compounds presented in the following sections are added to the wet end of the paper-making process. TWSR have chemical structures, which include weaker bonds: hydrolysable covalent or ionic bonds [40,41]. They are macromolecules that may or may not react with cellulose.

The ideal candidate as TWSR might be a cationic polymer (for retention), with pendant functional groups able to form weak bonds with the cellulose fibers or to self-cross-link. The benefits of this "reactive functionality" are double-fold: on the one hand such a resin is able to provide strength, but on the other hand the strength may decay after soaking in water. A challenge associated with such a resin is that its reactivity can shorten its shelf life.

The resin effectiveness depends on the number of reactive groups in one molecule and the specific reactivity of those groups. For a polymeric material, the number of functional groups per molecule is associated to its molecular weight. In order to cross-link cellulose fibers, a polymer must have at least two reactive functional groups, per one macromolecule. For higher cross-linking efficiency, the number of reactive functional groups needs to be above two. However, with more cross-links, the decay rate decreases [42]. Therefore, the number of reactive functional groups on a macromolecule must have an optimum value.

TWSR generate a "smart material" after cross-linking on paper [43]: they are able to respond to their own physical or chemical environment. In the presence of a stimulus (water in this case), they recognize, discriminate, and react to produce several useful effects. In this kind of system, chemical sensitivity arises from the presence of reversible chemical cross-links.

Polymeric candidates as TWSR must provide strength, which should be based on strong chemical bonds. In the presence of water, the same resin loses its strength due to hydrolysable bonds. That coexistence of strong and weak bonds in the same molecule is specific to the chemical structure of the TWSR.

4.2.1 STRONG BONDS AND WEAK BONDS IN ORGANIC CHEMISTRY

TWSR should be a cationic poly-functional macromolecular compound. An infinite network is developed by reacting TWSR with cellulose or by self-cross-linking. The network includes both strong and weak bonds. "Strong bond" means [44–46] a covalent bond that is stable in the presence of water. A chemical bond is considered weaker if it is breakable in the presence of water.

The general picture for a cross-linked system is a number of backbone macromolecules linked together by a cross-linker. The most common system is polystyrene cross-linked by divinylbenzene. In this case, both the backbone structure and the cross-linking bridges are formed from carbon–carbon bonds. The bonds are equally strong regardless of their position: in the backbone or in the cross-linking units (IV)1. That makes the network very chemically stable.

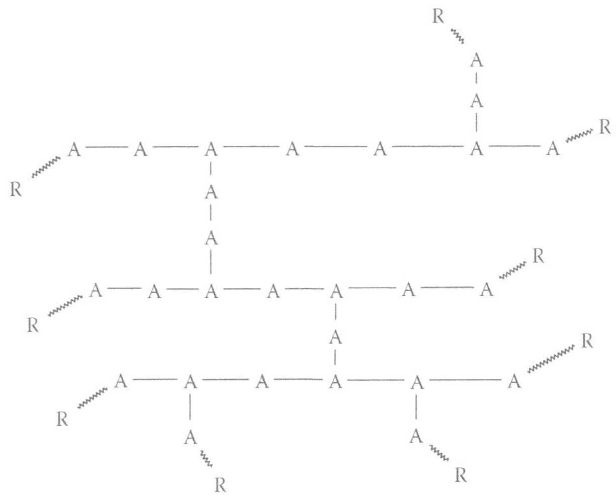

(IV)1

There is another system in which backbone bonds are different from those existing in the cross-linking units. It is the case of a polyacrylamide (PAAm) network made through acrylamide copolymerization with methylene bis-acrylamide (MBAA) [47]. MBAA [48] is a difunctional monomer. Its copolymerization with acrylamide results in a cross-linked structure. The backbone is made from carbon–carbon bonds but within the cross-linking units there are secondary amide bonds. Such a network structure can be hydrolyzed keeping unchanged the backbone molecular weight. In other words, this is a structure with stronger and weaker bonds. The weaker amide bonds are located in the cross-linking segments (IV)2.

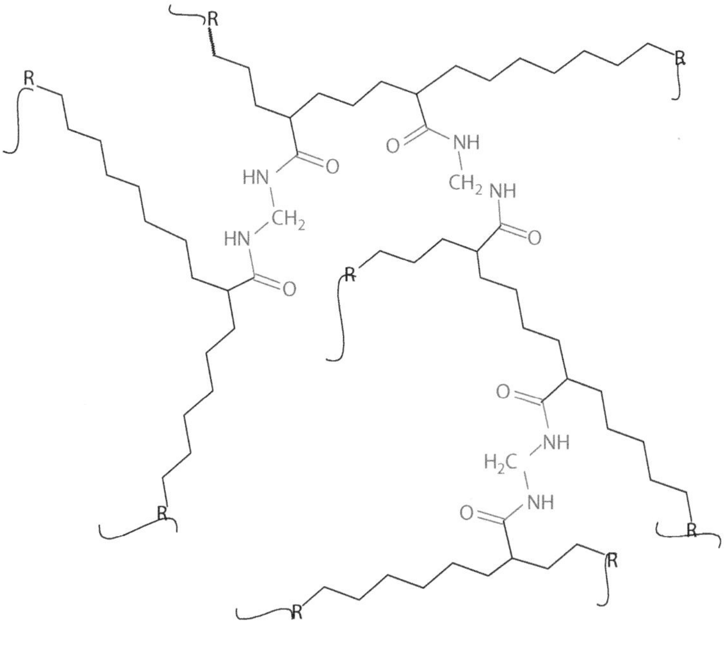

(IV)2

A cross-linked structure [49] can also be obtained by Michael addition of secondary amine from amino-polyamide to MBAA. Breakable bonds (amide: CO-NH, in red (IV)3) are present both in the backbone and in the cross-links as well.

This type of structure is disintegrated only at high pH. In neutral pH and at room temperature, a polyamide structure is quite stable.

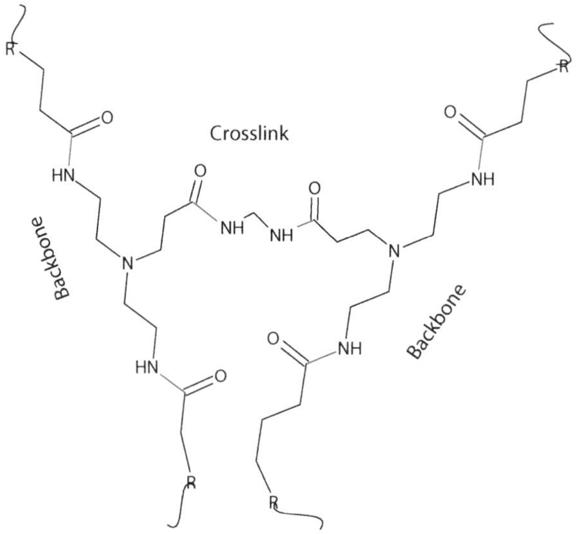

TWSRs must have bonds hydrolytically unstable (at neutral pH) or shear sensitive at room temperature. A hydrogel from a copolymer of isopropylacrylamide and a zirconium, alkoxide cross-linker degraded within days [50], while the degradation time for 2-hydroxyethyl acrylate and 5% cross-linker (N,N'-dimethacryloyloxy malonamide, (IV)4) ranged from 7 h to 31 days (at 37°C—body temp.—and pH = 7.4) [51,52].

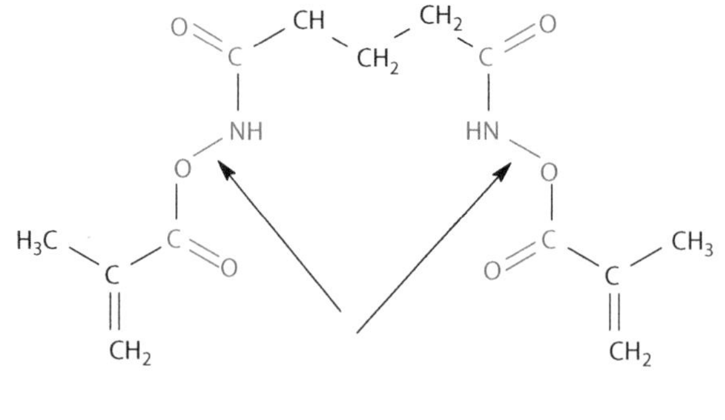

That degradation rate is too slow to be considered for a TWSR. However, the segment –CO–NH–O–CO– (in which the amide bond is connected to an ester group that will be mutually weakening) is a suggestion to be followed for faster decay. The nitrogen–oxygen bond is weaker due to those two carbonyl groups belonging to the system.

Aliphatic polyesters show quite a strong chemical bond: the hydrolysis in water at pH = 7.4 and at 37°C temperature takes about 20 h to reach a 50% conversion [53]. The ester function can be made weaker if the ether functionality is located in its neighborhood: hemiacetal ester [54–56] (IV)5 (see also acetoacetates [57]). This complex functionality can be built up through an addition of (meth) acrylic acid to ethylene glycol divinyl ether. The weakness of a chemical bond depends on the electron distribution, which, in its turn, depends on the type of other functional groups located next to that chemical bond.

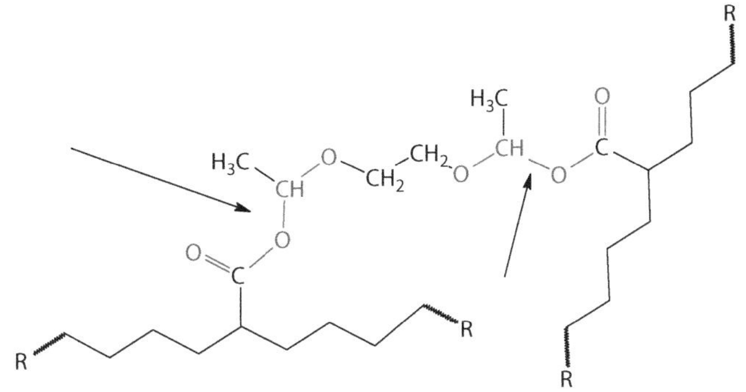

(IV)5

Poly-hydroxylic alcohols react with boric acid to reversibly form a much stronger monobasic acid (IV)6 [58,59].

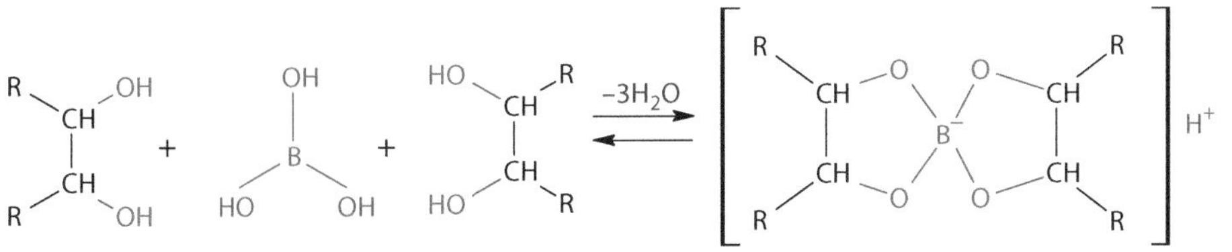

(IV)6

Carbohydrates and related substances (such as starch [60]) can also interact with boric acid [61,62], or boronic acid-containing compounds [63]. Polyvinyl alcohol (PVA) forms a gel with borate ions as well. A PVA with 46% of the diol groups in iso-tactic configuration will have about half of its coordinating sites available for complexation. Due to its high molecular weight, the local concentration of the diol groups does not show a linear dependence on the total PVA concentration [43]: some intramolecular complexation takes place. The gel is formed only above a critical concentration, which is a downside of that reaction because at the wet end the concentrations are very low.

PVA is typically used at up to 10% and borax at levels up to 2% based on the mass of cellulose. The complex made by boric acid shows a different stability at different temperatures, water concentration, and pH values [31,64–66]. The gel retains strength temporarily in the presence of limited quantities of liquid, but breaks down in the presence of a large volume of water [67].

Aldehyde functionalities react with phenols, primary and secondary amines [68], alcohols (to form acetals), amides, and other aldehydes within condensation reactions [69]. In order to make a hydrolysable structure, the most attractive reaction seems to be that between an amide and an aldehyde. Under neutral and mildly basic conditions, the aldehyde combines with an amide containing an N-H function (primary or secondary amide) [70] to make an amidol (IV)7:

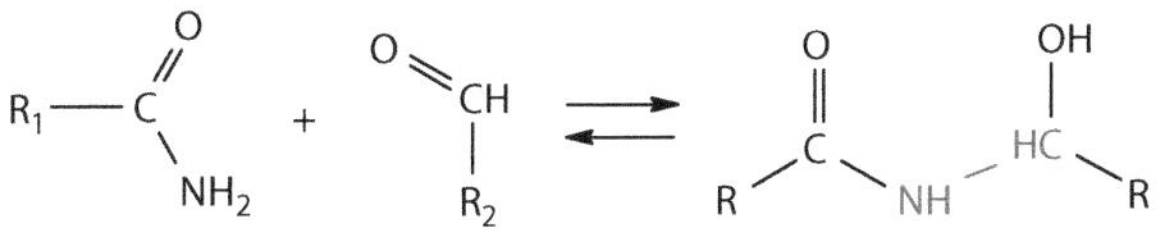

(IV)7

The amidol has an NH–CH weak bond formed within an equilibrium reaction. This is a key property of the new compound: by dilution in water, the equilibrium moves toward the left side. This equilibrium reaction is very similar to the one involved in a hemiacetal synthesis (IV)8:

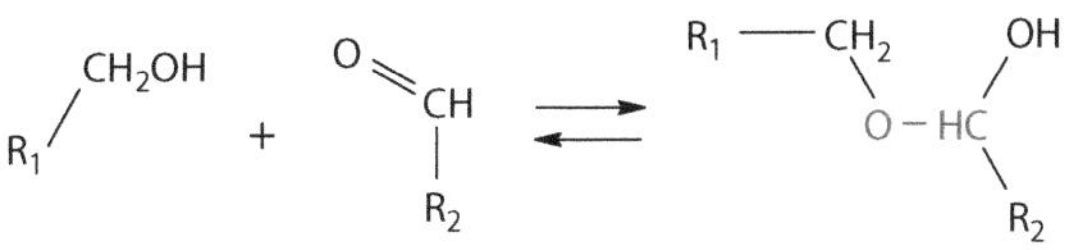

(IV)8

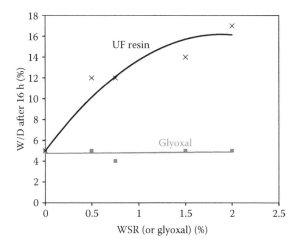

FIGURE 4.5   Wet-strength decay for paper strengthened with glyoxal or UF resin.

Cellulose has many hydroxyl groups and the hemiacetal bond formation is expected to occur in reaction with aldehydes. A di-aldehyde such as glyoxal is able to cross-link cellulose fibers [71]. Glyoxal is a nonionic compound and it should be added directly to the wet web [72]. The glyoxal strengthening capabilities were studied versus an UF or MF resin [73]. The UF resin (Figure 4.5) shows a much lower decay than glyoxal. After 16 h of soaking in water, the UF resin still shows a very high wet strength, while paper treated with glyoxal (regardless of the initial glyoxal concentration) has lost the entire wet strength down to the wet strength of water leaf (bleached sulfite pulp).

Glyoxal is significantly less toxic than formaldehyde. Because of the low volatility of its hydrate, high concentrations of the vapors are unlikely under normal use or mill conditions. Spraying the 30% aqueous solution of glyoxal into a flame failed to ignite it [73].

Straight glyoxal is hardly the ideal temporary wet-strengthening agent: it is not cationic and consists of a small molecule. Therefore, glyoxal is not retained at the wet end and is rather ineffective due to its low molecular weight. In other words, the aldehyde functionality should be attached to an ionic (co)polymer.

4.2.2   THE BACKBONE STRUCTURE FOR CARRIERS OF ALDEHYDE GROUPS

A potential TWSR is a cationic macromolecular compound with multiple pendant aldehyde groups. A cationic polymer with aldehyde moieties may also include other functional groups able to be involved in the cross-linking process and/or in the hydrogen bonding.

The strengthening performance and its decay depends on the polymer molecular weight, the number of functional groups, the cationic charge density, the ratio between aldehyde and the cofunctional groups, and the reactive group distribution.

A thermodynamic and a statistical factor are controlling the rate of network disintegration: the weakness and the number of cross-linking bonds, and the molecular weight of the backbone.

A macromolecular compound with multiple reactive functions may undergo "self-cross-linking" (intramolecular) reactions and/or "co-cross-linking" (intermolecular) reactions in which both TWSR and cellulose fibers are involved. For a temporary character of the paper strength, both processes should be reversible chemical reactions.

The ionic bonds are developed through electrostatic attractions and can occur over greater intermolecular distances than those necessary for hydrogen and/or covalent bond formation. However, the ionic bonds are also reversible and thus they suit the "temporary strengthening" concept.

For a given reactive group (such as aldehyde), the paper dry strength and the initial wet strength are higher as the molecular weight of the TWSR and the number of cross-linkable functionalities are higher. However, the wet-strength decay is faster when the polymer molecular weight and the number of breakable bonds are smaller. Actually, the balance between the initial strength and the wet-strength decay is a matter of optimum. Moreover, the distribution of functional groups (ionic and/or nonionic) and their availability for interactions seems to be another important parameter for both initial strength and its decay.

Aldehyde functionalities can be bonded to the carrier backbone through either strong or weak bonds. For retention purposes (capabilities to interact with a cellulose fiber), a TWSR has ionic charges (cationic or anionic).

4.2.2.1 Backbone with Aldehyde Functionality Bonded through Strong Bonds

Macromolecular compounds having aldehyde functionalities connected to the backbone through strong bonds (polyaldehyde) do not include the weak amidol or the hemiacetal bonds. The aldehyde moiety can be attached to a carbon–carbon macromolecular backbone or through a carbon–oxygen bond (ether). Ether bonds can be developed by a Michael type addition of an unsaturated aldehyde (acrolein or crotonaldehyde) to a macromolecular compound having hydroxyl groups.

The strength of the bond between aldehyde and the backbone, and the number of aldehyde functionalities impart high wet strength to paper and make it resistant to degradation from exposure to water over long period of time [74].

4.2.2.1.1 Aldehyde Incorporated by Copolymerization

Polymeric polyaldehydes are obtained by free radical copolymerization of unsaturated aldehydes (such as acrolein, methacrolein, crotonaldehyde, and their derivatives) with cationic comonomers (IV)9. If the aldehyde functionality (in red) is connected to a carbon–carbon double bond through strong covalent bonds, those strong bonds are preserved during the polymerization process and will be present in the final copolymer. To make cationic polyaldehyde, a cationic comonomer is involved in the copolymerization reaction.

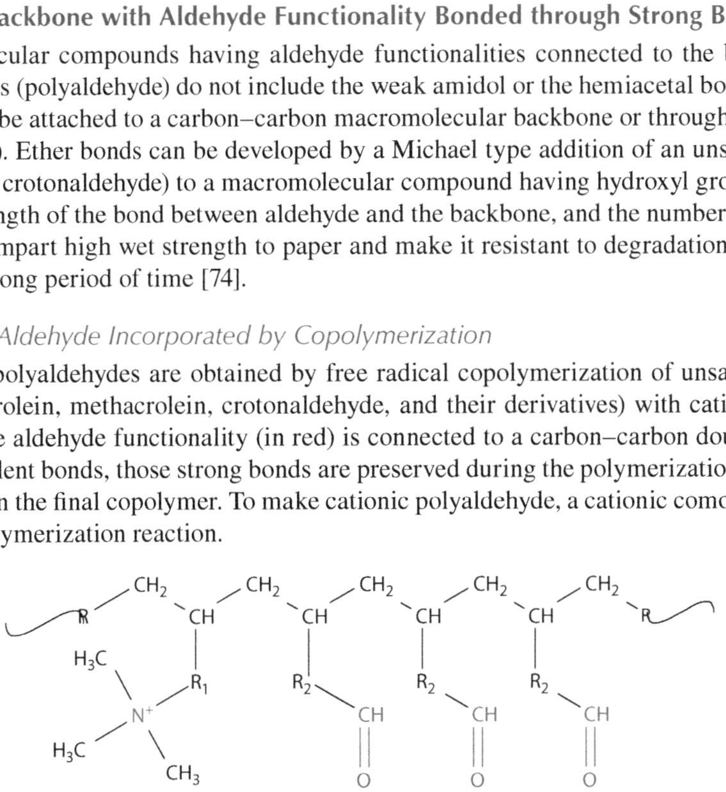

(IV)9

The use of TWSR requires balancing sufficient wet strength of the sanitary tissue during use with the ability of fast decay in an aqueous environment. For instance, the general structure of the polymer backbone for TWSRs may have the following general structure [44–46]:

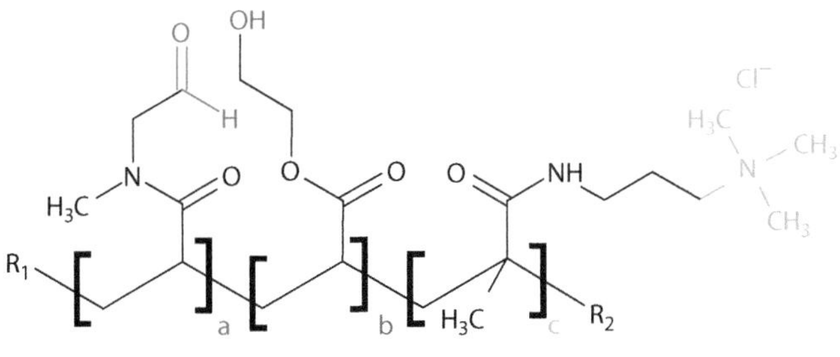

A co-cross-linking monomer with an aldehyde functionality (a is ranging from 5% to about 30% mol) is copolymerized with a cationic comonomer ($c = 5\%$–10%) and optionally with a diluter

(*N*-vinyl pyrrolidone or *N,N*-dimethyl acrylamide), and with a homo-cross-linking monomer (*b* ranging from about 30% to about 80% mol). The homo-cross-linking comonomer can develop a reaction and make a covalent bond (weak hemiacetal bond) between two temporary wet strength (TWS) macromolecules.

Molecular weight is preferably from about 100,000 to about 200,000. Higher molecular weights (>200,000) do not achieve a wet-strength decay of at least 35% after 5 min and at least 65% after 30 min. Lower molecular weights (<70,000) have a very low wet strength. A molecular weight of over 70,000 is needed because no other type of interaction with cellulose fibers (such as hydrogen bonds) is present.

A cationic reactive copolymer may include a variety of other comonomers (Table 4.1). It incorporates in its anionic version, acrylic acid [74] and needs an aluminum compound for retention. Amphoteric copolymers are also recommended as TWSR [75]. The aldehyde monomer is copolymerized with a small amount of cationic comonomer. The cationic binary copolymer (aldehyde + cationic comonomer) contains too high a concentration of the reactive aldehydes. To reduce their concentration, a third comonomer is added to make a ternary copolymer (Table 4.1) [76].

The third comonomer (a diluter) should be selected with a chemical structure able to avoid interaction with aldehyde groups. A potential reaction between the diluter (such as acrylamide) and aldehyde may result in self-cross-linking and shorter product shelf life. Depending on the amount and structure of the diluter comonomer, the final copolymer can be water soluble or water dispersible.

The reactivity of the aldehyde functionality is a permanent concern both during the polymer synthesis and its storage time. Several blocked aldehyde monomers were synthesized such as *N*-(2,2 dimethoxyethyl) acrylamide [77,78] and used to replace acrolein [79] in copolymerization with redox initiation [80] (Table 4.1). If the stable acetal group is within the structure of the main comonomer structure, acrylamide can be used as diluter.

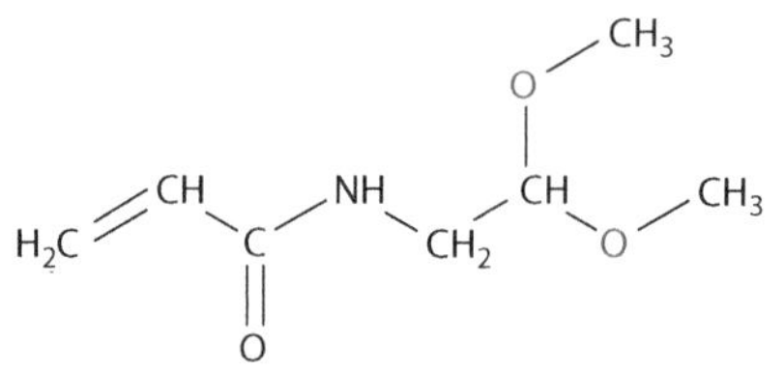

TABLE 4.1
Polyaldehydes Used as Temporary Wet-Strength Resins

Aldehyde Comonomer	Third Monomer (Diluter)	Cationic Comonomer	References
Acrolein (34%)	*N*-vinyl pyrrolidone (66%)	(Betaine hydrazide chloride)	[81,82]
Acrolein	Acrylonitrile	2-hydroxy-3-methacryloyloxy propyltrimethylammonium chloride	[83]
Acrolein	—	Tri-methyl allyl ammonium chloride	[84]
Acrolein (30%)	*N*-vinyl pyrrolidone (60%)	[3-(methacryloylamino)propyl trimethylammonium nitrate (10%)	[80,85]
Acrolein (15%)	*N,N*-dimethyl acrylamide (73%)	Methacrylamidoethyl trimethylammonium chloride (12%)	[86]
N-(2,2 dimethoxyethyl) acrylamide	Acrylamide (30%)	DADMAC (6%)	[77,87]
N-2,2-dimethoxyethyl)–*N*-methyl acrylamide (14%)	2-Hydroxyethyl acrylate (76%)	Methacrylamidoethyl trimethylammonium chloride (10%)	[44–46,88]
N-(2,2 dimethoxyethyl) acrylamide	Acrylic acid	3-Methacryloyloxyethyl trimethylammonium chloride	[75]

Unfortunately, the acetal functionality is much less reactive with the cellulose hydroxyl groups (curing time 3 min at 160°C) [89] than aldehyde. That is why, the acetal is hydrolyzed back to aldehyde before it is used on paper. In order to free the aldehyde group, a tedious procedure should be put in place: acetal is hydrolyzed with hydrochloric acid 0.5N for about 22 h to get an 88% conversion. The hydrolyzed copolymer is used to impart temporary wet strength to paper.

Monomers with "latent" aldehyde groups [90–92] such as graft copolymers of acrylamide, acrylic acid, DADMAC, and an acetal comonomer ([5-(dimethoxymethyl) furfur-2-yl] methyl acrylate, were used to synthesize grafted starch [93])—see below:

Blocked aldehyde groups allow for the synthesis of copolymers with higher molecular weights ($M_w = 90{,}000$) [44–46,80]. In the papermaking process, the acetal group comes right before the polymer application.

4.2.2.1.2 Aldehyde Groups Attached to Preexisting Macromolecular Compounds

Nonreactive polymers with strong bonds in their backbone can be converted to reactive macromolecular compounds by polymer analogous reactions. Cellulose oxidation results in aldehyde groups located on the cellulose fiber surface and improve the sheet performances [94,95]. That shows the capability of aldehyde groups to enhance paper wet strength through a cross-linking process (hemiacetal bond formation with cellulose fibers). The hemiacetal bond can also be formed with polysaccharides soluble in water [96] or with PVA [97].

Starch oxidation reactions are easier to perform [98,99] (see Section 2.4.2). As far as obtaining starch derivatives is concerned, three are the factors of interest: (1) the degree of substitution (how many units are affected by derivatization, expressed as mole of reacted D-glucopyranosyl units per total number of glucopyranosyl units—which can be $DS < 1.0$), (2) the reaction selectivity (derivatization vs. degradation), and (3) the product heterogeneity (how many starch molecule have zero derivatization).

During the oxidation of polysaccharides, the hydroxyl groups are converted to aldehyde functionality. Unfortunately, the aldehyde functionality is easily converted to carboxylic group by further oxidation [100]. The aldehyde groups formed during oxidation with hypochlorite were predominantly in the C-6 position (primary hydroxyl). The ratio between carboxyl and the carbonyl groups is roughly 2:1. It depends on the concentration of hypochlorite, time, temperature, and pH [98]. Oxidation with hypochlorite is nonselective. The scission of several glucosidic linkages also occurs, which results in a decrease in molecular weight.

In order to reduce the carboxylic group concentration in the final product, starch slurry was oxidized by sodium hypobromite at pH = 7.0, at a temperature of 20°C. With this procedure [101], the carboxyl/carbonyl ratio is reduced to 1:4. The degree of substitution is quite low: only about 10% of units are oxidized. A better selectivity is obtained when the oxidation of C_6 is performed with an enzyme (galactose oxidase) [102,103].

Besides the obvious reduction in the degree of aldehyde substitution [104], the presence of carboxylic groups in the aldehyde starches has several drawbacks, namely: the introduction of hydrophilic properties, the increase of anionic charges, and the possible cross-linking due to the presence of a multivalent metal cation.

A special technique is required to stop the oxidation at the aldehyde level. The oxidation of starch [105] or various gums [106] with periodates is characterized by a remarkably easier control and homogeneity of the product [107]. The reaction temperature 12°C–45°C, the yield was 85%–98% and only 0.03%–0.2% of carboxyl groups were present at the end of the reaction. About 98% of the repeating polymer units of starch have been converted to the di-aldehyde [108]. Peracetic acid can also be used [109] as an oxidant, but the reaction was run at pH < 2.0 (or even below 1.0), which makes that procedure difficult to use.

Potassium ferrate (K_2FeO_4) can oxidize the primary hydroxyl group only (C-6) from starch to an aldehyde [110]. The amount of oxidant exceeds the amount of starch, which makes that method impractical.

Certain polysaccharides (i.e., guar gum [111]) can be oxidized (by breaking the C–C bond between *cis*-hydroxyl groups) in the presence of ozone to form an aldehyde functionalized polysaccharide. Oxidation, which is performed at room temperature in a solution of only 1% polysaccharide, is not selective and the molecular weight is reduced (viscosity decreases from >10,000 cPs, to about 50 cPs). Oxidation by ozone was also performed on a starch [112] or a cellulose ester [113].

The reaction selectivity was much improved by performing the oxidation of polysaccharides [114–118] with a limited amount of oxidant mediated with a nitroxyl radical (TEMPO at low temperature: 5°C [119]). The limited amount of TEMPO is allowed because the catalyst is regenerated during the oxidation process [120].

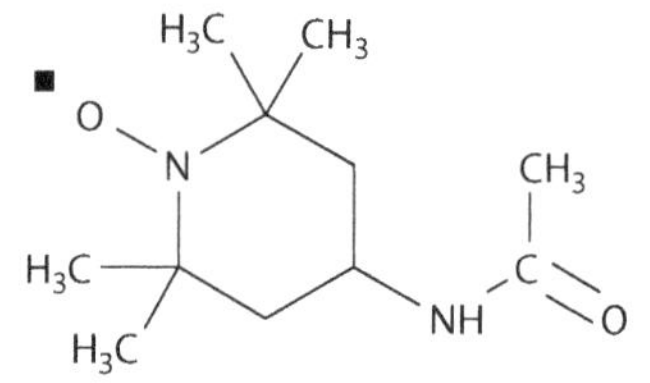

Selective oxidation of a ternary copolymer (2-hydroxyethyl acrylate, *N,N*-dimethyl acrylamide and a cationic comonomer) is also performed with TEMPO and sodium hypochlorite to convert the hydroxyl group to an aldehyde functionality [121–123]. The ternary cationic copolymer has a molecular weight of 70,000–200,000. The degree of substitution with an aldehyde group is about 2%–5% and the ternary copolymer is converted to a quaternary copolymer. Aldehyde may react both with the hydroxyl groups from the cellulose fibers and with the unreacted hydroxyl group from the TWSR.

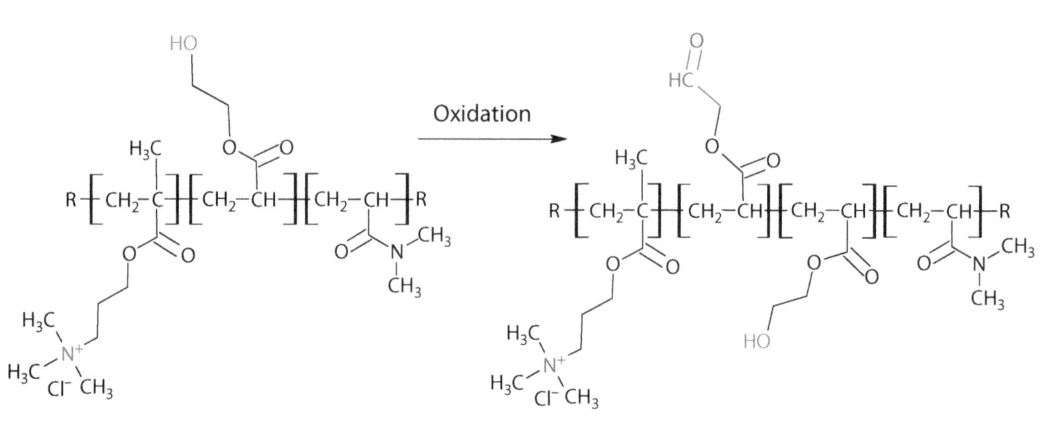

In order to improve the selectivity of the oxidation reaction, cellulose was functionalized with an easier to oxidize double bond. Cellulose is esterified with *cis*-1,2,3,6-tetrahydrophthalic acid (30–120 min at 120°C–180°C in the presence of sodium hypophosphite) and then the double bond is oxidized with ozone [113]:

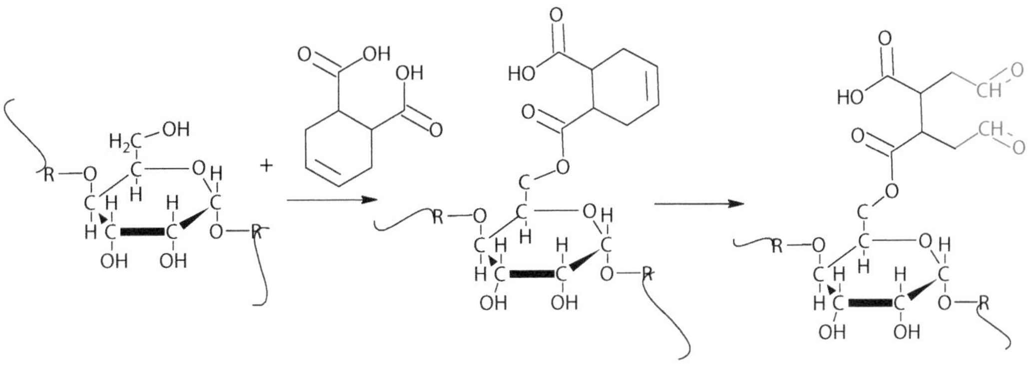

However, oxidative processes are not selective, and non-oxidative methods were developed for adding aldehyde groups to starch [124]. During that new process, any interference of the reactive aldehyde group is prevented by protecting it as acetal.

A substantially noncross-linked granular starch derivative is prepared through a reaction with an unsaturated aldehyde (methacrolein, acrolein and crotonaldehyde). This Michael addition is performed at 50°C and a pH=8.0. The reaction temperature is below the starch gel point. Therefore, the reaction develops only on the surface (heterogeneous process). This type of process results in a very broad compositional distribution: some starch is strongly derivatized and some will have no aldehyde functionality.

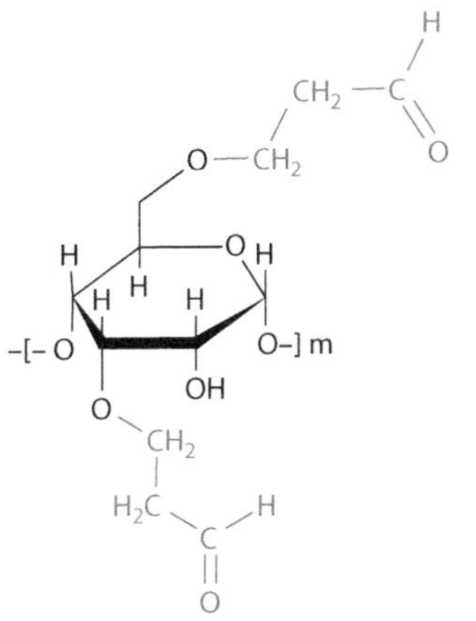

Aldehyde groups are designed to react with the hydroxyl group from cellulose (a co-cross-linking process) in order to impart strength to paper. The hydroxyl groups from starch display the same reactivity as those from cellulose and the intermolecular hemiacetal formation will result in starch homo-cross-linking, which is why the product has a short shelf life.

In order to improve the shelf life for functionalized starch with aldehyde groups, the aldehyde functionality is protected as acetal [104,125–129]. N-(2,2-dimethoxyethyl)-N-methyl-2-chloracetamide (an intermediate in starch derivatization) is prepared by adding chloracetyl chloride to methyaminoacetaldehyde dimethyl acetal:

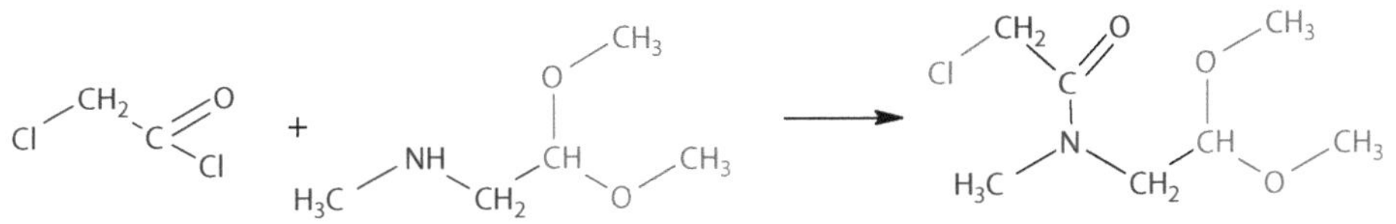

N-(2,2-dimethoxyethyl)-*N*-methyl acrylamide is prepared by adding acryloyl chloride to methyl-aminoacetaldehyde dimethyl acetal:

These compounds can react with starch and provide new functionalities differentiated by the structure of the R group.

Aldehyde-containing dextran [130] is obtained by hydrolysis of the corresponding diethyl acetal. The acetal synthesis starts from dextran and bromoacetaldehyde diethyl acetal. The dextran diethyl acetal is hydrolyzed and used immediately. Diethyl acetal has an indefinite shelf life as compared to aldehyde that has a very short shelf life.

The cellulose acetal is converted to the corresponding aldehyde by slurring the acetal in water and adjusting the pH to 2.5–3.0 with hydrochloric acid. Temperature has been set over the gelatinization point. The total cooking time was about 20 min. In order to prevent homo-cross-linking, the cook should be cooled down rapidly.

Aldehyde-containing polymers are not stable upon storage because they can undergo cross-linking and oxidation [130]. Starch with chemically blocked aldehyde groups is stable and the reactive aldehyde functionality is unblocked during cooking, before delivery to the wet end [131,132].

The procedure of the protected aldehyde groups offers some benefits but it has at least three reaction steps, the last of which (the conversion of acetal to aldehyde) should be performed at very low pH, where corrosion can be an issue.

Several polyaldehydes are used in a dosage of 2% based on cellulose fiber [111], and they should be spread on paper because no ionic group is present to help retention at the wet end. If ionic groups are present for retention on cellulose fibers at the wet end [75,77], the chemistry is even more difficult. A second approach can be considered: an anionic polymer, much easier to obtain (see oxidation reaction), is retained with a cationic retention aid [133,134].

Aldehyde starch provides a good initial wet strength [119]. The temporary character of the paper wet strength [131] is diminished due to the strong bond between the aldehyde group and the backbone of the TWSR. High molecular weight aldehyde starch has multiple aldehyde pendant groups and it is highly unlikely that all the hemiacetal groups formed with the cellulose fiber can be broken simultaneously. A decay of about 30%–50% is reached after more than 1 h soaking time in deionized (DI) water [135,136].

Both carbon–carbon and ether bonds are very stable toward hydrolysis. The weaker hemiacetal bonds coexist with strong bonds and the paper strengthened with that type of resin shows a slower decay. This is why a TWSR with an increased number of weak bonds is needed.

4.2.2.2 Carriers of Aldehyde Group through a Weaker Bond (Hemiacetal or Amidol)

Compounds with aldehyde groups form weak hemiacetal bonds with cellulose. Polymeric poly-aldehyde may develop multiple hemiacetal bonds and create a pseudo-irreversible link between polymer and fibers because hydrolysis is an equilibrium process unlikely to break all the hemiacetal bonds simultaneously. Hemiacetal bonds are responsible for the paper dry strength and initial wet strength, but the wet-strength decay depends on the number of such bonds and the polymer molecular weight. In order for TWSR to show a faster wet-strength decay, the polymer molecular weight should be lower and the connection between the aldehyde functionality and the backbone must include another weak bond.

In the case of cross-linking through multiple weak bonds, two other parameters are still important: the higher molecular weight of the backbone will provide a higher tensile but a slower decay, and the higher number of reactive aldehyde in the TWSR molecule results in a higher initial strength and a slower wet-strength decay. The backbone molecular weight and the concentration of the aldehyde groups are interrelated. The best TWSR must have an optimum ratio M_w/number pendant group.

4.2.2.2.1 TWSR with Two Identical Weak Bonds

The first logical attempt is to add a second hemiacetal bond between the aldehyde functionality and the polymer backbone. A symmetrical linkage with two hemiacetal bonds for each connection between the TWSR and the cellulose fiber is obtained in the final dry paper.

Chemical compounds prepared by glyoxal (or cyclic urea) and poly-hydroxylic compounds provide TWS for paper [137]. Since these poly-aldehydes do not have ionic charges, they are added to the wet web.

The hydroxyl groups from a poly-hydroxyl compound react with di-aldehydes and hemi-acetal bonds are obtained. At higher conversions of this reaction, a branched polymer can be formed and even a cross-linking process (gel formation) can take place, due to multiple functionality compounds.

Glyoxal (or blocked glyoxal) reacts with cationic starch and forms highly labile hemiacetal bonds (in red) along with aldehyde groups (in blue) [138,139]. The glyoxal units within the starch structure disrupt the crystallinity of amylose, thus significantly inhibiting retrogradation [138]. PVA and its cationic copolymers [140] follow the same reaction with glyoxal.

The multiple hydroxyl groups from the starch structure have virtually the same reactivity against glyoxal. But the pendant aldehyde group on the functionalized starch has the same reactivity with hydroxyl group, as the free glyoxal does. Because the glyoxalation of the hydroxyl group is an equilibrium process, a large amount of free glyoxal is needed to move the equilibrium in the right direction. Obviously, this would result in cross-linking, which continues even after the reaction has been quenched. A glyoxalated starch solution has a very short shelf life: a solution of 30% solids shows an increase in viscosity of about 100% after 5 days [139].

Cross-linked glyoxalated starch includes two symmetric hemiacetal bonds for each bridge between two starch molecules. The cross-linking reaction involving two different weak bonds will develop a nonsymmetrical bridge between macromolecules. The second weak bond must have a different bond strength and a different equilibrium constant.

4.2.2.2.2 Cross-linking Process through Two Different Weak Bonds

The second weak bond, along with the hemiacetal bond, is the amidol bond formed through a reaction between and aldehyde and an amide. In this case, the backbone must have multiple amide pendant groups.

The macromolecular carrier for amide groups can be generated through a reaction of the pre-existing polymer or by (co)polymerization. Free radical copolymerization results in a carbon–carbon backbone but with an important possibility of incorporating different pendant groups. The main amide monomer is acrylamide. Its copolymerization is discussed in a separate chapter (Section 4.3).

The preexisting polymer can be involved in a polymer analogous reaction to add amide groups. The backbone may have a variety of chemical structures, which may also include hydrolysable bonds (such as ester, amide, and acetal).

Carbamylethylation is a Michael addition reaction in which a compound having protons (hydroxyl, secondary or primary amine) reacts with the carbon double bond of acrylamide [141]. When the proton is located on a macromolecule, multiple amide functionalities are attached to the backbone.

Starch reacts with acrylamide in the presence of sodium hydroxide. Cationic starch [49,142] is the preferred support. Starch carbamylethylation results in a high degree of substitution. Some amide groups may hydrolyze to a carboxylic structure. Starch seems to be an attractive backbone, but the starch molecule is not easy to handle due to its high viscosity in solution, its sensitivity to hydrolysis, and to the uncontrolled reduction of the molecular weight. Carbamylethylated starch is subject to glyoxalation to make a wet-strength additive [28,143].

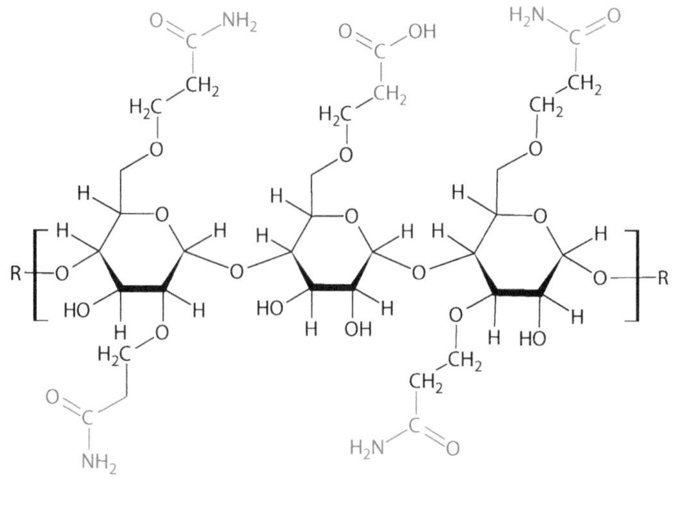

Other molecules subjected to carbamylethylation were an aminopolyamide [144], a polyamine or a polyvinyl amine [145], and a polyalkylene polyamine [146]. The aminopolyamide can be built up at any molecular weight, with any amino group concentration, with any charge density. During the Michael addition, tertiary amines are formed. They have potential cationic charges in a neutral or acid pH. Polyamidoamines obtained by the condensation of carboxylic acid (containing 16–20 carbon atoms) and DETA [147] are not soluble in water, but the addition of acrylamide makes them water soluble.

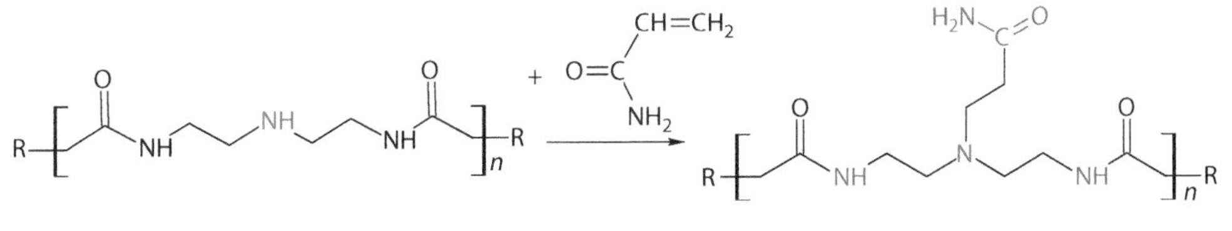

A sufficient amount of acrylamide will be used to react with substantially all of the amine groups in the polyamidoamine. A high pH during the reaction is favored since it frees the amino groups from their salts. Both primary and secondary amides (in blue) reacted with glyoxal to form an amidol bond (see Section 4.4).

4.3 THE SYNTHESIS OF POLYACRYLAMIDE

PAAm is a perfect candidate for a backbone able to react with di-aldehyde. Acrylamide can be polymerized through a free radical or an anionic mechanism. PAAm obtained through free radical polymerization has only carbon–carbon bonds in its backbone, while the backbone of PAAm obtained with an anionic catalyst contains multiple secondary amides (poly(β-alanine)).

PAAm is a polyamide with an amide functionality as a pendant group (primary amide). Both types of acrylamide polymers can react with glyoxal to make a TWSR. Poly(β-alanine) reacts with glyoxal [148,149] at the secondary amide atom.

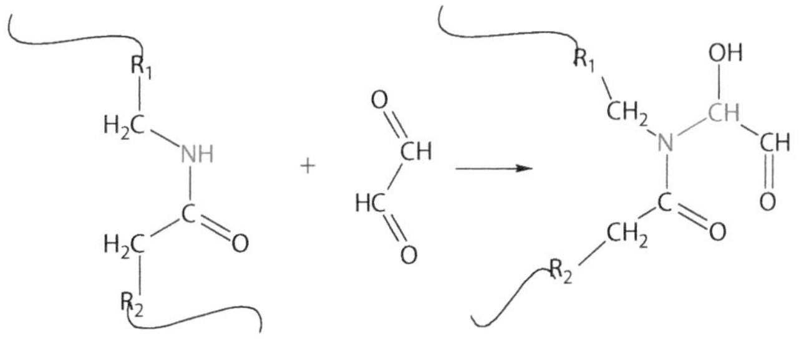

Poly β-alanine is not cationic and must be treated with caustic in order to make it anionic. The solution of glyoxalated anionic poly β-alanine (10% solids and pH=5.0) has a shelf life that exceeds 6 months [149].

4.3.1 CATIONIC POLYACRYLAMIDE THROUGH FREE RADICAL COPOLYMERIZATION

PAAm does not react with cellulose fibers at 105°C (drying temperature), but the amide pendant groups do develop hydrogen bonding with them, which may improve the paper initial wet strength. Cationic PAAm is synthesized through the free radical copolymerization of acrylamide with a

TABLE 4.2

Cationic Copolymers of Acrylamide

Carrier	Cationic Comonomer	References
Acrylamide	Di-allylamine	[26,155]
Acrylamide	DADMAC and allylamine	[156]
Acrylamide	Morpholino-alkylvinyl (or allyl) ethers	[157]
Acrylamide (70%–97.8%)	DADMAC (2.2%–30%)	[151–153,158]
Acrylamide (80%)	Methacrylamidoethyl trimethylammonium chloride (20%)	[152]
Acrylamide (99%–95%)	Cationic acrylate (<5%)	[159]

cationic comonomer. When the reaction is performed in the presence of starch, a cationic, graft copolymer is obtained [150]. During that process, it is possible to set the molecular weight, molecular weight distribution, the charge density, and the charge density distribution of the acrylamide copolymer.

The simplest structure for such a copolymer would be an acrylamide binary copolymer with a cationic comonomer [151–154] (Table 4.2). Acrylamide copolymerization is performed in the presence of 2-mercapto-ethanol as chain transfer agent. In order to prevent the Michael addition of mercaptan to the acrylamide double bond, pH is adjusted to 4.0 with a solution of sulfuric acid. The cationic comonomer concentration ranges from 2% to 30% by weight in copolymer. A TWSR should have a T_g less than 100°C [45].

The cationic comonomer or a third comonomer (diluter) should be water soluble and have functional groups that do not react with glyoxal. Tertiary amines, such as morpholino-alkylvinyl ether and/or morpholino-alkylallyl ether [157], do not react with aldehyde, but secondary and primary amines do.

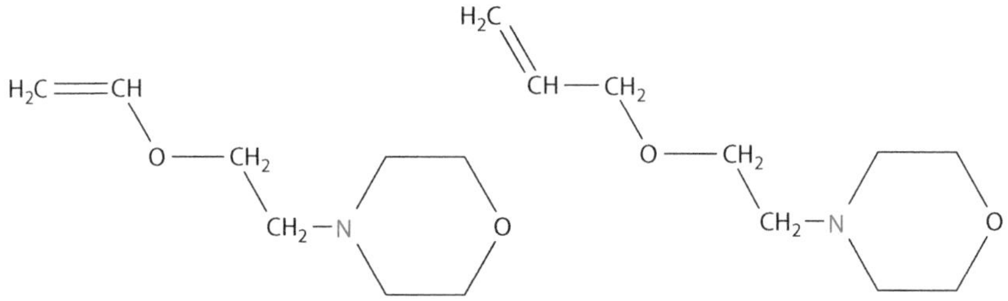

Comonomers with amine groups (such as allyl or di-allyl amine) need special attention during the free radical copolymerization and during the reaction with aldehyde. In order to convert an amino group to an ammonium salt, the copolymerization should be performed in acidic pH (pH=3–4) [26,155]. In acid pH, the amine is protonated and it cannot interfere with the initiation system or to be involved in the Michael addition to the double bond of acrylamide.

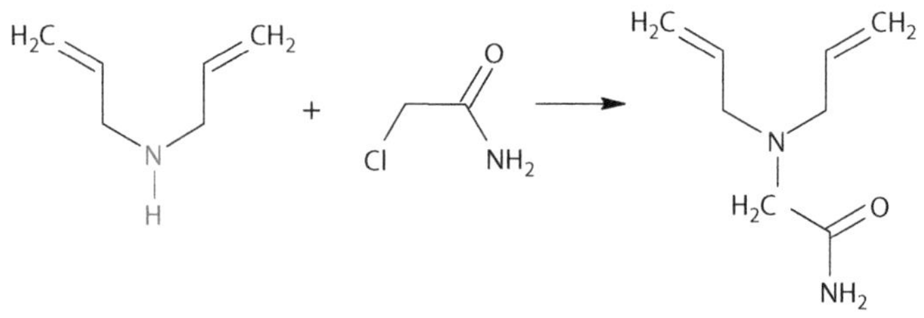

However, the best option to protect the amine is through alkylation. Di-allyl amine is alkylated to obtain DADMAC or is converted to a tertiary amine and a new cationic charge is added by reaction with cyanamide [160].

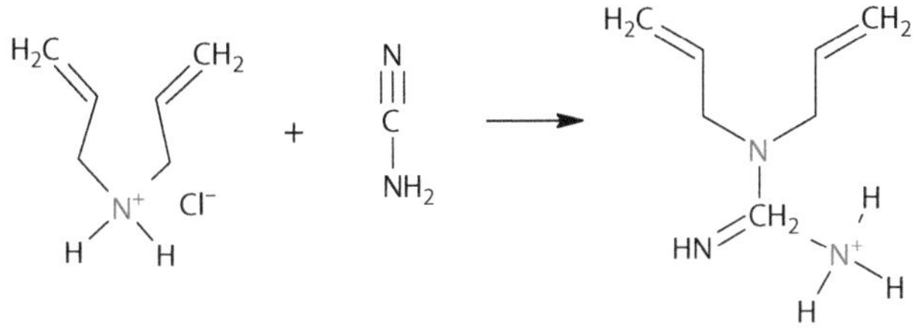

If chloroacetamide is used for alkylation, the final product is a multifunctional monomer with a double bond, and a tertiary amine, which is a potential carrier for cationic charges, and a new amide group [161]. This new monomer is polymerized after its conversion to a quaternary ammonium salt in acidic pH.

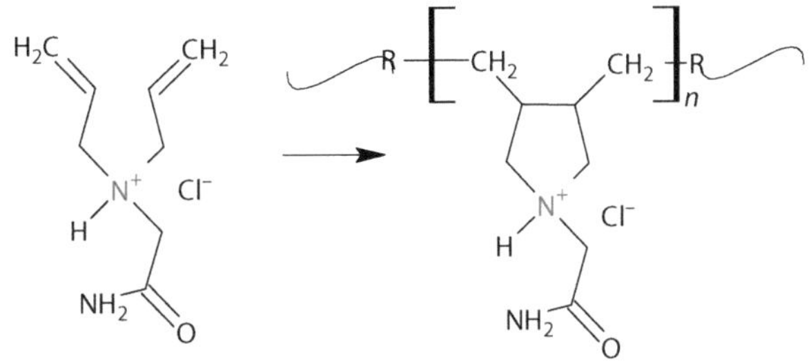

In the copolymerization with acrylamide, the double bonds in di-allyl derivatives are cyclo-polymerized: regardless of the multiple functionalities, no cross-linking reaction is recorded (a difunctional monomer behaves as a monofunctional compound). DADMAC as a cationic como-nomer [162] is converted through cyclo-copolymerization [163] to a five-member ring, and the residual double bonds (due to 1–2 allyl polymerizations) are less than 3%.

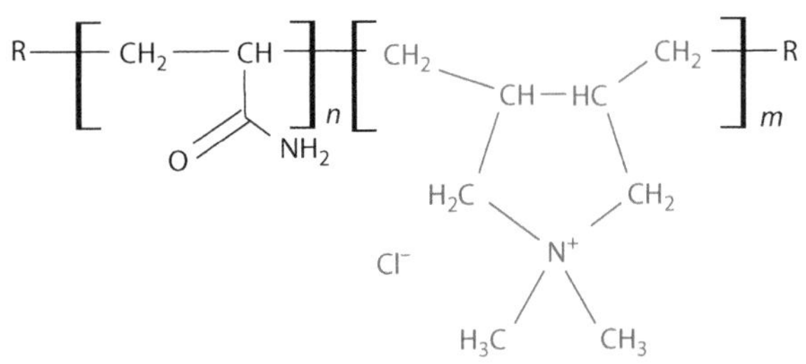

The kinetics of DADMAC cyclo-homopolymerization follows the known rule (Equation 4.1) of free radical processes [164]:

$$R_p = k_p[M]^{1.0}[I]^{0.5} \tag{4.1}$$

where
 M stands for the monomer concentration
 I stands for the initiator concentration

This makes the copolymerization of DADMAC with acrylamide a classic binary copolymerization system. Equation 4.2 describes the copolymer composition as a function of the composition of the comonomers and their reactivities [165,166] (see also Sections 2.4.1 and 9.1.2.2):

$$\frac{dM_1}{dM_2} = \frac{m_1}{m_2} = \frac{M_1}{M_2}\left[\frac{r_1M_1 + M_2}{M_1 + r_2M_2}\right] \tag{4.2}$$

where
  $r_1 = k_{11}/k_{12}$ and $r_2 = k_{22}/k_{21}$ are the monomer reactivity ratios
  m_1 and m_2 are the copolymer composition obtained starting from a particular monomer composition (M_1 and M_2)

Table 4.3 shows few reactivity ratios for binary copolymerizations of acrylamide with the most common ionic or nonionic comonomers used in the synthesis of TWSR. Nonionic comonomers are used as diluters.

The most common cationic comonomer is DADMAC. As shown in Table 4.3, different authors provide significantly different values for its reactivity or for the acrylamide–acrylonitrile copolymerization. The reactivity ratios for acrylamide–DADMAC copolymerizations are in the following ranges: for acrylamide (M_1) $2.6 < r_1 < 5.2$ and $0.05 < r_2 < 0.15$ [184]. Those reactivity ratios show that the cationic comonomer is less reactive (about 100 times) and less incorporated in the copolymer structure. In a batch-wise process, for a 100% conversion of acrylamide, DADMAC is incorporated in the copolymer up to 30%–40% only [185].

TABLE 4.3
Reactivity Ratios in Free Radical Copolymerization of Acrylamide

| | | Reactivity Ratios | | |
		r_1	r_2	References
Monomer 1	Monomer 2			
Acrylamide	DADMAC	6.4	0.06	[167,168]
Acrylamide	DADMAC	6.7	0.58	[169]
Acrylamide	Dimethyl aminoethyacrylate quat	0.29	0.34	[168]
Acrylamide	Dimethyl aminoethymethacrylate quat	0.61	2.52	[168]
		0.52	1.9	[170]
		0.25	1.71	[169,171]
Acrylamide	Dimethylaminopropyl acrylamide	0.95	0.47	[169]
Acrylamide	Acrylic acid	0.57	1.45	[172]
		0.47	1.3	[173]
Acrylamide	Itaconic acid	0.67	1.25	[174]
		1.18	0.60	[175]
Acrylamide	Acrylonitrile	0.55	1.91	[176]
		0.74	0.05	[177]
Acrylamide	2-Hydroxyethyl methacrylate	0.0002	2.33	[178]
Acrylamide	Vinyl acetate	9.3	0.09	[179]
Acrylamide	Ethylene sulfonic acid	11.1	0.02	[180]
Acrylamide	N-Vinyl pyrrolidone	1.2	0.8	[170]
n-Butyl acrylate	Allyl acetate	11.7	0.04	[181]
N-Antipyryl acrylamide	MMA	0.98	1.53	[182,183]
N-Antipyryl acrylamide	Acrylonitrile	0.82	1.14	[182,183]
N-Antipyryl acrylamide	Vinyl acetate	1.22	0.22	[182,183]

Based on the reactivity ratios shown in Table 4.3, and for a mole ratio of 0.95/0.05 acrylamide/DADMAC, for every 125 units of acrylamide, there will be one DADMAC unit. For a copolymer with a molecular weight of about 3000 (45 acrylamide repeating units), one DADMAC molecule is incorporated for every macromolecule only at a DADMAC concentration of 15% mol (26% by weight). This is valid for copolymers obtained at very low conversions.

Copolymerization is a kinetically controlled process and the copolymer composition is a function of conversion. Figure 4.6 shows the instantaneous copolymer compositions as a function of conversion up to 100% conversion. In the case of comonomers with a reactivity similar to acrylamide (such as acrylonitrile or N-vinyl pyrrolidone), the copolymer composition is almost constant up to high conversion values.

For a batch copolymerization process between acrylamide and DADMAC, using reactivity data from Table 4.3 (r_{AA}=6.4 and r_{DADMAC}=0.06), at a high concentration of DADMAC (25% mol), the copolymer composition (obtained even at low conversions) is richer in acrylamide (95%). That results in a faster consumption of acrylamide. Thus, at conversions of over 90% acrylamide is fully converted, but some DADMAC remains un-reacted (Figure 4.6). In other words, about 50% from DADMAC will not be incorporated in the acrylamide copolymer. It is not surprising that the final product (after dilutions and glyoxalation) still shows about 1.1% of free DADMAC [186] (more than 5% based on active glyoxalated polyacrylamide [G-PAm]). As a cationic compound with small molecule, DADMAC will compete with the cationic polymer for the anionic charges from cellulose fibers, and thus possibly damage polymer retention.

PAAm can be obtained by direct solution polymerization [158] or in an inverse-emulsion polymerization [187,188]. A cationic initiator (AIBA) + sodium persulfate were used at a temperature of 35°C and then 75°C.

The acrylamide–DADMAC copolymer obtained through a batch-wise process [152] will show a very broad chemical composition distribution. That is why that type of copolymerization should be run within a semi-continuous process [26,155,166] (a small amount of acrylamide and the entire amount of DADMAC as pre-charge, while the rest of acrylamide is added continuously [189]). The addition time for acrylamide should be correlated with the DADMAC conversion in order to get a uniform copolymer composition. To incorporate 2.2% DADMAC [190], a semi-continuous process (100 min acrylamide addition—at 82°C, reflux temperature for IPA [151]) is used. To incorporate larger amounts of DADMAC (up to 40% in copolymer), more complicated processes are used [153].

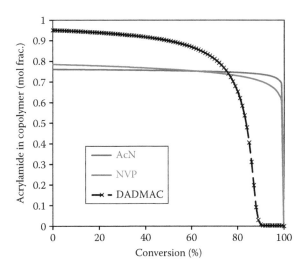

FIGURE 4.6 Composition/vs./conversion with acrylamide copolymers.

The composition of the acrylamide–DADMAC copolymer also depends on the total concentration of both comonomers [167] and the presence of a "template" (such as the polyacrylic acid), which may change the distribution of cationic units within the copolymer [191].

DADMAC is an ionic monomer; its reactivity depends on pH [152,158] and on the presence of inorganic salts (such as ammonium sulfate and/or sodium sulfate). A pH of 4.0 is suggested before starting the acrylamide addition. In the presence of an inorganic salt, the newly formed polyelectrolyte (acrylamide-DADMAC copolymer) can precipitate and the precipitate will be swollen with a comonomer composition different than the one used in the continuous phase. Cationic polyelectrolyte (such as poly-DADMAC) [192] is used to stabilize the precipitate of cationic polyacrylamide.

The batch-wise process always produces a broad distribution of the copolymer composition: the cationic acrylamide–DADMAC copolymer will coexist with nonionic PAAm. By manipulation of the semi-continuous reaction, the ratio between ionic and nonionic polymers can be adjusted.

The semi-continuous synthesis of the acrylamide–DADMAC copolymer is difficult, and other more reactive cationic comonomers are considered: cationic acrylamides and/or acrylates [159,192–194] (see Table 4.3).

Inverse-emulsion polymerization of acrylamide, in which each particle is a small reactor performing as a batch process, for a blend of 15% DADMAC and 85% acrylamide, after a conversion of 96% of acrylamide, DADMAC is only 40% incorporated [195]. That means that the inverse-emulsion polymerization, which cannot be handled as a semi-continuous process, does not improve the DADMAC incorporation.

Due to the difficulties in obtaining a homogeneous acrylamide copolymer, cationic charges are added to a nonionic acrylamide homopolymer through a polymer analogous reaction: trans-amidation of nonionic PAAm with dimethylaminomethyl amine [196] or 3-(dimethylamino) propylamine [151].

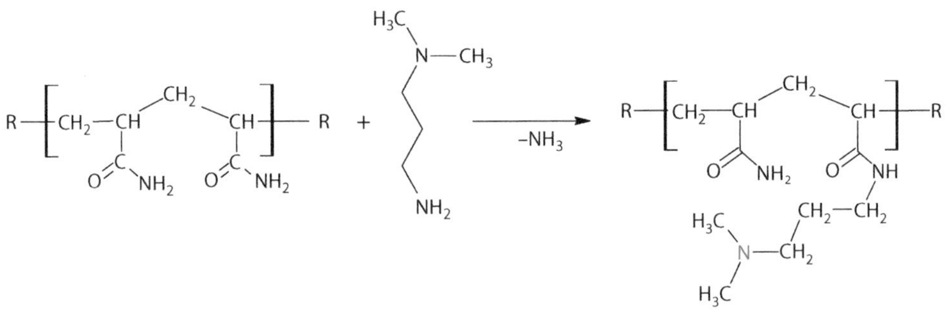

Anionic or amphoteric PAAms (copolymers with acrylic acid and—optionally—with a cationic comonomer) are also potential backbones for glyoxalated resins [133,134,159,197,198]. An anionic acrylamide copolymer (acrylamide-acrylic acid) is much easier to obtain [133,134] (see Table 4.3). The copolymerization of the acrylamide–acrylic acid depends also on pH [199]. Anionic polymers are retained by a cationic promoter [197].

4.3.2 ACRYLAMIDE COPOLYMERS WITH A "DILUTER"

The chemistry related to the temporary strengthening of the paper (glyoxalation rate, resin stability, strength performances, decay of paper strength when soaked in water) includes reversible processes. Thus, the synthesis of TWSR, the product stability, and the paper wet-strength decay, can be manipulated by changing the concentrations of chemical compounds and the coefficients of the reaction rate.

The amide group concentration can also be decreased by dilution in backbone with an inert monomer, in the glyoxalation step. However, diluter units (ethyl acrylate, acrylonitrile [190], methyl methacrylate and styrene [152], *N*-vinyl pyrrolidinone, *N,N*-dimethyl acrylamide [200]), etc.) are used in small amounts [151,152]. 2-Hydroxyethyl acrylate is a special case because its hydroxyl group reacts with glyoxal.

A diluter for acrylamide can reduce the glyoxalation rate and increase the stability with no impact on strength performances. A "control" backbone [151] (but with only 3%–5% glyoxalated acrylamide units) has a wet tensile decay of 54% after 30 min. With a diluter (46% *N,N*-dimethylacrylamide), the wet tensile decay reaches 65% for the same amount of glyoxalated units [75].

The presence of a diluter changes the binary copolymerization in a ternary copolymerization, which is more difficult to control. An easier dilution of the amide units can be obtained by a grafting process on a preexisting polymer. Unfortunately, the grafted copolymer will coexist with an un-grafted polymer backbone and an un-grafted PAAm (see grafting ratio in Section 2.4.2).

Acrylamide grafted copolymers on starch [201,202], PVA [150,203], or polyethylene oxide [204,205] are potential TWSRs. Cationic starches are obtained by grafting cationic acrylamide copolymers on starch backbones.

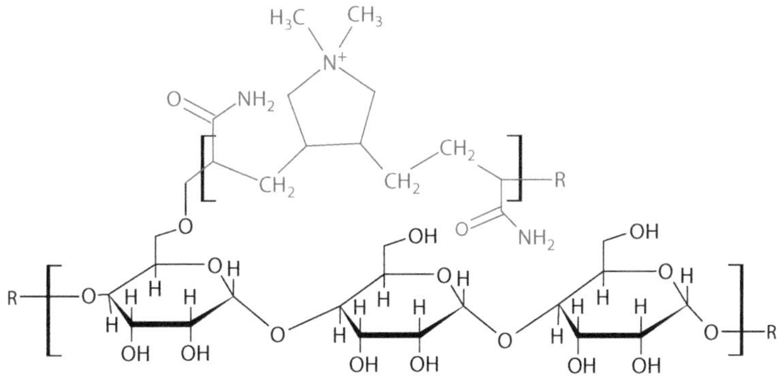

In order to improve the grafting degree, starch was derivatized with allyl glycidyl ether starch (starch has a pendant double bond); acrylamide and cationic comonomers are copolymerized in the presence of that starch.

The grafting reaction is also performed on polyamine obtained by condensation epichlorohydrin and methyl amine [206]. The tertiary amine was changed into a quaternary ammonium salt:

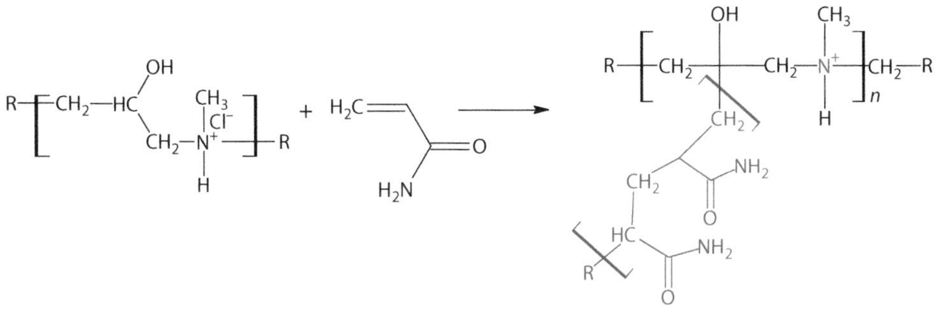

An important amount of un-grafted PAAm is expected along with some grafted copolymer. The glyoxalation step involves both the grafted polymer and the acrylamide homopolymer. All grafted copolymers (without any aldehyde functionality) [207] were tested as dry and wet-strength resins. The paper wet-strength decay is only 20% in 30 min.

Grafted polymers have, by definition, a molecular weight larger than the initial backbone. The wet-strength decay depends on the polymer molecular weight, on the presence of functional groups, and on the number and distribution of reactive groups. For faster wet-strength decay, the molecular weight of PAAm is a critical parameter.

4.3.3 POLYACRYLAMIDE MOLECULAR WEIGHT

In order to perform as a TWSR, the molecular weight of the PAAm should reach an optimum [208]: too high a molecular weight results in lower wet-strength decay, too low a molecular weight polymer is less effective as strength resin. Moreover, the backbone of the PAAm is partially cross-linked during the glyoxalation step and a higher molecular weight of the backbone may form a cross-linked polymer in the early stages of the glyoxalation.

During the glyoxalation reaction, glyoxal reacts at the same rate with all the amide groups, regardless if they are in a smaller or larger molecule. From the composition stand point, final glyoxalated PAAm is a homogeneous material. After retention on paper, after drying, and during paper testing by soaking in water, a G-PAm from a lower molecular weight PAAm imparts wet strength to paper in the manner of its higher molecular weight counterparts. A higher molecular weight due to the cross-linking process will increase the initial wet strength. However, paper treated with G-PAm from low molecular weight PAAm shows a faster wet-strength decay.

The estimation and control of the molecular weight of PAAm [209] are crucial for a good TWSR. During the acrylamide polymerization, the polymer molecular weight is controlled by a chain transfer agent (CTA). The polymer molecular weight is a function of the concentration of the CTA, and its chain transfer constant (Equation 4.3) [210]:

$$\frac{1}{n} = \frac{1}{n_o} + C_T \frac{[CTA]}{[M]} \qquad (4.3)$$

where

    n is the degree of polymerization in the presence of CTA
    n_o is the degree of polymerization in the absence of CTA
    C_T is the chain transfer constant for a particular CTA [211]
    [CTA] and [M] are the concentrations of CTA and monomer, respectively

The mercaptoethanol and isopropyl alcohol [212] are used as CTA. Isopropanol is much less effective in controlling the PAAm molecular weight; a larger amount of CTA is required, which makes the polymer precipitate at lower conversions.

PAAm used as a carrier for a potentially reactive aldehyde has a molecular weight lower than 25,000 [151,152]. The molecular weight is estimated according to the general rules for polyelectrolytes: intrinsic viscosity is measured in a solvent containing 1 M NH$_4$Cl [153].

Unexpectedly, it was found [158,213] that the addition of modest amounts of a high molecular weight PAAm did not adversely affect the decay of the initial wet strength, and also raised the initial values of wet and dry strength over those obtained by using the low molecular weight component alone. Both fractions with high molecular weight and low molecular weight are glyoxalated separately or within the same glyoxalation step [214].

4.3.4 POLYMER BLENDS AS TWSR

Paper used for towels, napkins, and tissues must have several attributes, such as good dry strength, high initial wet strength, and fast decay of the paper wet strength after soaking in water. All these attributes can hardly be achieved by a single TWSR. In order to find synergetic effects, blends of different resins were used as TWSR. To improve the decay of the paper wet strength, polyaldehydes are always present in these blends.

Many polymers that improve the dry strength of paper are anionic under normal papermaking conditions. A cationic compound may be used to retain these anionic substances to the pulp fibers, which are anionic by nature [75]. In the free radical copolymerization, the acrylic acid has a reactivity similar to that of acrylamide, and thus the anionic PAAm is easier to synthesize. In the presence of a high amount of buffer (15%–20% based on solids), glyoxalated nonionic PAAm is partially hydrolyzed before being blended with poly (DADMAC) [215] for retention purposes. In that mixture, polyDADMAC can be replaced by other cationic polymers such as the dimethylamine–epichlorohydrin copolymer [216].

Blends of polymers are obtained by the synthesis of a macromolecular compound in the presence of another polymer. The synthesis of polystyrene in emulsion is performed in the presence of cationic acrylamide [217]. PAAm copolymers are obtained in the presence of glyoxal and urea when polycondensation and polymerization take place simultaneously [218].

It was found that paper having improved strength can be obtained from recycled paper by mixing PAE resin and G-PAm. This blend was added to the wet end of the papermaking process [22,27]. Paper treated with it exhibits a significantly increased dry-strength performance as compared to the joint use of the individual resins. G-PAm and aminopolyamide–epichlorohydrin resins are present at a weight ratio of about 1:1 to about 5:1.

In a blend of the polyaldehyde polymer and the polyhydroxy polymer [219], hydroxyl groups react with the aldehyde groups of polyaldehyde to form covalent bonds (hemiacetals). Polysaccharides, as polyhydroxy polymers, provide a high level of TWS. The *cis*-hydroxyl groups may support neighboring group participation, which facilitates the formation of covalent bonds with poly-aldehyde. Additionally, the *cis*-hydroxyl groups may form hydrogen bonding with the cellulosic fibers.

4.4 POLYALDEHYDE COPOLYMERS FROM POLYACRYLAMIDE

It is well known that aldehydes react with cellulose or cellulose derivatives [220,221]. Formaldehyde and glyoxal are the most common aldehydes used in the paper industry. Glyoxal—the first compound in the poly-aldehyde series—represents a special case: it is nonionic; it has two aldehyde groups with the same reactivity (symmetric molecule). Glyoxal must be attached to a cationic compound in order to be incorporated in the paper web. The ideal attempt would use the first functionality for the attachment to the cationic backbone, and the second aldehyde group would be used in the reaction with cellulose fibers. Unfortunately, due to the same reactivity of those two aldehyde groups, the glyoxalation reaction of a cationic macromolecular compound will result in side reactions (cross-linking). As a consequence, the reaction conversion, which is measured by the increase in viscosity of the polymer solution, and the shelf life of the final product are hard to control.

4.4.1 GLYOXALATION OF POLYACRYLAMIDE

Glyoxal reacts with cationic PAAm, which is an ideal candidate for a glyoxalation reaction in slightly alkaline pH [161]. The amide group is attached to an acrylamide copolymer, which can be linear, branched, or grafted [140,150,151,222].

The glyoxalation of the PAAm is an equilibrium reaction. In that reaction, both aldehyde groups of glyoxal can react. The amidol bond (in red) is formed by the amide (weak base) addition to the C=O double bond in an acid and base catalysis [70,223].

Amidol

Although several other potential backbones (such as polyvinyl alcohol [140]) were studied as well, this chapter is focused on the polyamide–di-aldehyde reaction. The PAAm glyoxalation is an equilibrium reaction, given as

$$K = \frac{[CONH_2][Gly]}{[Amidol]} \qquad (4.4)$$

An equilibrium reaction will always show an important amount of un-reacted glyoxal. "Un-reacted" glyoxal is a complex concept: it involves both a certain amount of free glyoxal and an important amount of glyoxal reacted with only one aldehyde group. A glyoxal capped acrylamide is a unit substituted at the amide nitrogen with alpha-hydroxy acetaldehyde. As part of this equilibrium, the amidol bond (NH–CH) is sensitive to water in both acid and alkaline environments.

Higher concentrations of amide and glyoxal will result in higher concentrations of amidol bonds. However, free glyoxal and un-reacted amide will always exist along with the glyoxalated compound (see the analysis of free glyoxal [224]). A catalyst can increase the reaction rate to reach the equilibrium, but it cannot alter the equilibrium composition.

The cationic PAAm designed as a backbone for G-PAm has a low molecular weight (2000–5000). For a small molecular size, the average number of amide groups in one macromolecule is about 30–70. During the glyoxalation reaction, a cross-linked network (a gel) can be formed due to the high functionality of the polyamide and the glyoxal difunctionality. The cross-linking process shows a second-order kinetic [225]. PAAm glyoxalation supposes the skill to introduce as many aldehyde groups as possible into the PAAm molecule, before the gel formation.

The glyoxalation parameters are the following: the PAAm concentration and molecular weight, the glyoxal concentration, the catalyst concentration (pH), and the temperature. If a diluter is used for the synthesis of PAAm, the polymer concentration should be converted to the concentration of the amide group. This reaction reaches an equilibrium after a very long reaction time, which is why glyoxalation is performed at a high reaction rate (higher concentrations) and at a certain viscosity of the glyoxalated polymer solution. The reaction is quenched at pH = 2–4. Thus, the final composition is a system that tends to continue to react toward the equilibrium described by the concentration of the reactants.

Low molecular weight polymers and low concentrations are required to increase the amount of bonded glyoxal {–CONHCH(OH)CHO} to the polymer, without infinite cross-linking (gelling) [187,188]. When a lower molecular weight PAAm is used, the cross-linking process is slower and the amount of total solids can be increased to 20% [158]. For higher molecular weights (M_w about 10,000), the overall concentration reaches only about 11%. For Mw of 100,000, the final solids were reduced to 2% [226]. The glyoxalation of high molecular weight PAAm at a high polymer concentration is performed in inverse-emulsion polymerization and the glyoxalated polymer takes the form of micro gels [227].

Regardless of the cationic charge concentration in a macromolecule, the amide group has the same reactivity against glyoxal. In other words, the glyoxalation step can impart a more uniform distribution of the cationic charges.

In the absence of extensive information about a direct reaction between glyoxal and PAAm, we can learn more about this process by examining the reaction of formaldehyde with different amides [228]. Amides and imides react reversibly with formaldehyde in acid, neutral, or basic media, but, over a relatively wide pH range (2–12), the activation energy of the reverse reaction remains greater than that of the forward process by a nearly constant amount (5 kcal/mol) [229].

Groups of peptide bonds belonging to secondary amides or polyamides do not react with formaldehyde at room temperature [230]. That is why formaldehyde or glyoxal do not lower the water absorption of polyamides [231]. However, in agreement with recent findings [232,233], the

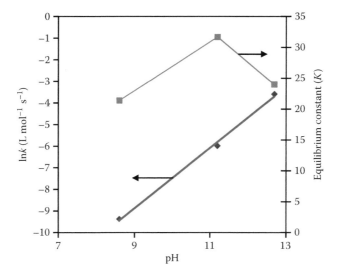

FIGURE 4.7   The effect of pH on the reaction of acetamide with formaldehyde.

methylolation of N-alkyl substituted amides is shown to occur. The secondary proton from the amide group can react with the methylol group, which explains the cross-linking of the UF resin.

In the reaction between acetamide and formaldehyde, the reaction rate increases more with the increasing pH (Figure 4.7) than with the increasing temperature (Figure 4.8) [234]. From pH = 8.6 to 11.2, the rate constant (k) increases up to 300 times, although the equilibrium constant (K) remains at about the same value. The pH value recommended for glyoxalation is 7.5–8.0 [151,152,158,235]. Buffer is used for a better control of the pH [201].

The reaction temperature has a smaller effect on the reaction rate (only about a three times higher reaction rate for an increase of 15°C in temperature—Figure 4.8). At higher temperatures, the equilibrium constant is smaller.

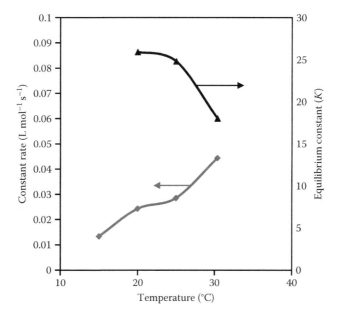

FIGURE 4.8   The temperature effect on the reaction of acetamide with formaldehyde.

The reaction of the amide group with glyoxal or to other polyaldehydes (such as that obtained by a condensation between isocyanuric acid and acrolein [236–239]) has received much less attention. The base-catalyzed addition proceeds best when a fresh glyoxal solution is used in the pH range 8–10 [240]. The condensation reaction competes with the Cannizzaro reaction [225]: an intramolecular redox process with the formation of glycolic acid. The solution slowly turns acid and more alkali should be added.

The condensation rate depends not only on the pH (and on temperature) but also on the chemical structures of the amide and aldehyde. The reaction rate of acrylamide with aldehyde [241] is about 2000 times slower than with urea [242].

The products of the reaction of glyoxal, or blocked glyoxal [77] with a PAAm [243,235] are excellent cross-linking agents. During glyoxalation, the pendant aldehyde group can react with an amide group from another molecule in a cross-linking process.

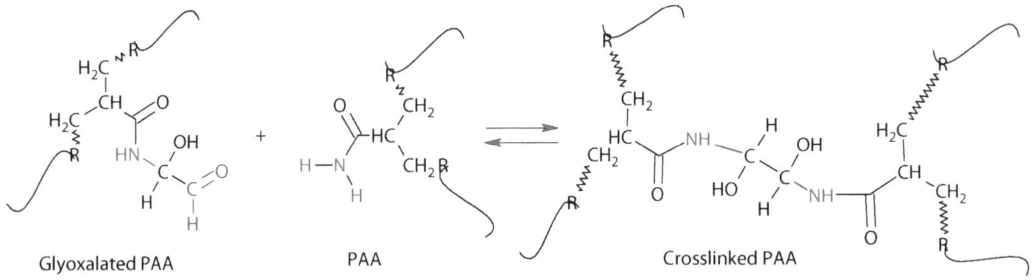

Glyoxalated PAA   PAA   Crosslinked PAA

The formation of both amidol bonds is part of an equilibrium reaction; the two bonds coexist with un-reacted glyoxal. If 25 mol% of glyoxal are used (based on acrylamide units), the glyoxalated polymer includes less than 12.5 mol% aldehyde functionalities [79].

According to the analysis of glyoxalated cationic PAAm [186], the copolymer has the following composition of the repeating units:

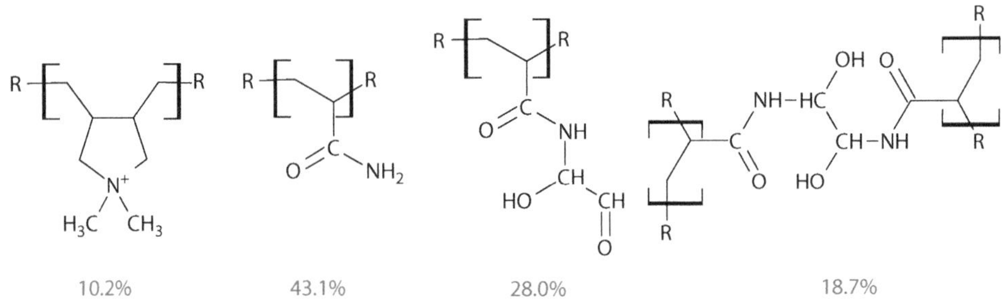

10.2%   43.1%   28.0%   18.7%

The acrylamide provides sites to which the –CH(OH)–CHO are attached and which will react with cellulose during the thermosetting reaction. About 10 mol% of acrylamide units (based on the total numbers of vinyl monomer units present) appear to be the minimum needed to provide the necessary number of sites [151].

The glyoxal/acrylamide ratio ranges from 0.2 to 0.4 [75,77,79,87,152,208], but large excess of glyoxal is also used: 3.4 g of 40% aqueous glyoxal for 1 g backbone [200]. Regardless of the amount of glyoxal added to the glyoxalation reaction, there is always an excess of free glyoxal in the TWSR solution. Free glyoxal is nonionic and it has a low molecular weight.

The effect of the concentration of glyoxal units [200] is still under debate (see also Section 2.5): if the initial wet strength is obviously improved as the concentration increases, the decay rate shows very scattered values (Figure 4.9).

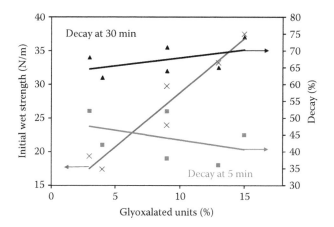

FIGURE 4.9   The effect of the concentration of glyoxalated units on polymer properties.

The number of pendant aldehyde groups can be increased by the glyoxalation of a copolymer with preexisting aldehyde functionalities (such as acrylamide–acrolein copolymers) [154].

4.4.2   THE GLYOXALATED POLYACRYLAMIDE STABILITY

The amidol bond formation is an equilibrium reaction. At the equilibrium point, the number of new amidol bonds is equal to the number of broken amidol bonds. A glyoxalation performed with a small amount of glyoxal reaches the equilibrium in several weeks. If the glyoxal/PAAm ratio is small enough, at the equilibrium point, the glyoxalated polymer is still soluble in water and has an indefinite shelf life.

In order to make the glyoxalation process economically affordable, the reaction time should be reduced to several hours. To speed up the glyoxalation process, an excess of glyoxal is used. At a desired molecular weight of the glyoxalated PAAm, estimated based on the viscosity measurements, the pH is dropped to 2–3 (quench) in order to slow down the cross-linking process. At the quench moment, the reaction mixture is far from the equilibrium conditions. Even at that low pH, the reaction is trying, albeit at a very low rate, to reach equilibrium. Advanced glyoxalation results in an increased viscosity of the G-PAm solution and a shorter shelf life.

Resin stability can be tested by measuring its viscosity versus time. In order to speed up the ageing process, the testing temperature is increased to 37°C [244]. The resin viscosity is increasing steadily [153]. Figure 4.10 shows a typical profile for the resin viscosity over time (20% resin concentration, Ambond 1600). The viscosity curve profile helps to predict gel formation.

For a final concentration of G-PAm of 8% [190], as the ageing progresses, the samples diluted to 2%–5% yield hazy solutions, which show that the polymer therein is in colloidal state.

Commercial products delivered at 10% solids [28,79,102] will gel in less than 30 days at room temperature. For a longer shelf life, they need to be stored under refrigeration [198]. In order to prevent the loss of material (cleaning the storage tank when a gel of G-PAm has been formed, is rather difficult), the glyoxalation reaction is performed at the paper mill in a continuous process [235].

There have been attempts at producing TWSR with a longer shelf life. Scavenger agents for glyoxal, such as glycols [245] or amide [246], can slow down the aldehyde group reaction with amides. Aldehyde groups can be involved in other reactions as well (with phenols in the presence of transition metal ion) [247,248].

The reaction of the G-PAm with sodium bisulfite has also been performed [151,218,249]. A much longer shelf life is reached when sodium bisulfate is added [151] to a G-PAm (Figure 4.11).

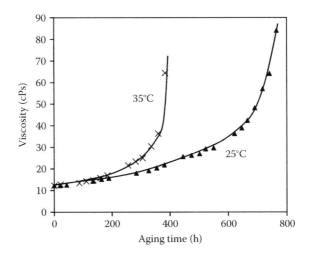

FIGURE 4.10   Self life for a G-PAm solution.

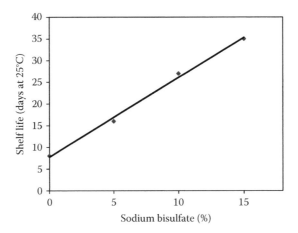

FIGURE 4.11   The effect of sodium bisulfate on the G-PAm shelf life.

Unfortunately, the solutions of sulfite-stabilized G-PAm lose nearly all of their wet strengthening effectiveness after storage for a few weeks under normal conditions: 10% concentration, at 25°C and pH = 7.0. The larger the quantity of reacted sulfite present in G-PAm and the higher the temperature at which the solution is stored, the more rapid and more severe is the loss in effectiveness.

The loss in effectiveness is due to the anionic charges developed by bisulfate, which make the polymer amphoteric with a lower retention [81,82]:

On the other hand, the aldehyde functionality is no longer available for the cross-linking process with the cellulose fiber. Water-soluble bisulfite adducts of polyacrolein, used in a comparative test, also show poor results when used as strengthening resins. Because glyoxal and formaldehyde are competing for the bisulfite molecules, the addition of formaldehyde can move the equilibrium back to the aldehyde pendant moiety [249].

The equilibrium reaction is manipulated to produce a resin with a longer shelf life. The post addition of glyoxal [222] increases the amount of free glyoxal, and, therefore, the cross-linking process is slowed down because the free glyoxal reacts first to provide pendant groups. The self-cross-linking process is also reduced by lowering the amid group concentration with a diluter (such as *N,N*-dimethyl acrylamide), or with a higher concentration of the cationic comonomer [244].

The fundamental change in the equilibrium of the glyoxalation reaction is to change the reactivity of the aldehyde by using a blocked aldehyde group [250]. Modified glyoxal (dimethoxyethanal) [208] react with the acrylamide–DADMAC copolymer to form pendant acetal groups ($m = n + p$).

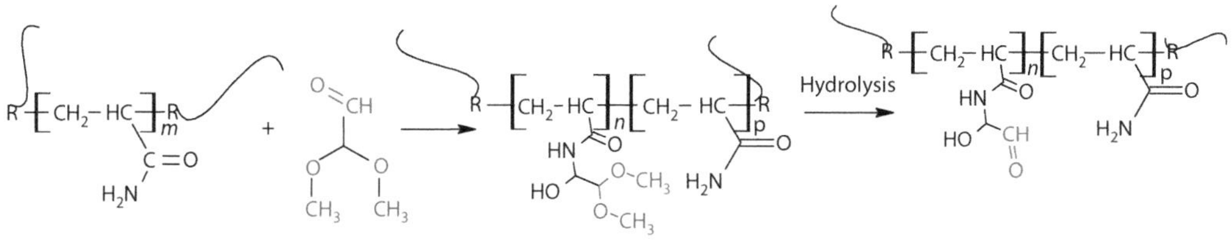

The polymer (10% solids solution) is kept for 20 h, at about 40°C, in the presence of 5% 2.5N hydrochloric acid solution (pH = 1, see [75]) to convert it back to the aldehyde functionality.

There is a possibility to glyoxalate the acrylamide monomer under mildly alkaline conditions before polymerization. The glyoxalated acrylamide is copolymerized with a cationic comonomer [251]. Now the final viscosity (for G-PAm) is controlled by controlling the monomer conversion (an inhibitor prevents any further polymerization).

4.5   PAPER WET STRENGTH AND ITS DECAY

The weakness of hemiacetal and amidol bonds are tested by measuring the decay of the paper wet strength. During the papermaking process, the TWSR, which has pendant aldehydes bonded through the amidol bond to the PAAm backbone, reacts with the hydroxyl groups from cellulose to form hemiacetals bonds [252–258]. The reaction is performed at 90°C–110°C [151]. Polyaldehyde makes a bridge between the two cellulose fibers. The potential ionic bonds [40,259,260] (bond energy of 10–20 kcal/mol) between the anionic charges on the cellulose fiber and the cationic charges of the TWSR must also be considered along with amidol and hemiacetal bonds. The amide functionality may be involved, as well as the hydrogen bond formation.

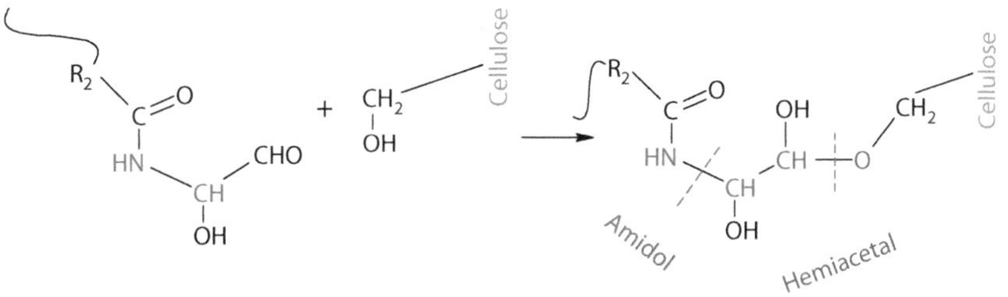

Due to the new covalent bond formation, both dry and wet strength will increase. The presence of hydrolysable, weak chemical bonds (amidol, hemiacetal and ionic bonds) is the reason for the temporary character of the paper wet strength. The paper sheet containing aldehyde starch shows a decrease in its wet strength after a prolonged period of soaking in water [261].

TWSRs are generally distinguished from permanent wet-strength agents in that they provide a certain degree of wet strength immediately (e.g., 5 s after the paper is wetted), but a good portion (e.g., 30%–75%) of this immediate wet strength is lost after 30–90 min of soaking in water [46].

The wet-strength decay rate is defined according to the following equation (Equation 4.5) [80,111]:

$$\%\text{Decay} = \frac{WT_i - WT_R}{WT_i - WT_0} \times 100 \tag{4.5}$$

where
WT_i is the initial wet tensile strength of paper treated with TWSR
WT_R is the wet tensile strength of paper treated with TWSR after 90 min soaking in neutral pH water
WT_0 is the wet tensile strength of paper without TWSR after a 90 min soaking in water at neutral pH

The testing method is equally critical for accurate results. The characteristics of water used for soaking paper samples are of great importance. The soaking water [75] should have a pH = 7.5 and a standard hardness (100 ppm hardness, 50 ppm alkalinity based on calcium carbonate).

In spite of the efforts to standardize wet-strength measurements, a large variability in paper wet-strength values is recorded (see also Section 2.5), which results in a higher standard deviation for WT_i and therefore, for the W/D ratio. An aggravating factor is that the WT_R is measured on a paper sample and the WT_i is estimated on another paper sample. The overall accuracy is poorer especially when the decay ratio is considered at shorter soaking time [200] (Table 4.4).

Wet-strength decay over soaking time is a nonlinear process. Figure 4.12 shows the paper wet-strength decay for two different TWSRs (resin concentration 2 lb/ton) [46]. The slope for the paper wet-strength decay is steeper during the first minute, i.e., within a short soaking time interval, a differentiation between resins is recorded.

In order to overcome that huge decay in the first minute, a very high initial wet strength (at 5 s) is needed. By changing the TWSR chemical structure with homo-cross-linking units in G-PAm [46] was possible to obtain higher initial wet strength and a faster decay (Figure 4.12, blue curve).

The hemiacetal bond formation is also a reaction sensitive to pH. The paper wet strength depends on the wet end pH [135] (Figure 4.13): paper made with 1% di-aldehyde gum at higher pH at the wet end shows a lower wet strength.

The paper wet-strength decay also depends on the access of water and on the local distribution of the polymer–polymer interactions. Those interactions between polyelectrolytes can be manipulated by the presence of inorganic electrolytes [262,263]. The loss of strength rate is an exponential decay curve and reflects water penetration quite well [264]. If a hydrophobic compound (ASA) is used,

TABLE 4.4
The Testing Data for Two Replicate (Handsheets Made with 0.5% TWSR)

	Initial Strength			Wet Tensile Decay (%)	
	Dry	Wet	Wet/Dry (%)	5 min	30 min
1-A	646	62	9.6	52	71
1-B	700	77	11.0	38	64

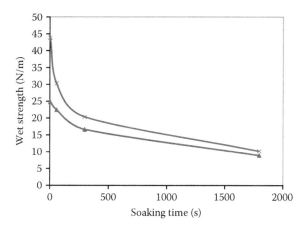

FIGURE 4.12   Paper wet strength vs. soaking time.

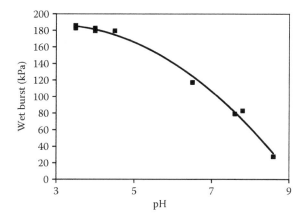

FIGURE 4.13   The pH effect on the wet burst of paper with 1% di-aldehyde gum.

water penetration is slower and the initial wet strength is higher [265]. In order to help water penetration, paper is impregnated [30] with a mixture of water-soluble (glycerol, sorbitol, polyethylene glycol, CMC) and water-insoluble compounds (liquid paraffin), or it is strengthened only in special areas located next to the dispersibility regions [266].

Due to the hemiacetal bond sensitivity to higher pH, the incorporation of an alkaline reagent into the tissue product results in the enhanced paper degradation [267]. Sodium bicarbonate is applied to the finished sheet (0.5% based on total sheet) [75,77,79,87,208] at the dry end [267]. The catalyst is located next to the TWSR to improve paper flushability [268]. Alkaline hydrolysis removes the entire amount of glyoxal being bound to the cellulosic material [245]. While the alkaline reagent improves flushability, it does not substantially affect the initial tensile strength of the tissue.

For a better understanding of the cross-linking process, numerous studies on the cross-linking of cellulose by glyoxal alone were performed [245,269,270]. Competitive hemiacetalizations of glyoxal (using various alcohols as models) show that vicinal diols (primary or secondary) [222] are much more reactive than isolated alcohols [269], either in aqueous solution or when water is removed (see the paper drying step).

Paper strengthened with G-PAm includes both amidol and hemiacetal bonds. The bond strength C–N or C–O depends on their vicinity [271,272]. Hemiacetal bonds contribute to a rate of decay in paper products, which is believed to be orders of magnitude faster than amidol bonds at neutral pH [79]. Therefore, the hydrolysis of strengthened paper with G-PAm takes place at the hemiacetal bond [273]. The rate of paper wet-strength decay will be proportional to the relative number of amidol and hemiacetal bonds [75,77,79,87,208].

Because the rate of decay of hemiacetal bonds is faster than that of amidol bonds (at neutral pH), the rate of the wet tensile decay increases when the relative proportion of amidol bonds decreases. The number of amidol bonds can be reduced by adding a diluter N,N'-dimethyl-acrylamide in the TWSR [200].

The decay mechanism presented above suggests that a faster decay is expected when the polymer molecular weight is lower, when a number of branches are present, when the chemical structure of the polymer reduces the possible interactions with cellulose, when homo-cross-linking through hemiacetal bonds is present.

Starch has a large molecular weight, which makes its potential wet-strength decay slower. Aldehyde starch was tested as wet-strength resin and as TWSR [104]. As expected, the aldehyde functionality has a big impact on both wet and dry strength, but the decay values are low. After 16h of soaking in water, paper made with aldehyde functionalized starch shows a much higher wet strength than the initial wet strength for paper made with regular cationic starch (Table 4.5).

Although its chemistry is almost identical with that of G-PAm, aldehyde starch displays a very slow wet-strength decay [42,274]. Dextran (2%) with a degree of substitution with aldehyde of about 0.22 [130] and a molecular weight of about 500,000, has a slower decay even at pH = 3 (Figure 4.14). Regardless of good final decay (about 56% after 2 days), aldehyde starch seems not to meet the recommended performance as "temporary" wet-strength resin. Figure 4.14 shows that G-PAm with low-molecular-weight backbone provides the fastest wet-strength decay.

However, all these parameters and especially the amount of reactive groups and the charge density must reach an optimum. A balance needs to be set between the backbone molecular weight, the

TABLE 4.5

The Effect of Cationic Aldehyde Starch on Paper Strength and Its Decay

Starch	Dry Strength (BL) (m)	Wet Strength (BL) (m)			Decay (%)	
		Initial	30 min	16 h	30 min	16 h
Cationic starch (no functionality)	1640	83	43	—	48	—
Cationic starch (aldehyde groups)	2140	382	260	122	32	68

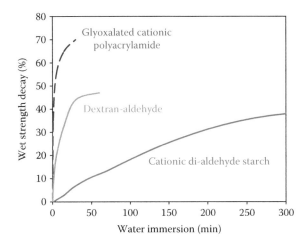

FIGURE 4.14 Wet-strength decay for high and low molecular weigh aldehyde polymers.

amide group concentration, the aldehyde pendant group concentration, the glyoxal bridges between the backbone segments and the molecular weight of the final glyoxalated polymer. Moreover, some other parameters may also play an important role: the molecular weight distribution and the cationic charge distribution.

The hydrolysis process (in excess of water) will release G-PAm, glyoxal, and/or PAAm in water, which will result in less cross-linking of cellulose fibers and in a lower strength (paper dispersibility in water [30]).

According to the above formula, the decay rate should be at least 80% after 90 min [80]. For a classic glyoxalated PAA, the best results are described as around 35% tensile decay at 5 min and around 65% at 30 min after saturation rewetting [79].

The concept of standard method for the decay rate measurement has a history of its own: in 1986 [152], it was suggested that the soaking time should be of 16 h, in 1991 [80], 90 min were recommended, and in 2001, 2 h [75]. During that lapse of time paper dispersibility in a sewer system is simulated (on a small scale). The test involves standardized paper size, curing temperature and time, water volume for soaking, soaking time and temperature, etc., and must also simulate the turbulence typically observed in a toilet bowl while flushing [268,275].

REFERENCES

1. A.A. Robertson, *Sven. Papperstidn.*, **66**, 477–497 (1963).
2. A.J. Stamm, *Tappi*, **33(9)**, 435–439 (1950).
3. H. Schott, *J. Macromol. Sci.-Phys.*, **B27(1)**, 119–123 (1988).
4. R.S. Seth, D.H. Page, M.C. Barbe, and B.D. Jordan, *Sven. Papperstidn.*, **87(6)**, R36–R43 (1984).
5. D.K. Ray-Chaudhuri, US 3,891,589 (1975).
6. P.M. Westfall, N.R. Eldred, and J.C. Spicer, US 3,372,086 (1968).
7. V.T. Stannett, *Surf. Coat. Relat. Paper Wood*, *Symp.*, 269–299 (1967).
8. N. Dunlop-Jones, Wet strength chemistry, in *Paper Chemistry*, J.C. Roberts (Ed.), Blackie Academic, London, U.K., 1996, pp. 98–119.
9. S.R.D. Guha, *Indian Pulp Paper*, **20(1)**, 33–37 (1965).
10. T. Vitale, *Materials Research Society Symposium Proceedings*, Vol. 267 (Materials Issues in Art and Archaeology III), 1992, pp. 397–427.
11. N. Gurnagul and D.H. Page, *Tappi J.*, **72(12)**, 164–167 (1989).
12. A.A. Robertson, *Tappi*, **53(7)**, 1331–1339 (1970).
13. G. Broughton and J.P. Wang, *Tappi*, **37(2)**, 72–79 (1954).
14. T.J. Senden and M.B. Lyne, *Nord. Pulp Paper Res. J.*, **15(5)**, 554–563 (2000).
15. A.H. Nissan, *J. Educ. Modules Mater. Sci. Eng.*, **1(4)**, 773–844 (1979).
16. M.N. Fineman, *Tappi*, **35(7)**, 320–324 (1952).
17. H.H. Espy, *Tappi J.*, **70(7)**, 129–133 (1987).
18. M. Sain and H. Li, *J. Wood Chem. Technol.*, **22(4)**, 187–197 (2002).
19. C.S. Maxwell and W.F. Reynolds, *Tappi*, **33(4)**, 179–182 (1950).
20. A. Jurecic, C.M. Hou, K. Sarkanen, C.P. Donofrio, and V. Stannett, *Tappi*, **43**, 861–865 (1960).
21. P. Pezzaglia, A. Debenedictis, W. Ewert, W.H. Houff, R.C. Hurlbert, A.C. Nixon, and I. Sobolev, *Tappi*, **48(5)**, 314–320 (1965).
22. R.T. Underwood, R.J. Jasion, S.P. Hoke, and G.G. Spence, US 5,674,362 (1997).
23. W.M. Thomas, US 2,394,273 (1946).
24. M. Goehler, F.H. Reimer, and E.F. Davis, US 2,423,097 (1947).
25. R.R. House and Y. Jen, US 2,872,313 (1959).
26. R.R. Staib, J.R. Fanning, and W.W. Maslanka, US 6,245,874 (2001).
27. W.B. Darlington and W.G. Lanier, US 5,427,652 (1995).
28. L.L. Williams and A.T. Coscia, US 3,740,391 (1973).
29. K.P. Mohammadi, L.O. Seward, and D.M. Rasch, US 6,149,769 (2000).
30. K. Taniguchi, US 5,935,384 (1999).
31. D.V. Duchane, US 3,654,928 (1972).
32. H. Böe, US 3,034,922 (1962).
33. Y. Mizutani and S. Sakamoto, US 3,431,166 (1969).

34. R.M. Krajewski and R.E. Erickson, US 4,186,233 (1980).
35. R.S. Yeo, US 5,509,913 (1996).
36. M. Komatsu and I. Toki, US 5,312,883 (1994).
37. D.B. Cole, V.K. Shah, K.J. Bevernitz, F.M. Chen, E.D. Johnson, F.J. Lang, J.D. Lindsay, L.A. Rivera, K.G. Schick, and K.D. Stahl, US 6,444,214 (2002).
38. F.J. Lang, Y. Chang, F.M. Chen, P.S. Mumick, W.S. Pomplun, and D.A. Soerens, US 6,548,592 (2003).
39. F.J. Lang, Y. Chang, F.M. Chen, P.A. Dellerman, E.D. Johnson, J.D. Lindsay, L.A. Rivera, K.G. Schick, W.T. Schultz, V.K. Shah, P.S. Mumick, W.S. Pomplun, D.M. Jackson, D.A. Soerens, K.Y. Wang, US 6,579,570 (2003).
40. D.C. Smith, US 5,338,406 (1994).
41. A.J. Allen, E. Echt, W.W. Maslanka, and J.C. Peters, US 6,294,645 (2001).
42. B.T. Hofreiter, H.D. Heath, A.J. Ernst, and C.R. Russell, *Tappi*, **57(8)**, 81–83 (1974).
43. E.T. Wise and S.G. Weber, *Macromolecules*, **28**, 8321–8327 (1995).
44. R.L. Barcus, K.P. Mohammadi, A.M. Leimbach, and S.R. Kelly, US 7,125,469 (2006).
45. R.L. Barcus, K.P. Mohammadi, A.M. Leimbach, and S.R. Kelly, US 7,258,763 (2007).
46. K.P. Mohammadi, R.L. Barcus, A.M. Leimbach, and S.R. Kelly, US 7,217,339 (2007).
47. H. Kasgoz, S. Ozgumus, and M. Orbay, *Polymer*, **44(6)**, 1785 (2003).
48. L.A. Lundberg, US 2,475,846 (1949).
49. D.E. Floyd and D. Potente, US 3,127,365 (1964).
50. A. Grosse-Sommer, M. In, and R.K. Prudhomme, J., *Polym. Sci. Part A: Polym. Chem.*, **34**, 1447–1453 (1996).
51. H. Zhang and A. Schwarz, US 6,713,646 (2004).
52. H. Zhang and A. Schwarz, US 7,135,593 (2006).
53. A.-C. Albertsson and M. Eklund, *J. Polym. Sci. Part A: Polym. Chem.*, **34**, 1395–1405 (1996).
54. E. Ruckenstein and H. Zhang, US 6,323,360 (2001).
55. E. Ruckenstein and H. Zhang, *Macromolecules*, **32**, 3979–3983 (1999).
56. E. Ruckenstein and H. Zhang, *J. Polym. Sci. Part A: Polym. Chem.*, **38**, 1848–1851 (2000).
57. J.S. Witzeman, *Tetrahedron Lett.*, **31(10)**, 1401–1404 (1999).
58. A.I. Vogel, *Vogel's Textbook of Practical Organic Chemistry*, 5th edn., Longman Scientific & Technical, London, U.K., 1989, p. 1224.
59. G. Carpeni, *Bull. Chem. Soc. France*, 1280–1294 (1950).
60. Y.C. Huang, K.M. Singh, Y. Hong, and M.B. Lyne, US 7,608,166 (2009).
61. H.B. Davis and C.J.B. Mott, *J. Chem. Soc. Faraday 1*, **76**, 1991–2002 (1980).
62. P.J. Antikainen and I.P. Pitkanen, *Suom. Kemistil. B*, **41(3)**, 65–69 (1968).
63. R.H. Pelton and C. Lu, US 2008/0099172 (2008).
64. D.V. Duchane, US 3,808,165 (1974).
65. D.V. Duchane, US 3,689,314 (1972).
66. D.V. Duchane, US 3,692,725 (1974).
67. E.G. Varona, US 4,309,469 (1982).
68. J.M. Kliegman and R.K. Barnes, *J. Heterocycl. Chem.*, **7(5)**, 1153–1155 (1970).
69. N. Platzer, B. Lefebvre, J. Temmen, and G. Mattioda, *Bull. Chem. Soc. Fr.*, **1–2(Pt. II)**, 40–42 (1981).
70. J. Zabicky, in *The Chemistry of the Carbonyl Group*, Interscience, New York, 1970, p. 45.
71. W.U. Day, H.F. Donnelly, and H.L. Rohs, US 3,096,228 (1963).
72. P.V. Luu, G. Worry, R.J. Marinack, H.S. Ostrowski, D.M. Bhat, US 6,059,928 (2000).
73. N.R. Eldred and J.C. Spicer, *Tappi*, **46(10)**, 608–612 (1963).
74. S.P. Malchick, W.G. James, and J.S. Munday, US 3,325,347 (1967).
75. M.T. Crisp and R.J. Riehle, US 6,197,919 (2001).
76. D.W. Bjorkquist, US 5,085,736 (1992).
77. R.H. Jansma and K.R. Sandberg, US 4,508,594 (1985).
78. R.K. Pinschmidt, G.E. Davidowich, W.F. Burgoyne, D.D. Dixon, and J.E. Goldstein, *Polym. Mater. Sci. Eng.*, **56**, 780–784 (1987).
79. R.H. Jansma, J. Begala, and G.S. Furman, US 5,401,810 (1995).
80. D.W. Bjorkquist, US 4,981,557 (1991).
81. G.T. Kekish, US 3,317,370 (1967).
82. G.T. Kekish, US 3,410,828 (1968).
83. A.R. Mills, US 3,347,832 (1967).
84. A.C. Nixon, P.J. Berrigan, and P.H. Williams, US 3,278,474 (1966).
85. D.W. Bjorkquist, US 5,138,002 (1992).

86. D.W. Bjorkquist, US 5,008,344 (1992).
87. R.H. Jansma and K.R. Sandberg, US 4,605,718 (1986).
88. R.L. Barcus, K.P. Mohammadi, A.M. Leimbach, and S.R. Kelly, US 7,691,233 (2010).
89. J.G. Frick and R.J. Happer, *J. Appl. Polym. Sci.*, **29**, 1433–1447 (1984).
90. R.K. Pinschmidt, D.D. Dixon, and W.F. Burgoyne, US 4,691,026 (1987).
91. R.K. Pinschmidt, D.D. Dixon, W.F. Burgoyne, and J.E. Goldstein, US 4,663,410 (1987).
92. A.F. Nordquist and R.K. Pinschmidt, US 4,959,489 (1990).
93. J.J. Tsai, P.G. Jobe, and R.L. Billmers, US 4,866,151 (1989).
94. B. Alince, *Sven. Papperstidn.*, **78(7)**, 253–257 (1975).
95. R.A. Young, *Wood Fiber.*, **10(2)**, 112–119 (1978).
96. D.J. Smith and M.M. Headlam, US 6,319,361 (2001).
97. A.L. Cimecioglu, J.S. Thomaides, K.A. Luczak, and R.D. Rossi US 6,368,456 (2002).
98. M.W. Rutenberg and D. Solarek, Starch derivatives: production and uses, in *Starch Chemistry and Technology*, R.L. Whistler, J.N. BeMiller, and E.F. Paschall (Eds.), Academic Press, New York, 1984, pp. 312–388.
99. R.L. Billmers, M.M. Tessler, D.M. Del Giudice, and C. Leake, US 4,788,280 (1988).
100. B.T. Hofreiter, I.A. Wolff, and C.L. Mehltretter, US 2,894,945 (1959).
101. D.H. Le Roy and S.M. Parmerter, US 3,553,193 (1971).
102. C.W. Chiu, R. Jeffcoat, M. Henley, and L. Peek, US 5,700,917 (1997).
103. F.J. Germino, US 3,297,604 (1967).
104. D.B. Solarek, P.G. Jobe, and M.M. Tessler, US 4,675,394 (1987).
105. W.E.C. Yelland, US 2,606,188 (1952).
106. R.A. Jeffreys, B.E. Tabor, and D.M. Burness, US 3,062,652 (1962).
107. W. Dvonch and C.L. Mehltretter, US 2,648,629 (1953).
108. J.E. Slager, US 3,086,969 (1963).
109. G.D. Fronmuller, US 2,803,558 (1957).
110. J.N. BeMiller and S.D. Darling, US 3,632,802 (1972).
111. D.J. Smith, US 5,760,212 (1998).
112. D.J. Smith and M.M. Headlam, US 5,656,746 (1997).
113. D.J. Smith and J.E. Ruth, US 5,698,688 (1997).
114. A.L. Cimecioglu and D.E. Harkins, US 6,228,126 (2001).
115. A.L. Cimecioglu and D.E. Harkins, US 6,562,195 (2003).
116. A.L. Cimecioglu, D.E. Harkins, M. Merrette, and R.D. Rossi, US 6,695,950 (2004).
117. A.L. Cimecioglu and J.S. Thomaides, US 6,872,821 (2005).
118. A.L. Cimecioglu and J.S. Thomaides, US 6,586,588 (2003).
119. A.L. Cimedoglu and J.S. Thomaides, US 7,247,722 (2007).
120. A.E.J. de Nooy, A.C. Besemer, and H. van Bekkum, *Tetrahedron*, **51**, 8023–8032 (1995).
121. R.L. Barcus and K.P. Mohammadi, US 7,598,331 (2009).
122. R.L. Barcus and K.P. Mohammadi, US 7,259,218 (2007).
123. R.L. Barcus and K.P. Mohammadi, US 7,625,989 (2009).
124. S.M. Parmerter, US 3,519,618 (1970).
125. D.B. Solarek, P.G. Jobe, M.M. Tessler, R.L. Billmers, D.L. Lamb, and J.J. Tsai, US 4,703,116 (1987).
126. J.J. Tsai, P.G. Jobe, D.L. Lamb, R.L. Billmers, and M.M. Tessler, US 4,983,748 (1991).
127. D.B. Solarek, P.G. Jobe, and M.M. Tessler, US 4,731,162 (1988).
128. D.B. Solarek, P.G. Jobe, M.M. Tessler, R.L. Billmers, D.L. Lamb, and J.J. Tsai, US 4,741,804 (1988).
129. D.B. Solarek, P.G. Jobe, M.M. Tessler, R.L. Billmers, D.L. Lamb, and J.J. Tsai, US 4,804,769 (1989).
130. N. Chen, S. Hu, and R. Pelton, *Ind. Eng. Chem. Res.*, **41**, 5366–5371 (2002).
131. L.R. Peel, *Pap Technol.*, **30(12)**, 20–23 (1989).
132. M. Laleg and I.I. Pikulik, *Nord. Pulp Paper Res. J.*, **8(1)**, 41–47 (1993).
133. W. Brevard and M. Ryan, WO 2004072376 (2004).
134. M. Ryan and W. Brevard, US 7,736,465 (2010).
135. J.W. Opie and J.L. Keen, *Tappi*, **47(8)**, 504–507 (1964).
136. B.T. Hofreiter, G.E. Hamerstrand, C.L. Mehltretter, W.E. Schulze, and A.J. Ernst, *Tappi*, **43(7)**, 639–643 (1960).
137. J.H. Dwiggins, R. Ramesh, F.D. Harper, A.O. Awfeso, T.P. Oriaran, G.A. Schulz, and D.M. Bhat, US 6,277,467 (2001).
138. W.C. Floyd, N. Thompson, and L.R. Dragner, US 6,303,000 (2001).
139. L.R. Dragner, W.C. Floyd, and J.W. Ramp, US 5,032,683 (1991).

140. L.L. Williams and A.T. Coscia, US 3,597,313 (1971).
141. L. Pardo, R. Osman, H. Weinstein, and J.R. Rabinowitz, *J. Am. Chem. Soc.*, **115**, 8263–8269 (1993).
142. L.H. Elizer, G.C. Glasscock, and J.M. Seitz, US 3,051,691 (1962).
143. J.R. Stephens and L. Rapoport, US 2,938,026 (1960).
144. H.H. Espy, US 3,607,622 (1971).
145. H.H. Espy, US 3,728,214 (1973).
146. H.H. Espy, US 3,728,215 (1973).
147. D.E. Floyd and D. Potente, US 4,060,507 (1977).
148. T.W. Rave, US 4,079,043 (1978).
149. T.W. Rave, US 4,035,229 (1977).
150. L.L. Williams and A.T. Coscia, US 3,772,407 (1973).
151. A.T. Coscia and L.L. Williams, US 3,556,932 (1971).
152. G.J. Guerro, R.J. Proverb, and R.F. Tarvin, US 4,605,702 (1986).
153. S. Cyr, H. Bakker, W.M. Stevels, and R.R. Staib, WO 2006/068964 (2006).
154. E.D. Kaufman, US 3,819,555 (1974).
155. R.R. Staib, J.R. Fanning, and W.W. Maslanka, US 6,103,861 (2000).
156. S.A. Fischer, N.S. Clungeon, and D.I. Devore, US 5,382,324 (1995).
157. T.-W. Lai, R.K. Pinschmidt, and W.F. Burgoyne, US 4,931,501 (1990).
158. D. Dauplaise and G.J. Guerro, US 5,723,022 (1998).
159. K. Moriya and I. Honda, US 4,122,071 (1978).
160. F.C. Schaefer and A.C. Wright, US 3,734,939 (1973).
161. R.A. Bankert, US 3,728,216 (1973).
162. W. Jaeger, M. Hahn, C. Wandrey, F. Seehaus, and G. Reinisch, *J. Macromol. Sci. Chem.*, **A21(5)**, 593–614 (1984).
163. G.B. Butler, US 3,288,770 (1966).
164. A.I. Martynenko, C. Wandrey, W. Jaeger, M. Hahn,, D.A. Topchiev, G. Reinisch, and V.A. Kabanov, *Acta Polym.*, **36(9)**, 516–517 (1985).
165. G.E. Ham, *Copolymerization*, Interscience, New York, pp. 2–9 (1964).
166. C. Hagiopol, Copolymerization: towards a systematic approach, Kluwer-Academic Press, New York, 1999.
167. Ch. Wandrey and W. Jaeger, *Acta Polym.*, **36(2)**, 100–102 (1985).
168. W. Baade, D. Hunkeler, and A.E. Hamielec, *Polym. Mat. Sci. Eng.*, **57**, 850–853 (1987).
169. H. Tanaka, *J. Polym. Sci. Polym. Chem. Ed.*, **24**, 29–36 (1986).
170. Y.V. Bune, A.I. Barabanova, Y.S. Bogachev, and V.F. Gromov, *Eur. Polym. J.*, **33(8)**, 1313–1323(1997).
171. H. Tanaka, A. Swerin, L. Odberg, and S.-B. Park, *J. Pulp Paper Sci.*, **25(8)**, 283–288 (1999).
172. S.M. Shawaki and A.E. Hamielec, *J. Appl. Polym. Sci.*, **23**, 3155–3166 (1979).
173. A. Chapiro, J. Dulieu, Z. Mankowski, and N. Schmitt, *Eur. Polym. J.*, **25(9)**, 879–884 (1989).
174. C. Erbil, S. Ozdemir, and N. Uyanik, *Polymer*, **41**, 1391–1394 (2000).
175. S.M. Mokhtar, *J. Macromol. Sci. Pure Appl. Chem.*, **A34(5)**, 865–879 (1997).
176. A. Chapiro and L. Perec-Spritzer, *Eur. Polym. J.*, **11**, 59–69 (1975).
177. S-M. Deng and F.-M. Meng, *J. Macromol. Sci.-Pure Appl. Chem.*, **A31(9)**, 1289–1301 (1994).
178. M. Kucharski and R. Lubczak, *J. Appl. Polym. Sci.*, **64(7)**, 1259–1265 (1997).
179. M. Mukherjee, S.K. Chatterjee, A.S. Brar, and K. Dutta, *Macromolecules*, **31**, 8455 (1998).
180. D.S. Breslow and A. Kutner, *J. Polym. Sci.*, **27**, 295–312 (1958).
181. F. Heatley, P.A. Lovell, and J. McDonald, *Eur. Polym. J.*, **29**, 255–268 (1993).
182. S.H. El-Hamouly, S.A. El-Kafrawi, and N.N. Messiha, *Eur. Polym. J.*, **28(11)**, 1405–1410 (1992).
183. S.H. El-Hamouly and O.A. Mansour, *J. Polym. Sci. Part A: Polym. Chem.*, **31**, 1335–1337 (1993).
184. T. Siyam, *Egypt. J. Chem.*, **37**, 69–78 (1994).
185. H.A. Gartner, US 5,110,883 (1992).
186. T.P. Oriaran, B.T. Burrier, H.S. Ostrowski, E.W. Post, and J.H. Propp, US 6,207,012 (2001).
187. D.L. Dauplaise, J.J. Kozakiewicz, and J.M. Schmitt, US 4,954,538 (1990).
188. D.L. Dauplaise, J.J. Kozakiewicz, and J.M. Schmitt, US 5,320,711 (1994).
189. J.R. Sanchez, US 6,315,866 (2001).
190. EP 0133699 (American Cyanamid) (1984).
191. L.X. Gong and X.F. Zhang, *eXPRESS Polym. Lett.*, **3(12)**, 778–787 (2009).
192. R. Nagarajan and J.B.W. Shing, US 6,007,679 (1999).
193. H. Takeda and M. Kawano, US 5,006,590 (1991).
194. H. Takeda and M. Kawano, US 4,929,655 (1990).

195. J.J. Kozakiewicz, D.L. Dauplaise, J.M. Schmitt, and S.-Y. Huang, US 5,037,863 (1991).
196. J.C. McClendon, US 3,478,003 (1969).
197. M. Ryan, W. Brevard, D. Dauplaise, M. Lostocco, R. Proverb, and D.W. Lipp, US 6,939,443 (2005).
198. R.P. Geer and R.R. Staib, WO 00/11046.
199. K. Plochocka, *J. Macromol. Sci.-Rev. Macromol. Chem.*, **C20(1)**, 67–148 (1981).
200. D.W. Bjorkquist and W.W. Schmidt, US 4,603,176 (1986).
201. D.N. Van Eenam, EP 147 380 (1984).
202. C.E. Farley, G. Anderson, and K.D. Favors, 6,787,574 (2004).
203. G. Mino and S. Kaizerman, US 2,922,768 (1960).
204. T.G. Shannon and D.A. Soerens, US 6,994,770 (2006).
205. J.H. Wang, D.M. Schertz, and D.A. Soerens, US 6,172,177 (2001).
206. A.T. Coscia and L.L. Williams, US 3,734,977 (1973).
207. J.J. Tsai and E.A. Meier, US 4,973,641 (1990).
208. R.H. Jansma, A.J. Begala, and G.S. Furman, US 5,490,904 (1996).
209. J. Klein and K.D. Conrad, *Makromol. Chem.*, **181**, 227–240 (1980).
210. G. Odian, *Principles of Polymerization*, Wiley-Interscience, New York, 2004.
211. J. Brandrup and E.H. Imergut (Eds.), *Polymer Handbook*, 4th edn., John Wiley & Sons, New York, 1999.
212. N. Huebner, W.R. Mueller, B.W. Peters, and L. Schleferstein, US 6,180,705 (2001).
213. R.J. Proverb and L.M. Pawlowska, US 2009/0283231 (2009).
214. R.J. Proverb and L.M. Pawlowska, US 7,608,665 (2009).
215. E.G. Ballweber, R.H. Jansma, and K.G. Phillips, US 4,217,425 (1980).
216. E.G. Ballweber, R.H. Jansma, and K.G. Phillips, US 4,233,411 (1980).
217. W.W. Maslanka and G.G. Spence, US 4,403,062 (1983).
218. R.P. Avis, US 3,773,612 (1973).
219. M.M. Haedlam and D.J. Smith, US 5,690,790 (1997).
220. A.E. Broderick, US 2,285,490 (1942).
221. A.E. Broderick, US 2,329,741 (1943).
222. C. Hagiopol, Y. Luo, D.F. Townsend, K.D. Favors, J.W. Johnston, C.E. Ringold, L.D. Saddler, and D.G. Jenkins, US 7,119,148 (2006).
223. J. Zabicky, *The Chemistry of Amides*, Interscience, New York, 1970.
224. R.E.J. Mitchel and H.C. Birnboim, *Anal. Biochem.*, **81**, 47–56 (1977).
225. A. Omari, *Polymer*, **35(10)**, 2148–2152 (1994).
226. M.D. Wright, US 2009/0126890 (2009).
227. D.L. Dauplaise, J.J. Kozakiewicz, and J.M. Schmitt, US 5,041,503 (1991).
228. M. Imoto and M. Kobayashi, *Bul. Chem. Soc. Jpn.*, **33**, 1651–1656(1960).
229. H.E. Zaugg and W.B. Martin, *Org. React.*, **14**, 52–269 (1965).
230. H.F. Conrat and H.S. Olcott, *J. Amer. Chem. Soc.*, **70**, 2673–2684 (1948).
231. H.S. Olcott and H.F. Conrat, *Ind. Eng. Chem.*, **38**, 104–106 (1946).
232. S.L. Vail, C.M. Moran, and H.B. Moore, *J. Org. Chem.*, **27**, 2067–2070 (1962).
233. J.P. Chupp and A.J. Speziale, *J. Org. Chem.*, **28(10)**, 2592–2595 (1963).
234. G.A. Crowe and C.C. Lynch, *J. Amer. Chem. Soc.*, **72**, 3622–3623(1950).
235. J. Schaffer, US 2008/0216979 (2008).
236. S.M. Cohen and J.R. LeBlanc, US 4,293,693 (1981).
237. S.M. Cohen and J.R. LeBlanc, US 4,321,375 (1982).
238. S.M. Cohen and J.R. LeBlanc, US 4,326,057 (1982).
239. S.M. Cohen and J.R. LeBlanc, US 4,375,537 (1983).
240. S.L. Vail, C.M. Moran, and R.H. Barker, *J. Org. Chem.*, **30(4)**, 1195–1199 (1965).
241. H. Feuer and U.E. Lynch, *J. Amer. Chem. Soc.*, **75**, 5027–5029 (1953).
242. J. Lalo, R. Garrigue, and F. Daydou, *Bull. Soc. Chim. France*, **3**, 474–477 (1988).
243. S.H. Hui, US 4,544,609 (1985).
244. C. Lu and J.D. Ward, US 2008/0308242 (2008).
245. C. Schramm and B. Rinderer, *J. Appl. Polym. Sci.*, **88**, 1870 (2003).
246. C. Hagiopol, Y. Luo, D.F. Townsend, and J.W. Johnston, US 7,034,087 (2006).
247. N.A.-K. Mumallah and A. Moradl-Araghi, US 4,822,842 (1989).
248. N.A.-K. Mumallah and A. Moradl-Araghi, US 4,884,636 (1989).
249. L.L. Williams and A.T. Coscia, US 3,556,933 (1971).
250. R. Epton, J.V. McLaren, and T.H. Thomas, *Polymer*, **15(9)**, 564 (1974).
251. R.W. Faessinger, US 3,709,857 (1973).

252. M.M. Strung and F.O. Guenther, *J. Amer. Chem. Soc.*, **73**, 1884 (1951).

253. F. Chastrette, M. Chastrette, and C. Bracoud, *Bull. Chem. Soc. France*, **5**, 822 (1986).

254. S.K. Gupja, Brit. 1,473,782 (1975).

255. A. Stambouli, F. Hamedi-Sangsari, R. Amouroux, F. Chastrette, A. Blanc, and G. Mattioda, *Bull. Soc. Chim. France*, **1**, 95 (1988).

256. F.S.H. Head, *J. Chem. Soc.*, 1036–1037 (1955).

257. J.M. Kliegman and R.K. Barnes, *J. Org. Chem.*, **38**, 556 (1973).

258. J.M. Kliegman and R.K. Barnes, *J. Org. Chem.*, **39**, 1772 (1974).

259. S.P. Dasgupta, US 5,338,407 (1994).

260. G.G. Allan and W.M. Reif, *Trans. Tech. Sect.* (Canadian Pulp and Paper Association), **1(4)**, 97–101 (1975).

261. A. Meller, *Tappi*, **41(11)**, 684–686 (1958).

262. M.S. Ghafoor, M. Skinner, and I.M. Johnson, US 6,031,037 (2000).

263. M.S. Ghafoor, M. Skinner, and I.M. Johnson, US 6,001,920 (1999).

264. E.T. Reaville and W.R. Hine, *Tappi*, **50(6)**, 262–270 (1967).

265. M. Ryan, D. Dauplaise, and W. Brevard, US 2009/0114357 (2009).

266. M. Ryan and D. Dauplaise, WO 2004/001127 (2003).

267. T.C. Shannon, M.J. Smith, P.P. Chen, and G. Jiminez, US 6,548,427 (2003).

268. T.G. Shannon, M.J. Smith, P.P. Chen, and G. Jimenez, WO 01/38638 (2001).

269. F.H. Sangsari, F. Chastrette, M. Chastrette, A. Blanc, and G. Mattioda, *Recl. Trav. Chim. Pays Bas*, **109**, 419–424 (1990).

270. K. Yamamoto, *Text. Res. J.*, **52(6)**, 357–362 (1982).

271. C.M. Hadad, P.R. Rablen, and K.B. Wiberg, *J. Org. Chem.*, **63**, 8668–8681 (1998).

272. V.E. Tumanov, E.A. Kromkin, and E.T. Denisov, *Russ. Chem. Bull.*, **51(9)**, 1641 (2002).

273. L.L. Chan and P.W.K. Lau, *Pulp Paper Can.*, **89(8)**, 57–60 (1988).

274. W.L. Kaser, J. Mirza, J.H. Curtis, and P.J. Borchert, *Tappi*, **48(10)**, 583–587 (1965).

275. M.J. Smith, J.M. Kaun, and V.R. Gentile, US 5,993,602 (1999).

5 Wet-Strength Resins

Paper structure refers to a geometric arrangement of cellulose fibers. Paper wet strength is a result of cellulose fiber strength [1] and interfiber interaction. Hydrogen bonds are involved in both the structure of cellulose fibers and the interfiber bonding. The number of hydrogen bonds is larger within the tight crystalline structure of the cellulose fiber which is less sensitive to water. A lower number of hydrogen bonds are developed within the amorphous regions; it is a structure less ordered and more sensitive to water [2].

The interfiber hydrogen bonds require an internuclear distance of only 2.7 Å. They are developed—supposedly—between the amorphous zones of two different fibers. Therefore, the effectiveness of interfiber hydrogen bonds depends on the apparent density of paper and on the residual moisture content.

Figure 5.1 shows the paper wet-strength decay as a function of the moisture content [3]. When ordinary (water leaf) paper has been completely rewetted, the residual paper strength is less than 10% of its original dry strength. The presence of a wet-strength resin (WSR) makes paper with a residual wet strength about 30% of its initial strength. Paper containing wet-strength resins shows a significant residual wet strength which is constant regardless of the moisture content and long-lasting (permanent) wet strength [4].

Paper tissues typically contain a blend of relatively long fibers (usually softwood fibers) and relatively short fibers (hardwood fibers) [5]. It is common practice in the industry to add strengthening agents. To increase the permanent paper wet strength, and make it less sensitive to water, the hydrogen bonds developed within the interfibers bonded area should be protected; they should be enhanced and/or new covalent (50–100 kcal/mol) and/or ionic bonds (10–30 kcal/mol) should be added.

If more than 15% of the original dry strength is retained after soaking in water, paper is normally regarded as being of wet-strength quality. This chapter presents chemical compounds able to deliver a "permanent" wet strength to paper [6]. Wet-strength resins must [7] (1) be water soluble, (2) have a high molecular weight (polymer), (3) have cationic charges, (4) have multiple reactive functional groups, and (5) be able to generate strong bonds during the curing process.

Any chemical compound which increases the number of interfiber bonds (covalent, ionic or hydrogen bonds) is a potential wet-strength resin. More hydrogen bonds at the point of contact between two cellulose fibers can be added by a "glue" [8] with a similar chemical structure (such as regenerated cellulose from xanthate [9,10], cellulose "hydrate" [11], or locust bean gum [12]).

More covalent bonds are added to the bonded area by chemical compounds able to react with two hydroxyl groups from cellulose, such as anhydride, aldehyde, or methylol [13–16].

We will see in what is to follow that all those five attributes are more or less flexible: there are WSR which are not soluble in water (dispersions or di-anhydrides); the reactive groups are not always present and the need for a macromolecular compound is debatable. For instance, compounds with small molecules, such as tris(2,3-epoxypropyl) amine [17] or formaldehyde [18], improve the wet strength of cellulosic materials. That diversity explains the difficulty in designing a general mechanism for the wet strengthening effect. However, the most accurate description for a WSR would be a macromolecular cationic compound with multiple functionalities, a "reactive, post-cross-linkable polymer" [19].

The effectiveness of a WSR is expressed through the wet strength of paper (Equation 5.1), which is much lower than its dry strength. The wet strength of a resin-containing sheet is commonly expressed as a percentage of the dry strength (W/D ratio) [20]:

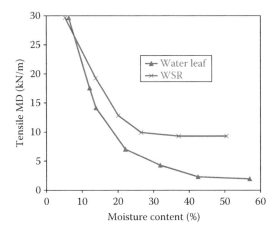

FIGURE 5.1   The moisture effect on paper wet strength.

$$\%WS = \frac{WS}{DS} \times 100 \tag{5.1}$$

It would be more accurate to indicate the increments of wet and dry strength (ΔW and ΔD) in relation to the strength values for waterleaf (WL), WS_o and DS_o, respectively (Equation 5.2):

$$\%WS = \frac{WS_o + \Delta W}{DS_o + \Delta D} \times 100 \tag{5.2}$$

The usual wet-strength resins improve both paper wet strength and dry strength [21]. However, a good WSR is expected to have little effect on the dry strength of the paper and a strong effect on its wet strength (as compared to WL). For a wide range of WSR (polyamidoamine epichlorohydrin (PAE)) concentration [22,23] (Figure 5.2), the wet strength (W/D ratio) increases faster than the dry strength.

Wet-strength resins may have other collateral effects: they can make the sheet stiffer or more difficult to repulp [24–26]. To address those issues, several changes in WSR chemical structure are made; the W/D ratio is increased by reducing the paper dry strength with a debonder [27] or by increasing the paper softness [28].

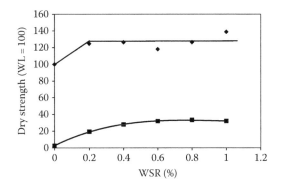

FIGURE 5.2   The effect of WSR concentration on the paper dry and wet strengths.

There are many criteria according to which WSR are classified: their effectiveness at a particular pH; their solubility in water (solutions and dispersions); their concentration, shelf life, viscosity, number of components, type of reactive functionality, etc. The present review considers the chemical structure to be the main criterion. The reasons for that choice are the understanding of the affinity for cellulose fibers, the feasibility of the synthesis, the way the cross-linking process is working, and how easy the process of repulpingis.

All the resins presented in this chapter belong to organic chemistry. However, there are inorganic compounds (silicic acid [29]) with unexpected wet strengthening capabilities. The organic compounds used as WSR can be ionic or nonionic. They can be added to the pulp at the wet end or to the web at the dry end [30,31]. At the dry end the absorption of WSR can be performed from a bath or can be "printed" with a special roll [32].

Because some WSR are reactive, for a good performance, the reaction time and the temperature at the dry end is taken into account. For instance, the effect of formaldehyde or urea–formaldehyde (UF) resins is boosted at higher temperatures [33].

That definition of WSR (cationic, high molecular weight, water-soluble reactive molecule) includes some potential conflicts: its reactivity reduces the shelf life or may react with water [34]; the higher molecular weight increases the viscosity of the solution and imposes the use of a diluted solution. These issues have implications on the WSR synthesis.

Because it is difficult to dissolve high molecular weight polymers at a reasonable concentration, the WSR synthesis implies, as a first step, the synthesis of a prepolymer with a lower molecular weight. The prepolymer concentration in water can reach 50% or higher. In the second step, the prepolymer chemical structure and molecular weight are adjusted to fit the final requirements for a WSR in the papermaking process.

5.1 PREPOLYMER SYNTHESIS

The number of cationic charges and reactive groups, and their reactivity and availability [35] are a problem which must be solved during the prepolymer synthesis. The prepolymer structure is designed by taking into account the second reaction, the final WSR capabilities and even the repulpability of paper.

Based on the backbone structure, there are two large categories of macromolecular compounds: (a) with at least a heteroatom included in the backbone, and (b) with only carbon–carbon bonds. The carbon–hetero-atom (nitrogen, oxygen, etc.) bond is weaker than a carbon–carbon bond and it is easier to hydrolyze them during the repulpable process.

5.1.1 Prepolymers with a Hetero-Atom in the Backbone

The macromolecular compounds having hetero-atoms in their backbone are obtained mainly through polycondensation between poly-acids and poly-alcohols, poly-acids and polyamine, polyamine and aldehyde, urea and formaldehyde, etc. Phosphorus-containing resins [36] or phenol–formaldehyde (PF) resins [37] are less involved in the papermaking process: the PF modified resins with amino acids (β-alanine or glycine [38]) are recommended for acid-resistant paper [39].

5.1.1.1 Urea–Formaldehyde Resins

Polycondensation reaction involves a compound with at least two reactive groups and one compound with three or more reactive groups in order to develop a three-dimensional network [40]. Urea is a tetra-functional condensation monomer, while formaldehyde is a difunctional condensation monomer. By changing the UF ratio and their functionality, a cross-linked structure can be built. A third compound, such as furfuryl alcohol [41], can be a part of the condensation reaction as well.

The first step of the condensation reaction is the synthesis of mono-methylol (UF mole ratio 1:1) and di-methylol urea synthesis (UF mole ratio 1:2). This is an equilibrium reaction: un-reacted urea and free formaldehyde are present at any value of the reaction conversion [42].

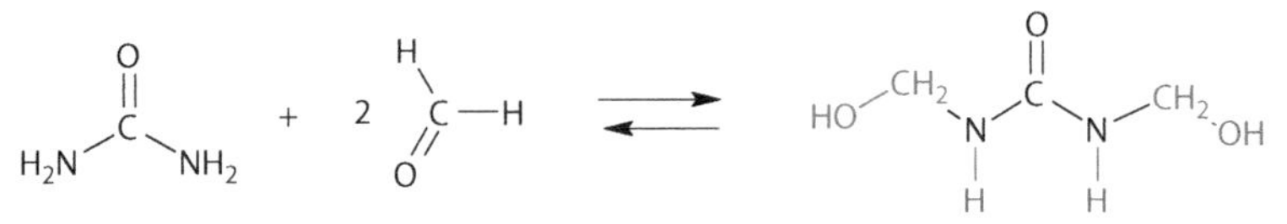

Due to its higher functionality (in blue and red), di-methylol urea has two other reactive positions left: the hydrogen atoms of the secondary amide. The di-methylol urea is still a tetra-functional compound. That higher functionality is the driving force for a cross-linking process through methylene groups (blue bridges).

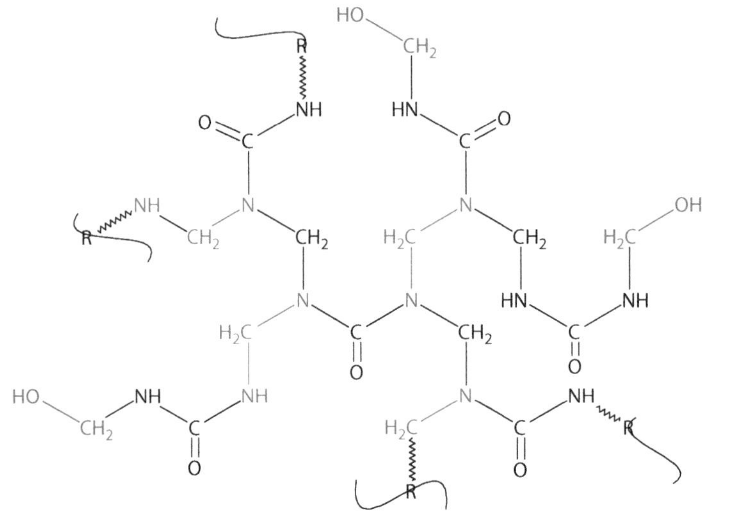

The branched UF resins still have multiple reactive methylol groups (in red). The reactive methylol can react with other amide groups (from urea units) and provide a more cross-linked structure. Resin is self-cross-linkable and a resin solution in water (50%–70% [43]) is not stable. Condensation is an equilibrium reaction: the methylol functionality can release formaldehyde from a UF resin at any moment during its application.

UF resins with higher molecular weight are better retained and show a higher effectiveness as wet-strength resins for paper [44]. The methylol group reacts with paper, making an ether bond, or with resorcinol [45] to develop a hybrid resin. The reaction with the hydroxyl group from cellulose (curing step) may provide wet strength for paper treated with UF resin [46] (Figure 5.3): higher temperatures and a lower pH will result in better strength developed in a shorter time. An acceptable increase in paper wet strength is obtained after a curing time exceeding 10 min.

Unfortunately, absorbent paper grades, such as toweling and facial tissues, are under a handicap at pH 4.5 because water absorbency tends to be lower. Corrosion problems, the fact that the paper results in a more brittle product [47], and difficulties in recovering the broke [48] (see Section 5.6) are other issues involved in using UF resins to increase paper wet strength.

Most of the UF resins are not soluble in water. A water-soluble version of the UF resins was obtained by a new condensation in which a cationic polymer reacts with UF resin. The cationic polymer is obtained by classic condensation between a polyamine and a dihaloalkane or a halohydrine [49,50].

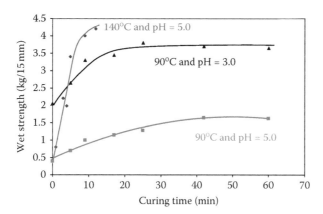

FIGURE 5.3 The effect of curing time, temperature, and pH on the paper strengthened with UF resin. (From Hazard, S.J. et al., *Tappi*, 44, 35, 1961. With permission.)

The presence of amine functionalities [51,52] (primary, secondary, or tertiary) makes the UF resin potentially cationic in acid or neutral pH. Amine functionalities are also available for further reactions, such as condensation with aldehyde or trans-amidation with urea. The reaction product is recommended as highly effective for improving wet strength [50].

5.1.1.2 Melamine–Formaldehyde Resins

Melamine reacts with formaldehyde in water or in isopropanol/water blend [53,54], under slightly alkaline conditions (pH = 7.7), within a reversible process resulting in methylol derivatives [40,55,56].

In this reaction, formaldehyde has a functionality of 2, and melamine a functionality of 6: there are six hydrogen atoms bonded to the amine nitrogen, all with the same reactivity. Even at a mole ratio of 1:1 melamine/formaldehyde (MF), a mixture of mono-, di-, tri-, etc. derivatives is formed along with un-reacted melamine and free formaldehyde. At an MF ratio of about 1:3, the methylol group from one melamine molecule will react with the un-reacted amine hydrogen from another melamine molecule, and a methylene (–NH–**CH₂**–NH–) or a methylene ether linkage (–NH–**CH₂**–**O**–**CH₂**–NH–) is formed. This process leads to a less soluble macromolecule and eventually in a cross-linked material [43]. The number of methylol groups and the probability to get a cross-linked polymer depend on the MF ratio. For an excess of formaldehyde (1/8 mol ratio), a hexamethylol melamine is obtained.

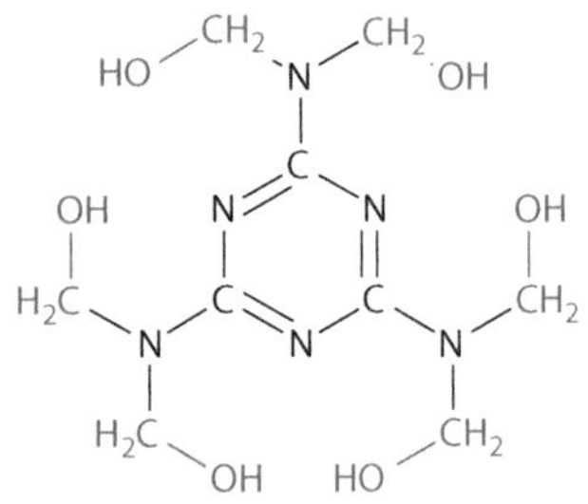

During the MF resin application to the wet end, resin is greatly diluted; changing concentrations may release formaldehyde and shift the resin composition [55]. The adsorbed polymer always contains less methylol groups than its parent solution. For that reason, the MF resin dispersion was doped with more formaldehyde (5%–10% mole per mole of melamine) to improve the wet strength by about 30% [57].

In spite of the difference in reactivity of an amine in melamine and an amide group in urea, a copolymer of urea–melamine and formaldehyde (hybrid copolymer) can be obtained for a wide range of urea/melamine ratios [58–60].

For high wet strength, paper is dried at 100°C and then cured at about 150°C for 3 min [54]. However, the application of UF and MF resins is impaired by the requirement for a low pH of the paper slurry (4.5 < pH < 5.5). The un-reacted formaldehyde is also a concern because formaldehyde is well known as a harmful chemical compound [61].

The paper made in acid systems has a lower dry strength and is more brittle than paper made under neutral or alkaline conditions [62]. Acid conditions cause corrosion of the metal parts of the papermaking machine. Moreover, the strength developed through the use of UF and MF resins is frequently temporary because acid-cured resin is readily hydrolyzed in the presence of acid [63] (see the equilibrium reaction). These issues pushed the research teams to look for a free formaldehyde resin, able to cure in neutral or alkaline pH.

5.1.1.3 Polyamines and Polyethylene Imines

Polyamines are synthesized by condensation or by polymerization. The polycondensation reaction involves amines and halogen derivatives, epoxy compounds, di-aldehyde, di-halides, etc. Epichlorohydrin is a difunctional compound: there is an epoxy ring and also a chlorine derivative. For instance, ammonia, which is considered as the simplest amine, reacts with epichlorohydrin and forms 1.3-diamino-2-propanol [64].

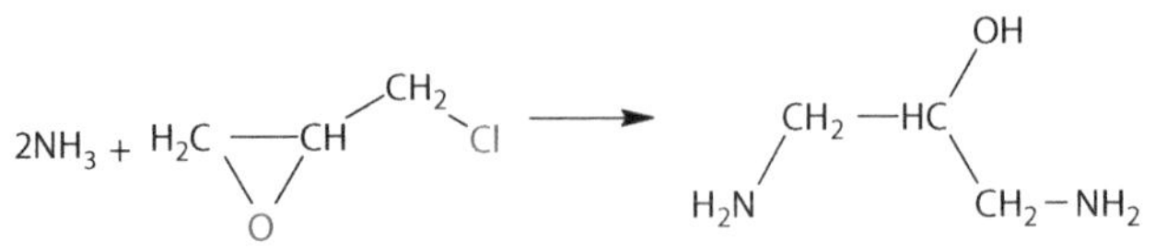

That equation is the most simplified presentation of the actual process. The reaction product is also an amine which can react with another molecule of epichlorohydrin and form a polymer (polyalkanolamine). The chemical structure of the polyamine depends on the molar ratio and on the order of addition [65]. There is a difference in the reactivity of epoxy and chlorine derivatives: epoxy reacts faster. The final compound is a water-soluble thermosetting cationic resin: it has both reactive amine and chlorine in the same molecule [65–70].

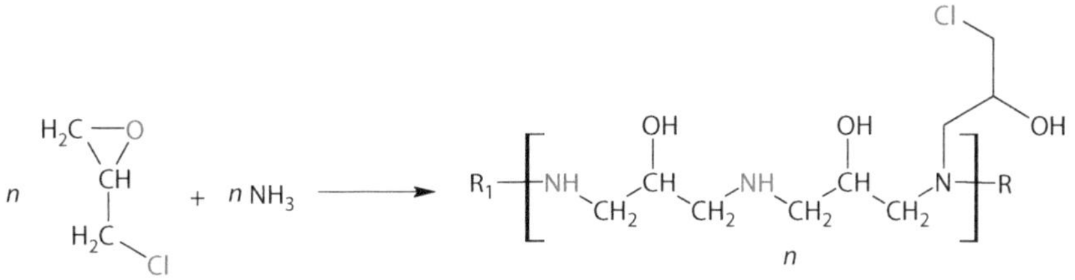

For better control of the process, the organic bound chlorine is adjusted by adding an excess of ammonia [71] and the amine functionality is reduced by replacing ammonia with secondary amines [72–74]. The reaction between the organic chlorine and the amine group is also controlled by lowering the pH in acid range (pH $<$ 5.0), when amines are protonated [75–76]. If the right molar ratio is used, a branched, high molecular weight poly-tertiary amine is obtained [77].

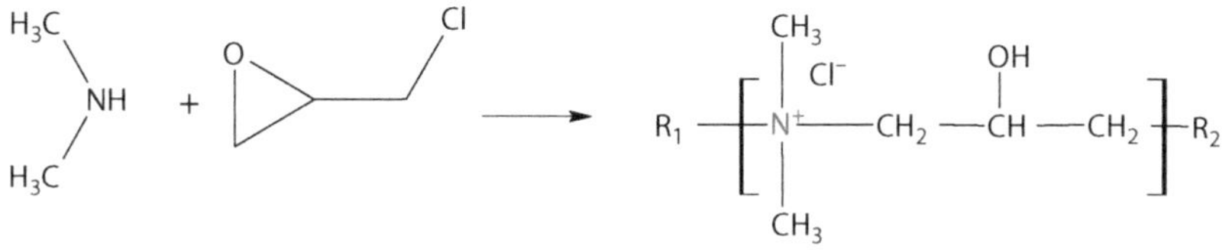

Any change in the molar ratio can lead to the formation of a thermosetting resin. In order to control the cross-linking reaction during the polyamine synthesis, an excess of amine or a mono-functional compound (such as trimethyl amine) is preferred. Due to the different types of condensation processes, the molecular weight distribution is bimodal [78].

DETA reacts with epichlorohydrin [79–81] in water or water–propylene glycol mixtures [82–84]. The molecular weight is controlled by acid addition.

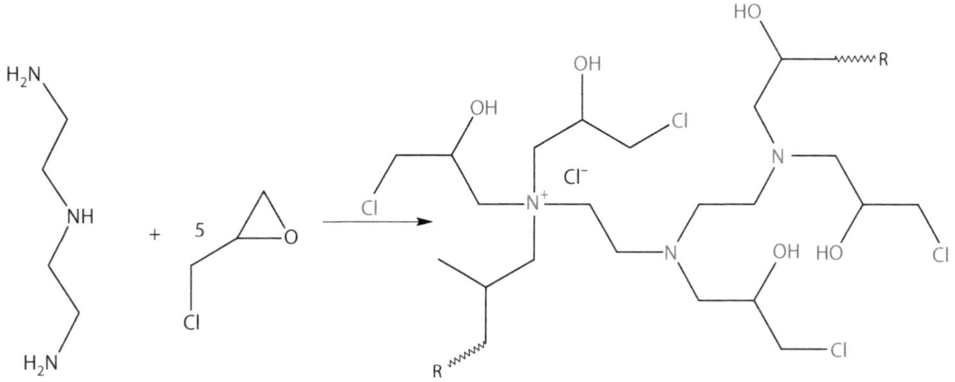

Primary, secondary, and tertiary polyamines react with halogen derivatives (at least a functionality of 2, such as 1,2-dichloroethane, 1,3-dichloro-2-propanol, or 3,4-dichlorotetrahydrothiophene-1,1-dioxide [85–90]. Hexa-methylene diamine is reacted with 1,2-dichloroethane to obtain a polyamine with secondary amine groups [87,91].

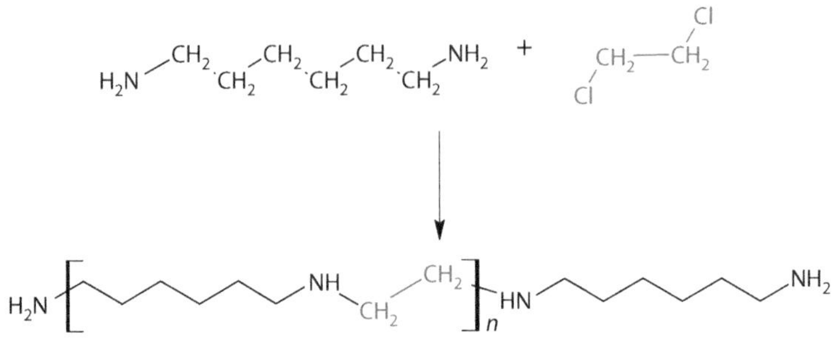

The reaction continues with a branching step because the organic bonded halogen reacts with secondary amine and the tertiary amine can also be alkylated to obtain a poly-quaternary ammonium salt. The progression of that reaction can reach the cross-linking point. In order to prevent further condensation, the amine functionality is reduced. Di-tertiary amine (N,N,N',N'-tetramethylethylenediamine [92]) forms a linear polymer. A Michael addition (ethylenediamine with acrylamide or N-methylol acrylamide, methacrylonitrile, acrylates) is a way to obtain tertiary amines, which are reacted with 1,4-dibromobutane [93]. The final product looks like a cationic polyacrylamide.

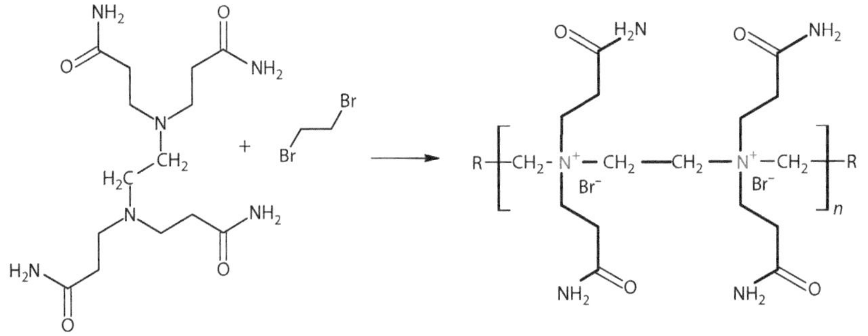

This type of condensation is characterized by the release of hydrochloric acid as a by-product. The acid is captured by the un-reacted amine and the protonated amines are no longer active in condensation.

In another type of condensation, polyamines react with a di-aldehyde (such DETA and glyoxal) [94] to obtain a poly-imine-amine [95]. The secondary amine functionalities are preserved and subject to a charge addition in a further step (polymer-analogous reaction with epichlorohydrin).

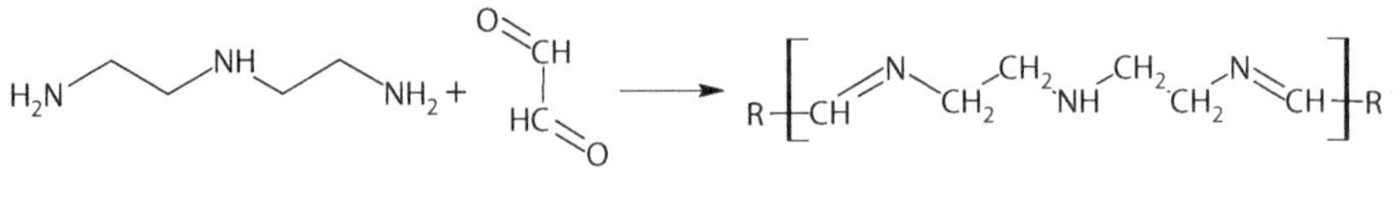

Guanidine is condensed with formaldehyde (and a ketone, such as acetone) to form a poly-amine used as wet-strength resin in alkaline pH [96]. Cyan-guanidine reacts with the aniline-formaldehyde condensation products to form a cationic poly-aryl-guanidine [97]. The new resin is modified with UF resins and used as wet- and dry-strength resins for paper.

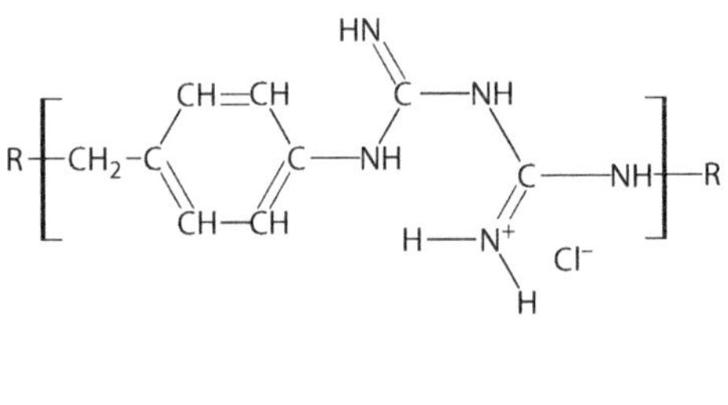

Unlike the condensation reaction, the polymerization process does not generate by-products. As early as 1888 it was discovered that ethylene imine polymerizes in aqueous solution [98–101], or in the presence of aliphatic halides as initiators [102]. The reaction product is a poly-secondary amine (polyethyleneimine (PEI)).

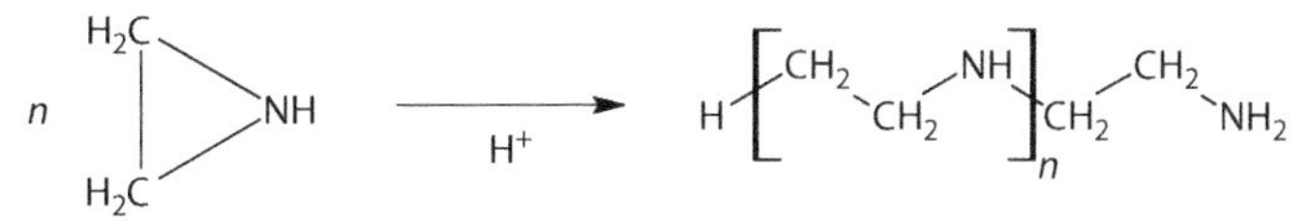

In spite of the difficulties to make linear PEI [103], PEI is produced in large amounts. As any other polyamine, PEI has cationic charges in water at acid or neutral pH and therefore is retained on anionic cellulose fibers.

The ethylene imine can also be involved in other reactions to generate polyamines: it can copolymerize with epichlorohydrin (less than 1% Epi) [104], or react with polysaccharides to make polyaminoethyl ether [105].

Macromolecular polyamines, obtained either through polycondensation or polymerization, have multiple reactive groups (mostly secondary amines). Those secondary amines allow for the addition of other groups to the backbone. Polyamines react with sodium cyanate and a poly-urea is obtained [90].

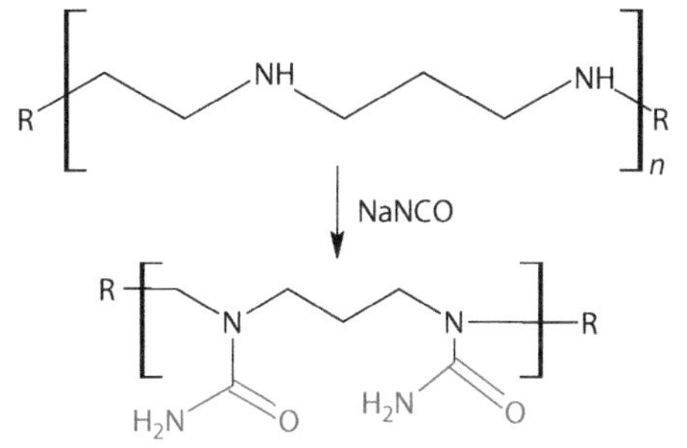

A linear polymer may react with fatty acid chloride, alkyl halide, chlorinated waxes, epichlorohydrin [87,106], or formaldehyde [101] to make branched or cross-linked materials. PEI reacts with urea in a trans-amidation and a cross-linking reaction [107].

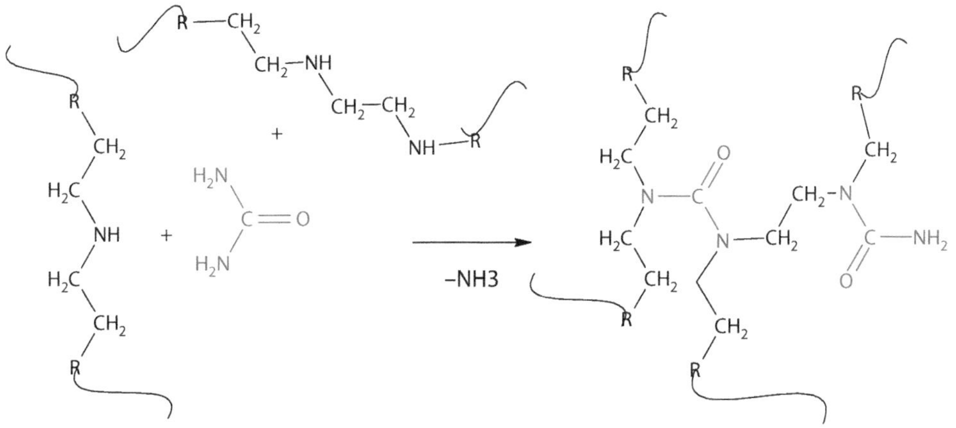

The amine groups are in excess, many secondary amines remain un-reacted, and the increase in the molecular weight is limited. It has an important potential as a wet-strength resin and its

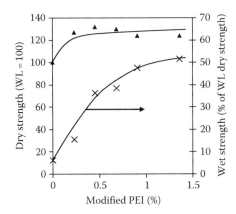

FIGURE 5.4   Modified PEI resin as wet-strength resin for paper.

repulpability should be an interesting experiment to perform. Figure 5.4 shows a much faster increase in the paper's wet strength than in its dry strength. The W/D ratio reaches good numbers for high PEI concentration in the case of unbleached Kraft pulp.

5.1.1.4   Polyamidoamine

UF and MF resins as wet-strength agents (and acid-curing processes) are used in old technologies. The modern technology uses WSR curable in neutral or alkaline pH. The backbone of those resins is a polyamidoamine [108,109].

In 1960, when research began, a patent was published which claimed a reaction between a polyamidoamine and formaldehyde [110] for curing in acid pH. This compound represents the transition from formaldehyde condensates to polyamidoamine WSR. The PAE resins, reaction products of polyamidoamines with epichlorohydrin, were the turning point in wet-strength resin synthesis: a curing reaction can be developed at alkaline pH [109].

Polyamidoamine is obtained from diacids and polyamines [111]. The polyamine must have both primary and secondary amine groups (such as diethylene triamine (NH_2–CH_2–CH_2–NH–CH_2–CH_2–NH_2). The secondary amine group can be replaced by a tertiary amine group [112–114] (such as N,N-bis(3-aminopropyl) methylamine). The mole ratio between polyamine and diacid is roughly 1:1 [67]. During the condensation reaction only the primary amine reacts with the carboxylic group of the diacid and forms an amide bond. The secondary amine remains available for further reactions.

$$R\left[\!\!\begin{array}{c} \underset{O}{\overset{O}{\underset{\|}{C}}}-R-CH_2-\overset{\overset{\displaystyle O}{\|}}{C}-NH-CH_2-\overset{\overset{\displaystyle H}{|}}{\underset{|}{N}}-CH_2-CH_2-NH \end{array}\!\!\right]_n R$$

There is a wide diversity of dicarboxylic acids, or their esters, used in reaction with polyamine to make a WSR: adipic, glutaric, succinic acids [113,115–117], oxalic acid [114]), itaconic acid [47,118,119], nitrilotriacetic acid [120], polyethylene glycol bis(carboxymethyl) ether [121]), polymeric fatty acids [122–124], aromatic acids (isophtalic acid [119,125–127]), or polysiloxane dicarboxylic acid [128,129]. Multi-carboxylic acid (such as ethylenediaminotetracetic, nitrilotriacetic acid) [130] or citric acid [118] are also used in that condensation.

Itaconic acid shows a triple functionality in the reaction with amines. Diamines (such as hexamethylene diamine which is a tetra-functional compound) react with itaconic acid first with a Michael addition, and then a polycondensation takes place [131]. The structure of that polyamide does not include a residual secondary amine.

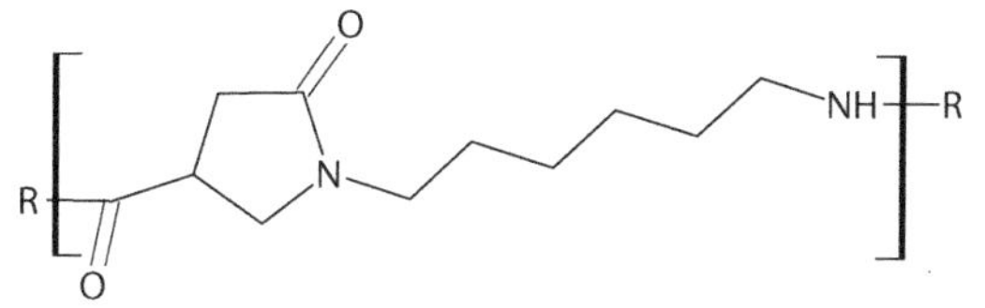

The same pyrrolidone structure is obtained from itaconic acid and an amino acid [132–134]. The new diacid can be involved in a condensation with a diamine.

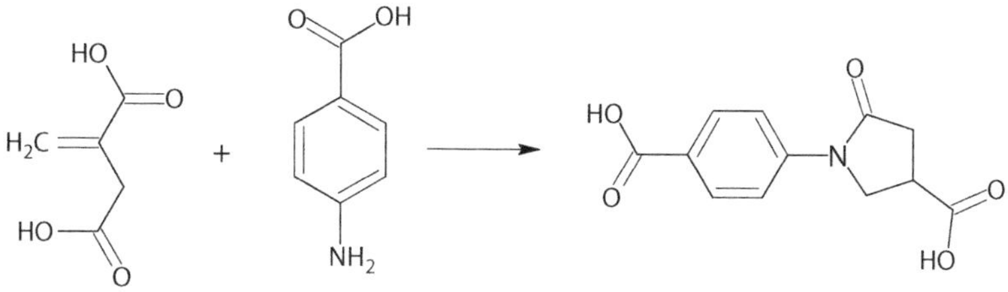

Polyamines are partners of diacids in the synthesis of the backbone of polyamidoamine. Their structure must have the general formula $NH_2-CH_2-CH_2-(NH-CH_2-CH_2)_a-NH_2$ where $1 \leq a \leq 6$ (two primary amines and at least one secondary or tertiary amine [113,135–138] are included in its molecule). Blends of different polyamines [[116,122,125–127,137,139–142] and blends of acids [143,144] are also recommended.

Changes in the acid and/or amine structure could bring important modifications in the backbone properties. Hydrophobic tails are added in both diacid [123] and polyamine [145] to make polyamidoamine a raw material for wet-strength resins and internal sizing agents. Polyamines may have a reactive double bond (such as bis(2-aminoethyl) allyl amine [146]) and polyamidoamine will have an allylic pendant functionality.

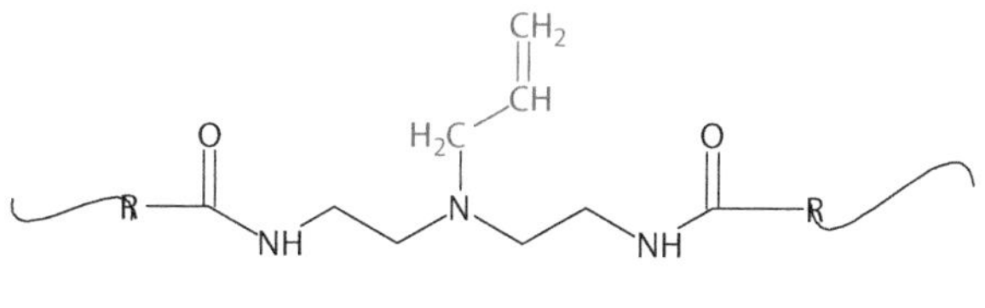

Polycondensation (adipic and DETA) is neatly performed with hydrochloric acid [147,148], sulfuric acid, or p-toluene sulfonic acid (0.03–0.04 mol for 1 mol of adipic acid) as a catalyst [149,150]. The presence of a catalyst shortens the reaction by about 40% [151].

The neutralization reaction between the diacid and the polyamine develops a large amount of heat, which is why water must be added in order to help the dissipation of neutralization heat. Condensation is an equilibrium reaction (see Section 2.4.1); water is released during synthesis. In order to move the equilibrium toward the polymer formation, the water should be distilled [152–154].

The polycondensation reaction is performed at temperatures of 160°C–200°C, and the final prepolymer solution (after dilution with water) has a concentration ranging from 20% [94] to over 50% [108,109].

The esters of diacid compounds can also react with polyamines [22,116,155,156] and the corresponding alcohol will be released. In that process no neutralization is involved and a low molecular weight alcohol is much easier to distil at lower temperatures (80°C–105°C). However, at lower temperatures viscosity is higher for a given molecular weight.

The most common prepolymer (adipic acid–DETA) has a secondary amine at every 10 carbon atoms. To increase the concentration of secondary amines (reactive group in the prepolymer), the acid must have a smaller number of carbons (oxalic acid replaces the adipic acid [137]), or an amino acid can be used. The ester of the amino acid is synthesized through a Michael addition [22,157–160] between methyl acrylate and DETA. However, secondary amines are also reactive in a Michael addition; therefore this reaction has a very low selectivity.

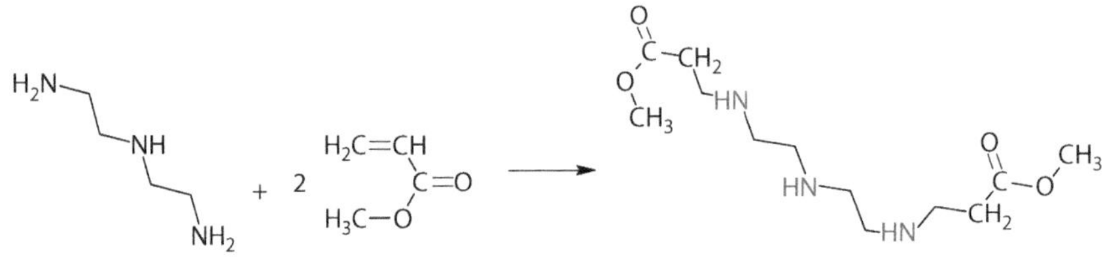

The amino acid ester is condensed with a new portion of DETA. The new polyamidoamine includes a higher number of secondary amines (in red) in its backbone.

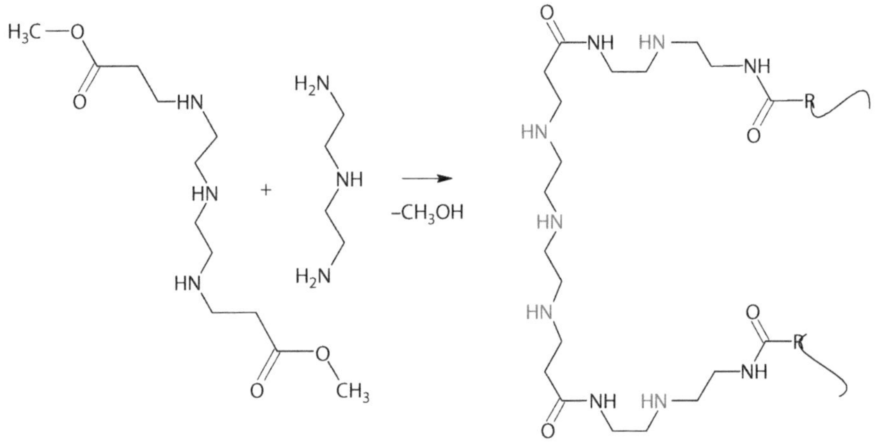

The really poor selectivity of the Michael addition is proven by the reaction between methyl acrylate and ethylenediamine [161].

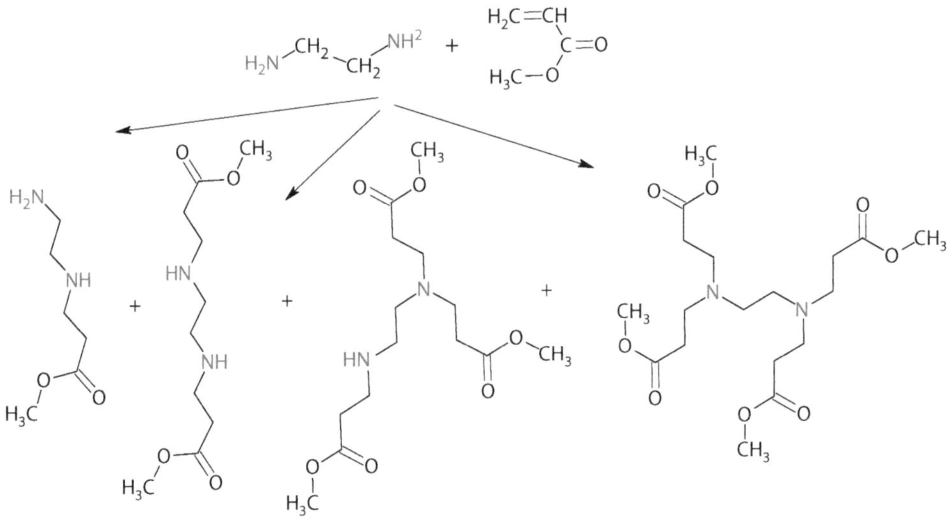

Four different amino acid esters are synthesized and after polycondensation with polyamine, a highly branched polyamidoamine (dendrimer) is obtained [161,162].

In order to obtain a co-polyamidoamine, small amounts of organic compounds with both amine and acid functionalities (lactame [22,142,163,164] or amino acids [22]) can be involved in the polycondensation process [165].

Urea, which is the diamide of the carbonic acid, may react with secondary and/or primary amines through a trans-amidation reaction when ammonia is released as small molecules. The reaction with diamines (such as N-methyl bis (3-aminopropyl) amine) is performed at 175°C–225°C [166] to obtain a new polyamidoamine (polyaminoureylene).

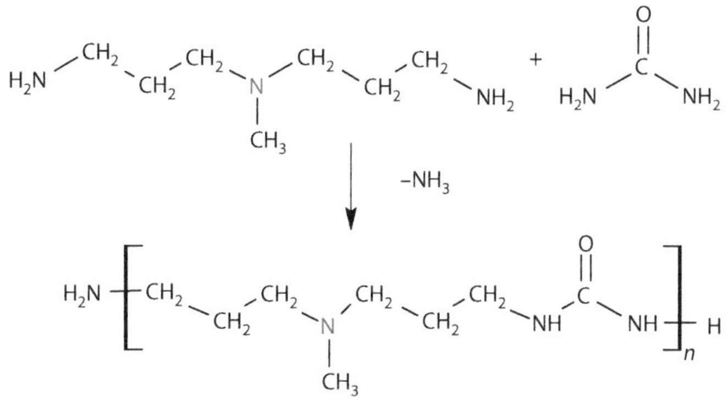

Condensation and trans-amidation can be performed simultaneously: N-methyl bis(3-aminopropyl) amine reacts, in one or two steps [114], with dimethyl oxalate and urea to make a hybrid polyamidoamine [115,167].

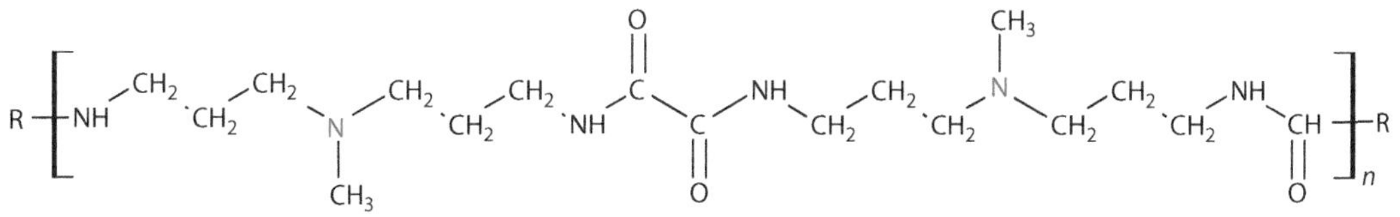

Polyaminoureylene has a tertiary amine in its backbone which can be transformed into a quaternary ammonium salt (epichlorohydrin reaction). The new resin has epoxy groups as reactive functionalities. The tertiary amine also helps the curing process during the drying step.

The molecular weight of the final polyamidoamine depends on the amount of water distilled. At high conversions, the distillation of a small amount of water results in an important increase in molecular weight. Because it is quite difficult to control molecular weight by water distillation, some other techniques are used. The prepolymer molecular weight can be tailored by changing the mole ratio between the polyamine and the diacid: for PA/DiAc>1.0, all the end groups are amines [168]. Molecular weight can also be adjusted by adding a mono-functional amine (i.e., mono-ethanol amine) [169,170] or a mono-functional acid or ester (such as methylbenzoate [171]).

In order to avoid an equilibrium reaction and the distillation of small molecules, a poly-addition reaction is used: polyamidoamine synthesis through multiple Michael additions between a polyamine (DETA) and N,N'-methylene-bis-acrylamide [172]. The lack of selectivity (see above) adds branches to the final product.

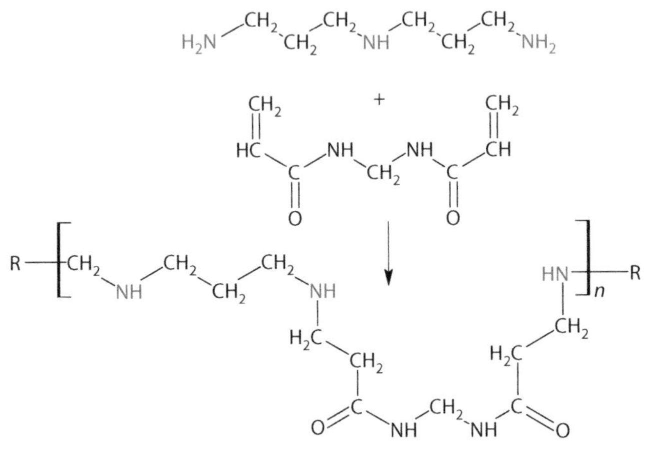

The Michael addition and polycondensation can be combined [47]. For that type of process, the diacid has in its structure a carbon–carbon double bond next to the carboxylic groups (such as maleic half-ester), which reacts with DETA within a Michael addition and a polycondensation [173–175].

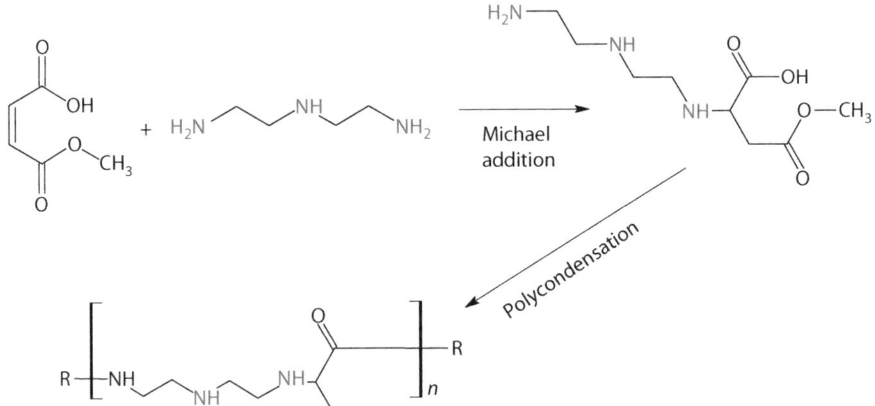

The resulting product is a polyamidoamine with a pendant ester group. Depending on the poly-condensation parameters, the ester group can also be involved in condensation and a branched structure can be obtained.

Di-isocyanates [176] react with polyamines through a poly-addition reaction and the poly-amide structure includes secondary amines, available for further reactions (polymer-analogous reactions).

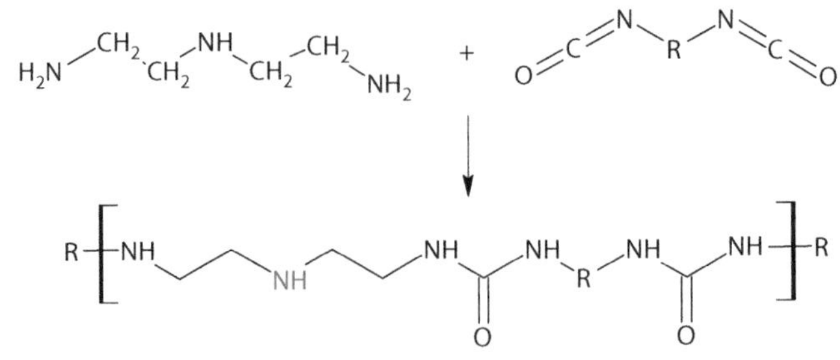

Secondary amines react with acrylates (Michael addition) [159], with the epoxide with long hydrocarbon tail (1,2-epoxyoctadecane), or with the acylchloride of fatty acids [177]. All these reactions reduce the number of secondary amines and make the polyamidoamine more hydrophobic.

Formaldehyde, which is a difunctional compound, adds methylol groups to the secondary amines [178] and then cross-links the polymer. Other difunctional compounds, such as dichloroethane, diglycidyl ether, N,N'-methylene bis-acrylamide, or glyoxal [138], are also cross-linkers for polyamidoamines.

The secondary amine reacts with potassium cyanate [110] and is converted to amide functionality (urea derivatives) which can react with formaldehyde and becomes WSR.

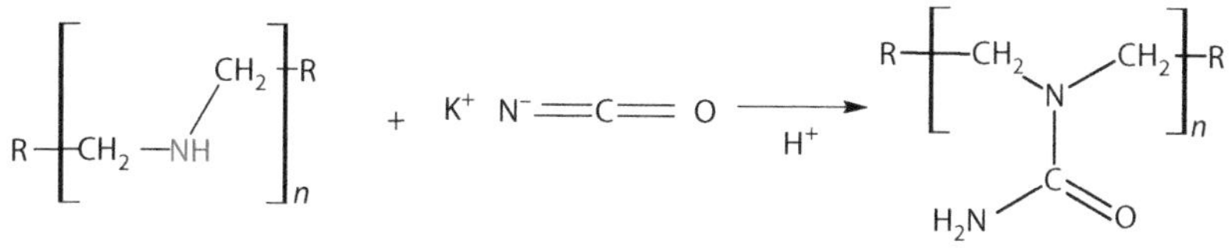

The diversity of chemical reactions performed on the polyamidoamine backbone shows the broad chemistries available to obtain WSR. However, all these potential technologies should consider the addition of the cationic charges (see Section 5.2). There is a possibility to build molecular weight through a condensation reaction and quaternize nitrogen atoms at the same time [179]. The adduct between 1,3-bis(dimethylaminopropyl)urea with 2 mol of epichlorohydrin is a di-halogen derivative quaternized with a di-tertiary amine. The reaction is actually a poly-quaternization.

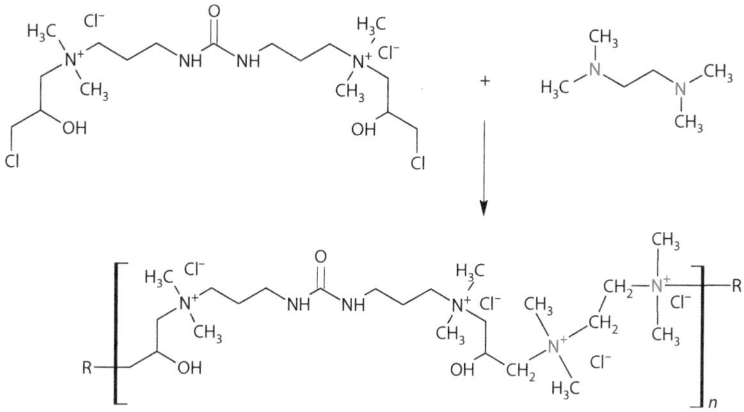

5.1.1.5 Polyamidoamine Esters

A polyamide-type structure and a polyester-type structure (from a diacid and a diol) can be built within one polycondensation step [180]. Actually, the diol (ethylene glycol or diethanolamine [141]) will replace about 25% mol of DETA. The copolymer (polyamidoamine ester) is water soluble and it has a corresponding residual secondary amine available for quaternization. From adipic acid, DETA and ethylene glycol result in a backbone with ester bonds (blue), amide bonds (in green), and a secondary amine (in red).

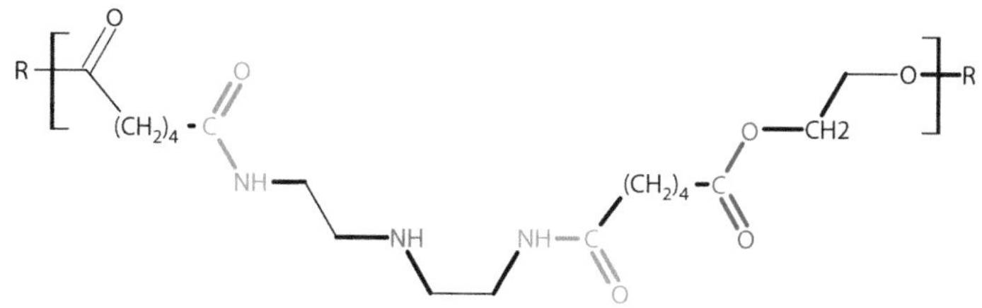

The ester functionality is also obtained by partially replacing DETA with about 10% molar mono-ethanol amine [181] or di-ethanol amine [141]. That type of process shows how versatile the condensation process is. However, the di-alcohol replaces the polyamine; consequently the polyamidoamine ester has a lower concentration of secondary amines.

A polyamine ester structure can be built on a polyester backbone (diethylene glycol and maleic anhydride polycondensation), and pendant amine groups are added by a Michael addition of difunctional amines [182].

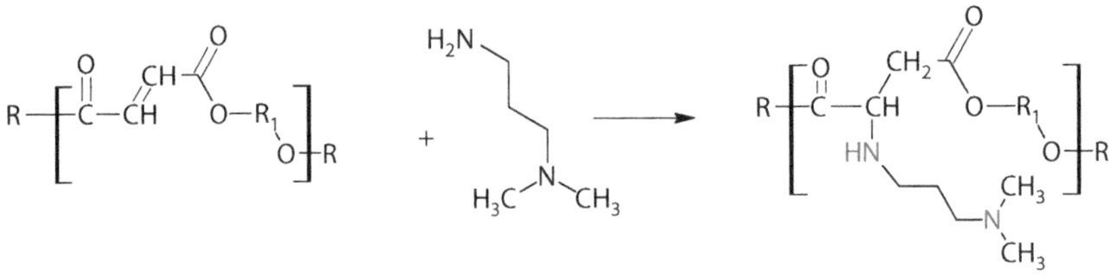

Bond angle and bond strength [183] make the difference between a polyamidoamine and a polyamidoamine ester in terms of solubility, chemical reactivity, repulpability, and properties as wet-strength resins.

5.1.1.6 Polysaccharides

Although cellulose, starch, and their derivatives display potentially important features as wet-strength resins, such as high molecular weight and chemical structure similar to that of pulp fibers, other important characteristics are missing: water solubility is limited, they are mostly nonionic and nonreactive with cellulose. However, their chemical structure allows for the addition of reactive groups and cationic charges. The addition of cationic charges may improve water solubility as well.

Chapter 4 describes the conversion of regular starch to cationic di-aldehyde starch and its effectiveness as temporary wet-strength resin. The slow decay of the paper wet strength after soaking in water invalidates cationic di-aldehyde starches as temporary wet-strength resins, and suggests that the combination of aldehyde functionalities, high molecular weight compounds (such as polysaccharides), and cationic charges may provide an adequate permanent wet-strength resin.

Aldehydes and their derivatives (such as formaldehyde, glyoxal, urea, and glyoxal [184] or methylol compounds of urea or thiourea [185]) react with cellulose fibers [186–188] and the resulting paper shows a good wet strength, dimensional stability, and reduced swelling capacity. Because glyoxal lacks specific substantivity to cellulose, it must be applied to the web after sheet formation [189].

The treatment with glutaraldehyde alone causes severe paper embrittlement and extensibility reduction. The presence of a reactive polymer (1% polyvinyl alcohol) reduces the effect of the downsides of low molecular weight aldehydes: treatment with 2% glutaraldehyde (at 140°C and $Zn(NO_3)_2$ as catalyst) [190] increases both the dry strength and the W/D ratio up to 50%.

The water-soluble hydroxyethyl cellulose modified with glyoxal [191] or pyruvic aldehyde [192] was used to increase the paper wet strength. Glyoxalated hydroxyethyl cellulose is used as wet-strength resin by impregnation (or in size press solution) because no charge is attached to it.

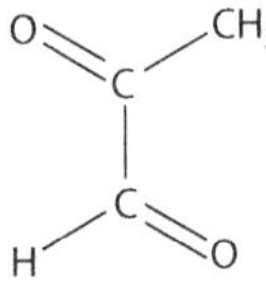

Pyruvic aldehyde

For retention at the wet end, a potential wet-strength resin must have both reactive groups and electric charges. The combination of the molecular weight of starch and the high number of aldehyde groups defines this product (cationic di-aldehyde starch [193–195]) as a wet-strength resin candidate especially for repulpable paper [196]. Due to the very low concentration of polymers used at the wet end, high molecular weight starches can be used without significantly increasing viscosity of the process water [197].

The aldehyde functionality is added to polysaccharide by oxidation [198,199], which can be performed at different conversions [200]. During that reaction some aldehyde functionalities are converted to carboxyl groups [201–203]. Oxidized compounds are anionic and the preferred aldehyde/carboxyl ratio is up to 10:1.

Oxidized starches [195,204–208] (see Section 2.3) as well as their bisulfite adducts [200,209], di-aldehyde galactomannan gum [210,211], starch with blocked aldehyde groups [212], and grafted starch with polyacrylamide are carriers for aldehyde functionalities [213–215] and anionic carboxylic groups. Oxidized polysaccharides improve the paper wet strength when used as a beater additive [216] or at the wet end in the presence of alum.

More anionic charges are added on starch by reaction with carbon disulfide in the presence of potassium hydroxide, when a xanthate derivative [217,218] is obtained.

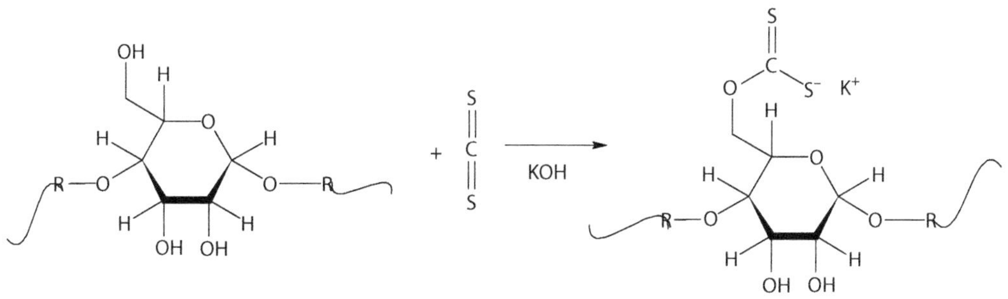

The di-aldehyde starch retention is lower than 100% [207] and depends on the amount of alum and the stock pH. Instead of alum as retention aid, cationic starch [206,219], PAE-type resin [220], zirconium salt [221], or borax [204] are used to retain anionic di-aldehyde.

Cationic charges can be added [222] before oxidation through a reaction at the hydroxyl group with diethyl-aminoethyl chloride hydrochloride in the presence of caustic [223,224] or sodium methoxide [225]. The addition of cationic charges to the oxidized starch (aldehyde starch) is performed by reacting the aldehyde groups with melamine [226], with PAE resins [220], with unsymmetrical dimethyl hydrazines, with betaine hydrazide hydrochloride [227] or with Girard's reagent in order to attach a cationic charge [128].

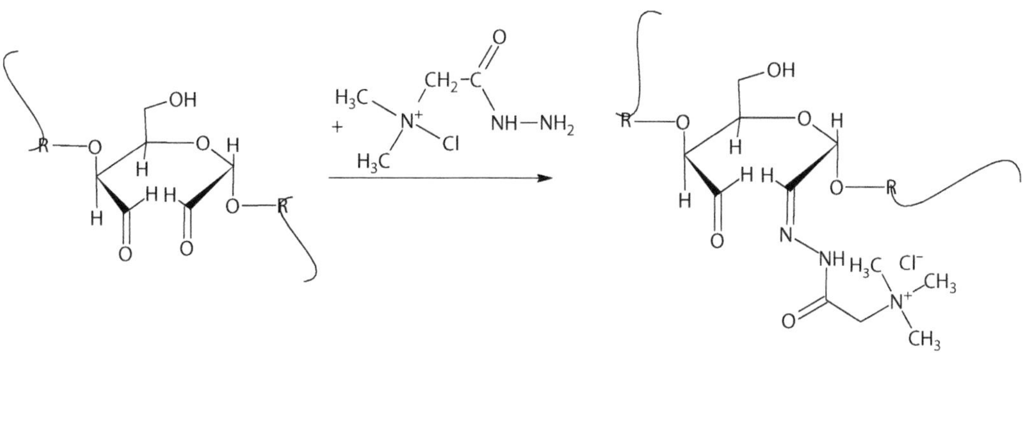

Anionic polysaccharides (such as xanthate derivatives) are changed into a cationic version through a reaction with PEI, which results in a starch polyethyleneiminothiourethane [229–231].

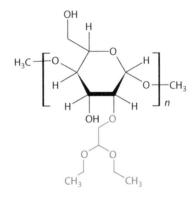

Starch xanthate also reacts with PAE [232]. It can be cross-linked (by oxidation [217] or with diepoxides [233]) for better wet-strength performance. Depending on the ratio between anionic groups (xanthate) and secondary amines, the resulting compound can be cationic or amphoteric.

The effect of the cationic di-aldehyde starch on the wet/dry ratio (Figure 5.5) is lower than that developed by PEI resin (Figure 5.4).

Aldehyde-containing polymers are not stable upon storage [195] because they can undergo cross-linking, oxidation [234], or a Canizzaro reaction [214] (see also glyoxalated polyacrylamide). In order to prevent any further reaction, the aldehyde group in the modified dextran or starch [235,236] is protected by acetalization [194].

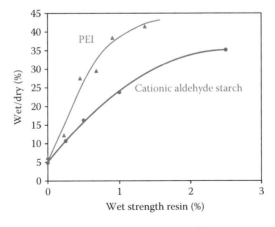

FIGURE 5.5   The effect of cationic di-aldehyde starch on the wet/dry ratio.

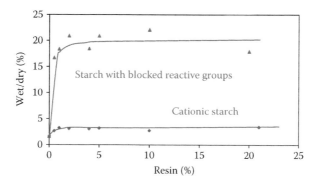

FIGURE 5.6   Cationic starch and reactive cationic starch as WSR.

Dextran acetal is converted to the corresponding aldehyde derivative by hydrolysis at pH 2.5–3 and 80°C–94°C for 20–120 min. The hydrolyzed polymer must be used immediately.

The paper wet tensile after treatment with aldehyde-containing dextran depends on the degree of substitution: a larger number of aldehyde groups per one macromolecule (higher degree of substitution) will provide a higher wet tensile. The covalent bond formation has an impact on the wet/dry strength ratio [212] (Figure 5.6): the reactive groups will make more covalent bonds, which will improve the paper wet strength.

The reactions performed to add reactive functional groups to starch can also be performed on cellulose (see Section 2.3). Functional cellulose fibers need lower amounts of paper chemicals. Oxidized pulp is self-cross-linkable and shows a better wet strength than regular pulp [237]. Di-aldehyde cellulose [237–240], obtained through cellulose oxidation, increases the paper wet strength from 5% to about 20% of the paper dry strength [241] and wet/dry ratio of 24% [242]. The presence of carboxylic groups along with aldehyde may have a catalytic effect on the hemiacetal bond formation.

Di-aldehyde cellulose is converted to cationic fibers by reaction with (carboxymethyl)trimethylammonium chloride hydrazide (Girard's reagent T). The paper wet strength is improved by adsorption of an anionic aldehyde starch on the cationic cellulose fibers [240].

For modified starch and cellulose with aldehyde groups, the curing mechanism involves hemiacetal formation within an equilibrium reaction [207] (see Chapter 4). Some other functionalities (aliphatic hydrocarbon tails or polysiloxane moiety) may be added to the cationic starch structure to make it a better wet-strength resin [243,244].

5.1.1.7   Polyisocyanates

A polyisocyanate has multiple isocyanate functional groups (hetero-cumulene double-bond system [245]).

The isocyanate group reacts easily with compounds containing "active hydrogen." The reaction is reversible and the equilibrium is moved in one direction or the other as a function of the active hydrogen compound and the resulting urethane.

Compounds containing active hydrogen are water, alcohols, phenols, acids, oximes (O–H bonds), mercaptans (S–H bond), amine (N–H bond), and malonates (C–H bond). The hydroxyl group in cellulose fibers is a candidate for that type of reaction [246–249]; polyisocyanate can develop a stronger interfiber bonding.

The main concern in this case is the side reaction of the isocyanate groups with water, which is present in large amounts around cellulose fibers [250]. Protecting their reactivity and preserving them for further reactions are key in the attempt to use polyisocyanates as WSR. As a general rule, aliphatic polyisocyanates are preferred insofar as prepolymers react more slowly with water than those prepared from aromatic polyisocyanates [251].

There are two different approaches to using polyisocyanates as WSR: the addition of a solution (or dispersion) of polyisocyanates as is (and lose an important amount of their functionalities by reacting with water), and the protection of the isocyanate group until the paper reaches the drying step.

5.1.1.7.1 Straight Polyisocyanates as Wet-Strength Resins

The use of the straight polyisocyanate for the cellulose fiber treatment involves the risk of losing some of the isocyante group in side reactions. There are strategies to limit those side reactions. Thus, the protection of the isocyanate group is ensured by incorporating it in a hydrophobic structure. Polyisocyanate was obtained by homopolymerization and copolymerization [252] with the styrene of 2-methacryloyloxyethyl isocyanate [246] in the absence of air and moisture (solvent tetrahydrofuran [THF]). In this case paper was soaked in the THF solution of those (co)polymers.

The infrared (IR) spectra of cured paper (at 100°C for 10 min) show some un-reacted isocyanate groups. This finding suggested that several isocyanate groups surrounded by hydrophobic phenyl groups (from styrene) are still available for further reactions, regardless of the residual moisture in paper. This is an indication of some sort of "physical protection" of the reactive groups by the hydrophobic parts of polyisocyanate. The reaction of isocyanates with water and the possible protection system are important for the commercial application of polyisocyantes as wet-strength resins.

The concept of straight "water-dispersible isocyanates" includes, after the contact with water, molecules with both reactive isocyanate groups and some urethane groups obtained by sacrificing the isocyanate group in the side reaction with water. The effectiveness of water-dispersible polyisocyanates is driven by the kinetic of reaction with water: how many isocyanate groups remain active in a certain amount of time?

The interaction between polyisocyanates and cellulose fibers depends on the size of the contact surface. For a large contact surface, the polyisocyanate must be dispersed in water with very small particle sizes. Aromatic di-isocyanates are emulsified in the presence of a polyvinyl pyrrolidinone (nonionic) type stabilizer [253]. A combination of nonionic (Propylene Oxide–Ethylene Oxide [PO–EO] copolymer) and anionic emulsifiers sodium lauryl sulfate (SLS) is also useful [254]. Those emulsifiers and stabilizers may create foaming issues and bring undesired chemicals on the fiber surface.

The preferred polyisocyanates are polymers having a hydrophobic part and a hydrophilic part that are able to self-emulsify. The hydrophobic part is the aliphatic or aromatic segment from polyisocyanate. In order to be self-retained on anionic cellulose fibers, the polyisocyanate must also have cationic charges. Both hydrophilic properties and cationic charges are obtained by converting some of the isocyanate groups [255–257]. Polyisocyanates partially react with hydrophilic polyethers

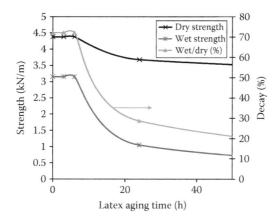

FIGURE 5.7   The strengthening properties of polyisocyanate dispersion versus aging time.

(mono-functional PEG) resulting in a water-dispersible polyisocyanate, which has a reduced number of isocyanate groups. In order to attach potential cationic groups to the hydrophilic polyisocyanate, other isocyanate groups react with the hydroxy-tertiary amine.

There is an unexpectedly low reaction rate for isocyanate with water. After storage of the aqueous emulsion for 6 h, the wet-strength action still reaches about 70% of the value on the immediate use of the emulsion. No significant loss in paper strength (dry or wet) occurs at up to 6 h of emulsion aging (Figure 5.7). The effectiveness of the emulsion diminishes in the first 24 h by 16% in terms of paper dry strength and by 67% in terms of wet strength [247].

The rate of disappearance of the isocyanate groups in emulsion depends on the hydrophobic character of polyisocyanate (Figure 5.8): the more hydrophilic it is, the higher the reaction rate with water [258]. For aromatic di-isocyanates the reaction rate with water is much slower [253].

Water-dispersible polyisocyanates have a stronger effect than the same concentration of PAE resin before curing (drying at 85°C for 8 min) [259]. After curing, the PAE resins are more effective than the polyisocyanate (Figure 5.9). Under those circumstances, a synergetic effect between PAE and polyisocyanates is not quite unexpected [259].

High molecular weight glycols react with an excess of di-isocyanates to obtain "reactive urethanes." "Reactive urethanes" are dispersed in water in the presence of an anionic emulsifier and the resulting emulsions are used to saturate the paper sheet [247]. There is a hydrophobic core with reactive isocyanate groups (isocyanuric trimer, in red) and a hydrophilic tail (in blue). Self-emulsified polyisocyanate (Bayer—ISOVIN) follows the same pattern [260].

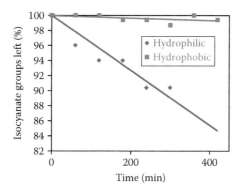

FIGURE 5.8   Chemical stability of polyisocyanate emulsion.

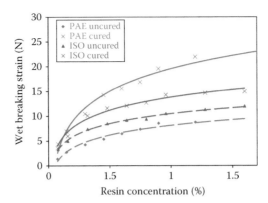

FIGURE 5.9   The effectiveness of polyisocyanate and PAE resins as WSR.

Difunctional polyether increases the molecular weight and functionality of reactive urethane. To make it cationic, an un-reacted isocyanate groups is converted to tertiary amine [261].

The new molecule has multiple isocyanate groups (in red) with an important hydrophilic part (in blue) and a potential cationic charge (in green). Because there is no differentiation between the reactivity of the isocyanate groups, the reaction with polyethers and hydroxylamine does not have any selectivity. As a result, a broad distribution of the chemical composition will be obtained: along with the target compound (cationic–hydrophilic polyisocyanate), compounds with no hydrophilic part and others with no cationic charges will coexist. For instance, the reaction of a di-isocyanate with a tri-ol (star-type polyether obtained from glycerol) [262,263] results in a higher molecular weight with an "average" composition.

The physical protection of the isocyanate groups can be better performed by encapsulation. The encapsulated material includes a polyisocyanate core confined to an envelope of protective substance which is inert to isocyanates and to aqueous media. One approach is to make a protective shell from deactivated polyisocyanate: a fraction of polyisocyanates react at the interface with polyamines (such as 4,4′-diamino-3,3′-dicyclohexylmethane [264]), polyamido-amines [265], and silanoates [266]. Encapsulation can also be performed on the blocked isocyanate [267]. The reaction is accompanied by protective film formation. Thus the progression of the isocyanate reaction with water is stopped and such systems have a longer shelf life (over 3 months) [268].

The protective shell can also be made from an inert material such as polystyrene. Naphthylene 1,5-di-isocyanate (average particle size of $5\,\mu$) is dispersed in a large volume of polystyrene solution in carbon tetrachloride (1.25%). A plasticizer for polystyrene, like chlorinated di-phenyl, can be added in order to choose the most convenient T_g for the protective shell. The suspension was atomized in a nitrogen current spray-dryer [269]. The average particle size was $30\,\mu$ and the di-isocyanate is well enveloped by the protective shell.

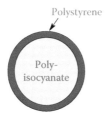

The partial consumption of the isocyanate groups will only slow down their reaction with water: the moment when the isocyanate conversion reaches 100% is only a matter of time. It is desirable that the polyisocyanate be released from the shell when it still has a functionality of at least 2. The protective shell can be destroyed by shear forces or through facilitating the diffusion of the poly-isocyanate through the poly-adduct layer or even through the dissolution of the poly-adduct layer by an appropriate solvent.

5.1.1.7.2 Blocked Polyisocyanates

The polyisocyanates presented in Chapter 4 have a relatively low molecular weight and their average functionality is about 3–4. Any reaction with water reduces the functionality down to a value where the cross-linking capabilities are almost lost. A higher number of reactive isocyanate groups can be incorporated into a polymeric structure. The synthesis and the handling of such polymeric polyisocyanates are more difficult. In order to make it easier to operate a polymeric polyisocyanate, the isocyanate groups are protected.

The protection of the isocyanate group converts the reactive groups to nonreactive functionalities (blocking reaction) [270–273]. A mono-functional blocking agent can be any compound capable of reacting with the isocyanate group so as to retard or prevent its reaction with water. The terms "blocked" and "latent" functionalities are interchangeable [274].

The sequence of the blocking process includes first the addition of the hydrophilic tail to a hydrophobic polyisocyanate. Then, the isocyanate-terminated prepolymers are blocked by heating with an excess (10%) of blocking agent, in dry toluene under nitrogen until the isocyanate IR absorption disappears. The solvent is removed under vacuum at temperatures below 50°C [275].

The blocking agents can be alcohols, phenols, amines, alkanol amines, or oximes [251,254,276–279]. The preferred blocking agents are of the lactame type, the oxime types and the hydroxamic acid ester, or the acyl hydroxamate type [280–282]. The blocking agent can also be difunctional and, in that case, a reaction with a difunctional isocyanate will result in a polymeric blocked poly-isocyanate [283]. The blocked polyisocyanate has "reactive" urethane type groups.

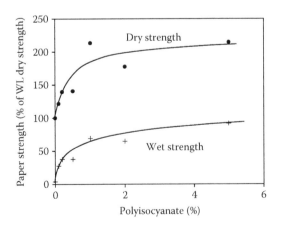

This type of reaction should be reversible; the blocking agent is released at higher temperatures [278] or replaced by a stronger hydrogen active compound (de-blocking). If the hydrogen-active compound has multiple functions, a cross-linking process can be developed. The curing by de-blocking of the blocked polyisocyantes with poly-nucleophiles depends on the chemical structure of the isocyanate, the blocking agent and nucleophile, the relative rate of reactions (nucleophile with the isocyanate compared to the reverse reaction rate of isocyanate with the blocking agent), the temperature, and the type and concentration of catalysts [271,277].

The cationic blocked polyisocyanate was tested in a hand sheet study [276]. For wet tensile, the samples were soaked in a solution (1%) of Aerosol OT (wetting agent) for 6h. Figure 5.10 shows that the strongest effect is obtained at very low concentrations of polyisocyanate. Blocked polyisocya-nate increases the WL dry strength up to 100%, which clearly separates them from other WSR. Wet strength is increased from less than 5% of the WL dry strength to about 90%.

The blocked isocyanate group can react with a hydroxyl group from a copolymer of 2-hydroxy-ethyl acrylate [284,285]. The polymer which has both reactive groups (blocked isocyanate and hydroxyl) is self-cross-linking [286]. Those two functionalities must be isolated from each other. The polymer structure should be rigid enough (such as solid particles) to prevent a reaction before the polymer is added to pulp). The emulsion copolymerization of butadiene, styrene, and amide-blocked isopropenyl-α-α-dimethyl benzyl isocyanate along with hydroxyethyl acrylate results in a reactive WSR [286–288].

FIGURE 5.10   Blocked isocyanate as wet-strength resin.

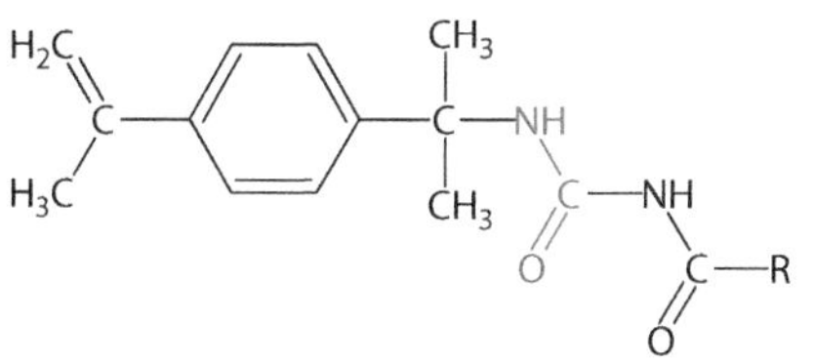

The de-blocking reaction (presumably during the paper drying) is the most important step when the strengthening capabilities of blocked isocyanate are evaluated. The thermal de-blocking step needs time and high temperatures: 15–60 min at 100°C–140°C [267,275]. The contact time in the drying step of the current papermaking process (<60 s at 105°C) is not sufficient to achieve quantitative de-blocking.

However, there is a broad range for the reactivity of blocked isocyanates. Blocked isocyanates with blocking agents as phenols, caprolactam, and alcohols were less reactive than those blocked with oximes, *N,N*-diethyl hydroxylamine, or hydroxypyridine. Prepolymers containing blocked aromatic isocyanates were more reactive than those containing blocked aliphatic isocyanates [275]. On the other hand, a "partial" de-blocking reaction may result in a reasonable increase in the paper wet strength.

The de-blocking process releases an organic compound into water, having no affinity for the cellulose and that may enter the white water of the paper machinery [261], which is another shortcoming of this method.

5.1.1.8 Polycarboxylic Acids

Polycarboxylic compounds, as potential WSR, deserve the attention of the scientist because carboxylic groups can react with the cellulosic hydroxyl (ester bond formation) and are stable in the presence of water.

A carboxylic functionality can be a pendant group attached either to a carbon–heteroatom backbone or to a carbon–carbon backbone (such as in acrylic acid copolymers [289,290]). Regardless of the backbone type, these compounds are located under the "carbon–heteroatom backbone" because they develop, after curing, an ester-type bond, which is breakable during the repulpable process.

The ester bonds between the cellulose and wet-strength resin will improve both its hydrophobicity and the paper strength. On the other hand, the un-reacted carboxyl group has a hydrophilic character, and thus it decreases wet strength by allowing more water around the cellulose fibers.

The following potential issues should be considered when a polycarboxylic compound is used as WSR: the anionic carboxylic groups alter the resin retention capabilities; the ester formation is an equilibrium reaction and water must be distilled to increase conversion; and esterification is usually performed at higher temperatures and in the presence of a catalyst.

Small molecules of polycarboxylic acids (such as 1,2,3,4-butantetracarboxylic acid [291], the succinic acid [292,293], the tricarballylic acid, the citric acid [294], and the poly-maleic acid with low molecular weight [295]) are very efficient as cross-linker agents for paper [295–298]. For citric acid the bridge is presented in the following figure (in red).

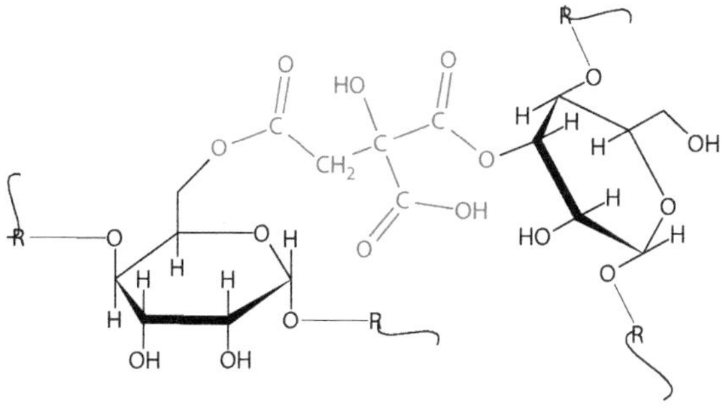

The distance between functional groups can be enlarged by reacting ethylene glycol (or polyethylene glycol) with trimellitic anhydride [299,300]; the resulting molecule still has a low molecular weight.

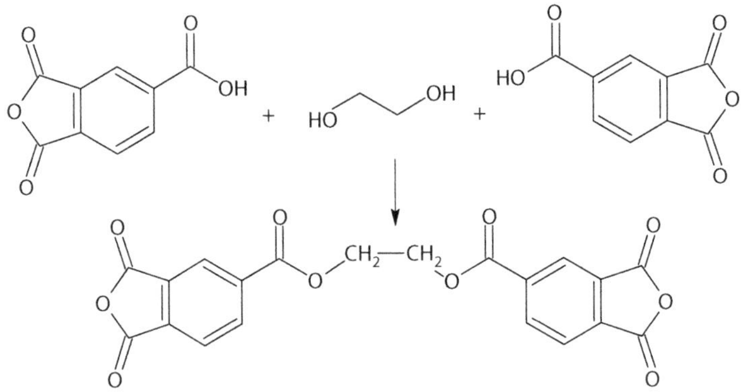

For those polycarboxylic compounds, the wet/dry ratio shows values ranging from 40% to 85%. The catalyst for the esterification reaction (such as sodium hypophosphite, NaH_2PO_2) [32,295,299,301] is also useful for the curing process.

The paper wet strength increases with the increasing content of poly-acids. Since ester linkages are stable in water and more hydrophobic, they prevent the swelling of the fibers and hold them together in the wet condition. Polycarboxylic compounds with medium molecular weights (less than 3000) are poly-maleic acid ($M_w = 800$ [190]) or copolymers of acrylic acid or maleic anhydride [32].

Polycarboxylic acid can be neatly obtained directly from cellulose fibers, by reaction with haloacetic acid [302,303] or by oxidation [304,305] (see also Section 2.3), or by methyl–vinyl ether copolymerization with maleic anhydride (1:1). That copolymer has a molecular weight of about one million [190]. A higher molecular weight of the maleic anhydride homopolymer ($M_w = 81,000$) is obtained at high temperatures (130°C–150°C) in the presence of large amounts of dibenzoyl peroxide [306].

Because the polycarboxylic compounds do not have cationic charges, they are added to the existing paper web [307] as water solution (carboxylalkyl cellulose [308]) or by printing with a selected pattern [32].

The ammonium salt of a carboxymethyl cellulose (CMC) (DS = 1.23) was used as a wet-strength resin [309] to treat paper. The wet strength of the final paper is a function of drying time and temperature. After 1 min soaking in water, the wet strength went to 62% from the initial dry strength (when paper was dried for 24 h at room temperature) to 82% when paper was dried at 280°F for 10 min, to 110% when paper was dried for 30 min, and 142% when it was dried for 2 h at the same temperature. The carboxylic groups of CMC or those developed on cellulose fibers can also be cross-linked with salts of polyvalent cations (zirconium, zinc, aluminum, etc.) with a view to improving the paper strength [305,310,311].

Mixed functionalities along with carboxylic groups are also useful in the cross-linking process. Poly-acids of the poly-maleamic type (half amide and half acid) are obtained from the copolymers of maleic anhydride [312] with ethylene, methyl–vinyl ether, etc. When treated with water solution of the copolymers dissolved in ammonia, paper needs 3 min curing time at 149°C.

High temperatures are always requested for the cellulose hydroxyl esterification [313]. The curing process based on the carboxyl–hydroxyl reaction asks for a higher temperature or a longer reaction time: at 250°C–320°C for less than 15 s [307], 90 s at 180°C [297], 2 min at 170°C [186,301] or 10–15 min at 130°C–150°C [289–291,296], or 20 min at 170°C [314].

The paper wet strength was measured after soaking in water for 12 h. After curing, paper (treated with 1,2,3,4-butanetetracarboxylic acid) shows an unchanged dry strength and a much higher wet

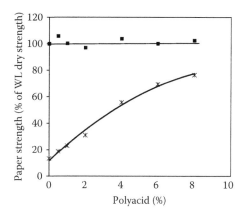

FIGURE 5.11   The effect of the 1,2,3,4-tetracarboxylic acid concentration on paper strength.

strength (Figure 5.11). The wet strength seems to be due to the ester bond formation: as the concentration of the carboxylic compound is higher, the concentration of ester carbonyl band is higher [301]. However, the significant effect was noted at high concentration of poly-acids. The covalent ester bond formation shows a differentiation in terms of dry strength properties from the urethane bond formation in the case of polyisocyanates as WSR (Figure 5.10).

In order to be used at the wet end, this strengthening material should be cationic (or amphoteric) and have a higher molecular weight. Water-soluble carboxymethyl hydroxyethyl celluloses [303] have both functionalities (hydroxyl and carboxyl) and can be applied either at the dry or the wet end (in the presence of alum).

5.1.1.9   Polyethers

Polyethers are synthesized by the ionic polymerization of cyclic ethers. Cationic polymerization is performed with boron trifluoride-etherate and ethylene glycol. If the di-ol is replaced by polyethylene oxide, a tri-block copolymer can be obtained. The solely polyether backbone has a nonreactive structure.

In order to perform as wet-strengthening agents, polyethers must have reactive functionalities as pendant groups. Cationic polymerization is very sensitive to other reactive groups (such as amine, hydroxyl, carboxyl, etc.), and therefore the cyclic ether structure must not include those functionalities. Other moieties, able to be reacted within a further polymer-analogous reaction, are incorporated in the cyclic ether structure. Epichlorohydrin fits that requirement [315,316]. Epichlorohydrin is (co)polymerized with boron tri-fluoride; the (co)polymer has pendant chlorine groups. After an amination (reaction with ethyl amine), a poly-secondary amine is obtained. During the last step, the secondary amine is reacted with epichlorohydrin to form an azetidinium cycle.

The curing mechanism depends on the chemical structure of the pendant group. Allylic pendant groups on a polyether backbone are obtained by cationic polymerization of allyl–glycidyl ether [317–322]. For allylic double bonds, the curing mechanism is based on an auto-oxidation process.

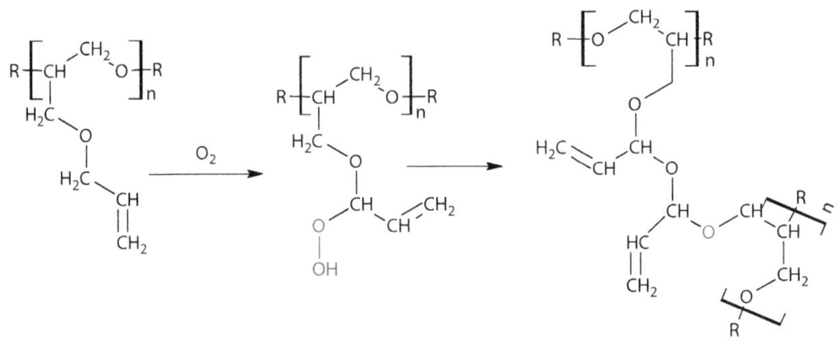

When the oxidation process has cross-linked the polymer, the paper wet strength reaches up to 50% of its dry strength. However, there are several shortcomings of this process: the final polymer cannot be added at the wet end (it is nonionic), the auto-oxidation process is slow, and the amount of resin added to paper is unusually high (20%).

5.1.2 BACKBONE WITH CARBON–CARBON BONDS ONLY

The carbon–carbon backbones are made mainly through condensation (phenolic rings are bonded through methylene groups [323], but that topic is not covered in this chapter) or though free radical (co)polymerization, a kinetically controlled process. The number of cationic charges, of reactive groups and their distribution within the macromolecular structure can be designed by choosing the comonomers molar ratio and the order of addition.

The simplest structure for carbon–carbon backbone is polyethylene, which is hydrophobic and not soluble in water. Some hydrophilic pendant groups are required in order to increase the polymer solubility in water. There are several potentially hydrophilic groups: amide, carboxyl (polyacrylic acid [298], hydroxyl, amine, and all the anionic and cationic functionalities known in organic chemistry. Those functionalities must provide not only the hydrophilic character, but also the reactivity during the curing step.

Some reactive and hydrophilic functionalities (such as amines, quinines, thiols, aldehydes, etc.) may interfere with the free radical process and must be avoided, which is why the monomers, initiators, and solvents should be carefully chosen. Any interference with the polymerization process reduces either the polymer molecular weight or the reaction rate.

The WSR can be synthesized in one or two steps. The one-step polymerization involves reactive monomers carrying functional groups and ionic charges. The two-step option includes the synthesis of a nonreactive (and/or nonionic) copolymer in the first step, and in a second step, the copolymer is modified by addition of cationic charges and/or reactive functionalities through polymer-analogous reactions.

5.1.2.1 Homopolymers as Wet-Strength Resins

Acrylic [324–328] or vinyl monomers [329] undergo faster polymerizations that make wet-strength resins in a one-step process. Acrylates or methacrylates have incorporated azetidine or azetidinium functions as pendant groups. The azetidinium cycle is both cationic and reactive.

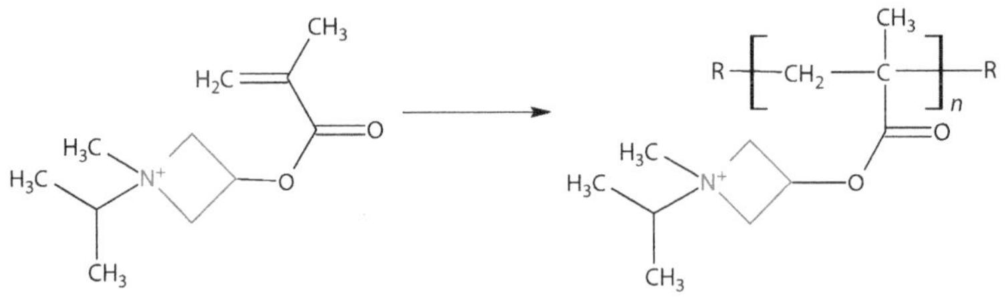

The polymerization was performed in water as a solvent (1 mol/L) at 40°C and the initiation was made by potassium persulfate (0.5 mol %). After a 12 h reaction time, the conversion was 100%. 3-(1-Cyclohexyl) azetidinyl methacrylate was polymerized at 70°C in 1,4-dioxane as solvent, in the presence of a free radical initiator [330]. The homopolymer can be conserved in dry form without showing any tendency to cross-link at room temperature [324].

Vinyl benzyl dimethyl ammonium chloride is polymerized in water [329,331] with ammonium persulfate. To make a WSR, in a second step, the poly-tertiary amine is partially reacted with epichlorohydrin (pH is 7.5) to obtain a quaternary ammonium salt with reactive epoxy functionality.

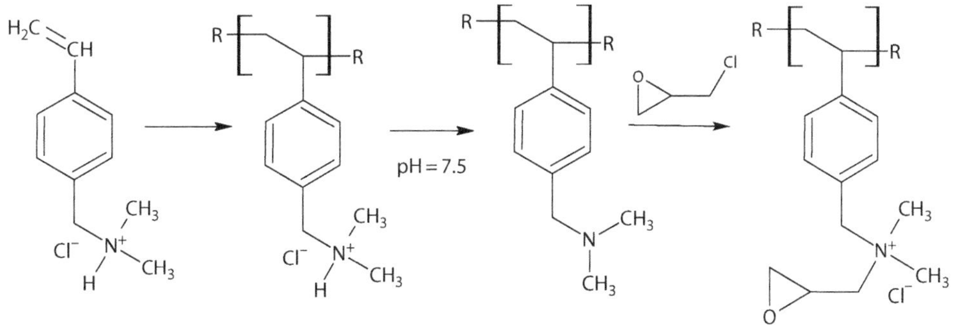

The aldehyde functionality is well known as a reactive group in the reaction with hydroxyl groups of cellulose fibers. Acrylic aldehyde (acrolein) is the most known aldehyde-type monomer [332]. Acrolein homo-polymerizes in the presence of free radical initiators [333] and the resulting polymer reacts with bi-sulfite. The soluble form (bi-sulfite adduct) is anionic. Cationic starch was used to retain it on the cellulose fibers [332].

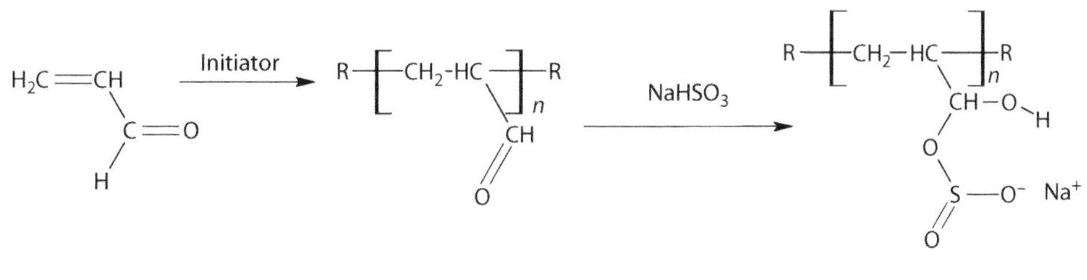

The simplest amine monomer, vinyl amine, is thermodynamically and kinetically unstable [334]. Allyl-amines show important chain transfers during free radical polymerization and produce mainly low molecular weight polymers and copolymers [335]. In order to obtain a carbon–carbon backbone, a different type of polymerization should be used (such as cyclo-polymerization) or a different structure of the initial monomer in which the amine group should be protected as a less reactive function (such as amide in N-vinyl formamide).

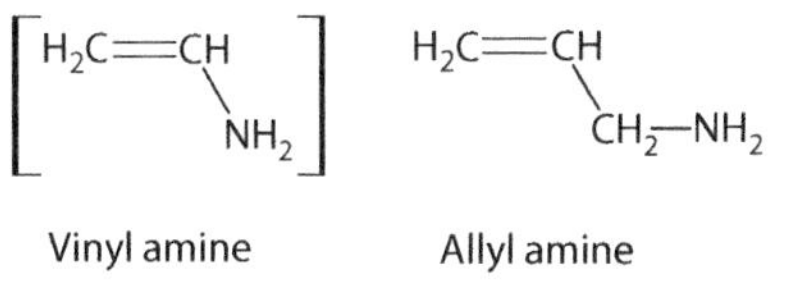

Vinyl amine   Allyl amine

To protect the amine group that can interfere with the initiator (oxidant), the free radical cyclo-polymerization of di-allyl amine should be carried out only after amine neutralization with hydrochloric acid [336–339].

Cyclo-polymerization (an internal copolymerization) results in a carbon–carbon backbone with a five-member cycle included in the propagation step. In other words, regardless of its functionality (a di-allyl monomer has two double bonds), a linear polymer is obtained.

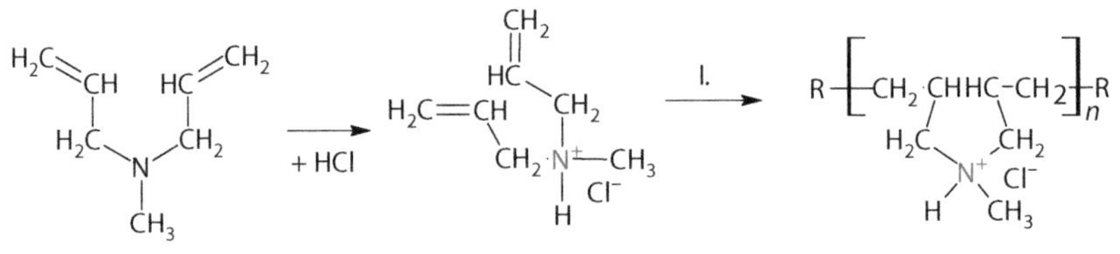

Allyl monomers are much less reactive than vinyl monomers and special initiators are required [340–342]. Anyway, the polymerization is slower and, for a reasonable reaction time, the conversion reaches a plateau at about 85% [336].

The cyclo-polymer and the un-reacted monomers go to the quaternization reaction step with epichlorohydrin [337,343–347]. Careful polymer purification is mandatory before quaternization (pH must be changed in the alkaline side and the water–monomer azeotropic mixture must be distilled). For quaternization, the preferred amount of epichlorohydrin ranges from about 1 mol to about 1.5 mol per mole of tertiary amine. Polymers of the N,N-diallyl amine quaternized with epichlorohydrin provide wet strength and a higher dry strength [344].

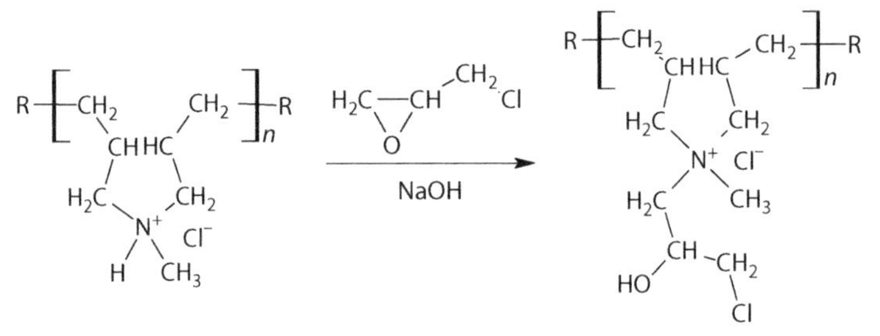

Allyl chloride, which is a precursor of epichlorohydrin, can also be used for quaternization [348–351] (the excess of allyl chloride is 1.2–1.3 mol based on the equivalent nitrogen).

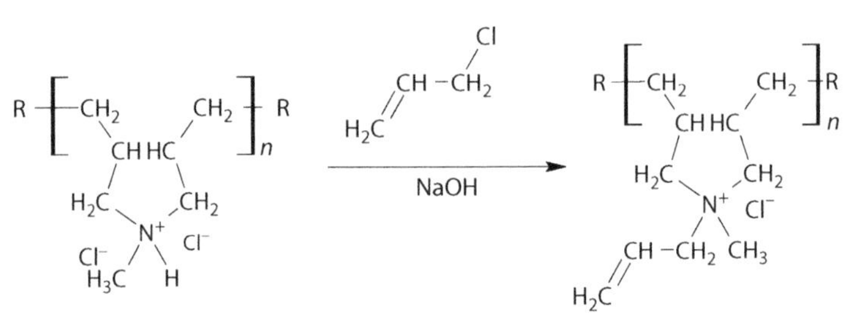

The quaternization reaction can be performed on the monomer before the polymerization: for instance, C-vinyl pyridine is quaternized with allyl chloride [351].

The amine functionality may be protected as amide and, after polymerization the polymer with amide as pendant groups is hydrolyzed to a polyamine. N-vinylformamide (or N-vinyl acetamide [352,353], or N-vinyl phthalimide [62]) is polymerized in water (solution, suspension [354], or in inverse emulsion polymerization [355,356]), with regular free radical initiators [357], such as cationic azo-derivatives [339]. The monomer purity is a key parameter for vinyl–amide polymerization [353,358].

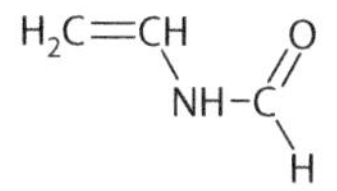

N-vinyl formamide 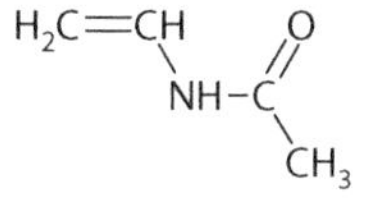   N-vinyl acetamide 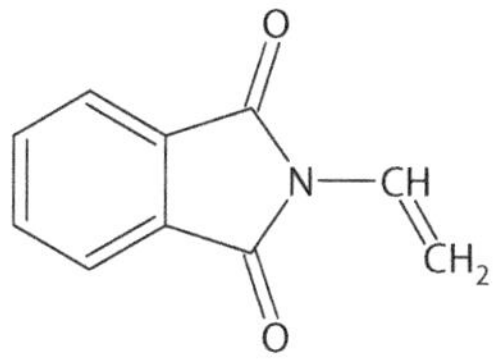   N-vinyl phthalimide

N-vinyl formamide has attractive physical and toxicological properties [359]: boiling point: 84°C at 10 mm Hg, melting point: 16°C, density: 1.014 g/cc, flash point: 102°C, viscosity: 4 cPs at 25°C, surface tension: 36.2 dyn/cm at 21°C, heat of polymerization: 19 kcal/mol, mutagenicity testing: negative, acute oral LD_{50} (rat): 1,444 mg/kg, dermal LD_{50}: >2000 mg/kg.

The resulting polymer is hydrolyzed to obtain the corresponding polyamine [360,361] in a slower reaction [362]. The hydrolysis is performed at temperatures ranging from 70°C to 90°C, in the presence of hydrochloric acid (1:1 mol).

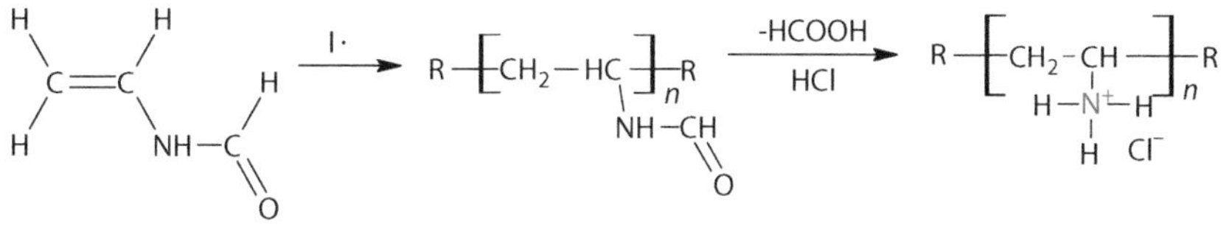

PVAm is a poly-primary amine. Because of the nearest neighbor effects, the PVAm is a weaker base than a primary amine with a small molecule [363]. It is assumed that under normal papermaking conditions (6.5 ≤ pH ≤ 7.0), the degree of protonation is 0.5 and because of its high cationic charge content, PVAm readily adsorbs onto fibers. The ability to change the reactivity and properties of amine groups by a simple change of pH provides numerous valuable options for viscosity control, polymer solubility, or formulating self-stable polymers (and making them reactive at another pH) [334].

Cross-linked polymers, recommended as wet-strength resins, are obtained from the polyamine-type polymers (such as PVAm) with polyphenols in the presence of an oxidizing agent [364].

A cationic resin with a right ratio between its hydrophilic and hydrophobic segments is a good WSR even if reactive groups are missing. Hexahydro-1,3,5-triacrylyl-s-triazine is condensed with 1–2 mol of fatty amine and protonated with acetic acid. The acrylic double bonds are polymerized through free radicals [365]; the new polycationic resin shows a good wet/*dry* strength ratio.

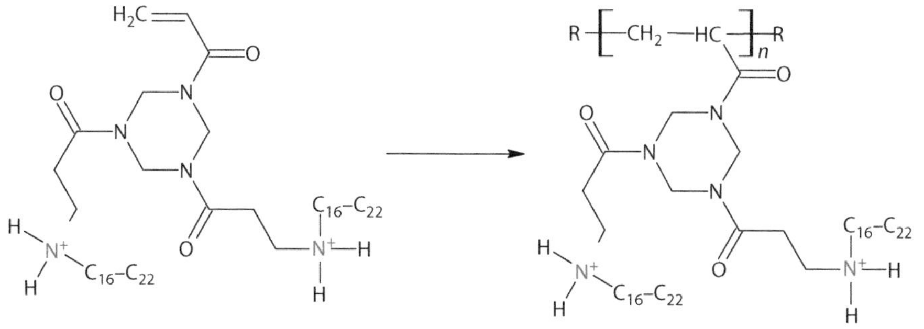

Homopolymers have a very high concentration of function groups (such as the homopolymer of N-(β-acrylamidoethyl)-ethyleneurea [366], or ureidoethyl vinyl ether [367]). It is also difficult to solve solubility, cationicity, and reactivity problems in one reaction step (homopolymerization). Some homopolymers were partly modified (such as the partial hydrolysis of polyvinyl formamide);

the resulting compound was a copolymer. That is why the involvement of a larger number of comonomers in the same process (copolymerization) opens the opportunity to make the right WSR in a one-step process.

5.1.2.2 Copolymers as Wet-Strength Resins

Two or more comonomers are involved in a copolymerization reaction. The simplest copolymerization is binary. It involves two comonomers and is governed by specific composition equations (Equation 5.3) concerning the relationship between the final copolymer composition (m_i), the initial monomer composition (M_i), and the reactivity of the comonomer (r_i) [368].

$$\frac{m_1}{m_2} = \frac{M_1}{M_2} \frac{r_1 + \dfrac{M_2}{M_1}}{1 + r_2 \dfrac{M_2}{M_1}} \tag{5.3}$$

For a larger number of comonomers, copolymerization equations are more complicated, but the process is not difficult to control [369].

The large diversity of comonomers generates, after copolymerization, an even larger diversity of copolymers. Their properties depend on the chemical structure of the monomers, and the ratio and distribution of comonomers. Thus, those copolymers can be ionic or nonionic, hydrophilic or nonsoluble in water, reactive or nonreactive. For instance, the acrylic copolymers with tertiary amine groups and a hydrophobic comonomer (ethyl acrylate) are dispersions in alkaline pH but become water soluble in acid pH [370]).

If the cationic comonomer is not a quaternary ammonium salt, the free radical copolymerization should be performed with special precautions. The acrylamide–dimethylaminomethyl methacrylates copolymerization [371] is performed in a mixture of solvents (water–acetone) and the pH is adjusted to 4. At that pH the tertiary amine is protonated and cannot interfere with the initiator (potassium persulfate). The copolymer of acrylamide and 20% 2-aminoethyl methacrylate hydrochloride [326] is a cationic polyamine able to make ionic bonds with anionic cellulose fibers.

Copolymers of acrylonitrile with 2-vinyl pyridine [372], of acrylamide with dimethyl-2-methylen-3-butenyl sulfonium chloride [373], or with trimethyl (2-methylene-3-butenyl) ammonium chloride [374] are wet-strength resins for paper.

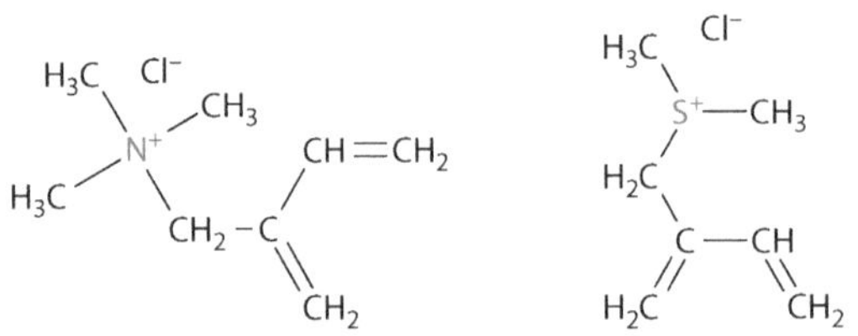

All those cationic copolymers are used as wet-strength resins even when functionalities able to develop covalent bonds with cellulose are missing. However, the ionic bonds are important. In order to develop stronger inter- and intramolecular ionic bonds, dimethylaminoethyl methacrylate (or its copolymers) is reacted with epichlorohydrin in acid pH to form a water-soluble cationic comonomer.

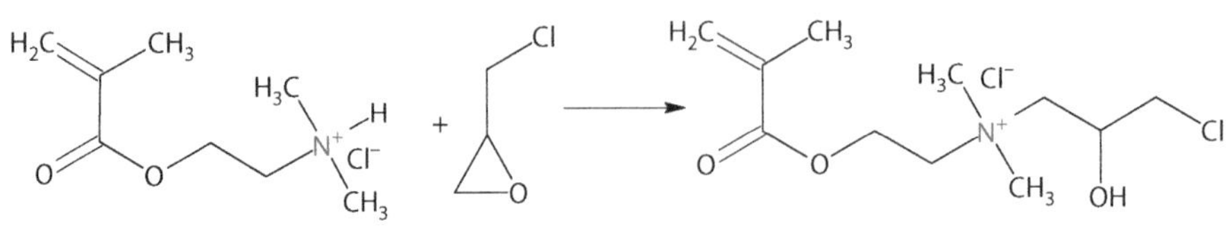

Copolymers of that cationic methacrylate with ethyl acrylate are partly hydrolyzed to obtain an amphoteric wet-strength resin, which performs better than PAE resin at the same dosage [327].

The copolymerization process is flexible and offers the possibility to combine, in the same macromolecule, cationic charges and reactive groups. The reactive group can preexist in the acrylic monomer (as in 2-(1-aziridinyl ethyl) methacrylate [375]), or can be developed into the final copolymer as in the copolymer of acrylamide with aminoethyl methacrylate (quaternized with epichlorohydrin), which is converted to a thermosetting resin in alkaline pH (epoxy ring formation) and used as a WSR [371].

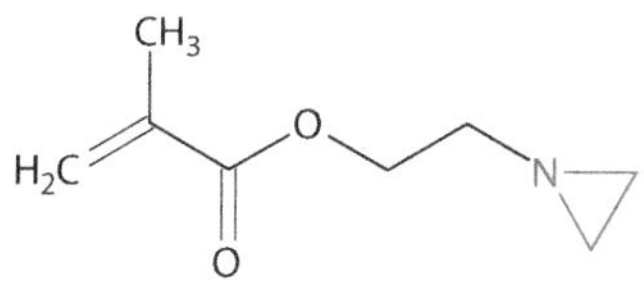

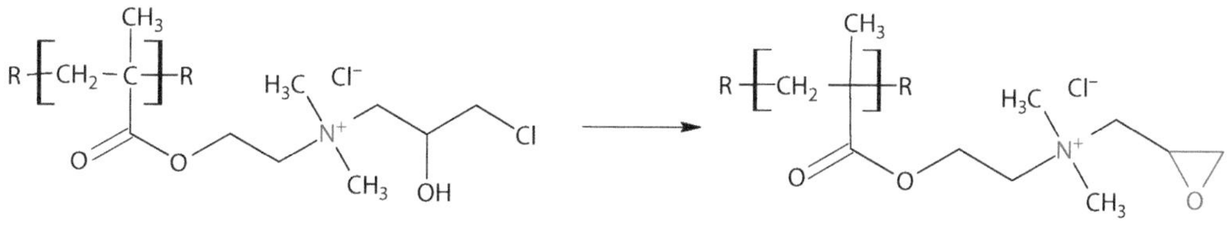

Different reactive groups can coexist in the same molecule. For instance, maleic acid and di-allyl amines can copolymerize to obtain an amphoteric copolymer [376] that is self-cross-linkable at higher temperatures.

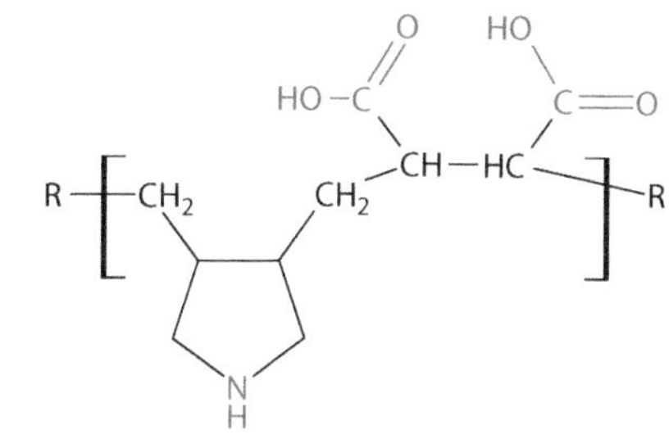

Reactive comonomers (such as vinyl isothiocyanate [377] or 2-methylene-3-butenyl isothiocyanate [378]) are copolymerized with acrylamide (12:88 weight ratio) and the resulting copolymer is self-cross-linkable in water solution in the presence of polyamines.

Acrolein is copolymerized with acrylonitrile and/or acrylic acid [379] or with 2-hydroxy-3-methacryloyloxy-propyl trimethyl ammonium chloride to make a wet-strength resin [380]. The aldehyde functionality can be blocked as acetal (such as acrylamidoacetaldehyde dimethyl acetal [19,381,382]) and the copolymers could still be reactive toward the hydroxyl groups.

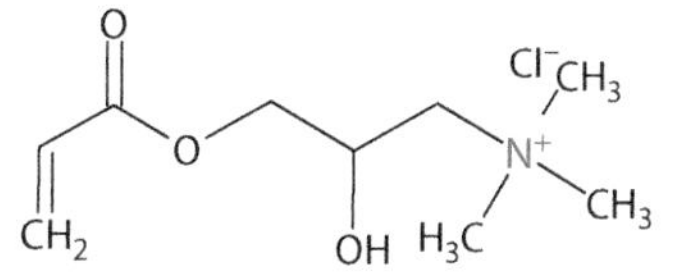

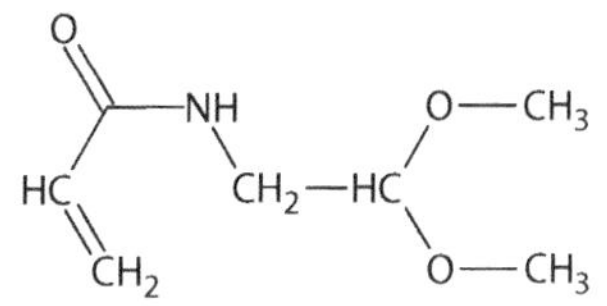

The copolymerization of acrylamide (cyclo-copolymerization with di-allyl amine comonomers) can be performed through a free-radical mechanism [326,337,343,344,383], where $n \gg m$. The copolymer is already a cationic polyelectrolyte (see also Section 3.3 for PAAm as retention aid). Quaternary ammonium salts with epichlorohydrin are obtained in the same way as poly-diallyl amines (see figure).

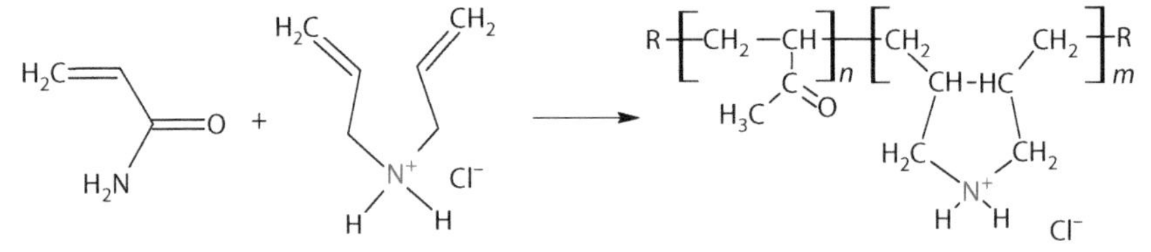

N,N-diallyl-3-hydroxyazetidinium chloride is a cationic monomer with a reactive azetidinium ring [384,385]. Its synthesis uses epichlorohydrin but the monomer is purified by vacuum stripping or distillation to substantially reduce the level of these by-products (chloropropandiol (CPD) and dichloropropanol (DCP), see Section 5.3.4).

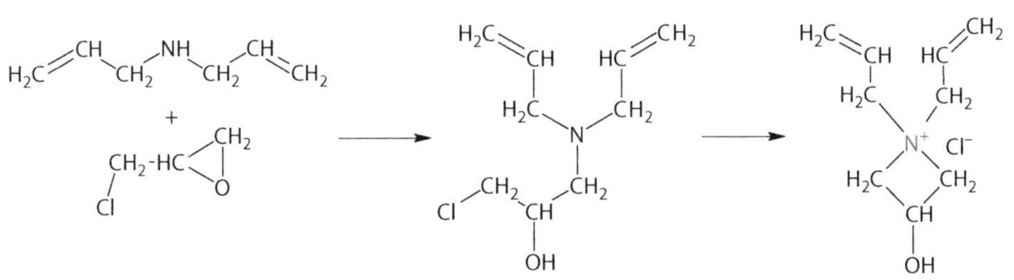

After the copolymerization of the purified new monomer, no by-products are present in the WSR solution. *N,N*-diallyl-3-hydroxyazetidinium chloride is involved in copolymerization reactions with comonomers such as acrylamide, *N*-vinyl formamide and *N*-vinyl 2-pyrrolidinone [384,385]. The total concentration of CPD and DCP falls below 200 ppm based on the polymer dry weight. The same type of comonomer is 1-diallylamino-2,3-epoxypropane, which copolymerizes with styrene [386].

The option of modifying the functional monomers before (co)polymerization was used for several other compounds. For instance, the reaction of *N*-vinylformamide—at the unusually acidic NH group—with isocyanate gives vinylacylureas. Methyl acrylates and acrylonitrile react with *N*-vinyl formamide according to the Michael addition to the same nitrogen atom [359].

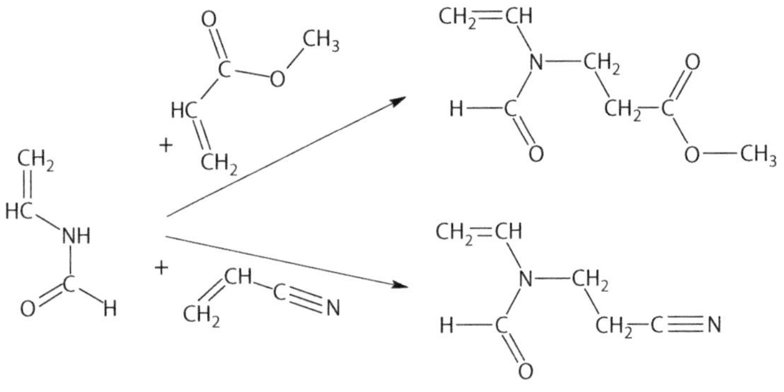

The new type of vinyl–amide monomers copolymerize and, after hydrolysis, a more hydrophobic poly-secondary amine is obtained.

The WSR synthesis through binary copolymerization is limited to a cationic comonomer and a reactive compound. However, a concentration of 5%–20% reactive groups per macromolecule is enough for the cross-linking step and the cationic charges, (needed for retention) must also be in a limited number. Therefore a third comonomer (nonionic, nonreactive, and less expensive) should be incorporated in a ternary copolymer. That ternary copolymer is obtained either through ternary copolymerization or through a polymer-analogous reaction on a binary copolymer (see Section 5.1.3).

A typical terpolymer used as wet-strength resin is the copolymer of acrylamide (nonionic)—N-methylol acrylamide (nonionic reactive)—dimethylaminoethyl acrylate (quaternized with dimethyl sulfate) which shows a wet-strength capability similar to the PAE resin [387]. The new WSR is soluble in water.

If the third comonomer (nonionic and nonreactive) is hydrophobic, the final copolymer can be either water soluble or an emulsion. This is the case for the ternary copolymer 2-hydroxy-3-methacryloyloxy propyltrimethylammonium chloride/acrolein/stearyl methacrylate (or acrylonitrile, or vinyl acetate) as wet-strength resins [388].

The (co)polymers with a carbon–carbon backbone and the possibility to form covalent bonds with cellulose fibers raise the concern about paper repulpability: a carbon–carbon bond is much stronger than an amide bond in PAE resins.

5.1.3 POLYMER-ANALOGOUS REACTIONS

Homopolymers of the common monomers (such as acrylamide) are easy to obtain, and a nonionic polymer can be converted to a wet-strength resin through a polymer-analogous reaction (see Section 2.4.1). The polymer-analogous reactions are selective and may add ionic charges and reactive functionalities to a preexisting backbone. The organic chemistry of high polymers is focused on changing different functionalities without any impact on the molecular weight or on the molecular weight distribution [389,390].

The amide group in polyacrylamide is converted to an amine functionality through a Hofmann degradation with sodium hypochlorite [391–393] (see Section 2.4.1). If the reaction conversion is lower than 100%, the final product is a binary copolymer of acrylamide and vinyl amine.

Polyacrylamide [394] ($M_w = 20,000$) undergoes a trans-amidation performed at 95°C–105°C. The amine must be difunctional (such as ethylene diamine) or polyamine (PEI [151]). They can carry a cationic charge (3-dimethylamino-propylamine) [395] or a hydrophobic tail [396,397]. For an excess of about 10:1 amine (ethylene diamine) to amide groups, only about 30% of amide groups are converted to amine pendant functionalities.

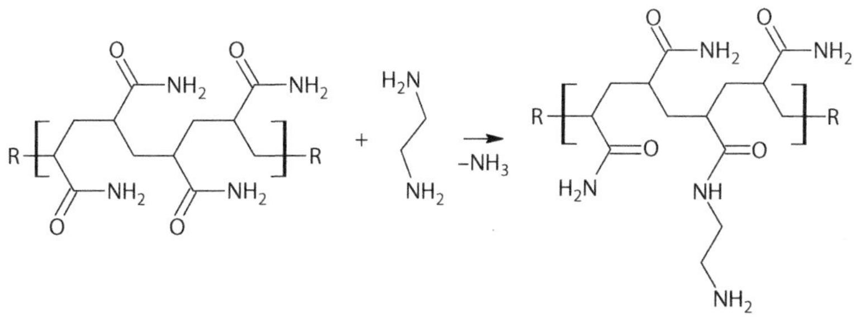

The new copolymer is a polyamidoamine having a carbon–carbon backbone. The amine will provide cationic charges by protonation in neutral or acid pH. Epichlorohydrin is used in excess to build up possible azetidinium functionalities, similar to a PAE resin.

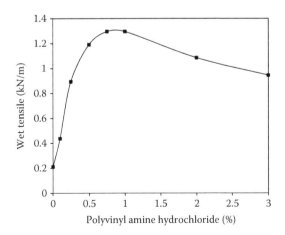

Cationic polyacrylamide can be functionalized with difunctional compounds such as di-alde-hyde (glyoxal) [398,399] and glycidyaldehyde [400] (obtained through the oxidation of acrolein with hydrogen peroxide).

Difunctional compounds, such as 3-glycidoxy-propyl trimethoxysilane [401], react with the polyamidoamine backbone to make a cross-linkable WSR.

Only 15%–50% of the secondary amines are converted during that reaction. Residual second-ary amines are used to add cationic charges through a reaction with glycidyltrimethyl ammonium chloride.

Polyvinyl amine hydrochloride, obtained through the hydrolysis of all the amide groups of poly-vinyl formamide (Section 5.1.2.1), is an attractive wet-strength resin at [62] (Figure 5.12).

FIGURE 5.12   The effect of PVAm concentration on the paper wet strength.

Paper treated with PVAm has a pronounced tendency to yellow [402]. In order to discontinue the colored compound formation, vinyl formamide copolymerization is a better choice. Actually, the synthesis of a WSR from PVAm involves three steps: the copolymerization of N-vinyl formamide, the hydrolysis of the copolymer (polymer-analogous reaction), and the cross-linking process on cellulose fibers.

The copolymers of N-vinylformamide [354] or foramidoethyl vinyl ether [403] are obtained either through the copolymerization or the partial hydrolysis (20%–90%) of polyvinyl formamide. The binary copolymer contains pendant amine and amide groups. Ternary copolymers with vinyl acetate, N-vinyl pyrrolidone, and acrylonitrile are an even better solution to the yellowish problem [339,402,404,405]. The whiteness made with this type of copolymer is identical to that obtained with PAE-type WSR.

N-vinyl formamide has a strong tendency to copolymerize (see the reactivity ratio values in Table 5.1, $r_{NVF} < 1.0$) [404–407]. However, the copolymerization of N-vinyl formamide [359] is not easy: the process implies a complicated addition procedure for the monomers and the initiator. Different reactivities in the copolymerization and hydrolysis processes are recorded when N-vinyl-tert-butyl carbamate is used instead of N-vinyl formamide [408].

When the comonomer is also a functional monomer (like maleic anhydride [359,409]) the process is even more difficult. However, reactive comonomers (such as boronic acid derivatives [410]) will react as cross-linkers for the cellulose fiber, starch, or guar [410,411] (see Section 4.2.1.).

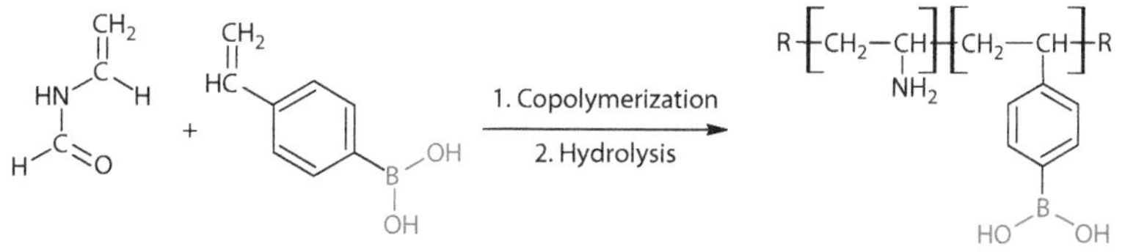

Polymers and copolymers of N-vinyl formamide grafted on a water-soluble polymer (such as polysaccharide [412], polyvinyl alcohol [413], or polyethylene glycol [414]) show good performances as wet-strength resins.

The synthesis of the NVF copolymers follows the same pattern as in the polyvinyl acetate (PVA) technology. Moreover, the most common comonomer is vinyl acetate. The base hydrolysis of the VAc-NVF copolymer almost favors ester cleavage, but in acid pH amid hydrolysis takes place alongside partial ester hydrolysis [334,354,415]. N-vinyl formamide copolymers with acrylamide [416], acrylate [417], and acrylonitrile [418,419] are also hydrolyzed in acid pH.

Copolymers N-vinyl acetamide–vinyl acetate show different T_g values as a function of the comonomer ratio in the final copolymer [358] and their hydrolysis requires much milder conditions than those used for the N-vinyl acetamide homopolymer [353]. The concentration of the vinyl–amide in those copolymers and the degree of hydrolysis generate multiple chemical microstructures. The

TABLE 5.1

Reactivity Ratios for Vinyl Formamide (M_1) Copolymerization

Comonomer (M_2)	r_{NVF}	r_2
Maleic anhydride	0.028–0.05	0.01–0.02
Vinyl acetate	4.62	0.31
Acrylamide	0.05	0.52
Sodium acrylate	0.25	0.58
n-butyl acrylate	0.06	0.54

vinyl–amide content is recommended to be over 1% but less than 15%. The vinyl acetate units are hydrolyzed in two steps: by alkali saponification using 0.1 mol Na methoxide in methanol (step one) and by concentrated hydrochloric acid at 90°C (step two), at over 75% while the amide hydrolysis is about 25% [420]. The final quaternary copolymer has hydrophilic units (in blue), hydrophobic units (in black), and potential cationic charges attached to the amine group (in red). That structure combines two different CCDs: the distribution generated during copolymerization, and the distribution generated during hydrolysis.

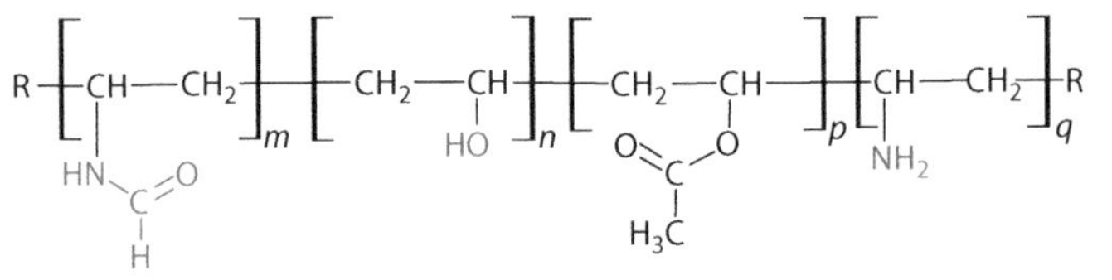

After the copolymer hydrolysis, the copolymerization of vinyl acetate with *N*-allyl formamide (or *N,N*-diallyl formamide made by trans-amidation of ethyl formate with the corresponding amine) [421] allows for the incorporation of primary amines, a cyclic secondary amine, and a hydroxyl group.

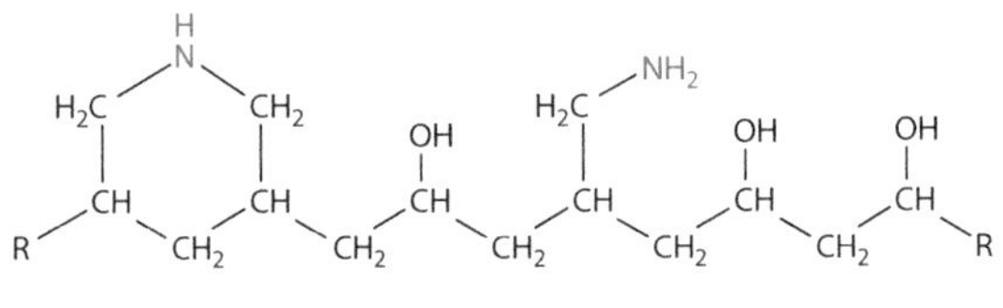

Polyvinyl amines undergo the well-known reactions of amines: acylation, alkylation, arylation, diazotation, oxidation, Michael addition to double bonds, etc. [422]. Most of these reactions which involve a macromolecular compound are quite different from those on small molecules [389]. When the polyamine is reacted with a difunctional compound (such as butandiol diacrylate, epichlorohydrin [423], or glyoxal [423,424]), the polymer-analogous reaction generates a cross-linked material even at room temperature.

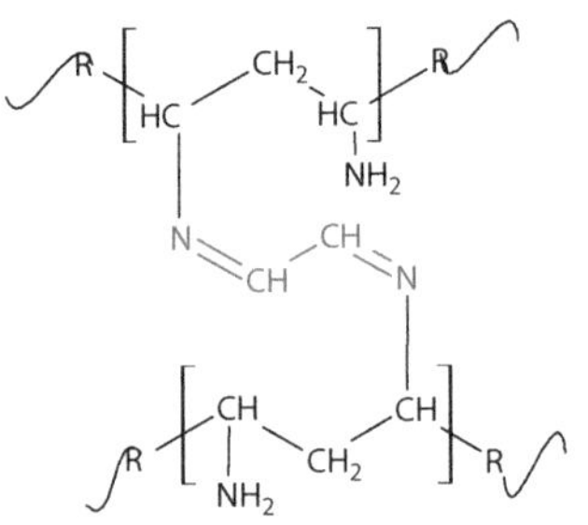

Aldehydes with only one functionality (such as butyraldehyde) react with vinyl amine–vinyl alcohol copolymer (6%/94%) in water solution to form a polyaminal [425,426].

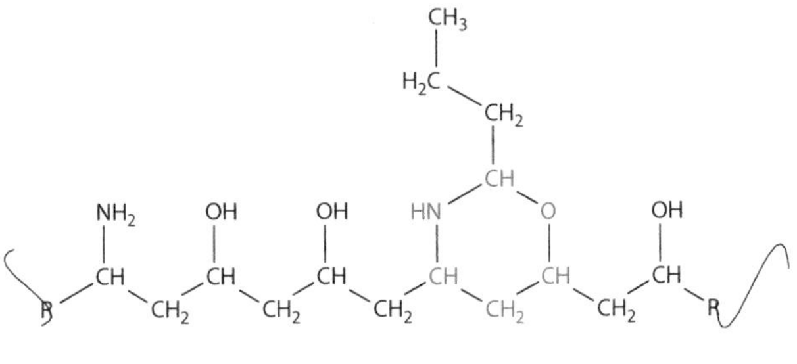

The long hydrocarbon tails [427] make the copolymer more hydrophobic. The hydrophobic part can also be added through a reaction with epoxides [428,429], with alkenyl ketene dimer (AKD) and/or alkenyl succinic anhydride (ASA) [177,430]. It seems that a synergetic effect is developed when the PVA–PVAmine is modified with AKD [431] because the polyvinyl alcohol (PVA)/AKD blend does not show any improvement in wet strength.

Maleic anhydride copolymers (with isobutylene, diisobutylene, styrene, and propylene) are very attractive as potential backbones for many reactions on the anhydride functionality. Thus the esterification with diethylaminoethanol [432] or the imidization [433] of the maleic unit with a difunctional amine (such as dimethyl aminopropylamine [434]) converts the polymer to a hydrophobic backbone with cationic charges. In a subsequent step, the tertiary amine is quaternized (with epichlorohydrin or any other alkylation agent [435]).

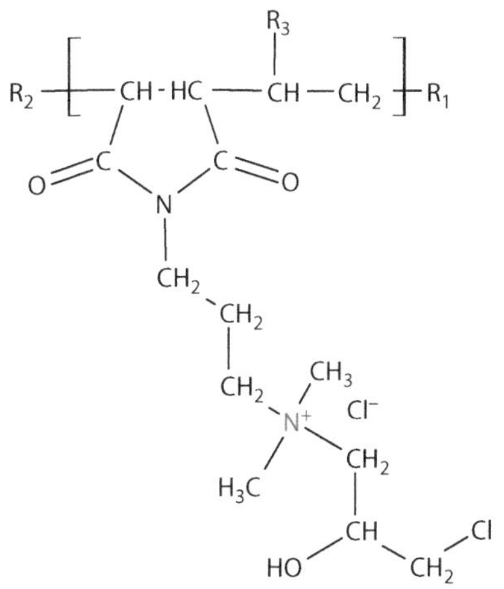

The grafting reaction may change the molecular weight of the backbone [436] but a hydrophilic macromolecular compound (such as polyethylene oxide) may be converted to a functional polymer. For instance, polyethylene oxide (molecular weight about 100,000) is reacted with vinyl functional monomers such as methacryloxypropyl trimethoxy silane in the presence of a free radical initiator [437].

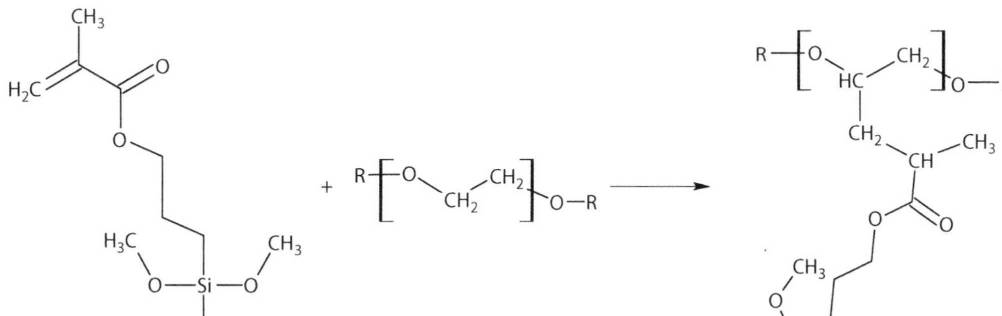

The reaction mechanism involves a chain transfer to PEO and the number of monomer molecules attached to the PEO macromolecule depends on the reaction parameters: initiator concentration, mole ratio between backbone and monomers, order of addition, etc. The final functional polymer is a WSR that still needs cationic charges for its retention at the wet end.

5.1.4 Polymer Latexes

Water-soluble polymers and copolymers are particularly effective because their cationic charges and reactive groups are wide open for interactions with cellulose fibers. Surprisingly, dispersions of polymers are also good WSR. Polymer particles may have ionic charges and reactive groups, but the interaction with the fibers is possible only for those groups that are distributed on the particle surface.

Polymers and copolymer latexes are obtained directly by emulsion (co)polymerization or by dispersing a solution of the (co)polymer in water (see Sections 2.4.1. and 9.1.2.2). Any latex used as WSR has three major characteristics: the type of (co)polymer (T_g and hydrophobicity seem to be the most important parameters), the particle size, and the type of charges and functional groups existing on the particle surface.

Styrene–butadiene copolymers are very hydrophobic. Paper treated with latex shows a very low water penetration: in order to achieve complete paper wetting (for a wet tensile-strength determination) a surfactant must be added to water. The effectiveness of water repellency depends on the T_g of the polymer, which is a function of the copolymer composition [438].

The properties of nonreactive neoprene latexes as WSR [439] differ significantly from those of UF or MF resins [440]. The neoprene particles are hydrophobic, with low T_g, nonreactive toward cellulose, and non-thermosetting. A poly-chloroprene concentration in paper of about 10% increases wet strength tremendously (up to about 80% of its dry strength).

The following latexes are used as wet-strength resins: butadiene–acrylonitrile and butadiene–styrene copolymers [441], polyacrylates, polyvinylidine chloride, paraffin wax emulsion [442–444], ethylacrylate–aconitic acid (10% acid) [445], vinyl acetate–butyl acrylate–maleamic acid copolymers [446] and blends of anionic latexes (styrene-butadiene rubber [SBR] and styrene–hydroxyethyl acrylate–itaconic acid) [447].

The emulsion polymerization of chloroprene results in a dispersion with anionic charges (rosin salt as stabilizer) on the particles' surface [448]. Thus, when the latex and the pulp are mixed, there should be no immediate co-coagulation (latex retention on the cellulose fibers). For better adsorption on cellulose fibers, aluminum sulfate [449,450], zinc white [451], and a cationic "deposition aid" [439,452] (such as polyvinyl pyridinium butyl bromide [444]) are involved. The best choice for the cationic "deposition aid" would be a cationic wet-strength resin (such as PAE) [451].

The particle size (the average diameter is about $0.2\,\mu m$) and the amount of emulsifier existing in the latex (see the surface tension of about 36 dyn/cm) are the key parameters for a proper co-coagulation [453,454]. In order to have a uniform deposition on fibers, the homo-coagulation (latex-to-latex flocculation) must be prevented [449].

The amount of neoprene needed to improve the W/D ratio is 2%–4% [449,450], but the curing process [454,455] still needs closer attention in terms of the time–temperature relationship.

Sticky polymer particles are able to improve wet strength in the absence of any new covalent bond formation. Moreover, the ionic bonds are less important and the *W/D* ratio is less sensitive to pH; the particles are too large to diffuse into existing fiber–fiber bonds [456]. What neoprene can do is to increase the hydrophobic character and glue the cellulose fibers through hydrophobe–hydrophobe interactions.

PVA latexes were also tested as nonreactive wet-strength resins [457]. Cationic modified PVAc latexes are self-retained, while the anionic PVAc latexes are retained with alum. The curing parameters (time–temperature) are related to the glass transition temperature of the PVAc. Plasticizers will lower the T_g but their type and their amount are still uncertain.

Due to their retention on the anionic cellulose fibers, the cationic latexes are preferred as dispersed WSR. Cationic charges are provided by a cationic surface active agent (such as trimethyl, hexadecyl ammonium chloride) [458]) and by a cationic comonomer (such as 2-vinyl pyridine copolymers [372]). The cationic charges can also be generated after particle formation by a polymer-analog reaction [459,460]. 2-Bromoethyl metacrylate and butadiene are copolymerized in emulsion and the resulting copolymer is quaternized with triethylamine. The quaternary ammonium salt is located on the surface and is chemically bonded to the polymer particle [461].

Anionic latexes (methacrylic acid as comonomer and sodium dodecylbenzene sulfonate as emulsifier and styrene maleic anhydride copolymer (SMA) as macromolecular stabilizer) are blended with a PAE-type resin. That blend is applied on the paper surface [462] or at the wet end. The cationic resin is in excess and provides a cationic character to the dispersion. The blend of hydrophobic latex and PAE resin shows a synergetic effect [463].

In order to make a reactive compound, the polymer particles are made from a soft copolymer (butadiene as a main comonomer) and with some reactive groups on the surface. Amide-blocked isocyanates [286], from N-methylol-acrylamides [464] or from acetoacetoxyethyl methacrylates [465], are examples of reactive groups.

In this case, the wet-strength mechanism seems to be of the hybrid type: a reactive group on the surface appears to help. However, the low T_g of the polymer seems to be the main reason for a cement-like behavior of polymer particles between the cellulose fibers.

The reactive latex (butadiene–styrene-2-isopropenyl-2-oxazoline [466]) is combined with reactive latex (carboxylated latex having itaconic acid as a comonomer [467]) for a blend of polymers curable at room temperature [468]. In order to avoid working with a highly toxic monomer (2-isopropenyl-2-oxazoline), a poly-methacrylic ester is converted to an oxazoline-containing polymer [469] through a polymer-analogous reaction. Carboxylic groups on the cellulose fibers may also be involved in the curing process. Self-curable latexes are obtained by a 2-isopropenyl-2-oxazoline functionality on top of the carboxylated latexes (seeded emulsion copolymerization) [470,471].

A special case is when the stabilizer is a wet-strength resin itself [159,472,473]. A PAE resin is modified 2-ethyl-hexyl acrylate (Michael addition). The acrylate part provides the hydrophobic character and the azetidinium ring provides the reactivity and retention capability (in red). The largest part of that stabilizer (in blue) will be distributed in water due to its hydrophilic character. In this case, there is no chemical bond between the stabilizer and the polymer which forms the particle core.

A chemical bond between the stabilizer and the polymer particle is formed when a macro-monomer is used in emulsion copolymerization. A kind of polyamidoamine (pre-polymer with secondary amine) reacted with acrylic acid (or glycidyl methacrylate [GMA]) resulted in a cationic macro-monomer.

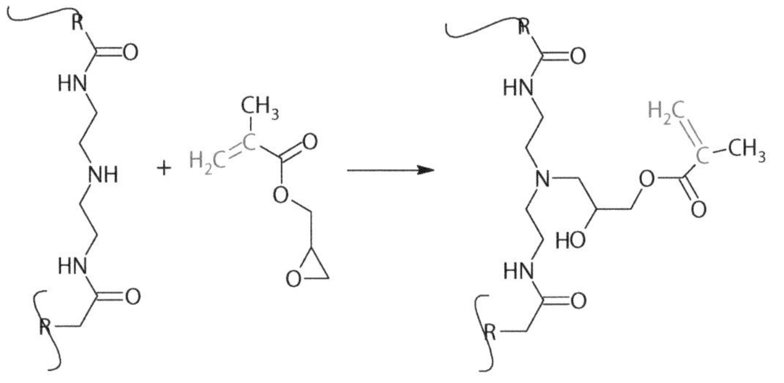

The cationic comonomer (10%) is compolymerized with styrene and butadiene to make a cationic latex (styrene-butadiene) [474].

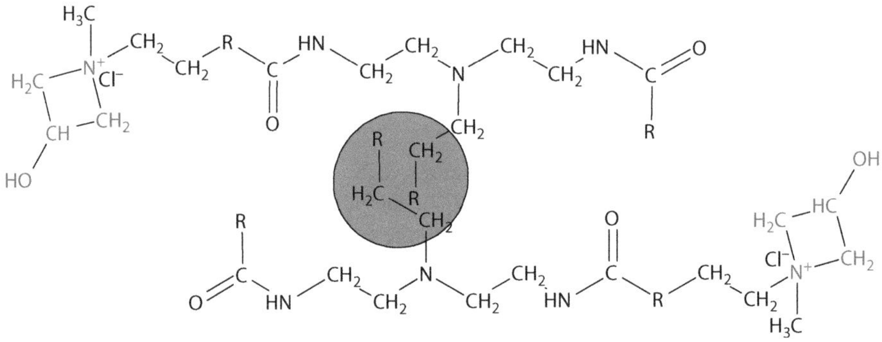

This type of resin can provide not only wet strength but also a higher hydrophobic character of the sheet along with an additional elasticity of the paper.

The reactivity of functional groups and their availability are equally important for latex as a wet-strength resin. Styrene–glycidyl methacrylates 1:1 copolymer [475,476] has a high T_g. Due to the reaction with water at the interface, about 5% of the glycidyl groups are lost during latex synthesis. Any attempt at involving the remaining 95% of reactive groups (located in the particle core) in a reaction with cellulose will require very high temperatures: 200°C for 20 min.

It is no longer a matter of glycidyl group reactivity but of how to make those groups available. Higher drying temperatures, which help the spreading process, will result in better paper wet strength [477] even for polymers with a low T_g number.

5.2   IONIC CHARGE ADDITION

Due to the absence of any ionic interaction between cellulose fibers and nonionic WSR at the wet end, the paper sheet must be tub-sized [33,445,478,479]. Nonionic compounds (such as dry formaldehyde monomer or dimethylolurea) are used for the direct treatment of paper articles [15,18,480]. When added to the wet end, UF resins show a poor retention on the cellulose fiber (less than 50%) [43].

Despite the fact that the UF resin may be precipitated on pulp fibers with the aid of alum [481], the presence of ionic groups would improve retention. The addition of ionic charges should be performed by avoiding cross-linking reactions. Thus, the pH, the concentration, and the reaction temperatures are controlled to prevent the reactive groups from developing side reactions.

5.2.1 ANIONIC GROUPS

For UF or MF resins, several methylol groups react with sodium bisulfate [482] (or sulfur dioxide [483,484], sodium lignin sulfonate [485], or hydroxyalkane sulfonate [486]) to add anionic charges to the macromolecule [487].

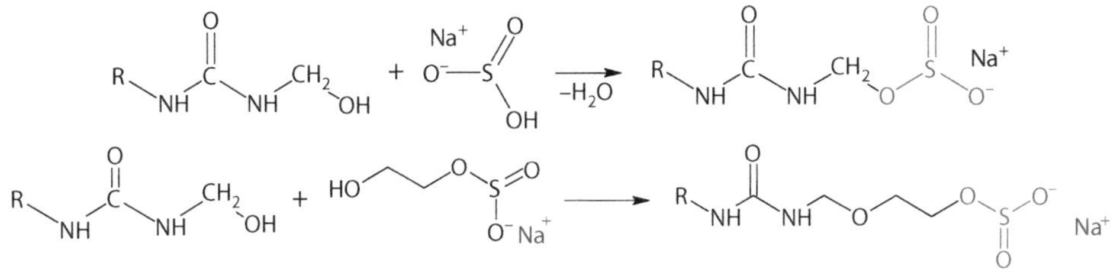

Bisulfite may be added at the beginning of the condensation [488] and uncontrolled polycondensation is reduced by adding an excess of butyl alcohol [489]. Anionic MF resins are obtained through a very similar procedure [490,491].

Anionic charges are added though a reaction with the methylol group or the resin; thus the anionic resin will have less reactive groups, but the ionic resin is still reactive because both UF and MF resins are branched, with multiple methylol groups. The new reactive anionic resins are retained in the presence of salts of a polyvalent cation (such as alum) [487].

5.2.2 CATIONIC GROUPS

The base strengths of the ionizable amine groups of the MF resin provide charges in the presence of an acid [55,58,492]. For MF resins (1:2.5 MF ratio) hydrochloric acid (0.5–3.5 mol of acid for 1 mol of melamine [493]) has been used.

Some of the preexisting functional groups in a nonionic WSR are sacrificed to make the polymer cationic. Cationic charges are located at a nitrogen or sulfur atom. For instance, cationic charges at a sulfur atom are in alkyl thioether [63].

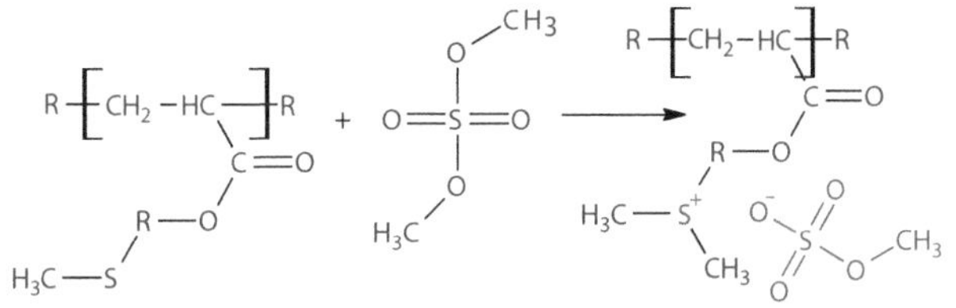

Cationic charges located at a nitrogen atom are easier to obtain. The methylol group reacts (at pH lowered to 5.5 with formic acid) with triethanolamine [53,54,494,495], ethylene polyamine [482,496], hexamethylene diamine [497], or dimethyl amino ethanol [498] and the new cationic charge is added to the polymer.

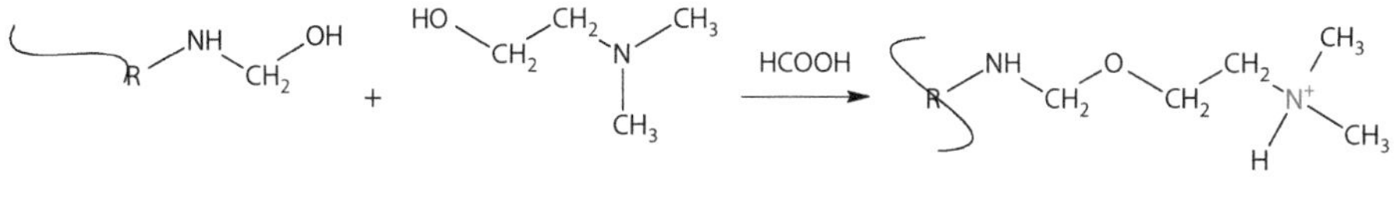

Triazone is a cyclic tertiary amine [499] which results from the condensation of formaldehyde, urea, and a primary amine. The amide functionalities are able to react with formaldehyde in a regular condensation. The final product is a di-methylol triazone. The tertiary amino group is potentially a cationic charge.

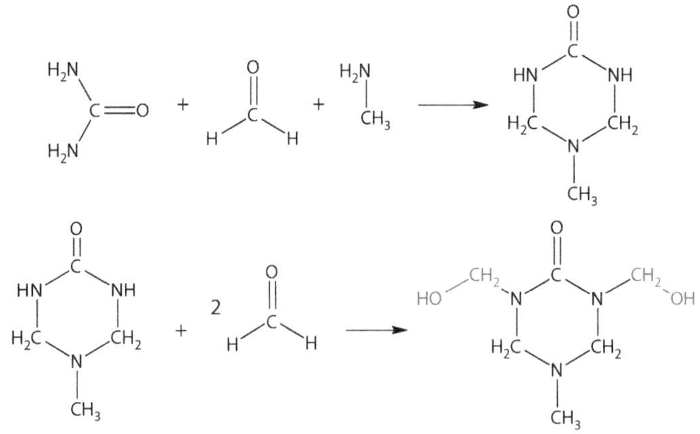

The methylol groups have a similar reactivity with those of UF resins and, therefore, di-methylol triazone can be added to the UF condensation [500] in order to obtain a cationic-modified resin.

Polyamines (such as triethylene tetramine) multiply the process of UF condensation–cationic charge addition [501]. The modified resin is used as WSR at a pH = 4.0 when residual amine groups are quaternary salts. Polyamines also reduce the content of free formaldehyde in the final resin [42,502].

The addition of cationic charges through methylol for a better retention of ionic WSR (UF and MF) on cellulose fibers also brings several other benefits: a better solubility in water, and a better control of the polymer reactivity for a longer shelf life.

The cationic charges are added by reaction of the aldehyde group of the backbone. The aldehyde moiety is introduced in the polymer structure by acrolein copolymerization [503] or by hydroformylation of polybutadiene [504] (in toluene as solvent at high pressure 600 psig).

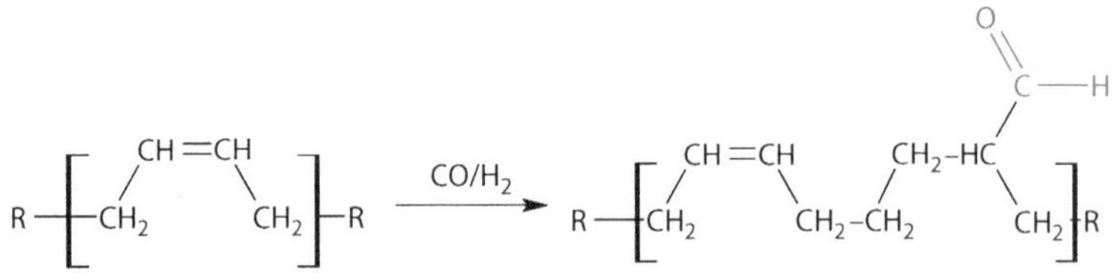

This is a synthesis in which the cationic charges are added before the reactive moieties. The aldehyde groups are converted to an amino group by reductive amination with an excess of amine [505] (see also the reductive amination performed on olefin–carbon monoxide binary copolymers [480]).

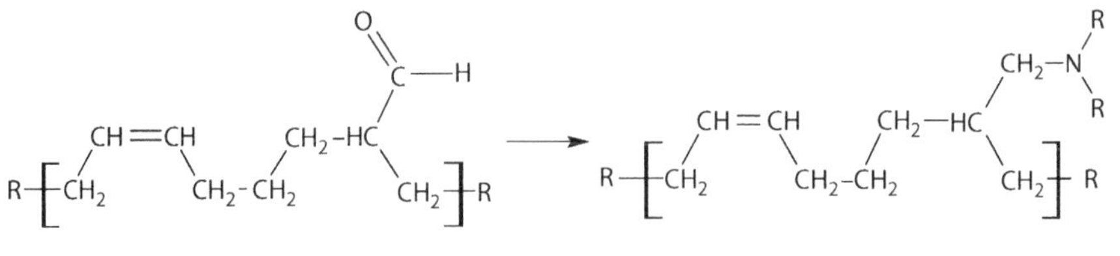

To make a reactive and cationic WSR, a new hydroformylation is performed. That compound is cationic at neutral pH and cross-linkable with epichlorohydrin [505].

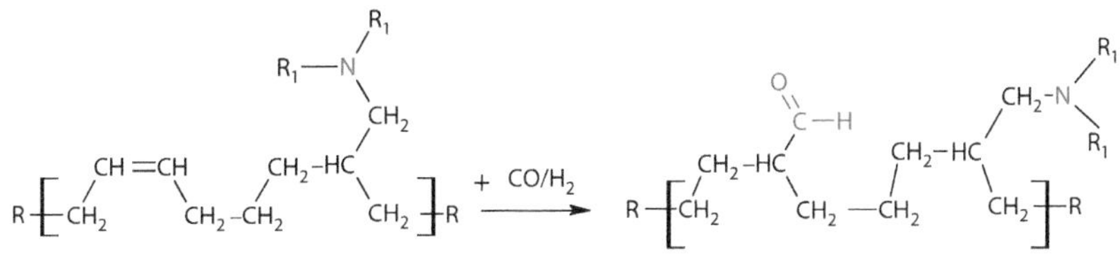

Acrolein copolymers react with diamine [506] to add cationic charges and the paper treated with them shows a *W/D* ratio of about 30% [507].

The amide moiety (as in polyacrylamide) is converted to a cationic group through Mannich reaction [400,508,509]. The polyacrylamide backbone changes into a copolymer of acrylamide with *N*-dimethylaminomethyl acrylamide ($n = m + p$).

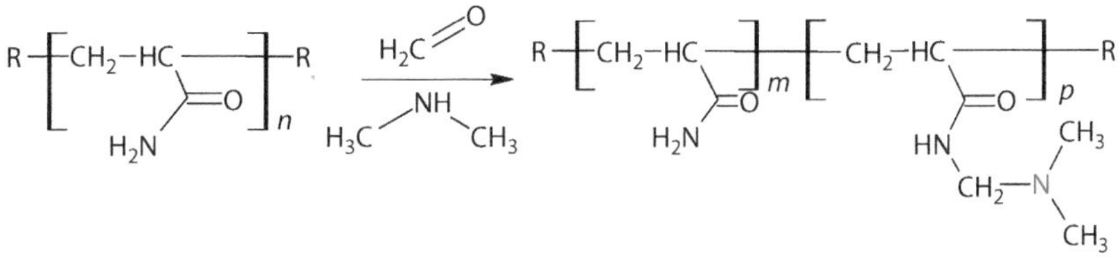

The m/p ratio can be changed by modifying the reaction parameters. The distribution of the cationic charges depends on the molecular coil shape in the solution. During these reactions the molecular coil shape can change due to different types of interactions with the solvent of the newly formed pendant groups.

Polyamines (PEI, polyvinyl amine, or polyallylamine [335]) in aqueous solution are cationic because the amine groups are protonated even in neutral pH [510] (equilibrium of neutralization). Quaternary ammonium salts provide a better charge density for the polyelectrolyte on a broader pH range, which makes it possible for the retention of the wet-strength resin to occur at neutral and slightly alkaline pH. The secondary amine is converted to a tertiary amine and quaternary ammonium salt by a reaction with glycidyl trimethyl-ammonium chloride [511].

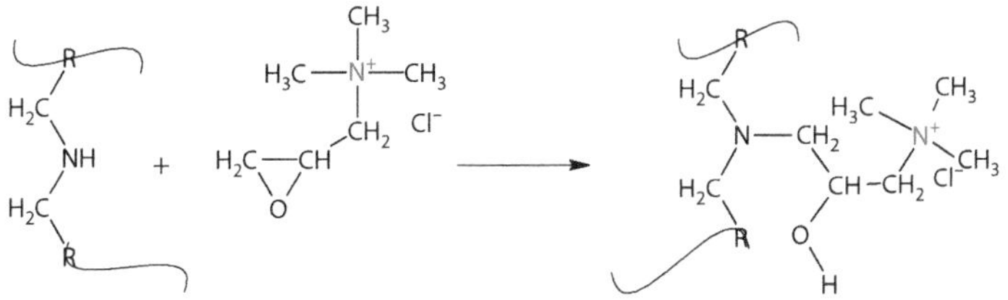

Polymers with tertiary amine groups may be quaternized by reagents such as *p*-toluenesulfonate [512] or with methyl chloride in the presence of methanol and sodium chloride [508].

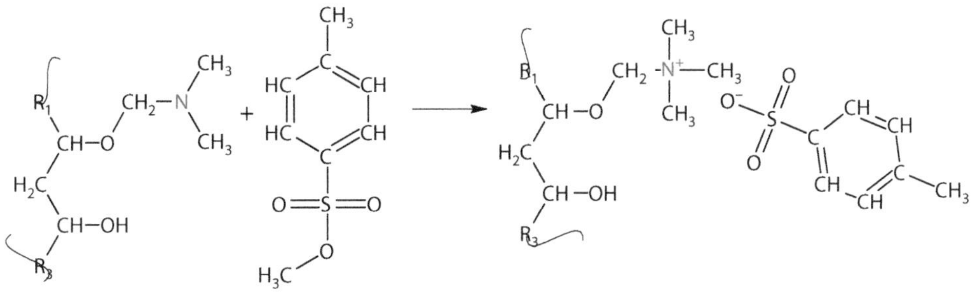

The cationic charge can be formed through intramolecular reaction: the primary amines in the hydrolyzed acrylonitrile–N-vinylformamide copolymer react with the nitrile groups (amidinization), in the presence of hydrochloric acid, to form the amidine cationic functionality [510].

5.2.3 PAE Resins Synthesis: The Epichlorohydrin Ability to Add Cationic Charges

Most backbones of the WSRs presented in previous chapters are polyamines [86], such as polyethylene imine [106,513]), polyamidoamine, the polyamidoaminester [180], or the polyester backbone with pendant secondary amine groups [182]. Amines react with epichlorohydrin to form tertiary amines and quaternary ammonium salts.

The reaction mechanism of amine and epichlorohydrin in the presence of water (or alcohol [145]) is supposed to involve an epoxy ring opening [514] and a ring re-formation [515] (the amine in excess will capture the released hydrochloric acid).

In order to reduce the hydrolysis of epichlorohydrin, the reaction is performed at low temperatures (5°C–30°C) [161] and high solids concentration (45%). During that stage no significant change in molecular weight was noticed.

The tertiary amine forms an internal quaternary ammonium salt (cyclization to 3-azetidinol [516]). The presence of the azetidinium cycle was identified by NMR [517] in PAE resins.

The linear chlorohydrin group was also identified in the same type of resin [518] and even in a concentration which exceeds the concentration of the azetidinium cation [519].

Epichlorohydrin is nonsymmetrically reactive. The two functionalities of epichlorohydrin react differently with amines: the ring opening of the epoxide does not release any small molecules, while organic chlorine reacts with an amine and the resulting hydrochloric acid is captured by another amine. There are thermodynamic and kinetic differences between those two reactions: the epoxy group reacts first. Thus, at a low epichlorohydrin/secondary amine ratio (0.25), only a branched structure is obtained in the absence of any azetidinium cycle [520].

Due to the epichlorohydrin involvement in hydrolysis reactions (see Section 5.3.4), the secondary amines are only partially converted. The un-reacted secondary amine reacts with the azetidinium cycle or with the chlorohydrin moiety within a cross-linking process. In order to control it, the secondary amine should be protonated with sulfuric acid [153].

Cationic charges increase (almost linearly) during the reaction with epichlorohydrin [519], meaning that the quaternization process is a unimolecular reaction. Viscosity follows a different pattern (Figure 5.13) showing that a certain amount of the azetidinium ion is needed to start the cross-linking process.

The chlorohydrin pendant groups are involved either in a unimolecular reaction (the azetidinium ring formation) or in a bimolecular reaction (cross-linking process). To prevent bimolecular reactions, the azetidinium cycle is synthesized in the second step performed at higher temperatures of 60°C–65°C but at a lower concentration (20% solids) [521].

The ratio between the cyclic form (internal quaternary ammonium salt) and the tertiary amine is the key of the conversion reaction and the final performances of the WSR [518]. The cyclization to form the azetidinium group is slower than the addition of the epoxide [517]. A slower process means a longer reaction time, and a longer reaction time will result in a partly cross-linked polymer. In other words, there are limitations in obtaining a high content of azetidinium cation in PAE resins. Changing the pH and the resin concentration are other tools for a better control of the cross-linking reaction [154].

Figure 5.14 shows that the azetidinium ring concentration reaches a maximum, after which its concentration is almost constant: the number of newly formed rings is equal with the number of the rings consumed in the cross-linking process. During this period of constant concentration of the azetidinium ring, viscosity increases very fast [521]. Apparently, the reason for cross-linking is the reaction of the azetidinium cycle [522]. The amount of azetidinium cation involved in the cross-linking reaction depends on its concentration and reactivity: the terminal groups are more reactive [523]. The increase in cationic charges over time (see Figure 5.13) suggests that the

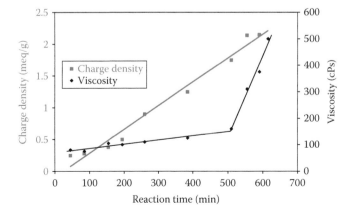

FIGURE 5.13 Epichlorohydrin reaction with PAA resin. (From Devore, D.I. and Fischer, S.A., *Tappi J.,* 76(8), 121, 1993. With permission.)

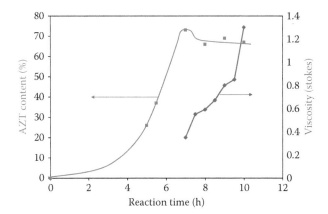

FIGURE 5.14   The azetidinium content (based on the repeating units) and the resin viscosity as a function of reaction time.

chlorohydrin pendant group reacts with the tertiary amine as well and generates a quaternary ammonium salt [519].

At high conversions, the newly formed polymer is self-cross-linkable: both reactive functionalities are on the same backbone. The azetidinium group is reactive with polyamide including proteins [524,525]. In order to reduce the functionality of the system and thus to prevent gel formation through bimolecular reactions, the epichlorohydrin reaction with polyamidoamine is performed in the presence of a polyamine with a small molecule (bis(3-aminopropyl) amine) [526], or of ammonia [67].

The cross-linking reaction is simplified by replacing the nonsymmetric reactive compound (epichlorohydrin) with a symmetric one (such as 1,3-Dichloro-2-propanol (DCP) [527]), or by controlling the backbone molecular weight (end-capped with hydroxyl amines). The shorter segments of the polyamidoamine backbone reduce the probability of gel formation.

Tertiary amines [113,370,528] can also be involved in a reaction with epichlorohydrin (or with any other halohydrine) to form a quaternary ammonium salt (pH=2–3) as WSR. The nitrogen atom of the quaternary ammonium salt is located in the backbone structure, or as a pendant group [338,371] when tertiary amines are within di-allylmethyl amine homopolymers or copolymers.

5.2.4   THE SYNTHESIS OF PAE-TYPE RESIN WITHOUT EPICHLOROHYDRIN AS RAW MATERIAL

The use of the PAE resin involves several drawbacks [166]: the hydrolysis of epichlorohydrin generates toxic by-products (see Section 5.3.4), their maximum wet-strength properties are gained after an aging period, the resins require a relatively low pH for stabilization, and corrosion problems are encountered in the handling and storage of PAE resins. Due to environmental and health hazards, the products obtained by using epichlorohydrin have become highly undesirable.

In order to rule out any by-products, other reaction routes are suggested [348,529]: water is eliminated from the synthesis involving epichlorohydrin, or an allyl group chemistry is used because the allylic structure is a precursor of epichlorohydrin. An allyl double bond can be added to the polyamidoamine directly from the condensation reaction, using bis(2-aminoethyl)allyl amine [146], or the quaternary ammonium salt is formed with an allyl chloride reacted with a polyamidoamine.

A polyamidoamine (as the prepolymer made from DETA and diacids) reacts with an excess of allyl chloride (1 mol secondary amine to 1.2–1.3 mol allyl chloride [348]). Due to that excess, both a quaternary ammonium and a tertiary amine are formed.

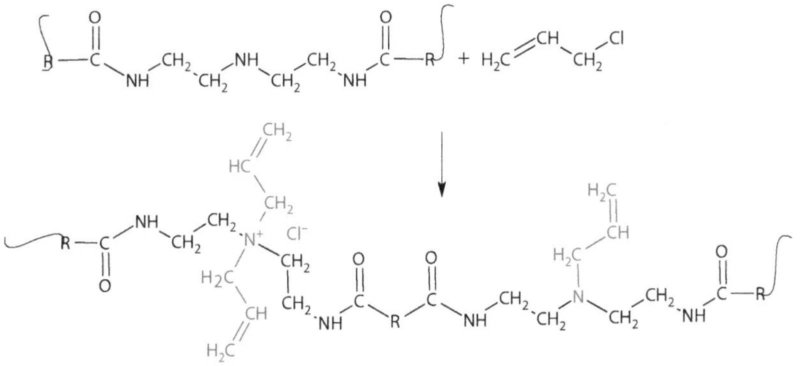

The quaternization of a tertiary amine [529] is done with a lower allyl chloride/polyamidoamine ratio. The reaction is performed at a temperature of 25°C–105°C in water and the allyl chloride is used in amounts ranging from 1.0 mol to about 1.05 mol per mole of tertiary amine [351].

The next step is to add the chlorohydrin functionality by a reaction of the allylic double bond with hypochlorous acid created in situ by sparging with chlorine.

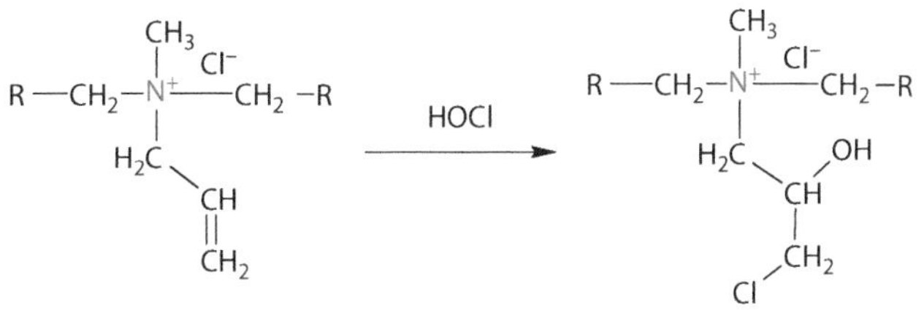

5.3 POLYAMIDOAMINE EPICHLOROHYDRIN POLYMERS AS WET-STRENGTH RESINS

The retention and the performances of PAE polymers as WSRs depend on their chemical structure: the molecular weight, the charge density, and the distribution of the azetidinium cycle. That chemical structure is set with the polymer synthesis, which involves three major steps: (1) the condensation of polyalkylene polyamine with a dicarboxylic acid to form a polyamidoamine (backbone synthesis), (2) the polyamidoamine reaction with epichlorohydrin at a temperature below 40°C (secondary amines are converted to tertiary amines), and (3) the tertiary amines are partially converted to quaternary ammonium salts (azetidinium cycle) at temperatures ranging from 50°C to 70°C.

While steps 1 and 2 show good selectivity, during step 3 the cyclization reaction competes with the cross-linking process. As a result, the PAE resin shows broad molecular weight and chemical composition distributions.

5.3.1 CHEMICAL STRUCTURE OF PAE RESINS

PAE-type resins contain a polyamidoamine backbone with a quaternary ammonium salt located in the backbone structure. Due to the excess of epichlorohydrin, there are virtually no secondary or primary amines left. At pH = 3, tertiary amines are protonated. The linear chlorohydrin conversion to the azetidinium cycle is developed in parallel with the cross-linking reaction. The number of linear units and the number of cyclic units depends on the epichlorohydrin concentration and reaction time [517]. Thus, there are four possible structures for the ionic units.

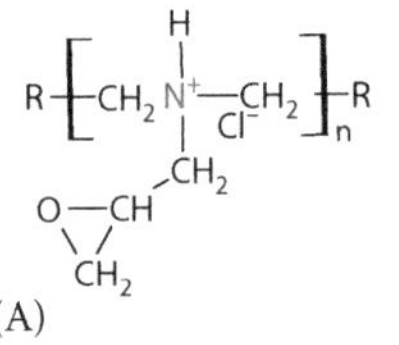

(A) 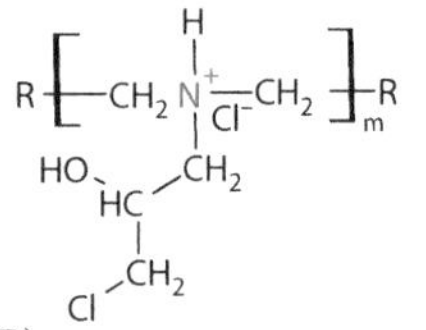   (B)

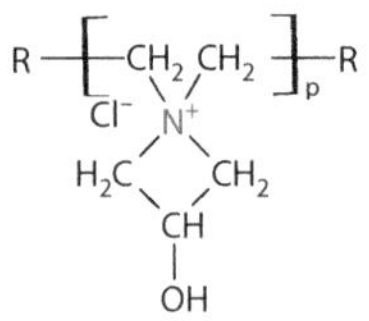

(C) 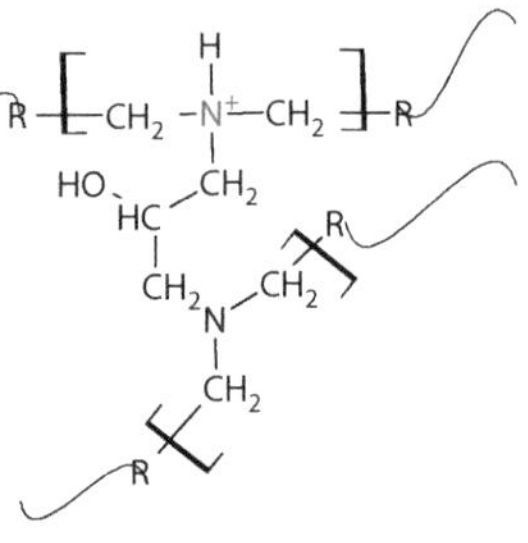   (D)

The structure, as illustrated above, is complex and shows a very reactive polymer. In order to prevent any chemical reactions (hydrolysis, cross-linking, or ring opening of the azetidinium cycle), the analysis of the PAE resin structure is performed in water, along with all the residuals from its synthesis [522].

There is no evidence for the presence of the epoxy structure (A), and the azetidinium cycle unit (C) is in a concentration of less than 70% mol [519,522]. During the synthesis of the PAE resin, hologenated alkyl amines (B structure) are converted (intramolecular reaction) to a cyclic quaternary ammonium salt [530–534] (structure C) or to a cross-linked macromolecule (structure D) through an intermolecular reaction.

Nonionic chlorine is converted to ionic chlorine through both intra- and intermolecular reactions. Intermolecular reactions result in higher molecular weights. According to the M_n measurements, there are about four bridges (D) between the polyamidoamine backbones in one big PAE macromolecule [521].

The PAE synthesis is a very delicate process during which the ideal polymer would have a higher concentration of (C) units with a reduced number of cross-links (D). Unfortunately, these two properties are interrelated (see Figure 5.15). Moreover, the presence of reactive functionalities determines the poor stability of the final product.

Tertiary amines (B and D) behave differently when pH changes from 3 to 8 [522]. Through 1H-NMR and ^{13}C-NMR on the PAE resin dissolved in water, it was possible to identify a differentiation in the hydrogen atoms bonded to the carbon atoms located next to the nitrogen from tertiary amines. These important changes can result in a different macromolecular coil conformation at different pH and electrolyte concentrations (see polyelectrolytes in solution).

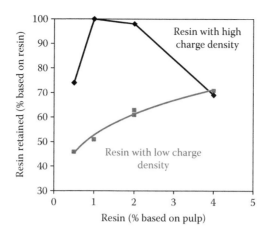

FIGURE 5.15   The effect of the charge density on the retention of PAE resin.

The type of ionic charges, the distance between cationic charges, and the characteristics of the environment (pH, ionic strength of the solution) will influence the shape of the molecular coil and its capability to adsorb on the cellulose fiber.

The retention of PAE-type resin [123,535] reaches a plateau at 1%–2% resin concentration. At higher resin concentrations, the resin adsorption (calculated based on pulp) increases but the adsorbed fraction of PAE resin starts decreasing after the 100% values has been reached.

The resin structure is responsible for the charge density and, thus, for the resin retention on cellulose. The charge density of the PAE resin is adjusted by the amine/epichlorohydrin ratio, by the diacid molecular weight, and by the polyamine functionality. A lower charge density is obtained with a higher diacid molecular weight and/or by substituting DETA with ethylenediamine [536].

Retention is a process driven by the charge density [122,123]: for a higher molecular weight of the diacid (dimer acid), the charge density (per gram of resin) is lower and the retention has lower values at the same resin concentration (Figure 5.15, in red).

PAE includes a broad distribution of macromolecules with a different potential as wet-strengthening material for paper. Different fractions (separated by dialyze) of polymers within a sample of PAE resin were identified as having a lower retention and much lower performances than average [537].

The retention values of PAE were higher when the pulp had a higher carboxyl content. However, small amounts of PAE were retained in the handsheets prepared from pulp with no carboxyl functionality detected by the TAPPI titration method [538].

5.3.2   MOLECULAR WEIGHT

The WSR molecular weight is a crucial parameter for its performances and shelf life. The polymer molecular weight is built up in two steps: during the pre-polymer synthesis (linear macromolecule) and during the epichlorohydrin reaction (branched macromolecules—a partially cross-linking process).

The attempt to use NMR for molecular weight measurements seems to lead to numerous inconsistencies [522]. Thus, data on the molecular weight of the PAE resins are obtained by working with diluted polymer solutions. The behavior of the PAE resin in solution becomes a key factor in understanding experimental data.

Polyamidoamines are polyelectrolytes: the amine groups are protonated in acid and even in neutral pH. The extent of protonation and the shape of the macromolecular coil depend on pH. That is why the molecular weight estimation for those polyamidoamine polymers must be performed in a standard solution of 1 N ammonium chloride [152]. The molecular coil size also depends on the presence of other electrolytes [77] (see Section 2.3.2), and on the rigidity of the polymer structure [521].

The gel permeation chromatography (GPC) method shows, for all cationic polyelectrolytes, a reduction in their apparent molecular weight when pH and/or the ionic strength are increased [539]. Intrinsic viscosity in $NaNO_3$ solution depends on the concentration of the inorganic electrolyte. Strong effect is noticed at very low concentrations of inorganic salt: from 1 dL/g in the absence of electrolyte to about 0.35 dL/g at 0.5 mol/L $NaNO_3$.

For consistent results, pH and the conductivity of the polyelectrolyte sample should be carefully managed. The molecular weight of the epichlorohydrin–dimethyl amine condensate was measured by high-performance liquid chromatography (HPLC) using an eluent of 0.3125 mol/L sodium acetate and 0.3125 mol/L acetic acid [78]. The same eluent was used for the molecular weight measurement of cationic polyelectrolytes by size-exclusion chromatography (SEC) [521,540]. Moreover, the column should be treated with specific reactants in order to add quaternary ammonium salt to the stationary phase and the eluent should be carefully chosen (0.1 N nitric acid for a cationic polyacrylamide-type copolymer) [541]. A better method (SEC with a multi angle laser light-scattering detector) has been developed using a solution of 0.5 M acetic acid and 0.2 M sodium nitrate as eluent, or acetic acid (0.3 M)-sodium acetate (0.3 M) [540].

During the PAE resin synthesis, parallel reactions develop (azetidinium cycle formation, alkylation, and cross-linking). The higher molecular weight fraction may undergo a faster cross-linking process and the higher cationic charge density in a molecule will change the shape of the coil (it extends due to the repulsive forces) for an easier action with epichlorohydrin. Therefore, the macromolecules start to be differentiated by their molecular weight and chemical composition.

The GPC of the PAE resin shows two peaks (bimodal distribution [542]) which shift positions and intensities when pH and/or the ionic strength is increased [517]. This suggests that there are at least two macromolecular species with different chemical structures and molecular weights. This polymer shows a polydispersity of both its molecular weight and chemical composition.

Different fractions are influenced differently by the electrolyte concentration: lower molecular weight (separated by fractionation) shows a lower impact of the electrolyte on its molecular shape [517].

A systematic study [543] reveals that there is no correlation between the molecular weight and the charge density of PAE resin (Figure 5.16).

For an ideal experiment aimed at studying the effect of the molecular weight on the resin performances, polymers with different molecular weights and with the same charge density should be

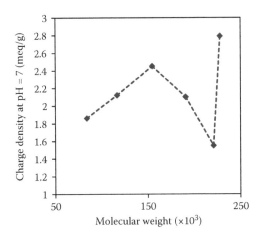

FIGURE 5.16   Charge density versus molecular weight of the PAE resin.

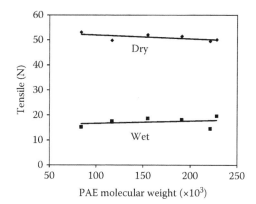

FIGURE 5.17   The effect of the molecular weight of the PAE resin on paper properties.

tested. Due to the lack of availability of such resins, apparently, the PAE resin molecular weight does not have any impact on the paper wet or dry strength (Figure 5.17). It is more reasonable to believe that although the molecular weight of the wet-strength resin is important, the experimental data cannot reveal it.

But when the molecular weight is associated with the concentration of the azetidinium cycles, during the PAE resin synthesis, the polymer molecular weight has a strong impact on the paper wet strength: the molecular weight should be over 10^4 and the azetidinium content over 65% [521].

The PAE resin solution viscosity is a function of the molecular size and shape of the molecular coil. For a given molecular weight, the polymer solution depends on the electrolyte added to the system: acids, buffers, and the electrolytes brought by water used for dilution. The presence of electrolytes reduces the size of the macromolecular coil, which also changes the molecule's vicinity. The environment, the distribution of azetidinium cycles, and the polymer molecular weight dictate not only the solution viscosity but also the resin stability.

5.3.3   RESIN STABILITY AND SHELF LIFE

Several patents claim a very stable PAE resin: 6 months at 25°C–30°C for 37% solids [22] or even 53% solids [141]. However, as a general behavior, for that type of resin, viscosity increases over time [544]. Figure 5.18 shows the viscosity of the PAE solution as a function of time. Regardless of

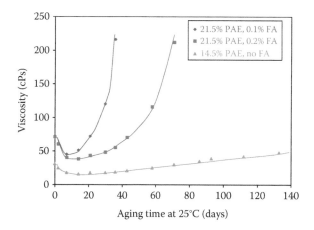

FIGURE 5.18   The stability of the PAE resin solution treated with formic acid (FA).

the polymer concentration, the curves profile is the same: with a shorter aging time, a decrease in viscosity is noted, while with a longer time, viscosity increases exponentially.

The cross-linking process is responsible for the increase in viscosity. There is no reason to believe that the cross-linking reaction starts later during the shelf life. It is more likely that the cross-linking reaction competes against the hydrolysis process from moment zero of the shelf life. Those two reactions are developed simultaneously.

The cross-linking mechanism is supposed to develop through the attack on the azetidinium cycle resulting in a bridge (in red) between two macromolecules (in blue).

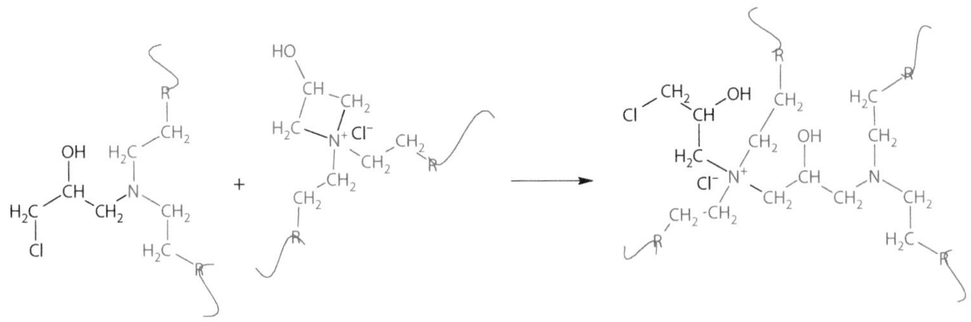

This cross-linking reaction keeps constant the number of tertiary amines and quaternary ammonium cations, but lowers the concentration of azetidinium. Higher–molecular weight polymers have a larger number of azetidinium and chlorohydrin groups and are more likely to involve them in the cross-linking step (statistical factor). The partially cross-linked polymer has an even higher molecular weight, which is why the process is developed exponentially (see Figure 5.18).

The macromolecule of the polyamide backbone can be broken [545], and hydrolysis was suggested as a reason for lower viscosity during the shelf life [517,546–548]. The amide bond is the weakest part of the backbone and its hydrolysis is supposed to take place at a low pH [549].

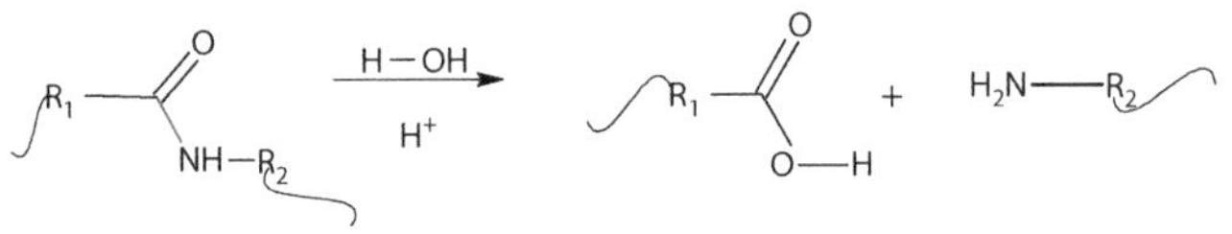

The signals in H-NMR spectra of 3-hydroxy-azetidinium groups of PAE decreased with aging time. The four-member cycle is hydrolyzed to 2,3-dihydroxypopyl groups [548,549].

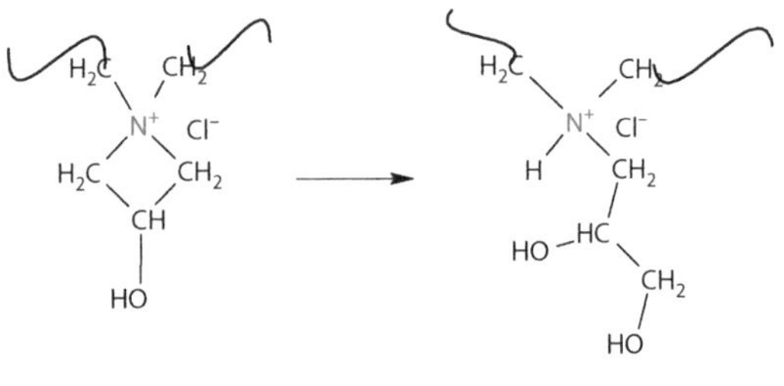

The hydrolysis and the cross-linking change the solution viscosity, reduce the azetidinium cycle concentration, and make the polymer shelf life shorter. From the commercial point of view the best WSR product would have a high concentration (lower costs for production, storage, and shipping), as well as high molecular weight and high azetidinium content for better performances. Thus, increasing resin stability is a common objective for scientists.

There are two approaches to improve the stability of the PAE resin solution: one is to use chemical reactions which interfere with the aging process and the other is to reduce the probability of developing higher molecular weight species.

At pH 8.0, even in a diluted solution, tertiary amines are less protonated, and the PAE resin shows an increase in molecular weight after 1 h [540]. Because secondary and tertiary amines seem to be involved in the cross-linking process, a chemical reaction used to improve resin stability was to convert all secondary amines to tertiary amines (reaction with ethylene oxide [550] or formaldehyde [551]) and to block the tertiary amine nitrogen by protonation with an acid [171,528,540]. Water-soluble acids are added to the resin solution in order to convert the reactive groups to quaternary ammonium salts (2.0<pH<3.0) [113,552].

Unfortunately, the quaternary ammonium salt is part of an equilibrium reaction and free tertiary amines are still present even at lower pH. That is why, even at a very low pH, the cross-linking process continues. The pH can be changed (pH = 7.0) either before [154] or after the reaction with epichlorohydrin. The pH of the PAE resin is adjusted with inorganic and organic acids and their blends [553], such as sulfuric and formic acids in different sequences. Those two acids can be added in a sequence (first formic acid [546,552]) or even better as a blend in only one shot [547]. A higher acid concentration provides better stability (see Figure 5.21). Stability is also improved by heating for 3.5 h at 70°C, and at pH = 2 [114], or by adding inorganic salts (calcium chloride, magnesium chloride, aluminum chloride, etc.) at pH = 4.5 [554].

The low pH and the potential corrosion concerns have led to an attempt to convert tertiary amines to quaternary ammonium salts. If all the nitrogen atoms were quaternary ammonium salts the resin stability would be much longer [347]. The quaternization reaction (with dimethyl sulfate) can be performed before [555] or after cross-linking with epichlorohydrin [536,556].

The cross-linking process is self-accelerating because the probability of developing a bridge between two molecules increases with their increasing molecular weight. Another chemical approach to increase the resin stability is to slow down the cross-linking reaction by adding a low molecular weight compound but with the same chemical structure [557]. The intermolecular reaction is developed at the same reaction rate, but the "cross-linked" polymer has a smaller partner and the increase in viscosity is not significant.

A reduction in the probability of developing a reaction can be obtained by decreasing concentration of the reactants. Therefore, good resin stability requires lower solid content, lower molecular weight, and lower azetidinium concentration [558–561].

The resin shelf life is longer in diluted solutions. However, dilution can be performed only down to a reasonable point. The first generation of PAE resins was delivered at 3% [94], 6% [562], or 9% concentration [108,109].

Resin stability can be improved by lowering the molecular weight of the resin and/or the concentration of the azetidinium cycle. A PAE resin made from a low molecular weight prepolymer (polyamidoamine obtained with an excess of amine [168]) will take more cross-links to build up molecular weight. This resin with a concentration of 40% is stable for 6 weeks at 90°F.

The prepolymer molecular weight is set during the adipic acid/polyamine condensation: a mole ratio close to 1.0 will result in a higher molecular weight for the polyamidoamine prepolymer [150]. At a mole ratio (adipic/DETA) lower than 1.0, the PAE resin stability was improved to over 30 days at 50°C [151].

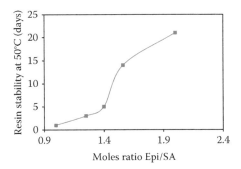

FIGURE 5.19   PAE resin stability at 50°C as a function of epichlorohydrin concentration.

The concentration of the azetidinium cycle and the corresponding tertiary amine concentration depend on the epichlorohydrin/secondary amine mole ratio [22]: the higher the mole ratio, the better is the resin stability (Figure 5.19). Although these experimental data are interesting, higher epichlorohydrin concentrations are prohibitive in terms of production costs and of by-products resulting from hydrolysis of epichlorohydrin (see Section 5.3.4. and Figure 5.20).

A PAE resin with a very low pH has good stability but loses its wet-strength efficiency with aging: both hydrolysis and cross-linking reduce the azetidinium cycle content. The "re-activation" of the low pH PAE resin is performed with a base [552], within a complicated process that requires an aging step of 4 h at room temperature before dilution from 25% to 2% and uses it promptly.

Due to the fact that resin stability is related to the concentration of the azetidinium cation, the reactive groups can be eliminated. The final polymer would no longer be a self-reactive, thermosetting resin [170]. The synthesis involves the same chemical compounds (adipic acid and DETA and epichlorohydrin) but the molecular weight of the prepolymer is controlled with mono-functional amines (mono-ethanol amine) and the epichlorohydrin concentration is kept below 0.5 mol per mole of secondary amine. The final polymer is highly branched, but the azetidinium cycle and the chlorohydrin pendant groups are missing. Actually, this branched compound is a poly-tertiary amine. The tertiary amine groups (in red) can easily be protonated to make a cationic resin.

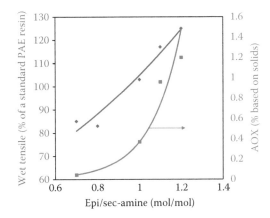

FIGURE 5.20   The effect of epichlorohydrin concentration on the AOX content and the PAE resin performances for mixed hardwood–softwood Kraft paper.

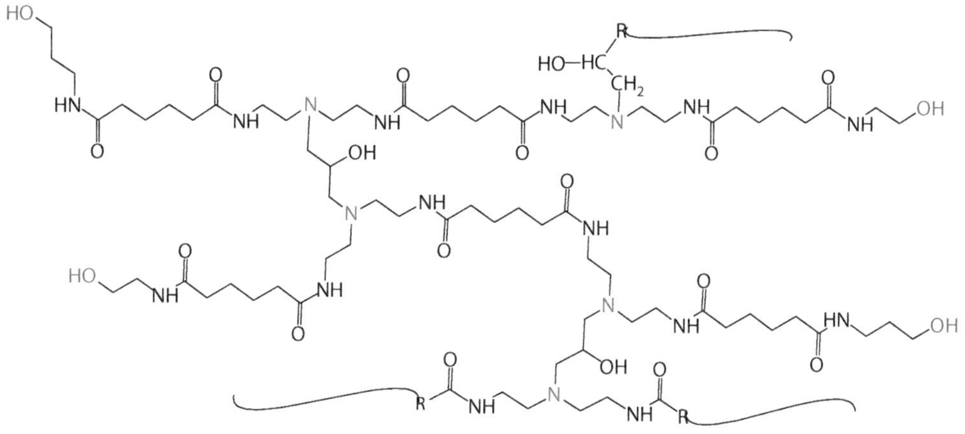

5.3.4   BY-PRODUCTS (DCP AND CPD) AND HOW TO LOWER THEIR CONCENTRATION

During the PAE resin synthesis, only a fraction of the epichlorohydrin reacts with the amine groups from the polyamidoamine prepolymer. Some of it remains un-reacted, and some reacts with the water or the chloride ion to form CPD and DCP respectively.

The amount of by-products (up to 40% based on the dry product [384]) depends mostly on the ratio between epichlorohydrin and amine. It is preferable to use sufficient epichlorohydrin to convert all secondary to tertiary amine groups or to add dimethyl sulfate to make quaternarium ammonium salt [152]. High Epi/amine ratio involves an important benefit: a higher conversion of the amino groups (a completely quaternized copolymer [563] and therefore a longer shelf life).

An excess of epichlorohydrin is involved in the other side reactions. Organic chlorine from epichlorohydrin is converted to anionic chlorine in quaternary ammonium salt. The ionic species (such as Cl⁻, H⁺, and HO⁻) may move the process toward the neutral organic compound formation (dichloro-2-propanol, DCP).

In an aqueous medium, about one-third of epichlorohydrin undergoes the reaction with hydrochloric acid to form DCP [64]. The consumption of hydrochloric acid changes the pH to alkaline [564], which favors the formation of 3-chloropropane-1,2-diol (CPD) and eventually glycerol.

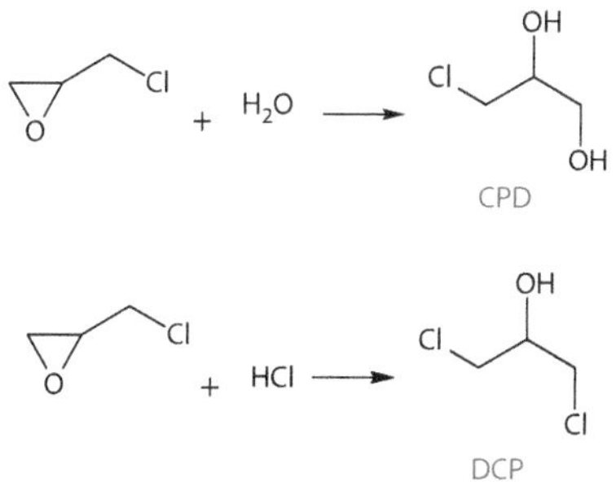

The presence of DCP and CPD is a specific characteristic of the synthesis of PAE resins. A PAE resin without those by-products can be obtained only by replacing epichlorohydrin with another cross-linker (such as 3-glycidoxypropyl trimethoxysilane) [401] or by using other anions in solution [331,346]. A change in chemistry through an "epichlorohydrin-free process" will also change the wet-strength resin application: dosages, pH, temperature, repulpability, etc. Because the epichlorohydrin technology is more affordable, the presence of DCP and CPD in PAE resins must be dealt with.

PAE resins with good performance as wet-strengthening agents should have a high concentration of the azetidinium group. That high concentration is obtained when a high ratio of Epi/secondary amine is used (1.3–1.5). A higher Epi concentration will generate a higher concentration of by-products: for the Epi/amine ratio ranging from 1.5 to 2.0, the DCP concentration is about 4%–6% based on solids [563]. Commercial PAE resins typically contain 1%–19% (dry basis) of epichlorohydrin, 1,3-dichloropropanol (DCP), and 3-chloropropandiol (CPD) [114,171,565,566].

Regardless of their nonionic character, about 50% of those by-products (absorbable organic halogen, AOX [261]) are retained on cellulose at the wet end [565] or are brought by water remaining in the wet web [567]. These organic chloride by-products are generally considered to be environmental pollutants. The environmental concerns have created an interest in wet-strength resins that have reduced levels of such by-products [568]. By 1990 [567], the residual AOX was reduced by 50%, while by 2006 the reduction was of 90% [562].

Both chemical and physical processes are used to reduce the DCP and CPD content in wet-strength resins [569]. But in order to reduce the AOX content in PAE resins, any changes in technology should be supported by very accurate analytical data. Gas chromatography (GC) or reverse-phase HPLC [570] are largely used for that purpose (GC on ethyl acetate extract with no derivatization [571,572]). This method can determine 10–80 ppm DCP and CPD based on PAE solution (the coefficient of variance is about 10 times higher for CPD than that for DCP). The CPD detection limit is 10 ppm.

The lack of accuracy in the estimation of CPD concentration is explained by its capability to react with other chemicals (such as ketones [573]). For that reason a derivatization of DCP and CPD is applied. After paper extraction with acetonitrile [574] a derivatization with N,O-bis(trimethylsilyl) trifluoroacetamide (or phenyl–boronic acid [575,576], 1-butan-boronic acid [577], hepta-fluorobutyric acid anhydride [578]) is performed; the internal standard being 1-fluoronaphtalene (or tetradecane, n-heptadecane, and d_5-3-monochloro-1,2-propandiol) [573].

One of the most important steps in AOX estimation from the water solution of resins is their extraction with an organic solvent (hexane [577], ethyl acetate, iso-octane). A high concentration of sodium chloride in water (12%–20%) is recommended for a better extraction [575]. Residual water from the AOX solution in an organic solvent is removed by the anhydrous sodium sulfate [578].

5.3.4.1 WSR with Low AOX by Adjusting the Synthesis Parameters

During the PAE resin synthesis, there is a competition between the reaction of epichlorohydrin with secondary amine and the involvement of epichlorohydrin in the side reaction (such as hydrolysis). In order to reduce the formation of by-products, the reaction temperature should be lower, and the concentration of polyamidoamine, higher [171]. An increase in the resin concentration from about 20% to about 40% and even 59% [579] reduces the total AOX by 50%. Epichlorohydrin is added in three portions [580] at a reaction temperature of 5°C [117,161,581,582].

Even at high concentrations of polyamine in water and at low temperatures, the ratio between epichlorohydrin and the secondary amine is still the most important parameter for the AOX concentration and the polymer quality as wet-strength resin for paper. If the Epi/secondary amine ratio ranges from 0.95 to 1.0, the DCP content is 0.13% [583]. For a ratio much lower than 1.0 [572,579,581,582], the final content of DCP may reach a value of less than 0.01% [149]. Figure 5.20 shows the increase in total chlorine by-products and in paper wet strength as the epichlorohydrin concentration goes up [565].

The by-product composition includes a larger amount of DCP: it is more likely a reaction between epichlorohydrin with hydrochloric acid. To reduce the amount of DCP, a specific amount of a halogen-free mineral acid is added to the reaction mixture after the major part of the epichlorohydrin charge has alkylated the polyamidoamine resin [584].

5.3.4.2   The Reduction of the Concentration of DCP and CPD by Their Hydrolysis

Lower concentrations of epichlorohydrin, DCP, and CPD are achieved by post-treatment of the aqueous solution of PAE with an inorganic base (2.8 mmol NaOH/g dry resin reduces the AOX from 2500 ppm to less than 20 ppm) [139,181,566,544] or a dibasic phosphate and an alkanolamine. The hydrolysis of epichlorohydrin [585] is an equilibrium reaction, and with a very diluted base (0.5 M NaOH) DCP is converted back to epichlorohydrin. The amount of DCP depends on the base concentration [564]. The reduction of the by-product concentration is based on the hydrolysis reaction.

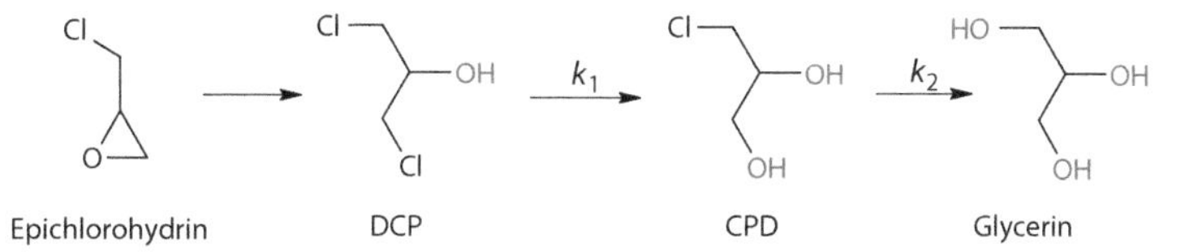

Epichlorohydrin   DCP   CPD   Glycerin

In order to simplify the picture, we can assume that $k_1 = k_2$. The reaction kinetic is a pseudo-unimolecular process (water in large excess): the residual concentration (C) of DCP (or CPD) depends (Equation 5.4) on their initial concentration (C_o) and time (t) [586]:

$$C = -kt + \ln C_o \tag{5.4}$$

and the rate constant (k) is a function (Equation 5.5) of pH and temperature (T) [570]:

$$k = 10^{7.73347 + 0.83775\,\mathrm{pH}}\, e^{-\frac{E_a}{RT}}\ (s^{-1}). \tag{5.5}$$

At a constant value for pH (pH = 6.8), the CPD hydrolysis follows Equation 5.6 (the activation energy is about 118 kJ/mol) [586]:

$$\ln k = -\frac{14,209}{T} + 34.683. \tag{5.6}$$

The inorganic base increases the pH up to 12.0, and after the hydrolysis of chlorinated compounds, the pH is reduced to stabilize the final resin [544]. The PAE resins with a longer shelf life and a lower covalent bonded chlorine are obtained [587] by hydrolysis at pH = 10 with sodium hydroxide. The mixture is then acidified to pH = 2.0 by a sulfonic acid (methansulfonic acid, benzenesulfonic acid, p-toluenesulfonic acid, formic acid, etc.).

The addition of organic bases (such as diethanolamine) can reduce the total organic chlorine from 0.8% to 0.07% [588]. The treatment with organic amines can also be combined with an inorganic base [72]. Ammonia, primary amines (monoethanol amine or ethylene diamine), secondary amines (dimethyl amine), and tertiary amines (trimethyl amine) can reduce the DCP concentration by about 50% [583]. Ammonia can be added even during the prepolymer–epichlorohydrin condensation process [67].

A bioprocess is also suggested [589–591] in which a reduction below 10 ppm DCP and CPD is obtained after about 20 min in a bioreactor. An enzyme is added to the composition in an amount of about 5×10^{-5} weight percent [592]. The four-component bacteria of community H10 are shown to completely convert DCP to glycerol [593,594].

The possibility to address the problem of by-product formation through bioprocesses suggests that the PAE synthesis could be revised and a higher ratio of Epi/secondary amine might be used. The higher substitution degree will result in a more stable resin (longer shelf life), while the DCP and CPD will be eliminated within the final step by a reaction with an enzyme [595]. In order to help the enzymatic process, a hybrid procedure is suggested [596,597]: the PAE-type resins are treated with sodium hydroxide or sulfuric acid to reduce the AOX content; the enzyme treatment is the final step.

5.3.4.3 Producing WSR with Reduced AOX via Physical Processes

The previous processes are based on chemical reactions. Physical processes, such as adsorption, distillation, and dilution, are also considered in order to lower the AOX content.

Selective adsorption is effective in reducing the AOX concentration from about 7000 ppm to less than 5 ppm total AOX [598,599]. Adsorbents are selected from a group consisting of ion-exchange resins (strongly basic ion-exchangers containing tertiary amino groups [600,601]), nonionic polymeric resins, activated carbon, zeolites, silica, clay, and alumina. Such adsorbents have a limited capacity; they need to be regenerated once they no longer remove the epichlorohydrin by-products efficiently. A dwell time of less than 1 h is preferred. The total chlorine (organic and inorganic) is reduced by more than 90% (the reduction in AOX exceeds 99%). The ion-exchanger regeneration (from the chlorine to the hydroxyl form) is needed before a new cycle. The effectiveness of treated wet-strength resin is identical to that of the original resin [600,601].

DCP and water form an azeotrope (11.3% DCP with a boiling point of 99.4°C), which can be distilled under reduced pressure (below 0.5 kg/cm^2) [69,602].

An attempt at reducing the AOX concentration by dilution with another cationic polymer (pDADMAC or cationic polyacrylamide [603]) has been made [604]. Due to the synergetic effect between PAE and pDADMAC, dilution is followed by a reduction in the total amount of WSR (see Section 5.4).

5.3.4.4 Epichlorohydrin-Free Resins as Paper Wet-Strengthening Agents

Epichlorohydrin is a cross-linker with two functionalities having different reactivities against the secondary amine. The cross-linking reaction of the polyamidoamine with epichlorohydrin takes place in two steps: the epoxy group reacts at lower temperatures and the newly formed chlorohydrin pendant groups make bridges between two different molecules or make an azetidinium cation by intramolecular reaction.

Ideally, a WSR free of chlorinated by-products would be obtained through a reaction in which epichlorohydrin is no longer involved. In that case the cross-linking agent can be a different epoxy-type compound. Because chlorine is missing from that picture, the new compound should have a poly-epoxy structure such as 1,3,5-triglycidylisocyanurate [605].

A poly-epoxy compound has three identical reactive groups and is used to cross-link the regular polyamidoamine backbone. The new WSR performs like the regular PAE resin, but residual DCP is less than 0.05 ppm and CPD less than 0.1 ppm.

Nonsymmetrical cross-linkers were also explored. 2,3-Epoxypropyl sulfonate (R_1 is an alkyl group) [606,607] or 3-glycidoxypropyl trimethoxysilane [608] can replace epichlorohydrin in the reaction with a polyamine. The resulting polymer doesn't show any residual organic halogen; the sulfonate and silane groups are reactive to cellulose.

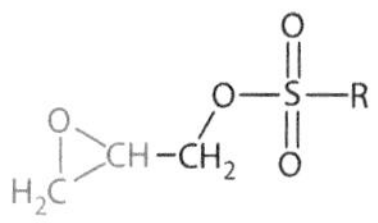

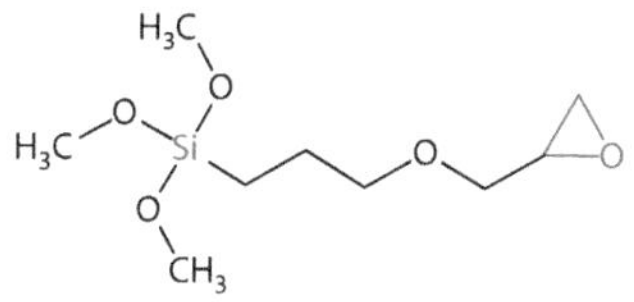

These compounds react based on the same concept as epichlorohydrin does: they are difunctional cross-linkers for polyamine, with a nonsymmetrical structure. The two reactive groups have a different reactivity and the cross-linking process is developed in two steps. The silyl-linked polyamidoamines are thermosetting and they are prepared with minimal levels of halide ion.

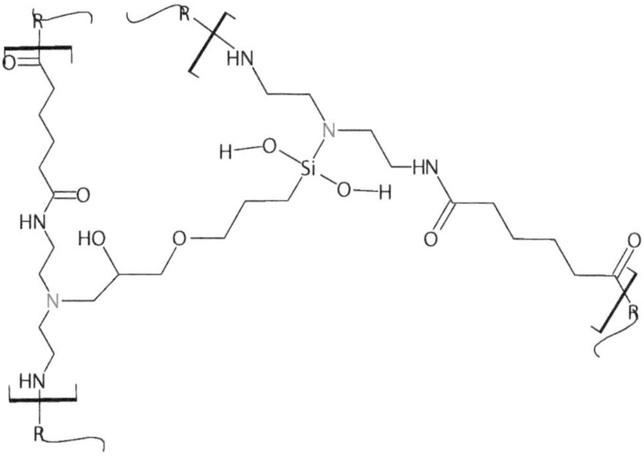

5.4 WSR MADE FROM BLENDS

WSRs are used in conjunction with other chemicals in order to improve their performances. There are several reasons for blending two or more compounds as wet-strength resins: the dilution (for reduced cost and lower AOX for PAE resins) [163,604], the manipulation of the charge density (WSR and/or cellulose charge density) in order to increase the WSR retention, the improvement of the cellulose chemical reactivity, and the synergetic effects [609].

Blends of wet-strength resins are obtained either by mixing the components or by synthesizing a new resin in the presence of another resin. Each component of the final mixture can interfere with cellulose in a separate way or a synergetic effect can be developed by component interaction before or during the curing step. The existence of a synergetic effect is recorded when a response (R) does not fit the simplest mixture rule [610].

5.4.1 BLENDS OF RESINS WITH SIMILAR CHEMISTRY AND NO SYNERGETIC EFFECT

No synergetic effect is expected between two cationic resins with similar chemistry or within interpolymers made from compounds with similar chemical structures. Blends of polyvinyl amine and PAE (or polyDADMAC—epi) [339,604] or cationic polyacrylamide (5% DADMAC) with PAE resins [603] are used as wet-strength resins. PolyDADMAC and PAE-type resins can be added in one shot or sequentially to the pulp fiber slurry [611,612]. During their retention, the two different cationic resins compete for the same anionic spot on fiber.

The above-mentioned blends are obtained by simply mixing the water solution of the preexisting polymers. As expected, the blend of a permanent wet-strength resin (such as PAE) and a cationic polyacrylamide provides wet strength to paper but at a lower value than the PAE resin by itself at the same dosage [603].

In order to improve the interactions between the two cationic resins, the synthesis of one resin is performed in the presence of the other resin or of its precursor. The precursor of the PAE-type resins is the polyamidoamine prepolymer. The reaction of the prepolymer with epichlorohydrin is performed in the presence of ammonia [67]. The final product is a blend of PAE resin and thermosetting polyamine.

5.4.2 Synergetic Effects Provided by Blends of Resins with Different Chemistries

Wet-strength resins are cationic polymers with reactive functional groups. The term "different chemistries" implies chemical or ionic interactions between two components of the blend, or between WSR and cellulose. The effect of grafted starch to improve compatibility between starch and wet-strength resins [412] can also be considered as a synergetic effect.

The (co)polymers presented in this chapter have their own mechanism to impart wet strength to paper. These curing mechanisms are based on their particular chemistry (organic chemistry of the functional groups), and blends of different wet-strength resins (MF, PAE, and acrylamide copolymers) are claimed to have a synergetic effect [613].

The synergetic effect of the blends of different WSRs can be explained by an ionic interaction between polyelectrolytes. Polyanionic and polycationic compounds interact in the continuous phase or at the interface of water–cellulose fibers. For those types of blends, a covalent bond between resins is not a must.

Blends of cationic and anionic compounds [614–616] result in supra-molecular aggregates able to impart wet strength to paper. The cationic polymers are thermosetting resins with quaternary ammonium salts as cationic moieties. The anionic polyelectrolyte is an acrylamide–acrylic acid copolymer, poly-methacrylic acid, or a styrene–maleic anhydride copolymer. Poly-anionic phosphorus compounds [87,617], poly-phosphonium, or poly-sulfonium compounds are also recommended [618,619].

The cationic copolymer of acrylamide with acryloyloxyethyl–trimethyl ammonium chloride (46% mol) is added first, followed (after 3 min) by an anionic acrylamide (24% mol anionic comonomer: 2-acrylamido, 2-methylpropane sulfonate). The effect on the paper wet strength reaches 85% of the standard PAE resin [615]. Still, the best performance is obtained at an almost double concentration of WSR (cationic plus anionic resins) (Figure 5.21), and both anionic and cationic resins perform poorer than the PAE resin at the same concentration. Figure 5.21 shows a curve with a maximum (the synergetic effect): the blend performs better as a wet-strength resin than each of its components.

The idea of "dual systems" as wet-strength agents is quite old: in 1946 [483] (with a continuation in 1952 [484]) anionic UF resins were blended with cationic MF resins, and in 1954 a blend of cationic MF resin and an anionic (oxidized) guar was suggested [620] in acid pH in the presence of alum.

A cationic macromolecular compound (PAE resin, cationic UF resin, or cationic guar gum) [621,622] is added first, and an anionic compound afterward: lignin sulfonic acid, anionic UF resin, soybean oil modified with fumaric acid, the sodium salt of CMC [621], the copolymer of ethylene with maleic anhydride modified with fatty amines [623], the glyoxalated anionic

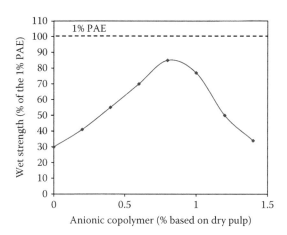

FIGURE 5.21   Paper wet strength after the addition of 1% cationic resin and different amounts of anionic copolymer.

polyacrylamide [624], or the ammonium salt of ethylene–maleic anhydride copolymer, to increase the wet strength for paper (up to 100% [625]) as compared to the cationic resins only [482,626].

A dual system can be prepared in situ at the wet end or in a separate reactor. At the wet end, the addition of the anionic polyelectrolyte (anionic guar) is followed by the addition of the cationic guar gum [627]. The supra-molecular aggregate is developed in the aqueous phase and is then retained on pulp (see also Section 3.3.3.5). Ternary systems are formed from two cationic resins (only one is thermosetting) which are added first to the pulp, and an anionic polymer (copolymer acrylamide–acrylic acid) added later [628].

Generally speaking, the addition in a sequence of cationic and anionic polymers is a ternary system: cellulose, as an anionic polyelectrolyte, is the third partner. A multilayer structure obtained by successive addition—directly to the pulp—of cationic, then anionic, and then cationic again [629–634] seems to be a new direction for better wet-strength resins.

Anionic–cationic macromolecular complexes (CMC—partially hydrolyzed polyvinyl formamide) [616] can also be prepared separately, before their addition to the pulp [614]. In this case, at the wet end, anionic cellulose is a part of a binary system and deals with an anionic–cationic polyelectrolyte aggregate. The complex is effective as wet adhesive only if the concentration of the cationic group exceeds the concentration of the carboxyl group. Those aggregates are mentioned as amphoteric retention aids as well (see Section 3.3.3.5).

Ionic interactions are critical within the aggregate and during the adsorption of the polyelectrolyte complex on the cellulose fibers. However, the covalent bonds are much stronger and a synergetic effect on paper wet strength is expected when chemical bonds are developed along with ionic bonds. The synergetic effect can be the result of the new chemical bond formation within the polyelectrolyte complex or during the papermaking process.

Two reactive polymers may react with each other to form an inter-polymer through covalent bonds. Inter-polymers are obtained between polyvinyl alcohol and PEI resin (modified with epichlorohydrin) [635], MF resins [636], or cationic polyacrylamide (functionalized with blocked glyoxal) [637]. Every time the polyvinyl alcohol is converted to a cationic wet-strength resin.

The epichlorohydrin reaction with the backbone of polyamidoamine is performed in the presence of polyvinyl amine [638] or starch, and the azetidinium cation formation takes place simultaneously with the formation of the inter-polymer. As a proof of the synergetic effect, the inter-polymer (about 50% PAE resin) has the same performance (wet/dry) as the straight PAE resin at the same polymer concentration [163].

Covalent bonds between two polymers are also formed when one macromolecular compound is obtained in the presence of the other. During the free-radical copolymerization of vinyl or acrylic monomers in the presence of PAE, a large fraction of grafted polymer is obtained. Acrylamide [639], N-vinyl pyrrolidone, vinyl triazole, or vinyl imidazoles are grafted [640] on PAE-type resins. The grafted copolymer brings more hydrogen bonding to the wet-strength resin. The grafted polymer is obtained along with a large amount of linear (co)polymers which are not retained on fibers.

These wet-strength resins (inter-polymers and graft-copolymers) show a synergetic effect but, at the same time, their synthesis is not selective, which may result in linear polymers less effective as wet-strength resins. Moreover, both inter-polymers and graft-copolymers have a molecular weight much higher than the individual compounds and therefore their solutions have higher viscosities. It seems a better idea to build the covalent bonds during the papermaking process.

The anionic groups (carboxyl) on cellulose fibers are responsible for wet-strength retention and (arguably) for covalent bond formation by reacting with the azetidinium cycles. In other words, higher wet strength is expected to be retained when more carboxylic groups are on the cellulose fibers. Due to the low concentration of carboxyl on cellulose fibers, carboxylated compounds (or amphoteric [641,642]) are used to improve the PAE resins (or the modified PAE resin [556]) capabilities as a wet-strength resin. Carboxylated compounds are latexes (ethylene–vinyl acetate [30,643] or butyl acrylate–methyl methacryclate (MMA)–methacrylic acid copolymers [463]), CMC [314,644,645], maleic anhydride copolymer, polyacrylic acid [153], carboxymethylated starch, oxidized starch [646,647], or starch xanthate [218,648,649]. The film cast from the solutions of PAE and carboxylated polymers is over 90% insoluble in water. In the case of starch xanthate, the cationic–anionic interactions are followed by ester bond formation during the papermaking process.

An ionic interaction takes place before a real covalent bond is formed. The overall concentration (as low as 2% [153]), the mixing temperature, the presence of electrolytes, the order of addition, and the polymer molecular weights are critical parameters for the intermolecular ionic bond formation. The charge density is also important: if CMC is in excess in the aggregate containing the PAE resin, the cellulose fibers should be pretreated with an additional amount of PAE resin for a better adsorption [632]. PAE develops better wet strength at an optimum amount of carboxymethyl cellulose [20,650]. The charge density can also be balanced by adding cationic surfactants [651,652].

The high reactivity of the PAE resin allows for a diversity of reactions with water-soluble gum [152], cationic starch grafted with acrylonitrile, butyl acrylate and styrene [653], glyoxalated polyacrylamide [651,652,654–656], and polyamines (such as the hydrolyzed copolymers of N-vinyl formamide) [657]. PAE resins can also be included in ternary systems with di-aldehyde starch and CMC [609], or anionic polyacrylamide and acrylic emulsion [658].

Other wet-strength resins are also involved in blends with synergetic effects. Polyvinyl amine (co)polymers are cross-linked with UF, PF, MF resins or glyoxal [659] or with anionic polyelectrolytes (ethylene–maleic anhydride copolymers) [660]. Cationic starch is reacted with glyoxalated resins [661] or other compounds able to generate free glyoxal [662] to form a cationic reactive wet-strength resin [663]. The effectiveness of the cationic aldehyde starch is enhanced by the presence of zirconium oxychloride ($ZrOCl_2$) or chitosan [664]).

Phenol–formaldehyde resins [323] can be associated with PVA latexes [457] in a ratio of 1:9: after curing for 2 h at 140°C; the wet-strength resin is about 30% of the initial dry strength. The very high concentration of the resins (10% based on dry paper), the high curing temperature, and a very long curing time make this process less attractive.

Due to the huge amounts of water used in the papermaking industry, "green chemistry" is an interesting concept [665]. Polyvinyl amines make a gel with saccharides (or polysaccharides) containing a reducible aldehyde function, based on the Maillard reaction [666,667].

5.5 PAPER WET-STRENGTHENING MECHANISMS

Ordinary paper retains less than 5% of its dry strength after it is fully wetted. The addition of a wet-strength resin increases the paper wet strength up to an average of about 25%. The previous chapters have shown the diverse chemical structure of wet-strength resins: reactive or nonreactive, able to provide hydrogen bonding or not, hydrophilic or hydrophobic, ionic or nonionic, flexible or rigid molecule, etc. This diversity has an impact on their adsorption and conformation on the cellulose fiber surface, on their interaction with water, on their involvement in the ion-exchange retention process, on their chemical reactions, and on hydrogen bonding. At first glance, these diverse structures would point to the difficulty to find a unique strengthening mechanism for wet paper.

Despite literature reviews published every decade (1965 [668], 1967 [669], 1979 [670], 1987 [20], 1995 [7]), in 2002 [234], arguably, "the origin of the wet-strength mechanism is not clear." And that, for a good reason: it is hard to incorporate that diversity in a unique, generally accepted, wet-strength mechanism.

There are data supporting a certain mechanism but that particular mechanism is not confirmed by other experimental data. Terms such as "perhaps," "probably," "the evidence must be questioned," "the proposed mechanisms are uncertain," "any present explanation is highly speculative," etc., are often used in relation to this topic. In 2003, Lars Wagberg wrote: "[T]here is still a debate regarding the molecular mechanism responsible for the wet strengthening action" [671].

The statement "One very common pitfall with paper systems is that of analyzing strictly chemical phenomena on the basis of physical effects produced" [672] made in 1959 is still valid today, i.e., after 60 years. On the other hand, the WSR concentration is below 2% based on dried pulp, and the reacted functional group concentration is much lower. A very low concentration of the potentially newly formed bonds makes it very difficult to identify them by current analytical methods. Based on the dry and wet-strength measurements on paper we can only speculate which mechanism is more likely.

Strictly accurate data concerning the WSR diffusion, adsorption, and distribution on the fiber surface, the nature of potential chemical reactions, and their conversions are still missing. The entire picture gets even more complicated due to the intrinsic properties of the macromolecular compounds used as WSR: the distribution of the molecular weight and of the chemical composition. The broad distribution of polymer molecules is set up against the distribution of the size, porosity, and anionic charge density of cellulose fibers. For instance, highly beaten pulp having low freeness shows increased wet strength [309].

5.5.1 THE STRENGTH OF WET AND DRY PAPER

There is a reciprocal interaction between water and cellulose: water association is changed in the presence of cellulose [673–675] and the presence of water disrupts the interfiber hydrogen bonding in paper [676]. The wet strength of untreated paper is less than 5% from its dry strength; therefore the water's fate in the wetted paper is crucial for the understanding of the wet-strength mechanism [677,678] (see Section 4.1).

The effect of water molecules on the paper structure depends on the distribution and the strength of existing bonds. Since the enthalpy of hydrogen bonds (4.75 kcal/mol) [679] is much lower than that of ionic bonds (5–20 kcal/mol) and covalent bonds (80 kcal/mol), the latter are more stable in water [314]. However, the presence of the hydrogen bonds in the paper structure may allow for the existence of other molecular forces [680] such as Van der Waals' bonds of about 1–2 kcal/g mol.

In terms of their sensitivity to water, there are two different types of hydrogen bonds in the paper sheet: one within the fiber architecture and the other between fibers (interfiber bonding). About one-third of the hydrogen bonds existing in paper are involved in intramolecular (cellulose) hydrogen bonding and are therefore not available for fiber-to-fiber bonding [679,681].

The water content in paper (w) ranges from zero to saturation. There is free water and about 5% of bonded water [682]. The critical water content (w_c) [683] is a limit up to which the hydrogen bonds which have been triggered to break have little opportunity to remain broken and are quickly re-formed. The general relationship is (Equation 5.7) [681]:

$$\ln \frac{E_i}{E_o} = f(T, w, t),\qquad(5.7)$$

where
E_o is Young's modulus for the standard conditions
E_i is Young's modulus for paper kept at temperature T, with a moisture w after the time t

If $E_o = 28.8$ GPa, then Young's modulus for wet paper can be calculated (for given temperature and time) and plotted, as in Figure 5.22. There is a change in the curve slope at a critical moisture value ($w_c = 4.5\%$) [684,685]: at lower moisture ($w \leq 0.045$) Equation 5.8 fits the experimental data:

$$E = E_o e^{-w},\qquad(5.8)$$

while at higher moisture content ($w \geq w_c$) a new value (Equation 5.9) governs the process [679]:

$$E = E_o e^{(0.2433 - 6.4074w)}\qquad(5.9)$$

When $w \geq 0.045$, the interfiber hydrogen bonds are cooperatively dissociated by water [686]: the dissociation of one hydrogen bond will induce the dissociation of seven other hydrogen bonds ("the zipper effect"). Interfiber hydrogen bonds are randomly distributed and in a much smaller number per macromolecule, which makes the "zipper effect" very effective. At higher moisture content ($w \geq 0.045$) the paper strength decays much faster (Figure 5.22).

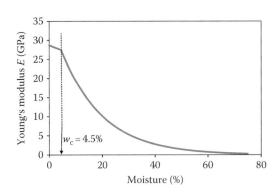

FIGURE 5.22   The effect of paper moisture on Young's modulus.

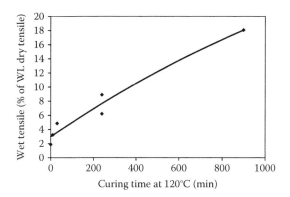

FIGURE 5.23   The effect of the curing time on paper wet strength.

The water effect on the paper strength should be reviewed starting with "zero" concentration for water: dried paper. The untreated paper, dried at 140°C for 4 h, shows an important increase (30%) in its dry strength [687]. When paper is heated under controlled conditions, its swelling in water can be reduced by 60% [187,688].

The curing process seems to help with hydrogen bonding formation. The higher number of hydrogen bonds may improve the paper wet strength as well. The heat cure increases the wet/dry ratio up to 25% at 170°C and up to 33% at 195°C [314]. If the WL (in the absence of any WSR) is heated at 245°C–260°C for a short period of time (1–2 s), then an important increase in wet strength is noticed [33].

Taylor [689] published a systematic study on the curing effect on paper without any resin. For a long curing time at 120°C it is possible to reach a wet strength higher than about 20% from the initial dry strength (Figure 5.23).

The curing process is a function of time: perhaps in the amorphous area cellulose fibers need time to reorient themselves to a new position so that they are able to provide more hydrogen bonds with the neighbors. The access of water to the hydrogen bonds may also be changed. In other words, in the absence of any wet-strength resin, hydrogen bonds alone are able to improve the paper wet strength. The availability and the "protection" of the existing hydrogen bonds can be a reasonable explanation for the wet-strength mechanism.

The boost in wet strength provided only by higher temperatures reinforces the idea that the paper wet-strengthening mechanism is related to hydrogen bonds. The rewetting process can be seen as the reverse of the drying step in the papermaking process: the swelling of paper in water is a combination of a reversible separation of fibers and reversible swelling of individual fibers [187,690]. Intramolecular hydrogen bonds are less sensitive to water: their number and distribution are associated with the macromolecule rigidity which blocks the zipper effect. Therefore, fibers are swollen in water but fiber integrity is kept within reasonable range.

The approach for solving the paper wet-strengthening mechanism must firstly include the quantitative dimension of the process: WSR retention. The adsorbed WSR should be then placed in the real circumstances of the papermaking technology: the water effect on paper and on cellulose fibers. Finally, the diversity of the chemical structure of WSR should be discussed in connection with their effectiveness in a particular mechanism. In its diversity, the chemical structure incorporates, along with cationic polymers, several ammonium salts of anionic copolymers used as WSR [691,692].

5.5.2   WSR Retention Mechanism

Any comments on physical or chemical interactions with cellulose should include the quantity of the wet-strength resin involved in the paper-strengthening process. The wet-strength retention will set the amount of adsorbed resin and its distribution on the cellulose fiber. The retained polymer is a part of the adsorption equilibrium because some of the wet-strength resins remain dissolved in water [693].

At the wet end, the cationic WSR macromolecular coil is dissolved in water and will adsorb on the anionic cellulose fibers swollen in water. Swollen fibers are different from fibers obtained after drainage and drying [694]. Water adsorption takes place at the same amount regardless of the "new external surface" developed during pulp beating [695]. Because the "internal surface" is about 3×10^6 cm^2/g, and the "external surface" is about 1000 times smaller (2×10^3 cm^2/g) [696] the beating increases external surface to a value below the detection limit of the water adsorption method. The absorbed water interacts with the hydroxyl groups. Water adsorption on the internal surface of the fibers generates the swelling process, which changes the distribution of hydroxyl groups and their availability for hydrogen bonding. Pulp oxidation reduces the number of hydroxyl groups and therefore reduces the capacity of cellulose to swell in water [239].

The fact that ionic interactions are crucial for resin adsorption has long been acknowledged [697]. In that dispersed system, the distance between individual cellulose fibers is much higher than in the final sheet and much higher than the WSR molecular coil diameter.

There are many parameters of the retention process [698]: pH, the carboxylic group concentration on cellulose fibers, the WSR ionic charge density, temperature, the ionic trash concentration, the degree of pulp refining, the time of contact between resin and pulp, the presence of an electrolyte, etc. [699,700] (see Chapter 3).

The retention of wet-strength resins on cellulose takes place in two steps: diffusion (resin transport from solution to the surface of the fiber) and adsorption [700,701]. Both diffusion and adsorption are time and temperature dependent. Besides polymer concentration, stirring rate, type of cellulose fiber, temperature, and presence of electrolytes, the M–F resin retention is a function of the time of contact [56,701] (Figure 5.24). The largest amount of resin is retained within the first 5 min of contact. A longer contact time improves the amount of retained resin only with a limited quantity. However, even after a very long contact time, a large amount of cationic resin is still in the aqueous phase. PEI retention on cellulose fibers [702] follows the same general pattern: for high resin concentration (2.5% based on dried fibers), the equilibrium is reached only after about 5 min.

The time needed for retention may be different from one resin to the other [537], but higher temperatures increase the retention rate. Figure 5.25 shows how higher temperatures improve the cationic MF resin adsorption on the sulfite pulp at pH = 7.0 [703]. The curves have the same profile as those in Figure 5.28 and, even at a higher retention rate, the amount of adsorbed resin is far from 100%. For instance, the PAE resin is retained 100% only at very low concentration of the resin: 0.3% based on the dry pulp [535]. In order to increase the amount of adsorbed resin, more of it should be added to the system, and more remains dissolved in water.

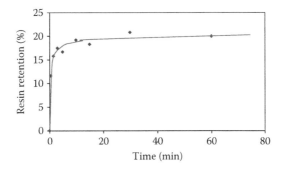

FIGURE 5.24   MF resin retention on sulfate pulp.

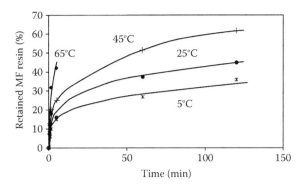

FIGURE 5.25 The temperature effect on the adsorption of cationic MF resin. (From Thode, E.F. et al., *Tappi*, 42(3), 170, 1959. With permission.)

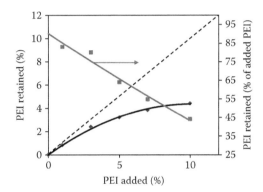

FIGURE 5.26 PEI retention on bleached Kraft pulp.

The PEI retention (black curve in Figure 5.26) on bleached Kraft pulp follows the same pattern: it is far from 100% retention (dotted line). As expected, at lower concentrations of PEI resin, the percentage of retained polymer is higher. But at 10% added PEI (based on pulp), about 65% [704] of the polymer remains in water (red curve). White water contains the non-retained material and a second pass (recirculation) will improve retention [705] based on a longer time provided for adsorption.

During the retention step, each WSR (UF, MF, or PAE) shows a specific behavior [706]: there are different adsorption profiles. At very high resin concentrations, the amount of resin adsorbed on paper reaches saturation point (the plateau in Figure 5.27).

Figure 5.31 shows a complex adsorption mechanism in which the excess of WSR and the presence of WSR in the white water seem to be the intrinsic characteristics of the papermaking process.

The ion-exchange mechanism involves the following equilibrium.

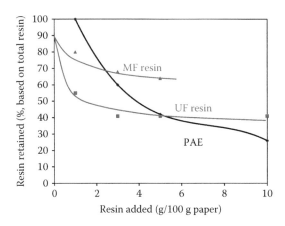

FIGURE 5.27　PAE, MF, and UF resins retention on bleached Kraft pulp.

The ion-exchange mechanism implies the reversibility of each ionic bond. As the molecular weight and the charge density are higher, the polymer adsorption becomes pseudo-irreversible: it is difficult to break "all" the ionic bonds at once.

Equilibrium depends on the pH and the basicity–acidity of the ionic species. PEI is retained under neutral or alkaline pH with a maximum retention at pH = 9.9. At this pH value carboxylic groups are largely dissociated and ready to form ionic bonds [101]. The steric availability of the ionic charges (pore size), the polymer molecular weight, and the shape of the macromolecular coil should also be considered as factors which influence WSR adsorption [707].

The retention mechanism reveals [707] that the positively charged sites (on the WSR) should be in close enough proximity to those negatively charged for electrostatic attraction to occur. Adsorption is also a matter of steric hindrance: branched PEI with high molecular weight (DP = 1165) has less than 0.2% of the cationic charges involved in electrostatic bonding. Therefore, a large amount of retained PEI provides the cationic character of the cluster and most of the anionic charges remain occluded on the cellulose fiber due to the PEI "cloud."

There is a big difference between the anion concentration of the pulp and the amount of cationic groups brought by the adsorbed PEI: about tenfold more cationic groups are present on cellulose fibers. Regardless of the large excess of cationic resin and the repulsive effects, cellulose fibers can adsorb increasing quantities of cationic resin.

The relation between the anion concentration (estimated by regular titration method involving a small molecule) and a poly-cationic macromolecular compound is a fascinating topic. The issues of how the anions and cations interact and how the "charge density" for the new aggregate changes are still under debate.

The position occupied by WSR in the paper structure was identified by staining with OsO_4 [537]. Crude data show that a major amount of the resin remains on the fiber surface, but the penetration of the wood cell wall cannot be rejected either. Some resin is only physically adsorbed [672].

The ion-exchange mechanism suggests the importance of the ion concentration (charge densities of anionic groups on cellulose, and cationic on resin) and the effect of other ionic species which might be present at the wet end (pH, electrolytes, anionic trash, other ionic paper chemicals).

The retention of wet-strength resins (such as PEI [101]) increases with the increasing content of carboxyl groups in the cellulose fiber [537]. The number of carboxylic groups on cellulose fibers can be increased by oxidation reaction [706,708,709], with reactive dyes [707,710], carboxymethylation [302], reaction with poly-acids (such as succinic acid [292,293], polyacrylic acid [711]), and by CMC irreversible adsorption [644,712–714].

Any other cationic compound (such as a debonder) will compete with WSR for an anionic spot on cellulose fiber. The WSR adsorption is reduced and, as a consequence, both the paper dry and wet strength are reduced [28,121]. Surprisingly, the debonder effect on the paper dry strength is relatively stronger.

In other words, the ratio between anionic and cationic species is crucial for wet-strength retention. Resin retention can be improved by increasing the anionic charges on cellulose fibers (see above) or by reducing the relative charge density of the wet-strength resin. That adjustment can be performed at the wet end by adding CMC [715], starch xanthate [218], or an amphoteric aggregate obtained separately (see Section 3.3.3.5).

Cationic wet-strength resins are polyelectrolytes. Their solubility in water, molecular coil shape, and, eventually, their adsorption on the cellulose fibers depend on the pH [716]. For instance, the stock pH changes the adsorption of the cationic MF resin [703].

There are multiple ionic bonds between the anionic groups of cellulose and the adsorbed cationic WSR; dilution with water does not reverse the adsorption process. The number of ionic bonds between the cellulose fibers and WSR depends on the shape of the polyelectrolyte coil. On the fiber surface, the adsorbed macromolecular coil takes a different shape as a function of the pH and/or the concentration of electrolytes in its vicinity (such as sulfate ion for MF resin [55,717]). The concentration of the divalent metal ion is particularly important [718].

PAE resin retention is strongly influenced by the presence of inorganic electrolytes (Figure 5.28). That is also valid for PEI resins adsorption [702]. Calcium is more effective than sodium in competing with PAE for the carboxyl group in the pulp [712,718]; as a result, paper strength decreases as the ionic strength and the fraction of divalent metal ions in the pulp increase.

The retention is far from 100% of the added resin, and even at very high concentrations of PAE in water, the saturation point is not reached. There is a competition for cellulose carboxylic groups and the polymeric cation shows a pseudo-irreversible adsorption. The calcium cation is replaced by changing the equilibrium conditions [719] (Figure 5.29).

Bleaching residues interfere with the retention of WSR. They are insoluble anionic particles, which adsorb some of the resin. Many of these fine particles with accompanying resin presumably pass through the wire into the white water [698].

The ion-exchange retention mechanism explains the resin retention up to the neutralization of the anionic charges onto the cellulose fibers. However, the resin retention exceeds that point. The amount of cationic resin adsorbed on cellulose fibers and the paper wet strength show two distinct slopes [720]. The break point is related to a zero streaming current, after which retention is most likely due to hydrogen bonding and/or van der Waals' forces. The excess of cationic resin changes the electrophoretic mobility of the pulp–resin complex [716].

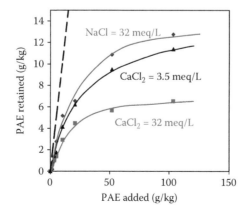

FIGURE 5.28   The effect of the electrolytes on the PAE resin retention (the dotted line is calculated based on 100% retention).

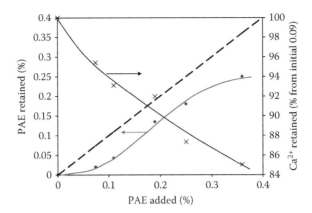

FIGURE 5.29 PAE resin adsorption in the presence of Ca^{2+} cation. (From Strazdins, E., *Tappi Papermakers Conference*, Atlanta, GA, 1974, pp. 59–65. With permission.)

Adsorbed WSR can migrate and the macromolecular coil can change its shape. The migration of cationic polymers from the fiber surface to its interior can diminish the effectiveness of the WSR [7].

Most of wet-strength resins are cationic polymers. The retention of the anionic wet-strength resins (such as UF resin) is performed with alum: more the alum, more the resin is retained within a faster process [698]. Anionic poly-acrolein needs retention aid but its retention shows a lower yield: less than 70% of the resin is retained on the pulp [332]. The retention process through a retention aid appears to be more complicated and more difficult to control.

The major portion of water is adsorbed (as multilayer) [721] in the amorphous region, which may have an effect on the interfiber bonding. However, hydrogen bonding is a reversible and dynamic process: even in the presence of water in large excess, segments of amorphous cellulose can rearrange and re-form an organized structure.

How the paper swelling process can be controlled is a major question when one attempts at describing the wet-strength mechanism. The swelling process should also be associated with the water "structure' [722], its availability for interactions with cellulose, and its effect as plasticizer for cellulose [723]. Plasticizer can greatly reduce the glass transition temperature, making the polymer chains more flexible.

At the wet end, the wet-strength resins are added to plasticized fibers, which have crystalline and amorphous zones. The wet-strength resins hold together a fiber mass, within which the fibers themselves have substantially no cohesion [724]. WSR retention takes place in the presence of a huge volume of water as continuous phase and involves ionic interactions with cellulose fibers. The drying step complicates the picture even further: WSR may be involved in the new interfiber hydrogen bond formation, in the protection of the existing hydrogen bonds and in the formation of new covalent bonds. In other words, the presence of wet-strength resins may change the paper rewettability and interfiber bonding by physical and/or chemical processes.

The forces required to bring fiber surfaces into close contact and thus consolidate the paper web are provided by water itself as it departs during drying [725] (the tensile index of the wet web starts to increase [726] at a concentration of fibers higher than 50%).

New covalent bonds between an additive and the cellulose fibers may have a strong impact on paper properties. For a given pulp type [727], water absorbency depends on the presence of cross-linkers (such as formaldehyde or methylolated urea) [728–730]: the increase in wet strength is correlated with a lower swelling index. Other potential cross-linkers are butane tetracarboxylic acid or PAE resins, which reduce the pore size of the cellulose fibers [291] and the water access to the hydroxyl groups.

5.5.3 Diverging Views on the Wet-Strength Mechanism

Water added to dried paper damages the cellulose fiber structure, which results in a dramatic decrease in paper strength. Retained WSR reduces this deterioration of the paper structure. The effect of the WSR is similar to that of the hemicellulose–lignin matrix on wood strength, in the presence of water, by holding the cellulosic fibrils together in the wood cell wall [731]. The degradation of this matrix may allow the fibrils to slide over one another [732].

The addition of a very small amount of WSR (less than 2%) increases the residual wet strength of paper from 5% to about 25% of its dry strength. However, even at high WSR concentrations, it is next to impossible to bring the paper wet strength to the same level as its dry strength.

It is rather difficult to explain the paper wet-strengthening mechanism. The first reason for that difficulty is the large number of parameters to take into consideration: the chemical structure of WSR, the charge density, the molecular weight, the resin concentration in the final paper, the temperature, the formation of ionic and covalent bonds, the ionic strength of water, the pH, the specific area of paper (2–$10\,m^2/g$ [733]), the carboxyl group concentration on fibers (2–$20\,meq/100\,g$ [734]), the curing time, etc. Most of these data are missing in studies developed for resins associated with any particular furnish.

The final paper (cellulose fibers and adsorbed macromolecular compounds) may be regarded as a composite [670]. After the drying step, the chemical composition and the relative distribution of the components have changed. After soaking in water the properties of the cellulose fiber are modified and the number of interfiber bonds is altered.

The loss in paper strength on rewetting is not related to the change in the degree of polymerization of cellulose but to the plasticization by water of cellulose fibers [723]. Therefore, changes in the fiber architecture (such as fiber size and distribution, porosity, etc.) and the interfiber bonds are responsible for the low wet strength of paper soaked in water.

Regardless of the hydrophilic character of "the repeating unit" of cellulose (glucose), cellulose, which still has three hydroxyl groups for each repeating unit, is no longer soluble in water. The association of macromolecules, within the cellulose fibers, through intermolecular hydrogen bonding allows for a lower number of possible interactions of those hydroxyl groups with water molecules [735]. Water cannot disrupt those hydrogen bonds and, in that case, the zipper effect does not function.

The failure of paper under wet stress shows that only 2% of the fibers are broken in the presence of the MF resin, while over 60% of the fibers are broken during the dry tensile measurements [679]. It is unlikely that the WSR can penetrate the fiber surface. It is much more reasonable to consider that WSR adsorbs on the fiber surface and improves only the interfiber bonded area.

It is generally accepted that different WSRs are effective through different mechanisms [671]. In Figure 5.30 the wet tensile values for three different resins are shown as a function of the retained resin [706]. Each resin involving a different chemistry provides a different wet

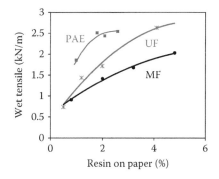

FIGURE 5.30   The effect of the amount of adsorbed WSR on the paper wet strength.

strength for paper. The PAE resin is more efficient than the UF and MF resins, especially at lower concentrations.

Wet strength depends both on resin adsorption and its reactivity. Different chemical structures (functional reactive groups), involved in a curing process, have different activation energies at a given temperature. A corona treatment also accelerates the curing reaction [736].

The possible modes of action of various wet-strength resins [23] are very diverse: ionic bond formation with fibers; formation of covalent bonds with fibers ("co-cross-linking" or "reinforcement"); prevention of water penetration into the fiber and the interfiber area ("protection" mechanism); formation of secondary valence bonds that are not disrupted by water; formation of an intimate physical mixture of additives and fibers by interdiffusion; development of a separate network by self-cross-linking reactions. It is quite likely that parallel mechanisms are operative and one mechanism can be dominant based on the wet-strength resin type.

The large number of parameters and the very low concentration of WSR in the final paper make it very difficult to study the paper-strengthening mechanism. That is why, in the absence of direct measurements, indirect data are considered valuable [535].

5.5.3.1 Are Cellulose Fibers Involved in New Covalent Bond Formation?

The most compelling mechanism would be the one involving covalent bond formation between different fibers through the reactive functional groups of the wet-strength resin. Cross-linking reactions with formaldehyde, in the presence of acid, strongly improve the paper wet strength [18] and reduce its specific surface [723]. That is an indication of the cellulose fiber involvement in the covalent bond formation [671]. Moreover, the C_6 hydroxyl oxidation of cellulose fibers to aldehyde [737] reduces the paper density and dry strength due to a reduction in hydrogen bonding. The hemiacetal bond formation, between the hydroxyl groups of one fiber and the aldehyde functionality of the other, increases the paper wet strength.

Several commercially available WSRs have a macromolecular structure and reactive functionalities, such as methylol in the UF and MF resins, or the azetidinium cycles in the PAE resins. This reactivity of the wet-strength resin supports the idea of covalent bond formation. However, the WSR concentration in paper is very low and only a small number of reactive groups may be involved in chemical reactions. The number of new covalent bonds is very small; common analytical methods can hardly identify them.

The covalent bond formation takes into account the chemical reactivity of cellulose fibers [738]. The number of hydroxyl groups available for reaction, the swelling degree of amorphous cellulose, and the reorientation of the WSR molecule after reactions could be part of the covalent bond formation mechanism.

The idea of interfiber bonding by covalent bonds developed within a reaction with the WSR is supported by the fact that WSR contributes to the paper dry strength as well [7]. Dry strength cannot be related to "the protection of existing hydrogen bonds against water interaction." On the other hand, those resins are also self-cross-linkable, and the interfiber covalent bonds can increase the number of hydrogen bonds. Thus, a mixed mechanism can also be possible [670].

In order to verify the homo-cross-linking versus the co-cross-linking mechanism, several model compounds were synthesized [739–743]. There are two different groups of potential WSR: one is designed to avoid chemical interaction with cellulose and the other prefers to react with the functional groups existing on cellulose fibers.

The polymer synthesis was performed either by direct polymerization of the appropriate (co) monomers or by polymer-analogous reactions. Ideally, the models should only differ from each other in terms of their chemical behavior during the papermaking process. However, it is hard to believe that different copolymers obtained through different reaction pathways will show the same retention, the same molecular weight, the same charge distribution, and the same physical interactions with the fiber surface.

1. Polymers intended for *homo-cross-linking* have both functionalities (amino and vinylsulfonyloxy) in the same molecule (self-cross-linkable).

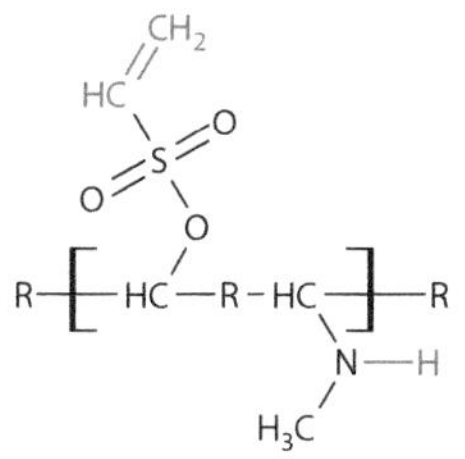

The homo-cross-linking reaction is an intermolecular Michael addition developed in neutral and alkaline pH. Paper strengthened with that WSR shows a wet strength of 11%–18% from the initial dry strength [744]. Auto-oxidation for poly-allyl glycidyl ether [317–322] is another reaction designed to improve the paper wet strength, in which the cross-linking process does not involve cellulose. Thus, the self-cross-linking reaction cannot be rejected as a possible part of the wet-strength mechanism. However, a higher concentration of polymer is required to improve the paper wet strength when both Michael addition and auto-oxidation reactions are used as self-cross-linking mechanisms.

2. *Polymers with pendant vinylsulfonyl groups for cross-linking cellulose fibers*
In order to get univocal reactions between cellulose fibers and WSR, the fiber surface has been modified by adding more reactive functional groups [314]. The WSR is designed to have only one type of reactive group, which prevents the self-cross-linking reaction. The homopolymer with vinylsulfonyloxy groups may react through a Michael addition with cellulose only. Other polymers intended for co-cross-linking were equipped with ketene or *N*-hydroxysuccinyl functionalities.

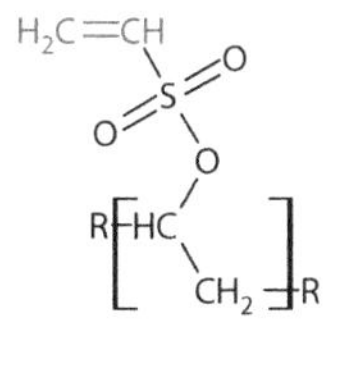

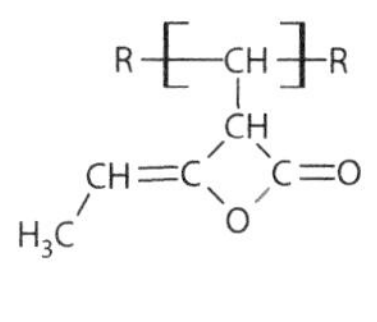

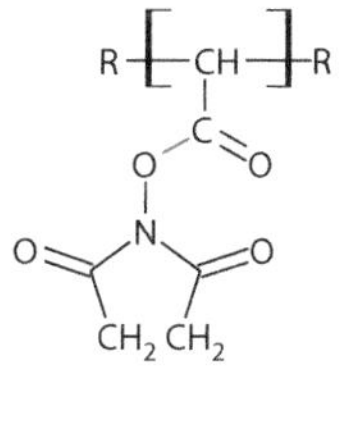

The reactive bonds (in red) are able to react with the hydroxyl groups. The expected covalent bonds formed between cellulose and the wet-strength resin are ester or ether bonds. The co-cross-linking reaction is more effective: the paper wet strength reaches up to 39% from its dry strength [744], which proves that the paper wet-strengthening mechanism works through covalent bonds within a co-cross-linking reaction.

3. *Wet-strengthening by the entanglement mechanism* requires that the originally soluble polymer could be rendered "insoluble" during the drying of the paper without the formation of any cross-links [739]. A macromolecular compound may become insoluble due to polar group interactions (see polyethers–polycarboxylic acids) or change the resin solubility through a unimolecular reaction, such as the elimination of hydrophilic moieties.

Polymers that support the third mechanism have multiple quaternary ammonium pendant groups located at the right distance from a carbonyl (or sulfonyl) group. The base-catalyzed elimination of the hydrophilic part (the quaternary ammonium group) is performed through a Hofmann degradation of the quaternary ammonium base.

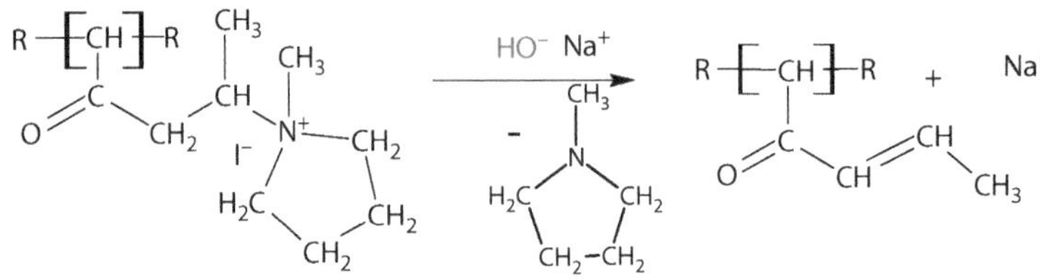

The entanglement mechanism resulted in fairly modest levels of wet strength: only 18% from the initial dry strength [744], which is still a confirmation of that concept.

The model compound study did not clearly rule out any of the possible strengthening mechanisms. The covalent bond formation with cellulosic support, the self-cross-linking of the WSR, and the entanglement mechanism are still valid explanations for how WSR might work to improve paper wet strength.

Formaldehyde (up to 7%) [745], polycarboxylic acids, or poly-aldehydes [295] develop covalent bonds with cellulose hydroxyl groups. Polycarboxylic acids react with the cellulose fibers (curing temperature 140°C–160°C) in the presence of a catalyst (sodium hypo-phosphite NaH_2PO_2) [190]. Paper with cross-linked fibers shows up to 60% lower water absorption and lower specific surface (from 1.8×10^6 cm^2/g to 0.7×10^6 cm^2/g) [723,746]. Due to the cross-linking reaction the paper wet and dry strengths are higher, and some other properties, such as stiffness, are altered as well.

The commercially available WSRs (UF, MF, and PAE) are different from the cross-linkers mentioned above: they are cationic and have a higher molecular weight. The behavior of these resins is somehow ambiguous: they may both cross-link cellulose fibers and homo-cross-link. During the paper drying step, the methylol groups of UF or MF resins react with cellulose or are dehydrated to form methylene and/or methylene ether bridges between different macromolecular units to generate a cross-linked structure which involves WSR only [7]. The curing process is developed at acid pH. Paper wet strength drops by about 40% when the stock pH is increased from 4 to 7 [747]. UF and MF resins boost both paper wet and dry strengths [21].

The chemistry of the azetidinium functionality is very interesting [558–561]. The reactive azetidinium cycle of PAE-type resins may polymerize (cationic mechanism) [748–753], or can react with carboxylic acids and amines [754,755]. Both tertiary amines and carboxylic groups are reactive and thus PAE resin may cross-link cellulose fibers and/or homo-cross-link to make an independent network.

PAE resin retention is based on the ionic interaction between azetidinium and the carboxyl anion from cellulose fibers and thus two reactive groups are close to each other. The carboxyl group reacts with the azetidinium cycle resulting in an ester bond formation [7]. It is well documented that PAE resins react with the carboxyl groups from the carboxymethylated cellulose fibers to form the ester bonds shown in the IR spectrum [542].

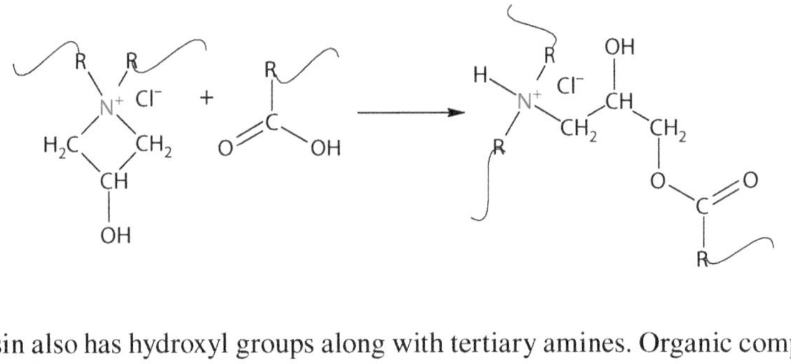

The PAE resin also has hydroxyl groups along with tertiary amines. Organic compounds with an azetidinium cycle [559] react with ammonia or with primary and secondary amines [756].

The azetidinium cycle is also reactive to the tertiary amine. For PAE wet-strength resins, a tertiary amine (like N-methyl bis-(3aminopropyl) amine) in the backbone provides a faster curing process [519] than PAE made from DETA [112].

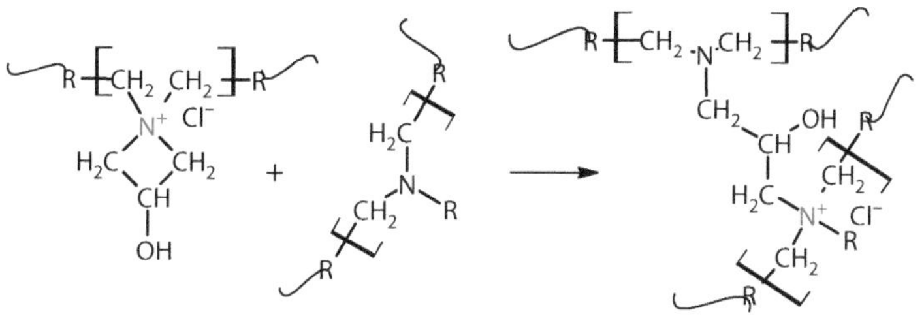

Based on this self-cross-linking mechanism, the quaternary ammonium concentration does not change. Another curing mechanism was suggested by Bogaert et al. [324] for the cross-linking reaction of poly-methacrylates of 1-alkyl-3-hydroxyazetidines. This mechanism involves two azetidinium ions and the resulting cross-linked polymer has a lower number of cationic charges. The self-cross-linking reaction simulates the evolution of the molecular weight (and viscosity) during the PAE synthesis.

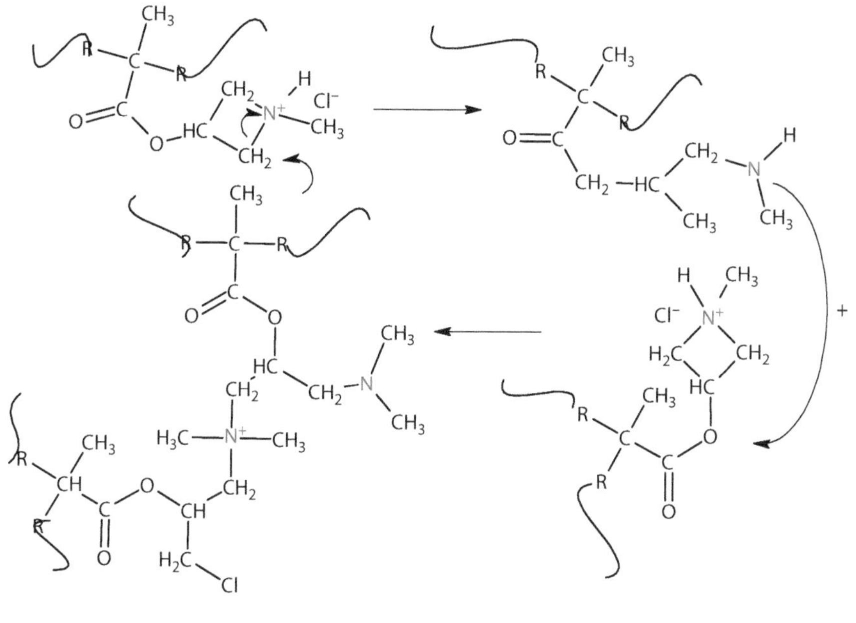

The azetidinium concentration decreases (while a steady-state tertiary amine concentration was maintained) during the thermosetting reaction of the PAE resin. The curing process follows a pseudo-first-order kinetic [519].

Starch with blocked reactive groups reacts with cellulose, resulting in higher paper wet strength, as compared with regular cationic starch [212]. That reaction can also be both a co-cross-linking and a homo-cross-linking process.

Cellulose fibers have two types of potentially reactive groups: hydroxyl and carboxyl. Changes in chemical structure take place during the curing step, but the nature of those new covalent bonds (co-cross-linked or self-cross-linked) seems to be undecided. The ideas presented so far leave open the topic of which wet-strength mechanism prevails. Another detail makes the picture even more complex: the chemical bonds developed in paper strengthened with UF, MF, and PAE resins are very strong, and strengthened paper is not repulpable.

There have been several attempts to solve the problem concerning which wet-strengthening mechanism is involved when WSR is added to paper. Kinetic and thermodynamic studies were performed along with the use of model compounds to simulate possible reactions.

These studies involve many experimental difficulties because the resin concentration in paper is very low. Moreover, the involvement of functional groups into covalent bond formation with cellulose is difficult to identify due to the presence of other potential reactions with water and other paper chemicals (starch, CMC, anionic trash, etc.).

Both WSR and cellulose have reactive groups in large numbers. However, it was found that only a minor fraction of the reactive groups of the PAE resin was consumed when the resin alone was cured to a water-insoluble film [670] and therefore few covalent homo-cross-links are very effective in imparting wet strength. Less than 20 out of a total of 70 reactive sites present per 100 repeating units are sufficient for full wet-strength development [23]. However, the absorbance due to the ester bond formation in the PAE-treated handsheets increased nearly equal to that in the case of the cured PAE films [538].

There is some other evidence for the PAE resins homo-cross-linking process: different pulp furnishes give similar percentages of wet/dry ratio when treated with PAE resin at the same concentration [20,650], which suggests that the pulp is not directly involved in the chemical reactions.

The self-cross-linking of the UF resins [671] is still an option because thermodynamic measurements show that the activation energy for the curing process is similar to that for resin cured in the absence of cellulose [323,687]. The average activation energy associated with the wet-strength development is about $22.6 \pm 0.8\,kcal/mol$ for the cellule fibers and UF resin, which is almost identical (within the experimental errors) with that for glass paper and UF resin ($23.8 \pm 0.6\,kcal/mol$) [687]. This is a strong indication that the reaction leading to wet strength is independent of the fiber.

In order to check the strengthening mechanism, the reaction was performed with model molecules and at higher concentrations of the reaction partners. It was found that N,N-dimethyl-N'-methylolurea, which has an active methylol group, did not react with the hydroxyl group of cellulose [687]. Model molecules of cellulose (such as sucrose or methyl-α-glucopyranoside [440]) did not react with the UF resin, with the PAE, or with the MF resins [757].

The PAE reactions with the model compounds (acetic acid, ethanol, dibutyl amine, methyl-β-glucoside) [23] were performed and, within the experimental errors, no covalent bond between the azetidinium ring and acetic acid or ethanol was identified [519]. There is negligible bonding of glucoside to the insolubilized resin either at pH=6 or under alkaline conditions [23]. The extent of the co-cross-linking process (if any) appeared to be rather limited [669].

However, the H-NMR data are not accurate enough to reach the final conclusion concerning the curing mechanism for PAE as WSR. If the chemical bond formation between WSR and cellulose is unlikely, and the separate cross-linked structure is formed based only on the very low polymer concentration, it is very interesting to explore the effect of the change in chemical structure on other properties of fibers.

5.5.3.2 To What Extent Does Hydrogen Bonding Explain the Paper Wet Strength?

The formation of covalent bonds would increase both the dry and wet strengths. Surprisingly, in the presence of wet-strength resins, the paper wet strength is much more improved than its dry strength. A little change in the paper dry strength [687] was found to accompany the substantial increase in wet strength. For instance, 1.2% PAE increases the dry strength by 35% while, for the same paper, the wet strength increases by about 20-fold [670]. In the case of the PEI resin, when the polymer content in paper increases from 0% to 1.54%, the dry strength increases from 18.57 to 22.18 lb/in. (20%), while the wet tensile strength ranges from 0.69 to 3.78 lb/in. (5.5-fold), and the W/D ratio changes from 4% to 17% [101]. Because the number of new covalent bonds (if any) is low, the effect of hydrogen bonding on the paper wet strength must be revisited.

If the paper dry strength is generated by the hydrogen bond formation, the loss of strength of paper on rewetting could be the result of breaking the interfiber hydrogen bonds by water. To accomplish this, water must penetrate and have access to the bonding area.

The importance of the hydrogen bond formation can also be explored by using polymers which do not have any chemical reactivity toward cellulose. This is the case of PEI [707], PVAm [363], or blends of nonreactive anionic and cationic resins [616]. Those WSRs can only be involved in ionic and hydrogen bond formation.

Ionic bonding is considered very important in the case of PEI [758]. The mechanism put forward by Trout in 1951 [101] was an ion exchange mechanism: anionic species from cellulose fibers exchange their cation with the cations delivered by PEI [704,707]. The fact that the paper needs to be dried in order to enforce this interaction is hard to explain.

PEI and polyvinyl formamide are by far the polymers with a much higher charge density per gram than pDADMAC. PEI and polyvinyl amine are "isomers" with the same general formula (C_2H_5N) and a charge density of 22 meq/g in acid pH (about 10 times higher than the regular PAE resin). The very short distance between the potential cationic charges in PEI and the molecular capabilities to rotate around the simple covalent bonds are also factors which must be taken into account. More frequent cationic charges will better fit the random distribution of anionic charges on cellulose fibers: it is more likely to couple an anionic group with a cationic group through the macromolecule re-conformation. The larger number of ionic bonds could be the reason for the pseudo-irreversible PEI retention. However, the transport and reorientation of macromolecules on the cellulose fiber surface is a time-consuming process [672].

Ionic bonds are 2–3 times stronger (10–20 kcal/mol) than hydrogen bonds (only about 5 kcal/mol). By consequence, a polymer with multiple electric charges is an interesting candidate as wet-strength resin, even in the absence of any reactive functionality. This idea led to the concept of modifying the cellulose fibers themselves with pendant groups carrying zwitterions (N6-(1,3,5-triazin-2-yl)-L-lysine) [725]. Modified fibers show a significant higher paper wet strength without the involvement of any other strengthening agent.

Polyvinyl formamide partially hydrolyzed (11% cationic units) is capable of providing both dry and wet strengths [361]. Figure 5.31 shows that the high wet strength is obtained only at higher polymer concentrations. This kind of graph is representative for wet-strength resin performances (see also Figures 5.4 and 5.5). The only difference is the amount of resin needed to reach a certain level of the *W/D* ratio.

The importance of ionic interactions is challenged by the fact that poly-DADMAC, a cationic polyelectrolyte, does not improve wet-strength. pDADMAC, which has a very rigid backbone, can interact with cellulose fibers through ionic bonds but cannot be involved in any hydrogen bonding.

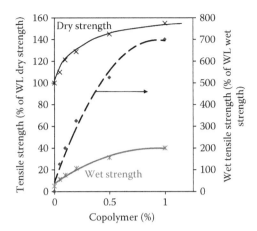

FIGURE 5.31   Copolymer vinyl amine–vinyl formamide as WSR. (From Wang, F. and Tanaka, H., *J. Appl. Polym. Sci.*, 78, 1805, 2000. With permission.)

Modified PEI with epichlorohydrin [513] shows the same performances as the PAE resin. The potential cross-linking definitely helps in the wet-strengthening process. The effect of the thermosetting resin on the paper wet strength can be explained either by the restriction on water access or by the increase in the glass transition temperature of cellulose [723]. The adsorbed PAE resins may change the macromolecular coil shape in the presence of electrolytes: higher conductivity (>2 mS/cm) reduced the performance of the resin [712].

The results obtained on UF resin (homo-cross-linking) are compatible with the hypothesis that wet strength is largely due to the protective action of resin on the interfiber hydrogen bonds. A necessary condition for the high degree of wet strength may be the penetration of resin into the amorphous cellulose of the interfibrillar regions [440].

The mechanism of sheet reinforcement by latex particles must take into account that more polymer is needed (up to 10%) to reach a reasonable wet strength. In that case no reaction takes place and no new ionic bond is formed. The hydrophobic polymer particles incorporate many macromolecules; their effectiveness is reduced because the macromolecule coil is smaller. Thus, a small cellulose fiber surface is covered by a thick polymer layer. However, a thicker polymer layer can increase the number of interfiber bonds: the interfiber bonds are substituted by polymeric bonds [759].

5.5.3.3   What Would a Protective Mechanism Look Like?

Paper is an aggregate of cellulose fibers. Cellulose fibers are formed from linear cellulose macromolecules which interact with each other through intramolecular and intermolecular hydrogen bonds. Paper disintegrates after soaking in water but fibers retain their shape because the "internal" hydrogen bonds are, somehow, not broken by the water molecule. There is something in the hydrogen bonds' organization within the fibers that makes them behave not as individuals but as an "association," or as a system of reciprocal protection against the zipper effect.

Due to a large number of intermolecular hydrogen bonds, there is an extended hydrophobic area resulting in the reorientation of the macromolecules (see crystalline structure of the cellulose fiber). The hydrophobic–hydrophobic interactions can improve the fiber strength and prevent water penetration into fiber structure. Swelling occurred in hydrophobic organic solvents, which proves the existence of a hydrophobic part in paper [760].

The failure of paper shows that only 2% of the fibers are broken under wet stress [687]. The intermolecular hydrogen bonds developed in interfiber bonding area are broken when paper is soaked in water. If the hydrogen bonds within the cellulose fibers are compared with those developed between fibers, it seems that the interfiber hydrogen bonds are not protected against water aggression.

The fiber internal strength can be a model of how to improve the interfiber strength. There are several ways for a wet-strength resin to interact with the existing interfibers bonds [667]: making the hydrogen bonding less sensitive to the swelling and dissolving action of the water; bringing hydrophobicity; increasing the number of hydrogen bonds; adding ionic and covalent bonds which restrict the freedom of the macromolecules involved in interfiber bonding will improve the paper wet strength. The higher molecular weight of WSR has the benefit of pseudo-irreversible ("partially irreversible" [735]) interactions as well.

The amorphous zone is the most likely interfiber bonding area because cellulose macromolecules from different fibers can be entangled. The entanglement will generate a large number of hydrogen bonds. The paper dry strength is based on those hydrogen bonds. The amorphous part is also highly susceptible to the swelling effects produced by water [706]. The structure swollen in water has a larger distance between molecules and, consequently, a lower number of intermolecular hydrogen bonds (see "zipper" effect [686]; see Figure 5.25). The increase in the paper wet strength can be due to the protection of hydrogen bonds against water aggression.

In has been shown [187] that impregnation with polyethylene glycols greatly reduces paper swelling. In that case no chemical reaction was developed. However, water penetration in the paper structure has been changed by the presence of a polyether (or polycarboxylic acid, poly-alcohols [761–763]) able to occupy the space around the hydroxyl groups. The macromolecular character of polyether makes the interaction pseudo-irreversible and may retard the dissociation of hydrogen bonds.

A dynamic process of breaking and re-formation of hydrogen bonds takes place within those polymeric aggregates [735], but the total number of intermolecular bonds is constant even in the presence of water. In these aggregates more than two polymers are involved. If two of those polymers are cellulose molecules and the third is WSR, then interfiber interaction can be reinforced.

Hemicellulose can be the third component. Hemicellulose is very much involved in the strengthening process and is responsible for the paper dry strength [764–766]. Water has very little impact on the T_g of the crystalline cellulose [767] but hemicellulose shows a much lower softening temperature when water—as a "solvent"—is present. The hemicellulose sensitivity to water is part of the explanation for the weakness of wet paper.

Ethylenediamine does not show any effect on the paper wet strength but a low molecular weight polyamine (pentaethylene–hexamine) does impact the paper wet strength [758]. Unlike most other WSR, PEI is essentially a nonreactive polymer, with high charge density and relies on its strong, largely irreversible, adsorption to the cellulose fibers [704,768]. Polysaccharides have been found to give precipitates (complex formation) when mixed with wet-strength resins [762] (see also the complex formation between cellulose and amines and polyamines [769–772]).

The tight aggregate between polyamine and cellulose may reduce the "zipper" effect by water molecules. Hydrogen bond formation between amino and hydroxyl group is well documented [773].

The number of bonds between fiber and resin is a function of charge density and molecular weight. The number of interactions between PEI and cellulose can be increased by increasing the number of anionic charges of the cellulose fibers (modified fibers) [758], or by increasing the PEI molecular weight (reaction with formaldehyde) [101]. With a higher number of ionic bonds, wet strength increases tremendously for the same amount of retained PEI with a molecular weight of 75,000 (Figure 5.32). That means that a higher number of ionic bonds is probably able to make a tighter structure, and the water effect ("zipper") is slowed down.

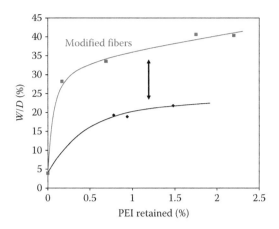

FIGURE 5.32 The interaction of PEI resin with modified cellulose fibers.

Water penetration into the interfiber bonding area can be controlled with small amounts of hydrophobic material [774]. Wet strength was improved by simply adding an internal sizing agent (rosin) [775], lecithin [776,777], or dextrane palmitate [778]. The acetalylation of the hydroxyl group makes the fiber more hydrophobic and less sensitive to water penetration; therefore the wet strength of acetylated paper is higher [728,779].

Polycationic wet-strength resins with no reactive groups, but with hydrophobic tails [365], are more efficient WSRs. A more hydrophobic PAE resin [177,430,431,780] provides a higher wet/dry tensile ratio. Hydrophobic polymers like PVA reduce the water adsorption of cellulose [781]. AKD along with temporary wet strength [654] also enhances the wet strength of board.

Latexes [782] can significantly improve wet strength without any new covalent bond formation, with no redistribution of the cationic charges, and no macromolecular penetration in the cellulose pores. Paper treated with neoprene latex shows a very low strength decay after soaking in water for about 120 h at 25°C [440]. Wet strength seems to be related to the hydrophobic interactions and the formation of a polymeric film sandwiched between the fibers [783].

The hydrophobic properties of the cellulose fiber surface are improved by ionic charge neutralization with another polymer. The fiber surface behaves like an anionic polyelectrolyte gel. After the adsorption of cationic wet-strength resins, a polyelectrolyte complex (anion–cation) is formed on the cellulose fiber surface. The aggregate becomes more hydrophobic (even before the drying step), water is expelled, and the aggregate shrinks [784]. By rejecting water molecules from the interfiber volume, more hydrogen bonds are formed. Their sensitivity to water is lower due to the more hydrophobic character of the polyelectrolyte complex.

The PAE resin can be made more hydrophobic in a complex with an anionic copolymer (ethylene–maleic anhydride) modified with N-methyl-N-octadecylamine [623].

The hydrophobic character of the wet-strength resin dual system increases the paper wet strength by 25%. Blends of PAE resin and a reactive anionic polymer [780] also enhance it. However, the aggregate formed between the cationic PAE resin and the poly-anionic reactive polymer (SMA reacted with *N*-methyl-*N*-octadecylamine) will add more homo-cross-links in which cellulose is not involved.

The additivity rule follows Equation 5.10 [538,610]:

$$DT_r = DT_b + (WT_r - WT_b),$$ (5.10)

where
  WT is wet tensile index
  DT is dry tensile index
  r stands for "resin-treated handsheet"
  b stands for "blank (untreated) handsheet"

The dry tensile index meets the additivity rule and therefore the wet-strength mechanism of the PAE resin can be explained by a "protection mechanism."

The curing process is a chemical modification reaction of the WSR: the hydrophilic methylol groups (or the azetidinium cycle) are converted to the more hydrophobic carbon–carbon bonds. The rate of water adsorption is reduced by acid-curing wet-strength resins, such as UF or MF resins [628,785]. The reduction of water adsorption can be due to a protection system that reduces the water access to the hydrogen bonds, or to the reaction between the hydrophilic groups and the wet-strength resin. It has been suggested [55] that wet strength is the result of the formation of MF resin bonds which are not sensitive to water, and of a protective "shielding" of the cellulose-to-cellulose hydrogen bonds.

The protection mechanism will allow the wet-strength resin to be adsorbed in the amorphous zone and will prevent the zipper effect of water molecules on the interfiber bonds by ionic, covalent, and hydrogen bonds. Physical and/or chemical interactions between fibers and WSR bring segments of macromolecules in short range where the hydrogen bonding is effective (<0.2 nm). The angle between a given hydrogen bond donor and acceptor is also restricted to a narrow range [786]. Therefore, the resin merely reinforces and protects preexisting hydrogen bonds and thus the water sensitivity decreases and the paper wet strength increases [706,787].

The aggregate stability depends on the molecular weight of the wet-strength resin. A polymer with a higher molecular weight is more likely to develop a larger number of hydrogen bonds with an increased probability for a rigid structure. However, the resin–cellulose cross-link formation cannot be completely ruled out, but if those cross-links do exist, they only help the protection effect by freezing the polymer coil shape.

The protective mechanism is a hybrid one: the new chemical bonds (ionic or covalent) can be formed but their effect consists in not only the increased number of links but also in the increasing rigidity of the polymeric aggregate and boosting of the protection of existing bonds. Homo-cross-linking [788] and/or co-cross-linking increase the molecular weight as well, which makes the structure even less flexible due to a larger number of possible interactions. The ionic, hydrogen, and hydrophobic bonds are all conducive to higher paper wet strength [712]. The cooperative effect of the "hydrophilic" strength (hydrogen bonding protection) and "hydrophobic" strength should be incorporated in order to have a better picture of the paper wet-strengthening mechanism.

5.6 PAPER REPULPABILITY

Paper wet strength is an important property for both paper application and production. The fact that the wet strength is more or less "permanent" raises the questions about broke recycling [789] and the ability of the final paper product to be reintroduced into the papermaking process (repulpability). In other words, the pulp and paper industry was concerned about sustainability long before that concept became popular.

In 2008 there was a 56% recovery rate for fiber, the goal being to reach 60% [790]. However, the quality for recovered fiber is lower and the amount of recycled paper in the blend with virgin fiber is only about 34%. Due to higher repulping temperatures and repeated drying operations, the strength properties of recycled fibers are dramatically decreased with recycling [791,792]. After recycling the fines, crystallinity is increased, their surface is richer in carboxylic groups [793], the water retention value is lower [794], and the interfiber bonding is weaker (small bonded area [795]).

A milder repulping process is key in preserving fiber characteristics. WL readily disintegrates in water. Paper chemicals slow down the disintegration process because most of them are hydrophobic and/or thermosetting compounds [21,716]. In other words, the recovery of the cellulose fibers during the repulping process has to do with the paper chemicals incorporated in paper and mainly with the wet-strength resins.

Paper chemicals are designed to make cellulose fibers more hydrophobic and to increase the interfiber bonding. What used to be positive contributions during the papermaking are now obstacles for fiber recovery.

The ideal paper chemical would provide properties to the web for an indefinite number of repulping cycles. Ideally, only the interfiber bonds should be broken during the repulping process but they must re-form during the further papermaking process. The real repulping process is far from ideal and the current goal is to extract the cellulose fibers from paper by sacrificing all the paper chemicals previously used to make that paper. It is like a cleaning process which tries not to harm the fiber.

Commercial wet-strength resins fall into two categories. Temporary wet-strength resins provide paper wet strength, which decays by 50% in 5 min and about 90% in 1 h after soaking in water. Permanent wet-strength resins impart long-lasting wet strength. The wet strength of a sheet soaked for 2 h will typically be 80%–90% of the value measured after a 10 s wetting and does not fall much further after longer soaking [795].

The structure of the paper wet-strengthened with permanent WSR is different from WL, which makes the reworking of broke difficult [797,798]. To be readily repulpable, paper should disintegrate easily at a temperature of 38°C–60°C and a pH of 6–8 [654,799,800]]. The mild conditions are needed to prevent any chemical degradation of the cellulose fibers. A reduction in chemicals discharged in water after repulping is also important [801].

However, regular repulping procedures involve mechanical energy, heat, and chemical treatment in a very large volume of water. Special equipment [25,802] is required for higher pH and higher temperatures. At 135°C (obtained with superheated steam [25]), repulping is performed under pressure.

In order to preserve the cellulosic substrate unchanged, the repulping process should be selective. Due to the chemical architecture developed during the papermaking process, there are two approaches for repulping paper: one is related to improving repulping chemicals and equipment, and the second to making readily repulpable paper.

5.6.1 Fighting the Chemicals that Yield Permanent Wet Strength

Water is the continuous phase for the pulping process, and chemicals used for pulping are supposed to be soluble in water and penetrate the paper structure. Due to the presence of hydrophobic and cross-linked compounds, water penetration is slow. Better wettability is obtained when surfactants (sodium oleate [803] or nonionic compounds [804]) are added to the repulper.

The structure of the wet-strengthened paper includes strong (carbon–carbon, ether) bonds and weaker bonds: amide, esters, ionic and hydrogen bonds. In order to liberate cellulose fibers from that structure, the weaker bonds can be broken by hydrolysis or by oxidation. The cross-linked structure is broken into short segments which are dissolved in water as a part of that "cleaning." In other words, the ideal repulping process would be able to separate individual fibers by selectively disintegrating only paper chemicals (or the bond between the fiber and the wet-strength resin, such as the ester bond formed between a functional latex and cellulose [446]). The art of the repulping process lies in its selectivity: chemical compounds used to repulp paper should be active at lower temperatures, lower pressure, and moderate pH.

Hydrolysis can be performed in acid or alkaline pH. UF or MF resins are hydrolyzed at pH<3.0 with regular acids [48,805] or with alum [48]. Glyoxalated polymers are hydrolyzed at pH > 9 [796,806] and paper made with a PAE resin was repulped in aqueous NaOH at pH = 12 (temperature of about 85°C for 60 min) [113,135–137]. Carboxylated polymers incorporated in the coating formulas [807] are also hydrolyzed at higher pH.

Paper mills use chemicals available on their facility for the repulping process. Hypochlorous acid is obtained by continuous addition of chlorine to the repulper [808] and can oxidize the PEI resins. Sodium hypochlorite (NaOCl) is active in alkaline pH [789]. However, hypochlorites generate absorbable organic halides in paper mill effluents [809]. These substances are environmentally undesirable [804].

Alkali metal persulfates (oxidizing agents) have been used as reagents to replace hypochlorites [804] in repulping the paper strengthened with the PAE resin, but the process is slower than with hypochlorite [809]. As persulfate repulping proceeds, the pH may fall and a base [810,811] or buffer systems are recommended [812] (acetic acid/sodium acetate for 5.0<pH<5.5 [813]). Shorter repulping times were required at a pH of about 10 with sodium carbonate. The persulfate activity is enhanced by the presence of a copper salt [814]. For paper made with UF and MF resins as WSR, a mixed oxidizing agent (sodium hypochlorite and ammonium persulfate) is recommended [26].

The oxidizing agents, such as hydrogen peroxide or peracetic acid, are activated as repulping agents by the presence of ferrous sulfate [815]. Chemical treatment may combine caustic and oxidizing agents in order to help the repulping process [802]. The oxidizing process is performed in two steps: the first in acid pH (pH = 3 for less than 60 min) and the second at higher pH (pH = 11 for 30 min at 70°C) [804,816–818]. That procedure increases the repulpability yield up to 90%.

The effect of the repulping process parameters (temperature, pH, time, chemicals, etc.) must be evaluated within an adequate standard method [805] applied to both filtrate (for residual chemicals) and pulp. A testing method for repulpability is presented by Darlington and Lanier [654]: 110 g board, 2895 g DI water, temperature 120°F, pH = 6–7 in Maelstrom pulper. A 15 min soak time is recommended before turning on the pulper motor. After 15 min of agitation, a sample of pulp containing 11 g of dry fiber was taken from the pulper and screened through a vibrating plate, Somerville screen with 0.006 in. slots. The fiber retained on the screen and that passing through the screen were collected separately, dried, and weighted after equilibrating at 50% relative humidity and 23°C. The percent of the fiber fed to the screen that passed through was the repulping yield.

5.6.2 THE REPULPING MECHANISM

The wet strength of paper treated with UF and MF resins decreases at high temperatures and different pH [6]. The acid hydrolysis mechanism of UF resins [819] is a reversed condensation reaction.

The mechanism of the alkaline hydrolysis of PAE resins involves a cleavage of amide bond [716,819].

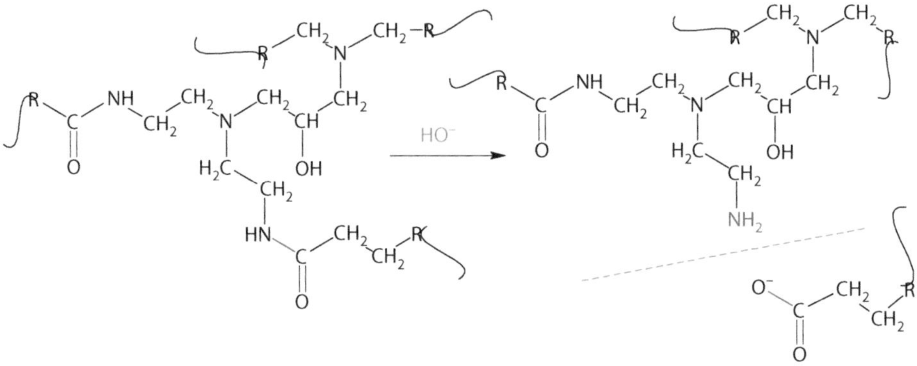

The hydrolysis process can be performed under milder conditions (pH = 9.0 and temperature 40°C) in the presence of enzymes [820]. The resin molecular weight was reduced and the presence of adipic acid showed an accelerated degradation of the polyamide structure after 45 min.

5.6.3 REPULPABLE PAPER

An analysis of the paper repulpability process is actually a review of the entire paper chemistry in a new perspective: what kind of paper chemicals (or binders in coatings [821–825]) are required, given the need for them to be recovered by the process as part of the recycled paper.

In order to set an easier protocol for paper recycling, the wet end chemistry was changed. The wet-strength paper seems to be developed based on the increased number of hydrogen bonds and fewer covalent and ionic bonds. The addition of the weaker bond (within WSR structure or in within the interconnection between WSR and cellulose fibers) can help the recycling process [654]. It should be no surprise that a larger number of amide bonds per gram of resin (due to the partial replacement of adipic acid with succinic or oxalic acid in polyamidoamine resin [143,144]) makes the repulping process easier [113,135–137].

A weaker PAE chemical structure was obtained by the reaction of the adipic–DETA prepolymer (large excess of DETA) with glyoxal to introduce weaker bonds. After the reaction with epichlorohydrin, the mixed cross-links and di-aldehyde bridges (double-base Schiff in red) are weaker during the repulping process [826,827].

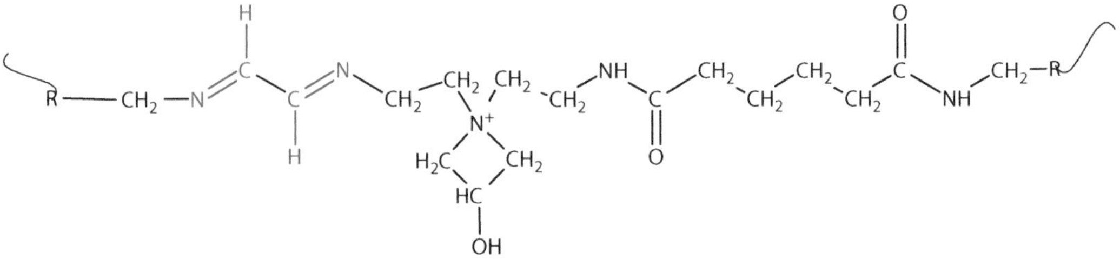

The sensitivity of WSR to hydrolysis depends on the weaker bond position. Because the azetidinium cycle is considered the reactive functionality in the reaction with cellulose fibers, it was located in a pendant group [161]. Now paper is easier to recycle because the weak bond (in red) is more accessible.

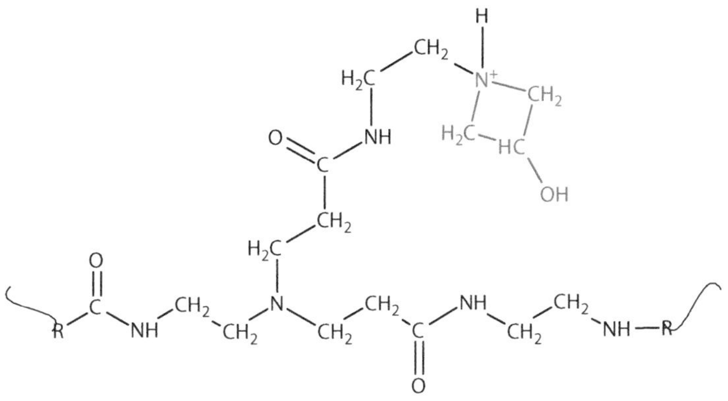

An example of weak bond is the hemiacetal bond presumably formed between aldehyde starch and cellulose (193–195). Broke and waste paper containing di-aldehyde gum can be readily recycled by subjecting them to an alkaline repulping system [214].

The weaker bonds are introduced in the MF resin through a reaction between the MF resin and starch [828]. The new resin provides a better dry strength without any significant effect on wet strength over the paper treated only with starch.

Blends of different resins as WSR lower the cost of energy for its recovery from recycled paper [196]. Some polymer blends (such as PAE and cationic PAAm) show the paper strength but allow for a reduction in the PAE dosage, which results in better repulpability [603].

The repulpability of the linerboard and newsprint paper obtained with blends of starch xanthate and PAE [218] shows better repulpability. An inter-polymer between PAE and ammonia–epichloro-hydrin resins seems to be easier to repulp as well [716,829].

Temporary wet-strength resins contain a large amount of un-reacted glyoxal and are also hydro-lytically unstable. Thus they open the possibility to obtain 100% repulpable paper. However, paper (or paperboard) strength should be high enough when wet for certain applications. It is suggested that a combination of wet-strength resin (0.01%, with a low repulpability rate) with temporary wet-strength resin (0.05%, with a high rate of hydrolysis), in the presence of internal sizing agents [799], can reach a reasonable balance between the wet strength and the rate of disintegration [655,656].

Blends of PAE resins with glyoxalated polyacrylamide [654,655,830] can be made before their addition at the wet end or can be added in a sequence. Any other carrier for glyoxal (such as poly-di-allyl amine or its copolymers with acrylamide [4,831]) is also usable to make the paper easier to recycle. The chemical interaction between PAE and G-PAm seems to be a must. Aldehyde groups (from G-PAm) react with secondary amines or amide groups (from PAE resin) [832], and an inter-polymer is obtained. That inter-polymer has a higher molecular weight and contains functional groups which may react with the cellulose fibers to form hydrolysable bonds. In order to opti-mize that interaction, the resin concentration and the contact time should be considered. Those two parameters are interrelated: a higher resin concentration asks for a shorter reaction time.

To show the balance between strength and repulpability, a relationship (Equation 5.11) between the repulpability index (RI), the paper wet tensile strength (WTS), and fiber yield (FY) [4,831] was put forward:

$$RI = \frac{WTS \times FY(\%)}{100}.$$ (5.11)

According to this formula, the PAE resin has a repulping index of 2.0, and a blend of 5.7.

Another type of weak bond is the ionic interaction: anion–cation. Blends of the PAE resin and anionic SMA copolymers provide a reasonably fast reworkability of the paper [833].

An interesting concept is that the cationic starches used at the size-press are able to improve the repulped fiber quality: improved filler retention along with higher strength were noted [834].

5.6.4 Improved Recycled Fibers

The use of recycled paper (in board grades and even in tissue manufacture [835]), which remains cheaper than virgin fibers, will continue to support the demand for a wide variety of paper chemicals to improve the quality of recycled fibers [836,837].

In order to regain the initial strength properties of untreated pulp, the treatment of recycled pulp with different chemicals (ammonia, caustic, organic solvent) was performed but with very little success [791].

One attempt is to increase the strength of the bonds between the chemicals and the cellulose fibers. Regenerated cellulose fibers need a treatment with high molecular weight polyacrylic acid to improve the properties of their blend with virgin fibers [838].

Another attempt is to incorporate chemicals that are able to go trough the repulping process and be retained in the recycled fibers when used to form a new paper web. Thus, cationic starch (made from oxidized starch cooked in the presence of cationic resins Epi + ammonia) [839] and a dual system (anionic + cationic polymers) [840] are tightly bound to paper and are still present during the repulping of broke.

Some paper chemicals (such as those from recycled coated broke) will form a separate phase (white pitch). Cationic resins (polyacrylamide with more than 20% DADMAC) coagulate the white pitch on the repulp fibers [841]. Thus, recycled fibers bring some old chemicals adsorbed on their surface in the mix with virgin fibers.

REFERENCES

1. P. Somboon and H. Paulapuro, *Tappi J.*, **8(5)**, 4–10 (2009).
2. A.J. Stamm, *Tappi*, **33(9)**, 435–439 (1950).
3. K.W. Britt, *Tech. Assoc. Papers*, **31**, 594–596 (1948).
4. R.R. Staib, J.R. Fanning, and W.W. Maslanka, US 6,245,874 (2001).
5. M.J. Smith, J.M. Kaun, and V.R. Gentile, US 5,993,602 (1999).
6. C.S. Maxwell and W.F. Reynolds, *Tappi*, **33(4)**, 179–182 (1950).
7. H.H. Espy, *Tappi J.*, **78(4)**, 90–99 (1995).
8. M.O. Schur, US 2,076,599 (1937).
9. G.A. Richter, M.O. Schur, and R.H. Rasch, US 1,745,557 (1930).
10. G.A. Richter and M.O. Schur, US 1,971,274 (1934).
11. G.A. Richter and M.O. Schur, US 2,096,976 (1937).
12. M.O. Schur, US 2,215,136 (1940).
13. S. Zakaria and J.C. Roberts, *Prepr. Int. Paper Coat. Chem. Symp. 5th* (2003), pp. 241–243.
14. M.A. Hubbe, D.G. Wagle, and E.R. Ruckel, US 5,958,180 (1999).
15. Scott Paper Company, GB 502,724 (1939).
16. K.W. Britt, US 2,325,302 (1943).
17. J.B. McKelvey, R.R. Benerito, R.J. Berni, and B.G. Burgis, *Textile Res. J.*, **33**, 273–281 (1963).
18. J. Russell, W.E. Carlson, and C.H. Mayhood, US 3,310,363 (1967).
19. R.K. Pinschmidt, G.E. Davidowich, W.F. Burgoyne, D.D. Dixon, and J.E. Goldstein, *ACS Symposium Series 367 Cross-Linked Polymers*, American Chemical Society, Washington, DC, 1988, pp. 467–478.
20. H.H. Espy, *Tappi J.*, **70(7)**, 129–133 (1987).
21. C.S. Maxwell, Dry strength from wet strength agents, in *Dry Strength Additives*, W.F. Reynolds (Ed.), TAPPI Press, Atlanta, GA, 1980, pp. 119–123.
22. D.K. Ray-Chaudhuri, US 3,891,589 (1975).
23. N.A. Bates, *Tappi*, **52(6)**, 1162–1168 (1969).
24. M.T. Crisp and R.J. Riehle, US 6,197,919 (2001).
25. M. Goehler, F.H. Reimer, and E.F. Davis, US 2,423,097 (1947).
26. R.R. House and Y. Jen, US 2,872,313 (1959).

27. B.J. Kokko and S.L. Edwards, US 7,585,392 (2009).
28. B. Freimark and R.W. Schaftlein, US 3,755,220 (1973).
29. K.W. Britt, US 2,399,981 (1946).
30. M.T. Goulet, T.H. Mathews, S.L. Pomeroy, and M. Tirimacco, US 7,297,231 (2007).
31. W.D. Harrison and R.E. Matthews, US 2,503,267 (1950).
32. T. Sun and J.D. Lindsay, US 6,610,174 (2003).
33. M.O. Schur, US 2,116,544 (1938).
34. J.C. Decroix, J.M. Bouvier, R. Roussel, A. Nicco, and C.M. Bruneau, *J. Polym. Sci. Symp.*, **52**, 299–309 (1975).
35. S.H. Rider and E.E. Hardy, *Adv. Chem. Ser.*, **34**, 173–190 (1962).
36. H.P. Wohnsiedler, US 3,305,436 (1967).
37. R.P. Foulds, J.T. Marsh, and F.C. Wood, US 1,734,516 (1929).
38. G.F.D. Alelio, US 2,247,772 (1941).
39. T.J. Suen US 2,639,242 (1953).
40. C.P. Vale and W.G.K. Taylor, *Aminoplastics*, Lliffe Books Ltd., London, U.K., 1964, pp. 20–60.
41. E.S. Valentine, US 3,455,860 (1969).
42. A.H. Conner, Urea-formaldehyde adhesive resins, in *Polymeric Materials Encyclopedia*, Vol. 11, 2nd edn., Joseph C. Salamone (Ed.), CRC Press, Boca Raton, FL, 1996, pp. 8496.
43. L. Henrikson and B. Steenberg, *Sven. Papperstidn.*, **50(23)**, 547–556 (1947).
44. E.E. Morse, US 2,757,086 (1956).
45. H.W. Nackinney and S.J. Schultz, US 2,642,360 (1953).
46. S.J. Hazard, F.W. O'Neil, and V. Stannett, *Tappi*, **44**, 35–38 (1961).
47. P.M. Westfall, N.R. Eldred, and C.S. Maxwell, *Tappi*, **49(1)**, 56A–58A (1966).
48. W.M. Thomas, US 2,394,273 (1946).
49. P.J. McLaughlin, US 2,765,229 (1956).
50. R.S. Yost and R.W. Auten, US 2,616,874 (1952).
51. J.B. Davidson and E.J. Romatowski, US 2,729,617 (1956).
52. A.M. Schiller and T.J. Suen, US 2,698,787 (1955).
53. R.P. Hofferbert, US 2,291,079 (1942).
54. R.P. Hofferbert, US 2,291,080 (1942).
55. W.F. Linke, T.F. Ziegler, E.C. Eberlin, and R.R. House, *Tappi*, **45(10)**, 813–819 (1962).
56. B. Steenberg, *Sven. Papperstidn.*, **49(14)**, 311–323 (1945).
57. C.S. Maxwell, US 2,986,489 (1961).
58. F.W. Boughton, US 2,548,513 (1951).
59. H.P. Wohnsiedler and W.M. Thomas, US 2,485,079 (1949).
60. H.P. Wohnsiedler and W.M. Thomas, US 2,485,080 (1949).
61. G.E. Myers, *Forest Prod. J.*, **36(6)**, 41–51 (1986).
62. C.A. Weisgerber, US 2,721,140 (1955).
63. N.R. Eldred, G.W. Buttrick, and J.C. Spicer, US 3,207,656 (1965).
64. J.B. McKelvey, B.G. Webre, and R.R. Benerito, *J. Org. Chem.*, **25**, 1424–1428 (1960).
65. J.H. Ross, D. Baker, and A.T. Coscia, *J. Org. Chem.*, **29**, 824–826 (1964).
66. V.R. Gaertner, US 3,247,048 (1966).
67. D.C. Babcock, US 3,224,990 (1965).
68. J.M. Baggett, US 3,947,383 (1976).
69. J.M. Baggett, US 3,655,506 (1972).
70. J.B. McKelvey, R.R. Benerito, R.J. Berni, and B.G. Burgis, US 3,351,420 (1967).
71. O. Stallmann, US 1,977,251 (1934).
72. H. Bachem, J. Muszik, W.-D. Schroder, and C. Suling, US 4,857,586 (1989).
73. A.T. Coscia, US 3,248,353 (1966).
74. R.H. Jansma, US 3,954,680 (1976).
75. K.W. Dixon and S.M. Menchen, US 4,575,527 (1986).
76. J.C. Bolger, H.E. McCollum, and R.W. Hausslein, US 3,577,313 (1971).
77. K.J. McCarthy, C.W. Burkhardt, and D.P. Parazak, *J. Appl. Polym. Sci.*, **34**, 1311–1323 (1987).
78. L. Wagberg and T. Lindstrom, *Nord. Pulp Paper Res. J.*, **2(2)**, 49–55 (1987).
79. J.H. Daniel and C.G. Landes, US 2,595,935 (1952).
80. J. Green, US 2,969,302 (1961).
81. W. Lehmann and O. Bayer, US 2,849,411 (1958).
82. S.A. Fischer and R.H. Grinstein, US 5,116,887 (1992).

83. S.A. Fischer and R.H. Grinstein, US 5,120,773 (1992).
84. N.S. Clungeon, and S.A. Fischer, US 5,492,956 (1996).
85. H.E. Fritz and R.P. Yunick, US 3,306,912 (1967).
86. J.P. Petrovich and D.L. Taylor, US 3,855,158 (1974).
87. J.P. Petrovich and D.L. Taylor, US 3,899,388 (1975).
88. J.P. Petrovich and D.L. Taylor, US 4,147,586 (1979).
89. Y. Jen and S.T. Moore, US 2,834,675 (1958).
90. T.J. Suen, Y. Jen, and S.T. Moore, US 2,834,756 (1958).
91. J.P. Petrovich and D.N. Van Ecnam, US 4,129,528 (1978).
92. S.J. Buckman, J.D. Pera, F.W. Raths, and G.D. Mercer, US 3,784,649 (1974).
93. R.J. Schaper, US 4,166,894 (1979).
94. P.M. Westfall, N.R. Eldred, and J.C. Spicer US 3,372,086 (1968).
95. J.M. Kliegman and R.K. Barnes, *Tetrahedron*, **26**, 2555–2560 (1970).
96. B. McDonnell and H.E. Jackson, US 3,002,881 (1961).
97. J.R. Dudley and J.A. Anthes, US 2,596,014 (1952).
98. W.F. Tousignant and C. Moore, US 3,200,088 (1965).
99. A.L. Wilson, US 2,553,696 (1951).
100. H. Ulrich and W.H. Dormagen, US 2,182,306 (1939).
101. P.E. Trout, *Tappi*, **34(12)**, 539–544 (1951).
102. B.W. Wilson, US 3,203,910 (1965).
103. D.S. Zhuk, P.A. Gembitsky, and A.I. Chmarin, US 4,032,480 (1977).
104. A. Goldstein and J.W. Brook, US 3,294,723 (1966).
105. R. Nordgren, US 3,303,184 (1967).
106. H. Ulrich, US 2,272,489 (1942).
107. G.W. Strother, US 3,617,440 (1971).
108. G.I. Keim, US 2,926,154 (1960).
109. G.I. Keim, US 2,926,116 (1960).
110. Y. Jen and R.R. House, US 2,948,652 (1960).
111. W.H. Carothers, US 2,149,273 (1939).
112. G.I. Keim, US 3,332,901 (1967).
113. W.W. Maslanka, US 4,388,439 (1983).
114. B.K. Bower, US 6,111,032 (2000).
115. G.I. Keim, US 4,501,862 (1985).
116. W.W. Maslanka, US 4,605,709 (1986).
117. D.I. Devore and S.A. Fischer, US 5,364,927 (1994).
118. S.A. Lipowski, US 3,793,279 (1974).
119. S. Loshaek and E.A. Blommers, US 3,125,552 (1964).
120. N.W. Dachs and G.M. Wagner, US 3,565,754 (1971).
121. T.G. Shannon, M.T. Goulet, D.A. Clarahan, and W.Z. Schroeder, US 6,896,769 (2005).
122. J.W. Hyland, US 3,434,984 (1969).
123. J.W. Hyland, US 3,248,280 (1966).
124. D.E. Floyd, R.J. Ess, J.L. Keen, and J.W. Opie, US 3,138,473 (1964).
125. A.T. Coscia and L.L. Williams, US 3,813,362 (1974).
126. L.L. Williams and A.T. Coscia, US 3,778,339 (1973).
127. L.L. Williams and A.T. Coscia, US 3,733,290 (1973).
128. W.Z. Schroeder, D.A. Clarahan, M.T. Goulet, and T.G. Shannon, US 6,235,155 (2001).
129. W.Z. Schroeder, D.A. Clarahan, M.T. Goulet, and T.G. Shannon, US 6,465,602 (2002).
130. S.A. Lipowski and A. Christiansen, US 3,535,288 (1970).
131. E.F. Morello, US 4,418,189 (1983).
132. F.S. Munjat and S.H. Kim, US 3,869,342 (1975).
133. F.S. Munjat and S.H. Kim, US 3,869,343 (1975).
134. F.S. Munjat and S.H. Kim, US 3,761,350 (1973).
135. W.W. Maslanka, US 4,708,772 (1987).
136. W.W. Maslanka, US 4,515,657 (1985).
137. W.W. Maslanka, US 4,487,884 (1984).
138. E. Strazdins and R.R. House, US 3,329,657 (1967).
139. H. Bachem, J. Muszik, J. Reiners, C. Suling, and W.-D. Schroer, US 4,975,499 (1990).
140. G.I. Keim, US 3,039,889 (1962).

141. J.-C. Bonnet and G. Tesson, US 4,075,177 (1978).
142. J.S. Conte, R. Park, and R.W. Faessinger, US 3,250,664 (1966).
143. R.R. Staib, US 6,171,440 (2001).
144. R.R. Staib, US 6,355,137 (2002).
145. V.R. Gaertner, US 3,314,897 (1967).
146. O.G. Weaver, T.W. Glass, and W.G. Buckner, US 5,278,255 (1994).
147. W.W. Maslanka, US 5,644,021 (1997).
148. W.W. Maslanka, US 5,668,246 (1997).
149. T. Hasegawa, H. Takagishi, and H. Horiuchi, US 5,017,642 (1991).
150. H. Takagishi and N. Kondo, US 4,287,110 (1981).
151. H. Takagishi and N. Kondo, US 4,336,835 (1982).
152. G.I. Keim and A.C. Schmalz, US 3,058,873 (1962).
153. R.W. Butler and G.I. Keim, US 3,224,986 (1965).
154. R.L. Juhl, J.L. Potter, and S.C. Polemenakos, US 4,714,736 (1987).
155. L.-L. Chan and J.E. Amy, US 3,887,510 (1975).
156. J.C. Spicer, P.M. Westfall, and N.R. Eldred, US 3,351,520 (1967).
157. Q.-M. Gu, A. Michel, H.N. Cheng, W.W. Maslanka, and R.R. Staib, US 6,667,384 (2003).
158. W.D. Emmons, US 3,305,493 (1967).
159. H. Wallenius, S. Sandberg, M. Gorzynski, and O. Struck, US 6,852,197 (2005).
160. L. Pardo, R. Osman, H. Weinstein, and J.R. Rabinowitz, *J. Am. Chem. Soc.*, **115**, 8263–8269 (1993).
161. G.R. Killat and L.R. Wilson, US 4,416,729 (1983).
162. X.C. Peng, X.H. Peng, S.M. Liu, and J.Q. Zhao, *eXPRESS Polym. Lett.*, **3(8)**, 510–517 (2009).
163. D.K. Ray-Chaudhuri and C.P. Iovine, US 3,962,159 (1976).
164. J.S. Conte and R.W. Faessinger, US 3,320,215 (1967).
165. Y. Terada, M. Urata, and K. Yamamoto, US 3,645,954 (1972).
166. R.H. Earle, US 3,240,664 (1966).
167. G.I. Keim, US 4,537,657 (1985).
168. W.W. Maslanka, US 6,908,983 (2005).
169. A.J. Allen, E. Echt, W.W. Maslanka, and J.C. Peters, US 6,294,645 (2001).
170. A.J. Allen, US 5,786,429 (1998).
171. B.J. Kokko and E.W. Post, US 6,222,006 (2001).
172. L.A. Lundberg and L.H. Wilson, US 3,019,156 (1962).
173. W.F. McDonald, Z.H. Huang, S.C. Wright, M. Danzig, and A.C. Taylor, US 6,495,657 (2002).
174. W.F. McDonald, Z.H. Huang, S.C. Wright, M. Danzig, and A.C. Taylor, US 2005/0043506.
175. Z.H. Huang, W.F. McDonald, S.C. Wright, and A.C. Taylor, US 6,399,714 (2002).
176. P.M. Westfall, N.R. Eldred, and J.C. Spicer, US 3,372,085 (1968).
177. B.J. Kokko, US 6,673,205 (2004).
178. F. Horowitz, US 3,914,155 (1975).
179. J.G. Fenyes and J.D. Pera, US 4,506,081 (1985).
180. R.R. House, J.H. Rassweller, and J.M. Schmitt, US 3,086,961 (1963).
181. M. Marten and W. Kamutzki, US 5,019,606 (1991).
182. R. Backai, US 3,715,335 (1973).
183. C.M. Hadad, P.R. Rablen, and K.B. Wiberg, *J. Org. Chem.*, **63**, 8668–8681 (1998).
184. J. Kamlet, US 2,624,686 (1953).
185. C. Heuck and P. Esselmann, US 1,737,760 (1929).
186. G.G. Xu and C.Q. Yang, *Tappi J.*, **84(6)**, 68 (2001).
187. A.J. Stamm, *Tappi*, **42(1)**, 44–50 (1959).
188. W.E. Woods and W.A. Schenck, US 2,622,960 (1952).
189. R.A. Young, *Wood Fiber*, **10(2)**, 112–119 (1978).
190. G.G. Xu, C.Q. Yang, and Y. Den, *J. Appl. Polym. Sci.*, **101**, 277–284 (2006).
191. A.E. Broderick, US 2,285,490 (1942).
192. A.E. Broderick, US 2,329,741 (1943).
193. M. Laleg and I.I. Pikulik, *J. Pulp Paper Sci.*, **17(6)**, J206–J216 (1991).
194. M. Laleg and I.I. Pikulik, *J. Pulp Paper Sci.*, **19(6)**, J248–J255 (1993).
195. M. Laleg and I.I. Pikulik, *TAPPI Proceedings—Papermakers Conference*, Atlanta, GA, 1991, pp. 577–590.
196. B.T. Hofreiter, H.D. Heath, A.J. Ernst, and C.R. Russell, *Tappi*, **57(8)**, 81–83 (1974).
197. J. Baas, P.H. Brouwer, and T.A. Wielema, *PTS-Symposium*, Atlanta, GA, 2002, pp. 38/1–38/30,.

198. D.H. Le Roy and S.M. Parmerter, US 3,553,193 (1971).
199. J.N. BeMiller and S.D. Darling, US 3,632,802 (1972).
200. B.T. Hofreiter, G.E. Hamerstrand, and C.L. Mehltretter, US 3,062,703 (1962).
201. J.W. Thornton, D.L. Van Brussel-Verraest, A. Besemer, and S. Sandberg, US 6,896,725 (2005).
202. J.W. Thornton, D.L. Van Brussel-Verraest, A. Besemer, and S. Sandberg, US 6,582,559 (2003).
203. J.W. Thornton, A. Besemer, D.L. Van Brussel-Verraest, and S. Sandberg, WO 0183887 (2001).
204. J.W. Swanson, E.J. Jones, and C.L. Mehltretter, US 3,184,333 (1965).
205. J.L. Keen, W.J. Ward, R.R. Swanson, and H.N. Dunning, US 3,239,500 (1966).
206. G.E. Hamerstrand, B.T. Hofreiter, C.L. Mehltretter, W.E. Schulze, and D.J. Kay, *Tappi*, **44**(6), 430–433 (1961).
207. B.T. Hofreiter, G.E. Hamerstrand, C.L. Mehltretter, W.E. Schulze, and A.J. Ernst, *Tappi*, **43**(7), 639–643 (1960).
208. C.L. Mehltretter, US 2,713,553 (1955).
209. J.L. Keen and H.G. Simmerman, US 3,231,560 (1966).
210. J.W. Opie and J.L. Keen, US 3,205,125 (1965).
211. J.W. Opie and J.L. Keen, US 3,236,832 (1966).
212. L.R. Peel, *Paper Technol.*, **30**(12), 20–23 (1989).
213. D.N. Van Eenam, EP 147 380 (1984).
214. J.W. Opie and J.L. Keen, *Tappi*, **47**(8), 504–507 (1964).
215. P.B. Davidson, US 2,549,177 (1951).
216. A. Meller, *Tappi*, **41**(11), 684–686 (1958).
217. C.R. Russell, R.A. Buchanan, and C.E. Rist, US 3,160,552 (1964).
218. M.E. Carr, B.T. Hofreiter, G.E. Hamerstrand, and C.R. Russell, *Tappi*, **57**(10), 127–129 (1974).
219. B.T. Hofreiter, G.E. Hamerstrand, and C.L. Mehltretter, US 3,067,088 (1962).
220. P.J. Borchert, W.L. Kaser, and J. Mirza, US 3,269,852 (1966).
221. J.H. Curtis, US 3,236,721 (1966).
222. E.J. Barber, R.H. Earle, and G.C. Harris, US 3,219,519 (1965).
223. B.T. Hofreiter, G.E. Hamerstrand, and C.L. Mehltretter, US 3,087,852 (1963).
224. C.G. Caldwell and O.B. Wurzburg, US 2,813,093 (1957).
225. R.J. Berni, R.R. Benerito, J.B. McKelvey, T.L. Ward, and D.M. Soignet, US 3,644,082 (1972).
226. P.J. Borchert, US 3,021,329 (1962).
227. C.L. Mehltretter, US 3,251,826 (1966).
228. C.L. Mehltretter, T.E. Yeates, G.E. Hamerstrand, B.T. Hofreitter, and C.E. Rist, *Tappi*, **45**(9), 750–752 (1962).
229. G.C. Maher, US 3,436,305 (1969).
230. M.E. Carr, B.T. Hofreiter, and C.R. Russell, *J. Polym. Sci. Polym. Chem. Ed.*, **13**, 1441–1456 (1975).
231. G.G. Maher, A.J. Ernst, H.D. Heath, B.T. Hofreiter, and C.E. Rist, *Tappi*, **55**(9), 1378–1384 (1972).
232. M.E. Carr, W.M. Doane, G.E. Hamerstrand, and B.T. Hofreiter, *J. Appl. Polym. Sci.*, **17**, 721–735 (1973).
233. G.G. Maher, US 3,730,829 (1973).
234. N. Chen, S. Hu, and R. Pelton, *Ind. Eng. Chem. Res.*, **41**, 5366–5371 (2002).
235. L.L. Williams and A.T. Coscia, US 3,740,391 (1973).
236. D.B. Solarek, P.G. Jobe, and M.M. Tessler, US 4,675,394 (1987).
237. A. Meller, *Tappi*, **41**(11), 679–683 (1958).
238. B. Alince, *Sven. Papperstidn.*, **78**(7), 253–257 (1975).
239. M.O. Schur and R.M. Levy, *Paper Trade J.*, **124**(20), 43–46 (1947).
240. A.C. Besemer, A.M.Y.W. Verwilligen, H.J. Thiewes, and D.L. van Brussel-Verraest, US 6,849,156 (2005).
241. P. Luner, K.P. Vemuri, and B. Leopold, *Tappi*, **50**(3), 117–120 (1967).
242. A.L. Cimecioglu, J.S. Thomaides, K.A. Luczak, and R.D. Rossi US 6,368,456 (2002).
243. W.Z. Schroeder, D.A. Clarahan, M.T. Goulet, and T.G. Shannon, US 6,398,911 (2002).
244. T.G. Shannon, D.A. Clarahan, M.T. Goulet, and W.Z. Schroeder, US 6,596,126 (2003).
245. A.A.R. Sayigh, H. Ulrich, and W.J. Farrissey, Diisocyanates, in *High Polymers*, Vol. 27, (Condensation Monomers), J.K. Stille and T.W. Campbell (Ed.), Wiley, New York, 1972, pp. 369–476.
246. X. Zhang and H. Tanaka, *J. Wood Sci.*, **45**, 425–430 (1999).
247. R.L. Berger and C.H. Gelbert, *Tappi*, **44**(7), 467–472 (1961).
248. W.E. Hanford and D.F. Holms, US 2,339,913 (1944).
249. F.R. Hunter, US 5,008,359 (1991).
250. A.R. Adams, US 3,007,763 (1961).

251. I.E. Isgur and W.D. Delvecchio, US 4,442,259 (1984).
252. K.-C. Liu, US 7,625,462 (2009).
253. W. Reuther, A. Segnith, O. Wittmann, and H. Schatz, US 4,260,532 (1981).
254. G. Daude, J.M. Lasnier, and J. Pijselman, EP 0 017 698 (1983).
255. B. Jensen, J. Konig, P. Nowak, and J. Reiners, US 5,718,804 (1998).
256. B. Jensen, J. Konig, B. Thiele, J. Reiners, and R.V. Meyer, US 6,080,831 (2000).
257. J. Reiners, J. Kopp, J. Konig, H. Traubel, E. Wenderoth, B. Jensen, and J. Probst, US 6,090,871 (2000).
258. H.J. Laas, T. Hassel, W. Kubitza, R. Halpaap, and K. Noll, US 5,252,696 (1993).
259. H. Traubel, H.J. Laas, H. Reiff, J. Konig, J. Reiners, and H. Faika, US 6,143,132 (2000).
260. Bayer Product Information Bulletin: "ISOVIN"—New wet strength additive for paper based on isocyanate chemistry (1998).
261. J. Reiners, H.J. Laas, J. Konig, H. Reiff, J. Probst, B. Bomer, R. Halpaap, F. Puchner, and H. Traubel, US 5,503,714 (1996).
262. I.E. Isgur, A.B. Holmstrem, and N.J. Hayes, US 4,182,649 (1980).
263. I.E. Isgur, A.B. Holmstrem, and N.J. Hayes, US 4,110,508 (1978).
264. W. Drouve, P. Hohlein, and W. Wieczorrek, US 5,185,422 (1993).
265. G. Uhl, R. Blum, and H. Belde, US 4,849,262 (1989).
266. W.H. Breen and G.F. Sirine, US 3,551,346 (1970).
267. H. Yang, S.K. Mendon, and J.W. Rawlins, *eXPRESS Polym. Lett.*, **2(5)**, 349–356 (2008).
268. G. Grogler, H. Hess, and R. Kopp, US 4,483,974 (1984).
269. W. Mehlo, R. Titzmann, and R. Zinsmeister, US 3,409,461 (1968).
270. Z.W. Wicks, *Prog. Org. Coat.*, **9**, 3–28 (1981).
271. D.A. Wicks and Z.W. Wicks, *Prog. Org. Coat.*, **36**, 148–172 (1999).
272. D.A. Wicks and Z.W. Wicks, *Prog. Org. Coat.*, **41**, 1–83 (2001).
273. A.J. Morak, US 3,492,081 (1970).
274. T.C. Duvall and J.L. Carpenter, US 7,025,904 (2006).
275. G.B. Guise, G.N. Freeland, and G.C. Smith, **J.** *Appl. Polym. Sci.*, **23**, 353–365 (1979).
276. I.E. Isgur, W.D. Delvecchio, J.L. Ohlson, and N.J. Hayes, GB 2,068,034 (1981).
277. P.A. Longo, I.E. Isgur, W.D. Delvecchio, and J.L. Ohlson, GB 2,048,289 (1980).
278. E. Querat, L. Tighzert, and J.P. Pascault, *Angew. Makromol. Chem.*, **219**, 185–203 (1994).
279. K. Sato, S. Wada, N. Sainai, and N. Yamaji, JP 5-51896 (1993).
280. W.J. Mijs and J.B. Reesink, US 4,008,192 (1977).
281. A.M. Brownstein and J.P. Sibilia, US 3,465,024 (1969).
282. S.J. Mels, B.C. Petrie, and R.J. Pokusa, US 4,444,954 (1984).
283. R. Swidler, US 3,954,718 (1976).
284. S.J. Mels, B.C. Petrie, and R.J. Pokusa, US 4,444,955 (1984).
285. S.J. Mels, B.C. Petrie, and R.J. Pokusa, US 4,561,952 (1985).
286. P.C. Hayes, US 5,326,853 (1994).
287. P.C. Hayes, US 5,425,999 (1995).
288. P.C. Hayes, US 5,494,963 (1996).
289. R.L. Barcus and D.W. Bjorkquist, US 5,698,074 (1997).
290. R.L. Barcus and D.W. Bjorkquist, US 5,443,899 (1995).
291. M. Haggkvist, D. Solberg, L. Wagberg, and L. Odberg, *Nord. Pulp Paper Res. J.*, **13(4)**, 292–298 (1998).
292. R.A. Jewell, US 6,471,824 (2002).
293. R.A. Jewell, US 6,579,414 (2003).
294. S.A. Naieni, US 5,873,979 (1999).
295. C.Q. Yang, Y. Xu, and D. Wang, *Ind. Eng. Chem. Res.*, **35**, 4037–4042 (1996).
296. Y.J. Zhou, P. Luner, and P. Caluwe, *J. Appl. Polym. Sci.*, **58**, 1523–1534 (1995).
297. D.F. Caulfield, *Tappi J.*, **77(3)**, 205–212 (1994).
298. C.T. Arkens, S.L. Egolf, R.D. Gleim, O.H. Hsu, and K.J. Weisinger, US 5,427,587 (1995).
299. R.L. Anderson, W.W. Cattron, V.F. Smith, and D.J. Fenoglio, US 6,248,879 (2001).
300. R.L. Anderson, W.W. Cattron, V.F. Smith, and D.J. Fonoglio, US 5,981,739 (1999).
301. C.Q. Yang and Y. Xu, *J. Appl. Polym. Sci.*, **67**, 649–658 (1998).
302. D. Barzyk, D.H. Page, and A. Ragauskas, *J. Pulp Paper Sci.*, **23(2)**, J59–J61 (1997).
303. J.L. Azorlosa and D.P. Hug, US 2,635,972 (1953).
304. T. Jaschinski, S. Gunnars, A.C. Besemer, and P. Bragd, US 6,635,755 (2003).
305. T. Jaschinski, US 6,409,881 (2002).
306. A. Merijan, US 3,385,834 (1968).

307. T. Sun and J.D. Lindsay, US 6,322,665 (2001).
308. P.K. Chatterjee and M.C. Kwok, US 3,723,413 (1973).
309. W.H. Ashton and J.N. Masci, US 2,766,137 (1956).
310. J.W. Swanson, Mechanism of paper wetting, in *The Sizing of Paper*, 2nd edn., W.F. Reynolds (Ed.), Tappi Press, Atlanta, GA, 1989, pp. 133–154.
311. S.A. Weerawarna and S. Bing, US 7,591,891 (2009).
312. J.E. Drach, US 4,391,878 (1983).
313. J.A. Cuculo, US 3,555,585 (1971).
314. A.N. Neogi and J.R. Jensen, *Tappi*, **63(8)**, 86–88 (1980).
315. S. Evani and G.R. Killat, US 4,198,269 (1980).
316. S. Evani and G.R. Killat, US 4,156,775 (1979).
317. D.N. Van Eenam, US 4,145,248 (1979).
318. D.N. Van Eenam, US 4,333,971 (1982).
319. D.N. Van Eenam, US 4,308,187 (1981).
320. D.N. Van Eenam, US 4,289,864 (1981).
321. D.N. Van Eenam, US 4,346,027 (1982).
322. D.N. Van Eenam, US 4,412,017 (1983).
323. H.L. Marder, S.E. Church, and V. Stannett, *Tappi*, **40(10)**, 829–832 (1957).
324. Y. Bogaert, E. Goethals, and E. Schacht, *Makromol. Chem.*, **182**, 2687–2693 (1981).
325. A.C. Nixon, P.J. Berrigan, and P.H. Williams, US 3,278,474 (1966).
326. D.P. Sheetz and C.G. Humiston, US 3,255,072 (1966).
327. S.N. Lewis, R.F. Merritt, and W.D. Emmons, US 3,702,799 (1972).
328. B.F. Aycock and E.M. Hankins, US 2,831,833 (1958).
329. D.N. Van Eenam, US 4,226,957 (1980).
330. M. Coskun, H. Erten, K. Demirelli, and M. Ahmedzade, *Polym. Degrad. Stab.*, **69**, 245–249 (2000).
331. D.N. Van Eenam, US 4,250,212 (1981).
332. P. Pezzaglia, A. Debenedictis, W. Ewert, W.H. Houff, R.C. Hurlbert, A.C. Nixon, and I. Sobolev, *Tappi*, **48(5)**, 314–320 (1965).
333. W.H. Houff and A.C. Nixon, US 3,079,296 (1963).
334. R.K. Pinschmidt and T.W. Lai, US 5,194,492 (1993).
335. L. Gardlund, J. Forsstrom, and L. Wagberg, *Nord. Pulp Paper Res. J.*, **20(1)**, 36–42 (2005).
336. D.N. Van Eenam, US 4,298,715 (1981).
337. G.I. Keim, US 3,772,076 (1973).
338. G.I. Keim, US 3,840,504 (1974).
339. K. Yoshitani, WO 2005/059251.
340. K. Kageno, T. Ueda, and S. Harada, US 4,812,540 (1989).
341. K. Kageno, H. Sakuro, S. Kiyoshi, and S. Harada, EP 196 588 (1986).
342. T. Kato, Y. Fujita, I. Hayashi, H. Ohta, and K. Kitamura, US 6,395,849 (2002).
343. G.I. Keim, US 3,700,623 (1972).
344. G.I. Keim, US 3,833,531 (1974).
345. D.N. Van Eenam, US 4,233,417 (1980).
346. D.N. Van Eenam, US 4,222,921 (1980).
347. D.N. Van Eenam, US 4,340,692 (1982).
348. W.W. Maslanka, US 4,520,159 (1985).
349. R.A. Bankert, US 4,419,500 (1983).
350. R.A. Bankert, US 4,419,498 (1983).
351. R.A. Bankert, US 4,354,006 (1982).
352. R.D. Gless, D.J. Daeson, and R.E. Wingard, US 4,018,826 (1977).
353. R.W. Stackman and R.H. Summerville, *Ind. Eng. Chem., Prod. Res. Dev.*, **24**, 242–246 (1985).
354. R.K. Pinschmidt and K. Yacoub, US 5,519,093 (1996).
355. T.W. Lai and B.R. Vijayendran, US 4,952,656 (1990).
356. T.W. Lai and B.R. Vijayendran, US 4,804,793 (1989).
357. F. Brunnmueller, R. Schneider, M. Kroener, H. Mueller, and F. Linhart, US 4,421,602 (1983).
358. R.H. Summerville and R.W. Stackman, *Polym. Prepr.*, **24(2)**, 12–13 (1983).
359. R.K. Pinschmidt, W.L. Renz, W.E. Carroll, K. Yacoub, J. Drescher, A.F. Nordquist, and N. Chen, *J. Macromol. Sci. Pure Appl. Chem.*, **A34(10)**, 1885–1905 (1997).
360. F. Linhart and W. Auhorn, *Papier*, **46(10A)**, 38–45 (1992).

361. F. Wang and H. Tanaka, *J. Appl. Polym. Sci.*, **78**, 1805–1810 (2000).
362. L. Shi, K.A. Boduch-Lee, J.T. Henssler, E.J. Beckman, and T.M. Chapman, *J. Polym. Sci., Part A: Polym. Chem.*, **42**, 4994–5004 (2004).
363. R. Pelton and J. Hong, *Tappi J.*, **10(1)**, 21–25 (2002).
364. T.T. Nguyen, US 6,146,497 (2000).
365. L.A. Lundberg and W.F. Reynolds, US 2,794,737 (1957).
366. E.M. Hankins and S. Melamed, US 2,727,016 (1955).
367. S. Melamed, US 2,689,844 (1954).
368. G.E. Ham, *Copolymerization*, Interscience, New York, 1964, pp. 2–9.
369. C. Hagiopol, *Copolymerization: Towards a Systematic Approach*, Kluwer-Academic Press, New York, 1999.
370. S.N. Lewis, R.F. Merritt, and W.D. Emmons, US 3,694,393 (1972).
371. G.I. Keim, US 3,842,054 (1974).
372. J.L. Azorlosa, US 2,654,671 (1953).
373. G.D. Jones, G.R. Geyer, and M.J. Hatch, US 3,494,965 (1970).
374. G.D. Jones, G.R. Geyer, and M.J. Hatch, US 3,544,532 (1970).
375. D.H. Klein, US 3,846,828 (1974).
376. M. Hahn, W. Jaeger, R. Schmolke, and J. Behnisch, *Acta Polym.*, **41(2)**, 107–112 (1990).
377. G.D. Jones and R.L. Zimmerman, US 2,757,190 (1956).
378. G.D. Jones, C. Kleeman, and D. Villani-Price, Polymer science and technology, in *Water Soluble Polymers*, Vol. 2, N.M. Bikales (Ed.), Plenum, New York, 1973, pp. 255–266.
379. S.P. Malchick, W.G. James, and J.S. Munday, US 3,325,347 (1967).
380. I. Sobolev, US 3,428,617 (1969).
381. R.K. Pinschmidt, *Polym. Mater. Sci. Eng.*, **59**, 1156–1160 (1988).
382. R.K. Pinschmidt, W.F. Burgoyne, D.D. Dixon, and J.E. Goldstein, *ACS Symposium Series 367, Cross-Linked Polymers*, American Chemical Society, Washington, DC, 1988, pp. 453–466.
383. G.I. Keim, US 3,686,151 (1972).
384. A.J. Allen, US 5,510,004 (1996).
385. R.S. Buriks and E.G. Lovett, US 4,341,887 (1982).
386. F.W. Michelotti, US 3,316,225 (1967).
387. S.A. Fischer and D.I. Devore, US 5,408,024 (1995).
388. A.R. Mills, US 3,347,832 (1967).
389. R.W. Lenz, *Organic Chemistry of Synthetic High Polymers*, John Wiley & Sons, New York, 1967, pp. 687–770.
390. Y. Gnanou and M. Fontanille, *Organic and Physical Chemistry of Polymers*, Wiley-Interscience, New York, 2008, pp. 357–376.
391. R.R. House, S.T. Moore, and A.M. Schiller, US 2,729,560 (1956).
392. H. Tanaka, A. Swerin, L. Odberg, and S.-B. Park, *J. Pulp Paper Sci.*, **25(8)**, 283–288 (1999).
393. H.-K. Lee and W.-L. Jong, *J. Polym. Res.*, **4(2)**, 119–128 (1997).
394. L.L. Williams and A.T. Coscia, US 3,507,847 (1970).
395. C.S. Scanley and H.P. Panzer, US 3,503,946 (1970).
396. D.W. Fong, US 5,084,520 (1992).
397. D.W. Fong, US 5,075,390 (1991).
398. D.L. Dauplaise, J.J. Kozakiewicz, and J.M. Schmitt, US 5,041,503 (1991).
399. J.J. Kozakiewicz, D.L. Dauplaise, J.M. Schmitt, and S.-Y. Huang, US 5,037,863 (1991).
400. D.R. Cosper, US 4,135,969 (1979).
401. A.J. Allen and A.C. Sau, US 6,315,865 (2001).
402. S. Pfohl, M. Kroener, H. Hartmann, and W. Denzinger, US 4,774,285 (1988).
403. S. Melamed, US 2,871,203 (1959).
404. S. Pfohl, M. Kroener, H. Hartmann, and W. Denzinger, US 4,880,497 (1989).
405. S. Pfohl, M. Kroener, H. Hartmann, and W. Denzinger, US 4,978,427 (1990).
406. J. Brandrup and E.H. Imergut (Eds.), *Polymer Handbook*, 4th edn., John Wiley, New York, 1999.
407. R.K. Pinschmidt and T.-W. Lai, EP 0339371 (1990).
408. W.M. Brouwer, P. Piet, and A.L. German, *J. Polym. Sci. Polym. Chem. Ed.*, **22**, 2353–2362 (1984).
409. Y. Chang and C.L. McCormick, *Macromolecules*, **26**, 4814–4817 (1993).
410. R.H. Pelton and C. Lu, US 2008/0099172.
411. C. Bonnet-Gonnet, J.C. Castaing, and P. Le Cornec, WO 98/55694.

412. H. Hartmann, W. Denzinger, M. Kroener, C. Nilz, F. Linhart, and A. Stange, US 5,334,287 (1994).

413. W. Denzinger, M. Rubenacker, C. Nilz, P. Lorencak, D. Monch, R. Schuhmacher, and A. Stange, US 6,060,566 (2000).

414. W. Denzinger, M. Rubenacker, C. Nilz, P. Lorencak, D. Monch, R. Schuhmacher, and A. Stange, US 6,048,945 (2000).

415. R.K. Pinschmidt and T.W. Lai, US 5,300,566 (1994).

416. T. Itagaki, M. Shiraga, S. Sawayama, and K. Satoh, US 4,808,683 (1989).

417. T. Itagaki, M. Shiraga, S. Sawayama, and K. Satoh, US 5,037,927 (1991).

418. T. Itagaki, M. Shiraga, S. Sawayama, and K. Satoh, US 5,064,909 (1991).

419. T. Itagaki, M. Shiraga, S. Sawayama, and K. Satoh, US 4,957,977 (1990).

420. L.M. Robeson and T.L. Pickering, US 5,380,403 (1995).

421. R.K. Pinschmidt, T.W. Lai, and L.K. Wempe, US 5,155,167 (1992).

422. V. Barboiu, E. Streba, C. Luca, and C.I. Simionescu, *J. Polym. Sci. Part A; Polym. Chem.*, **33**, 389–398 (1995).

423. R.K. Pinschmidt and B.R. Vijayendran, US 4,973,410 (1990).

424. M. Aoyama and M. Tomida, *Polym. Prepr.*, **45(1)**, 853–854 (2004).

425. R.K. Pinschmidt and T.W. Lai, US 5,086,111 (1992).

426. J.G. Smigo, T.P. McAndrew, R.K. Pinschmidt, and A.F. Nordquist, US 5,185,083 (1993).

427. T.P. McAndrew, A.F. Nordquist, R.K. Pinschmidt, and D.P. Eichelberger, US 5,270,379 (1993).

428. R.K. Pinschmidt, J.G. Smigo, A.F. Nordquiust, US 5,324,787 (1994).

429. J. Utecht, M. Niessner, D. Monch, and M. Rubenacker, US 6,057,404 (2000).

430. L.M. Robeson, G. Davidowich, and R.K. Pinschmidt, US 5,397,436 (1995).

431. L.M. Robeson, G. Davidowich, and R.K. Pinschmidt, GB 2,268,758 (1994).

432. E.S. Blake, US 2,723,195 (1955).

433. W. Lobach, P. Haas, G. Kolb, G. Sackmann, and J. Probst, US 4,423,194 (1983).

434. E. Valton, J. Schmidhauser, and M. Sain, *Tappi J.*, **3(4)**, 25–30 (2004).

435. W. von Bonin, P. Mummenhoff, and H. Baumgen, US 4,381,367 (1983).

436. J.H. Wang, D.M. Schertz, and D.A. Soerens, US 6,172,177 (2001).

437. T.G. Shannon and D.A. Soerens, US 6,994,770 (2006).

438. B. Alince, *Tappi*, **60(12)**, 133–136 (1977).

439. J.R. Lindquist, US 2,668,111 (1954).

440. A. Jurecic, C.M. Hou, K. Sarkanen, C.P. Donofrio, and V. Stannett, *Tappi*, **43**, 861–865 (1960).

441. I. Noda and D.F. Hager, US 4,835,211 (1989).

442. K.D. Vinson, J.A. Ficke, and H.T. Deason, WO 98/51862 (1998).

443. N. Pourahmady and A. Williamson, *Tappi Proceedings—Nonwovens Conference*, Charlotte, 1996, pp. 185–198.

444. C.L. Weiduer and I.R. Dunlap, US 2,745,744 (1956).

445. P.J. McLaughlin and W.W. Toy, US 3,017,291 (1962).

446. J.E. Drach, US 4,420,368 (1983).

447. D.P. Stack, US 5,198,492 (1993).

448. R.H. Welsh and W.W. Pockman, US 2,657,991 (1953).

449. R.H. Walsh, H.H. Abernathy, W.W. Pockman, J.R. Galloway, and E.P. Hartsfield, *Tappi*, **33(5)**, 232–237 (1950).

450. G.W. Wheeler, A.M. Borders, J.W. Swanson, and G.R. Sears, *Tappi*, **34(7)**, 297–301 (1951).

451. M. Amano and T. Koike, US 4,121,966 (1978).

452. F.W. Meisel and K.C. Larson, US 3,844,880 (1974).

453. E.K. Thommen and V. Stannett, *Tappi*, **41(11)**, 692–695 (1958).

454. W.R. Abell and C.H. Gelbert, *Tappi*, **48(8)**, 97A–100A (1965).

455. C.H. Gelbert, *Tappi*, **43(2)**, 207A–209A (1960).

456. R. Pelton, *Appita J.*, **57(3)**, 181–190 (2004).

457. W.J. Andrews, R.P. Barber, R.W. Reiter, and V. Stannett, *Tappi*, **40(9)**, 744–749 (1957).

458. I. Noda and D.F. Hager, US 4,785,030 (1988).

459. W.A. Foster, D.M. Pckelman, and R.A. Wessling, US 4,189,345 (1980).

460. R.A. Wessling, W.A. Foster, and D.M. Pickeman, US 4,178,205 (1979).

461. C.Y. Huang, S. Shimizu, and H. Adachi, US 3,926,890 (1975).

462. M.T. Goulet, S.L. Pomeroy, and M. Trimacco, US 7,189,307 (2007).

463. L. Leibler, A. Schroder, I. Betremieux, and I. Silberzan, US 2002/0096282 (2002).

464. D.F. Diehl and P.C. Hayes, US 5,523,345 (1996).

465. P.C. Hayes, US 5,362,798 (1994).
466. J.E. Schuetz and W.H. Keskey, US 4,460,029 (1984).
467. W.H. Keskey, J.E. Schuetz, and A.D. Hickmann, US 4,644,032 (1987).
468. W.H. Keskey, J.E. Schuetz, Do I. Lee, and J.E. Scwartz, US 4,474,923 (1984).
469. D.R. Swanson and D.A. Tomalia, US 5,705,573 (1998).
470. W.H. Keskey, J.E. Schuetz, and Do I. Lee, US 4,508,869 (1985).
471. D.B. Rice, J.G. Galloway, and W.H. Keskey, US 5,629,396 (1997).
472. M. Gorzynski, C. Biermann, H.J. Macherey, and A. Anderson, US 6,702,923 (2004).
473. I. Noda, US 5,200,036 (1993).
474. I. Noda, US 5,342,875 (1994).
475. X. Zhang and H. Tanaka, *J. Appl. Polym. Sci.*, **81**, 2791–2797 (2001).
476. X. Zhang and H. Tanaka, *J. Appl. Polym. Sci.*, **80**, 334–339 (2001).
477. B. Alince, M. Inoue, and A.A. Robertson, *J. Appl. Polym. Sci.*, **20**, 2209–2219 (1976).
478. M.O. Schur, US 2,035,024 (1936).
479. H. Corten, R.P. Foulds, J.T. Marsh, and F.C. Wood, GB 557,182 (1943).
480. M.E. Cupery, US 2,526,638 (1950).
481. M.O. Schur, GB 523,185 (1940).
482. C.J. Allison, P.Y. Jone, and P.E. Jacobson, US 3,448,005 (1969).
483. C.S. Maxwell, US 2,407,376 (1946).
484. C.S. Maxwell, US 2,582,840 (1952).
485. G.I. Keim, US 2,622,979 (1952).
486. R.W. Auten and J.L. Ratney, US 2,444,802 (1948).
487. R.W. Auten and J.L. Rainey, US 2,407,599 (1946).
488. T.J. Suen, US 2,559,578 (1951).
489. B.E. Sorenson, US 2,213,921 (1940).
490. A. Aignesberger and P. Bornmann, US 3,941,734 (1976).
491. T.J. Suen and A.M. Schiller, US 2,730,516 (1956).
492. C.S. Maxwell and C.G. Landes, US 2,559,220 (1951).
493. H.P. Wohnsidler and W.M. Thomas, US 2,345,543 (1944).
494. R.W. James and W.B. Pings, US 2,626,251 (1953).
495. J.H. Daniel, C.G. Landes, and T.J. Suen, US 2,657,132 (1953).
496. J.B. Davidson and E.J. Romatowski, US 2,683,134 (1954).
497. S. Kono, H. Mouri, M. Tobita, M. Ishida, N. Eto, T. Akiya, S. Toganoh, M. Higuma, M. Sakaki, and R. Arai, US 4,801,497 (1989).
498. R.W. Auten, US 2,471,188 (1949).
499. T.A. Martone, US 2,641,584 (1953).
500. G.I. Keim, US 2,826,500 (1958).
501. J.W. Eastes and R.W. Faessinger, US3,275,605 (1966).
502. K. Flory, A. Stange, M. Kroener, and N. Sendhoff, US 5,478,656 (1995).
503. E.D. Kaufman, US 3,819,555 (1974).
504. M.P. McGrath, E.D. Sall, and S.J. Tremont, *Polym. Prepr.*, **41(2)**, 1513–1514 (2000).
505. P.E.A. Hansson and F. Jachimowicz, US 4,566,943 (1986).
506. G.T. Kekish, US 3,317,370 (1967).
507. G.T. Kekish, US 3,410,828 (1968).
508. M.L. Zweigle, US 4,096,133 (1978).
509. K. Fujimura and K. Tanaka, US 3,790,529 (1974).
510. S. Sawayama, K. Satoh, S. Sato, and T. Sakakihara, US 5,389,203 (1995).
511. I.A. Pudney, B.M. Stubbs, and M.J. Welch, US 5,912,306 (1999).
512. D.D. Reynolds, US 3,345,346 (1967).
513. H.H. Roth, US 3,520,774 (1970).
514. S. Swier and B. Van Mele, *J. Polym. Sci. Part B Polym. Phys.*, **41**, 594–608 (2003).
515. J.B. McKelvey, R.R. Benerito, and T.L. Ward, *Ind. Eng. Chem. Prod. Res. Dev.*, **6(2)**, 115–120 (1967).
516. V.R. Gaertner, *Tetrahedron Lett.*, **39**, 4691–4694 (1966).
517. G.B. Guise and G.C. Smith, *J. Appl. Polym. Sci.*, **30**, 4099–4111 (1985).
518. H.R. Kricheldorf, *J. Polym. Sci. Polym. Chem. Ed.*, **19**, 2195–2214 (1981).
519. D.I. Devore and S.A. Fischer, *Tappi J.*, **76(8)**, 121 (1993).
520. A.J. Allen, US 5,902,862 (1999).
521. T. Obokata, M. Yanagisawa, and A. Isogai, *J. Appl. Polym. Sci.*, **97**, 2249–2255 (2005).

522. T. Obokata and A. Isogai, *J. Appl. Polym. Sci.*, **92**, 1847–1854 (2004).
523. H. Oike, T. Yaguchi, and Y. Tezuka, *Macromol. Chem. Phys.*, **200(4)**, 768–773 (1999).
524. A.T. Coscia and J.H. Ross, US 3,634,399 (1972).
525. A.T. Coscia and J.H. Ross, US 3,494,775 (1970).
526. H. Zieman and W. Lehmann, US 3,640,840 (1972).
527. O. Hertel, F. Linhart, and E. Scharf, US 4,450,045 (1984).
528. R.H. Earle, US 3,311,594 (1967).
529. R.J. Riehle and R.R. Staib, US H1629 (1997).
530. C.F. Gibbs and C.S. Marvel, *J. Am. Chem. Soc.*, **57**, 1137–1139 (1935).
531. M.R. Lehman, C.D. Thompson, and C.S. Marvel, *J. Am. Chem. Soc.*, **55**, 1977–1981 (1933).
532. C.F. Gibbs and C.S. Marvel, *J. Am. Chem. Soc.*, **56**, 725–727 (1934).
533. E.R. Littmann and C.S. Marvel, *J. Am. Chem. Soc.*, **52**, 287–294 (1930).
534. C.F. Gibbs, E.R. Littmann, and C.S. Marvel, *J. Am. Chem. Soc.*, **55**, 753–757 (1933).
535. T. Yano, H. Ohtani, and S. Tsuge, *Tappi J.*, **74(2)**, 197–201 (1991).
536. G.I. Keim and A.C. Schmalz, US 3,240,761 (1966).
537. N.A. Bates, *Tappi*, **52(6)**, 1157–1161 (1969).
538. T. Obokata, A. Isogai, and F. Onabe, *Proceedings of 2nd ISETPP*, Guangzhou, 2002, pp. 613–617.
539. G.B. Guise and G.C. Smith, *J. Chromatogr.*, **235**, 365–376 (1982).
540. A. Swerin and L. Wagberg, *Nord. Pulp Paper Res. J.*, **9(1)**, 18–25 (1994).
541. C.P. Talley, G.M. Bradley, and R.T. Guliana, US 4,118,316 (1973).
542. L. Wagberg and M. Bjorklund, *Nord. Pulp Paper Res. J.*, **8(1)**, 53–58 (1993).
543. S.A. Fischer, *Tappi J.*, **79(11)**, 179–186 (1996).
544. R.J. Riehle, US 7,081,512 (2006).
545. R.F. Moore, *Polymer*, **4(4)**, 493–513 (1963).
546. A.C. Schmalz, US 3,197,427 (1965).
547. J.F. Miller, US 4,853,431 (1989).
548. T. Obokata and A. Isogai, *J. Polym. Environ.*, **13(1)**, 1–6 (2005).
549. T. Obokata and A. Isogai, *Appita J.*, **57(5)**, 411–414 (2004).
550. H. Asao, F. Yoshida, K. Tomihara, M. Akimoto, and G. Kubota, US 3,609,126 (1971).
551. G.I. Keim, US 3,227,671 (1966).
552. R.H. Earle, US 3,352,833 (1967).
553. M. Wei, V.A. Grigoriev, and G.S. Furman, US 7,291,695 (2007).
554. A.T. Coscia and L.L. Williams, US 3,259,600 (1966).
555. J.B. Llosas and R.P. Subirana, US 5,955,567 (1999).
556. W.W. Maslanka, US 6,352,613 (2002).
557. Y. Luo, C.F. Ringold, D. Johnson, and C. Hagiopol, US 7,868,071 (2011).
558. V.R. Gaertner, *J. Org. Chem.*, **35(11)**, 3952–3959 (1970).
559. V.R. Gaertner, *J. Org. Chem.*, **33(2)**, 523–530 (1968).
560. V.R. Gaertner, *J. Org. Chem.*, **32(10)**, 2972–2976 (1967).
561. V.R. Gaertner, *Tetrahedron Lett.*, **4**, 343–347 (1967).
562. R. Lacher, *Wochenblatt fuer Papierfabrikation*, **134(5)**, 192–195 (2006).
563. D.N. Van Eenam, US 4,298,639 (1981).
564. J.B. McKelvey, R.R. Benerito, R.J. Berni, and C.A. Hattox, *Textile Res. J.*, **34(9)**, 759–767 (1964).
565. D.I. Devore, N.S. Clungeon, and S.A. Fischer, *Tappi J.*, **74**, 135–141 (1991).
566. R.J. Riehle, US 6,429,267 (2002).
567. M. Gorzynski and J. Emanuelsson, *TAPPI Proceedings—Papermakers Conference*, Atlanta, GA, 1996, pp. 459–473.
568. L. Webb, *Pulp & Paper Industry*, January 1998, pp. 33–39.
569. M. Ortolani, *Industria della Carta*, **1(4)**, 41–47 (2001).
570. C.G. Hamlet and P.A. Sadd, *Eur. Food Res. Technol.*, **215**, 46–50 (2002).
571. T.W. Mueller and S.A. Fischer, *Tappi J.*, **75(4)**, 159–162 (1992).
572. D.I. Devore, S.A. Fischer, and N.S. Clungeon, *TAPPI Proceedings—Papermakers Conference*, Atlanta, GA, 1991, pp. 377–390.
573. E. Kissa, *J. Chromatogr.*, **605**, 134–138 (1992).
574. L. Boden, M. Lundgren, K.E. Stensio, and M. Gorzynski, *J. Chromatogr.*, **A 788**, 195–203 (1997).
575. W.J. Plantinga, W.G. Van Toorn, and G.H.D. Van der Stegen, *J. Chromatogr.*, **555**, 311–314 (1991).
576. L.E. Rodman and R. Dudley Ross, *J. Chromatogr.*, **369**, 97–103 (1986).
577. R.L. Pesselman and M.J. Feit, *J. Chromatogr.*, **439**, 448–452 (1988).

578. W.-C. Chung, K.-Y. Hui, and S.-C. Cheng, *J. Chromatogr.*, **A 952**, 185–192 (2002).
579. B.K. Bower, US 5,714,552 (1998).
580. L.L. Chan and P.W. Lau, CA 2,041,444 (1992).
581. D.I. Devore and S.A. Fischer, US 5,189,142 (1993).
582. D.I. Devore and S.A. Fischer, US 5,239,047 (1993).
583. B.K. Bower, US 5,614,597 (1997).
584. A.J. Miller and B.M. Stubbs, US 5,171,795 (1992).
585. G.N. Merrill, *J. Phys. Org. Chem.*, **17**, 241–248 (2004).
586. X. Xing, Y. Cao, and L. Wang, *J. Chromatogr.*, **A 1072**, 267–272 (2005).
587. H. Bachem, C. Suling, J. Muszik, and W.D. Schroer, US 4,929,309 (1990).
588. M.A. Dulany and C.E. Ringold, US 5,256,727 (1993).
589. A. Bull, D.J. Hardman, B.M. Stubbs, and P.J. Sallis, US 5,470,742 (1995).
590. D.J. Hardman, M. Huxley, A.T. Bull, J.H. Slater, and R. Bates, *J. Chem. Tech. Biotechnol.*, **70**, 60–66 (1997).
591. R.J. Riehle, R. Busink, M. Berri, and W. Stevels, US 7,303,652 (2007).
592. D.J. Hardman, B.M. Stubbs, and A. Bull, EP 0 510 987 (1992).
593. A. Bull, D.J. Hardman, B.M. Stubbs, and P.J. Sallis, US 5,871,616 (1999).
594. A. Bull, D.J. Hardman, B.M. Stubbs, and P.J. Sallis, US 5,843,763 (1998).
595. R. Bates, H.J. Branton, D.J. Hardman, and G.K. Robinson, US 5,972,691 (1999).
596. R.J. Riehle, A.J. Allen, M. Hofbauer, A.J. Haandrikman, R. Busink, M.T. Crisp, J.J. Hoglen, H.N. Cheng, F.J. Carlin, J.A. Lapre, and H. Jabloner, US 6,554,961 (2003).
597. R.J. Riehle, A.J. Allen, M. Hofbauer, A.J. Haandrikman, R. Busink, M.T. Crisp, J.J. Hoglen, H.N. Cheng, F.J. Carlin, J.A. Lapre, and H. Jabloner, US 7,175,740 (2007).
598. R.L. Amey, US 6,057,420 (2000).
599. R.L. Amey, US 6,056,855 (2000).
600. M. Gorzynski and A. Pingel, US 6,376,578 (2002).
601. M. Gorzynski and A. Pingel, US 5,516,885 (1996).
602. D.N. Van Eenam, US 4,269,951 (1981).
603. G.J. Guerro and L.A. Lawrence, US 6,429,253 (2002).
604. D.I. Devore, N.S. Clungeon, and S.A. Fischer, US 5,575,892 (1996).
605. C.D. Walton and J.F. Warchol, US 2003/0114631.
606. H. Bachem, G. Schroder, C. Suling, J. Reiners, J. Muszik, D. Arit, M. Jautelat, and W.-D. Schroer, US 5,082,527 (1992).
607. H. Bachem, G. Schroder, C. Suling, J. Reiners, J. Muszik, D. Arit, M. Jautelat, and W.-D. Schroer, US 5,093,470 (1992).
608. A.J. Allen and A.C. Sau, US 5,990,333 (1999).
609. L.G. Garth, US 3,320,066 (1967).
610. L.E. Nielsen, *Predicting the Properties of Mixtures*, Marcel Dekker, New York, 1978.
611. D.I. Devore, N.S. Clungeon, and S.A. Fischer, US 5,503,713 (1996).
612. D.I. Devore, N.S. Clungeon, and S.A. Fischer, US 5,350,796 (1994).
613. W.H. Griggs, US 3,592,731 (1971).
614. K.P. Kehrer, S.M. Atlas, V.A. Kabanov, A. Zenin, and V. Rogachova, US 6,755,938 (2004).
615. J.P. Lallier, J.F. Argillier, S. Fouquay, and D. Yanhoye, WO 94/29523.
616. X. Feng, K. Pouw, V. Leung, and R. Pelton, *Biomacromolecules*, **8**, 2161–2166 (2007).
617. T.M. King, US 3,629,124 (1971).
618. R.W. Morgan and M.J. Hatch, US 3,146,157 (1964).
619. J.P. Lallier, J.F. Argillier, S. Fouquay, and D. Vanhoye, FR 2,706,496 (1994).
620. D.R. Spear, US 2,685,508 (1954).
621. R.P. Avis, US 4,076,581 (1978).
622. R.P. Avis, US 4,557,801 (1985).
623. B.J. Kokko and D.W. White, US 7,041,197 (2006).
624. M. Ryan, W. Brevard, D. Dauplaise, M. Lostocco, R. Proverb, and D.W. Lipp, US 6,939,443 (2005).
625. L. Gardlund, L. Wagberg, and R. Gernandt, *Colloids Surf. A Physicochem. Eng. Aspects*, **218**, 137–149 (2003).
626. R. Gernandt, L. Wagberg, L. Gardlund, and H. Dautzenberg, *Colloids Surf. A Physicochem. Eng. Aspects*, **213**, 15–25 (2003).
627. S.P. Dasgupta, US 5,318,669 (1994).
628. H.H. Espy, US 5,316,623 (1994).

629. R. Lingstrom and L. Wagberg, *J. Colloid. Interface Sci.*, **328**, 233–242 (2008).
630. M. Eriksson, G. Pettersson, and L. Wagberg, *Nord. Pulp Paper Res. J.*, **20(3)**, 270–276 (2005).
631. L.E. Enarsson and L. Wagberg, *Nord. Pulp Paper Res. J.*, **22(2)**, 258–266 (2007).
632. L. Gardlund, M. Norgren, L. Wagberg, and A. Marklund, *Nord. Pulp Paper Res. J.*, **22(2)**, 210–216 (2007).
633. G. Pettersson, H. Hoglund, and L. Wagberg, *Nord. Pulp Pap Res. J.*, **21(1)**, 115–121 (2006).
634. G. Pettersson, H. Hoglund, and L. Wagberg, *Nord. Pulp Paper Res. J.*, **21(1)**, 122–128 (2006).
635. P. Lagally and J.W. Brook, US 3,348,997 (1967).
636. R.L. Adelman, US 4,461,858 (1984).
637. W.C. Floyd and L.R. Dragner, US 5,147,908 (1992).
638. W.W. Maslanka, US 5,994,449 (1999).
639. Y. Wang and J. Liu, *China Synth. Resin Plastics*, **19(6)**, 19–22 (2002).
640. J. Detering, C. Shade, K. Oppenlaender, M. Zirnstein, W. Trieselt, and V. Schwendemann, US 5,677,384 (1997).
641. A.J. Miller and B.M. Stubbs, US 5,525,664 (1996).
642. T. Maki, S. Hori, Y. Sakamoto, M. Yoshimoto, and T. Manabe, US 3,949,014 (1976).
643. M.T. Goulet, T.H. Mathews, S.L. Pomeroy, and M. Tirimacco, US 7,299,529 (2007).
644. J. Laine and T. Lindstrom, *IPW*, **1**, 40–45 (2001).
645. J. Laine, T. Lindstrom, G.G. Nordmark, and G. Risinger, *Nord. Pulp Paper Res. J.*, **17(1)**, 57–60 (2002).
646. P.J. Borchert and J. Mirza, *Tappi*, **47(9)**, 525–528 (1964).
647. W.L. Kaser, J. Mirza, J.H. Curtis, and P.J. Borchert *Tappi*, **48(10)**, 583–587 (1965).
648. G.E. Hamerstrand and M.E. Carr, US 3,763,060 (1973).
649. G.E. Hamerstrand and M.E. Carr, US 4,152,199 (1979).
650. H.H. Espy and T.W. Rave, *Tappi J.*, **71(5)**, 133–137 (1988).
651. W. Brevard and M. Ryan, WO 2004/072376.
652. M. Ryan and W. Brevard, US 7,736,465 (2010).
653. K. Luukkonen, US 7,045,563 (2006).
654. R.T. Underwood, R.J. Jasion, S.P. Hoke, and G.G. Spence, US 5,674,362 (1997).
655. W.B. Darlington and W.G. Lanier, US 5,427,652 (1995).
656. R.P. Geer and R.R. Staib, WO 00/11046.
657. E. Kruger, M. Wendker, S. Frenzel, and C. Bottcher, US 2009/0008051.
658. M. Howle and K.B. Himmelberger, US 2003/0079847.
659. J.G. Smigo, L.M. Robeson, G. Davidowich, G.D. Miller, and W.E. Carroll, US 5,281,307 (1994).
660. J. Lindsay, T.G. Shannon, M. Goulet, M. Lostocco, T. Runge, K. Branham, L. Flugge, J. Foster, F. Lang, T. Sun, and G. Garnier, US 6,824,650 (2004).
661. W.C. Floyd and S.H. Hui, US 4,695,606 (1987).
662. D.A. Capwell, US 2005/0155723 (2005).
663. W.C. Floyd, N. Thompson, and L.R. Dragner, US 6,303,000 (2001).
664. M. Laleg and I.I. Pikulik, *Nord. Pulp Paper Res. J.*, **8(1)**, 41–47 (1993).
665. S. Ritter, *Chemical Engineering News*, May 2, 2005, p. 38.
666. W.E. Carroll, C.J. Rendu, and E.J. Beckman, US 7,494,566 (2009).
667. E.J. Beckman, W.E. Carroll, T. Chapman, K.E. Minnich, D. Sagl, and R.J. Goddard, US 7,090,745 (2006).
668. V.T. Stannett, *Wet Strength in Paper and Paperboard*, Tappi Monograph Serie No 29, J.P. Weidner (Ed.), 1965, pp. 85–104.
669. V.T. Stannett, *Surf. Coat. Relat. Paper Wood, Symp.*, 269–299 (1967).
670. L. Westfelt, *Cellul. Chem. Technol.*, **13**, 813–825 (1979).
671. L. Wagberg, *5th Intrnational Paper and Coating Symposium*, Montreal, Quebec, Canada, 2003, pp. 281–284.
672. E. F. Thode, *Tappi*, **42(12)**, 983–985 (1959).
673. T.C. Maloney, H. Paulapuro, and P. Stenius, *Nord. Pulp Paper Res. J.*, **13(1)**, 31–36 (1998).
674. D.A.I. Goring, *Fibre-Water Interact. Paper-Making, Trans. Symp.* Oxford, September 1977, vol. 1, 1978, pp. 43–62.
675. D.F. Caulfield, *Fibre-Water Interact. Paper-Making, Trans. Symp.* Oxford, September 1977, vol. 1, 1978, pp. 63–82.
676. H.W. Haslach, *Mech. Time-Depend. Mater.*, **4(3)**, 169–210 (2000).
677. R.W. Hoyland, *Paper-Making, Trans. Symp.* 1977, **2**, 557–579 (1978).

678. R.H. Marchessault, M. Dube, J.St. Pierre, and J.F. Revol, *Fibre–Water Interact. Paper-Making, Trans. Symp.*, Vol. 2, Oxford, September 1977, pp. 795–818 (1978).
679. A.H. Nissan and G.L. Batten, *Tappi J.*, **80(4)**, 153–158 (1997).
680. J.A. Van Den Akker, *Tappi*, **42(12)**, 940–947 (1959).
681. A.H. Nissan, *Macromolecules*, **10(3)**, 660–662 (1977).
682. S. Yoshioka, H. Yamada, H. Sato, and M. Kamei, US 4,882,087 (1989).
683. S. Zauscher, D.F. Caulfield, and A.H. Nissan, *Tappi J.*, **79(12)**, 178–182 (1996).
684. A.H. Nissan, *Nature*, **263**, 759 (1976).
685. S. Zauscher, D.F. Caulfield, and A.H. Nissan, *Tappi J.*, **80(1)**, 214–223 (1997).
686. A.H. Nissan, *Fibre–Water Interact. Paper-Making, Trans. Symp.*, 1977, **2**, 609–640 (1978).
687. A. Jurecic, T. Lindh, S.E. Church, and V. Stannett, *Tappi*, **41**, 465–468 (1958).
688. A.A. Robertson, *Sven. Papperstidn.*, **66**, 477–497 (1963).
689. D.L. Taylor, *Tappi*, **51(9)**, 410–413 (1968).
690. A.J. Stamm and W.E. Cohen, *Aust. Pulp Paper Ind., Tech. Assoc. Proc.*, **10**, 346–365 (1956).
691. H. Robinette and C.W. Pheifer, US 2,621,169 (1952).
692. T.J. Drennen and L.E. Kelley, US 2,999,038 (1961).
693. J.W. Swanson, *Wet Strength in Paper and Paperboard*, Tappi Monograph Serie No 29, J.P. Weidner (Ed.), 1965, pp. 74–84.
694. R.C. Howard, *Pap Technol. Ind.*, **28(2)**, 425–427 (1987).
695. C.O. Seborg and A.J. Stamm, *Ind. Eng. Chem.*, **23**, 1271–1275 (1931).
696. A.J. Stamm and M.A. Millett, *J. Phys. Chem.*, **45**, 43–54 (1941).
697. J.W. Swanson, *Tappi*, **39(5)**, 257–269 (1956).
698. C.S. Maxwell, W.F. Reynolds, and R.R. House, *Tappi*, **34(5)**, 233–238 (1951).
699. J.-F. Lafaye, *Peintures Pigments Vernis*, **43(9)**, 594–611 (1967).
700. S.F. Kurath, C.Y. Chu, and J.W. Swanson, *Tappi*, **42(3)**, 175–179 (1959).
701. J.J. Becher, G.R. Hoffman, and J.W. Swanson, *Tappi*, **44(4)**, 296–299 (1961).
702. B. Alince, A. Vanerek, and T.G.M. van de Ven, *Berichte der Bunsen-Gesellschaft*, **100(6)**, 954–962 (1996).
703. E.F. Thode, J.W. Swanson, S.F. Kurath, and G.R. Hoffman, *Tappi*, **42(3)**, 170–174 (1959).
704. K.V. Sarkanen, F. Dinkler, and V. Stannett, *Tappi*, **49(1)**, 4–9 (1966).
705. D. Horn and F. Linhart, in *Paper Chemistry*, J.C. Roberts (Ed.), Blackie Academic, London, 1996, pp. 64–82.
706. R.J. Kennedy, *Tappi*, **45(9)**, 738–741 (1962).
707. G.G. Allan and W.M. Reif, *Sven. Papperstidn.*, **74(18)**, 563–570 (1971).
708. R.A. Jewell, J.L. Komen, Y. Li, and B. Su, US 6,379,494 (2002).
709. A.A. Weerawarna, J.L. Komen, and R.A. Jewell, US 7,135,557 (2006).
710. T. Sun, WO 98/24974 (1998).
711. J.A. Westland, US 5,755,828 (1998).
712. T. Kitaoka and H. Tanaka, *J. Wood Sci.*, **47**, 322–324 (2001).
713. J. Laine, T. Lindstrom, G.G. Nordmark, and G. Risinger, *Nord. Pulp Paper Res. J.*, **17(1)**, 50–56 (2002).
714. J. Laine, T. Lindstrom, G.G. Nordmark, and G. Risinger, *Nord. Pulp Paper Res. J.*, **15(5)**, 520–526 (2000).
715. E. Strasdins, *TAPPI Proceedings—Papermaking Conference*, Atlanta, GA, 1992, pp. 503–505.
716. L.L. Chan, Wet and dry strength short course, in *TAPPI Seminar Notes*, Atlanta, GA, 1988, pp. 25–30.
717. C.S. Maxwell, US 2,559,221 (1951).
718. R.S. Ampulski and C.W. Neal, *Nord. Pulp Paper Res. J.*, **4(2)**, 155–163 (1989).
719. E. Strazdins, *Tappi Papermakers Conference*, Atlanta, GA, 1974, pp. 59–65.
720. D.L. Kenaga, W.A. Kindler, and F.J. Meyer, *Tappi*, **50(7)**, 381–387 (1967).
721. H. Hatakeyama, T. Hatakeyama, and K. Nakamura, *J. Appl. Polym. Sci., Appl. Polym. Symp.*, **37**, 979–991 (1983).
722. H.S. Frank and W.-Y. Wen, *Discus. Faraday Soc.*, **24**, 133–140 (1957).
723. D.F. Caulfield and R.C. Weatherwax, *Tappi*, **59(7)**, 114–118 (1976).
724. S.M. Parmerter, US 3,519,618 (1970).
725. E. Delgado, F.A. Lopez-Dellamary, G.G. Allan, A. Andrade, H. Contreras, H. Regla, and T. Cresson, *J. Pulp Paper Sci.*, **30(5)**, 141–144 (2004).
726. G. Szeiffova and B. Alince, *Proceedings of the international conference*, Chemical Technology of Wood, *Pulp and Paper*, Bratislava, Slovakia, September 17–19, 2003, pp. 225–228.

727. F. Chen, M.A. Burazin, M.A. Hermans, D.H. Hollenberg, R.J. Kamps, B.E. Kressner, and J.D. Lindsay, US 6,808,790 (2004).
728. A.J. Stamm and W.E. Cohen, *Aust. Pulp Paper Ind., Tech. Assoc. Proc.*, **10**, 366–393 (1956).
729. L.J. Bernardin, US 3,224,926 (1965).
730. P.A. Graef and F.R. Hunter, US 5,225,047 (1993).
731. J.-F. Lafaye, *Peintures, Pigments, Vernis*, **43(5)**, 321–324 (1967).
732. N. Gurnagul and D.H. Page, *Tappi J.*, **72(12)**, 164–167 (1989).
733. J.-F. Lafaye, *Peintures, Pigments, Vernis*, **43(6)**, 375–384 (1967).
734. J.-F. Lafaye, *Peintures, Pigments, Vernis,* **43(8)**, 517–528 (1967).
735. H.G. Higgins and A.W. McKenzie, *Appita*, **16(6)**, 145–164 (1963).
736. J. Jardeby, A. Nihlstrand, T. Chihani, and S. Sandberg, US 6,423,179 (2002).
737. P. Luner, E. Eriksson, K.P. Vemuri, and B. Leopold, *Tappi*, **50(1)**, 37–39 (1967).
738. A. Meller, *Tappi*, **36(6)**, 264–268 (1953).
739. B. Fredholm, B. Samulesson, A. Westfelt, and L. Westfelt, *Cellul. Chem. Technol.*, **15**, 247–263 (1981).
740. I. Hagland, B. Samulesson, A. Westfelt, and L. Westfelt, *Cellul. Chem. Technol.*, **15**, 295–303 (1981).
741. A. Westfelt and L. Westfelt, *Cellul. Chem. Technol.*, **17**, 165–177 (1983).
742. B. Samulesson and L. Westfelt, *Cellul. Chem. Technol.*, **17**, 179–184 (1983).
743. A. Westfelt and L. Westfelt, *Cellul. Chem. Technol.*, **17**, 49–54 (1983).
744. B. Fredholm, B. Samuelsson, A. Westfelt, and L. Westfelt, *Cellul. Chem. Technol.*, **17(3)**, 279–295 (1983).
745. W.E. Cohen, A.J. Stamm, and D.J. Fahey, *Tappi*, **42(12)**, 934–940 (1959).
746. R.C. Weatherwax and D.F. Caulfield, *Tappi*, **59(8)**, 85–87 (1976).
747. C.G. Landes and C.S. Maxwell, *Paper Trade J.*, **121(6)**, 37–46 (1945).
748. E.J. Goethals, E.H. Scacht, P. Bruggeman, and P. Bossaer, *ACS Symp. Ser.*, **59**, 1–12 (1977).
749. S. Hashimoto and T. Yamashita, *J. Macromol. Sci. Chem.*, **A23(5)**, 597–603 (1986).
750. A.K. Banthia, E.H. Schacht, and E.J. Goethals, *Makromol. Chem.*, **179**, 841–844 (1978).
751. E.H. Schacht, P.K. Bossaer, and E.J. Goethals, *Polym. J.*, **9**, 329–336 (1977).
752. E.J. Goethals, E.H. Scacht, Y.E. Bogaert, S.I. Ali, and Y. Tezuka, *Polym. J.*, **12(9)**, 571–581 (1980).
753. E.J. Goethals, EP 0454226 (1991).
754. V.R. Gaertner, *J. Heterocycl. Chem.*, **6(3)**, 273–277 (1969).
755. Y. Tezuka and E.J. Goethals, *Eur. Polym. J.*, **18**, 991–998 (1982).
756. H.H. Espy and S.T. Putnam, US 3,966,694 (1976).
757. N.A. Bates, *Tappi*, **49(4)**, 1984 (1966).
758. G.G. Allan and W.M. Reif, *Trans. Tech. Sect. (Can. Pulp Paper Assoc.)*, **1(4)**, 97–101 (1975).
759. B. Alince, *J. Appl. Polym. Sci.*, **23**, 549–560 (1979).
760. J.-F. Lafaye, *Peintures Pigments Vernis*, **43(4)**, 245–246 (1967).
761. D. Lath and M. Sivova, *Makromol. Chem. Macromol. Symp.*, **58**, 181–187 (1992).
762. J. Russell, O.J. Kallmes, and C.H. Mayhood, *Tappi*, **47(1)**, 22–25 (1964).
763. C. Hagiopol, M. Georgescu, T. Deleanu, and V. Dimonie, *Colloid Polym. Sci.*, **257(11)**, 1196–1202 (1979).
764. F. Mobarak and A.E. El-Ashmawy, *Cellul. Chem. Technol.*, **11(1)**, 109–113 (1977).
765. A.E. El-Ashmawy, F. Mobarak, and Y. Fahmy, *Cellul. Chem. Technol.*, **7**, 315–323 (1973).
766. S.R.D. Guha and P.C. Pant, *Indian Pulp Paper*, **25(1–6)**, 385–388 (1970).
767. D.A.I. Goring, *Pulp Paper Mag. Can.*, **12**, 515–527 (1963).
768. K.G. Phillips, US 3,468,818 (1969).
769. L. Segal, L. Loeb, and C.M. Conrad, US 2,955,014 (1960).
770. L. Segal and J.J. Creely, *J. Polym. Sci.*, **50**, 451–465 (1961).
771. L. Segal, *J. Polym. Sci.*, **B 1(5)**, 241–244 (1963).
772. K. Ward, C.M. Conrad, and L. Segal, US 2,580,491 (1952).
773. J.F. Harrod, *J. Polym. Sci.*, **A1**, 385–391 (1963).
774. D. Choi, J. Gast, E. Echt, K. Niskanen, and H. Tattari, *Pulp Paper*, **75(10)**, 59–67 (2001).
775. H.L. Jones, US 3,421,976 (1969).
776. E.D. Martinez and T.F. Duncan, US 4,970,250 (1990).
777. E.D. Martinez and T.F. Duncan, US 5,034,097 (1991).
778. L.J. Novak and J.T. Tyree, US 2,786,786 (1957).
779. A.J. Stamm and J.N. Beasley, *Tappi*, **44(4)**, 271–275 (1961).
780. B.J. Kokko and D.W. White, US 2006/0124264.
781. B. Alince, L. Kuniak, and A.A. Robertson, *J. Appl. Polym. Sci.*, **14**, 1577–1590 (1970).

782. E.F. Horsey and W.D. Thompson, US 2,650,163 (1953).

783. B. Alince, *Paper Technol.*, **46(6)**, 14–17 (2005).

784. R. Pelton, *Nord. Pulp Paper Res. J.*, **8(1)**, 113–119 (1993).

785. J.W. Swanson, *Tappi*, **44(1)**, 142–181 (1961).

786. J. Zhang, R. Pelton, L. Wagberg, and M. Rundlof, *Nord. Pulp Paper Res. J.*, **15(5)**, 400–405 (2000) 837.

787. M.N. Fineman, *Tappi*, **35(7)**, 320–324 (1952).

788. D.I. Devore, N.S. Clungeon, and S.A. Fischer, *ACS Sym. Ser.*, **548**, 394–405 (1994).

789. H.R. Miller, US 3,427,217 (1969).

790. G. Rodden, *Pulp Paper*, **9**, 26–30 (2008).

791. N. Wistara and R.A. Young, *Cellulose* (Netherlands), **6(4)**, 291–324 (1999).

792. P. Howarth, *Fibre–Water Interact. Paper-Making, Trans. Symp.*, Vol. 2, Oxford, September 1977, pp. 823–833 (1978).

793. N. Wistara, X. Zhang, and R.A. Young, *Cellulose (Netherlands)*, **6(4)**, 325–348 (1999).

794. A.A. Robertson, *Pulp Paper Mag. Can.*, **65(3)**, T161–T168 (1964).

795. S. Khantayanuwong, *Kasetsart J. (Nat. Sci.)*, **36**, 193–199 (2002).

796. H.H. Espy, *Progress in Paper Recycling*, August 1992, pp. 17–23.

797. A.T. Coscia and L.L. Williams, US 3,734,977 (1973).

798. H.H. Espy and S.T. Putnam, US 3,966,684 (1976).

799. H.N. Cheng, Q.M. Gu, and R.G. Nickol, US 6,159,721 (2000).

800. J.S. Michelman and D.M. Capella, *Polymer, Laminates & Coatings Conference Tappi Proceedings*, Atlanta, GA, 1991, pp. 197–201.

801. J.C. Rankin, A.J. Ernst, B.S. Phillips, B.T. Hofreiter, and W.M. Doane, *Tappi*, **58(1)**, 106–108 (1975).

802. M.P. Bouchette, W.F. Winkler, and H.C. Thomas, US 6,458,240 (2002).

803. J.J. Magda and J.Y. Lee, *Tappi J.*, **82(3)**, 139 (1999).

804. B. Balos, N.S. Clungeon, J.V. Patterson, J.M. Rodriguez, and S.A. Fischer, US 5,593,543 (1997).

805. C. Njaa, A. Renders, and P. Walsh, *Progress in Paper Recycling*, November 1994, pp. 26–31.

806. R.H. Jansma, J. Begala, and G.S. Furman, US 5,401,810 (1995).

807. A.L. Berzins, T.C. Ma, and C.J. Davis, US 5,626,945 (1997).

808. C.S. Maxwell, US 3,407,113 (1968).

809. H.H. Espy and G.W. Geist, *Tappi J.*, **76(2)**, 139–142 (1993).

810. F.E. Caropreso, D.S. Thorp, and R.H. Tieckermann, US 5,718,837 (1998).

811. F.E. Caropreso, D.S. Thorp, and R.H. Tieckermann, US 5,972,164 (1999).

812. H.H. Espy, US 5,674,358 (1997).

813. R.A. Gelman and J.S. Hyunh-Ba, US 5,904,808 (1999).

814. R.H. Tieckelmann and D.S. Thorp, US 5,888,350 (1999).

815. J.S. Sajbel, J.R. Heyward, P.E. Share, and S.A. Fischer, US 5,447,602 (1995).

816. S.A. Fischer, *TAPPI Proceedings—Papermakers Conference*, Atlanta, GA, 1996, pp. 403–409.

817. J.S. Sajbel, J.R. Heyward, P.E. Share, S.A. Fischer, B. Balos, N.S. Clungeon, J.V. Patterson, and J.M. Rodriguez, WO 95/06157.

818. S.A. Fischer, *Tappi J.*, **80(11)**, 141–147 (1997).

819. L.L. Chan and P.W.K. Lau, *Pulp Paper Can.*, **89(8)**, 57–60 (1988).

820. M.A. Johnson, A.R. Pokora, and J.B. Henry, US 5,330,619 (1994).

821. H.J.F. van den Abbeele and T. Kimpimuki, US 7,244,510 (2007).

822. H. Gotoh, A. Igarashi, R. Kobayashi, and K. Akiho, US 4,117,199 (1978).

823. A. Takahira and Y. Yoshii, US 5,527,623 (1996).

824. R.E. Locke, G.N. Prentice, and C.M. Vitori, US 6,053,439 (2000).

825. S. Berube, US 5,929,155 (1999).

826. M A. Dulany, C.E. Garvey, C.E. Ringold, and R. Srinivasan, US 5,585,456 (1996).

827. M A. Dulany, C.E. Garvey, C.E. Ringold, and R. Srinivasan, US 5,567,798 (1996).

828. N.T. Woodberry, US 3,594,271 (1971).

829. L.-L. Chan and P.W. Lau, US 4,722,964 (1988).

830. R.T. Underwood, US 5,783,041 (1998).

831. R.R. Staib, J.R. Fanning, and W.W. Maslanka, US 6,103,861 (2000).

832. J.M. Kliegman and R.K. Barnes, *J. Heterocycl. Chem.*, **7(5)**, 1153–1155 (1970).

833. T.W. Rave, US 4,158,595 (1979).

834. B.T. Hofreiter, H.D. Heath, M.I. Schulte, and B.S. Phillips, *Starch/Staerke*, **33(1)**, 26–30 (1981).

835. J.C. Roberts, Applications of paper chemistry, in *Paper Chemistry*, 2nd edn., J.C. Roberts (Ed.), Blackie, London, U.K., 1996, pp. 1–8.

836. E. Strazdins, *Pulp & Paper*, March 1984, pp. 73–77.
837. The Freedonia Group, *Pulp & Paper*, April 2008, p. 21.
838. J.H. Filling, US 3,093,534 (1963).
839. P.D. Buikema, US 4,029,885 (1977).
840. D.R. Cosper and P.R. Stoll, US 4,268,352 (1981).
841. M.R.St. John, US 5,131,982 (1992).

6 Dry-Strength Resins

Failure, during the breakage of dry paper, occurs either at the point of contact between the fibers (bonded area [1,2]) or as a complete rupture of the fiber wall [3]. The dry strength of paper is derived in part from the intrinsic strength of cellulose fibers [4–6]. It is also derived from the papermaking parameters, which include the pulping process, the presence of surfactants that can modify the water's surface tension [7,8], the beating process [9–11], the retention of fines [12], the sheet formation [13], and the wet and dried sheet compression [14,15]). The paper dry strength typically decreases with higher hardwood content [16] or recycled fibers, but the cellulose fiber selection may improve the paper strength [17]. When the sheet is heavily wet pressed, a boost in paper dry strength is noted [18–20]. This boost is at least partly due to the Campbell effect, which involves a drawing together of the fibers and increase in the number of hydrogen bonds during drying, giving rise to forces in excess of 100 atm) [21].

Some wet end additives can interfere with the bonding between cellulose fibers. Such effects can either enhance or reduce the ability of those fibers to form a strongly bonded sheet of paper. However, the dry strengthening effect of chemical additives is generally smaller than the effects of pulp beating [22]. Moreover, even the effect of the dry-strength resin (DSR) depends on the degree of beating. In the case of cationic polyacrylamide [23], not only does the unbeaten pulp show a lower tensile index, but also a plateau, meaning that polymer adsorption is limited (Figure 6.1). Beaten pulp provides a higher dry strength for paper even in the absence of resin due to due to the tremendous increase in the conformability of the fiber surface.

Dry strength resins are "specialized compounds" because they impart strength to paper "only when the paper is dry". In this respect they are different from the wet-strength resins that were described in Chapter 5. The latter are paper chemicals with reactive functions capable of crosslinking the cellulose fibers. Paper chemicals with reactive functions crosslink the cellulose fibers and that may result in some undesired side effects: formaldehyde [24] increases the embrittlement of the paper, acrylamide–acrolein copolymers [25,26], acrylamide–glycidyl methacrylate copolymer [27] polyacids [28], epichlorohydrin [29–34], dimethylol urea [35] and polyisocyanate [36–37] increase the paper dry strength, which makes it difficult to repulp. Formaldehyde-based resins are also involved the issue of formaldehyde emission [38].

By contrast, a dry strength resin (DSR) should increase inter-fiber bonding, but the paper must remain repulpable: that means that when the paper is soaked, the effects of the DSR should disappear.

6.1 INVOLVEMENT OF CHEMICALS IN THE DRY STRENGTH MECHANISM OF PAPER

The precise meaning of "bonding" in paper is still unclear [39]: the bonds involved in ordinary dry paper products are drastically reduced (by 95%) after soaking in water. The extension of the bonding area in dry paper, the hydrogen bonding, the molecular entanglement in the amorphous phase, the ionic bonding, the van der Waals interactions, and the hydrophobic interactions are still under scrutiny in an effort to clarify the mechanism of paper dry strength.

Arguably, if the fiber mechanical entanglement and covalent bonding are rejected [5] (especially because the mechanical entanglement and the covalent bonds should not be water sensitive), the hydrogen bonds are responsible for inter-fiber bonding.

The entire papermaking process can be called "the process within which hydrogen bonding is managed". The beating, the web formation, the wet press and the drying steps are milestones for the formation and activation of hydrogen bonding [15]. The number of inter-fiber hydrogen bonds is

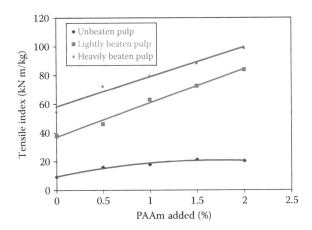

FIGURE 6.1   The effect of beating the pulp on the paper dry strength.

also related to the pulping process when carboxylic groups are added to the cellulose structure. The number and the distribution of carboxylic groups may enhance the swelling and the plasticization of fibers leading to more extensive bonding during sheet formation [40,41].

The high dry strength of paper indicates that interfiber bonding involves the hydrogen bonding between the hydroxyl groups of adjacent cellulose molecules [42]. The hydrogen bonding capabilities depend on the availability and reactivity of hydroxyl groups. Any damage to those interfiber bonds results in lower paper dry strength. The paper dry strength is reduced by the addition of filler particles [39,43], hydrophobic compounds, such as paraffin wax [44], hydrophobic resins [45,46], or hydrophobic pendant groups [47].

The paper dry strength is also diminished by changes in the chemical structure of the hydroxyl groups such as acylation [8,48] and cyanoethylation [49,50], C_6 hydroxyl oxidation [51], or by the presence of a compound (such as polyether [52]) able to compete with the hydroxyl groups in hydrogen bond formation.

Hydrogen and ionic bonds (as well as van der Waals forces) are active only over a short distance (of about 2.5 Å), and, hence, the bonding surface must be in molecularly close contact. It is difficult to bring the crystalline area of the cellulose fiber to that distance. Rough cellulose fiber surfaces have larger asperities, which physically prevent the formation of hydrogen bonds [53].

It is more likely to have a macromolecular entanglement in the amorphous zone [54], which is swollen by water. Segments of macromolecules of cellulose from the amorphous zone located at the contact area between two fibers need water as a solvent to allow for the hydroxyl groups to reach a proper position and generate the hydrogen bonds. Swollen fibers have more flexible surfaces and are able to develop a higher proportion of bonded area under a given set of conditions for forming, pressing, and drying of the paper sheet [55].

The morphology of the fibers changes during water evaporation. The water-swollen fibers and the movable fibrils attached to them lose their flexibility during drying. The air dried pulp has more intra-fiber hydrogen bonding and fewer positions are left for interfibers interactions. As a result, the once-dried pulp (see also the recycled pulp) produces paper with a lower water retention and lower strength than never-dried pulp [8,56].

It is interesting to see how a paper chemical may fit in the fiber geometry and what structures are involved in the hydrogen bonding process. DSR may fill out the asperities with more flexible molecules [56], and so can hemicellulose [54], starch [57], acidified starch [58], protein modified with PAAm [59], or modified starch with formaldehyde [60–62]. Although a difunctional small molecule (such as maleic anhydride) provides higher dry strength for the pulp cured at 65°C–115°C [63], dry-strength resins are ionic macromolecular compounds. Because it is hard to believe that polymers can penetrate the fiber structure, the dry-strength resins are designed to improve the interfiber bonding.

Hemicellulose shows a particular propensity to improve the paper dry strength [64–67]. Hemicellulose is partly extracted during pulping, but residual hemicellulose remains on the cellulose fibers (about 10% in hardwood pulp) [68].

The structure of hemicellulose includes units of arabinose (pentosan) (>10%), xylose (55%–85%), hexoses (glucose) [64], and uronic acid (3%–7%) [65]. The presence of hemicellulose on the fiber surface and its chemical structure (hydroxyl and carboxyl groups) open the possibility of an interaction with the paper chemicals used as dry-strength resins.

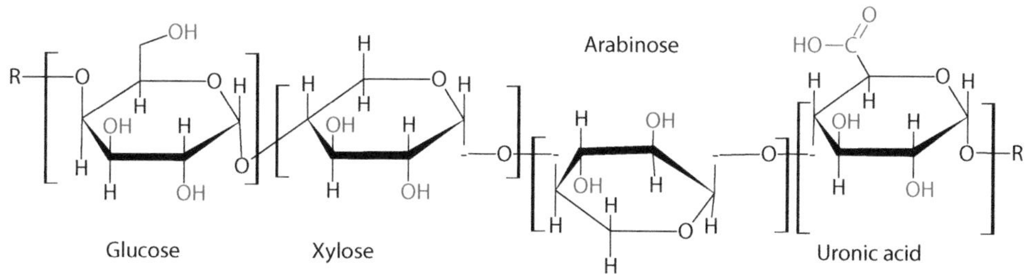

The involvement of paper chemicals in enhancing hydrogen bonds with the hydroxyl groups in cellulose is a key part of the papermaking process. This type of interaction has been well demonstrated for polyethylene oxide (PEO), grafted N-vinyl formamide on PEO [69] or polyvinyl alcohol (PVA) [70]. The hydroxyl at the C-6 position was engaged in a strong hydrogen bonding with the ether oxygen in PEO. PVA is involved in a diversity of hydrogen bonds with both hydroxyl and the ether group (C5) of the cellulose.

The "pseudo-irreversible" character of this bonding is provided by the number of intermolecular hydrogen bonds, which depends on the resin molecular weight, the frequency of the active groups in hydrogen bonding (hydroxyl, ether, carboxyl, and amine), the macromolecular coil flexibility, the polymer–water interactions, etc.

A macromolecular compound can be formed on the surface of the cellulose fiber by polycondensation: pulp treatment with γ-aminopropyl trimethoxy-silane has improved the paper strength by about 20% [71]. Dry-strength resins can also be added by spraying on the wet web [72], at size press [73–74], by impregnation (glyoxal and hydroxyethyl cellulose [75]) or by application on the Yankee dryer [76]. A solution of nonionic dextran is almost 100% retained at the dry end [74].

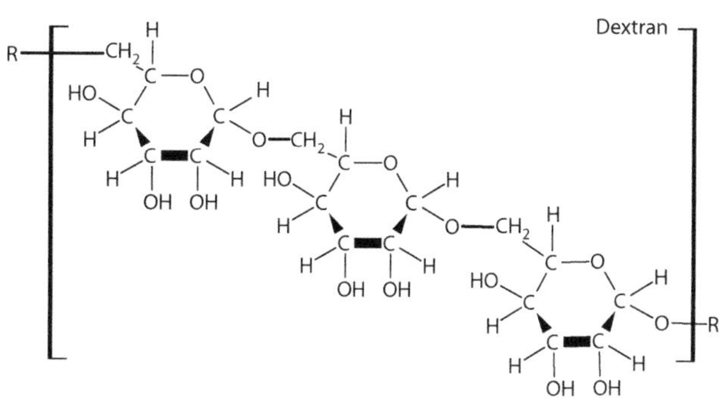

The paper dry tensile is significantly improved if the dry-strength resin is added at the wet end, such that the sheet is strengthened throughout its structure. The DSR added at the wet end should have ionic charges: the vinyl or acrylic monomers are copolymerized with ionic monomers (acryl-amide–acrylic acid copolymers) [77]. Preexisting polymers (such as starch or PVA) are chemically modified by the addition of ionic charges. Dry-strength resins are polyelectrolytes involved in the web formation as well [78,79]. Dry-strength additives increase the paper dry strength by positively influencing the sheet formation, enlarging the bonded area, and increasing the fiber to fiber bonding [79,80]. An increase in paper density is also noticed [81].

The amount of ionic material that can be retained is limited, due to the limited fiber charge density [82]. Retention capabilities should be considered along with strengthening properties for each dry-strength resin. Because the nonionic compounds added to the wet end (such as retention aid resin) make nearly no contribution to paper strength [83], the paper chemical contribution to the paper dry strength is focused on the ionic compounds.

The paper strength is developed in several steps during the papermaking process. Dry-strength resins are ionic macromolecular compounds and are involved in retention and web formation processes. The large majority of chemical structures recommended for dry-strength resins are identical to those of retention aids (see Chapter 3). At the wet end, the dry-strength resin may adsorb onto cellulose fibers (probably in the amorphous area) and large loops and tails of the polymer remain in the water phase.

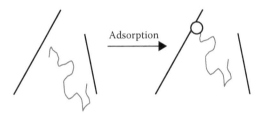

The increase in the number of fiber-to-fiber bonds (relative bonded area) plays a major role in the enhancement of the dry tensile strength of paper. Suitable strength additives enhance the bond strength per unit of bonded area [84–86]. However, the size of the bonded area and the relative position of fibers to each other are set during the sheet formation. Cationic starch seems to be very effective as a dry-strength resin due to its preferential retention on the most likely potential bonded area [21]. The flocculation process occurs when a loop or the tail of the adsorbed polymer is adsorbed on the other cellulose fiber (probably on the amorphous area as well).

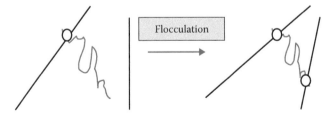

The dry-strength resin must diffuse into the area where macromolecules of hemicellulose or segments of cellulose preexist. In the presence of water (a good solvent for each polymer), an actual solution can be formed at the fiber surface. The paper strength depends on the compatibility of polymers at the fiber–water interface [47] and their capability to develop hydrogen bonds.

It is obvious that the amount and the type of retained resin are important for the paper dry strength. The amount of retained resin depends not only on the number of anionic charges existing on the cellulose fibers, but also on the resin charge density: more resin is retained when the resin charge density is lower [46].

Although the polymer solution is diluted at the wet end, the polymer concentration at the interface is higher. At that moment the "concentrated solution" may behave as an adhesive between fibers. Different polymers on the fiber surface, with different chemical structures, which do not mix, give a weaker interfiber bonding [47].

During the sheet formation, the polymer coil shape shrinks, the fibers are brought closer to each other, and the adsorption areas may overlap; thus the future bonded area is prefigured (see also the effect of the surface tension [10]).

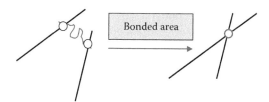

Up to this point (a large amount of water is still present), the strength between fibers is still weak and the stress due to the shear rate may break the bond. The crucial final step of the sheet formation is the pressure applied under high temperature (drying step [87]) when the bonded area is fortified by fibers brought within very close proximity to one another and by water evaporation. Thus, the dry-strength additives intensify bonding within the bonded area at each fiber–fiber crossover [92].

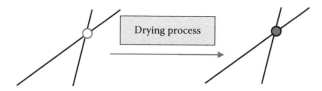

The tendency of bonding within areas where the fibers contact each other seems to be valid even for the DRS applied at the size press [16].

A dry strength agent compensates for the tendency of fillers and low-quality fibers [62,88–90] to reduce interfiber bonding [43]. Thus, DSRs allow for a reduction in grammage and increase in the filler content while maintaining the paper strength [91,92].

Nonionic polymers (such as PVA, starch [61,93], and PAAm) show great potential as dry-strength resins, but their adsorption on anionic cellulose fibers is very limited. Native starches (basically nonionic or slightly anionic) show a retention of less than 40% [53], while a slightly cationic starch (degree of substitution [DS] = 0.023) is retained more than 90% [79].

An ionic macromolecular compound used as a paper dry-strengthening agent must be able to improve the paper dry strength by at least 20%–40% over the normal dry strength of paper [94,95], impart a low degree of wet strength, and have no adverse effect on the drainage rate [88].

6.2 ANIONIC DRY-STRENGTH ADDITIVES

In the absence of a retention aid, the retention mechanism for anionic polymers looks like a mechanical entanglement. Adsorption of most anionic or nonionic starches is a fairly reversible process. In order to improve the amount of retained starch, the slightly anionic starch is used along with a cationic retention aid [96].

The paper dry strength is not improved by low molecular weight polycarboxylic acids (such as butane tetra carboxylic acid [97]). The dry-strength resin must have a high molecular weight [74]: the pulp treated with low molecular weight carboxy methyl cellulose (CMC) shows less strength gains than that treated with high molecular weight CMC [98].

Anionic polymers that improve the paper dry strength may be synthetic macromolecular compounds or modified natural polymers, such as anionic starch [93], starch and boric acid [99], sodium carboxymethyl cellulose [100], carboxymethyl guar, starch xanthate (cross-linked with di-epoxides [101]), or modified lignin [102–104]. Anionic copolymers can also be grafted on a polysaccharide backbone [105,106].

Synthetic copolymers with an anionic functionality are of a wide diversity, with different types of anionic groups, with or without other reactive functionalities (such as anionic acrolein copolymers [107]). The most frequently used anionic dry-strength resins are the acrylamide copolymers [77,108,109]. The anionic comonomers incorporate phosphonic acid [110,111], sulfonate [112] or sulfonic groups [111], as well as carboxyl groups from the itaconic acid [113,114]. Although some other comonomers may be involved in the copolymerization process (such as modified acrylonitrile [115]), the most popular anionic dry-strength resin is the copolymer of the acrylamide–acrylic acid [116–118].

The acrylamide-acrylic acid copolymer is obtained either by the hydrolysis of PAAm or by binary copolymerization. The amide hydrolysis from acrylamide copolymers [119–123] is a difficult process: upon heating the acrylamide-acrylic acid copolymer (30% acrylic acid) in water at pH = 4–7, a conversion to 90% acrylic acid content has been reached after 16 days [121]. There is a slowing down of the process as the conversion increases ("an auto-retarded" process [120]) due to the nearest neighbor effect [122] and the accumulation of the anionic charges on the copolymer macromolecular coil. The repulsive forces developed by the anionic charges existing on the macromolecular coil prevent the anionic hydroxyl group from performing the amide hydrolysis. Therefore, the final anionic polyelectrolyte has a different carboxylic group distribution depending on the concentration of caustic used for hydrolysis [119].

Due to the shortcomings of hydrolysis, the synthesis of the copolymer of the acrylamide–acrylic acid through binary copolymerization is preferred [109,124] (especially by inverse water-in-oil emulsion copolymerization [125]). Copolymerization is performed in water and the reactivity ratios depend on the pH [124] and the nature of the cations present in the system (Na, Ca, Ba, and/or Sr) [126]. In other words, there is a difference in reactivity of the acrylic acid and the acrylate salts, and the copolymer composition can be controlled not only by the comonomer ratio but also by the pH of the comonomer solution and the electrolyte concentration.

Difunctional monomers (e.g., N,N'-methylene-bis-acrylamide-MBA) can be used to increase molecular weight and to form branched structures of ternary copolymers. The comonomer ratios were 92% acrylamide, 0.018% MBA, and about 8% acrylic acid [127].

Natural compounds can be combined with synthetic polymers, as in the case of blends of CMC and anionic latexes [128]. Starch and PAAm both are capable of increasing the hydrogen bonding. A PAAm grafted on starch is therefore a good fit as a dry-strength resin [95,129,130].

Acrylamide–acrylic acid copolymers are water soluble and able to develop hydrogen bonding. The addition of more hydrophobic comonomers, such as acrylonitrile, methyl acrylate, or methyl methacrylate (MMA), reduces the hydrogen bonding capability. If the amount of hydrophobic comonomer (ethyl acrylate, MMA) is low enough, the copolymers are still soluble in water [131] or in an ammonia solution [73] (such as the styrene–maleic anhydride copolymer (SMA) copolymers [132,133]).

The hydrophobic part of the molecule has a significant sizing effect and can also reduce the dry-strength capabilities of the copolymer. The interaction of an anionic rosin salt with an anionic dry-strength resin in the presence of alum [134,135] imparts hydrophobic properties to paper.

At higher concentrations of hydrophobic comonomers (>10%), the copolymers may not be soluble in water even in alkaline pH. They are as dispersions [136]. Carboxylated latexes are retained on pulp by a cationic polyelectrolyte or alum [137]. The latex concentration on paper can reach up to 40%, but the final paper still keeps its porosity [138].

In the absence of an anionic comonomer, but in the presence of the anionic emulsifier as stabilizer, dispersions of nonionic polymers (such as N-vinyl pyrrolidone/MMAs/acrylamide copolymer) [139] are retained with alum and function as dry-strength resins. When both the polymer (such as ethylene–vinyl acetate copolymers) and the stabilizer are nonionic, particle retention is performed with a cationic polyelectrolyte, which interacts with the residual sulfate ions resulting from the decomposition of the initiator [140].

The use of an aluminum salt such as alum ($Al_2(SO_4)_3 \cdot 14H_2O$) or poly-aluminum chloride (for anionic starch) [92,93,103] is one approach to increase the retention of anionic polymers on the cellulose fibers. The alum should be added first at a concentration of 2% based on dry pulp. The optimum pH is related to the alum chemistry (see Section 7.1) and the ionization degree of the polyanionic dry-strength resin. The maximum dry-strength improvement is achieved at the stock pH between 4.2 and 5.0 [77,141,142].

However, for high-tensile-strength paper, the higher valence (+3) cation concentration should be kept at a minimum [143]. Moreover, the presence of alum, a very common paper chemical, can adversely affect the solubility, viscosity, and effectiveness of anionic polymers [144] (Figure 6.2). The acrylamide–maleic acid copolymer is less sensitive to the stock pH than the acrylamide–acrylic acid copolymer [145].

That is why some other options should be considered as potential dry-strength resins. When alum is replaced by a cationic anchoring resin [146], the cationic wet-strength resins can be used to retain potato starch phosphate, which has a negative charge [92]. Alum and a cationic resin can also be used along with an anionic dry-strength resin (such as anionic PAAm) [147]. The optimum ratio of cationic resin to anionic dry strength and the optimum proportion of alum depend on the acidity of pulp, the resin charge densities, and the pH of the aqueous phase. In the presence of a resin with quaternary ammonium cations in its structure, the retention of the anionic dry-strength resin is less pH sensitive.

The DSR efficiency depends on the type of cellulose (Figure 6.3) and has a tendency to reach a plateau at the higher resin dosages [77]. For a broad distribution of cellulose fibers (large amount of fines), the anionic PAAm has a stronger effect when added to the long fiber fraction [81].

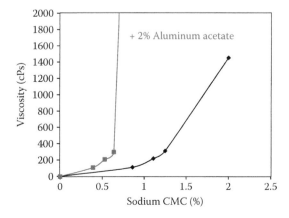

FIGURE 6.2 The effect of inorganic salt on viscosity of CMC solution.

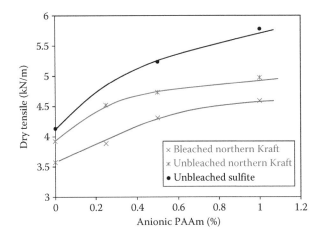

FIGURE 6.3 Anionic PAAm as dry-strength additive for different pulp types in the presence of the same amount of alum.

The idea of using dry-strength agents (anionic copolymer of acrylamide–acrylic acid together with alum) in papermaking dates back prior to 1951 [109]. One of its shortcomings is the fact that that paper is produced at a pH of about 4.5, which needs to be very closely controlled. Paper produced at pH = 4.5 is significantly acidic and undergoes acid tendering on aging. Moreover, the system operating at low pH results in a significant amount of corrosion of the apparatus.

6.3 CATIONIC DRY-STRENGTH ADDITIVES

Cationic dry-strength resins are self-retained; alum is no longer needed. The chemical structure of the cationic dry-strength resins include, besides the units with cationic charges, segments of groups able to be involved in hydrogen bonding or hydrophobic soft parts designed to interact with cellulose fibers through hydrophobic interactions. Chemical compounds with multiple hydroxyl and/ or amide groups and hydrophobic latexes (low T_g values are preferred) will perform as good dry-strength resins.

6.3.1 CATIONIC STARCH AS A DRY-STRENGTH ADDITIVE

Cationic starch is the most common cationic additive and the third most frequently applied wet end additive [148]. Starch has a more flexible macromolecule and can accommodate the uneven fiber surface and act like "cement" between fibers. Starch has a chemical structure very similar to cellulose and can bond with fines and fibrils to enhance strength [149] by performing the same function as hemicellulose[21]. Even lower molecular weight cationic starch shows a strong improvement of the paper dry strength [150–152].

Cationic starch is generally obtained by reacting starch (or starch ethoxylated [153]) with acrylonitrile and hydroxylamine [154], with diethyl amino-ethyl chloride hydrochloride [155,156], with 4-chloro-2-butenyl trimethyl ammonium chloride [157] (more reactive chlorine in the allylic position), or with 2,3-epoxypropyl trimethyl ammonium chloride [158] (see Section 2.3.1). Nonionic starch is also modified with cationic polymers (such as melamine-formaldehyde-guanidine resin [159]) in alkaline conditions [160,161] or in the presence of an oxidizing agent [162].

The addition of cationic charges is performed in a heterogeneous system and it may result in a broad chemical composition distribution. The degree of substitution (0.03–0.1) is an average value, some starch molecules (on the particle surface) show a higher concentration of cationic charges, while some other macromolecules remain nonionic. In order to improve the cationic charge distribution, the polysaccharide structure is modified before the addition of cationic charges. The linear structure is changed through successive degradation (hydrogen peroxide and potassium permanganate) and cross-linking steps [163–165].

Cationic starch added at the wet end is a dry-strength resin, which is also involved in web formation [166]. Cationic starch (up to 2.5% based on dry fibers) is completely retained by pulp [167]; its adsorption is virtually irreversible [168]. The interactions between cationic starch and cellulose fibers are ionic, hydrogen type, and van der Waals bonds that can be disrupted only by acid treatment [168]. Therefore, the retained cationic starches increase the bonded areas and the bonding strength [54,78,79].

Cationic starch increases the dry strength of paper [169] up to 50% (for 2% additive) but does not improve the wet strength of the web [167]. In order to increase the paper dry and wet strength, chemical bonds between starches and cellulose are required [148,167,170]. The starch effectiveness is strongly increased by reactive groups (such as aldehyde [171,172]) attached to the starch structure [148,167,170,173–179].

Cationic aldehyde starch (CAS) is obtained through starch modification (see Sections 2.4.2 and 5.1.1.6) or by ionic interaction between aldehyde starch (slightly anionic due to the oxidation reaction) and polyvinyl amine [180] or with urea-formaldehyde resin (cationic groups by polyamine [181]). For instance, corn starch can be reacted with MF resin in acid pH at room temperature [182].

The aldehyde groups react with the cellulose hydroxyl groups and form the hemiacetal bonds (see Chapters 4 and 5), which is why CAS [183] is a more effective dry-strength resin than cationic starch (Figure 6.4). The new covalent bonds have a strong effect on the dry strength of the paper.

CAS (or starch with blocked aldehyde groups [183,184]) is assumed [148] to form covalent bridges between fibers, and thus the strength of the rewetted paper [170] is higher (Figure 6.5). The effect of di-aldehyde gum is evident for a wide pH range [185], but the increase in paper wet strength is significant at high dosages of CAS resin only and the *W/D* is still inferior to that provided by polyaminoamine epichlorohydrin polymer (PAE) resin at much lower concentrations.

In order to improve the hydrogen bonding capability of starch (or other polysaccharides such as chitin [106]), free radical grafting reactions were developed [94]. The retrogradation process (supramolecular organized gel) for grafted cationic starch is slower due to the cationic pendant groups and the grafted branches.

Cationic charges can preexist on starch or can be brought about by grafting copolymerization in which a cationic comonomer is involved. The following cationic comonomers are used along with nonionic comonomers (such as acrylamide [94,186]): chloride of trimethyl amino ethyl methacrylate and diallyl dimethyl ammonium chloride (DADMAC).

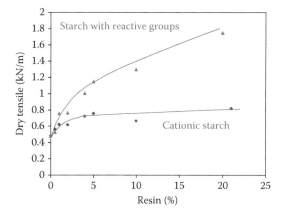

FIGURE 6.4   Cationic starch and CAS as dry-strength resins.

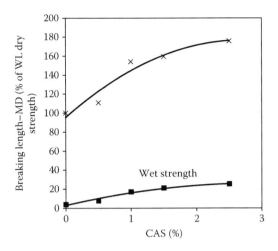

FIGURE 6.5 The effect of CAS on paper dry and wet strength.

The grafting degree is less than 100%, which means that some (co)polymer is linear and some starch may remain un-grafted. Grafted nonionic starches show lower performances because the un-grafted nonionic starch is not retained on cellulose fibers.

6.3.2 Cationic Polyvinyl Alcohol as Dry-Strength Additive

PVA may replace starch as dry strength. Nonionic PVA is obtained by hydrolyzing the homopolymer of vinyl acetate. The cationic PVA is obtained through the hydrolysis of the cationic copolymers of vinyl acetate [187,188]. Cationic comonomers are derivatives of acrylamide or methacrylamide such as N-(3-dimethylaminopropyl) acrylamide, and the copolymerization reactions are performed after neutralization of the tertiary amine with acetic acid. The cationic PVA is obtained in alkaline hydrolysis of the ester bonds when the amide bond (in red) shows a good stability.

The copolymerization reaction has a higher number of steps some of which are rather difficult. That is why it is preferable to modify the commercially available, nonionic PVA. Cationic charges are added to the PVA backbone through a polymer analogous reaction (an acetalization with 2,2,6,6-tetramethyl-4-piperidone) [189]

or a *trans*-acetalization with 4-amino butyraldehyde dimethyl acetal [190]).

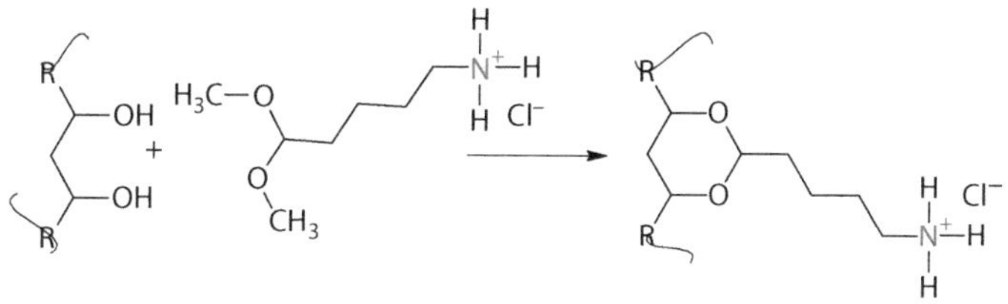

The cationic group is also added by a reaction with an epoxy compound [190–192] (such as glycidyl trimethyl ammonium chloride, see also the cationic starch synthesis)

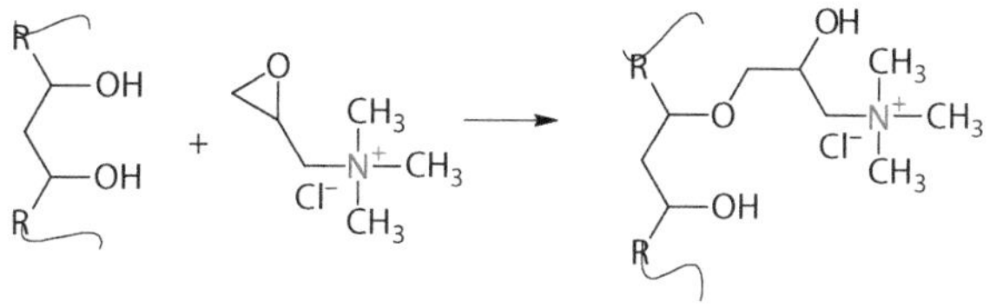

or with formaldehyde and a secondary amine [193]

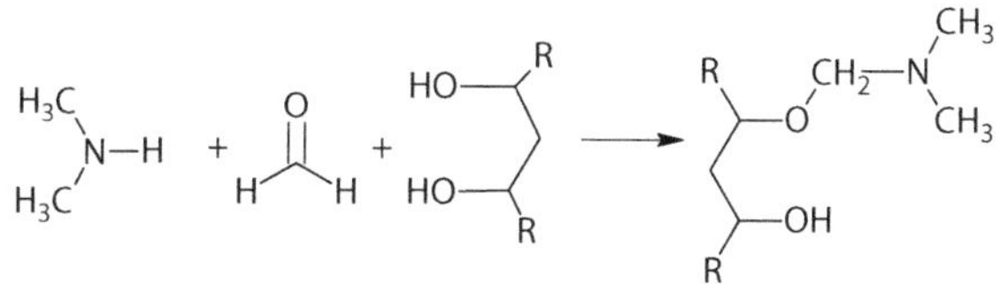

As in the case of starch, cationic charges are added to PVA by interaction with a cationic polymer. The inter-polymer PVA-melamine-formaldehyde resins [194] or the hydrolyzed polyvinyl formamide grafted PVA [195] improve the paper dry strength.

The dry strength of the paper made with those compounds is significantly higher than the water leaf. However, due to the lack of chemical reactivity, the wet strength provided by cationic PVAs is less than 12% of their dry strength.

6.3.3   CATIONIC POLYACRYLAMIDE AS A DRY-STRENGTH RESIN

PAAm is a good dry-strength resin but its effectiveness depends on the type of pulp [12]. cPAAm is easy to synthesize. Because acrylamide is a nonionic monomer, cPAAms are obtained either by copolymerization [196] or by polymer analogous reactions [197].

High molecular weight cPAAms, suitable as dry-strength resins, are obtained by copolymerization of acrylamide (or methacrylamide) with cationic acrylamide (or methacrylamide) comonomers [198], diallyl-melamine [199], DADMAC, or with vinyl pyridine [200] (see also Section 4.3). The acrylamide copolymerization has several drawbacks: cationic comonomers are expensive, the difference in reactivity in the free radical polymerization may lead to complicated technologies, the un-reacted cationic comonomer poses environmental problems, etc. Moreover, the water solution of high molecular weight cPAAm (a solution of 0.045% has a reduced viscosity of 5.0 dL/g) shows high viscosity at low concentrations. Due to the high viscosity, PAAm is precipitated to dispersion [196]. Inorganic electrolytes (ammonium sulfate) are used to partially precipitate the cationic polymer, and the dispersion of 20% has a viscosity of 1500 cPs.

In order to obtain a dry-strength resin, a polymer analogous reaction involves a nonionic PAAm (or its copolymer with styrene) and the amide group is converted to a cationic functionality through a Mannich reaction [201–203] or by a sequence of three reactions [204,205] (see also Section 5.1.3).

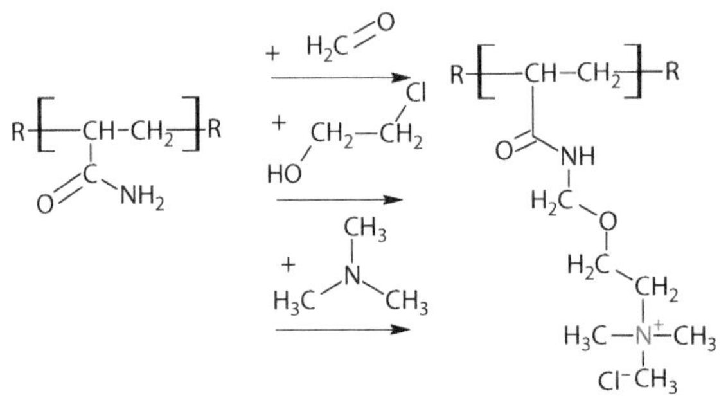

Only less than 10% of the amide units are converted to cationic groups. A cationic copolymer can also be obtained by partial conversion of the amide group to amine (Hofmann degradation) [206,207]. Amines are protonated in aqueous solution even at neutral pH.

cPAAm with higher molecular weight can be functionalized by glyoxalation in order to obtain a dry-strength resin [208–210]. Due to the higher molecular weight, the wet strength decay is much slower. The reactive functionality can be methylol in the copolymer with N-methylolacrylamide and a cationic comonomer [211].

Both reactive groups and cationic charges can be incorporated in the structure of PAAm by reaction with a cationic polymer obtained through polycondensation (dicyandiamide, formaldehyde, urea with or without polyamine and/or epichlorohydrin) [212,213]. PAAm is glyoxalated first to make it reactive to polyamidoamine-epichlorohydrin or to di-methylamine-epichlorohydrin resins [214], which add cationic charges to the new inter-polymer.

The capabilities of hydroxyl groups and amide groups to form hydrogen bonds can be combined in one molecule (inter-polymer) through an intermediate reaction involving asymmetrically blocked glyoxal [215] (such as dimethoxyethanal [216]). The reactivity of glyoxal is controlled by blocking with urea derivatives [217]: the inter-polymer is carrying blocked aldehyde functionalities available to react with cellulose during the paper-drying step. However, due to the high functionality of PVA and PAAm, gel formation during reaction and a short shelf life product should be expected. Efforts to explore the diversity of inter-polymer structures suggest the need to explore the polymer blends as well.

6.3.4 Blends of Cationic Resins as Dry-Strength Additives

Dry-strength additives are made from high molecular weight cPAAm (Mw = 100,000), which is glyoxalated at very low polymer concentration (2%); the polymer is blended with a glyoxalated polyvinyl amine [218].

cPAAm and cationic starch help the web formation. Blends of those two polymers show a synergetic effect on the sheet formation [166]. The PAE-type resin has a strong tendency for thermosetting and enhances the paper dry and wet strength [219]. The thermosetting effect is due to the presence of the azetidinium cation. The dilution with cPAAm [220] increases the dry strength and keeps the wet strength at the level set for PAE or lower. In order to reduce the thermosetting effect and to increase the hydrogen bonding capabilities, the azetidinium cation should be in a lower concentration (almost zero azetidinium groups [221]) and more hydroxyl groups should be added to the DSR.

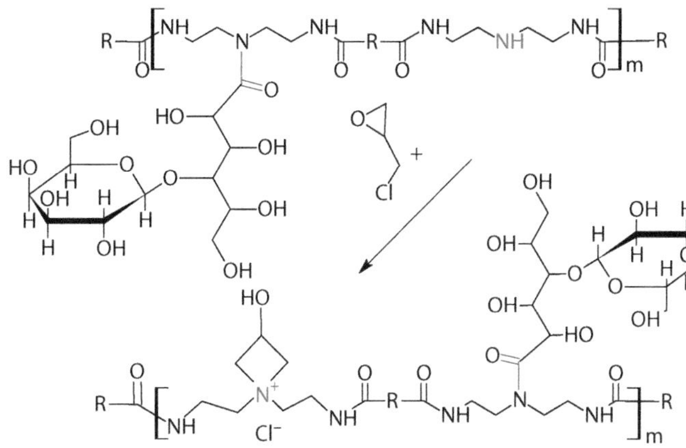

There are two approaches [222,223] to building a new macromolecule: (1) to reduce the number of secondary amines in the polyamidoamine prepolymer by a reaction with a hydroxyl–carboxylic acid (lactobionic acid) and (2) to convert the azetidinium cation in PAE resin with hydroxyl–amino compounds, such as glucosamine [223].

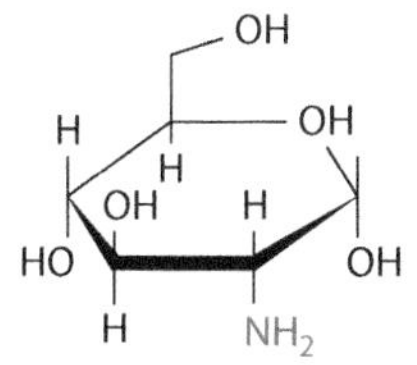

6.3.5 POLYAMINE

Polyamine homopolymers (such as the poly-allyl amine [224,225]), copolymers of vinyl amine [226–229], or branched polyamines (condensation with epichlorohydrin or glyoxal [230] or grafted with *N*-vinylpyrrolidone [231]) are protonated in neutral or acid pH and are effective dry-strength resins. A poly-tertiary amine is obtained through a Mannich reaction: DETA + TETA (>3 mol), acetone (0.5 mol) and formaldehyde (1.5 mol) in water at 92°C.

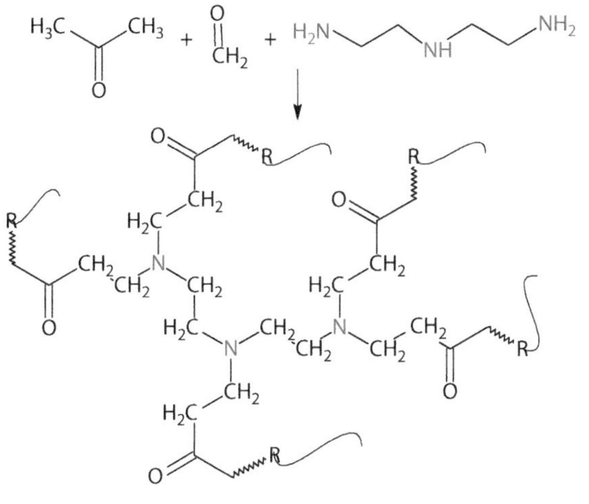

Poly-quaternary ammonium salts (such as poly-ionene) are effective on a broader pH range [232]. Most of the polyamines are cooked with starches first [233–236] to combine the starch effect and make it cationic (the cationic resin is seen as "retention aid for starch" [237]). Polyethylene imine (PEI) (poly-secondary amine) has a strong impact on the dry stiffness and dry tensile strength in the presence of *tris*-(1-aziridinyl) phosphine oxide [238].

Macromolecular compounds presented so far in this chapter are all water-soluble materials, which explains why their interaction with cellulose is supposed to be at the molecular level—hydrogen bonding.

6.3.6   CATIONIC LATEXES AS DRY-STRENGTH ADDITIVES

Certain polymeric dry-strength resins [239] are not soluble in water, for example cationic latexes. Cationic charges are added to polymer latex by copolymerization or by reactions with the functional groups located on nonionic latex [240]. In other words, a latex particle contains copolymers in which at least one comonomer is hydrophobic (styrene, acrylic, or methacrylic esters) and at least one has cationic charges or reactive functionalities.

The very hydrophobic homopolymer of styrene is converted to a cationic copolymer by successively performing a chloromethylation and then a reaction with trimethyl amine [241]. This copolymer does not have any capability to develop hydrogen bonds. But an increase of over 30% in the paper dry strength was noticed when cellulose pulp was treated with it.

The particles of the styrene–acrylamide copolymer (89:11) are synthesized through emulsion copolymerization and reacted with formaldehyde and di-methyl amine (Mannich reaction) to develop cationic groups on the acrylamide units located on the particle surface [242–244]. The dispersion shows high viscosity because a large amount of the acrylamide homopolymer has remained in the water phase. Thus, DSR combines cPAAm and a cationic dispersion of the styrene–acrylamide copolymer.

The cationic polystyrene latex ($T_g > 105°C$) needs high drying temperatures (1 h at 150°C) to reach its effectiveness as a dry-strength resin. A strong increase in breaking length was noticed [245] at regular drying temperature only for copolymers with low glass transition points [246,247] (such as styrene–butadiene copolymers). Cationic styrene-butadiene latexes were prepared with different T_g [240] and used as dry-strength agents (Figure 6.6) [248]. The paper dry strength increases to over 100% the water leaf dry strength. At about 5% latex concentrations

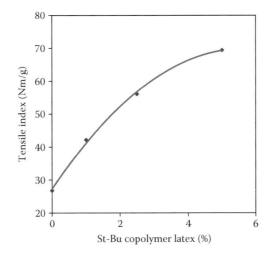

FIGURE 6.6   The effect of St-Bu latex concentration on the paper strength.

(based on dry paper), the adsorption plateau was not reached yet. The latex composition was 65% styrene and 110 nm particle size.

The amount of cationic resin retained on fiber is, unfortunately, limited to the number of existing anionic charges on the cellulose pulp. In other words, there is a limit for the concentration of the potential cement able to link together the cellulose fibers. The amount of retained polymer can be increased by using blends of anionic and cationic polymers (see Chapter 3 and Section 6.5).

The nature of the mechanism of paper dry strengthening through hydrophobic polymers is still undecided. Cellulose fibers are smoother after latex deposition and film formation but whether or not that will increase the hydrogen bonding is still under debate [246]. A more likely explanation for this effect is that additional interfiber bonding arises due to hydrophobic interactions [249] involving the cellulose surface and hydrophobic molecular segments of the polymer. These types of interactions are not sensitive to water (see wet strengthening by latexes, Section 5.1.4). Therefore, the repulpability of paper treated with latexes as dry-strength resins should be a concern.

6.4 AMPHOTERIC DRY-STRENGTH RESINS

Amphoteric compounds display both anionic and cationic charges in the same molecule. There is a new opportunity to manipulate the charges for an easier way to use dry-strength resins at the wet end. Better dry-strength properties provided by amphoteric resins can be attributed to the higher amount of resins adsorbed on cellulose fibers (their superior bonding ability), due to their ionic groups of opposite charge (making a significant contribution to the overall bonding mechanism), and due to their capability to increase the amount of surface-associated water (possibly helping the hydrogen bond formation during the drying step) [250]. Amphoteric resins also tend to self-associate when the pH conditions favor ionization of both positive and negative groups [250].

Amphoteric polymers are efficient dry-strength resins on a broader pH range and more robust on a wider electrolyte concentration [251–253]. A straight anionic dry-strength resin is able to increase the paper dry strength at a pH ranging from 4.0 to 4.7, after which the dry-strength values steadily declined. The resin became completely inactive at pH = 9.0. In the case of the amphoteric polymer, the dry strength increased steadily up to about pH = 7.0 [254].

Amphoteric dry-strength resins are synthesized by copolymerization process [255], grafting, and polymer analogous reactions. The copolymerization involves at least three comonomers: a nonionic monomer (such as acrylamide, which is the major component of the final ternary copolymer), an anionic monomer (the acrylic acid), and a cationic monomer, such as DADMAC [221,254]. Acrylic acid can be replaced by itaconic acid and DADMAC by dimethyl aminopropyl acrylamide or acryloyloxyethyl trimethyl ammonium chloride [256,257] to obtain a polyampholyte [250,252].

Amphoteric dry strength resins are made by grafting a ternary system (acrylamide, cationic and anionic comonomers) on a nonionic backbone (starch) [258] and by grafting an anionic PAAm on cationic backbones (polyamine or polyamidoamine) [259–261]. The grafting reaction is performed with cerium nitrate [262] (see also Section 2.3.2).

Amphoteric polymers are also obtained through a polymer-analogous reaction: an anionic or cationic copolymer is converted to an amphoteric compound by adding cationic or anionic groups, respectively. Hofmann and Mannich reactions add cationic charges to an anionic PAAm [253], cationic groups are added to anionic potato starch [263], cationic starch is xanthated with CS_2 [264]. Amphoteric starch is also obtained by reacting starch with functionalized amino acids (such as 2-chloroethyl, N-methylaminosuccinic acid) [265–267]. The presence of a pair of anion–cation charges in the same pendant group (zwitterions [167]) significantly increases the starch adsorption on the anionic pulp.

From a polyamine (or polyamidoamine) [268–270] the ionic charges are added in two successive steps: reaction with sodium chloroacetate (or acrylic acid for a Michael addition [269]) and then a reaction with epichlorohydrin [270].

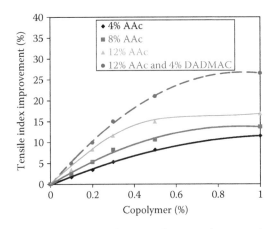

The amount and the chemical composition of the amphoteric dry-strength resin used at the wet end depend on the pH, the anionic/cationic charges ratio, the fiber anionicity, and the presence of another polyelectrolyte. If another cationic polyelectrolyte is present, a higher concentration of anionic charges in the amphoteric compound is required [257].

Figure 6.7 shows the effect of adding DADMAC [271] to an anionic PAAm, at a given copolymer molecular weight. A higher acrylic acid content (AAc) results in higher dry strength; DADMAC further enhances the paper strength.

The reduced bonding ability of recycled fibers requires an effort to restore the interfiber bonding by using the right dry-strength resin [56]. The large number of dry-strength resins should be associated with their blends. Thus, the effectiveness of dry-strength resins is enhanced if the resin structure (macromolecular polyelectrolytes) matches the cellulose fiber structure (which is a macromolecular polyelectrolyte as well).

6.5   BLENDS OF ANIONIC AND CATIONIC RESINS

The effectiveness of a dry-strength resin depends on the amount of resin retained on the anionic cellulose fibers. Cationic DSR is retained directly, while anionic DSR is retained with a cationic retention aid. Regardless of the adsorption method there is a limit of the amount of cationic polymer adsorbed on pulp. That limit (saturation point) depends on the level of fiber anionicity, the cationic charge density, the molecular weight of the cationic additives [158,272], and pH [273].

More anionic charges added to pulp by oxidation decrease the dry strength of the paper [274]. Thus, in order to increase the cationic polymer retained on fibers, the anionic charges are supplied by an anionic polyelectrolyte. The anionic macromolecular compound can be either a water-soluble polymer or a precipitated form (such as "fibers" copolymers of the styrene-maleic anhydride (92/8) with molecular weights ranging from 50,000 to 200,000 [275]).

An excess of cationic resin added at the wet end may change the electrophoretic mobility of the fibers to cationic, and a large amount of cationic resin will be lost in white water. However, the cationic charges of the modified cellulose fibers open the possibility to adsorb an anionic dry-strength resin on top of the preexisting cationic polymer. Afterward, more cationic polymers can

FIGURE 6.7   The effect of the amphoteric copolymer on the paper dry strength.

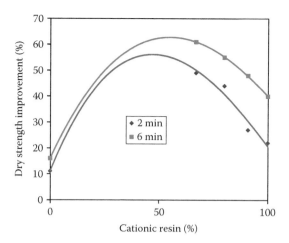

FIGURE 6.8   The synergetic effect of the anionic–cationic resin blends (drying time 2 or 6 min at 105°C).

be adsorbed by the last added anionic layer. By that sequence of addition, the total amount of resins adsorbed on the cellulose fiber is much higher. The ratio between the cationic and the anionic resins becomes a tool in designing the final dry strength of the paper [272].

Blends of anionic PAAm and cationic poly-DADMAC develop a higher dry strength than poly-DADMAC alone used in the same concentration [276]. In other words, it is not only a matter of the amount of resin retained, but also an increase in hydrogen and ionic bonding between the components present in the bonded area.

The adsorption of the cationic polymer follows the general rule: some cationic charges interact with the randomly distributed anionic charges on the cellulose fibers, and, thus, large loops and long tails are formed. Some cationic charges are available on the fiber surface, which are more "hairy." The anionic polymer will be retained based on ionic interactions. This new architecture on the fiber surface is a typical opportunity for a synergetic effect.

Such a synergistic effect on the paper dry strength [110,183,277–279] is developed by using blends of cationic polysulfonium chloride and anionic PAAm (Figure 6.8) [118]. Any blend ratio improves the paper dry strength more than either resin alone. Regardless of the fact that the maximum effect appears to be observed at 1:1 ratio, to ensure a good retention, an excess of cationic resin is preferred (see Section 3.3.3.5). For the blend of cationic starch and anionic PAAm [158], cationic starch is added in excess: starch 0.5%–1.5% based on dry pulp and anionic resin only 0.1%–0.2%.

Cellulose fibers are polyelectrolytes as well. The way macromolecules (poly-cations and poly-anions) are organized on the fiber surface depends on the order of addition of the polyelectrolytes. Since cellulose fibers have anionic charges, common sense suggests to add the cationic polymer first and the anionic component second. An optimum ratio between the cellulose charge density, the amount of the cationic anchor, and the amount of anionic polymer should be considered in this case [56]. Finding the pulp saturation point is crucial for the dual system effectiveness [56]. If the cationic component (CAS) is added in insufficient amounts (after adsorption of the cationic additive, the pulp is still anionic), the anionic resin is not retained and does not have any impact on the paper strength [148,170].

A multilayer structure can be obtained by successively adding a cationic polymer and an anionic polyelectrolyte, followed by another layer of cationic polymer and so on [280–284]. The cationic poly-allyl amine layer was followed by a polyacrylic acid layer. The cationic polymer was added in the third step on top of the anionic polymer, and after five successive layers the paper dry strength was increased over 100%. The 10 layers were about 56 nm thick [285].

In order to avoid flocculating the pulp with cationic polymers [286], the anionic polymer may be added prior to the cationic polymer (for instance CMC first and PAE second [287], caboxymethyl guar and then cationic guar [278], or anionic lignosulfonate and then PAE resin [288–293]). Preferably, the cationic resin should be added at a point as close to the forming section of the paper machine as possible [279].

Sequential addition is a difficult task for the wet end process. It would be much easier to handle if the two resins were blended before being added to the pulp dispersion [294]. For instance, polyelectrolyte complexes are formed from cationic PVA and anionic copolymers having sulfonate (or carboxyl) groups [187], PAE and CMC (or carboxylated polymers) [268,295], starch xanthate and PEI [296], polyethylene polyamine and lignosulfonate [297], anionic starch (modified with maleic anhydride) and cationic starch [298–300], cationic starch and amphoteric PAAm [253], or anionic starch and starch richer in cationic amylopectin [301].

The additives used to make anionic–cationic complexes are made through polycondensation [146], polymer analogous reactions [88,277], or free radical polymerization [277,302].

Table 6.1 includes several blends used at the wet end. In order to present the difficulties encountered in that process, the sequence of addition is also presented. Regardless of understandable

TABLE 6.1
Blends of Anionic and Cationic Polymers as Dry-Strength Resins

Addition Sequence	Anionic (A)	Cationic (C)	References
Cationic first	Anionic PAAm	Cationic starch	[303]
Cationic first	Anionic PAAm	Cationic latex	[304]
Cationic first	Anionic starch	Cationic PAAm	[305]
Cationic first	CMC	Poly-DADMAC	[272]
Cationic first	Ethylene–MA copolymer (maleamic acid)	Cationic guar gum	[306]
Cationic + tri-sodium citrate (as moderator)	CMC	Cationic PAAm poly-DADMAC	[144]
Alum + C + A	Anionic PAAm	PAE	[147]
Alum + C + A	Black liquor	AAm–DADMAC copolymer	[307]
Alum + A + C	Anionic PAAm	PAE (low Mw)	[221]
CaCl₂ + CMC + PAE	Carboxymethyl cellulose	PAE	[295,308]
(A + C) and ionization suppressor	Anionic PAAm	PAE (low Mw)	[309]
A first and C second (complex)	Anionic PAAm	Polyvinyl amine	[110]
A first and C second	Anionic PAAm	Poly-DADMAC	[276]
Anionic first	Carboxymethyl guar	Cationic guar gum, cationic PAAm, or PAE	[310–312]
Anionic first	Polyacrylic acid, SMA	p-DADMAC, cationic polyacrylates	[275,313]
Anionic first	Acrylamide–maleic acid copolymer	Polyamine-formaldehyde resins	[279]
Cationic first	Anionic PAAm copolymer with styrene (latex)	Polyamine and/or modified PAE	[302]
Simultaneously (complex)	Solubilized lignin, hemicellulose, synthetic polymers	Cationic guar or AAm–DADMAC copolymers	[286]
Premixed	Anionic starch	Cationic starch	[298,299,314]
Premixed	Di-aldehyde starch	PAE	[315]
Precipitation	Starch xanthate	PAE	[316]
C & A	Starch xanthate	PAE	[317]
Anionic + Al³⁺ salt (moderator) + cationic	Carboxymethyl cellulose	Cationic PAAm poly-DADMAC	[144]

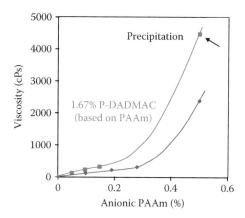

FIGURE 6.9   Viscosity of aqueous blends of anionic PAAm and p-DADMAC.

difficulties, the addition of two resins (anionic and cationic) brings more flexibility in getting the right paper properties.

The properties and stability of the polyelectrolyte complex depend on the charge density, the molecular weight, the presence of a moderator compound [144], pH [318], and the protocol for making it (rate of addition, shear rate, temperature, time, etc.). The complex stability depends on the number of "active sites" [249]: higher molecular weight and higher charge density ensure a higher complex stability.

Cationic polyelectrolytes interact in water with the anionic polyelectrolytes, resulting in the formation of a precipitate. Precipitation is anticipated by an increase in viscosity of the blend showing the presence of the strong interaction between those two polyelectrolytes. Figure 6.9 shows (in blue) the viscosity of the solution of anionic PAAm in the absence of a cationic polymer. The addition of anionic PAAm to a solution of cationic p-DADMAC [144] increases the solution viscosity (Figure 6.9, in red). At 0.5% PAAm, precipitation takes place.

In order to control the precipitation and the viscosity associated with it, a moderating agent or an ionization suppressor is used [144,302]: anionic CMC is moderated with aluminum acetate (Al^{3+}) before adding the cationic resin. Cationic resin is moderated by tri-sodium citrate before an anionic polyelectrolyte is added. Tall oil soap (1.5% based on dry pulp [307]) added at the pulp dispersion can be seen as a moderator for the cationic polyelectrolyte and an enhancer of the pulp anionic charges.

Another strategy to solve compatibility issues is to divide the pulp in two parts, and treat each part with cationic resin and anionic resin, respectively. The blend of two different charged fibers show an improvement both in fiber retention and paper strength [319].

The charge density of the premixed anionic–cationic polyelectrolyte complex [309,320,321] depends on the polymer molecular weight, the relative charge concentration, and the charge distribution. The polysalt shape is also controlled by an ionization suppressor like sodium sulfate, sodium chloride, hydrochloric acid, etc. Alum is also involved for retention when the overall charge is anionic.

Polyelectrolyte complexes (polysalt interpolymer) are supramolecular aggregates with larger molecular weights. After adsorption, they bring more material onto cellulose fibers before charge neutralization. Thus, superior dry-strength properties are obtained (45%–70% higher) [281,317,322] as compared with one dry-strength component able to provide only 20%–30% increase in the paper strength. The strengthening mechanism involves a larger number of both ionic and hydrogen bonds, broken by water, and is able to ensure a good repulpability.

PAE resins are largely used in papermaking as cationic wet-strengthening agents. They are also involved in polyelectrolyte complex formation. The PAE resin is blended with anionic or amphoteric

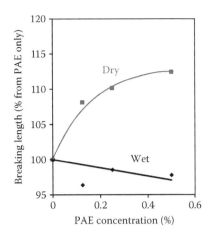

FIGURE 6.10   The effect of sodium starch xanthate (1%) on the wet and dry paper strengthened with PAE resin.

PAAm [323] or glyoxalated polyacrylamide (G-PAm) [324]. In the blend with sodium starch xanthate [325], the PAE resin improves only the dry strength [316], keeping the wet strength at the same level (Figure 6.10).

The structural similarities between retention aids and dry-strength resins open the possibility of using one single compound for multiple applications, which is a step toward simplifying the complex papermaking process.

REFERENCES

1. D.H. Page, *Nord. Pulp Paper Res. J.*, **17(1)**, 39–44 (2002).
2. R.W. Perkins and R.E. Mark, *J. Appl. Polym. Sci. Appl. Polym. Symp.*, **37**, 865–880 (1983).
3. L. Wagberg, *5th International Paper and Coating Symposium*, Montreal, Quebec, Canada, 2003, pp. 281–284.
4. P. Karenlampi, *Tappi J.*, **81(11)**, 137–147 (1998).
5. R.W. Davison, Theory of dry strength development, in *Dry Strength Additives*, W.F. Reynolds (Ed.), TAPPI Press, Atlanta, GA, 1980, pp. 1–31.
6. J.A. van den Akker, A.L. Lathrop, M.H. Voelker, and L.R. Dearth, *Tappi*, **41(8)**, 416–425 (1958).
7. A. Brucato, US 4,347,100 (1982).
8. H.G. Higgins and A.W. McKenzie, *Appita*, **16(6)**, 145–164 (1963).
9. C.H. Mayhood, O.J. Kallmes, and M.M. Cauley, *Pulp Paper Mag. Can.*, **62**, T479–T484 (1961).
10. W.B. Campbell, *Paper Trade J.*, **95(8)**, 29–33 (1932).
11. S. Khantayanuwong, *Kasetsart J. Nat. Sci.*, **36**, 193–199 (2002).
12. G. Carlsson, T. Lindstrom, and C. Soremark, *Sven. Papperstidn.*, **80(6)**, 173–177 (1977).
13. M.J. Korteoja, A. Lukkarinen, K. Kaski, and K.J. Niskanen, *J. Pulp Paper Sci.*, **23(1)**, 18–22 (1997).
14. B. Alince, A. Vanerek, M.H. de Oliveira, and T.G.M. van de Ven, *Nord. Pulp Paper Res. J.*, **21(5)**, 653–658 (2006).
15. A.K. Vainio and H. Paulapuro, *BioResources*, **2(3)**, 442–458 (2007).
16. I. Reinbold and H. Wilfinger, *Pulp. Paper Mag. Can.*, **68(4)**, 110–112 (1967).
17. K.D. Vison and J.P. Erspamer, US 5,228,954 (1993).
18. R.W. Davison, *Tappi*, **55(4)**, 567–573 (1972).
19. I.B. Strohbeen and J.H. Dinius, US 4,377,543 (1983).
20. R.I. Cole, D.E. Holcroft, E.A. Milligan, and S. Greenhalgh, US 3,432,936 (1969).
21. H.W. Moeller, *Tappi*, **49(5)**, 211–214 (1966).
22. T. Lindstrom, L. Wagberg, and T. Larsson, On the nature of joint strength in paper, in *Proceedings of 13th Fundamental Research Symposium*, Cambridge, U.K., 2005, pp. 457–562.
23. D. Tatsumi, T. Yamauchi, and K. Murakami, *Nord. Pulp Paper Res. J.*, **10(2)**, 94–97 (1989).
24. W.E. Cohen, A.J. Stamm, and D.J. Fahey, *Tappi*, **42(12)**, 934–940 (1959).

25. E.D. Kaufman, US 3,819,555 (1974).
26. D.N. Van Eenam, US 4,226,957 (1980).
27. M. Singh and R. Cockcroft, US 2009/0120601.
28. T. Jaschinski, US 6,409,881 (2002).
29. J.A. Harpham and H.W. Turner, US 3,069,311 (1962).
30. C.H. Lesas and M. Pierre, US 4,204,054 (1980).
31. I.L.-A. Croon and N.V. Blomqvist, US 3,700,549 (1972).
32. F.H. Steiger, US 3,658,613 (1972).
33. C.M. Herron and D.J. Cooper, US 5,183,707 (1993).
34. C.M. Herron, D.J. Cooper, T.R. Hanser, and B.S. Hersko, US 5,190,563 (1993).
35. J.A. Westland, US 5,755,828 (1998).
36. D.F. Hiscock, J.A. Jacomet, and D.H. Carter, US 4,617,223 (1986).
37. E. Wenderoth, J. Kopp, H. Traubel, and H.A. Burkhart, *Wochenblatt fuer Papierfabrikation*, **124(9)**, 379–383 (1996).
38. G.E. Myers, *Forest Prod. J.*, **36(6)**, 41–51 (1986).
39. A. Tanaka, E. Hiltunen, H. Kettunen, and K. Niskanen, *Nord. Pulp Paper Res. J.*, **16(4)**, 306–312 (2001).
40. S. Katz, N. Liebergott, and A.M. Scallan, *Tappi*, **64(7)**, 97–100 (1981).
41. S. Katz and A.M. Scallan, *Tappi J.*, **66(1)**, 85–87 (1983).
42. A.W. McKenzie and H.G. Higgins, *Aust. J. Appl. Sci.*, **6**, 208–217 (1955).
43. Y. Xu, X. Chen, and R. Pelton, *Tappi J.*, **4(11)**, 8–12 (2005).
44. R.G. Capell and P.R. Templin, US 2,898,293 (1959).
45. R. Pelton, *Appita J.*, **57(3)**, 181–190 (2004).
46. J. Zhang, R. Pelton, L. Wagberg, and M. Rundlof, *Nord. Pulp Paper Res. J.*, **15(5)**, 440–445 (2000).
47. J. Zhang, R. Pelton, L. Wagberg, and M. Rundlof, *Nord. Pulp Paper Res. J.*, **15(5)**, 400–405 (2000).
48. A.J. Stamm and J.N. Beasley, *Tappi*, **44(4)**, 271–275 (1961).
49. A.J. Stamm and W.E. Cohen, *Aust. Pulp Paper Ind. Tech. Assoc. Proc.*, **10**, 366–393 (1956).
50. R.F. Schwenker and E. Pacsu, US 3,312,642 (1967).
51. P. Luner, E. Eriksson, K.P. Vemuri, and B. Leopold, *Tappi*, **50(1)**, 37–39 (1967).
52. A.J. Stamm and W.E. Cohen, *Aust. Pulp Paper Ind. Tech. Assoc. Proc.*, **10**, 346–365 (1956).
53. J. Marton, Dry-strength additives, in *Paper Chemistry*, 2nd edn., J.C. Roberts (Ed.), Blackie Academic & Professional, London, U.K., 1996, pp.83–97.
54. H.W. Moeller, *Paper Technol.*, **6(4)**, 279–284 (1965).
55. M.A. Hubbe and O.J. Rojas, *Chem. Eng. Educ.*, **39(2)**, 146–155 (2005).
56. M.A. Hubbe, T.L. Jackson, and M. Zhang, *Tappi J.*, **2(11)**, 7–12 (2003).
57. B.W. Rowland, US 2,207,555 (1940).
58. L.H. Rees, US 2,635,068 (1953).
59. M. Singh, F. Herren, C. Northfleet, R. Cockcroft, K.C. Symes, S. Donnelly, A. McCann, and V. Reynolds, WO 2006/050865.
60. B.W. Rowland and J.V. Bauer, US 2,113,034 (1938).
61. P.H. Brouwer, J. Beugeling, and B.C.A. Ter Veer, *PTS-Symposium*, 1999, 1-15/28.
62. P.R. Shildneck and M.L. Cushing, US 3,151,019 (1964).
63. M.A. Hubbe, D.G. Wagle, and E.R. Ruckel, US 5,958,180 (1999).
64. F. Mobarak and A.E. El-Ashmawy, *Cellulose Chem. Technol.*, **11(1)**, 109–113 (1977).
65. S.R.D. Guha and P.C. Pant, *Indian Pulp Paper*, **25(1–6)**, 385–388 (1970).
66. J.O. Thompson, J.W. Swanson, and L.E. Wise, *Tappi*, **36(12)**, 534–541 (1953).
67. M.N. Fineman, *Tappi*, **35(7)**, 320–324 (1952).
68. A.E. El-Ashmawy, F. Mobarak, and Y. Fahmy, *Cellulose Chem. Technol.*, **7**, 315–323 (1973).
69. W. Denzinger, M. Rubenacker, C. Nilz, P. Lorencak, D. Monch, R. Schuhmacher, and A. Stange, US 6,048,945 (2000).
70. T. Kondo, C. Sawatari, R.St.J. Manley, and D.G. Gray, *Macromolecules*, **27**, 210–215 (1994).
71. M. Sain and H. Li, *J. Wood Chem. Technol.*, **22(4)**, 187–197 (2002).
72. R.S. Ampulski and P.D. Trokhan, US 5,246,545 (1993).
73. P.J. McLaughlin and W.W. Toy, US 3,017,291 (1962).
74. R. Pelton, J. Zhamg, N. Chen, and A. Moghaddamzadeh, *Tappi J.*, **2(4)**, 15–18 (2003).
75. A.E. Broderick, US 2,285,490 (1942).
76. M.J. Smith and S.J. McCullough, US 5,853,539 (1998).
77. W.F. Reynolds, L.H. Wilson, W.M. Thomas, N.T. Woodberry, J.C. Barthel, and C.G. Landes, *Tappi*, **40(10)**, 839–846 (1957).

78. J.C. Roberts, C.O. Au, G.A. Clay, and C. Lough, *J. Pulp Paper Sci.*, **13(1)**, J1–J5 (1987).
79. J.C. Roberts, C.O. Au, and G.A. Clay, *Tappi J.*, **69(10)**, 88–93 (1986).
80. F.M.K. Werdouschegg, Natural gums as dry strength additives, in *Dry Strength Additives*, W.F. Reynolds (Ed.), TAPPI Press, Atlanta, GA, 1980, pp. 67–93.
81. E. Strazdins, *TAPPI Proceedings, Papermakers Conference*, Atlanta, GA, 1982, pp. 145–152.
82. D. Gilttenberg, R.J. Tippett, and P. Leonhardt, *Paper Technol.*, **45(6)**, 27–33 (2004).
83. J.A. Van Den Akker, and W.A. Wink, *Tappi*, **52(12)**, 2406–2420 (1969).
84. F. Tiberg, J. Daicic, and J. Froberg, Surface chemistry of paper, in *Handbook of Applied Surface and Colloid Chemistry*, K. Holmberg (Ed.), John Wiley, New York, 2001, pp. 123–173.
85. P.H. Brouwer, M.A. Johnson, and R.H. Olsen, Starch and retention, in *Retention of Fines and Fillers during Papermaking*, J.M. Gess (Ed.), TAPPI Press, Atlanta, GA, 1998, pp. 199–242.
86. S. Dasgupta, *Tappi J.*, **77(6)**, 158–166 (1994).
87. H. Corte, H. Schaschek, and O. Broens, *Tappi*, **40(6)**, 441–447 (1957).
88. L.L. Chan and A.H. Guitard, US 4,007,084 (1977).
89. C.E. Farley, *TAPPI Proceedings, Alkaline Papermaking*, Atlanta, GA, 1985, pp.73–78.
90. M.A. Hubbe, Pulp and paper chemical additives, in *Encyclopedia of Forest Sciences*, J. Burley, J. Evans, and J. Youngquist (Eds.), Elsevier, Amsterdam, the Netherlands, 2005.
91. B. Alince, R. Lebreton, and S. St-Amour, *Tappi J.*, **73(3)**, 191–193 (1990).
92. J. Baas, P.H. Brouwer, and T.A. Wielema, *PTS-Symposium*, 2002, 38/1–38/30.
93. T.A. Wielema and P.H. Brouwer, *Paper Technol.*, **44(9)**, 27–40 (2003).
94. G.L. Deets and W.G. Tamalis, US 4,744,864 (1988).
95. F. Ide, T. Kodama, and Y. Kotake, US 3,785,921 (1974).
96. G. Greenwood, US 5,876,563 (1999).
97. M. Haggkvist, D. Solberg, L. Wagberg, and L. Odberg, *Nord. Pulp Paper Res. J.*, **13(4)**, 292–298 (1998).
98. J. Laine, T. Lindstrom, G.G. Nordmark, and G. Risinger, *Nord. Pulp Paper Res. J.*, **17(1)**, 50–56 (2002).
99. Y.C. Huang, K.M. Singh, Y. Hong, and M.B. Lyne, US 7,608,166 (2009).
100. J.L. Azorlosa and D.P. Hug, US 2,635,972 (1953).
101. G.G. Maher, US 3,730,829 (1973).
102. E. Bobu and V.I. Popa, *Cellulose Chem. Technol.*, **23**, 723–731 (1989).
103. V.I. Popa, E. Bobu, C. Beleca, and V. Negreanu, *Cellulose Chem. Technol.*, **28**, 445–455 (1994).
104. K.G. Forss, A.G.M. Fuhrmann, and M. Toroi, US 5,110,414 (1992).
105. R.C. Slagel and G. Di Marco, US 3,709,780 (1973).
106. R.C. Slagel and G. DiMarco Sinkovitz, US 4,056,432 (1977).
107. S.P. Malchick, W.G. James, and J.S. Munday, US 3,325,347 (1967).
108. R.W. Davison, US 3,049,469 (1962).
109. J.L. Azorlosa, CA 477265 (1951).
110. X. Chen, R. Huang, and R. Pelton, *Ind. Eng. Chem. Res.*, **44**, 2078–2085 (2005).
111. D.K. Drummond and P.C. Wernett, US 6,090,242 (2000).
112. C.L. McCormick, G.-S. Chen, and B.H. Hutchinson, *J. Appl. Polym. Sci.*, **27**, 3103–3120 (1982).
113. S.M. Mokhtar, *J. Macromol. Sci. Pure Appl. Chem.*, **A34(5)**, 865–879 (1997).
114. N. Uyanik and C. Erbil, *Eur. Polym. J.*, **36**, 2651–2654 (2000).
115. N.T. Woodberry, US 2,959,514 (1960).
116. M.T. Crisp and R.J. Riehle, US 6,197,919 (2001).
117. A.M. Swift, *Tappi*, **40(9)**, 224–227 (1957).
118. R.W. Morgan and M.J. Hatch, US 3,146,157 (1964).
119. K. Yasuda, K. Okajima, and K. Kamide, *Polymer J.*, **20(12)**, 1101–1107 (1988).
120. H. Kheradmand, J. Francois, and V. Plazanet, *Polymer*, **29**, 860–870 (1988).
121. G. Muller, J.C. Fenyo, and E. Selegny, *J. Appl. Polym. Sci.*, **25**, 627–633 (1980).
122. S. Sawant and H. Morawetz, *Macromolecules*, **17**, 2427–2431 (1984).
123. H. Morawetz, in *Chemical Reactions of Polymers*, J.L. Benham and J.F. Kinstle (Eds.) (ACS Symposium Series, 364), 1988, American Chemical Society, Washington, DC, pp. 317–325.
124. W.R. Cabaness, T.Y.-C. Lin, and C. Parkanyi, *J. Polym. Sci.*, **A-1(9)**, 2155–2170 (1971).
125. B. Walchuk, F. Zhang, J.C. Harrington, W.S. Carey, and R.L. Brady, US 7,250,448 (2007).
126. K. Plochocka and T.J. Wojnarowski, *Eur. Polym. J.*, **8**, 921–926 (1972).
127. W. Rodriguez, US 5,543,446 (1996).
128. W.J. van der Geer and P. Wietsma, US 3,622,447 (1971).
129. D.N. Van Eenam, US 4,575,528 (1986).
130. D.N. Van Eenam, US 4,604,163 (1986).

131. T.J. Drennen and L.E. Kelley, US 2,999,038 (1961).
132. J.E. Drach, US 4,391,878 (1983).
133. J.A. Cuculo, US 3,555,585 (1971).
134. R. Peppmoller and F. Koschier, US 4,773,967 (1988).
135. R. Peppmoller and F. Koschier, US 4,711,919 (1987).
136. E. Strazdins, US 3,840,489 (1974).
137. P.W. Pretzel, US 2,375,245 (1945).
138. R. Bartelloni, US 4,510,019 (1985).
139. H.S. Killam, US 4,167,439 (1979).
140. F.J. Kronzer, US 5,895,557 (1999).
141. E. Strazdins, *Tappi J.*, **67(4)**, 110–113 (1984).
142. E. Strazdins, *Nord. Pulp Paper Res. J.*, **4(2)**, 128–134 (1989).
143. A.M. Scallan and J. Grignon, *Sven. Papperstidn.*, **80(2)**, 40–47 (1979).
144. C.L. Burdick, US 6,359,040 (2002).
145. L.L. Chan and J.F. Schaffer, US 4,088,530 (1978).
146. P.J. McLaughlin, US 2,765,229 (1956).
147. W.F. Reynolds, US 3,332,834 (1967).
148. M. Laleg and I.I. Pikulik, *Papermakers Conference*, 1991, pp. 577–590.
149. D.E. Boardman, *Tappi J.*, **76(12)**, 148–152 (1993).
150. T. Vihervaara, K. Luukkonen, and V. Niinivaara, US 7,094,319 (2006).
151. T. Vihervaara, K. Luukkonen, and V. Niinivaara, US 6,398,912 (2002).
152. T. Vihervaara, K. Luukkonen, and V. Niinivaara, US 6,716,313 (2004).
153. E.J. Barber, R.H. Earle, and G.C. Harris, US 3,219,519 (1965).
154. Y. Matsunaga, T. Sugiyama, and E. Maekawa, US 4,632,984 (1986).
155. C.G. Caldwell and O.B. Wurzburg, US 2,935,436 (1960).
156. C.G. Caldwell and O.B. Wurzburg, US 2,813,093 (1957).
157. R.N. DeMartino and A.B. Conciatori, US 4,031,307 (1977).
158. J.D. Van Handel, C.K. Klaisner, and J.C. Gast, US 6,723,204 (2004).
159. M.J. Jursich and T. Aitken, US 3,424,650 (1969).
160. T. Altken, D.R. Anderson, and M.J. Jursich, US 3,674,725 (1972).
161. T. Altken, US 3,854,970 (1974).
162. T. Aitken and W.D. Pote, US 4,097,427 (1978).
163. M.M. Merrette, J.J. Tsai, and P.H. Richardson, US 6,843,888 (2005).
164. A.P. Kightlinger, E.K. Crosby, and E.L. Speakman, US 3,884,909 (1975).
165. A.P. Kightlinger, E.K. Crosby, and E.L. Speakman, US 3,778,431 (1973).
166. C. Gaiolas, M.S. Silva, A.P. Costa, and M.N. Belgacem, *Tappi J.*, **5(6)**, 3–8 (2006).
167. M. Laleg and I.I. Pikulik, *J. Pulp Paper Sci.*, **19(6)**, J248–J255 (1993).
168. J. Marton and T. Marton, *Tappi*, **59(12)**, 121–124 (1976).
169. R.C. Howard and C.J. Jowsey, *J. Pulp Paper Sci.*, **15(6)**, J225–J229 (1989).
170. M. Laleg and I.I. Pikulik, *J. Pulp Paper Sci.*, **17(6)**, J206–J216 (1991).
171. P.J. Borchert and J. Mirza, *Tappi*, **47(9)**, 525–528 (1964).
172. W.L. Kaser, J. Mirza, J.H. Curtis, and P.J. Borchert, *Tappi*, **48(10)**, 583–587 (1965).
173. D.B. Solarek, P.G. Jobe, and M.M. Tessler, US 4,675,394 (1987).
174. D.B. Solarek, P.G. Jobe, M.M. Tessler, R.L. Billmers, D.L. Lamb, and J.J. Tsai, US 4,703,116 (1987).
175. J.J. Tsai, P.G. Jobe, D.L. Lamb, R.L. Billmers, and M.M. Tessler, US 4,983,748 (1991).
176. D.B. Solarek, P.G. Jobe, and M.M. Tessler, US 4,731,162 (1988).
177. D.B. Solarek, P.G. Jobe, M.M. Tessler, R.L. Billmers, D.L. Lamb, and J.J. Tsai, US 4,741,804 (1988).
178. R.L. Billmers, M.M. Tessler, D.M. Del Giudice, and C. Leake, US 4,788,280 (1988).
179. D.B. Solarek, P.G. Jobe, M.M. Tessler, R.L. Billmers, D.L. Lamb, and J.J. Tsai, US 4,804,769, p. 3157 (1989).
180. K.R. Anderson, A. Esser, L.E. Fosdick, K.O. Hwang, N. Mahr, J.T. McDonald, D.S. Sivasligil, A. Stange, S. Veelaert, and M. Wendker, US 6,710,175 (2004).
181. K. Flory, A. Stange, M. Kroener, and N. Sendhoff, US 5,478,656 (1995).
182. N.T. Woodberry, US 3,594,271 (1971).
183. L.R. Peel, *Paper Technol.*, **30(12)**, 20–23 (1989).
184. M. Laleg and I.I. Pikulik, *Nord. Pulp Paper Res. J.*, **8(1)**, 41–47 (1993).
185. J.W. Opie and J.L. Keen, *Tappi*, **47(8)**, 504–507 (1964).
186. G.L. Deets and W.G. Tamalis, US 4,684,708 (1987).

187. T. Morinati and J. Yamauchi, *Polymer*, **39(3)**, 559–572 (1998).
188. T. Moritani, J. Yamauchi, and M. Shiraishi, US 4,311,805 (1982).
189. J.J. Rabasco, E.H. Klingenberg, G.P. Dado, R.K. Pinschmidt, and J.R. Boylan, US 6,096,826 (2000).
190. L.M. Robeson and T.L. Pickering, US 5,380,403 (1995).
191. R.I. Davis, J.C. Phalangas, G.R. Titus, and A.J. Restaino, EP 0141269 (1989).
192. A.J. Restaino, C.J. Phalangas, and G.R. Titus, US 4,689,217 (1987).
193. D.D. Reynolds, US 3,345,346 (1967).
194. R.L. Adelman, US 4,461,858 (1984).
195. W. Denzinger, M. Rubenacker, C. Nilz, P. Lorencak, D. Monch, R. Schuhmacher, and A. Stange, US 6,060,566 (2000).
196. J.R. Sanchez, US 6,315,866 (2001).
197. A.T. Coscia and L.L. Williams, US 3,556,932 (1971).
198. F. Poschmann, US 3,256,140 (1966).
199. S.T. Moore, US 3,077,430 (1963).
200. J.J. Padbury, W.M. Thomas, and W.H. Schuller, US 2,963,396 (1960).
201. K. Fujimura and K. Tanaka, US 3,790,529 (1974).
202. O. Grimm and H. Rauch, US 2,328,901 (1943).
203. W.A. Foster and J.E. Stout, US 3,323,979 (1967).
204. D.E. Nagy, US 3,980,800 (1976).
205. D.E. Nagy, US 3,957,869 (1976).
206. N.T. Woodberry, S.T. Moore, and Y. Jen, US 2,890,978 (1959).
207. R.R. House, S.T. Moore, and A.M. Schiller, US 2,729,560 (1956).
208. S. Cyr, H. Bakker, W.M. Stevels, and R.R. Staib, WO 2006/068964.
209. D.L. Dauplaise, J.J. Kozakiewicz, and J.M. Schmitt, US 5,041,503 (1991).
210. J.J. Kozakiewicz, D.L. Dauplaise, J.M. Schmitt, and S.-Y. Huang, US 5,037,863 (1991).
211. S.A. Fischer and D.I. Devore, US 5,408,024 (1995).
212. S.A. Lipowski and J.C. Queen, US 3,703,563 (1972).
213. F.E. Woodward, US 3,509,021 (1970).
214. R.T. Underwood, US 5,783,041 (1998).
215. W.C. Floyd and L.R. Dragner, US 5,147,908 (1992).
216. R.H. Jansma, J. Begala, and G.S. Furman, US 5,401,810 (1995).
217. B.F. North, US 4,284,758 (1981).
218. M.D. Wright, US 2009/0126890.
219. L. Wagberg and M. Bjorklund, *Nord. Pulp Paper Res. J.*, **8(1)**, 53–58 (1993).
220. G.J. Guerro and L.A. Lawrence, US 6,429,253 (2002).
221. A.J. Allen, E. Echt, W.W. Maslanka, and J.C. Peters, US 6,294,645 (2001).
222. B.K. Bower, US 6,165,322 (2000).
223. B.K. Bower, US 6,346,170 (2002).
224. M.S. Rathi and C.J. Biermann, *Tappi J.*, **83(12)**, 62 (2000).
225. L. Gardlund, J. Forsstrom, and L. Wagberg, *Nord. Pulp Paper Res. J.*, **20(1)**, 36–42 (2005).
226. H.J. Degen, S. Pfohl, V. Weberndoerfer, G. Rehmer, M. Kroener, and A. Stange, US 4,818,341 (1989).
227. F. Wang and H. Tanaka, *J. Appl. Polym. Sci.*, **78**, 1805–1810 (2000).
228. A. Esser, H.J. Hahnle, and T.A. Von Vadkerthy, US 2009/0272506.
229. C. Lu and J. Tan, US 2009/0314446.
230. N.T. Woodberry and W.F. Reynolds, US 3,258,393 (1966).
231. H. Hartmann, W. Denzinger, and C. Nilz, US 5,753,759 (1998).
232. R.J. Schaper, US 4,166,894 (1979).
233. A. Stange, H.J. Degen, W. Auhorn, V. Weberndoerfer, M. Kroener, and H. Hartmann, US 4,940,514 (1990).
234. M. Niessner, C. Nilz, P. Lorencak, M. Rubenacker, and R. Ettl, US 6,160,050 (2000).
235. F. Linhart, A. Stange, R. Schuhmacher, H. Hartmann, W. Denzinger, M. Niessner, C. Nilz, W. Reuther, and H. Meixner, US 5,851,300 (1998).
236. M. Niessner, C. Nilz, P. Lorencak, M. Rubenacker, and R. Ettl, US 6,235,835 (2001).
237. R. Dyllick-Brenzinger, P. Lorencak, H. Maixner, P. Baumann, E. Kruger, A. Stange, and M. Rubenacker, US 6,616,807 (2003).
238. H. Osborg, J.W. Brook, and A. Goldstein, US 3,298,902 (1967).
239. N. Pourahmady and A. Williamson, *Tappi Proceedings—Nonwovens Conference*, Charlotte, NC, 1996, pp. 185–198.

240. R.A. Wessling, W.A. Foster, and D.M. Pickeman, US 4,178,205 (1979).
241. L.H. Wilson, W.M. Thomas, and J.J. Padbury, US 2,884,057 (1959).
242. J.A. Sedlak, US 3,875,098 (1975).
243. J.A. Sedlak, US 3,875,097 (1975).
244. J.A. Sedlak, US 3,874,994 (1975).
245. B. Alince, *Tappi J.*, **74(8)**, 221–223 (1991).
246. B. Alince, P. Arnoldova, and R. Frolik, *J. Appl. Polym. Sci.*, **76(11)**, 1677–1682 (2000).
247. B. Alince, *Tappi*, **60(12)**, 133–136 (1977).
248. B. Alince, *Tappi J.*, **82(3)**, 175–187 (1999).
249. D. Lath and M. Sivova, *Makromol. Chem. Macromol. Symp.*, **58**, 181–187 (1992).
250. M.A. Hubbe, O.J. Rojas, N. Sulic, and T. Sezaki, *Appita J.*, **60(2)**, 106–111 (2007).
251. J. Song, Y. Wang, M.A. Hubbe, O.J. Rojas, N. Sulic, and T. Sezaki, *J. Pulp Paper Sci.*, **32(3)**, 156–162 (2006).
252. M.A. Hubbe, O.J. Rojas, D.S. Argyropoulos, Y. Wang, J. Song, N. Sulic, and T. Sezaki, *Colloids Surf. A Physicochem. Eng.*, **301**, 23–32 (2007).
253. A. Sato, M. Ogawa, T. Oguni, and M. Fujii, Amphoteric PAMS as dry strength resins, in *Practical Papermaking Conference*, Milwaukee, WI, 2005.
254. W.H. Schuller, W.M. Thomas, S.T. Moore, and R.R. House, US 2,884,058 (1959).
255. C. Hagiopol, *Copolymerization: Towards a Systematic Approach*, Kluwer-Academic Press, New York, 1999.
256. T. Kiyosada, A. Endoh, S. Iwata, and M. Ogawa, US 7,482,417 (2009).
257. E. Echt and R.P. Geer, WO 98/06898.
258. T. Morijiri, Y. Nagashima, and M. Kawamura, JP 11302992 (1999).
259. D.E. Nagy, US 3,697,370 (1972).
260. D.E. Nagy, US 3,567,659 (1971).
261. V.N.G. Kumar and P.G. Jobe, US 5,294,301 (1994).
262. G. Mino and S. Kaizerman, US 2,922,768 (1960).
263. D. Glittenberg, *Tappi J.*, **76(11)**, 215–219 (1993).
264. M.E. Carr, B.T. Hofreiter, and C.R. Russell, *J. Appl. Polym. Sci.*, **22**, 897–904 (1978).
265. K.A. Bernard, J. Tsai, R.L. Billmers, and R.W. Sweger, US 5,455,340 (1995).
266. K.A. Bernard, J. Tsai, R.L. Billmers, and R.W. Sweger, US 5,500,087 (1996).
267. M.M. Tessler, US 4,017,460 (1977).
268. A.J. Miller and B.M. Stubbs, US 5,525,664 (1996).
269. W. Lehmann, G. Troemel, K. Ley, and F. Muller, US 3,932,363 (1976).
270. W. Lehmann, G. Troemel, K. Ley, and F. Muller, US 4,166,002 (1976).
271. P. Marimuthu, A. Gopalan, and G. Venkoba Rao, *Ippta J.*, **16(3)**, 65–74 (2004).
272. M.C. Lofton, S.M. Moore, M.A. Hubbe, and S.Y. Lee, *Tappi J.*, **4(9)**, 3–8 (2005).
273. H.-K. Lee and W.-L. Jong, *J. Polymer Res.*, **4(2)**, 119–128 (1997).
274. R.A. Young, *Wood Fiber*, **10(2)**, 112–119 (1978).
275. T.W. Rave, US 4,158,595 (1979).
276. S. Chen and H. Tanaka, *J. Wood Sci.*, **44**, 393–309 (1998).
277. L.L. Chan and A.H. Guitard, CA 1,555,340 (1979).
278. S.P. Dasgupta, US 5,318,669 (1994).
279. L.-L. Chan and A.H. Guitard, US 4,036,682 (1977).
280. M. Eriksson, G. Pettersson, and L. Wagberg, *Nord. Pulp Paper Res. J.*, **20(3)**, 270–276 (2005).
281. L.E. Enarsson and L. Wagberg, *Nord. Pulp Paper Res. J.*, **22(2)**, 258–266 (2007).
282. G. Pettersson, H. Hoglund, and L. Wagberg, *Nord. Pulp Paper Res. J.*, **21(1)**, 115–121 (2006).
283. G. Pettersson, H. Hoglund, and L. Wagberg, *Nord. Pulp Paper Res. J.*, **21(1)**, 122–128 (2006).
284. R. Lingstrom and L. Wagberg, *J. Colloid. Interface Sci.*, **328**, 233–242 (2008).
285. L. Wagberg, S. Forsberg, A. Johansson, and P. Juntti, *J. Pulp Paper Sci.*, **28(7)**, 222–228 (2002).
286. D.C. Smith, US 5,338,406 (1994).
287. G.I. Keim and A.C. Schmalz, US 3,058,873 (1962).
288. B.A. Owens, D.I. Collias, and A.J. Wnuk, US 6,114,471 (2000).
289. B.A. Owens, D.I. Collias, and A.J. Wnuk, US 6,211,357 (2001).
290. B.A. Owens, D.I. Collias, and A.J. Wnuk, US 6,620,461 (2003).
291. B.A. Owens, D.I. Collias, and A.J. Wnuk, US 6,281,350 (2001).
292. B.A. Owens, D.I. Collias, and A.J. Wnuk, US 6,623,806 (2003).
293. B.A. Owens, D.I. Collias, and A.J. Wnuk, US 6,458,419 (2002).

294. K.P. Kehrer, S.M. Atlas, V.A. Kabanov, A. Zenin, and V. Rogachova, US 6,755,938 (2004).

295. R. Gernandt, L. Wagberg, L. Gardlund, and H. Dautzenberg, *Colloids Surf. A Physicochem. Eng. Aspects*, **213**, 15–25 (2003).

296. G.G. Maher, A.J. Ernst, H.D. Heath, B.T. Hofreiter, and C.E. Rist, *Tappi*, **55(9)**, 1378–1384 (1972).

297. J.W. Adams, US 3,180,787 (1965).

298. D. Glittenberg and P. Leonhardt, US 7,022,174 (2006).

299. D. Glittenberg and P. Leonhardt, US 7,147,753 (2006).

300. R.A. Jewell, US 5,667,637 (1997).

301. T.A. Wielema, J. Hendriks, R.P.W. Kesselmans, and J. Terpstra, US 6,767,430 (2004).

302. E. Strazdins, US 4,002,588 (1977).

303. T. Lindstrom, L. Wagberg, and H. Hallgren, *Industria della Carta*, **25(5)**, 227–232 (1987).

304. D.M. Pickeman, W.A. Foster, and R.A. Wessling, US 4,187,142 (1980).

305. D. Owen, WO 96/05373.

306. R.P. Avis, US 4,557,801 (1985).

307. J.A. Dickerson, H.J. Goldy, D.C. Smith, and R.R. Staib, US 6,228,217 (2001).

308. J. Laine, T. Lindstrom, G.G. Nordmark, and G. Risinger, *Nord. Pulp Paper Res. J.*, **17(1)**, 57–60 (2002).

309. P. Economou, US 3,677,888 (1972).

310. S.P. Dasgupta, US 5,338,407 (1994).

311. S.P. Dasgupta, US 5,502,091 (1996).

312. S.P. Dasgupta, US 5,633,300 (1997).

313. K.P. Kehrer, S.M. Atlas, V.A. Kabanov, A. Zenin, and V. Rogachova, US 6,716,312 (2004).

314. S. Gosset, P. Lefer, G. Fleche, and J. Schneider, EP 0 282 415 (1988).

315. P.J. Borchert, W.L. Kaser, and J. Mirza, US 3,269,852 (1966).

316. G.E. Hamerstrand and M.E. Carr, US 3,763,060 (1973).

317. G.E. Hamerstrand and M.E. Carr, US 4,152,199 (1979).

318. B. Philipp and N. Lang, *Tappi*, **52(6)**, 1179–1183 (1969).

319. W.I. Lyness, R.A. Gloss, and N.A. Bates, US 3,998,690 (1976).

320. P. Economou, US 3,660,338 (1972).

321. P. Economou, US 3,790,514 (1974).

322. L. Gardlund, L. Wagberg, and R. Gernandt, *Colloids Surf. A Physicochem. Eng. Aspects*, **218**, 137–149 (2003).

323. R. Winiker, US 5,032,226 (1991).

324. R.P. Geer and R.R. Staib, WO 00/11046.

325. M.E. Carr, B.T. Hofreiter, G.E. Hamerstrand, and C.R. Russell, *Tappi*, **57(10)**, 127–129 (1974).

7 Internal Sizing Agents

The penetration of liquid into paper is due to the fiber wettability with respect to the imbibing liquid (water, ink, milk, etc.) and the pore structure of the fiber network [1]. When an aqueous liquid contacts the surface of paper, it must first wet the fibers. The liquid then tends to spread out laterally along the fiber surface (feathering) and can also move transversely through the paper (penetration) [2].

Most often, a reduction in wettability against aqueous liquids is desired. In order to prevent the liquid feathering and/or penetration, the sizing agent should provide a hydrophobic character to the cellulose fibers. The term "sizing" is generally limited to the process whereby a chemical additive provides paper products with resistance to wetting, penetration, and absorption by aqueous liquids (ink, milk, etc.) [2–4].

Sized paper is obtained by both internal sizing (at wet end [5]) and surface sizing (at dry end) [6]. Surface sizing (see Chapter 9) is designed for a reduced number of applications in which the paper surface is mainly involved (such as printing). There are other applications that require internal sizing in order to prevent water penetration through paper: from packaging (for milk or juices) to cigarettes production.

The rate of liquid penetration (dl/dt) through a capillary (Washburn equation) [3,7] is a function of the capillary radius (R), the liquid surface tension (γ), the liquid viscosity (η), and the contact angle between the liquid and the capillary wall (θ):

$$\frac{dl}{dt} = \frac{1}{4}\left[\frac{\gamma}{\eta}\right]\left[\frac{R}{l}\right]\cos\theta \qquad (7.1)$$

The radius and the length of the capillaries are governed by the bulk density, the porosity, and basis weight of the paper. The sizing process aims at changing the fiber surface characteristics (cos θ). The liquid can penetrate both through the z-direction and the x–y plan of the paper [8].

However, data are indicating that the Washburn equation is inadequate to describe the penetration of aqueous solution into paper [9] because the paper rewetting is a "nonequilibrium" process [10]. Water penetration through paper starts with some swelling of the fibers before the liquid front penetrates a significant fraction of the sheet thickness [11]. Cellulose fibers swell in water within a dynamic process; the interface tension between wet fibers and water (or ink) changes over time. The Washburn equation is valid only for non-swelling liquids and in the absence of any chemical reaction (such as acid effect on rosin–alum system [12]).

The commercial requirement for a sizing agent is that it should provide paper with reasonable resistance to wetting and penetration by aqueous liquids. This protection is recognized and accepted as time limited because sizing retards but does not totally prevent liquid movement.

There are several approaches for the sizing measurement [6,13]: ink penetration, water absorption, the break of wet paper, water (or ink) contact angle, etc. It is difficult to correlate the results obtained through different methods (i.e., contact angle and water adsorption). However, a higher contact angle may be translated in a lower water adsorption. But the accurate water adsorption (Cobb) value cannot be estimated based on a contact angle measurement. A logarithmic graph describes Cobb values as a function of HST numbers [14].

Based upon the process that imparts hydrophobicity to fibers [15], there are two kinds of internal sizing materials:

1. A very hydrophobic compound (wax with no polar group [16,17]) that is dispersed in water, the hydrophobicity of which is transferred to fibers without any involvement of a chemical reaction.
2. The hydrophobic character is developed through chemical and/or physical processes (esterification, imidization, migration, reorientation, etc.), and the sizing effect is noticed after the drying step. Sometimes, those processes need lower moisture contents, higher temperatures, and longer reaction times. Included in this category are amphiphilic molecules with reactive groups (such as rosin, alkenyl ketene dimer—AKD and alkyl succinic anhydride—ASA).

The hydrophobic character of the internal size is essential but not sufficient. Many other properties should be added: molecular weight, a polar group to help molecular orientation, an appropriate balance between the polar group and the hydrophobic tail, an appropriate T_g to flow within the drying temperature range, vapor pressure [18], etc. A small molecule, with higher mobility, seems to be a better choice [3], but high molecular weight compounds are recommended as well [19].

The molecular characteristics of the sizing agent are also related to the pH of the application point, to the retention process [20], to the amount of extractive [21] and the coefficient of friction of paper [22].

The effectiveness of internal sizing agents depends on their retention [23]. Cationic compounds or cationic dispersions are self-retained on the cellulose pulp, but anionic sizing agents need a cationic retention aid. Sizing chemistry is inexorably linked to the pH of the papermaking system at the wet end [24]. The pH at the wet end has a strong impact on the paper quality as well. Librarians are concerned about acid paper deterioration. In 1992, the Canadian government decided to print on alkaline-based paper only, all publications it expects to be retained for information or historical purposes [25].

7.1 THE CHEMISTRY OF ALUM IN THE PAPERMAKING PROCESSES

Aluminium salts have had a transformative effect on the papermaking industry: everything became easier after alum started to be used. Most of the formulas for rosin size applications use alum and an acid pH. Despite the fact that many aspects of the aluminum chemistry in papermaking process remained to be clarified, the possibility of an inorganic-type polyelectrolyte formation is made clearer in this chapter. Arguably, the role of polyelectrolytes as retention aids (see Chapter 3) is important when aluminum compounds with multiple cationic charges, soluble or precipitated, able to form complexes with polycarboxylic acids, are involved in the same type of retention process.

The primary source of aluminum in most papermaking systems is alum [26] (aluminum sulfate: $Al_2(SO_4)_3 \cdot 14H_2O$). In water solution, the aluminum cation is hydrated with six molecules of water [27].

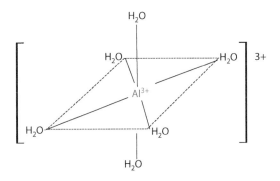

The papermaking process involves a broad range of pH. For both retention and sizing processes, two aspects are of importance: the aluminum species as a function of the pH and the chemical structure of paper chemicals designed to interact with aluminum compounds.

As an amphoteric element, aluminum forms both aluminum salts and aluminates. The metal reacts with the common strong acids [28–30]:

$$2\,Al + 6\,H_3O^+ + 6\,H_2O \rightarrow 2\left[Al(H_2O)_6\right]^{3+} + 3\,H_2$$

The aqueous aluminum ion (and ferric ion as well [31]) is a weak polyfunctional acid. For $AlCl_3$ dissolved in water, a wide distribution of aluminum species is noticed for a very narrow pH range ($4 < pH < 5.5$) [32]. Below $pH = 4.6$, the Al^{3+} is the major form, while at $pH > 5.0$ the major component is $Al(OH)_3$. In between ($4.6 < pH < 5.0$), there are both Al^{3+} and $Al(OH)_3$, along with $[AlOH]^{2+}$ and $[Al_8(OH)_{20}]^{4+}$.

The aqueous equilibrium of aluminum sulfate is considerably more complicated than aluminum chloride [33] because of the possible formation of mixed sulfato-hydroxo-aluminum complexes (for instance, $Al(OH)_{2.44}(SO_4)_{0.28}$ [27]).

The single trivalent cation Al^{3+} is present only at pH below 3.0. At higher pH, hydroxyl becomes a ligand [28,29,34]:

$$\left[Al(H_2O)_6\right]^{3+} + H^+/OH^- \leftrightarrow \left[Al(H_2O)_5(OH)\right]^{2+} + H^+$$

$$\left[Al(H_2O)_5(OH)\right]^{2+} + H^+/OH^- \leftrightarrow \left[Al(H_2O)_4(OH)_2\right]^{+} + H^+$$

$$\left[Al(H_2O)_4(OH)_2\right]^{+} + H^+/OH^- \leftrightarrow \left[Al(H_2O)_3(OH)_3\right] + H^+$$

$$\left[Al(H_2O)_3(OH)_3\right] + H^+/OH^- \leftrightarrow \left[Al(H_2O)_3(OH)_4\right]^{-} + H^+$$

and poly-aluminum compounds are also formed $(Al_2(OH)_2(H_2O)_8^{4+}$ [33]:

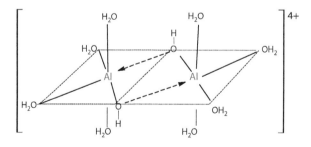

The extension of this process leads to the formation of aggregates containing more than two aluminum atoms (polymeric ions [35]), for instance $[Al_8(OH)_{20}]^{4+}$ [36,37]. At $4.1 < pH < 4.7$, the following aluminum compound is considered $[Al_8(OH)_{10}(SO_4)_5]^{4+}$ [30,38] in equilibrium with Al^{3+}.

Precipitation is noticed at pH higher than 4.1 [32,39]. A gelatinous precipitate begins to form at pH 4.8 and becomes essentially the sole species present above pH 5.0 [29]. The growth of these flocs depends on the presence of other inorganic or organic ions [40]. Freshly precipitated aluminum hydroxide is readily soluble in acids and solutions of strong bases [28].

The hydrolysis of Al^{3+} produces a complex mixture of products [35,41], which may include mononuclear and polynuclear species and most likely $[AlO_4Al_{12}(OH)_4(H_2O)_{12}]^{7+}$. The formation of $[Al(OH)_4]^-$ ion begins at pH 8, while the concentration of $Al(OH)_3$ starts to decrease [30]. However, cationic charges are still present up to pH 8.5 [29] and therefore it is of interest to obtain high basicity poly-aluminum sulfates such as poly-aluminum hydroxyl sulfate [42], poly-aluminum hydroxyl silicate sulfate [43], or phosphate-stabilized polyaluminum sulfate [44].

The shape of the aluminum compounds precipitate can be modified by using both alum and sodium aluminate [45]. At pH = 10.0, the major component is $Al(OH)_4^{1-}$, which can change in AlO_2^{1-} (aluminate) through the loss of two water molecules. Since aluminate is part of the aluminum compounds conversion during pH variation, the sulfate ion (from alum) can be removed from that scheme by using sodium aluminate ($NaAlO_2$) and changing the pH to get the same type of poly-cationic aluminum aggregates.

Poly-aluminum chlorides are polymers made from aluminum chloride and sodium carbonate to produce a complex molecule with the following formulas [46]:

$$\left[Al_{13}O_4(OH)_{24}(H_2O)_{12}\right]^{7+}[Cl^-]_7 \quad \text{or} \quad \left[Al6(OH)_{12}(H_2O)_{12}\right]^{6+}[Cl^-]_6$$

Due to its high charge (+3), for a small ion radius (0.57 Å) (compared to Na^{1+}, which has an ion diameter of 0.95Å [29]) the aluminum cation can form complexes with anionic and/or neutral molecules (electron donors). Polycarboxylic compounds (such as citric acid) or weak acids (like phenols) can generate strong complexes with aluminum salts [26]. It is possible to prepare solutions of an aluminum compound at a pH value at which aluminum hydroxide would ordinarily be precipitated by adding a precipitation inhibitor (citric acid or sodium citrate) [47].

There are other many carboxylic compounds involved in the wet end chemistry and interactions with aluminum cations may result in changes in the shape of the aluminum polycation flocs. Those anion–cation interactions help in paper chemicals retention, improvement of drainage, developing dry strength, or increasing the paper size.

However, the titratable charge density of the aluminum compound in precipitated form is less than 10% of the calculated trivalent Al^{3+} [33]. Only few cationic charges are available, the other cationic charges are inter-compensated or trapped within the flocs. Those charges are called "effective charges" or "apparent charge of the aluminum" [41].

The sequence of the polyelectrolyte addition and the pH adjustment are critical for the availability of the cationic charge on aluminum compounds [41,48]. There are very scattered results for the polyvinyl sulfonate, potassium salt (PVSK) titration with p-DADMAC in the presence of aluminum compounds: the aluminum compound interacts with the anionic polyelectrolyte in a way that is hard to control [41]. For each macromolecular shape, the local ion concentration—around the macromolecular coil—is different and therefore the aluminum compound precipitation will result in a different structure. On the other hand, the anionic charges on the PVSK macromolecular coil can interact with aluminum compounds, trap them, and prevent their interaction with other anionic charges.

In the presence of cellulose fiber as anionic polyelectrolyte, the alum chemistry can be developed on different pathways depending on the relative concentration of cellulose and aluminum salt, the pH value, and the presence of other ionic species (such as anionic trash). Every stock and every papermaking process may have a different composition and different shape for the aluminum precipitate.

The increase in pH creates more negative charges through carboxyl dissociation. At pH = 4.1, the cationic aluminum compound starts to precipitate and adsorbs onto the cellulose fibers [29]. More aluminum is adsorbed with increasing pH in the acid range with a maximum at about pH 4.5 [27]. After adsorption, cellulose fibers change their electrophoretic mobility from a negative to a positive range (excess of cationic charges over the anionic carboxylic groups) [49].

The retention of aluminum compounds depends on pH, temperature, shear rate, on the anionicity of the cellulose fiber, on the chemical structure and concentration of the aluminum compound, on

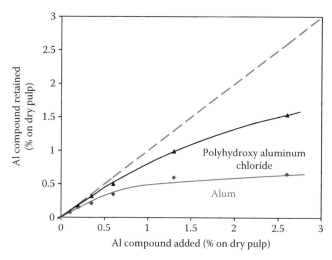

FIGURE 7.1   Aluminum compounds retention.

the presence of other cations, and also on the amount of fine and anionic trash [50–53]. Aluminum sulfate is less retained than polyhydroxyaluminum chloride (Figure 7.1) [53].

Figure 7.1 shows that only a fraction of aluminum compounds (regardless of its chemical structure) is retained on cellulose fibers, and the amount retained depends on the pH value [30]. There is a big jump in the adsorption curve at around pH 4.5 [54]. The alum adsorption seems to be an equilibrium process: the cellulose fibers are not saturated yet, but the alum is only partially retained. The presence of other polyelectrolytes (such as anionic dry strength), or dispersed particles (internal size) at the wet end make the aluminum chemistry even more complex.

The amount of alum is important not only for the retention, but also for the final paper properties. Alum contributes to sheet conductivity and dissipation of static charge [55]. The control of the sulfate ion concentration in the paper sizing process with rosin can be performed by using barium chloride [56].

7.2   ROSIN IS BACK ON THE CELLULOSE FIBERS

Before the nineteenth century, paper sheets were sized by impregnation with animal glue or vegetable gums, an expensive and tedious process. In 1807, in Germany, Moritz Friedrich Illig discovered that paper could be sized with rosin and alum [57].

Rosin is a natural product existing in wood and now—after extraction during the pulping process—it is added to the paper as sizing agent. Thus, a cycle is being closed: the extracted rosin gets back on the cellulose fibers.

Rosin is separated from the black liquor. Even without any further purification, the black liquor is considered a potential internal sizing agent [58] due to the presence of rosin. Rosin is a blend of hydrophobic compounds with only one carboxylic group per molecule, such as abietic acid ($C_{20}H_{30}O_2$):

The rosin molecule has that polar group (carboxylic acid) but the hydrophobic part (19 hydrocarbon atoms distributed in condensed cycles with conjugated double bonds) is quite rigid. The balance between the hydrophobic and hydrophilic parts can be changed by reactions applied to the double bonds and/or to the carboxylic group. In the following chapter, the organic chemistry of rosin is presented in its relationship with the new compounds that are also sizing agents.

7.2.1 Exploring the Organic Chemistry of Rosin: Rosin Derivatives as Sizing Agents

7.2.1.1 Reactions at Double Bonds

The Diels–Alder reaction is performed at the conjugated double bonds of rosin (which is the diene compound) with dienophile compounds: coumarone, furane, thiophene (in the presence of boron trifluoride as catalist) [59], acrylonitrile, acrylic acid, tetracyanoethylene [60]. The most used dienophiles are maleic anhydride [61], acrylic acid [62,63], fumaric acid [64–66], or aconitic acid (dehydrated citric acid) [67] (or blends thereof). The reaction is performed at high temperatures (200°C) in the absence of any catalyst. Less than 1% p-toluenesulfonic acid as rosin crystallization inhibitor [68] and 0.75% phenothiazine as antioxidant [69], are recommended for the Diels-Alder reaction.

Fumaric acid   Maleic anhydride   Aconitic acid   Acrylic acid

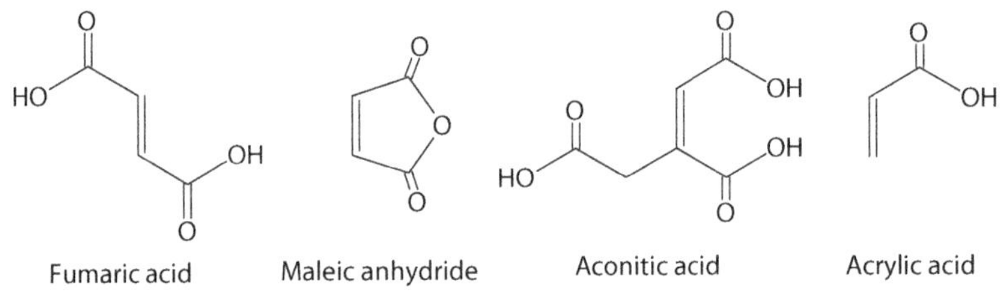

The reaction product with fumaric acid is called "rosin adduct." While rosin can theoretically react mole for mole with maleic anhydride, "fortified" rosin represents only 10%–25% in a blend with the excess of rosin [70,71].

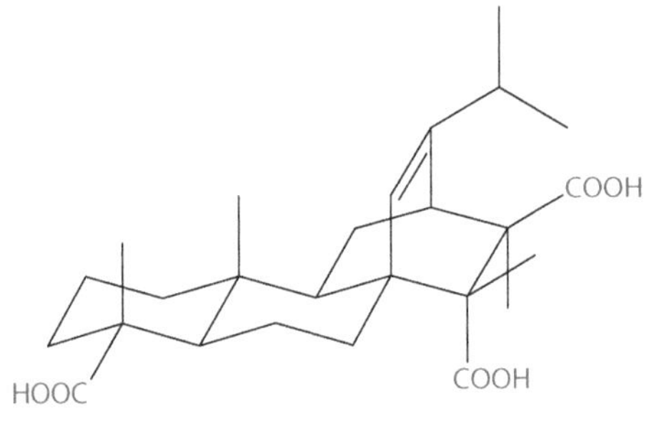

The new molecule has three carboxylic groups, which change the ration between the hydrophobic and the hydrophilic groups. Because the rosin conversion is always lower than 100%, that adduct is used as dispersed sizing agent for the un-reacted rosin [72].

The rosin double bonds "polymerize" [73] with sulfuric acid as catalyst [74]. Based on the molecular weight estimation, dimers, trimers and tetramers are obtained.

Rosin also reacts with formaldehyde (para-formaldehyde or trioxan) [74–77] at 120°C–300°C [78,79] in the presence of a strong acid or acid-treated clay [80] as catalyst. It is a substitution; the double bonds remain unchanged. The rosin/aldehyde ratio ranges from 1% formaldehyde to 2 mole of formaldehyde for one mole of rosin [77]. The common modified rosin may include only 3% of formaldehyde [63]. The conversion of formaldehyde depends on the reaction temperature and time.

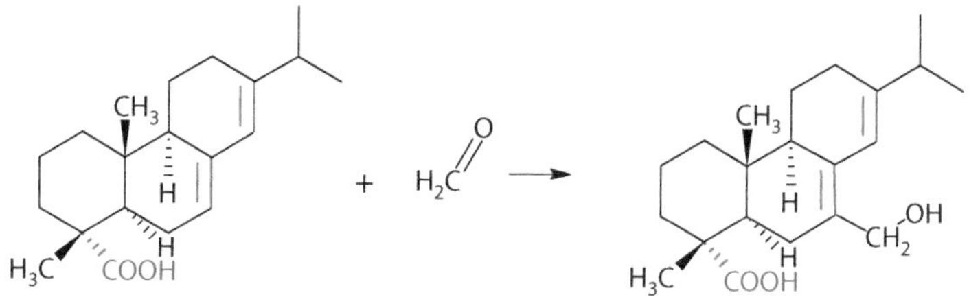

The modified rosin with formaldehyde is more hydrophilic and shows a reduced tendency to crystallize as compared to rosin. In order to improve the rosin "stability" (with reduced tendency to crystallize), both the reaction with formaldehyde and the one with fumaric acid are performed [70,76].

The modified rosin (with formaldehyde) has two conjugated double bonds and a hydroxyl group (allylic-type alcohol). At high temperatures, the hydroxyl group etherifies and the new compound can still react with maleic anhydride in a Diels–Alder reaction [81]:

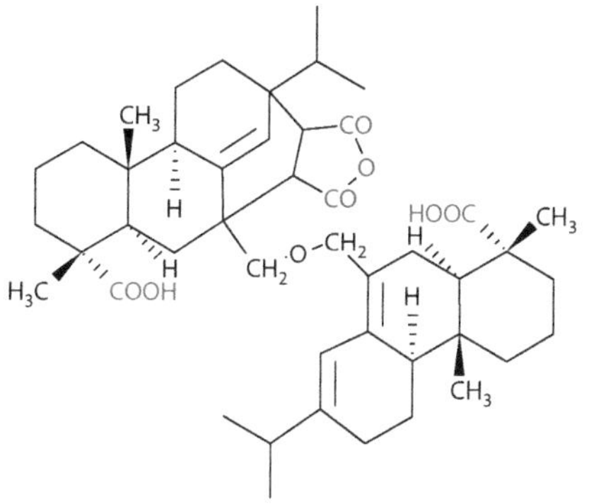

The new molecule has a higher molecular weight with a higher number of carboxylic groups.

The chemical modification can continue with other addition reactions at double bonds. For instance, hydrogenated rosin (hydrogen addition to the double bonds) was used as sizing agent in the presence of sodium aluminate [82].

7.2.1.2 Reactions at Carboxyl Group

The rosin ester is prepared from rosin and poly-hydroxyl alcohols (glycerin, trimethylol propane, pentaerythritol [83], etc.) or epoxy compounds [84]. The reaction temperature is 150°C–300°C and, if necessary, water can be carried out under azeotropy by using benzene or toluene as carrier [85]. Rosin esters are also blended with partially fortified rosin [86] as anionic dispersion (styrene—acrylic acid copolymer as stabilizer) for sizing in alkaline pH [87]. Cationic polymeric stabilizers are recommended for that blend [84].

The hydrophilic–hydrophobic balance should be carefully designed as a sizing agent. The fortified rosin is esterified with one mole of methoxy polyethylene glycol (hydrophilic tail) and 2 mol of glycidyl neodecanoate (hydrophobic tails) [88]. The ester chemical structure can improve the rosin compatibility with a variety of other resins (such as high-density polyethylene wax).

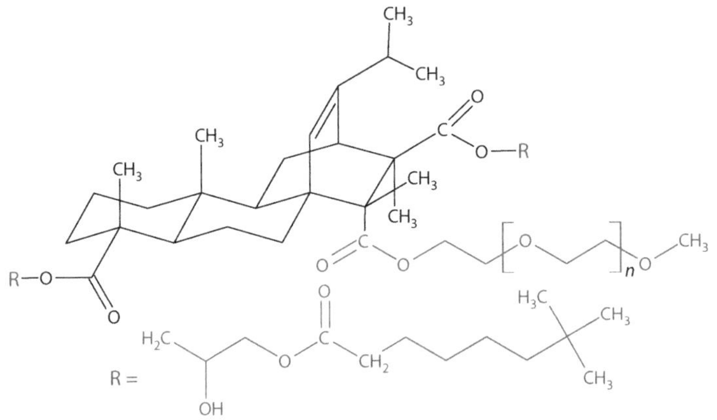

Amides (and polyamide) of rosin (and/or fortified rosin) are obtained by reacting at 185°C with polyalkylene polyamine [63]. The primary amine groups are converted to amides and the secondary amines are protonated with hydrochloric acid to make a rosin derivative as cationic size.

7.2.1.3 Rosin Neutralization

Rosin and its derivatives are high molecular weight carboxylic acids, which can be neutralized with any inorganic or organic base: caustic [69,89,90], sodium carbonate [91], with sodium aluminate [92], ammonia or low molecular weight amine [93,94]. Rosin potassium salt in solution shows a significantly lower viscosity than sodium salt at the same concentration [95].

The neutralized rosin (sodium or potassium salt) shows only a limited solubility in water (about 15%). In the presence of urea, the rosin soaps can reach over 35% concentration [79]. The sizing capability of the modified rosin soap [65] is enhanced by urea, urea-sulfamic acid adduct [68,96–98] and urea-thionyl chloride reaction product [99]. The rosin soaps behave like fatty acid soaps: they are good emulsifiers.

Rosin is used as sizing agent in both saponified and un-saponified forms [100]. Based on the degree of neutralization and the concentration of rosin in water, rosin sizing agents are available in two forms: a dispersion (acid size) in which only 5% of rosin was neutralized (and acts as an emulsifier), and the "neutral size" in which 80%–100% of rosin is sodium (or potassium) salt. Normally, both of these forms of size are chemically modified [15,18].

The viscosity (at 71°C) of 15% concentration in water of saponified rosin decreases from 2500 to 1000 cPs when 0.7% of sodium chloride was added [101]. The 77% solid rosin (mainly as potassium salt) is a paste containing less than 15% free rosin [102]. In order to make pumping easier, a water-soluble alcohol (C_1–C_4) is added (up to 20% based on rosin).

7.2.2 Anionic Rosin Size

The sodium, potassium, and ammonium salts of the rosin anion show a reduced solubility in water and a high viscosity for concentrated solution. Un-saponified rosin can be dispersed in water, in the presence of anionic stabilizers. A high content of organic phase carries a larger amount of potential anionic sizing agent. The dispersion is obtained by simply blending molten rosin with the water solution of stabilizer, by using a solution of rosin in an organic solvent (instead of molten rosin) at room temperature and by the phase-inversion method.

Molten rosin is dispersed in the aqueous solution of an anionic emulsifier (such as alkyl benzene sulfonate) at 160°C–190°C and a stable emulsion (35% solids) is obtained after its rapid cooling [103]. For a better control of the coagulation process, tannic acid [104] or tetrasodium

N-(1,2-dicarboxyethyl)-*N*-octadecylsulfosuccinamate [105] are use to improve the emulsion stability [106]. Those emulsifiers have a much larger hydrophilic part that can protect the particle even at lower pH (see sulfonate group).

Because rosin performs as an internal sizing agent for paper in the presence of alum, the anionic emulsifier may interfere with the sizing process. The neutralized rosin is also a good emulsifier and it can replace the other anionic stabilizer for the anionic rosin dispersion. Rosin neutralization and emulsification can be performed in one step: incomplete rosin neutralization at high rosin concentration in water results in anionic rosin dispersion (the system is self-emulsified). When rosin neutralization is performed with a volatile base (ammonia or low molecular weight amine [93,94]), less than 10% of the rosin carboxylic groups are in the ammonium salt form. The blend includes saponifiable rosin and preponderantly unsaponified material (free rosin) [107–109].

That dispersion is diluted with hot water and added to the pulp just before the latter is formed into a web. High free rosin dispersions (or colloidal rosin) are known to produce better sizing [110,111]. The presence of phosphate salt enhances the rosin performances [109].

Stable emulsions are obtained through the phase-inversion method [112,113]. The aqueous solution of an anionic emulsifier is added to the molten mixture of rosin and fortified rosin to obtain a dispersion of water-in-oil [101]. Hot water is added to inverse the phases and to make an oil-in-water dispersion. Foam formation can be reduced with an inorganic salt (such as $CaCl_2$) added in the surfactant solution [114].

In order to avoid the high temperatures of molten rosin, the rosin anionic emulsions are obtained by dissolving rosin in an organic solvent (benzene, toluene), dispersing the solution in water in the presence of an anionic emulsifier, and distilling the solvent [106,115]. The anionic stabilizer can be the rosin (or fortified rosin) partially neutralized with a small amount of potassium hydroxide. After homogenization, the solvent is distilled under reduced pressure [116].

During the papermaking process, active agents foam heavily with low molecular weight anionic surface. High molecular weight anionic (or amphoteric [117]) polymers are suggested as rosin emulsion stabilizers. Acrylamide copolymers with sodium styrene sulfonate, or styrene-acrylic acid copolymers [86,118,119] (which is a sizing agent as well) are stabilizers for an emulsion of rosin, which are stable and are used in the range of pH 6–8 [85].

Fortified rosin with fumaric acid contains an important amount of un-reacted rosin. That blend is dispersed (using a homogenizer) in water in the presence sodium hydroxide [64]. The fortified rosin is relatively more effective in sizing than rosin at low sizing agent concentration and low paper density [120]. At higher concentrations, that is no longer the case.

In order to reduce the melting point, the rosin dispersed phase may contain (as extenders [121]) a variety of other hydrophobic compounds: paraffin [45,122], hydrocarbon resins, extracts of petroleum oils [123], petroleum resin adduct with maleic anhydride [124], waxes, ethylene-bis-stearamide [125], indene-coumarone resins [126], styrene-butadiene block copolymers [127], and styrene-vinyl toluene copolymer (Mw = 1400). Anionic blends of rosin (80% fortified rosin and 20% rosin ester) and alkyl ketene dimer (AKD) show a synergetic effect for the alkaline (PCC) papermaking process [128].

The hydrophobic resins and rosin are melted together and dispersed in an alkali solution [129]. The dispersed phase contains free rosin and hydrocarbon resin, while in the continuous phase rosin soap is dissolved. Rosin soap is a sizing agent and a stabilizer for the dispersion at the same time. Alum helps to retain rosin particles that carry the petroleum resin on cellulose fibers.

7.2.3 Cationic Rosin Dispersions and Amphoteric Stabilizers

To be retained directly on the anionic cellulose fibers at the wet end, the sizing agent must have cationic charges. Molecular rosin retains its anionic character, which is why alum is required for retention and sizing (see Chapter 3).

Since 1932, the anionic rosin dispersion has been stabilized with protective colloids such as casein (1%–3% based on rosin) [111]. Cationic dispersed of rosin size contains particles with a positive charge [130,131]. The sizing is active at higher pH [132,133] but the presence of alum is still required (alum is added prior to the rosin addition [131]).

The anionic rosin particle includes only about 40% of the rosin in acid form because the anionic rosin salt is used in the stabilizing system along with casein. In order to increase the free rosin content (up to 100%), an anionic dispersant agent [134] (sodium alkyl-aryl sulfonate [135], styrene-acrylic acid copolymer [136], ethylene-acrylic acid copolymer [137]) is used to help the casein or cationic starch as stabilizers.

Cationic macromolecular compounds can replace casein as stabilizer. Polyaminoamine epichlo-rohydrin polymer (PAE)-type resins (1:1 to rosin) are used [138,139,140] to make a rosin emulsion with pH 6.8–7.2. Nonionic stabilizers (such as hydrophobic polyethoxylated alkylphenol) [141] can be associated with a cationic polyelectrolyte.

Fumaric acid-fortified rosin is dissolved in benzene (or methylene chloride [141]) and the solution is emulsified and homogenized (twice at 3000psi) in the presence of starch inter-polymer with PAE resin [62,142]. The resulting emulsion is stable and the solvent das been distilled [121,139,143]. The final solids are 35% and the particle size is less than 1 μm.

The hydrophilic stabilizer reduces the overall hydrophobic properties of the sizing agent. In order to compensate for that, the organic phase is made more hydrophobic by blending with hydrocarbon resins [144], or a more hydrophobic stabilizer is used (such as the cationic copolymer of acrylamide and iso-butyl methacrylate or styrene [145], or the cationic [146,147], or the amphoteric copolymer of styrene) [148].

Rosin and fortified rosin are anionic compounds and they are involved in the protection system. The anionic charges of the rosin derivatives interact with cationic charges of the stabilizer to make a larger aggregate that provides a better stability for the dispersed phase. The amphoteric compounds or the polyelectrolyte complexes have anionic groups available to interact with aluminum compounds for retention on the cellulose fibers. However, due to the presence of cationic polyelectrolyte, a smaller amount of alum is involved [149], and—sometimes—the alum is no longer needed [150].

7.2.4 Rosin Sizing Mechanism

The rosin sizing mechanism includes rosin precipitation, rosin interaction with aluminum compounds in co-precipitation of alum and rosin. The adsorption on cellulose fiber is a decisive step, but not the last to be involved in developing sizing properties: the rosin molecule migrates on the fiber surface where it is anchored and reoriented.

The first step in developing a more hydrophobic cellulose fiber is the rosin migration through water and its adsorption at the wet end. The rosin dispersions follow the general rules of colloidal systems: they are sensitive to pH, to electrolyte concentration [151], to dilution, and to shear rate. At the wet end, the rosin dispersion is diluted (from about 30% to less than 0.01%) and undergoes pH changes [107].

At the regular addition rate, the completely saponified rosin is soluble in water if the pH is in alkaline side. The completely neutralized sodium rosinate of 0.01 mol/L concentration (well bellow the critical micelle concentration—CMC) shows a pH of about 9.0 [152]. The alum addition changes the pH of the sodium abietate from about 8 to about 4 [153], a pH value at which rosin is protonated, loses its solubility, and a precipitate ($0.1\,\mu$) is formed [38,154].

The electrophoretic mobility changes from *negative* (abietic anion in alkaline pH) to *positive* values for colloidal aluminum hydroxide, which is composed of an equimolar mixture of aluminum di-rosinate and free rosin acid. The isoelectric point is at a pH value of about 6.5. The precipitation of both alum compounds and rosin takes place almost simultaneously. The rosin acid can make a complex molecule: aluminum di-rosinate mono-hydroxide, which is insoluble in water. Some free rosin acid can also be incorporated in that precipitate [36].

Therefore, the rosin retention on cellulose fibers shows a strong dependence on the stock pH [33,121,155–157]. A polynomial regression for the experimental data collected for rosin sizing properties as a function of temperature ($^\circ$C), dosage (%), and pH—as main parameters—shows a stronger positive effect of the increased dosage and a negative effect of the increasing temperature and pH [158–160]:

$$HST = C_0 - A \cdot Temp + B \cdot Dosage - C \cdot pH \qquad (7.2)$$

As expected, rosin esters are less sensitive to pH and more sensitive to temperature. Only about 20% of rosin is retained at pH = 4.3. At pH 4.6, rosin retention reaches the level of about 55% [154]. A higher pH reduces rosin retention to 40% (neutral pH = 7), and even down to 20% (alkaline pH = 8) [38].

The sizing shows a sharp maximum at a pH of about 4.5 [110]. Above pH = 5.0, the sizing drops and, for a long period of time, rosin has not been considered to be an effective sizing material above pH 6.5 (poor retention and inappropriately orientation [3]).

Alum chemistry is complex in itself. The addition of rosin changes the aluminum compounds precipitation. The right precipitation of rosin in the presence of precipitated aluminum depends on the aluminum/rosin ratio, the pH values, the sequence of acid addition, and the shear rate [161]. The anion of the aluminum salt is also important [4,153,155,157]: $AlCl_3$ behaves differently than $Al_2(SO_4)_3$, $Al(NO_3)$, or polyaluminium chloride (PAC).

The amount of alum added to the stock should be limited to prevent a too low acid pH [25]. Chelating agents (such as oxalic, succinic, citric acids or low molecular weight anionic polyacrylamide) are used to bond the excess of aluminum ion [74,162].

The rosin and the aluminum compound precipitation take place in the presence of cellulose fibers. The low overall concentration makes hetero-coagulation with cellulose fibers more likely [153]. The rosin sizing capabilities depend on the number of anionic charges existing on the cellulose fibers: more carboxylic groups (added by cellulose oxidation) improve the size by distributing rosin more evenly [163]. Due to the broad distribution of fiber dimensions and anionic charges, the hetero-coagulation of aluminum on cellulose fibers results in a nonuniform distribution of aluminum [164]. Consequently, the rosin is also uneven distributed.

The treatment of pulp with hydrochloric acid, prior to the addition of the sizing system, significantly improves the effect of the rosin size [165] while the conversion of carboxylic groups to non-ionic methylamide groups results in a size reduction [166] (or even zero size [118]).

It has been assumed that the alum forms a special bridge between anionic rosin and anionic cellulose fiber [36,120,153,167,168]. However, the aluminum–carboxyl bond is weak and rosin (as free acid) can be extracted with dioxane–water at 100°C [118]. The weakness of that bond may help the redistribution of rosin on the cellulose fiber surface.

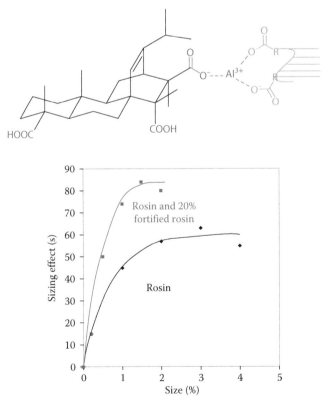

The aluminum rosinate is concentrated on the aluminum flocs with an uneven distribution on the fiber surface [169]. Due to its high melting point (120°C–250°C, depending on the moisture content) [170], the aluminum rosinate redistribution in the drying section is unlikely. Programmed drying (high temperature along with high moisture) can improve the redistribution of rosin.

The anchoring mechanism for the fortified rosin seems to be similar to that for rosin [171,172]. It is believed that fortified rosin reduces the rosin particle size and improves the bonding with aluminum compounds [170]. As a result, fortified rosin (20%) in a blend with rosin dispersion (70% neutralization) is more effective than rosin alone [173] (Figure 7.2).

Alum seems to interact with the carboxylic group because rosin esters ("neutral rosin") perform as a sizing agent both in the presence of alum (even at pH=8) [164] and in its absence [174]. Regardless of the amount of alum added for retention, the anionic rosin ester–fortified rosin emulsion is retained less than 40% [86]. The melting point of the rosin ester is about 80°C; the curing temperature appears to be the turning point in increasing the paper sizing effect [87]. After heating at 160°C, the SEM micrographs show only empty shells (amacromolecular stabilizer has a much higher melting point): the rosin ester has left the particle.

FIGURE 7.2   The sizing effect of rosin and blend of rosin with 20% fortified rosin.

In an attempt to separate the retention and the sizing processes, the surface sizing of paper having alum added at the wet end and rosin emulsion in the size press solution was performed. More alum added at the wet end results in better sizing. The anchoring effect of alum and probably an electrostatic interaction with rosin acid cannot be ruled out [175].

The complex formed between aluminum ions and pulp fibers is pulp-COOAl^{2+}, which can partly be converted (in acid-base equilibrium) to pulp-COOAl(OH)$^+$ and finally to nonionic pulp-COOAl(OH)$_2$ [156]. Therefore, longer contact times between alum and stock reduces the sizing performances [168]. At pH = 6.7, that process is much faster than that at pH = 4.5 and reaches zero sizing in less than 30 min [174].

The aluminum compounds help rosin retention and orientation processes [130,131,176]. The aluminum cation anchoring reduces the degree of freedom of rosin molecules, and thus, the hydrophobic part is rigidly oriented away from the surface of the cellulose fiber. The sodium rosin salt reaches that stage at the wet end. The dispersed free rosin needs to be heated in the drying step to develop the size [110]. However, the bridge between rosin and the cellulose surface is prone to alkaline hydrolysis and the paper shows no resistance to alkaline penetration [132].

When retained on the cellulose fiber, the size begins to react by producing either a "strong" or a "weak bond" [18]. In order to understand the rosin movement from a sized to an unsized sheet, the weak bond concept has been introduced. Some rosin is bonded to cellulose through strong bonds, which cannot be broken during the drying step. The weak bonds [95,177] are breakable in the drying section and resin can migrate. The rosin size migration is also pH dependent [18].

Arguably, the bonds between rosin and cationic polymers and between free rosin and aluminum compounds are weak. More than 80% of the rosin size molecule in the handsets (prepared by the anionic rosin emulsion size with alum) is as free acid [118,178]. After curing at 105°C, the free rosin acid particles have disappeared and empty shells of stabilizer are observed in SEM. However, the type of bond that is formed between rosin, the alum, and the fiber is uncertain [3] and after 200 years of rosin application in the papermaking process, the rosin mechanism is still undecided [95].

This sizing mechanism is indirectly supported by the desizing effect of formic acid and sodium citrate, which has a stronger affinity for aluminum ions [12] (the aluminum salt is attacked by acids and bases [179]).

7.2.5 TECHNOLOGICAL CONSEQUENCES OF THE ROSIN SIZING MECHANISM

Changes in the type of interactions as well as molecule reorientation take place during each step. Each step is a function of the rosin form (salt of free acid), the anionic charge density of the fibers, the rosin/alum ratio, pH, the sequence of addition, the temperature, or the contact time between alum and stock. Any failed step in that sequence will result in a sizing failure.

The "forward sizing" uses the first addition of the rosin soap and the alum is added later. That sequence may result in large amounts of foam and lost rosin in side reaction with divalent calcium ions, before the alum addition [168]. "Reverse sizing" is a better sequence: the aluminum compound flocs are added first on the cellulose fibers, which become cationic and retain rosin size [167,174].

The third option would be to make a premixed blend (rosin and alum) [180–182]. Anionic rosin emulsion becomes a cationic dispersion in the presence of alum (particle size lower than one micron [183]). The adsorption of cationic particles onto the anionic cellulose surface proceeds by a mechanism of opposite charge attraction [154].

The preexisting rosin–aluminum compound interactions make the process (aluminate formation, retention, orientation) more effective, and thus the premixed rosin–alum performs better than the conventional procedure ("forward" or "reverse" sizing, see Figure 7.3) [181]. The alum–rosin premixing (one-shot rosin emulsion [184]) greatly improves rosin size efficiency in a neutral-alkaline papermaking system [183].

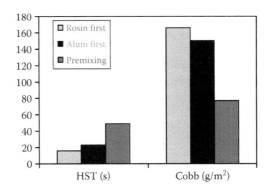

FIGURE 7.3 The effect of the order of addition of the sizing agent.

In order to make a stable dispersion (longer shelf life), the premixed fortified rosin (partially neutralized with tertiary amine and emulsified in the presence of casein [71]) and another stabilizer (cationic starch or degraded cationic starch [184]) is treated with alum [185]. Modified urea with sulfamic acid is incorporated in this blend [71] and the final pH is adjusted with caustic to 3.7. The dispersion keeps its properties for 3 months.

In the presence of an alkaline filler (calcium carbonate), the rosin sizing effect has been significantly affected. It was found that the addition of acidic compounds (alum, ferric chloride, sulfuric acid, or sodium bisulfate) after the rosin and filler retention and during the dilution step may restore the effectiveness of the rosin sizing [186–188].

Unfortunately, the reaction of rosin with polyvalent cations is not limited to aluminum. The size redistribution is further impacted if the stock contains Ca^{2+} and/or Mg^{2+}. In the presence of calcium (or magnesium) ions, the rosin acid will precipitate and that rosin is lost for sizing [120]. Calcium and magnesium (common constituents of hard water) compete with aluminum, react with rosin (to produce insoluble, non-sizing deposits [168]) and, therefore, decrease the effectiveness of rosin [189]. Soluble salts of calcium and/or magnesium can also precipitate the aluminum rosinate [92]. Aminopolycarboxylic acids (such as EDTA) are chelating agents for calcium and magnesium and thus, the rosin sizing agent is resistant to hard water precipitation.

The sizing effect of the rosin-type molecule is related to the hydrophobic/hydrophilic ratio. The hydrophobic segments added by the molecules of rosin reduce the interfiber bonds, which results in lower dry strength. In order to compensate for the loss of dry strength, an anionic polyacrylamide is added to the alum–rosin system [190,191]. Aluminum cations bring both anionic rosin and anionic polyacrylamide on the cellulose fiber.

Rosin without any alum shows only a little sizing effect, even in the presence of cationic resins [118,156]. However, the cationic macromolecular compounds are less sensitive to pH variation. Cationic resins used as anchoring compounds for rosin are epichlorohydrin–dimethyl amine condensates [18], polyalkylene polyamine (non-thermosetting) [192], polyvinyl amine [193–195], the product of a modified Mannich reaction (ammonium chloride reaction with formaldehyde and acetaldehyde) [196,197], poly-2-vinyl imidazoline [198], cationic starch, cationic or amphoteric polyacrylamide [195,199]. The charge density of those polyamines decreases in alkaline pH and consequently a decrease in sizing is noticed.

Those polymers are also used in conjunction with alum to retain and anchor the sodium salt of the fortified rosin. Alum with a post addition of cationic starch or cationic polyacrylamide improves rosin retention up to 80% [38,154].

7.2.6 Other Carboxylic Acids as Sizing Agents

The evaluation of the carboxylic (or polycarboxylic) acids as sizing agents (myristic [200] or stearic acids [201,202]) is an interesting opportunity to clarify the rosin sizing mechanism. Their chemical structure is quite similar to rosin: a carboxylic group with a hydrophobic tail and (for C_{18}–C_{22} fatty acids) the hydrophobic/hydrophilic ratio is very close to that for rosin. However, there is still a difference: the aliphatic tails of the carboxylic acids have more degrees of freedom.

The hydrophobic properties of the sizing agent are not sufficient to enhance the sizing performances of the sheet: the stearyl (or dehydroabietyl) group shows no sizing when chemically bonded through an amid bond to the cellulose pulp [118]. But, as aluminum complexes (or PAE resin and aluminum chloride), fatty acids behave in a way that is very similar to rosin [203,204]. The sizing mechanism should involve those processes enhanced by aluminum compounds, such as migration and reorientation of the hydrophobic tail.

Fatty acids are even better sizing agents than rosin [178], especially when they are in a mixture (see also the tall fatty acids fortified with 5% fumaric acid [205]). Unfortunately, only a very low concentration of fatty acid solutions can be obtained. A solution of 1% concentration of fatty acid soap is a solid mass at room temperature. A solution of 2%–3% sodium stearate is obtained at temperatures over 80°C [201].

The polydispersity of the carboxylic compounds molecular weight may prevent the molecular organization in water. A broader distribution of the molecular weight is obtained by mixing different compounds or by oligomerization. Maleated hydrocarbon resins in a blend with rosin [206], salts of oligomers of propylene [207], C_5 diolefins [208] (fortified with maleic anhydride) are dissolved in water at 30%–40% solids content.

Higher molecular weight compounds may include a broader molecular weight distribution and a higher number of carboxylic groups. A compound with multiple carboxylic groups is more likely to be immobilized on the fiber surface. Acrylic or methacrylic acid copolymers [209–211] provide the sizing material with a desired ratio between the hydrophilic (in blue) and hydrophobic parts $(n > m)$.

The hydrophobic comonomer is ethylene ($R_1 = H$), styrene ($R_1 = C_6H_5$), butadiene, α-methyl styrene, acrylic esters, etc. The preferred molecular weight of the copolymer is 4000–7000 for a composition of ethylene 87% and 13% acrylic acid.

The anionic copolymer of the ethylene-acrylic acid (EAA) [212] or of the styrene-maleic anhydride (SMA) are dissolved in ammonia solution (final pH is about 8 [213]) and sodium hydroxide [214]. If the base amount is less than 1:1 molar ratio to carboxylic acid, then dispersion is obtained [14,210]. Those resins are retained on the cellulosic pulp with iron salts ($FeCl_3$ as mordant) or cationic retention aids (PAE resins, polyamine [213,215], cationic starch, or alum). Water dispersible (co)polyester resins (polyethylene terephthalate with long hydrophilic chains) [216] are used in the presence of PAC.

The sizing capabilities of the acrylic acid copolymers depend on the base used to neutralize the carboxylic functionality: ammonia is better than sodium or potassium [209]. The ammonium salt will release ammonia during the drying step and a less hydrophilic material remains on the fiber. Sodium (or potassium) salts are still on fiber after drying and make the surface more hydrophilic than the acid form.

The acrylic acid copolymers are anionic polyelectrolytes; their retention at the wet end is performed with polycationic compounds. In this dual system, the order of addition is important. The anionic copolymer (EAA) is added first [209], the retention aid afterward. In order to make a cationic aggregate, the amount of the cationic retention aid is almost double as compared to that of the sizing agent. The cationic complex is retained on the anionic cellulose fibers and is effective in sizing over a wider range of pH: 4–10.

SMA copolymers (ammonium salt) are blended with cationic starch and cooked at 120°C–150°C. The blend is used as internal sizing agent in the presence of alum [217] and performs better than SMA at the same concentration.

7.3 REACTIVE INTERNAL SIZE (1): ALKYL KETENE DIMER

The generally accepted picture for the rosin sizing mechanism (still under discussion) is that ionic bonds are established between aluminum ions rosin and cellulose fibers, followed by the redistribution and reorientation of the anionic sizing molecules. A real covalent bond between cellulose and the sizing agent was considered a better option. After 1935, new options became available when a fatty acid anhydride was reacted with cellulose for a covalent ester bond formation and a good sizing effect [218].

The anhydride, ketene, acid chloride, lactone, and epoxide are potential reactive functionalities toward cellulose. However, a reaction of a reactive size with cellulose will always compete with the reaction of the reactive size with water.

Cellulose-reactive sizes were introduced in the papermaking process in the 1950s and they are usually used at neutral to slightly alkaline pH in an alum-free system [176]. The new neutral/alkaline process allows for a higher concentration of filler (calcium carbonate) and reduces corrosion [132,219]. The most common reactive sizing agents are the AKD and alkenyl succinic anhydride (ASA).

In spite of the similarity of their general application processes (synthesis, emulsification, and wet end application), AKD and ASA show different chemical reactivities, which will be discussed separately.

7.3.1 AKD Synthesis

The first ketene was isolated in 1905 [220]. Since then it has became a strong candidate for many reactions in organic chemistry due to its high reactivity. In 1946 [221], alkyl ketenes are mentioned as water-repellent compounds used for textile and paper.

Acyl halides undergo a reaction in which hydrochloric acid is released (intra-molecular dehydrohalogenation). That reaction is performed, under anhydrous conditions, at room temperature (or higher), using solvents (such as ethyl ether, benzene, methyl ethyl ketone, or xylene) in the presence of a tertiary amine (1:1–1.1:1 mol ratio [222,223]), able to capture the hydrochloric acid [224,225]. A large amount of tertiary amine is also used as a solvent [226].

The resulting ketene is unstable and will make a more stable dimer (β-lactone). The ammonium salt is easy to filtrate or separate from the dimerization slurry by liquid–liquid aqueous extraction. The AKD purification can be performed by recrystallization from acetone. The dimerization reaction involves the formation of a new, strong carbon–carbon bond, which extends the hydrophobic chain of the molecule. The new molecule is also a four ring ester (lactone), which is the reactive part of that new molecule.

R stands for any alkyl (the preferred alkyl contains from 12 to about 30 carbon atoms), straight and/or branched fatty acids (such as iso-stearic [227,228]), cyclohexyl [229], aryl [230], or fluoro-alkyl radicals [231,232]. In order to make a hydrophobic compound, the alkyl radical must have at least 10 carbons atoms.

The ketene dimer can be symmetrical (resulting from the dimerization of two ketenes with the same chemical structure) or asymmetrical (from ketenes with a different R alkyl). The reaction selectivity related to R is very poor and therefore, when different ketenes are involved in the dimerization process, a very complex mixture is obtained. The common mixture includes palmitic and stearic acids [227,233]. The AKD synthesis from C[14] labeled fatty acid in a mixture with unlabeled fatty acid [234] may result in a mixture of double labeled, mono-labeled, and unlabeled AKD in equal amounts.

The melting point of the dimer depends on its molecular weight and (for blends) on the molecular weight distribution. AKD made from stearic acid has a melting point of about 50°C [235]. Liquid AKD [236–239] comprises a mixture of the reaction products of the acid chloride dimerization of saturated and unsaturated fatty acids [240,241]: oleic (73%), linoleic (8%); the remainder being a mixture of saturated, unsaturated, and/or branched fatty acids. The complex mixture of acids may contain a multimer with the following general formula ($0 \leq n \leq 6$) [239]:

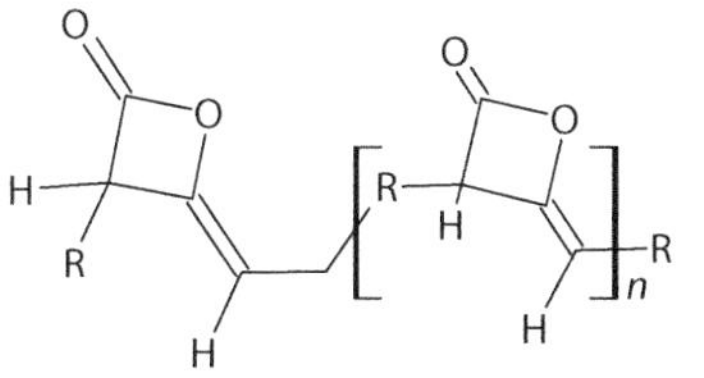

Liquid AKD provides a level of sizing similar to that of regular AKD but improves the handling performances in printing operations [242,243].

7.3.2 THE EMULSIFICATION OF AKD

Alkyl ketenes are not soluble in water and the first attempt to use them as sizing agents was made with a ketene solution in an organic solvent. In order to be added at the wet end [230], AKD must be dispersed in water. The emulsification process is important because it will set not only the emulsion stability but also the retention capabilities and the fraction of the lost reactive material as a result of hydrolysis with water.

Due to its reactivity with water, the main goal of the emulsification process is to obtain AKD dispersion able to preserve the β-lactone function. This can be achieved by emulsification at lower temperatures, by using a diluter and the right stabilizer. The mixing temperature should be above the AKD melting point but at the lowest possible temperature to avoid AKD hydrolysis [244].

The stability in water of the AKD particle is threaded from both outside (collision) and inside (crystallization). Those two processes are interrelated: AKD crystallization may result in a huge increase in particle surface (a much larger specific area), which requires more stabilizers.

The protection against effective collisions is achieved by high molecular weight water-soluble compounds used as stabilizers. In order to shift their adsorption equilibrium toward the "adsorbed" form, another stabilizer with opposite charges is suggested (sodium lignosulfonate and cationic starch, for instance).

7.3.2.1 Stabilizers for AKD Emulsion

The AKD emulsions are prepared by the supplier and are shipped to the mill at solids contents ranging from about 5% to 15% [132]. The AKD retention and efficiency should be developed in neutral-alkaline pH and the stabilizer (anionic, nonionic, cationic, or cationic–anionic complex) is a key component of the retention step. AKD is a nonionic molecule and its retention on the cellulose fiber involves the stabilizer, which should be substantive to the pulp [245] (such as cationic potato starch [246]).

On the other hand, the stabilizer should not react with the β-lactone function of AKD. For this reason, nonionic compounds are the preferred stabilizers: such as polyoxyalkylene sorbitol mono-palmitate[247], acetylated polyoxyethylene sorbitol mono-oleate [248], or alkyl-phenyl polyetoxylated [249].

Nonionic stabilizers with nonreactive molecules may provide a good stability of the AKD dispersion but do not help retention in the absence of ionic species. That is why, anionic and/or cationic compounds are added to the stabilization package.

Glycosides (natural compounds like saponine and its derivatives) are compounds containing a carbohydrate (in blue) and a noncarbohydrate residue (hydrophobic) in the same molecule.

They are effective in AKD emulsion stabilization (by themselves or in a mixture with phosphate esters of mono and di-glycerides) [250]. To increase the AKD concentration up to 40%, water-soluble macromolecular compounds (xanthan gum, polyvinyl pyrrolidone, methyl cellulose) are also part of the formulation. The nonionic emulsifier can also be associated with an anionic surfactant (calcium dodecyl benzene sulfonate) and a monohydric alcohol (dodecyl alcohol) to obtain AKD emulsions with 35% solids and about 28% active AKD [251]. These emulsions are used in the presence of cationic starch that helps their retention [249].

The anionic stabilizers help the interaction with a cationic retention aid. Due to their hydrophobic properties, fatty acids as stabilizers for AKD dispersion do not make any reduction in the sizing performance of AKD [252]. Fatty acids and their derivatives (amides, anhydrides, or acid chlorides) are also recommended as extenders and size enhancers for AKD [253,254]. A lower concentration of fatty acids has a negative impact on the emulsion stability and the AKD sizing performances [255], but at concentrations exceeding 12% (50% is preferred) in blends with AKD, they show a synergetic effect (see the AKD sizing mechanism).

Sodium lignosulfonate [256], colloidal anionic aluminum-modified silica particles [257], carboxymethyl cellulose [247,258], anionic starch [259], or fatty amide-modified carboxymethyl cellulose [260]) are macromolecular stabilizers for AKD dispersions. In those cases, the retention aid is a cationic condensate of dicyanamide and formaldehyde or cationic potato starch.

Low molecular weight [261] or macromolecular cationic compounds help both the dispersion stability and the AKD retention. The AKD emulsification with "gemini surfactants" (5%–15% based on AKD) is performed at 55°C by sonication [262]. These stabilizers (which are sizing agents by themselves [263]) are obtained by condensation of stearic (or hydroxystearic) acid with TETA (2:1 mol ratio). The hydrophobic aminoamide is protonated with acetic acid or quaternized with epichlorohydrin [264].

Cationic polymers, such as melamine-formaldehyde-imino-bis-propylmine resins ($M_w > 10,000$), acrylamide–DADMAC copolymers [265], polyvinyl amine hydrochloride [266], polyamines (di-amine–epichlorohydrin) [267], PEI, cationic starch, PAE, [268] are recommended emulsion stabilizers along with cationic starch for the AKD dispersions. Cationic resins are also sizing accelerators for AKD [229,269–271].

The adsorbed cationic stabilizer (such as cationic starch) is in equilibrium with its dissolved molecules in the continuous phase. The ratio between adsorbed and dissolved molecules depends on the AKD concentration and the particle size. Due to multiple dilutions during the addition of AKD emulsions at the wet end, the cationic stabilizer is depleted from the particle surface. In order to keep a reasonable stabilizer at the interface, the cationic starch (DS = 0.14–0.27) [272,273] was used in a large amount (2:1 starch/AKD). That amount of cationic stabilizer impacts both the general retention process at the wet end and the effectiveness of the AKD sizing.

In order to increase the concentration of the stabilizer at the interface, dual systems (anionic and cationic compounds) are used. This procedure decreases the amount of free cationic stabilizer present in continuous phase and increases the amount of stabilizer at the interface for a better protection of the AKD particle.

Two steps emulsification processes are performed. The first step is an emulsification of AKD in the presence of an anionic macromolecular stabilizer (sodium lignosulfonate, formaldehyde–sodium naphtalenesulfonate resin, 5% based on AKD). During the second step, a cationic polymer in large excess is added [274]. The two components, anionic sodium lignin sulfonate and poly-DADMAC may also be mixed before emulsification [275] and the final emulsion pH is adjusted with acetic acid or alum [276].

Anionic and cationic resin compounds with higher molecular weights (up to 20,000 Da) help the emulsion stability at higher AKD concentrations (30%) [277]. The effectiveness of cationic starch is enhanced by cross-linking with epichlorohydrin [278].

A better protection of the particle and a higher AKD concentration in emulsion (40% dimer) is obtained by adding a nonionic stabilizer (methylcellulose) along with anionic compounds (zinc stearate, sodium dioctyl sulfosuccinate, sodium naphthalene sulfonate–formaldehyde condensate) and cationic starch [279].

The AKD emulsion [280] is obtained with excess of cationic corn starch over sodium lignin sulfonate (6:1) [249,281,282], (or phenol/bisulfite/urea/formaldehyde condensate [283]). The excess of cationic resin is able to interact with the anionic trash [284,285], especially in the case of mechanical pulp [286] and improves the AKD retention. Unfortunately, the efficiency of high molecular weight cationic starch is reduced (and the sizing effect as well) by the presence of strong oxidizing agents used in bleaching and which can break down the retention aid [287].

7.3.2.2 AKD Dispersion with Higher Solids Content

"High solids AKD emulsion" does not always mean high AKD concentration. For instance, 15% AKD dispersion will contain only 5.4% AKD. The rest is cationic starch and PAE resin [143]. Therefore, even a higher amount of stabilizer should be used for higher solids content emulsions with a longer shelf life. But too much stabilizer will reduce the hydrophobic effect of the AKD and will increase the emulsion viscosity. To avoid that, a higher solids AKD emulsion must be associated with a more hydrophobic stabilizer obtained by reducing its hydrophilicity, or by chemical reactions.

The hydrophobicity of AKD is combined with the higher hydrophobicity of the stabilizer. Condensed naphtalene sulfonate (anionic) is combined with the hydrophobic cationic compound (di(hydrogenated tallow) dimethyl ammonium chloride) to make an emulsion containing 30% AKD [288,289].

The following hydrophobic cationic stabilizers are used for higher solids AKD emulsions: synthetic polymers with aromatic groups [290,291], cationic copolymers of acrylamide end capped with a long hydrophobic tail [292], or PAE resin modified with AKD, ASA or rosin isocyanate [293]. A powder of hydrophobic SMA has been suggested [294] to make 40% concentration AKD dispersion. Methyl cellulose can be a co-stabilizer along with SMA and the AKD retention is performed with cationic starch.

Attempts were made to change the structure of the dispersion by to locating AKD in the continuous phase and the stabilizer and water in the dispersed phase [295]. With this technique, a dispersion containing more than 60% active AKD was obtained and phases were reversed upon dilution at the paper mill.

7.3.2.3   AKD Emulsion Stability

Most of the AKD dispersions are unstable. Changes in the emulsion structure are noticed even in the first 2 weeks of storage time [296]. The AKD particles change their physical properties (crystallization) and chemical composition (alkyl ketone hydrolysis) [297].

Crystallization, which takes place in a very short period of time, results in the separation of AKD particles from the dispersion [297,298], and generates a larger surface area, which should be protected with more stabilizer, which will increase the rate of AKD hydrolysis.

The AKD crystallization can be stopped or delayed by dilution with a hydrophobic amorphous compound or by changing the AKD chemical structure (molecular weight distribution and the branching degree). Changes in both dilution and chemical structure may produce blends with a lower melting point.

The AKD is diluted (before emulsification) with an inert hydrocarbon resin (4%–60%) [298,299] or paraffin wax [250]. The hydrocarbon resin can be a synthetic resin (from olefins polymerization) or a natural resin (terpenes polymerization). Emulsions are obtained in the presence of cationic starch (25% based on AKD). The final product (13.5% solids content) shows an improved stability and no significant change in the paper coefficient of friction.

An AKD multimer obtained based on a regular procedure, from fatty acids in a blend with diacids, can prevent the AKD crystallization [300,301]. The polyfunctional compound makes a longer chain and the number of repeated units depends on the ratio between monoacids and diacids (1:1.5–1:4 fatty acid to dicarboxylic acid). The fatty acid is selected from the group consisting of oleic, linoleic and palmitoleic or a mixture thereof, while the dicarboxylic acid can be a dimer of an azaleic, sebacic, or fatty acid.

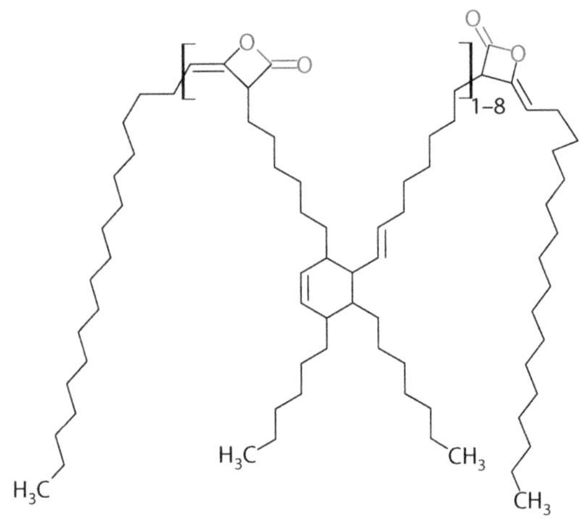

The new structure shows the same chemical reactivity and almost the same hydrophobic/hydrophilic ratio but with a broader molecular weight, which makes it less likely for the AKD molecules to organize into a crystalline structure, and reduces the melting point. The emulsification of liquid AKD produces the same type of emulsion as the regular AKD [237].

The liquid AKD shows sizing capabilities at lower temperature than the regular AKD [237]. However, PCC reduces the sizing efficiency of liquid AKD [238] and the presence of double bonds increases the risk of auto-oxidation.

The AKD dispersion shelf life is also related to the AKD hydrolysis [297,302]. The ester ring hydrolysis depends on the specific structure of the lactone AKD. The rate of hydrolysis of acyl phenyl esters increases as the chain length of acyl portion increases. The distortion of the C–O–C bond angle from 116° in a regular ester (such as ethyl acetate) to 94 in β-propiolactone increases tremendously the rate of hydrolysis by water [303].

Higher temperatures and a higher pH accelerate the AKD hydrolysis [304]. Alum has no significant effect on the hydrolysis of AKD in acid pH (pH = 4.8) but it accelerates the hydrolysis in alkaline pH (pH = 8) [305] and in the presence of PCC [306]. The size reversion also increases at higher PCC concentration when the AKD-cellulose bond is supposed to be hydrolyzed [307].

Emulsion stability and AKD hydrolysis are inter-related: the AKD hydrolysis results in lower emulsion stability, lower sizing effectiveness, and difficulties in running the papermaking equipment. To increase the emulsion stability, the type of AKD should be carefully selected. The presence of stearic anhydride in blends with AKD improves the emulsion stability in the presence of alum or at high alkaline pH [254].

The cationic starch with higher molecular weigh and higher charge density is a better stabilizer [308] and colloidal silica may enhance the dispersion stability [309], while sizing accelerators (such as PAE resins) increase the rate of AKD hydrolysis [302].

Since the presence of water is the main reason for the lower stability of the AKD dispersion, an extreme approach is to make a dry blend of AKD and a stabilizer in the absence of water. A cationic nonreactive surfactant (di(hydrogenated tallow) dimethylammonium chloride) is incorporated in molten AKD in the presence of a nonionic stabilizer (alkyl polyethylene oxide) to make a composition that is substantially water-free [310]. That composition has a long shelf life and is easily dispersed in water before its addition onto the paper stock.

7.3.3 AKD Retention

The AKD retention is the first step in the improvement of the paper hydrophobicity. The AKD sizing features depend on papermaking conditions such as pH, time, dryer temperature, kinds of pulp and fillers, etc. [235,311–313]. However, the amount of "active" AKD (non-hydrolyzed) retained on the cellulose fibers is the most important factor for the sizing process [314,315]. The AKD retention depends on the stabilizer type and concentration, on the electrolyte concentration, on the shear rate, and on the presence of alum and/or retention aids [316].

The AKD dispersion is retained due to the anionic charges on the cellulose fiber (nonionic pulp shows no sizing when AKD is added up to 0.2%). No coagulation was detected before the AKD particle retention onto mechanical pulp [286]. The amount of retained AKD was measured by pyrolysis-gas chromatography [317,318].

The AKD retention is far from 100%, it is even less than 50% [319]. The retention of AKD dispersion stabilized with cationic starch is about 10% [320] and reaches a maximum of 35% at pH = 8 [321]. The presence of PCC significantly enhanced the AKD retention [306]. A polyamine as promoter and an anionic polyacrylamide as retention aid improve the AKD retention to about 65% [322].

The interactions between pulp fibers and particles of AKD emulsion are not strong enough to prevent the partial removal of particles of AKD emulsion from the wet web, during drainage under reduced pressure, and/or during pressing. A longer waiting time between agitation and drainage

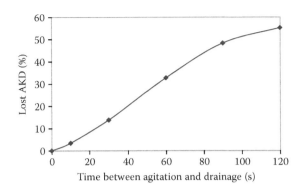

FIGURE 7.4 The amount of AKD lost before drainage. (From Isogai, A., *J. Pulp Paper Sci.*, 23(6), 276, 1997. With permission.)

lowers the AKD retention (Figure 7.4) and the loss of fines (about 75% of AKD is retained on fines) may increase the loss in AKD [311].

The sizing effectiveness of AKD dispersion made with cationic starch [319,323] or cationic starch and sodium lignosulfonate [324] is optimized by adjusting the cationic demand [325]. Anionic trash and other anionic compounds (such as calcium lignosulfonate, sodium oleate, or sodium abietate [326]) reduce the AKD retention down to about 5%.

The presence of Ca^{2+} in water (hardness) changes the charge density of the cellulose fibers. In order to prevent the failure of the adsorption of the cationic AKD particles, an organic complexing agent (amino-carboxylic acid) is used [327,328].

Highly charged cationic polyelectrolytes (such as PAE resins [317,318,325,329], N,N'-dimethylaminoethyl methacrylate/styrene/MMA/acrylonitrile/methyl acrylate copolymer, modified isobutylene–maleic anhydride copolymer with 3-dimethylaminopropylamine, or copolymers of DADMAC and diallyl amine [330,331] or chitosan [332]) are the efficient retention aids for the AKD dispersions.

This dual system (poly-cationic stabilizer and high molecular weight anionic polyelectrolyte) increases the AKD retention up to about 90% [333]. The pre-flocculation of the cationic dispersion of AKD with an anionic polyelectrolyte (CMC) increases the amount of retained AKD and reduces the Cobb values [334].

The AKD retention and distribution on the cellulose fibers is an important part of the sizing process. Therefore, the interaction between AKD and retention systems on the one hand, and the AKD redistribution on the other, should be carefully evaluated. AKD pretreated pulp shows a synergetic effect when the second treatment with traditional rosin is applied [335]. A synergetic effect is noticed even when blends of emulsions (cationic rosin and cationic AKD) are used [143]. Blends of AKD with ASA [278] or rosin are also recommended to decrease the friction coefficient of paper at high levels of AKD content in paper [336].

Some cationic polymers function not only as retention aids but also as "sizing enhancers" [331,337]. Cationic polymers (I and/or II) are able to accelerate the AKD off-the-machine sizing [281] (Figure 7.5).

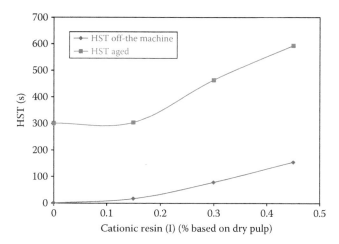

FIGURE 7.5   The effect of cationic resin (I) on the AKD sizing capability.

Cationic promoters (such as MF, dicyandiamide–formaldehyde resins [338] and their condensates with polyamidoamine [339]) are added to the wet end [340] or during the AKD emulsification
step [341].

The AKD effectiveness is reduced by shear, low pH, and high temperature. A carrier was used in
order to separate the sizing adsorption at the wet end: cellulose fibers were heavily sized in a separate process [342,343]. Only about 5% of AKD modified fibers provide uniform water repellency to
the remaining 95% of the fiber furnish.

Due to the large amount of stabilizer, the chemical compounds used to improve emulsion stability may have an important effect on the wet end chemistry. The number of cationic charges brought
by the sizing promoter may inhibit the retention and the effectiveness of the other paper chemicals.
For that reason, several changes in the order of addition can be made. For instance, the AKD internal size and wet-strength resin are added in one package [344] or the sizing promoter for AKD is not
a cationic compound but an amphoteric DADMAC–acrylic acid copolymer [345].

7.3.4   AKD Sizing Mechanism

The rosin size (neutral size and alum) also develops sizing by the time it leaves the wet end. By contrast, AKD is a pretty stable sizing agent and the AKD emulsion shelf life is several weeks [346].
The chemical stability of AKD may also result in a slow reaction with cellulose. The lower reactivity translates in the very slow development of sizing, with little or no size press holdout or sizing at
the reel. For mechanical pulp sized with AKD, the handsheets dried at 20°C show no sizing feature
up to 0.4% AKD addition level [286]. AKD develops the sizing as the paper is heated in the dryer
section [18].

For a wide range of nominal AKD dosage (up to 0.5%), the retained material is a linear process
but the sizing response (fiber contact angle) is not linear: it reaches a plateau for an AKD concentration of about 0.3% [1,347]. Therefore, the AKD addition rate is 0.04%–0.2%. In contrast, the rosin

ester sizes (nonreactive size) need an addition levels above 0.3%. The higher effectiveness of AKD was attributed to the sizing mechanism through a chemical bond formation.

Since AKD has chemical structures with a potential reactivity toward the hydroxyl groups of cellulose, the mechanism for efficient paper-sizing by AKD has been explained—in 1956—in terms of covalent bond formation between AKD and the hydroxyl group of cellulose [169].

Since 1956, the question of whether a covalent bond is established during the drying step has gotten different answers. There are two major points of controversy: the way AKD molecules spread on the fiber surface, and the existence of a real covalent bond between the sizing agent and the cellulose substrate [320,348].

The reactivity of AKD with the cellulose hydroxyl group is similar to that of water and water is present in an overwhelming amount even in the dried paper. The probability of the chemical reaction between AKD and cellulose should be evaluated taking into account the competition with the reaction with water (AKD hydrolysis).

AKD does not easily hydrolyze and the sizing reaction does not proceed before the drying step [132]. However, to hydrolyze 0.3% AKD retained on paper, only 0.012% moisture is needed [349]. At any point of the papermaking process, the water concentration is in excess of over 100 times the AKD concentration. Moreover, hydrolysis depends on pH (higher at higher pH) and it is well known that the optimum sizing effect of AKD is at pH 7.5–8.0 (the bicarbonate ion alkalinity is essential for AKD to function as internal size [350]).

The AKD molecule may also react with the amine groups from the cationic retention aids and thus the sizing agent can be involved in the following chemical reactions:

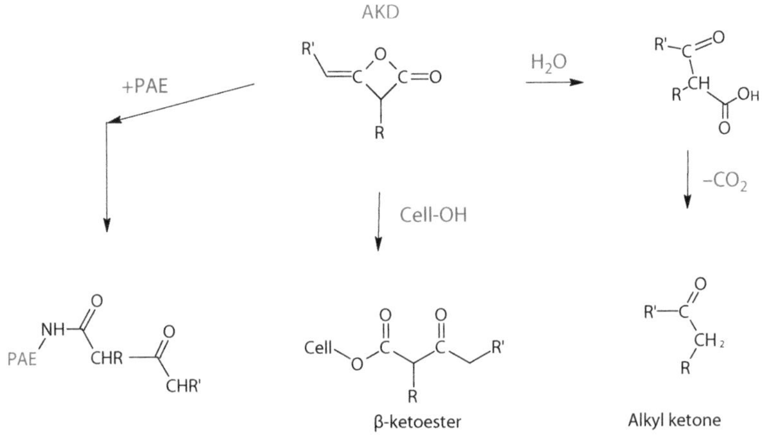

AKD is hydrolyzed to β-ketoacid (which is unstable) and, after decarboxylation, an alkyl ketone is obtained [3,4]. Fillers used in papermaking process [351,352] accelerate the hydrolysis of AKD [320]. AKD may also react with cellulose to make a β-ketoester or with PAE resin to make a hydrophobic cationic polymer (β-ketoamide).

If a reaction between AKD and cellulose results in a covalent bond formation, the bonded AKD would not be extractable with a nonreactive solvent. However, any evaluation of the extracted and/or "reacted" AKD should take into account the exact amount of retained AKD. The retention of AKD is lower than 100%. In the absence of any retention aid, the cationic AKD particles are retained in an amount of less than 20% [1,347].

7.3.4.1 Investigating the Formation of Covalent Bond between AKD and Cellulose

Any study attempting to solve the problem of the AKD sizing mechanism must deal with the potential presence of bonded AKD along with un-reacted AKD and a hydrolyzed by-product (β-ketone). Each compound must be quantitatively separated and qualitatively identified. Unfortunately, the published experiments designed to clarify the AKD sizing mechanism are not consistent: the AKD deposition from a solvent on a dry paper is different from an AKD dispersion retained on pulp at the wet end. Moreover, pulp used for scientific studies is of widely diverse types: different soft-hard wood ratios, bleached, unbleached, or mechanical pulp. That diversity is extended over to the type of fillers, starches, retention aids, etc. The goal of a published study is often in conflict with the complexity of the experiment.

An explanation of the AKD sizing mechanism would need to involve the measurement of the progression of each of the reactions that occur in a sized paper, and also the identification the reaction products. Separation was performed by extractions; the amount of extractable and the paper properties after extraction were evaluated. Several analytical techniques (such as IR or NMR) were employed for β-ketoester bond identification in the sized paper [312]. The possible chemical reactions were also simulated by using model compounds.

The supposed reaction between AKD and cellulose shows a higher rate at a higher pH and higher temperatures [353]. The concept of "reacted AKD" as different from that of "retained AKD" was largely propagated [354]. By using C^{14} [234,313,321,355,356] (or C^{13} [357,358]) labeled AKD, it was possible to identify the amount of "irreversibly bound" AKD left after extractions [321,355].

Less than 50% of retained AKD was identified as "reacted" [1,347]. The "reacted AKD" should be responsible for sizing because the hydrolyzed AKD product exhibited no sizing [3,4,357,358]. The amount of "un-reacted AKD" was estimated by extraction with THF and chloroform at room temperature [357], or for 50 h in chloroform [321], or 4 h THF [355] (both in Soxhlet).

The paper maintains its hydrophobic properties after extraction even if the amount of "bonded AKD" was 22%–50% [355] or only about 3% from the retained AKD [321]. IR spectroscopy shows the presence of β-ketoester (1741 cm^{-1}) in the paper sheet *after the extraction* at a concentration of about 0.05% [359] (but the reading was based on an absorbance multiplied by 10^3). The confidence in the existence of a chemically bonded AKD was built up on a simple balance calculation based on extraction processes. To what extent the extraction was completed and the analytical methods were accurate in not known.

Actually, the entire debate on the AKD sizing mechanism can be translated in how accurate the extractions and/or the analytical methods were [360] and how strong the covalent bond is (β-keto ester made from AKD and cellulose) in order to survive during extractions. That structure shows an internal hydrogen bond for the enolic form, which makes it more stable [359]. The stability of the presumed β-ketoester was tested in a potential hydrolysis reaction and that ester seems to be stable in water in alkaline pH [361].

The effectiveness of the extraction should always be a concern. In other words, the extraction procedure might not have really removed all un-reacted AKD. The solvent interaction with cellulose must also be considered as a potential replacement of the adsorbed material (such as the hydrogen bonds between dioxane and cellulose [362]). On the other hand, some components of the extraction system (such as Tween 20) may react with AKD [12].

There is a diversity of solvents used for extraction, at different temperatures for different periods of time, such as methylene chloride [325], refluxing chloroform for one hour [363], 18 h [364], or 50 h [321], THF for 4 h [326,353,355,365], 12 h [234], or 28 h [357,358] (Soxhlet), or water-dioxane (1:1) at 80°C (reflux, at equilibrium) for 8 h [312]. Regardless of the diversity of extraction protocols and the fact that lower molecular weight compounds are easier to extract [366]: the authors' conclusions are based on their confidence that **all** the un-bonded material was extracted. If the extraction failed to remove **all** AKD, the paper will show some size, which may lead to a wrong conclusion concerning the reaction between AKD and cellulose.

Eventually, a solvent and a protocol able to extract all the retained AKD has been found. The original handsheets with 1.4 mg/g AKD content were first defibrated in water and then dried fibers were extracted by chloroform. About 70% of AKD was extracted in that step [318,367]. 15% more AKD was extracted in water with 1% Tween 80 at room temperature and almost all AKD was removed when the defibrated handsheets were stirred in 1% Tween 80 at 60°C. Tween 80 has a neutral molecule and is also used for protein extractions.

If all AKD is solubilized and no cleavage of β-ketoester bonds occurred during that extraction even at 80°C for 24 h [318,361], the presence of bonded AKD is questionable and the kinetic model and reaction order calculated for the cellulose-AKD reaction [353,365] based on the unreliable extraction data seems to be too venturesome.

If no AKD is chemically bonded, some other species must provide sizing. The extracted (un-reacted) AKD has also an important effect on sizing [364]: regardless of the curing temperature (70°C, 90°C and 110°C), as more AKD is extracted ("un-reacted AKD"), lower HST values are obtained. In other words, the presence of un-reacted AKD contributes to the hydrophobicity of the paper. The effectiveness of un-reacted AKD depends on the amount of "reacted" AKD [234].

A direct identification of reacted AKD in sized paper should quantitatively measure the β-ketoesters concentration. The AKD concentration in paper can be estimated by pyrolysis–gas chromatography (GC) [286,311,368], but that method is not able to separate between reacted and un-reacted AKD. The solid-state ^{13}C NMR spectra of cellulose treated with ^{13}C labeled AKD shows no resonance peak due to the presence of that ester [317,318]. For both uncured and cured samples, the ^{13}C NMR spectrum indicates only the presence of un-reacted AKD or hydrolyzed AKD.

Paper sized with AKD may lose sizing over time [369]. The de-sizing effect is explained by the AKD desorption and vapor phase transfer [315]. If the "covalent bonds" mechanism is accepted, then the "sizing reversion" (the degradation over time of sizing performance) can also be explained by the covalent bond hydrolysis [350]. However, the β-ketoester is stable in an alkaline environment, in the presence of water at high temperatures. Therefore, size reversion must be explained in terms of some mechanism other than the cleavage of the β-ketoester bonds [361], for instance, by AKD molecular reorganization [363]. This seems to be a strong proof for no covalent bond formation.

The covalent bond formation raises the question related to the reactivity of the AKD lactone cycle with cellulose. IR spectra for the pulp modified with 0.5% stearoyl chloride show an absorption band at 1740 cm^{-1}. Cellulose also reacts with isocyanates (the paper was first dried in a desiccator over phosphorus pentoxide to prevent the parallel reaction with water [200]).

Unlike stearoyl chloride and isocyanates, AKD seems to be quite stable because after curing (30 min at 100°C [355]), about 95% of un-reacted AKD was extracted (chloroform as solvent for 50 h [321]) as un-reacted AKD. AKD lactone functionality is much more stable and can survive for a long period of time in the presence of water. Therefore, it was no surprise that no β-ketoester bond (expected to appear at 1730 cm^{-1}) was detected [370] in paper treated with AKD (detection limit 0.5%). Thus, the new data does not confirm the earlier findings [359]. Resonance peaks at about 205 ppm are generally due to the presence of carbonyl carbons of β-ketoesters. Such peaks were not detected in either paper dried at 20°C, or at 105°C for 20 min [367].

A DSC technique was used to identify a reaction between AKD and D-glucopyranoside. The possible reaction was performed in a calorimeter; no additional heat was developed or consumed by the mixture (except for melting) [370]. No changes in melting points were noted after heating a blend of these two compounds up to 175°C.

Methyl β-D-cellobioside (modified disaccharide) was used as another model for cellulose in a reaction with AKD, which is less reactive toward the hydroxyl groups of methyl cellobioside than it is to water [371].

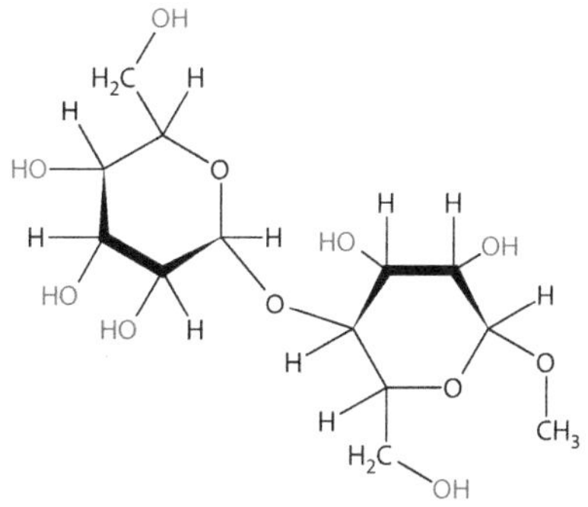

These results suggest that a reaction between cellulose and AKD would be most unlikely to occur, particularly in an aqueous system. However, some extreme conditions are requested for the reaction between cellulose and AKD to take place: a common solvent (dimethyl acetamide and 3%–15% lithium chloride [372]) and the presence of a catalyst (1-methyl imidazole) [675], or N-methyl morpholine [373]). The presence of β-ketoesters of cellulose (carbonyl group infrared absorption at 1742–1752 cm^{-1}) but the author [373] acknowledged that the reaction conditions are not papermaking conditions. Moreover, the methyl ester of AKD, as a model compound, was prepared in chloroform (in the absence of water) and with sodium hydroxide as a catalyst [359].

If there is no direct chemical bond between cellulose and AKD, the possibility to use an inter-mediate compound (such as PAE resin or alum) to link the ketene dimer to the cellulosic substrate was also evaluated. That option was chosen because the effectiveness of AKD is enhanced by the presence of other chemical compounds and their position seems to be between the AKD molecule and the cellulose fiber.

PAE resin improves the AKD retention up to 55%–60% (as compared to 20%–30% in the absence of PAE). An AKD emulsion is added first and the PAE solution second [286]. PAE resin also has an effect on the AKD sizing properties (Figure 7.6) [311]. It was suggested that the mechanism of the sizing enhancement is an irreversibly bonded formation between AKD and PAE resins [325]. Then the hydrophobic polyelectrolyte is pseudo-irreversibly adsorbed on the cellulose fiber.

Only a low level of alum enhances the initial sizing of AKD [219] but at higher concentrations, it has a detrimental effect on the AKD sizing capabilities [169,350,374]. The AKD molecule is not directly attached to fiber and an aluminum ion is the connection between AKD and cellu-lose [363,374,375], which makes the reversion process more effective [363]. Since the assumed

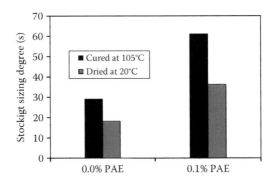

FIGURE 7.6   The effect of PAE resin on AKD sizing capability.

β-ketoester AKD-cellulose is a stable molecule, the reversion process shakes the confidence in strong covalent bond formation.

7.3.4.2   Alternative Suggestions for an AKD Sizing Mechanism

Extraction studies and the unsuccessful attempts at identifying the presence of β-ketoesters via different analytical methods, leave open the question of the AKD sizing mechanism. However, AKD is currently called "reactive sizing agent." Due to the uncertainty of the sizing mechanism, other ambiguous descriptions such as "anchored" or "direct size" have been suggested [12]. That was done to differentiate AKD from rosin, which is called "indirect size."

Regardless of the lack any definitive proof, the sizing mechanism by chemically bonded AKD is already called "traditional mechanism" [375]. Still, in order to explain the AKD sizing mechanism in the absence of a covalent bond, we have to consider the stronger adsorption of AKD molecule, as well as its migration and reorganization.

After extraction, water adsorption diminishes significantly. Therefore, the paper hydrophobicity can be provided by un-reacted AKD. The difficulty to extract the free AKD can be an indication of its stronger adsorption or trapped molecule into small pores on the fiber surface. The AKD extraction is as difficult as is the extraction of a nonreactive (hexadecyl haxadecanoate), which is not 100% extracted either [321].

The extraction (with chloroform) of AKD from precipitated aluminum oxides (in alkaline pH) [305] was hard to achieve as well. In that case, a covalent bond formation would be unreasonable, but a complex formation, physical entrapment and stronger adsorption on the surface, could be considered.

The adsorbed AKD molecules become effective only in the drying section where temperature may help the migration and/or the orientation of the sizing molecule. In order to check the migration concept on a really bonded molecule, strearoyl chloride was reacted with cellulose; the ester bond was identified at $1740\,cm^{-1}$. The un-reacted portion was washed off afterward. This chemically modified cellulose pulp exhibits a high degree of sizing. However, the sizing effect is not transferred to unsized fibers [370].

Two interesting experiments were designed to prove the AKD migration. One was the capability of 5% AKD-modified fibers to provide uniform water repellency to the remaining 95% untreated furnish [342]. The treated fibers are effective even when the fiber surface is not completely covered by AKD [343].

It is hard to believe that the covalent bonded AKD are mobile, it is more likely that the free AKD molecules are migrating from the sized to the unsized fibers. The second experiment is based on the same concept: this time 20 sheets of sized paper were placed on top of 20 sheets of unsized paper. The sizing effect on both sides was tested immediately, as well as after several days of storage at 25°C and 50% RH [376].

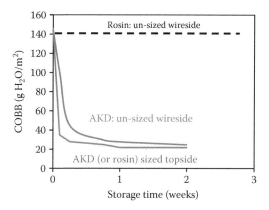

FIGURE 7.7 AKD migration vs. storage time. (From Isogai, A., *J. Pulp Paper Sci.*, 23(6), 276, 1997. With permission.)

The sizing effect of AKD was identified on the unsized paper (Figure 7.7), which indicates a significant migration of AKD molecules.

The same experiment was conducted with rosin as sizing agent, which makes insoluble, high molecular weight precipitate. The unsized side did not show any sizing properties over the same period of time. It might be concluded that, in the case of AKD, at least some of the AKD molecules do migrate. The mobile AKD molecules are not hydrolyzed and can impart sizing [377].

The AKD migration follows the general rules for that type of process: the temperature and the presence of a solvent favor the migration of AKD molecules. Paper sized with AKD shows higher waterproofing properties after the treatment with vapors of *n*-hexane (or any other good solvent for AKD) at room temperature [316].

The molten AKD migrates through porous paper due to interfacial tension and capillary force [378], or via vaporization and redeposition [348]. Migration through the vapor phase to impart sizing in unsized paper was also proven for stearic acid [200], or alkyl silicone halide [379–382]. In the case of the vapor-phase deposition of AKD, the hydrolysis reaction should also be considered [349].

In order to spread on the cellulose surface, the surface free energy of the liquid AKD should be lower than the surface free energy of the cellulose fiber [383], which explains why AKD does not a undergo spontaneous spreading on the smooth cellulose film [378]. The amphiphilic AKD molecules are adsorbed on the cellulose fiber surface and are organized with their hydrophobic end pointing into the air. The surface area occupied by one AKD molecule oriented perpendicularly is $24\,\text{Å}^2$ [355]. This stops the liquid from wetting the solid surface. However, the spreading process is assumed to have a lower activation energy than the chemical reaction [353].

The SEM show the retained AKD particle in the paper sheet dried at 20°C [367]. After curing at 105°C, only empty shell-like structures are observed on the pulp surface. The SEM observation and the result of solid-state ^{13}C NMR spectrum show that the increase in the sizing degree after curing is primarily due to the melting and spreading of the AKD molecules on the fiber surface [317]. By-products resulting from AKD hydrolysis do not migrate and they are located in spots [384].

The monolayer formation is still being discussed [315] and the fact that polymer particles are also internal sizing agents [385] but do not migrate should be a concern when the AKD sizing mechanism is evaluated. However, the amphiphilic molecules (like AKD) have a slower tendency to organize and this diffusion process explains the time needed for "curing" of AKD sized paper. To make this process faster, higher temperatures or liquid AKD are recommended [237]. The vapor diffusion plays an important role as well.

The contact angle measurement reaches a plateau at a fiber area of only 20% covered by AKD [347]. Therefore, the complete coverage of the fiber surface by the size is not necessary, but the reorientation of the hydrophobic tail is important [234,386]. In an attempt at explaining the inability of chloroform to extract all the AKD, the authors [321] suggested that the adsorbed AKD is highly oriented and difficult to extract or that the dimer has polymerized during the heat treatment. The AKD polymerization is a way to "organize" AKD molecules, to make them very effective in sizing and hard to extract.

The diketene polymerization is well documented [387,388]. Oligomers were identified within the reaction product of slightly alkaline hydrolysis of AKD [357]. AKD can polymerize by ring opening polymerization of the lactone cycle and develop linear polyester with branches of hydrophobic tails:

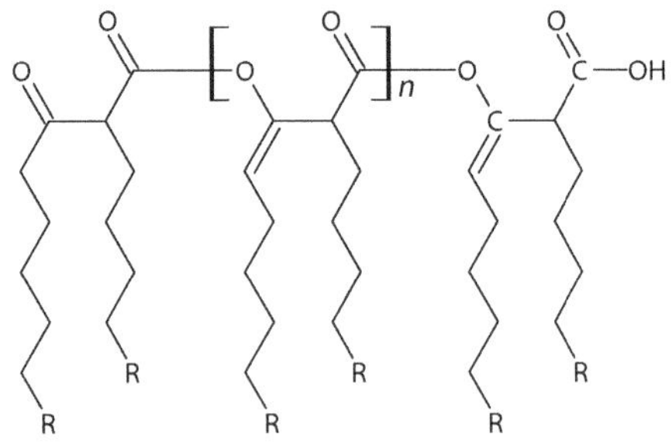

AKD oligomers (8% in AKD [255]) have a sizing effect similar to AKD. It seems reasonable that the oligomer may be formed under alkaline papermaking conditions. It is also plausible that cellulose initiates such an oligomerization of AKD [357]. Sodium bicarbonate and PAE resin show a synergetic effect on AKD sizing performances [365], which can also be related to the AKD polymerization.

Due to the presence of a template (cellulose fiber), the polymerization of adsorbed molecules may be developed in a particular way [389], with a special orientation of the hydrophobic tails, which may explain the sizing effect on paper. It is reasonable to think that heat is needed for migration, alignment, and polymerization.

If the covalent bond formation is not a must and the hydrophobic molecules are more efficient after their migration and organization, many other chemical structures can be effective in paper sizing. In order to demonstrate whether the chemically bonded sizing agent is needed, nonreactive molecules were tested. The effectiveness of AKD was compared with that of several derivatives, which are not expected to react with cellulose.

Polyamines are promoters for AKD [322]. The reaction products of AKD with polyamines (DETA, TETA) are neutral, waxy solids of no reactivity [169,370].

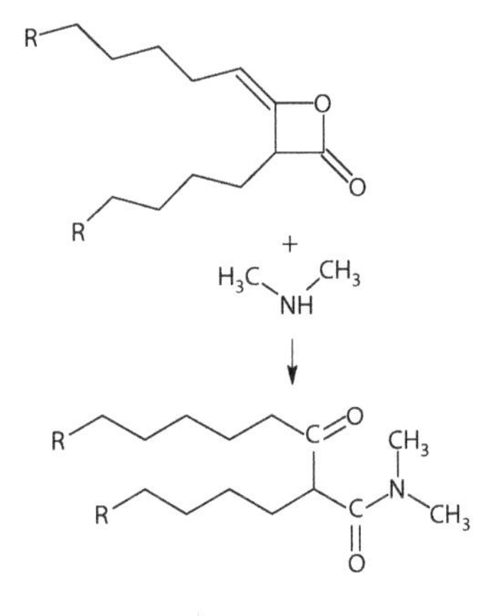

The β-keto amides are powerful sizing agents [370] with effectiveness comparable to that of AKD (Figure 7.8) (see the nonreactive internal sizing agents in Section 7.5.2).

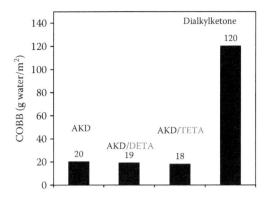

FIGURE 7.8   The sizing effect of some nonreactive sizing agents.

AKD reacts with polymeric polyamines to make cationic polyelectrolytes carrying hydrophobic pendant tails [390]. About 15% of the secondary amines have hydrophobic tails. The rest of the secondary amines are converted to cationic charges. The new hydrophobic cationic polymer is dispersed in water and it is a strong internal sizing agent. This can explain the effectiveness of the PAE resin (or poly-DADMAC-Epi resin [233]) when used along with the AKD emulsion: called an "accelerator."

The dispute over the existence of a covalent bond between AKD and cellulose extends the discussion over much weaker bonds developed during the adsorption and organization of AKD on the fiber surface. Thus a new theory was put forward, which takes into account both strong and weak bonds in an attempt to explain the AKD sizing mechanism [18]. Unlike the covalent bonds, the weak bonds can be broken by the energy available in the drying section.

According to that theory, the AKD is primarily bonded through a weak bond and the strong covalent bond (β-ketoester formation) is a minor side reaction at best [375,377]. There is a large majority of weak AKD–cellulose bonds, which are readily broken in the dryer section. The AKD molecules can migrate through a paper web via a sorption/desorption vapor phase mechanism.

The existence of covalent bonds between AKD and cellulose is under debate, but a very low concentration of β-ketoester cannot be definitively ruled out. Thus a hybrid mechanism can also be possible: a low concentration of β-ketoester molecules can anchor (through nucleation) un-bonded molecules (free AKD, alkanes [391], or akyl ketones [392]) and build up a hydrophobic layer. Both alkanes and un-reacted AKD (or alkyl ketone) are involved in this "adsorption by trapping." That can also explain why the hydrolyzed AKD (up to about 25% alkyl ketone in a blend with fresh AKD [318,392], or lanolin [393,394]) may enhance the AKD sizing effect. Moreover, blends of alkyl ketones with different melting points provide a noticeable sizing effect [314].

7.4   REACTIVE INTERNAL SIZE (2): AKENYL SUCCINIC ANHYDRIDE

ASA is a chemical compound that has an anhydride functionality and a hydrophobic tail. The anhydride group in ASA is more reactive than the lactone functionality in AKD and it is the most reactive neutral size. ASA is not soluble in water and, due to its reactivity, it must be emulsified at the paper mill before using [219,369].

There are linear (symmetric or nonsymmetric) anhydrides and cyclic anhydrides. Rosin anhydride [395] and stearic anhydride [218] are linear symmetric anhydride, which were tested as sizing agents. They are applied at the wet end as dispersions in water with cationic stabilizers [324,396–399]. The partial hydrolysis of the anhydride (to an acid) may play a role in the retention step [400].

The symmetric anhydrides react with a hydroxyl group of cellulose to form an ester and hydrophobically modified cellulose. Unfortunately, a fatty acid, which can migrate and interact with other paper chemicals is a by-product of that reaction. That may further complicate the papermaking process.

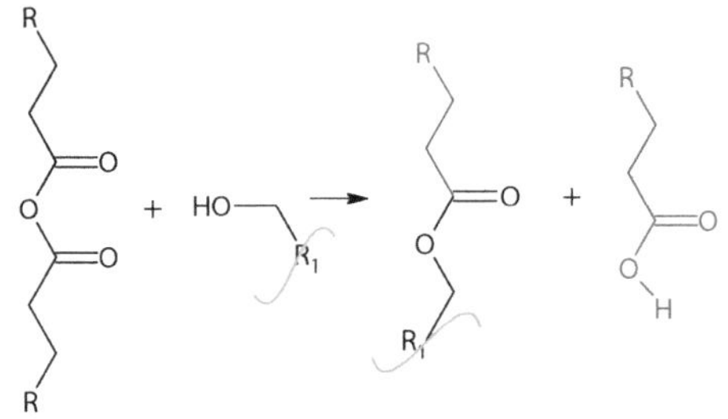

ASA (such as iso-octadecenyl succinic anhydride [401]) is a cyclic anhydride; its esterification reaction does not result in any by-product. The hydrophobic molecule is bonded to cellulose through an ester bond (in red) and a carboxylic group (in blue) now attached to cellulose. Both linear and cyclic anhydrides react easily with water (hydrolysis) and the hydrolysis products may interfere with the sizing process.

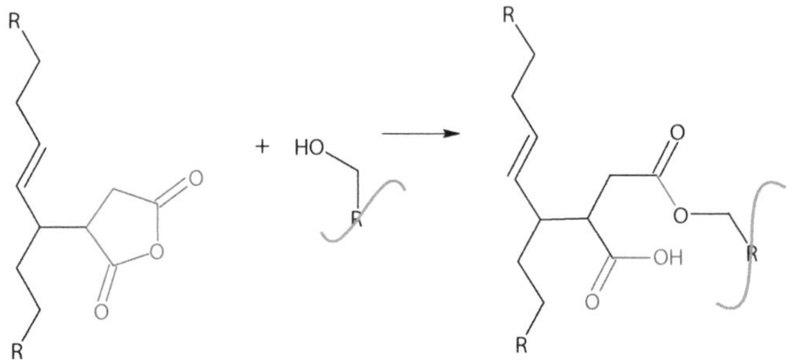

ASA is more reactive than stearic and abietic anhydrides [176]. It can be used under a wider range of pH (4.5–8.0) than AKD [350], and shows a better sizing than the rosin–alum system for the same amount of size [401]. Another benefit of the ASA is that the sizing effect is achieved while on the machine [132]. For those reasons, in recent years, there has been a trend toward replacing AKD with ASA in several paper grades [402]. However, due to its higher reactivity, ASA is more difficult to handle.

7.4.1 The Synthesis of ASA-Type Compounds

ASA has a molecular weight of about 350 and a very low solubility in water [403]. ASA is synthesized from maleic anhydride and olefins. A mono-olefin is first isomerized (acidic catalyst such as tungstosilicic acid carried on silica gel [404,405]) by randomly moving the double bond from the alpha position [406]. There are many isomers of the olefin based on the position of the double bond. Symmetric olefins perform better as internal sizing after reaction with maleic anhydride [403,407].

The ASA synthesis is an addition reaction of the allylic hydrogen from olefin to the maleic double bond. The presence of a stabilizer (hindered phenols) increases the reaction yield [408]. This gives an ASA, which is a liquid at room temperature.

During the ASA synthesis, the number of isomers is multiplied by the formation of two new asymmetric carbons (in red). About 300 isomers are likely to be present in a commercial ASA [3].

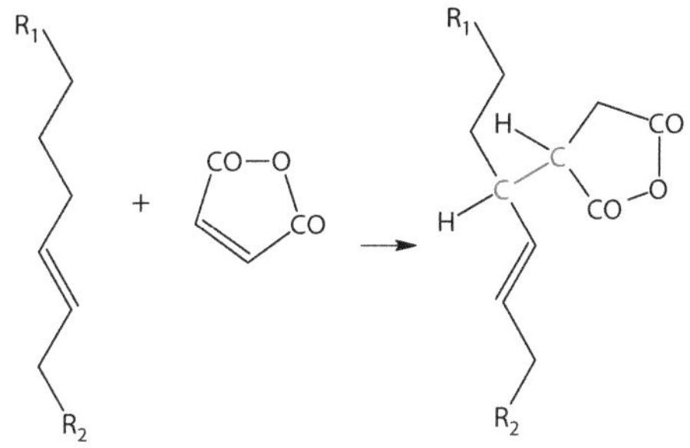

There are many sources of olefins: mixture of olefins [409], vinylidene olefins [410], isobutylene oligomers [411], oligomers of propylene (C_{12}–C_{18}) [207], oligomers of C_5 acyclic diolefins [208], or low molecular weight polybutadiene: $750 < M_w < 1500$ [412]. The olefin with an internal double bond can also be obtained by small molecular weight olefin oligomerization [413,414] with catalysts like $AlCl_3$, BF_3—donor (ethyl ether), phosphoric acid [415], etc. The oligomerization of n-octene-1 results in 89% C_{16} and 11% C_{24} of branched internal olefins. The branched olefins are blended with linear olefins (3:1) and reacted with maleic anhydride for a better sizing performance [405].

The olefin double bonds may be in an internal or terminal position [416]. Poly-butene has a terminal double bond and reacts with the maleic anhydride at 240°C and 40 psi for 5 h [414,417] to obtain an alkyl anhydride (where $5 \leq n \leq 16$).

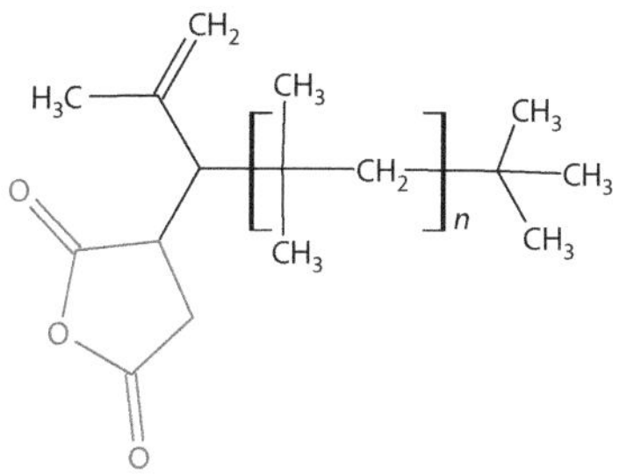

Paraffin (no allylic hydrogen) and alkyl-aromatic hydrocarbons (with a benzyl hydrogen) react in the same way but at higher temperatures (>300°C) and in the presence of iodine [418]. For instance, from ethyl benzene the addition takes place at the benzyl hydrogen.

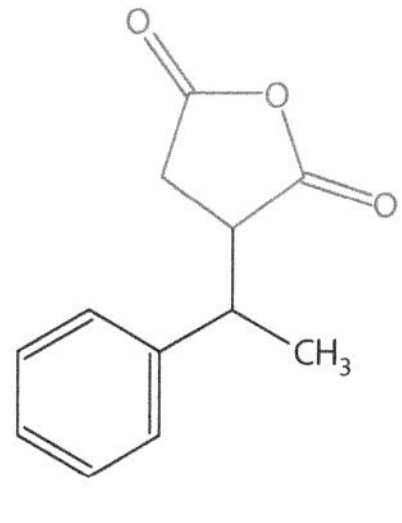

Because the reaction between olefin and the maleic anhydride is performed at elevated temperatures (200°C–300°C), secondary reactions take place (polymerization and decarboxylation), which may result in a darker product. In order to prevent these reactions, the following catalysts are recommended: p-toluene sulfonic acid, calcium bromide, aluminum acetylacetonate [419], tri-orthoalkylphenyl phosphate [420], or arylfluorophosphite [421]. Shorter reaction time and/or lower reaction temperatures are needed when a free radical initiator is used (di-t-butyl peroxide) [422,423].

The darker by-products can be eliminated by washing with water at 60°C–70°C when the anhydride is hydrolyzed. The anhydride ring is closed again by heating in xylene at 135°C [424]. The purified product is light amber in color.

7.4.2 ASA Emulsification

Because ASA is not soluble in water, it should be dispersed in water. Due to its reactivity with water, the emulsification should be performed on the site, fast, and at an optimum particle size. A homogenizer (or an unltrasonic processor [425]) is accommodated to the continuous papermaking process.

The preferred particle size ranges from 0.5 to 3 μ. Larger particle size leads to agglomeration and inefficient sizing. Smaller particles expose more surface area of the emulsion to the aqueous environment. This will result in a faster hydrolysis rate [426]. The hydrolyzed ASA changes the emulsion stability.

In order to obtain a stable emulsion, the following materials are used [427,428]: a surface active agent (*Activator*) that allows for the effective emulsification even under conditions of low mechanical energy, a cationic compound for retention (*Promoter*), and a high molecular compound (*Stabilizer*), able to inhibit the coalescence process. Besides, specially designed equipment for mixing may improve the emulsion quality [429–431].

Because ASA is easy to hydrolyze [432], the emulsification must take into account its sensitivity to water. The process should be fast, under lower share rate, in neutral pH, using a lower concentration of emulsifier, at room temperature, under normal pressure, and the emulsifier should not react with ASA.

Nonionic, anionic, cationic, or amphoteric [433] macromolecular compounds are used in high concentrations (1:1) as stabilizers for ASA emulsions at a total solids concentration of about 10% [434,435]. It would be beneficial for the papermaking process if that large amount of stabilizer could enhance other properties of the paper, like dry strength.

Cationic starch remains the most important stabilizer for the ASA dispersions because it functions as both a promoter (for retention) and a stabilizer [401,427,428,436,437]. The hydrolyzed ASA is an anionic compound, which may interact with the cationic starch [346,428]. Therefore, the cationic starch/ASA ratios must reach high values (1:1 up to 5:1) [438]. Higher starch concentrations result in smaller particle size and better sizing.

The molecular weight and the hydrophobicity of the cationic starch are critical for the dispersion quality. Cross-linked starch [278], grafted cationic starch [439], and more hydrophobic cationic starch perform better. In order to make it hydrophobic, starch was acetylated with acetic anhydride [440,441] or reacted with ASA [442,443]. The esterification of starch with octenyl succinic anhydride can prevent the ASA hydrolysis for more than 90 h. Hydrophobic tails attached to the starch macromolecule are supposed to align onto the ASA particle surface and thus prevent water molecules from approaching and causing the hydrolysis of ASA [444].

Starch can be made both cationic and hydrophobic in one step [445] with epoxy quaternary ammonium salt [446]. The newly modified starch is a good emulsifying agent (two parts starch for one part ASA) [446] and contributes to the sizing process.

Cationic grafted starch is obtained in one step by the free radical grafting copolymerization of acrylamide with DADMAC in the presence of hydroxylated corn starch [447]. A branched cationic starch can also be made through the glyoxalation of starch in the presence of cationic copolymer [448,449].

Tertiary amines [450], or the cationic synthetic polymers or their blends [451] can replace 100% the cationic starch [452]. The cationic polymer with molecular weights over 40,000 are obtained by the copolymerization of acrylamide with cationic comonomer (such as dimethylaminomethyl acrylate).

Cationic stabilizers [453–455] are better retained on the ASA particle surface by an anionic emulsifier (sodium di-octyl sulfosuccinate), anionic inorganic micro particles [456], or an anionic polyoxyethylene nonylphenylether phosphoric ester. The ratio between ASA and the cationic polymer is 2:1 and the anionic surfactant is less than 1% in the final emulsion [451]. Some of those formulations may use alum for the application ASA emulsions.

Nonreactive stabilizers protect ASA and the emulsion shows a longer shelf life [411]. The non-reactive, nonionic emulsifiers are used along with blends of cationic starch and cationic resins to stabilize the ASA particle [457].

In the absence of any ionic stabilizer, alkyl-aryl polyethoxylated compounds [458–462] (such as nonyl-phenol ethoxylated with 10 units of ethylene oxide) are dissolved in ASA (1:9) and dispersed in water at 2% concentration of organic phase. The mixtures between polyethoxylated compounds and ASA show self-emulsifying properties.

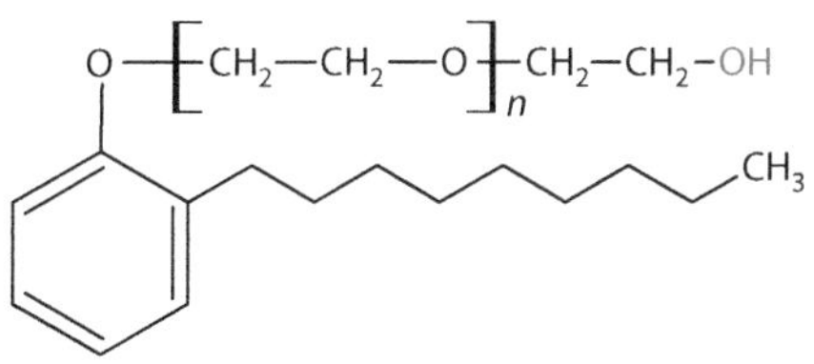

The tertiary hydroxyl (in polyoxyethylene–polyoxypropylene-type emulsifiers) does not react with ASA [463]. However, the primary hydroxyl group may react with ASA; the mixture is not stable. In order to improve the emulsion stability, the primary hydroxyl is esterified with fatty acids [464,465], or phosphoric acids [453], etc.). The esterification with an alkenyl anhydride is followed by decarboxylation (at 185°C) [466] for nonionic stabilizers. Those dispersions are nonionic and can be used only for sizing the paper surface.

The polyethylene glycol esterification is also performed with ASA in excess [455,467–469]. The final product is a blend of un-reacted ASA and an anionic stabilizer and it is easy to disperse in water. The new stabilizer has a carboxylic group. Cationic starch is used as retention aid on the cellulose fibers.

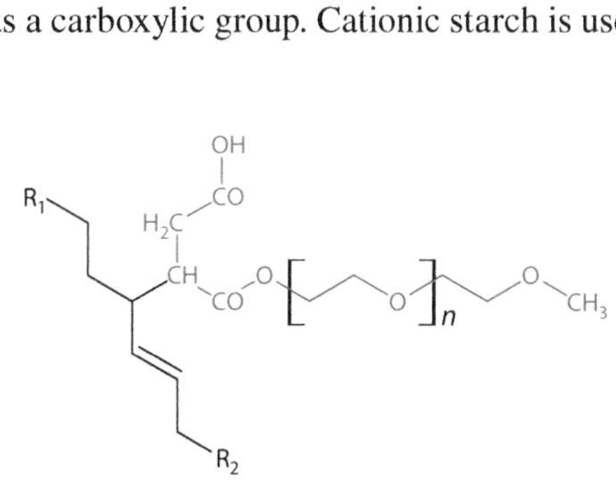

The performances of ASA are directly related to the quality of the emulsion produced prior to the addition of the size to the paper furnish [438,470,471]. The quality of an ASA emulsion is judged by [346] the particle size, the mechanical stability, the charge density, and the degree of hydrolysis

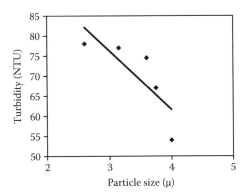

FIGURE 7.9 Emulsion turbidity vs. ASA particle size.

of ASA. The knowledge of the particle size—sizing properties relationship allows for a relatively easy fine-tuning of the emulsification process and for an optimized emulsion quality. An accurate method for measuring the particle size (such as light scattering) should be used. The turbidity method does not show a good correlation with the actual particle sizes (Figure 7.9) [438].

Besides large experimental errors [427], dependence does not even cover the entire range: the most likely range $0.5–1.5\,\mu$ is missing. Based on these experimental data, dependence cannot be extended over to the smaller particle size range (see Section 2.5). However, the particle size evaluation with a microscope is even worse [428].

7.4.3 EFFECTS OF ASA HYDROLYSIS ON ITS APPLICATION

An ASA emulsion has a limited shelf life [430]. That is why liquid ASA is supplied in a 100% active form and is emulsified at the paper mill. Liquid ASA must be protected from moisture at all time.

The hydrolysis of ASA [472] is the primary cause of any runnability problems such as spots, press picking, and holes that many mills using ASA are still facing. High temperatures during the emulsification of ASA, a higher alkalinity emulsion, long emulsion storage time, and poor first pass retention [473] result in a higher degree of hydrolysis [346]. The acid resulted from the ASA hydrolysis is a catalyst for the hydrolysis of the rest of the ASA molecule. In other words, the ASA hydrolysis process is auto-catalytic.

Since hydrolyzed ASA is anionic and nonreactive, it decreases the ASA retention because the anionic charge will neutralize the cationic charge of the retention aid, will reduce the ASA sizing effectiveness, and will create deposits on the paper machine.

An ASA emulsion can be stored for only several hours [474]. At room temperature, the hydrolysis rate for the emulsified ASA is a few percents per hour with complete hydrolysis occurring within 24 h [475]. The ASA hydrolysis is accelerated [428] at higher temperatures [470,471475] (Figure 7.10), in alkaline pH (Figure 7.11), in the presence of a surfactant or calcium carbonate as filler [476].

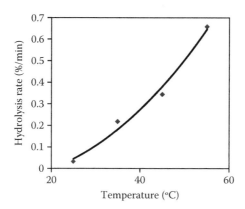

FIGURE 7.10 ASA hydrolysis rate vs. temperature at pH = 3.5. (From Chen, G.C.I. and Woodward, T.W., *TAPPI Proceedings, Papermakers Conference*, Atlanta, GA, 1986, pp. 37–40. With permission.)

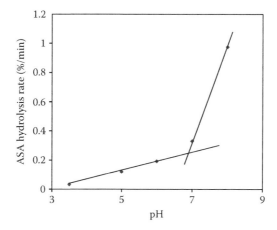

FIGURE 7.11 The pH effect on the ASA hydrolysis rate.

The pH effect on the ASA hydrolysis rate has an inflection point in the neutral range (pH = 7.0) (Figure 7.11), and the reaction rate in alkaline pH is much higher.

The effect of the particle sizes on the ASA hydrolysis rate is also important: after 3 h at 40°C and pH = 3.5, the degree of hydrolysis is 25% for large particle ($D = 8 \mu$) and 40% for small particles ($D = 1 \mu$) [475].

Hydrolysis can be minimized by keeping the emulsion temperature at 25°C and pH = 3.5 during and after emulsification, and a short time between preparation and use [477]. The holding time [472] and the emulsion temperature should be reduced in order to prevent a fast ASA hydrolysis (Figure 7.12).

ASA must be added to stock that is alkaline and relatively hot. ASA is added at the thick stock (to maximize its retention) or as close as possible to the headbox (to minimize its hydrolysis) [430]. Hydrolysis can occur after the emulsion has been added to the paper machine wet end [472]. Longer contact time will result in poorer sizing properties.

The cationic starch/ASA ratio is important to maximize ASA retention, to reduce the contact time and the amount of hydrolyzed ASA. More cationic starch (calculated in the blend with ASA) results in a better first pass retention [319], and a higher efficiency [472] (Figure 7.13).

Due to the large amount of stabilizer (Figure 7.13), the chemical compounds used to improve ASA emulsion stability may have an important effect on other paper properties. For instance,

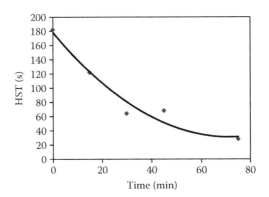

FIGURE 7.12 ASA performance vs. emulsion storage time.

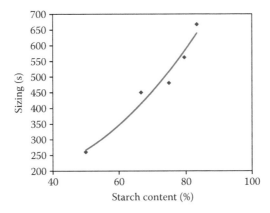

FIGURE 7.13 The effect of cationic starch concentration on the ASA sizing effectiveness.

internal size and wet-strength resin (polyvinyl alcohol (co)vinyl amine) are added at the wet end in one package [344].

The addition of the alum significantly increases sizing with an ASA emulsion and cationic starch [219,478]. The ASA concentration shows the same nonlinear effect on the hydrophobicity of the paper both in the absence [479] and in the presence of alum (Figure 7.14). The alum concentration should be correlated with the ASA concentration.

The hydrolysis rate of ASA is reduced at lower pH and alum helps that process. Alum also has been shown to act as a scavenger for hydrolyzed ASA, it forms a salt, which is less tacky than calcium salts and controls the deposit caused by hydrolyzed ASA [350]. The hydrolyzed ASA is still a hydrophobic material able to provide size only in the presence of a sufficient amount of alum [132,477].

In order to slow down the ASA hydrolysis, a good management of the pH, temperature, and storage time is recommended. There are some other approaches to minimize the ASA hydrolysis. One is to use a nonreactive diluter for ASA [480]. Another would be to make tablets of blends of ASA and cationic starch in the absence of water and stored in the absence of any moisture. Those pastilles are dispersed in water just before application [481].

7.4.4 ASA SIZING MECHANISM

The ASA emulsion particles are retained on the cellulose fibers at the wet end. The anionic carboxyl groups present in pulp fibers are the predominant retention sites of ASA particles stabilized with cationic compounds [482].

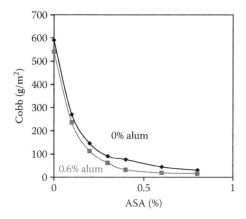

FIGURE 7.14 The effect of alum (0.6%) on the ASA sizing capability. (From Cousart, F.H., *TAPPI Proceedings, Papermakers Conference*, Atlanta, GA, 1989, pp. 99–102. With permission.)

In order to estimate the amount of ASA retained on fibers, the measurement of the ASA concentration in paper was performed by pyrolysis-GC [368,483]. The ASA retention on cellulose fibers was lower than 65%. The un-retained ASA increases the amount of deposits under forming fabrics and in the hoods of the dryer section [369].

The maximum retention of ASA emulsion (made with cationic starch) was only of about 20% (Figure 7.15). The addition of cationic PAE and alum contribute to a better ASA retention [483], but the maximum efficiency is reached at low amounts of ASA added to pulp. At higher ASA concentrations, the "excess" of ASA is lost in white water.

A reaction with cellulose is the generally accepted sizing mechanism [470] ("once retained in the sheet, ASA must react with the cellulose to give sizing" [346]). Arguably, the reaction rate depends on temperature, time, and sheet moisture. But the ASA sizing mechanism is much more complicated and still under study [430].

The sizing mechanism was studied by attempts at identifying the ester bond, using a reaction with model compounds, extracting for a balance of the reacted/un-reacted ASA in paper, and evaluating the ASA migration and sizing reverse process.

The evidence for chemical bond formation is the presence of the covalent bond (ester) between ASA and cellulose [470,484]. The presence of such strong bonds (ester) is also evaluated based on its stability. Paper sized with ASA loses its sizing over time [369]. A roll of paper treated with 2 lb

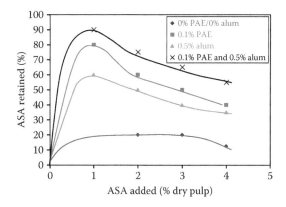

FIGURE 7.15 ASA retention performed with different retention aids.

ASA/ton paper was evaluated 15 months after the paper was manufactured: inside the roll, the degree of sizing was much higher than on the outside of the roll.

An ester bond was identified in the ASA reacted with a cotton linters pulp [484] after the reaction was performed in *N,N*-dimethyl formamide as solvent and triethyl amine as a catalyst. In the presence of water and nonreactive cationic retention aids, both ester bonds and carboxyl (from hydrolyzed product) are present in the IR spectra of the sized material with a high concentration of ASA (1.5%). However, after extraction with nonreactive solvent (toluene), the ester peak in the IR spectra decreases, which makes the presence of a strong, covalent ester bond doubtful.

The handsheets sized with ASA, after extraction with chloroform, still contain a certain amount of ASA and show a certain sizing degree. However, after the extraction of the defibrated handsheet with Tween 80 at 70°C, virtually no ASA or hydrolyzed ASA remained on the cellulose fibers [485]. The results of pyrolysis–GC analyses indicated that nearly no cellulose–ASA ester bonds were present in ASA-sized paper [486].

In order to study the possible reaction with ASA, low molecular weight cellulose was used as a model compound [487]. ASA does not react with cellulose even in the absence of water (dioxane solution). In the presence of water, the ASA hydrolysis proceeded predominantly during the drying process rather than the ester formation. Most of the hydrolyzed ASA was removed by extraction with acetone.

Due to the abundance of hot water and steam in the drying section of the papermaking process, and to the limited surface area of the fiber, the ASA is more likely to hydrolyze. Then, the hydrolyzed product is partially lost through vaporization [484].

If the real covalent bond between ASA and cellulose is unlikely, another mechanism must be involved in the sizing process. Chemical compounds (such ASA and its derivatives) able to provide sizing properties should be identified. As in the case of AKD, circulation and orientation of the molecule must be considered along with another type of weak bonds between the sizing agent and the cellulose fiber.

Blends of ASA and controlled amounts of hydrolyzed ASA show a decrease in sizing with the increasing amount of hydrolyzed compound [488]. Still, the aluminum salts of hydrolyzed ASA perform better than rosin as sizing agent [406,489]. Calcium salt may also contribute to the sizing [490]. Therefore, the saponified material has a potential as internal sizing agent.

The circulation of ASA within the paper thickness would be a proof of the mobility of the free molecule. At high temperatures, the ASA sprayed on one side of the paper migrates to the unsprayed side [491]. The presence of other hydrophobic compounds (such as lanolin [393,394]) may help that process.

ASA spreads easier before its hydrolysis. The hydrolyzed ASA (the acid form) has a melting point of 40°C, while the liquid ASA has a melting point of −20°C [485]. In order to spread on the cellulose fiber, the acid form needs higher temperatures in the drying step [484].

In the absence of ester bond formation between ASA and cellulose, the hydrolyzed ASA should be converted to a more "flexible" structure able to migrate on cellulose fibers. PAE and alum play an important role in developing different types of interaction between ASA (and its derivatives) and cellulose fibers. The aluminum salt of hydrolyzed ASA may interact with anionic charges on cellulose fibers, azetidinium cycle (from PAE resin) may react with the carboxylic group of the hydrolyzed anhydride [492], and/or strong hydrogen bonding between the hydrolyzed ASA and the cellulose hydroxyl groups are developed [487].

The absence of ester bonds opens the question whether a reaction between a sizing agent and cellulose is necessary for the sizing development. A long list of other chemical compounds used as internal sizing agents for paper confirms the existence of many possible mechanisms for the sizing process (see Section 7.5). Accordingly, a nonreactive compound (reaction product of ASA with hexadecylamine) shows a synergetic sizing effect in blends with ASA [393,394].

Not even the de-sizing mechanism involves a broken ester bond. Auto-oxidation is the predominant mechanism of size reversion for ASA-sized paper [493]. Hydroperoxide is formed in the allyl position and a new hydrophilic group is generated. This hydroperoxide can decompose in alcohol and/or ketone. A reduction in the molecular weight may also take place.

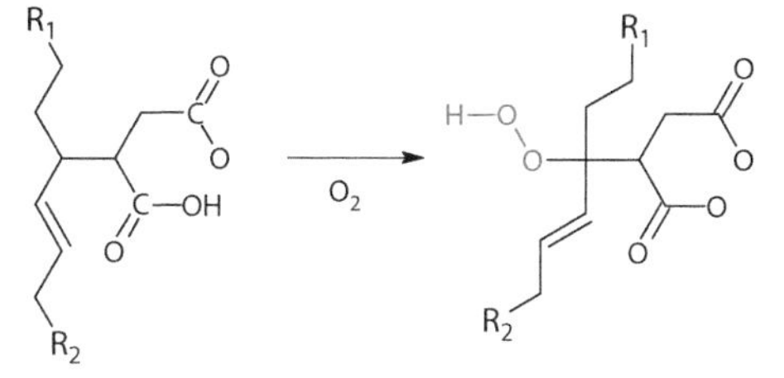

7.5 OTHER CHEMICAL COMPOUNDS ABLE TO FIT THE GENERAL CONCEPT FOR AN INTERNAL SIZING AGENT

From the previous chapters, we learned that rosin size is effective mainly at low pH. AKD and ASA perform at higher pH but their emulsions show a limited stability. A more versatile internal size with a lower reactivity in the presence of water is still needed [494].

Internal size compounds are small molecules or polymers, but always hydrophobic substances with a cationic charge for retention (or any other option for their interaction with a retention aid). This chapter is a collection of other chemical structures tested as internal sizing agents for paper.

From the perspective of the sizing mechanism, it is interesting to divide them in "potentially reactive" and "nonreactive" compounds. The reactive internal sizing agents underline the general characteristic of the papermaking process: the reactive group faces the presence of water and the possibility to be involved in side reactions. The nonreactive compounds bring new evidence for a sizing mechanism based on the physical interaction with cellulose fibers, migration, and reorientation.

7.5.1 OTHER "POTENTIALLY REACTIVE" COMPOUNDS AS INTERNAL SIZING ADDITIVES

The reactivity of an internal sizing agent refers to potential reactions with cellulose, with itself or to a preliminary reaction performed before the final product is applied to the papermaking machine as internal size [495,496]. The sizing compound is nonionic or cationic. Cationic compounds are self retained on fibers but nonionic compounds must be dispersed in water with a cationic stabilizer.

AKD as a "potentially reactive" compound can be blended with nonreactive materials, or used to prepare a hydrophobic material through a reaction performed separately. Liquid AKD is dispersed (2:1) along with polymer latexes stabilized by starch [497] to make a hybrid sizing agent.

In order to add hydrophobic tails to a water-soluble polymer, AKD reacts with polydiallylammonium chloride (with up to 50% mol of secondary amine) [498,499]. AKD is reacted with polyamine in two steps of polymer analogous reactions; the rest of secondary amines are reacted with an excess of epichlorohydrin. The new sizing agent has a reactive azetidinium cycle (in red):

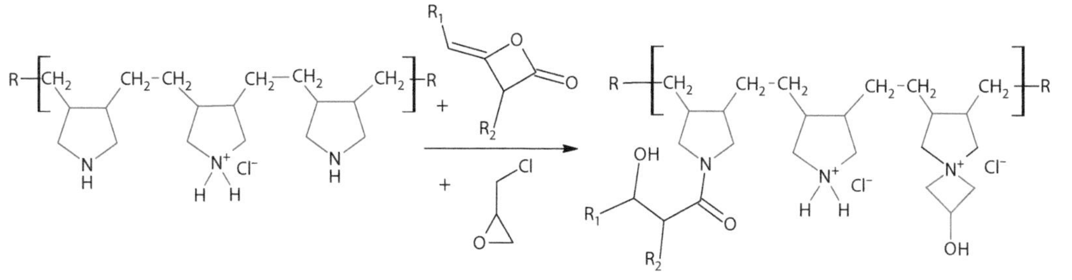

Macromolecular compounds with reactive groups are well known as strengthening agents (see Chapters 4 through 6). A hydrophobic tail may increase the water repellency but the amount of internal sizing compound is limited to the level at which the change in paper (dry and/or wet) strength is acceptable [500]. Copolymers of styrene and acrylic esters having cationic charges and epoxy reactive groups [501,502] are able to balance the effect of the sizing and the repulpability of paper.

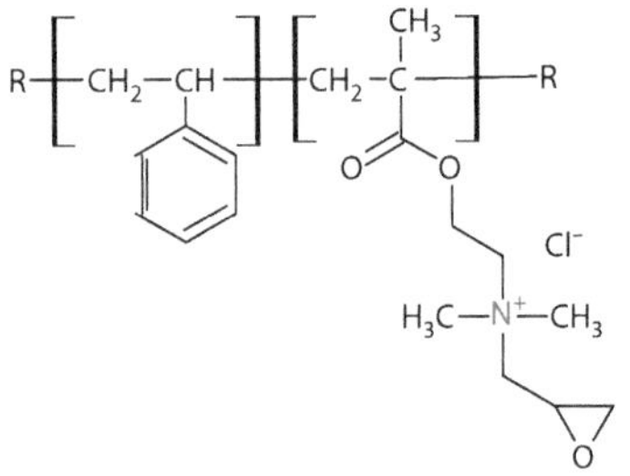

An epoxy group is also attached to maleic anhydride copolymers with olefins [503]. The anhydride functionality can be converted, through polymer analogous reactions, to a hydrophobic half ester, to epoxy ester, and cationic ammonium salt.

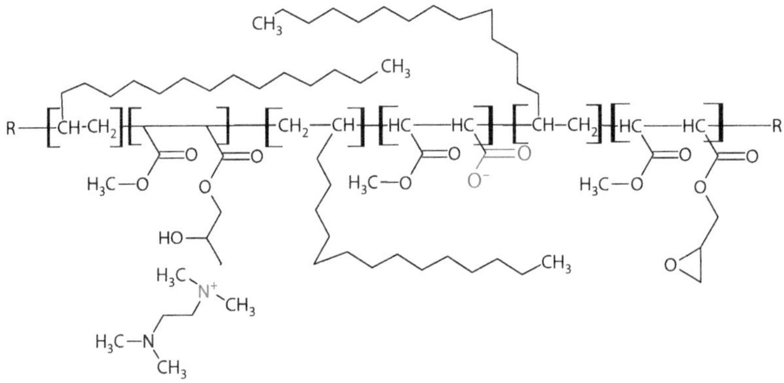

Olefin is present in 1:1 ratio to maleic anhydride (alternating copolymer) and the final compound has an amphoteric character.

The epoxy ring is supposed to be reactive with cellulose hydroxyl, and the N,N-bis(2,3-epoxypropyl) alkyl anilines are very effective sizing agents [504,505]. An aliphatic amine is converted to a cationic hydrophobic compound with an epoxy group [506,507].

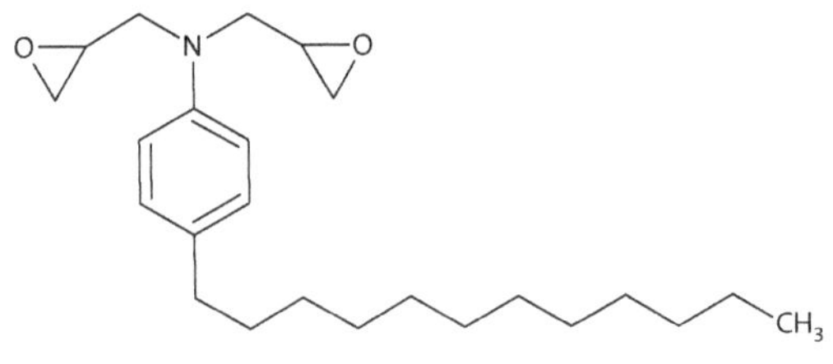

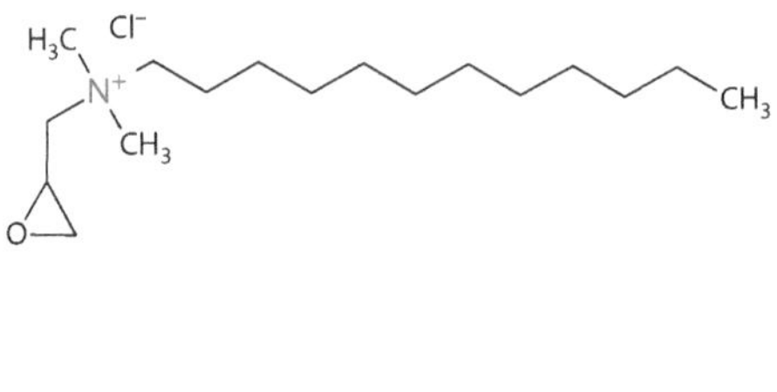

The reaction between octadecyl isocyanate and ethylene imine [508] is performed in the presence of a nonionic emulsifier. The stable dispersion of N-octadecyl-N'-N'-ethylene-urea [509] is used as internal sizing agent after retention with the PEI resin.

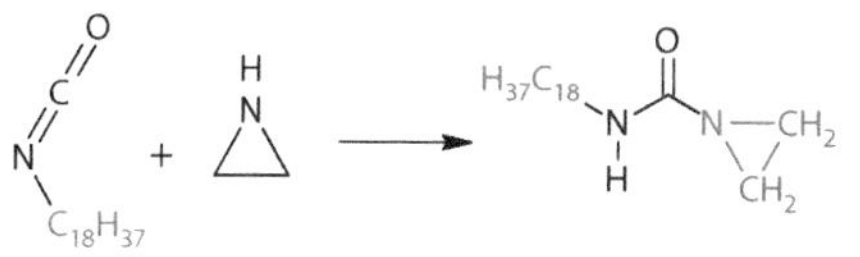

The same concept and reactions were extended to di-isocyanates [510]. N-substituted N',N'-alkylene ureas are recommended not only as sizing materials but also as cross-linkers [511], which shows their reactivity with cellulose.

The reactive imine ring (in red) makes the N-substituted aziridine a powerful sizing agent when connected to a hydrophobic structure [512]. The synthesis is a reaction of the Michael addition type. The new compound is dispersed in water in the presence of cationic starch.

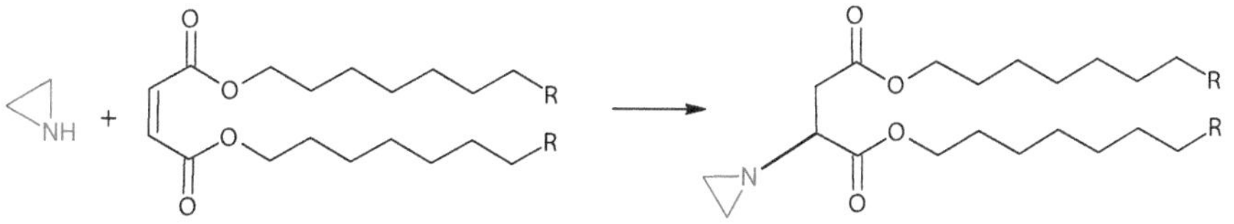

Anhydride groups are well known as reactive functionalities with cellulose (see ASA). Mono-alkyl or mono-alkenyl esters (or amides [513]) of trimellitic or pyromellitic anhydrides (obtained through a *trans*-acidolysis) are emulsified (cationic starch as stabilizer) and used as internal sizing agent in the presence of alum [179,514].

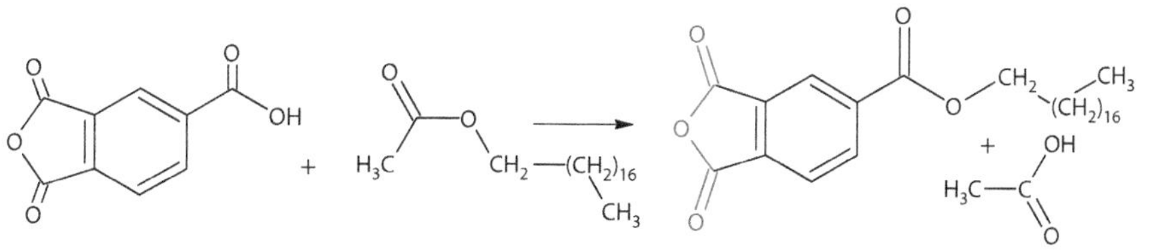

N-searoyl succinimide [515] is another reactive sizing agent: the imid cycle is reactive if an electron-withdrawing group is bonded to the imid nitrogen.

A combination of AKD and ASA structures can be obtained within a Friedel-Crafts reaction between an aromatic hydrocarbon (such as benzene, toluene or any alkyl benzene) and an alkyl succinic anhydride [516]. From an alkyl succinic anhydride (ASA, for instance), a lactone (such as AKD) is obtained.

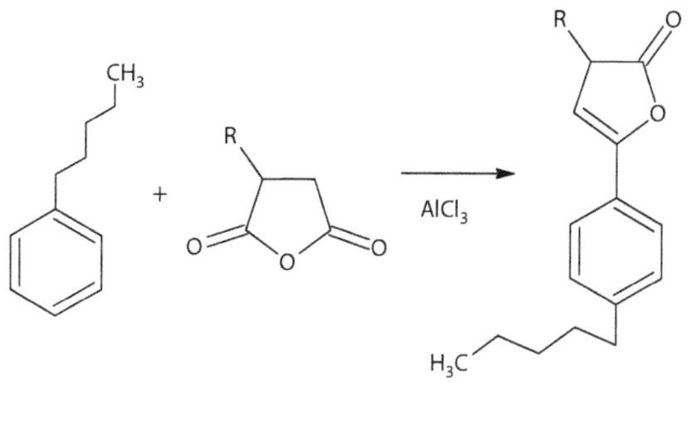

AKD has the disadvantage of a prolonged curing time. For a faster reaction with cellulose, highly reactive functionalities were investigated. Alkyl isocyanates (C_{18}) are very good sizing agents [4,200,517]. Stearyl [281] or octadecyl isocyanate [518] are emulsified with cationic starch and a cationic polymer (such as PAE resin) [385].

In stable dispersions in water, in the presence of cationic starch and sodium lignosulfonate [519], some isocyanate groups are lost due to the reaction with water. Molecules with just one isocyante functionality lose their reactivity. A multi-functional isocyanate will keep its reactivity even if some reactive groups are reacted with water. The linoleic acid dimerization results in hydrophobic diacid [520]. The carboxylic functionalities are converted to isocyanate groups and the di-isocyante is dispersed in water in the presence of a nonionic emulsifier (t-octylphenol polyethoxylated, $n = 7$).

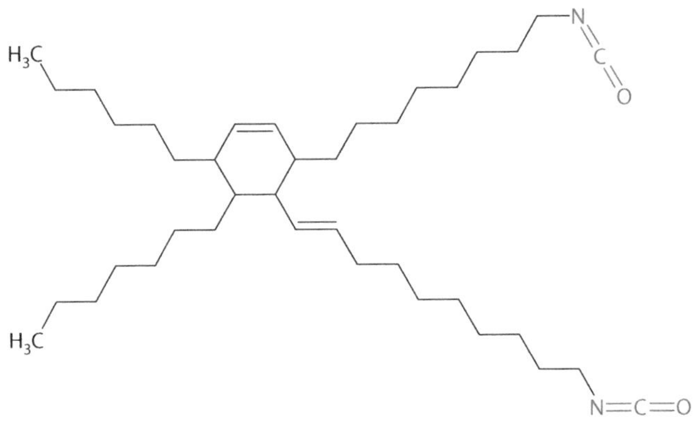

To make the organic phase self-emulsified, the isocyanate groups are partially reacted with polypropylene glycol mono-ether [521]. The hydrophobic compound is retained on the cellulose fibers by using cationic gum.

The organic chemistry of reactive small molecule has been brilliantly explored [503,504,522–525]. 2,4,6-Trichloro-sym-triazine (cyanuryl chloride) reacts with epoxides in the presence of aluminium chloride [522].

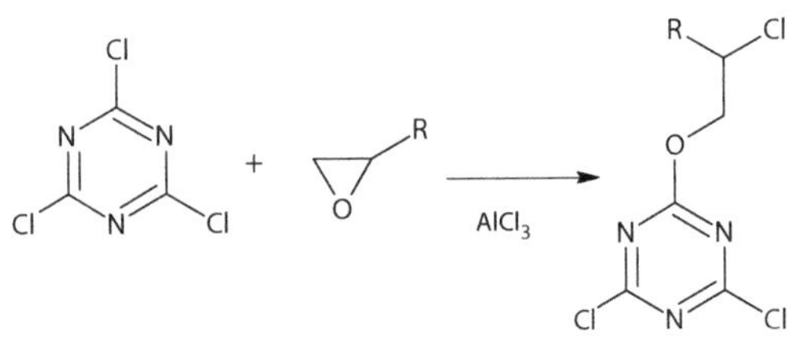

R stands for an alkyl with 12–28 carbon atoms. Epichlorohydrin can also be used as an epoxide. The chlorine atoms are important because they can form a covalent bond with cellulose. The triazine molecule is symmetrical: all chlorine atoms have the same reactivity. Di- and even trisubstituted will coexist with mono-substituted and with un-reacted cyanuryl chloride. Amines or polyamines are catalysts for that reaction.

Hydrophobic polyamines with N-3-chloro-2-hydroxypropyl pendant groups are quenched with a mineral acid in order to prevent the cross-linking process. Due to their cationic character, those polyamine are adsorbed on paper. On paper and neutral pH, the hydrophobic polyamines are self-cross-linked [525].

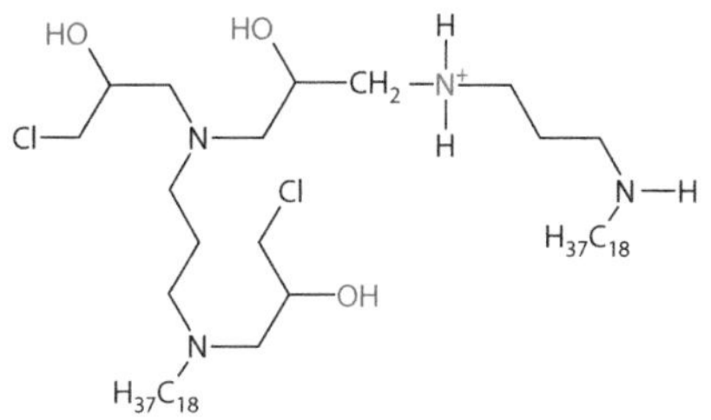

There are many versions of that reaction for changing the ratio between the hydrophobic tails and the amine groups, between amine groups and the N-3-chloro-2-hydroxypropyl pendant groups. For instance, epichlorohydrin reacts with ethylene di-amine (4:1) and the resulting adduct is reacted with one mol of octadecyl-1,3-propanediamine. The final chemical structure depends on the mole ratio, on conversion at the time the mineral acid was added, and on the final treatment with caustic to reclose the epoxy ring [524].

Alkyl chloroethyl sulfides or alkyl sulfones are effective sizing agents in alkaline pH [523].

Alkyl and di-alkyl carbamoyl chlorides are obtained through a reaction between alkyl amines and phosgene. Carbamoyl chlorides are as reactive as isocyanates with similar chemical structures. They are reactive to cellulose, and show a certain cationic activity, which is why they can be retained on cellulose fibers (at pH = 8). The alkyl groups provide the hydrophobic character that makes them effective internal sizing agents [526].

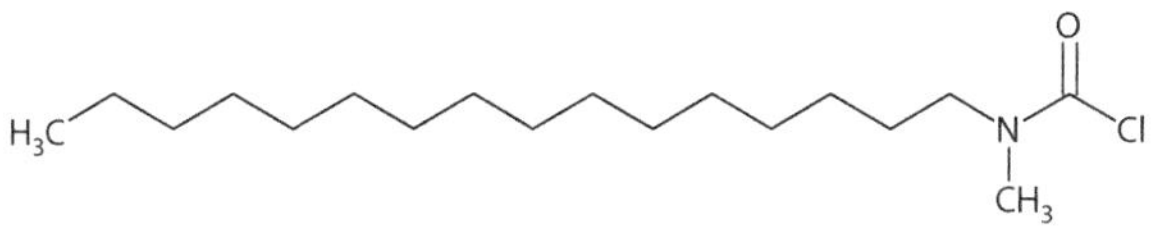

The reactivity of the hydrophobic compound should be taken into account during its emulsification. If the reactivity is too high, a solution in organic solvent (toluene) has been used [527].

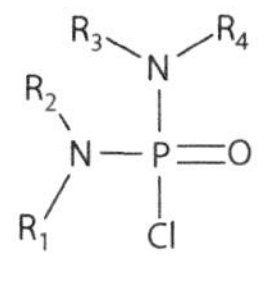

The distearyl acetal of stearyl aldehyde is much more stable than ASA or AKD and provides sizing when used as an emulsion (with cationic starch as stabilizer) at the wet end [528].

7.5.2 OTHER NONREACTIVE COMPOUNDS AS INTERNAL SIZING AGENTS

The reactive internal sizing agents have the following drawbacks: they are not water-soluble; stabilizers are required to disperse them in water. Most of them are not retained on the cellulose fibers because they are neutral, sensitive to water, and rapidly hydrolyzed to unusable chemicals. There is a continuous effort to overcome these drawbacks.

Although the internal sizing mechanism is still under study (reaction with cellulose, spreading on fiber, molecule reorientation, etc.), several experimental data are adding a new possible direction: the dispersion of nonreactive hydrophobic compounds. They cannot develop a chemical reaction with cellulose; some have very high molecular weights and are not distributed as molecules but as particles. The polymers have a very low solubility in water (if any) and their migration and "reorientation" would be hard to explain as a possible sizing mechanism. The chemical structures presented below are all proofs that a covalent bond is not necessary to develop size. However, for an imidized SMA both HST and Cobb values are improved only at a very high concentration (3%) of cationic polymer [529].

Nonreactive sizing agents are cationic or nonionic compounds with low, medium, and high molecular weights. Dispersions are obtained through emulsification of the preexisting hydrophobic material or by emulsion polymerization when the hydrophobic material is built up within the dispersed particle. If the sizing agent is not self-emulsified, the emulsification is performed in the presence of an emulsifier.

The cracked petroleum distillate (fraction C_5), butadiene–styrene mixture or turpentine is polymerized, and the polymer is modified with maleic anhydride. The fortified polymer is self-emulsified in alkaline pH before [530] or after blending with microcrystalline wax [531]. These anionic polymers, anionic latexes, or unmodified lignin [532] are retained with alum [533].

Cationic latexes are easily retained on fibers, which is why they are of special interest. Cationic polymers are synthesized by copolymerization or through a polymer analogous reaction (the modification of a preexisting nonionic polymer). Macromolecular compounds are copolymers with low T_g, such as vinylidene chloride-2-methyl-5-vinyl pyridine [534], styrene-butyl acrylate-cationic comonomer [535,536], polyvinyl acetate emulsion obtained in the presence of a cationic emulsifier [537], or of the copolymer of acrylonitrile butadiene. These latexes also increase the paper dry [538] and wet strength [533,539,540]. That is why the amounts of sizing agent are limited by their side effects.

Internal sizing agents are obtained from preexisting polymers modified by adding hydrophobic tails and/or cationic charges. A polymer analogous reaction can be performed even at the wet end to make a hydrophobic PEI through a Michael addition of octadecyl propiolate or acrylate [541].

Maleic anhydride copolymers with C_{16}–C_{18} olefins [542] are reactive hydrophobic compounds. In order to attach cationic charges, the anhydride groups undergo successive reactions; the first reaction is with methanol and forms half an ester. The carboxylic functionality from the half ester reacts with Epi in the presence of sodium methoxide. Eventually, the epoxide ring is reacted with polyamine and the final product is treated with hydrochloric acid to develop cationic charges. Cationic charges can also be added to the maleic unit in SMA copolymer by imidization [529].

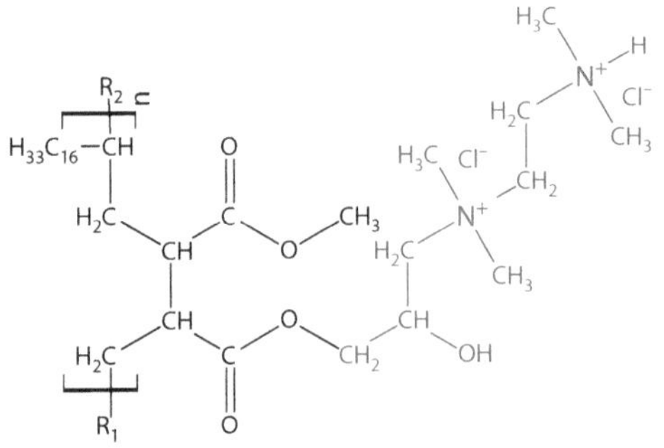

A cationic internal sizing agent is a pre-polymer obtained from a mixture of tallow N-alkenyl and N-alkyl-1,3-propanediamines (the alkenyl and alkyl radicals being C_{16}–C_{18}) with an excess of epichlorohydrin in methanol [543]. The hydrophobic tail is connected to a potential cationic center.

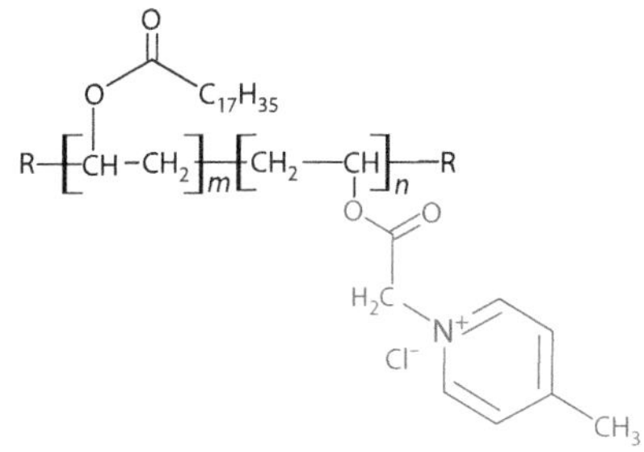

Copolymers of vinyl stearate and vinyl chloroacetate, or the polymer of *p*-chloromethyl styrene are reacting with tertiary amines (such as pyridine and picoline) [544–546]. The copolymerization reaction may set the m/n ratio and thus the balance between the hydrophobic (in blue) and the hydrophilic part. The new copolymer is a polyelectrolyte with a significant hydrophobic part. It is not soluble in water but it can be dissolved in an aqueous solution of *N*-octadecyl pyridinium chloride [547].

Lower molecular weight polymers are an intermediate category between high molecular weight compounds and small molecules. The hydrophobic material can be from paraffin, low molecular weight hydrocarbon resins, or fractions obtained by selective extraction of petroleum oils. Hydrocarbon resins with a molecular weight lower than 1000 and a melting point below 100°C are dispersed in water in the presence of a PAE-type resin a passed through a homogenizer. Small particles are retained on the cellulose fiber at the wet end and provide a good sizing for paper [548].

Wax emulsions of paraffin [549,550] are stabilized with methyl cellulose ethers, or gum arabic [551]. The solids content is about 15%–45% of which about 90% is the wax. Paraffin wax is also emulsified with casein and ammonium oleate [552–555], with gum and sodium rosinate [556], or ammonium oleate [550]).

Extract of aromatic and naphthenic compounds [123] are emulsified with salts of naphthenic acids in the presence of polyphosphoric acid compounds [557]. The anionic dispersions are retained with alum [550,558].

Small molecules with hydrophobic parts and ionic charges are either water-soluble or readily dispersible in water. There are nonreactive anionic or cationic compounds.

Anionic sulfonylcarbonylimide needs a retention aid such as polyalkylene polyamine [559]. The anionic group may result from the reaction of a distearylamine with phthalic anhydride [560] or as alkyl urea derivatives [20]. *N*-Methylstearoylhydroxamic acid and stearoylhydroxamic acid are recommended as sizing agents [18]. Anionic compounds ask for a cationic retention aid (such as alum or PEI).

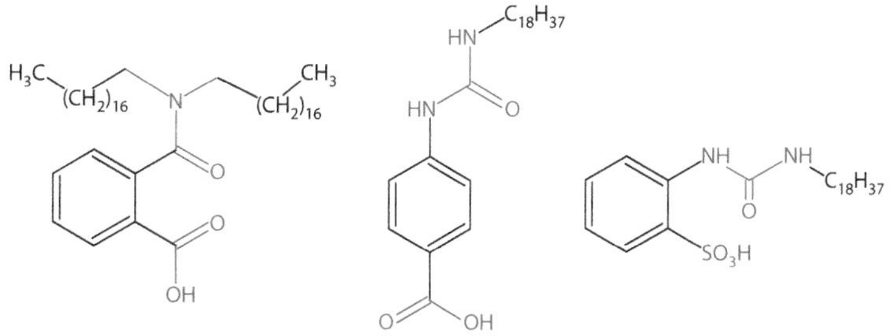

The ASA esterification with allylic alcohol results in half ester and the carboxylic group is neutralized with triethanolamine to make an internal sizing [561].

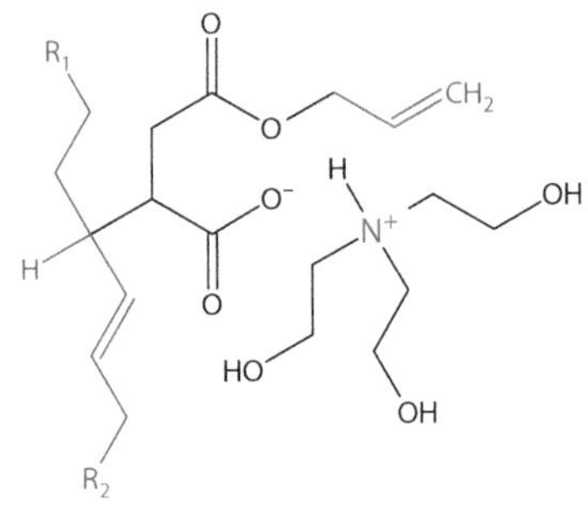

Cationic nonreactive sizing agents are self-retained on the anionic cellulose fibers. The organic chemistry can be manipulated to synthesize chemical compounds with long hydrophobic tails, polar groups, and potential cationic charges [20,559,562]. Fatty acids (such as stearic acid) react with DETA within a condensation reaction in which the primary amines are converted to amide groups and the secondary amines are subsequently converted to cationic charges [563]. The cationic compound is dispersed in water and used as an internal sizing agent [564].

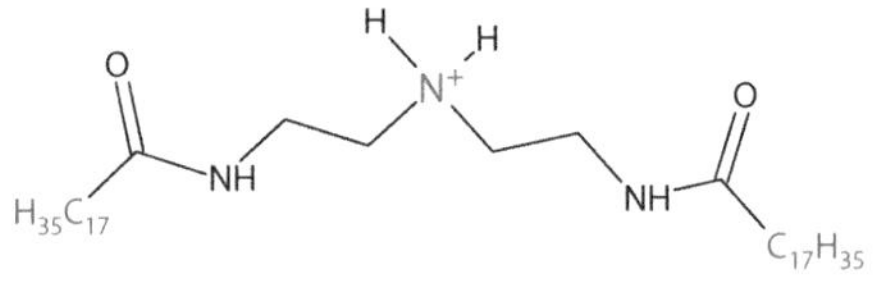

The esterification of triethanolamine with fatty acids (stearic and/or palmitic) results in a hydrophobic compound with a tertiary amine and therefore is potentially cationic [562]. The hydrocarbon tail can also have one or more double bonds [506,507].

Alkyl amines react with maleic anhydride (at temperatures below 100°C) and the resulting amide is reacted with another amine (or di-amine) at much higher temperatures (150°C–200°C) [494]. The resulting aspartiamide is emulsified in the presence of cationic emulsifier (hexadecyltrimethyl ammonium chloride and cationic starch and/or PAE resin).

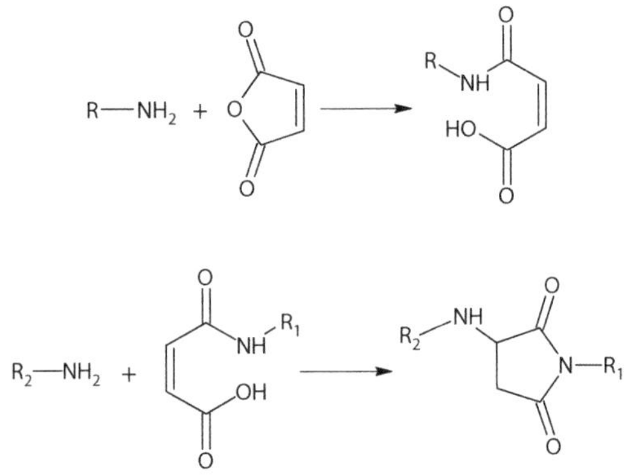

The hydrophobic tails of stearic acid are added by condensation to a polyalkylene polyamine (TETA, PETA), which are the potential cationic part after protonation with acetic acid [263].

The product of the Michael addition of fatty amines to hexahydro-1,3,5-triacrylyl-s-triazine (protonated with acetic acid) [565] is used as an internal sizing agent. The molecular weight depends on the ratio between the amine and carbon–carbon double bond.

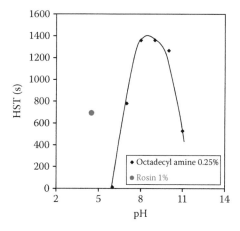

Fatty amines, such as octadecyl amine [566], are very attractive compounds as potential sizing agents: they are hydrophobic, nonreactive, and show a maximum effectiveness in sizing at pH = 7–10 (Figure 7.16).

The primary amine can be converted to secondary, tertiary, or quaternary ammonium salts. The fatty amine reacts with N,N'-methylene-bis-acrylamide through Michael addition to obtain a secondary amine [567]. That reaction is performed between two compounds with multiple functions (a primary amine has two hydrogen atoms able to react with the two equally reactive double bonds). The final product is a waxy material that can be self-dispersed in a solution of acetic acid.

The fatty primary amine (such as octadecylamine) reacts with epichlorohydrin [568,569], or with the di-glycidyl ether of 2,2-bis-(4′-hydroxyphenyl)-propane [570] to obtain secondary or tertiary amine. These products show an enhanced activity when they are in the presence of polyamine [570,571].

Fluorinated chemicals are nonreactive, water repellent, and also oil repellent [572]. Anionic fluorinated compounds are used as a solution in ethanol at the wet end [200] or as dispersions. The fluorine-containing organic compounds (difluoroalkyl phosphate diethanolamine salt [573–576], polyurethanes including a fluorinated segment [577], or perfluoropolyethers [582]) are anionic and

FIGURE 7.16 The pH effect on octadecyl amine sizing effectiveness.

they are retained almost 100% on paper with a cationic retention aid. As far as the sizing capabilities are concerned, the PAE resins performed better than poly-DADMAC [573] as retention aids. The addition of alum has a negative effect on the sizing behavior [575]. Cationic copolymers of fluoro-alkyl monomers (solution or latexes [578–580]) are recommended as internal sizing agents because they are self-retained [581].

REFERENCES

1. J.J. Krueger and K.T. Hodgson, *Tappi J.*, **77(7)**, 83–87 (1994).
2. R.W. Davison, *TAPPI Proceedings, Papermakers Conference*, Atlanta, GA, 1986, pp. 17–27.
3. J.C. Roberts, *Proceedings of SPIE—The International Society for Optical Engineering*, Bellingham, WA, 1997, 3227 (Interactive Paper), pp. 20–39.
4. J.C. Roberts, in *Fundamentals of Papermaking Materials*, C.F. Baker (Ed.), *Transaction of the Fundamental Research Symposium, 11th*, Cambridge, U.K., September 1997, Vol. 1, pp. 209–263.
5. M.A. Hubbe, *BioResources.*, **2(1)**, 106–145 (2006).
6. P. Wilson, Surface sizing, in *The Sizing of Paper*, J.M. Gess and J.M. Rodriquez (Eds.), Tappi Press, Atlanta, GA, 2005, pp. 211–235.
7. J.W. Swanson, Mechanism of paper wetting, in *The Sizing of Paper*, 2nd edn., W.F. Reynolds (Ed.), Tappi Press, Atlanta, GA, 1989, pp. 133–154.
8. B.W. Crouse and C.L. Warner, *TAPPI Papermaking Conference*, Atlanta, GA, 1986, pp. 7–14.
9. R.W. Hoyland, *Paper-Making Trans. Symp.* 1977, **2**, 557–579 (1978).
10. M.A. Hubbe, Wetting and penetration of liquids into paper, in *Encyclopedia of Materials Technologies*, Elsevier, Oxford, U.K., 2000, pp. 6735–6739.
11. J.A. Van Den Akker and W.A. Wink, *Tappi.*, **52(12)**, 2406–2420 (1969).
12. M. Chen and C.J. Biermann, *Tappi J.*, **78(8)**, 120–126 (1995).
13. E.T. Reaville and W.R. Hine, *Tappi.*, **50(6)**, 262–270 (1967).
14. M.F. Finlayson, K.T. Hodgson, J.L. Cooper, J.J. Gathers, and K.E. Spring, *TAPPI Papermaking Conference*, Atlanta, GA, 1996, pp. 309–314.
15. L. Westfelt, *Revue A.T.I.P.*, **28(4)**, 2007–2011 (1974).
16. P. Delcroix, US 2,061,935 (1936).
17. R.G. Capell and P.R. Templin, US 2,898,293 (1959).
18. G.T. Tiedeman and J.M. Gess, US 4,857,149 (1989).
19. M.P. O'Toole and O.S. dePierne, US 6,114,417 (2000).
20. M. Bernheim, D. Strasilla, B. De Sousa, and P. Rohringer, US 4,956,049 (1990).
21. N. Gurnagul, M.D. Ouchi, N. Dunlop-Jones, D.G. Sparkes, and J.T. Wearing, *J. Appl. Polym. Sci.*, **46**, 805–814 (1992).
22. R.W. Hoyland, M.P. Neill, and T. Keenan, *Paper Technol.*, **41(10)**, 50–52 (2000).
23. D. Johnson, Sizing in acid, neutral and alkaline conditions, in, *Applications of Wet-End Paper Chemistry* 2nd Edn., I. Thorn and C.O. Au (Eds.), Springer, Berlin, Germany, 2009, pp. 73–112.
24. J.C. Roberts, *Paper Technology.*, **34(3)**, 27 (1993).
25. N. Gurnagul, R.C. Howard, X. Zou, T. Uesaka, and D.H. Page, *J. Pulp Paper Sci.*, **19(4)**, J160–J166 (1993).
26. D. Eklund and T. Lindstrom, *Paper Chemistry*, DT Paper Science, Grankulla, Finland, 1991, pp. 135–144.
27. W.F. Reynolds and W.F. Linke, *Tappi.*, **46(7)**, 410–415 (1963).
28. K.S. Siefert, Aluminum compounds, in *Encyclopedia of Chemical Technology*, K. Othmer (Ed.), vol. 2, 1992, pp. 252–345.
29. W.E. Scott, *Principles of Wet End Chemistry*, Tappi Press, Atlanta, GA, 1996, pp. 91–98.
30. T.R. Arnson, *Tappi*, **65(3)**, 125–130 (1982).
31. J. Zhuang and C.J. Biermann, *Tappi J.*, **76(12)**, 141–147 (1993).
32. R.D. Crow and R.A. Stratton, *Tappi Proceedings*, Atlanta, GA, 1985, pp. 183.
33. E. Strazdins, *Nord. Pulp Paper Res. J.*, **4(2)**, 128–134 (1989).
34. J. Burgess, *Ions in Solution: Basic Principles of Chemical Interaction*, Halsted Press, John Wiley & Sons, New York, 1988.
35. M.J. Jaycock and D.K. Swales, The chemistry of aluminum, in *The Sizing of Paper*, J.M. Gess and J.M. Rodriquez (Eds.), Tappi Press, Atlanta, GA, 2005, pp. 27–55.
36. E. Strazdins, *Tappi J.*, **67(4)**, 110–113 (1984).

37. E. Matijevic, G.E. Janauer, and M. Kerker, *J. Colloid Sci.*, **19**, 333–346 (1964).
38. J. Marton and F.L. Kurrle, *Papermakers Conference Proceedings*, Atlanta, GA, 1985, pp. 197–207.
39. F.S. Potter, *TAPPI Papermaking Conference*, Atlanta, GA, 1996, pp. 315–324.
40. P. Lagally and H. Lagally, US 3,074,843 (1963).
41. J. Chen, M.A. Hubbe, J.A. Heitmann, D.S. Argyropoulos, and O.J. Rojas, *Colloids Surf. A.*, **246(1–3)**, 71–79 (2004).
42. G.M. Lindahl, US 4,536,384 (1985).
43. D. Haase, N. Spiratos, and C. Jolicoeur, US 5,149,400 (1992).
44. A.B. Gancy and C.A. Wamser, US 4,284,611 (1981).
45. R.W. Kumler and H. Sinclair, US 2,343,065 (1944).
46. B.E. Doiron, Retention aid systems in *Retention of Fines and Fillers during Papermaking*, J.M. Gess (Ed.), TAPPI Press, Atlanta, GA, 1998, pp. 159–176.
47. H.R. Rafton, US 1,904,251 (1933).
48. E. Strazdins, *Tappi J.*, **69(4)**, 111–114 (1986).
49. E. Strazdins, *Tappi Papermakers Conference*, Atlanta, GA, 1974, pp. 59–65.
50. M. Kato, A. Isogai, and F. Onabe, *J. Wood Sci.*, **46**, 310 (2000).
51. M. Kato, A. Isogai, and F. Onabe, *J. Wood Sci.*, **46**, 75–78 (2000).
52. M. Kato, A. Isogai, and F. Onabe, *J. Wood Sci.*, **45**, 154–160 (1999).
53. M. Kato, A. Isogai, and F. Onabe, *J. Wood Sci.*, **44**, 361–368 (1998).
54. T.R. Arnson and R.A. Stratton, *Tappi J.*, **66(12)**, 72–75 (1983).
55. C.L. Brungardt, R.J. Richle, and J.J. Zhang, US 6,048,392 (2000).
56. D. Lurie, US 2,711,370 (1955).
57. E. Stazdins, Paper sizes and sizing, in *Naval Stores*, D.F. Zinkel and J. Russell (Eds.), Pulp Chemical Association, New York, 1989, pp. 575–624.
58. E.M. Choy, US 4,517,052 (1985).
59. A.L. Rummelsburg, US 2,477,655 (1949).
60. E.J. Soltes, Chemistry of rosin, in *Naval Stores*, D.F. Zinkel and J. Russell (Eds.), Pulp Chemical Association, New York, 1989, pp. 261–345.
61. G.I. Keim, US 2,776,277 (1957).
62. P.H. Aldrich, US 4,263,182 (1981).
63. C.T. Leffler, US 4,219,382 (1980).
64. P.H. Aldrich, US 3,817,768 (1974).
65. R.W. Emerson, US 4,437,894 (1984).
66. W.D. McDavid, US 3,390,046 (1968).
67. R. Hastings, E.K. Drechsel, and E. Strazdins, US 2,771,464 (1956).
68. R.W. Emerson, US 4,483,744 (1984).
69. E. Strazdins, US 2,846,328 (1958).
70. E.T. Reaville, J.P. O'Brien, and L.P. Russe, US 2,994,635 (1961).
71. W.R. Bussell and N.S. Morgan, US 5,192,363 (1993).
72. R.W. Emerson, US 4,606,791 (1986).
73. J.N. Borglin, US 2,326,610 (1943).
74. A.L. Rummelsburg, US 2,773,859 (1956).
75. E. Strazdins, US 2,934,468 (1960).
76. S.H. Watkins, US 2,985,537 (1961).
77. A.L. Osterhof, US 2,084,213 (1937).
78. E. Strazdins, US 3,132,127 (1964).
79. S. Abell and T.L. Peltier, US 4,605,445 (1986).
80. A.L. Rummelsburg, US 2,720,514 (1955).
81. O.S. Eckhardt and I.E. Knapp, US 2,818,412 (1957).
82. F.W. Boughton, US 2,548,513 (1951).
83. A. Bender and A. Giencke, US 5,708,078 (1998).
84. K. Nakamura, K. Nagao, and M. Monobe, US 5,393,337 (1995).
85. Y. Sasaki, N. Tani, and D. Arai, US 5,817,214 (1998).
86. K. Ito, A. Isogai, and F. Onabe, *J. Wood Sci.*, **45**, 46–52 (1999).
87. K. Ito, A. Isogai, and F. Onabe, *J. Pulp Paper Sci.*, **25(6)**, 222–226 (1999).
88. G.D. Puckett and B. Woodworth, US 2007/0079730.
89. G.H. Foster, US 2,265,941 (1941).

90. A.C. Dreshfield and H.A. Johnstone, US 2,134,912 (1938).

91. H.F. Chappell, US 694,728 (1902).

92. W.J. Menzies, US 405,269 (1889).

93. E. Dowthwaite and I.R. Hiskens, US 3,906,142 (1975).

94. E. Dowthwaite and I.R. Hiskens, US 4,323,425 (1982).

95. J.M. Gess, Rosin, in *The Sizing of Paper*, J.M. Gess and J.M. Rodriquez (Eds.), Tappi Press, Atlanta, GA, 2005, pp. 57–73.

96. R.W. Emerson and J.R. Shattuck, US 4,025,354 (1977).

97. R.W. Emerson and W.L. Unterberger, *Tappi Proceedings*, *1989 Papermakers Conference*, pp. 97–98.

98. R.W. Emerson and J.R. Shattuck, US 4,022,634 (1977).

99. R.W. Emerson and J.L. Martin, US 4,141,750 (1979).

100. H.R. Rafton, US 2,056,209 (1936).

101. W.C. Hopkins and J.G. Senese, US 2,873,203 (1959).

102. M.J. D'Errico and R.J. Kulick, US 3,433,658 (1969).

103. E.D. Clemons and J.T. Daust, US 4,157,982 (1979).

104. J.A. De Cew, US 1,099,168 (1914).

105. K.L. Lynch, US 2,438,092 (1948).

106. R.J. Kulick and S.T. Moore, US 4,148,665 (1979).

107. H.R. Rafton, US 1,886,120 (1932).

108. S.R. Wagg, W.L. Wagg, and J.B. Wagg, US 1,032,973 (1912).

109. M.J. D'Errico and F.L. Wagner, US 3,433,659 (1969).

110. E. Strazdins, *TAPPI Papermakers Conference*, Atlanta, GA, (1977), pp. 149–155.

111. B. Wieger, US 1,882,680 (1932).

112. T. Okumichi and K. Kawatani, US 4,309,338 (1982).

113. T. Okumichi and K. Kawatani, US 4,267,099 (1981).

114. J.W. Gowan, US 4,461,646 (1984).

115. S. Ishibe, T. Okumichi, Y. Ishihara, and H. Naka, US 4,071,375 (1978).

116. R.W. Davison, US 3,565,755 (1971).

117. M. Nakajima, I. Sakai, and N. Tani, US 5,288,782 (1994).

118. T. Kitaoka, A. Isogai, and F. Onobe, *Nord. Pulp Paper Res. J.*,**12(1)**, 26–31 (1997).

119. K. Ohta, M. Yoshimura, and M. Takahashi, US 4,681,909 (1987).

120. E. Strazdins, *Tappi.*, **60(10)**, 102–105 (1977).

121. P.H. Aldrich, US 3,966,654 (1976).

122. O.F. Neitzke, US 2,198,289 (1940).

123. H.J. Tadema, US 2,602,739 (1952).

124. M. Arakawa, K. Hirooka, T. Kosugi, and M. Kawano, US 3,211,683 (1965).

125. D.W. Lovering and A.B. Poole, US 3,085,040 (1963).

126. O. Nashimura, S. Mano, I. Marikawa, and K. Yakahashi, US 3,299,034 (1967).

127. S. Kamiya, K. Satake, T. Sone, and T. Teraoka, US 4,199,490 (1980).

128. A.R. Colasurdo, I.R. Hiskens, N.S. Morgan, and K.J. Smith, US 5,510,003 (1996).

129. E.S. Fenelon and E.J. Pavilonis, US 2,502,080 (1950).

130. J.W. Sztuka, *J. Korea Tappi.*, **28(4)**, 66–75 (1996).

131. S.M. Ehrhardt and J.C. Gast, *Papermakers Conference*, 1988 pp. 181–187.

132. B.W. Crouse and D.G. Wimer, *Tappi J.*, **74**, 152 (1991).

133. B.W. Rowland, US 2,116,768 (1938).

134. W.-S. Schultz and U. Beyer, US 4,983,257 (1991).

135. R.T. Mashburn, US 2,393,179 (1946).

136. A.F. Nitzman and J.D. Reeves, US 2005/0090566.

137. Y.C. Huang, M.B. Lyne, and J.H. Stark, US 6,048,439 (2000).

138. I.A. Pudney, B.M. Stubbs, and M.J. Welch, US 5,393,338 (1995).

139. I.A. Pudney, B.M. Stubbs, and M.J. Welch, US 5,912,306 (1999).

140. F.B. De Young, US 3,186,900 (1965).

141. S.M. Ehrhardt and D.B. Evans, US 6,228,219 (2001).

142. P.H. Aldrich, US 4,374,673 (1983).

143. D.H. Dumas, US 4,522,686 (1985).

144. S.M. Ehrhardt and D.B. Evans, US 6,273,997 (2001).

145. T. Ikeda, K. Iwai, K. Ohta, S. Hyuga, and M. Hotta, US 5,438,087 (1995).

146. E. Strazdins, US 3,840,489 (1974).
147. M. Takahashi, T. Ikeda, and Y. Yamaguchi, US 4,943,608 (1990).
148. J.L. Azorlosa, US 2,592,107 (1952).
149. S.M. Ehrhardt and D.B. Evans, US 6,033,526 (2000).
150. J.W. Gowan, US 4,878,999 (1989).
151. E. Strazdins, Chemistry and application of rosin, in *The Sizing of Paper*, 2nd edn., W.F. Reynolds (Ed.), Tappi Press, Atlanta, GA, 1989, pp. 1–31.
152. R.W. Davison, *Tappi.*, **47(10)**, 609–616 (1964).
153. T. Lindstrom and C. Soremark, *Sven. Papperstidn.*, **80(1)**, 22–28 (1977).
154. R.W. Davison, *J. Pulp Paper Sci.*, **14(6)**, J151–J159 (1988).
155. E. Strazdins, *Tappi.*, **48(3)**, 157–164 (1965).
156. T. Kitaoka, A. Isogai, and F. Onabe, *Nord. Pulp Paper Res. J.*, **10(4)**, 253–260 (1995).
157. E. Strazdins, *Tappi.*, **46(7)**, 432–437 (1963).
158. A.F. Nitzman and A.T. Royappa, *Tappi J.*, **2(4)**, 8–11 (2003).
159. H. Zhang, A.F. Nitzman, and A.T. Royappa, *Tappi J.*, **3(9)**, 3–8 (2004).
160. M. Gyaneshwar, D. Hart, and W.E. Scott, *TAPPI Papermaking Conference and Trade Fair*, Atlanta, GA, 2000, pp. 413–443.
161. H.L. Jones, US 3,540,980 (1970).
162. F. Suyama, T. Yamamoto, K. Shimono, H. Kuwahara, M. Ishii, and I. Watanabe, US 3,725,195 (1973).
163. T. Kitaoka, A. Isogai, and F. Onabe, *Nord. Pulp Paper Res. J.*, **15(3)**, 177–182 (2000).
164. F. Wang, H. Tanaka, T. Kitaoka, and M.A. Hubbe, *Nord. Pulp Paper Res. J.*, **15(5)**, 416–421 (2000).
165. H.L. Jones, US 3,112,242 (1963).
166. M. Kato, A. Isogai, and F. Onabe, *J. Wood Sci.*, **45**, 481–486 (1999).
167. K. Ohno, A. Isogai, and F. Onabe, *J. Wood Sci.*, **45**, 238–244 (1999).
168. G.L. Batten, Rosin size, in *The Sizing of Paper*, J.M. Gess and J.M. Rodriquez (Eds.), Tappi Press, Atlanta, GA, 2005, pp. 75–133.
169. J.W. Davis, W.H. Robertson, and C.A. Weisgerber, *Tappi.*, **39(1)**, 21–23 (1956).
170. E. Strazdins, *Tappi.*, **64(1)**, 31–34 (1981).
171. R. Exner, *Paper Technol.*, **43(6)**, 45–51 (2002).
172. J. Bung, *Ippta J.*, **16(3)**, 29–31 (2004).
173. W.S. Wilson and A.H. Bump, US 2,628,918 (1953).
174. T. Kitaoka and A. Isogai, *Sen'i Gakkaishi.*, **59(9)**, 353–357 (2003).
175. T. Kitaoka, A. Isogai, and F. Onobe, *Nord. Pulp Paper Res. J.*, **16(2)**, 96–101 (2001).
176. D.H. Dumas, *Tappi.*, **64(1)**, 43–46 (1981).
177. J.M. Gess, *Tappi J.*, **72(7)**, 77–80 (1989).
178. T. Kitaoka, A. Isogai, and F. Onobe, *Nord. Pulp Paper Res. J.*, **12(3)**, 182–188 (1997).
179. M.E. Bateman and J.T. Palmer, US 4,065,349 (1977).
180. G. Katz, L.W. House, D.E. Alexander, S.L. Sowers, and J. Marton, US 6,540,877 (2003).
181. B.A. Kraus, F. DeStefano, and G. Katz, US 6,368,457 (2002).
182. I.R. Hiskens, EP 0 333 368 (1989).
183. Y. Zou, J.S. Hsieh, T.S. Wang, E. Mehnert, and J. Kokoszka, *Tappi J.*, **3(9)**, 16–18 (2004).
184. M.J. Jaycock, M.D. Phillipson, C. Zetter, and T. Vihervaara, US 6,290,765 (2002).
185. P.C. Street, US 4,333,795 (1982).
186. H.R. Rafton, US 1,803,643 (1931).
187. H.R. Rafton, US 1,803,647 (1931).
188. H.R. Rafton, US 1,914,526 (1933).
189. G.F. Feeney, US 4,699,663 (1987).
190. R. Peppmoller and F. Koschier, US 4,773,967 (1988).
191. R. Peppmoller and F. Koschier, US 4,711,919 (1987).
192. R.J. Kulick, US 3,526,524 (1970).
193. F. Wang, T. Kitaoka, and H. Tanaka, *Tappi J.*, **2(12)**, 21–25 (2003).
194. F. Wang and H. Tanaka, *J. Appl. Polym. Sci.*, **78**, 1805–1810 (2000).
195. F. Wang and H. Tanaka, *J. Pulp Paper Sci.*, **27(1)**, 8–13 (2001).
196. W. Neubert, H. Krizikalla, and R. Armbruster, US 2,492,702 (1949).
197. H. Krzikalla and R. Armbruster, US 2,296,211 (1942).
198. E. Strazdins, CA 1,008,609 (1977).
199. J.A. Sedlak, US 3,874,994 (1975).

200. K. Shimada, D. Dumas, and C.J. Biermann, *Tappi J.*, **80(10)**, 171 (1997).
201. T. Arnson, B. Crouse, and W. Griggs, Internal sizing with stearic acid, in *The Sizing of Paper*, J.M. Gess and J.M. Rodriquez (Eds.), Tappi Press, Atlanta, GA, 2005, pp. 143–150.
202. T. Arnson, B. Crouse, and W. Giggs, Internal Sizing with stearic acid, in *The Sizing of Paper*, 2nd edn., W.F. Reynolds (Ed.), Tappi Press, Atlanta, GA, 1989, pp. 79–86.
203. K. Ohno, A. Isogai, and F. Onabe, *J. Wood Sci.*, **48**, pp. 197–203 (2002).
204. W.H. Griggs and R.D. Zaffrann, US 3,096,231 (1963).
205. K.R. Andersson and N.E. Lyrmalm, WO 86/02677 (1989).
206. K. Funaoka and T. Miwa, US 3,804,788 (1974).
207. Y. Takahashi and T. Shoji, US 4,514,544 (1985).
208. H. Sato, S. Saitoh, M. Mohri, A. Miyahara, and H. Hayashi, US 4,212,783 (1980).
209. W.L. Vaughn and R.J. Beam, US 3,872,039 (1975).
210. M.F. Finlayson, K.E. Springs, J.J. Gathers, J.L. Cooper, and S.M. Oliver, US 5,993,604 (1999).
211. F.E. Carrock, US 3,839,308 (1974).
212. R.D. Jenkinson, US 3,677,989 (1972).
213. W.L. Vaugham and R.J. Beam, US 3,899,389 (1975).
214. T. Wang, J. Simonsen, and C.J. Biermann, *Tappi J.*, **80(1)**, 277–282 (1997).
215. W.L. Vaugham and J.A. Allen, US 4,181,566 (1980).
216. P. Pelletier and T.S. Wang, US 2003/0127210.
217. M. Nurminen and M. Niinikoski, US 6,939,441 (2005).
218. A. Nathansohn, US 1,996,707 (1935).
219. E. Bobu and G. Cimpoesu, *Cellulose Chem. Technol.*, **27**, 225–232 (1993).
220. A. Yarnell, *Chemical & Engineering News*, January 9, 2006, p. 46.
221. W.H. Rodleben and R. Hueter, US 2,411,860 (1946).
222. T.F. Nolan and B.M. Stubbs, US 5,484,952 (1996).
223. J.J. Zhang, US 5,525,738 (1996).
224. J.C. Sauer, US 2,238,826 (1941).
225. J.C. Sauer, US 2,369,9191 (1945).
226. P.S. McIntosh, US 5,399,774 (1995).
227. O. Malmstrom, M. Nurminen, R. Savolainen, A. Teijo, and C. Zetter, US 6,749,722 (2004).
228. M. Nurminen, K. Sundberg, and C. Zetter, US 7,232,503 (2007).
229. R.A. Bankert and D.H. Dumas, US 4,478,682 (1984).
230. W.E. Osberg, US 2,785,067 (1957).
231. D.C. England and C.G. Krespan, US 3,362,965 (1968).
232. K.J. Bottorff, US 5,252,754 (1993).
233. D.H. Dumas, US 4,295,931 (1981).
234. H.L. Lee and P. Luner, *Nord. Pulp Paper Res. J.*, **20(2)**, 227–231 (2005).
235. R.E. Cates, D.H. Dumas, D.B. Evans, and J.M. Rodriguez, Alkyl ketene dimer size, in *The Sizing of Paper*, J.M. Gess and J.M. Rodriquez (Eds.), Tappi Press, Atlanta, GA, 2005, pp. 193–209.
236. K.J. Bottorff, C.L. Brungardt, D.H. Dumas, S.M. Ehrhardt, J.C. Gast, and J.J. Zhang, US 5,685,815 (1997).
237. A. Isogai, *Nord. Pulp Paper Res. J.*, **16(2)**, 103–107 (2001).
238. A. Isogai, *Nord. Pulp Paper Res. J.*, **16(2)**, 108–112 (2001).
239. K.J. Bottorff, C.L. Brungardt, D.H. Dumas, S.M. Ehrhardt, J.C. Gast, and J.J. Zhang, US 6,007, 906 (1999).
240. C.L. Brungardt, J.C. Gast, and J.J. Zhang, US 5,725,731 (1998).
241. K.J. Bottorff, US 5,879,814 (1999).
242. K.J. Bottorff, US 6,197,417 (2001).
243. D.K. Black, K.J. Bottorff, C.L. Brungardt, D.H. Dumas, S.M. Ehrhardt, J.C. Gast, and J.J. Zhang, US 6,316,095 (2001).
244. C.H. Chapman, US 3,130,118 (1964).
245. C.A. Weisgerber, US 2,865,743 (1958).
246. D. Glittenberg, *Tappi J.*, **76(11)**, 215–219 (1993).
247. W.F. Downey, US 2,627,477 (1953).
248. C.A. Weisgerber, US 2,856,310 (1958).
249. E.D. Mazzarella, L.J. Wood, and W. Maliczyszyn, US 4,214,948 (1980).
250. D.H. Craig, US 5,403,392 (1995).
251. K. Homberg, H. Loijon, and K. Mohlin, US 6,692,560 (2004).

252. H.G. Arlt, US 2,901,371 (1959).
253. R.J. Kulick, E. Strazdins, and A.R. Savina, US 3,311,532 (1967).
254. J.M. Floyd, US 4,859,244 (1989).
255. K. Asakura, M. Iwamoto, and A. Isogai, *J. Wood Chem. Technol.*, **25(1–2)**, 13–26 (2005).
256. I.R. Hiskens, Brit. 1,457,428 (1976).
257. H. Johansson, US 5,876,562 (1999).
258. G.I. Keim and W.D. Thompson, US 2,762,270 (1956).
259. N.H. Yui and L.R. Cohen, US 3,524,796 (1970).
260. S. Frolich, E. Lindgren, and R. Sikkar, US 6,306,255 (2001).
261. C.A. Weidgerber and J.W. Davis, US 2,986,488 (1961).
262. H. Conner, T. Lin, G. Tuin, and H.G.M. van de Steeg, US 6,183,550 (2001).
263. W.F. Reynolds and L.A. Lundberg, US 2,772,969 (1956).
264. S. Yoshioka, H. Yamada, and A. Asakura, US 4,405,408 (1983).
265. H.G. Arlt, US 3,046,186 (1962).
266. C.A. Weisgerber, US 2,961,366 (1960).
267. A.T. Coscia, US 3,248,353 (1966).
268. C.O.A. Lundin, US 3,931,069 (1976).
269. D.H. Dumas, US 4,243,481 (1981).
270. E. Poppel and E. Bobu, *Cellulose Chem. Technol.*, **19**, 719–728 (1985).
271. S. Zauscher, D.F. Caulfield, and A.H. Nissan, *Tappi J.*, **79(12)**, 178–182 (1996).
272. G.C. Harris and C.A. Weisgerber, US 3,070,452 (1962).
273. Hercules Inc., Brit. 903,416 (1962).
274. D.H. Dumas, US 4,240,935 (1980).
275. R.V. Lauzon, US 6,315,824 (2001).
276. D.W. Eduards and D.F. Townsend, US 4,861,376 (1989).
277. H. Hallstrom, S. Frolich, E. Lindgren, and R. Sikkar, US 6,165,259 (2000).
278. P. Hakansson, US 6,585,859 (2003).
279. E.A. Jakaitis, F.W. Littler, and E.A. Roblendano, US 3,432,319 (1969).
280. A.R. Savina, US 3,223,544 (1965).
281. D.H. Dumas, US 4,279,794 (1981).
282. R. Ettl, W. Reuther, P. Lorencak, and J. Bonn, US 6,001,166 (1999).
283. W. Kortmann, W.D. Schroer, and K.H. Passon, US 5,028,236 (1991).
284. S. Frolich, E. Lindgren, and R. Sikkar, US 6,093217 (2000).
285. S. Frolich, E. Lindgren, and R. Sikkar, US 5,969,011 (1999).
286. B.Y. Kim and A. Isogai, *J. Korea Tappi.*, **32(5)**, 1 (2000).
287. R.G. Johnson, *TAPPI Papermaking Conference*, Atlanta, GA, 1985, pp. 85–88.
288. E. Lindgren, S. Frolich, M. Persson, and B. Magnusson, CA 2,418,400 (2002).
289. E. Lindgren, S. Frolich, M. Persson, and B. Magnusson, US 6,818,100 (2004).
290. E. Lindgren, S. Frolich, and M. Persson, US 6,846,384 (2005).
291. E. Lindgren, S. Frolich, M. Persson, and B. Magnusson, US 7,318,881 (2008).
292. H. Oikawa, M. Ogawa, K. Iwai, and M. Narushima, US 5,013,775 (1991).
293. P.H. Aldrich and D.H. Dumas, US 4,087,395 (1978).
294. D.H. Craig, US 6,156,112 (2000).
295. P. Flesher, J. Langley, N. Rosier, and D. Farrar, US 4,859,720 (1989).
296. P.D. Kaplan, A.G. Yodh, and D.F. Townsend, *J. Colloid Interface Sci.*, **155**, 319–324 (1993).
297. A. Isogai, R. Taniguchi, F. Onabe, and M. Usuda, *Nord. Pulp Paper Res. J.*, **7(4)**, 205–211 (1992).
298. T. Okumichi, O. Oseto, K. Matsumoto, S. Thuzimoto, and H. Sanda, US 4,296,012 (1981).
299. P.H. Aldrich, US 4,017,431 (1977).
300. C.L. Brungardt, R.J. Riehle, and J.J. Zhang, US 5,846,663 (1998).
301. C.L. Brungardt, R.J. Riehle, and J.J. Zhang, US 6,325,893 (2001).
302. J. Marton, *Tappi J.*, **73**, 139–143 (1990).
303. C.E. Stauffer and E. Zeffren, *J. Biol. Chem.*, **245(13)**, 3282–3284 (1970).
304. J. Marton, *TAPPI Papermaking Conference*, Atlanta, GA, 1995, pp. 97–100.
305. H. Jiang and Y. Deng, *J. Pulp Paper Sci.*, **26(6)**, 208–213 (2000).
306. A. Karademir, Y.S. Chew, R.W. Hoyland, and H. Xiao, *Can. J. Chem. Eng.*, **83(3)**, 603–606 (2005).
307. R.W. Novak and D.S. Rende, *Tappi J.*, **76(8)**, 117–120 (1993).
308. Y.S. Chew, G. Peng, J.C. Roberts, H. Xiao, K. Nurmi, and K. Sundberg, *Prepr. Int. Paper Coat. Chem. Symp.*, (2003) p. 331–337.

309. S.P. Dasgupta, US 5,433,776 (1995).
310. E. Lindgren, S. Frolich, and R. Sikkar, US 6,485,555 (2002).
311. A. Isogai, *J. Pulp Paper Sci.*, **23(6)**, 276 (1997).
312. A. Isogai, R. Taniguchi, F. Onabe, and M. Usuda, *Nord. Pulp Paper Res. J.*, **7(4)**, 193–199 (1992).
313. D.H. Dumas and D.B. Evans, *TAPPI Papermaking Conference*, Atlanta, GA, 1986, pp. 31–35.
314. R. Taniguchi, A. Isogai, F. Onabe, and M. Usuda, *Nord. Pulp Paper Res. J.*, **8(4)**, 352–357 (1993).
315. R. Seppanen, F. Tiberg, and M.-P. Valignat, *Nord. Pulp Paper Res. J.*, **15(5)**, 452–458 (2000).
316. E. Poppel and E. Bobu, *Cellulose Chem. Technol.*, **19**, 707–718 (1985).
317. A. Isogai, Fundamentals of papermaking materials in *Transactions of the Fundamental Research Symposium 11th*, Cambridge, U.K., September 1997, pp. 1047–1071.
318. A. Isogai, *Proceedings of 52nd APPITA Annual General Conference*, vol. 1, Carlton, Victoria, Australia, 1998, pp. 71–76.
319. J.E. Maher, *TAPPI Proceedings, Alkaline Papermaking*, Tappi, Atlanta, GA, 1985 pp. 89–92.
320. A. Esser and R. Ettl, Fundamentals of papermaking materials, in *Transactions of the Fundamental Research Symposium, 11th, Cambridge*, U.K., September 1997, Vol. 2, 997–1020.
321. J.C. Roberts and D.N. Garner, *Tappi J.*, **68(4)**, 118–121 (1985).
322. C. Cooper, P. Dart, J. Nicholass, and I. Thorn, *Paper Technology.*, **36(4)**, 30–34 (1995).
323. P.H. Aldrich, US 3,483,077 (1969).
324. D.H. Dumas, US 3,840,486 (1974).
325. E. Bobu, *Wochenblatt fur Papierfabrikation*, 14/15, 976–981 (2000).
326. T. Lindstrom and G. Soderberg, *Nord. Pulp Paper Res. J.*, **1(2)**, 31–38 (1986).
327. A. Juppo and A. Paren, EP 0 984 101 (2000).
328. A. Juppo and A. Paren, US 6,187,143 (2001).
329. G. Kemme, US 4,504,576 (1985).
330. W.D. Schroer, J. Probst, I. Kolb, P. Mummenhoff, and H. Baumgen, US 4,784,727 (1988).
331. K.J. Bottorff, US 5,853,542 (1998).
332. M. Hasegawa, A. Isogai, and F. Onabe, *J. Pulp Paper Sci.*, **23(11)**, J528–J531 (1997).
333. R.W. Davison and A.S. Hirwe, *TAPPI Papermaking Conference*, Atlanta, GA, 1985, pp. 7–16.
334. L. Odberg, R. Mattsson, and P. Barla, US 6,869,471 (2005).
335. C.A. Weisgerber, US 3,212,961 (1965).
336. C. Zetter, O. Malmstrom, and M. Nurminen, US 5,961,708 (1999).
337. R.A. Bankert, US 4,382,129 (1983).
338. T.E. Aderson, US 3,957,574 (1976).
339. R.A. Bankert, US 4,380,603 (1983).
340. M.O. Schur, US 3,006,806 (1961).
341. D.H. Dumas, US 4,317,756 (1982).
342. M.K. Gupta, *Tappi.*, **63(3)**, 29–31 (1980).
343. A. Ozersky, *Pulp & Paper*, February 2007, pp.46–50.
344. L.M. Robeson, G. Davidowich, and R.K. Pinschmidt, US 5,397,436 (1995).
345. D.F. Vernell, US 7,270,727 (2007).
346. C.E. Farley, *TAPPI Proceedings, Alkaline Papermaking*, Atlanta, GA, 1985, pp. 73–78.
347. J.J. Krueger and K.T. Hodgson, *Tappi J.*, **78(2)**, 154–161 (1995).
348. W. Shen, N. Brack, H. Ly, I.H. Parker, P.J. Pigram, and J. Liesegang, *Colloids and Surface.*, **A176**, 129–137 (2001).
349. U.D. Akpabio and J.C. Roberts, *Tappi J.*, **70(12)**, 127–129 (1987).
350. K.T. Hodgson, *Appita*, **47(5)**, 402–406 (1994).
351. T.H. Werner, W.H. Marra, and W.B. Gilman, US 2,992,964 (1961).
352. G.R. Evers, US 5,270,076 (1993).
353. T. Lindstrom and H. O'Brian, *Nord. Pulp Paper Res. J.*, **1(1)**, 34–42 (1986).
354. R.E. Cates, D.H. Dumas, and D.B. Evans, Alkyl ketene dimer sizes, in *The Sizing of Paper* 2nd edn., W.F. Reynolds (Ed.), Tappi Press, Atlanta, GA, 1989, pp. 33–50.
355. T. Lindstrom and G. Soderberg, *Nord. Pulp Paper Res. J.*, **1(1)**, 26–33 (1986).
356. T. Lindstrom, *TAPPI Papermaking Conference*, Atlanta, GA, 1986, pp. 29–30.
357. K.J. Bottorff and M.J. Sullivan, *Nord. Pulp Paper Res. J.*, **8(1)**, 86–95 (1993).
358. K.J. Bottorff, *Tappi J.*, **77(4)**, 105–16 (1994).
359. L. Odberg, T. Lindstrom, B. Liedberg, and J. Gustavsson, *Tappi J.*, **70(4)**, 135–139 (1987).
360. R. Seppanen, M. Von Bahr, F. Tiberg, and B. Zhmud, *J. Pulp Paper Sci.*, **30(3)**, 70–73(2004).
361. A. Isogai, *J. Pulp Paper Sci.*, **26(9)**, 330–334 (2000).

362. H.L. Lee and P. Luner, *J. Colloid Interface Sci.*, **146(1)**, 195–205 (1991).
363. P.A. Patton, *Proceedings Papermakers Conference*, Atlanta, GA, 1991, pp. 415–423.
364. M.J. Jaycock and J.C. Roberts, *Paper Technology.*, **35(4)**, 38–42 (1994).
365. T. Lindstrom and G. Soderberg, *Nord. Pulp Paper Res. J.*, **1(2)**, 39–45 (1986).
366. C.L. Brungardt and J.C. Gast, *TAPPI Papermaking Conference*, Atlanta, GA, 1996, pp. 297–308.
367. A. Isogai, *J. Pulp Paper Sci.*, **25(7)**, 251–256 (1999).
368. T. Yano, H. Ohtani, S. Tsuge, and T. Obokata, *Analyst.*, **117**, 849–852 (1992).
369. J.M. Gess, *Tappi J.*, **75(4)**, 79–81 (1992).
370. P. Rohringer, M. Bernheim, and D.P. Werthmann, *Tappi J.*, **68(1)**, 83–86 (1985).
371. J.C. Roberts and D.N. Garner, *Cellulose Chem. Technol.*, **18**, 275–282 (1984).
372. A.F. Turbak, A. El-Kafrawy, F.W. Snyder, and A.B. Auerbach, US 4,302,252 (1981).
373. S.H. Nahm, *J. Wood Chem. Technol.*, **6(1)**, 89–112 (1986).
374. U.D. Akpabio and J.C. Roberts, *Tappi J.*, **72(7)**, 141–145 (1989).
375. J.M. Gess and R.C. Lund, *Tappi J.*, **74(1)**, 111–113 (1991).
376. F. Tiberg, J. Daicic, and J. Froberg Surface chemistry of paper, in *Handbook of Applied Surface and Colloid Chemistry*, K. Holmberg (Ed.), John Wiley, New York, 2001, pp. 123–173.
377. R.C. Lund, *TAPPI Proceedings, Alkaline Papermaking*, Atlanta, GA, 1985, pp. 1–5.
378. W. Shen and I.H. Parker, *J. Colloid Interface Sci.*, **240**, 172–181 (2001).
379. E. Robbart, US 3,856,558 (1974).
380. E. Robbart, US 4,554,215 (1985).
381. E. Robbart, US 4,339,479 (1982).
382. W.I. Patnode, US 2,306,222 (1942).
383. J.A. Irvine, D.E. Aston, and J.C. Berg, *Tappi J.*, **82(5)**, 172–174 (1999).
384. L.C.A. Barbosa, A.J. Demiuner, C.R.A. Maltha, M.P. Cruz, and G. Ventorim, *Papel*, **66(9)**, 68–73, (2005).
385. J. Reiners, J. Kopp, J. Konig, H. Traubel, E. Wenderoth, B. Jensen, and J. Probst, US 6,090,871 (2000).
386. Y. Deng and M. Abazeri, *Wood Fiber Sci.*, **30(2)**, 155–164 (1998).
387. R. Oda, S. Munemiya, and M. Okano, *Makromol. Chem.*, **43**, 149–151 (1961).
388. J. Furukawa, T. Saegusa, N. Mise, and A. Kawasaki, *Makromol. Chem.*, **39**, 243–245 (1960).
389. L.X. Gong and X.F. Zhang, *eXPRESS Polym. Lett.*, **3(12)**, 778–787 (2009).
390. D.H. Dumas, US 3,992,345 (1976).
391. H.L. Lee and P. Luner, *Nord. Pulp Paper Res. J.*, **4(2)**, 164–172 (1989).
392. A. Isogai, R. Taniguchi, and F. Onabe, *Nord. Pulp Paper Res. J.*, **9(1)**, 44–211 (1994).
393. K.C. Dilts, R.J. Proverb, and D.L. Dauplaise, US 6,666,952 (2003).
394. K.C. Dilts, R.J. Proverb, and D.L. Dauplaise, US 6,576,049 (2003).
395. P.H. Aldrich, US 3,582,464 (1971).
396. E. Strazdins and R.J. Kulick, US 3,409,500 (1968).
397. R.J. Kulick and E. Strazdins, US 3,666,512 (1972).
398. A.M. Bills, US 3,821,075 (1974).
399. A.R. Savina, US 3,223,543 (1965).
400. R.J. Kulick and E. Strazdins, US 3,445,330 (1969).
401. O.B. Wurzburg and E.D. Mazzarella, US 3,102,064 (1963).
402. P. Pruszynski, Recent developments in papermaking chemicals, in *Wood, Pulp and Paper Conference*, Bratislava, Slovakia, 2003.
403. W.J. Ward, K. Andruszkiewicz, R.T. Gray, T.P. McGinnis, and R.W. Novak, US 2006/0231223.
404. W.A. Sweeney, US 4,431,826 (1984).
405. A. Sato, Y. Murai, M. Goto, and K. Mochizuki, US 4,514,229 (1985).
406. S.M. Blitzer and H.D. Wilder, US 4,302,283 (1981).
407. O.B. Wurzburg, US 3,821,069 (1974).
408. E.F. Zaweski and A.H. Filbey, US 3,476,774 (1969).
409. J.J. Zhang and S.M. Lai, US 6,348,132 (2002).
410. O.B. Wurzburg, US 3,968,005 (1976).
411. K. Kawatani, T. Fujikawa, and E. Watanabe, US 4,544,414 (1985).
412. I.R. Hiskens, D.C. Johnson, and M.J. Caudwell, US 4,222,820 (1980).
413. K. Kawatani, T. Fujikawa, and E. Watanabe, US 4,576,680 (1986).
414. F.A. Stuart, R.G. Anderson, and A.V. Drummond, US 3,361,673 (1968).
415. W. Pinkernelle, US 2,182,178 (1939).
416. W.M. Le Suer and G.R. Norman, US 3,172,892 (1965).

417. D. Shephard, US 4,207,142 (1980).

418. J. Binapfl, US 2,121,183 (1938).

419. H.E. Fried, US 4,761,488 (1988).

420. K.H. Shin and P.S. Hale, US 5,012,169 (1991).

421. P.S. Hale and K.H. Shin, US 4,958,034 (1990).

422. J.J. Harrison and W.R. Ruhe, US 5,286,799 (1994).

423. J.J. Harrison and W.R. Ruhe, US 5,319,030 (1994).

424. C.M. Selwitz and H.I. Thayer, US 4,158,664 (1979).

425. A. Tudorovic and T. Jacobson, US 2009/0139676.

426. R.J. Proverb and D.L. Dauplaise, *Papermakers Conference*, Atlanta, GA, 1989, p. 91.

427. G.C.I. Chen and T.W. Woodward, *Tappi J.*, **69(8)**, 95–97 (1986).

428. M.J. Lindstrom and R.M. Savolainen, *J. Disp. Sci. Technol.*, **17(3)**, 281–306 (1996).

429. D.G. Pardikes, US 6,207,719 (2001).

430. J.M. Gess and D.S. Rende, *Tappi J.*, **4(9)**, 25–30 (2005).

431. D.G. Pardikes, US 5,653,915 (1997).

432. J. Lindfors, J. Salmi, J. Laine, and P. Stenius, *BioResources.*, **2(4)**, 652–670 (2007).

433. S. Yoshioka, H. Yamada, K. Goto, Y. Adachi, and K. Miyahana, US 4,533,434 (1985).

434. D.R. Dostie and R.J. Nowicki, US 6,346,554 (2002).

435. P.R. Lakshmanan, US 4,355,125 (1982).

436. C.G. Caldwell and O.B. Wurzburg, US 2,813,093 (1957).

437. D.L. Dauplaise, R.J. Proverb, and K. Komarowska, US 6,210,475 (2001).

438. G.C.I. Chen and T.W. Woodward, *TAPPI Proceedings, Papermakers Conference*, Atlanta, GA, 1986, pp. 37–40.

439. H. Yamada, Y. Adachi, and K. Nishida, JP 9–111692 (1997).

440. J.J. Tsai and W. Maliczyszyn, US 5,658,378 (1997).

441. J. Tsai and W. Maliczyszyn, US 5,595,631 (1997).

442. C.G. Caldwell and O.B. Wurzburg, US 2,661,349 (1953).

443. P.T. Trzasko, M.M. Tessler, R. Trksak, and W. Jarowenko, US 4,721,655 (1988).

444. H.L. Lee, J.S. Kim, and H.J. Youn, *Tappi J.*, **3(12)**, 3–6 (2004).

445. E.F. Paschall, US 2,876,217 (1959).

446. P.T. Trzasko, M.M. Tessler, R. Trksak, and W. Jarowenko, US 4,687,519 (1987).

447. C.E. Farley, G. Anderson, and K.D. Favors, 6,787,574 (2004).

448. M.F. Ruiz, *Pulp & Paper International*, November 2009, pp. 35–36.

449. D.H. Denpwski, M.F. Ruiz, and W.B. Hill, US 2008/0277084.

450. R.B. Wasser, US 5,759,249 (1998).

451. D.E. Glover, US 5,962,555 (1999).

452. D.S. Rende and M.D. Breslin, US 4,657,946 (1987).

453. S. Yoshioka, A. Honma, and H. Sato, US 4,849,055 (1989).

454. S. Yoshioka, H. Yamada, H. Sato, and M. Kamei, US 4,882,087 (1989).

455. S. Yoshioka, T. Yoshida, H. Sato, H. Yamada, and Y. Adachi, US 4,810,301 (1989).

456. P. Peutherer, M. Waring, and L. Collett, US 6,284,099 (2001).

457. R.W. Novak, US 4,606,773 (1986).

458. E.D. Mazzarella, L.J. Wood, and W. Maliczyszyn, US 4,040,900 (1977).

459. E.D. Mazzarella, W. Maliczyszyn, and J. Atkinson, US 4,711,671 (1987).

460. E.D. Mazzarella, W. Maliczyszyn, and J. Atkinson, US 4,728,366 (1988).

461. E.D. Mazzarella, W. Maliczyszyn, and J. Atkinson, US 4,747,910 (1988).

462. E.D. Mazzarella, W. Maliczyszyn, and J. Atkinson, US 4,832,792 (1989).

463. K. Yokota, A. Ichihara, Y. Tanakamaru, and Y. Takahashi, US 4,666,523 (1987).

464. W.A. Sweeney, US 4,545,855 (1985).

465. W.A. Sweeney, US 4,545,856 (1985).

466. W.A. Sweeney, US 4,849,131 (1989).

467. W.A. Sweeney, US 4,695,401 (1987).

468. W.A. Sweeney, US 4,915,786 (1990).

469. R.J. Mills, J. Hoffmann, R. Kruckel, and A. Van Den Berg, US 6,444,024 (2002).

470. C.E. Farley and R.B. Wasser, Sizing with alkenyl succinic anhydride, in *The Sizing of Paper* 2nd edn., W.F. Reynolds (Ed.), Tappi Press, Atlanta, GA, 1989, pp. 51–62.

471. J.M. Gess and D.S. Rende, Alkenyl succinic anhydride, in *The Sizing of Paper*, J.M. Gess and J.M. Rodriquez (Ed.), Tappi Press, Atlanta, GA, 2005, pp. 179–192.

472. F.H. Cousart, *TAPPI Proceedings, Papermakers Conference*, Atlanta, GA, 1989, pp. 99–102.
473. H.A. Goldsberry, K.C. Dilts, C.R. Hunter, M.P. O'Toole, R.J. Proverb, L. Pawlowska, G.E. Baikow, K. Komarowska, D.L. Dauplaise, and M.J. Scanlon, US 2006/0049377.
474. H. Shigeto and H. Umekawa, US 5,391,225 (1995).
475. R.B. Wasser, *TAPPI Seminar Notes, Alkaline Papermaking*, Atlanta, GA, 1985, pp. 17–20.
476. R.M. Savolainen, *TAPPI Papermaking Conference*, Atlanta, GA, 1996, pp. 289–295.
477. R.B. Wasser, *TAPPI Proceedings. Papermakers Conference*, Atlanta, GA, 1986, pp. 1–6.
478. E. Strazdins, *TAPPI Proceedings, Alkaline Papermaking*, Atlanta, GA, 1985, pp. 37–42.
479. J.C. Winters and R.W. Best, *TAPPI Seminar Notes, Alkaline Papermaking*, Atlanta, GA, 1985, pp. 51–54.
480. Koyu Nikoloff and E. Takacs, US 5,176,748 (1993).
481. J.C. Winters, R.A. Mooth, and F.T. Orthoefer, US 4,629,655 (1986).
482. A. Isogai, M. Nishiyama, and F. Onabe, *Sen'i Gakkaishi*, **52(4)**, 195–201 (1996).
483. A. Isogai, *Sen'i Gakkaishi*, **56(7)**, 328–333 (2000).
484. W.R. McCarthy and R.A. Stratton, *Tappi J.*, **70(12)**, 117–121 (1987).
485. A. Isogai, *Sen'i Gakkaishi.*, **56(7)**, 334–339 (2000).
486. T. Sato and A. Isogai, *Appita J.*, **56(3)**, 204–212 (2003).
487. M. Nishiyama, A. Isogai, and F. Onabe, *Sen'i Gakkaishi*, **52(4)**, 180–188 (1995).
488. A. Isogai, M. Nishiyama, and F. Onabe, *Sen'i Gakkaishi*, **52(4)**, 189–194 (1996).
489. R.W. Liggett, US 3,139,373 (1964).
490. R.B. Wasser and J.S. Brinen, *Tappi J.*, **81(7)**, 139–144 (1998).
491. E.L. Back and S. Danielsson, *Tappi J.*, **74(9)**, 167–174 (1991).
492. A. Isogai, *J. Pulp Paper Sci.*, **25(6)**, 211–215 (1999).
493. T. Sato, A. Isogai, and F. Onabe, *Nord. Pulp Paper Res. J.*, **15(3)**, 172–176 (2000).
494. H. Beck, US 3,900,335 (1975).
495. D.H. Dumas, US 3,968,317 (1976).
496. D.H. Dumas, US 3,923,745 (1975).
497. R.V. Lauzon, US 6,414,055 (2002).
498. P.H. Aldrich and D.H. Dumas, US 3,922,243 (1975).
499. P.H. Aldrich and D.H. Dumas, US 3,990,939 (1976).
500. S.N. Lewis, R.F. Merritt, and W.D. Emmons, US 3,694,393 (1972).
501. S. Yoshioka and K. Okada, US 4,200,562 (1980).
502. S.N. Lewis, R.F. Merritt, and W.D. Emmons, US 3,678,098 (1972).
503. V.R. Gaertner, US 3,459,715 (1969).
504. V.R. Gaertner, US 3,391,018 (1968).
505. V.R. Gaertner, US 3,278,561 (1966).
506. J.P. Koskela and O.E.O. Hormi, *Appita J.*, **56(4)**, 296–300 (2003).
507. O.E.O. Hormi and J.P. Koskela, CA 2,265,432 (1999).
508. A. Goldstein and J.W. Brook, US 3,310,460 (1967).
509. H. Bestian and G. von Finck, US 2,314,968 (1943).
510. G.W. Sprenger, US 3,627,631 (1971).
511. Y. Morimoto, M. Saotome, and A. Komai, US 4,696,760 (1987).
512. G.H. Brown and M.M. Skoultchi, US 3,575,796 (1971).
513. M. Bernheim, H. Meindl, and P. Rohringer, US 4,735,685 (1988).
514. M.E. Bateman and J.T. Palmer, US 4,127,418 (1978).
515. K.J. Pickard and A. Smith, US 3,993,640 (1976).
516. V.R. Gaertner, US 3,345,252 (1967).
517. H.G. Arlt, US 3,050,437 (1962).
518. H.G. Arlt, US 3,084,092 (1963).
519. E. Strazdins, R.J. Kulick, and A.R. Savina, US 3,499,824 (1970).
520. M.R. Kamal and J.L. Keen, US 3,589,978 (1971).
521. J.R. Robertson, US 4,505,778 (1985).
522. V.R. Gaertner, US 3,214,325 (1965).
523. V.R. Gaertner, US 3,345,251 (1967).
524. V.R. Gaertner, US 3,278,560 (1966).
525. V.R. Gaertner, US 3,269,890 (1967).
526. K.U.E. Helmer and A.R. Reuterhall, US 3,887,427 (1975).
527. J.A.C. Bjorklund, T. Ekengren, US 4,123,319 (1978).
528. E.P. Pauley and K.S. Neighbor, US 6,165,321 (2000).

529. E. Valton, J. Schmidhauser, and M. Sain, *Tappi J.*, **3(4)**, 25–30 (2004).
530. P.H. Aldrich and H.I. Enos, US 3,193,449 (1965).
531. M. Arakawa, K. Hirooka, T. Kosugi, and M. Kawano, US 3,211,681 (1965).
532. K.G. Forss, A.G.M. Fuhrmann, and M. Toroi, US 5,110,414 (1992).
533. E.F. Horsey and W.D. Thompson, US 2,650,163 (1953).
534. J.H. Daniel, US 3,022,214 (1962).
535. H.S. Killam, US 4,152,201 (1979).
536. D.R. Spear and W.F. Fowler, US 2,748,029 (1956).
537. J.E. Smith, US 2,343,095 (1944).
538. B. Alince, *Tappi J.*, **74(8)**, 221–3 (1991).
539. W.T. Driscoll, W.F. Fowler, and R.J. Hellmann, US 2,887,380 (1959).
540. D.K. Pattilloch, US 2,694,633 (1954).
541. W.R. Hine and M.J. Holm, US 3,461,029 (1969).
542. V.R. Gaertner, US 3,562,102 (1971).
543. V.R. Gaertner, US 3,314,897 (1967).
544. W.F. Reynolds and D.R. Sexsmith, US 3,212,962 (1965).
545. J.H. Daniel, US 2,914,513 (1959).
546. Y. Jen and R.R. House, US 3,015,605 (1962).
547. E.H. Sheers and D.R. Sexsmith, US 3,268,470 (1966).
548. P.H. Aldrich, US 4,109,053 (1978).
549. H.R. Rafton, US 2,040,878 (1936).
550. H.R. Rafton, US 1,819,441 (1931).
551. H.R. Rafton, US 1,803,648 (1931).
552. O. Kress and C.E. Johnson, US 2,172,392 (1939).
553. O. Kress and C.E. Johnson, US 2,058,085 (1936).
554. O. Kress and C.E. Johnson, US 2,059,465 (1936).
555. O. Kress and C.E. Johnson, US 2,059,464 (1936).
556. H.R. Rafton, US 1,803,652 (1931).
557. A.B. Bakalar and R.D. Sullivan, US 2,665,983 (1954).
558. M.W. Mickell, US 2,993,800 (1961).
559. M. Bernheim, H. Meindl, P. Rohringer, H. Wegmuller, and D. Werthemann, US 4,623,428 (1986).
560. M. Bernheim, H. Meindl, and P. Rohringer, US 4,737,239 (1988).
561. A. Maeda, US 4,826,570 (1989).
562. R. Topfl, M. Bernheim, H. Meindl, H. Wegmuller, P. Rohringer, and D. Werthemann, US 5,201,998 (1993).
563. Y. Nakamura, T. Tamai, and K. Asakura, US 7,344,621 (2008).
564. K. Asakura and A. Isogai, *Nord. Pulp Paper Res. J.*, **18(2)**, 188–193 (2003).
565. L.A. Lundberg and W.F. Reynolds, US 2,794,737 (1957).
566. Q. Jing, M. Chen, and C.J. Biermann, *Tappi J.*, **81(4)**, 193–197 (1998).
567. L.A. Lundberg and W.F. Reynolds, US 2,801,169 (1957).
568. C.G. Landes and W.F. Reynolds, US 2,694,630 (1954).
569. C.G. Landes and W.F. Reynolds, US 2,694,629 (1954).
570. H. Tlach, K.D. Leifels, and W. Mischler, US 4,299,654 (1981).
571. C.G. Landes and W.F. Reynolds, US 2,698,793 (1955).
572. R.M. Chad and C.A. Schwartz, Fluorochemical sizing, in *The Sizing of Paper*, 2nd edn., W.F. Reynolds (Ed.), Tappi Press, Atlanta, GA, 1989, pp. 87–101.
573. S. Fukuda, A. Isogai, and F. Onabe, *Proceedings of 2nd ISETPP*, 2002, pp. 608–612.
574. S. Fukuda, A. Isogai, F. Onobe, M. Ishikawa, and T. Masutani, *Sen'i Gakkaishi*, **58(5)**, 170–175 (2002).
575. S. Fukuda, A. Isogai, M. Ishikawa, and T. Masutani, *Sen'i Gakkaishi*, **59(6)**, 239–242 (2003).
576. L. Yang, R. Pelton, F. McLellan, and M. Fairbank, *Tappi J.*, **82(9)**,128–135 (1999).
577. T. Trombetta, P. Iengo, and S. Turri, *J. Appl. Polym. Sci.*, **98**, 1364–1372 (2005).
578. C.A. Schwartz and M.M. Lynn, US 4,529,658 (1985).
579. P.C. Hupfield, T. Masutani, S. Minami, and I. Yamamoto, US 2010/0018659.
580. C.A. Schwartz and M.M. Lynn, US 4,579,924 (1986).
581. J.J. Fitzgerald, US 5,674,961 (1997).
582. P. Iengo and P. Gavezotti, US 7,534,323 (2009).

8 Creping Adhesives and Softeners

Sanitary tissue products (facial tissues, toilet tissues, and absorbent towels) are used in high quantities. The basis weight, thickness, strength, and dispersing medium of these products often differ widely, but they are all linked together by the common process through which they originate: the creped papermaking process (drying the paper web on a Yankee dryer and then scraping the sheet with a creping blade or air jets [1]).

The wet sheet travels to the Yankee by means of a felt, and it is transferred to the dryer surface at the pressure roll nip. At this point, the sheet is at 35%–80% consistency. The wet pressing step significantly increases the paper dry strength. By adjusting the pressure, the desired paper dry strength is obtained.

The sheet is further dried on the hot Yankee dryer to 90%–98% consistency [2]. The dryer temperature is about 100°C and the machine speed ranges from 900 to 2000 m/min [3]. Although that step is designed to help evaporate water from the wet sheet, the creping chemicals sprayed on the dryer surface are aqueous solutions. Thus, water circulation is also involved: we are dealing with evaporation not only from the adhesive solution and the wet web but also from the wet web to the dry adhesive film.

The wet web naturally adheres to the dryer surface: a thin layer of hemicellulose is transferred from the web to the hot dryer surface. The wetted hemicellulose, which shows a lower softening point [4], is a strong adhesive [5]. However, parameters such as the amount of hemicellulose and its distribution are hard to control. In order to supplement and control adhesion, creping adhesives with a controlled structure are added to the drier surface in specific amounts.

The adhesive helps the wet web to stick to the dryer surface; the dried paper is then scraped off by a blade. During that process, cellulose fibers are partially pulled out and many interfiber bonds are broken as a result. That process is aimed at imparting web properties to paper, such as softness, strength, and bulk. The wet creping [6,7] is a similar type of process and the need for an adhesive is greatly increased with the advent of the through-drying process [8].

Paper creping is a multiparameter process: the furnish quality (fiber length [9], hemicellulose concentration, Eucalyptus stock, etc.), paper chemicals added at the wet end [10], web consistency and water quality [5], basis weight [11], dryer temperature, type of creping machines [12], type of adhesive, debonder concentration, the adhesive solution concentration [13], the Yankee dryer speed, and the doctor blade angle are the most important [5,14–18]. Those parameters are instrumental in meeting the current need for changing the properties of the creping adhesive and optimize the creping process [19].

Creping aids interact with cellulose fibers on the Yankee dryer surface; the paper softness, strength, and bulkiness are a result of those interactions. The quality of the final sheet depends, to a large extent, on the quality of chemicals used at the wet end and during the creping process.

Creping aids are a complex mixture of small molecule (organic and inorganic) compounds and polymers, which must fit in the complexity of the papermaking process: special creping adhesives are designed for each furnish and each particular piece of equipment. For instance, the surface of the Yankee dryer may have a specific temperature profile; as a result, the paper is creped differently in different zones. Therefore, it is a specific area where the doctor blade is excessively worn out. In order to prevent that, different creping compositions (more or less humectant [20]) may be applied on different zones of the Yankee dryer.

The adhesion force should reach an optimum in terms of heat transfer and adhesion. When insufficiently adhered, webs tend to cause poor control of the sheet and poor creping while the drying process is incomplete. Higher levels of adhesion improve the heat transfer (are more energy efficient) and allow for a higher-speed operation of the creping process. For a more efficient heat transfer, the Yankee dryer surface is highly polished in order to increase the contact between the tissue web and the dryer.

The papermaking process actually involves a series of locations where chemicals are added and the creping step may not be the final addition point. Chemicals are added at the wet end [21] (ionic compounds), sprayed on the wet web [22] (especially nonionic compounds), incorporated in creping adhesives, sprayed on the dry (creped) paper sheet, or added with a printing roll in a predetermined pattern [23]. The addition of chemicals can also be performed simultaneously in multiple points: dispersion of paraffin wax at the wet end and a cationic resin sprayed on the Yankee surface [24].

The creping operation runs optimally when the adhesive is free of interference from non-creping-related chemicals such as wet-strength resin added at the wet end [25]. Therefore, some additives (softeners, humectants, emollients, etc.) may be added to the paper web after the creping step.

There are nonionic chemicals targeting specific paper properties (such as softness), which cannot be retained at the wet end. That is why, for better retention, those chemicals should be added after the creping step (see the lotion including skin pH balancing compounds [26]). The application of chemicals only to the paper surface ensures the effective use of the material since consumers only tactically interact with the surface. The chemicals distributed throughout the web are, in a large proportion, lost for the final application.

The paper web can have one or more layers [27]. In the case of the multilayer tissue paper, the distribution of layers and the properties of the layer, which comes into contact with the Yankee dryer (Eucalyptus layer), must also be considered.

After creping, several ingredients from the adhesive composition are distributed on paper and some paper components (especially hemicellulose, cellulose fibers, and inorganic fillers) are accumulated into the adhesive layer that remains on the Yankee dryer [3,5]. A cleaning blade is used to remove materials accumulated within the adhesive layer; the amount of recycled material should also be taken into account.

8.1 CREPING ADHESIVES

During the creping process, the cellulose fibers are partly pulled out; the adhesive disrupts many of the interfiber bonds formed during the drying step, thus helping in developing the desired properties of a tissue sheet, such as softness, bulk, and stretch [3]. An attempt to define paper creping in mathematical terms has been made [28]:

$$\%\text{crepe} = \frac{\dfrac{L_Y}{BW_Y} - \dfrac{L_w}{BW_w}}{\dfrac{L_Y}{BW_Y}} \tag{8.1}$$

where
  BW_Y is basis weight at the Yankee
  BW_w is basis weight at the rewinder
  L_Y is the length at the Yankee
  L_w is the length at the rewinder

The adhesive solution in water is sprayed onto the Yankee dryer surface. The coating should provide adhesion to both the paper and the Yankee dryer. The polymer film improves heat transfer by allowing for a more effective drying of the sheet [29], and it also provides lubrication between the doctor blade and the dryer surface [30].

A good creping adhesive should be water soluble or dispersible, must adequately adhere the sheet to the Yankee dryer surface, and it must produce a soft sheet, which is non-blocking when rolled.

All of these properties are hard to attain with a single chemical [31–34] and sometimes complicated mixtures are used as creping adhesives. Unfortunately, compatibility wise, many polymer mixtures pose quite a lot of problems (complex formations, chemical reactions, phase separation, etc.). If the interaction between components is too strong in solution, one component is added to the sheet and another to the Yankee. When the two components are brought into contact at the pressure roll nip, a complex adhesive might be formed [35].

In order to set an interface failure [36], an optimum adhesion between the creping aid and the paper should be set. If adhesion is too weak the sheet will lift off very easily and undergo very little creping. Greater adhesion causes increased softness, although,—generally—with some loss of strength [30] but a too strong adhesion causes the sheet to slip under the doctor blade. To control the adhesion, a release aid is added to the formula. The release aid is soluble or dispersible in the adhesive solution and it should be evenly distributed onto the adhesive film surface.

Ionic and/or hydrogen bonds between the cellulose fibers and the creping adhesive are key parts of the interactions on the Yankee dryer surface. Cationic resins (such as polyvinyl amine [34]) are more effective as adhesives when the cellulose web contains more carboxylic groups. Due to those interactions, the chemicals transferred from the adhesive layer to the paper sheet interfere with the chemicals added to the wet end, and, thus, the paper wet strength, its dry strength and softness may change.

Attempts to restore the paper strength lost by reduction of hydrogen bonding have included the addition of bonding materials (such as latexes of elastomers) [37]. In order to obtain the densified zone, pressure can be applied with a patterned rotogravure roll, along with bonding materials [38,39]. Pattern-densified tissue paper has a relatively low fiber density and an array of densified zone of relatively high fiber density [25,40,41].

8.2 COMPOSITION OF CREPING ADHESIVES

The hemicellulose from the cellulose fibers helps paper to stick on a clean dryer surface but the first creping aids added to the Yankee dryer surface were animal glues made from gelatin. These glues have an irregular molecular structure; the dry film is an irregular, disordered mass. Moreover, in order to achieve the necessary tack, animal glues require large amounts of water in the wet web [42]. A synthetic adhesive with adjustable properties was needed to improve the creping process.

The synthetic creping adhesive can have a single component, such as acrylonitrile-grafted copolymer onto cellulose [40], polyvinyl alcohol (PVA) [43,44], modified polyamidoamine with di-aldehyde [45,46], modified polyaminoamine epichlorohydrin polymer (PAE) resins [3], aromatic polyamidoamine [21], polyamine (polyionene) [47], or it can be a copolymer of the styrene–methacrylic acid (TEA salt) [48]. All those polymers are polar and some of them (such as PVA) may develop hydrogen bonding as well [49].

Due to their polarity, condensation polymers are water soluble and good adhesives. The adhesive chemical structure may be very diverse with hydroxyl groups (or ether group) and cationic charges in the same molecule (ionene polymers) [47,50]. These polymers are not thermosetting and the low molecular weight ($M_w < 5000$) materials are tacky.

Addition polymers obtained by free radical polymerization have the disadvantage of a hydrophobic carbon–carbon backbone. In order to make them more polar and water soluble, polar pendant groups and cationic charges should be attached to the hydrophobic backbone [28,51]. Copolymers of methyl methacrylate (MMA) with 30% dimethylaminoethyl methacrylate are still water insoluble (emulsions) and they become water soluble only after acetic acid addition. The copolymer molecular weight can be changed by using a chain transfer agent during the synthesis. A quaternization can be performed in order to obtain a cationic creping aid. Due to their low molecular weight, cationic creping aids do not alter the paper wet strength.

The paper web, which reaches the Yankee dryer may include a diversity of furnishes, a large variation of the residual moisture, and concentration of the paper chemicals used at the wet end. Creping adhesives must fit that diversity and, according to papers published in 2002 and 2005 [19,33], no single adhesive has provided a satisfactory combination of doctorability, rewettability, and level of adhesion.

In order to fit a particular paper mill, the creping adhesives may have a very sophisticated composition. They usually contain one or more polymeric adhesives, release aids, humectants, plasticizers, stabilizers, and softeners. Sometimes the chemical reaction between components is used to reach performances that the individual components or their physical mixtures are lacking. The optimum adhesive composition should have synergetic effects and control all possible chemical reactions (such as cross-linking and/or thermal degradation).

8.2.1 ADHESIVES FOR THE YANKEE DRYER

In order to fit different furnishes, different moisture profiles, web speeds, etc., the ideal adhesive must be a very complex blend of hydrophilic and hydrophobic compounds with ionic or nonionic structure, to which other plasticizers, humectants, release aids, and softeners should be added.

Blends of polymers used as creping adhesives are dispersions or homogeneous mixtures in water, and their constituents may or may not chemically interact with each other. The robustness of the adhesive film is obtained by controlling the potential chemical reactions (branching, cross-linking, and thermal degradation). Table 8.1 shows the most common adhesive mixtures used in the creping process.

8.2.1.1 Nonreactive Creping Adhesives

Water-dispersible polymers are hydrophobic (latexes); a film is formed after drying. In order to adjust the T_g of that film, polymeric latexes with high T_g are combined with low-T_g polymeric latexes [88]. The adhesive film must be water sensitive to stay tacky. For a good rewettability of the film (by absorption of water from the wet web), the hydrophobic character is balanced with a more hydrophilic compound (such as carboxymethyl cellulose [31], polyamine, or polyamide [32]).

TABLE 8.1
Adhesives for Yankee Dryer

Component A	Component B	References
Heterogeneous systems		
Polyvinyl acetate ($T_g < 30°C$)	PVA	[52,53]
EVA emulsion	PVA (10%–30%)	[54]
E-VA-methylol acrylamide copolymer (latex)	Anionic–nonionic surfactant	[55]
Acrylates copolymer latex	Polyamine, polyamide	[32]
Homogeneous, nonreactive systems		
PVA	Non-thermosetting polyamidoamine	[42,56]
PVA	EO-PO block-copolymer or stearyl-PPO	[8,57]
PVA (87% hydrolysis)	PEO ($M_w = 400$) and ammonium lignin sulfonate	[58]
PVA	Poly(vinyl-amine)/vinylformamide	[59]
PEI	Modified PEI	[60]
Homogeneous, reactive systems		
PVA	Cationic starch and PAE	[61–63]
PVA	PAE	[44,64–68]
PVA	Cationic (PAE) and anionic (styrene–maleic anhydride copolymer [SMA]) resins	[69]
PVA	PAE and AZC	[70]
PVA	AZC	[44,57,71–73]
PVA	G-PAm	[74]
Polyacrylic acid	Poly 2-ethyl-2-oxazoline or PVA, PEO	[35][a]
PVA	Poly 2-ethyl-2-oxazoline	[75][a]
Poly 2-ethyl-2-oxazoline	PAE-type resin	[76,77]
Poly 2-ethyl-2-oxazoline	PEI modified with Epi	[78]
Animal glue	PAE	[79]
PAE-1	PAE-2	[80,81]
PPG	PAE	[57]
PAE	Salt of multivalent metal ion	[19]
Vinyl alcohol-vinyl amine copolymer	AZC	[46,82–86]
Vinyl alcohol-vinyl amine copolymer	Polyphenols and oxidizing agents	[56,87]
Vinyl alcohol-vinyl amine copolymer	Glyoxal	[84]

[a] One to the sheet and another to the creping cylinder.

The ethylene-vinyl acetate copolymer (EVA) latex copolymer ($0°C < T_g < 10°C$) is blended with PVA, but its composition is impacted by the buildup of water-insoluble residues on process fabrics [75]. PVA also tends to coat the dryer with a hard and uneven film that accumulates during the drying and creping steps and results in uneven creping. Additionally, to remove the hard film, a cleaning blade is frequently used causing dryer surface wear [54].

The film made from blends of PVA and other water-soluble polymers (polyamidoamines as poly-secondary amines) [56] shows a better stability and take up water to a greater extent. In order to make the polyamidoamine film more hydrophilic, phosphoric acid is used to add cationic charges to that backbone [21,89].

Poly-secondary amines are reactive compounds incorporated in blends that can be lightly cross-linked and—in a separate step—make the blend more hydrophobic (such as polyethylene imine [PEI] reaction with alkenyl ketene dimer [AKD]) [60]. Epichlorohydrin is a nonsymmetrical cross-linker, which reacts in two steps: the epoxide ring and the chlorine pendant groups. The epoxide ring reacts much faster and its conversion is very high. The resulting polymer has chlorine pendant groups. The chlorine derivative reacts with secondary (in red) and tertiary amines within a slower reaction. If the chlorine pendant groups are 100% converted, the lightly cross-linked polymers are nonreactive. In order to obtain a lightly cross-linked adhesive with no chlorine pendant groups, the epichlorohydrin concentration should be lower; the reaction temperature is higher for a longer reaction time [90]. During the cross-linking reaction, hydrochloric acid is released and captured by the un-reacted amines as ammonium salt. In the new quaternary ammonium salt, the covalent-bonded chlorine (in blue) was converted to ionic chlorine.

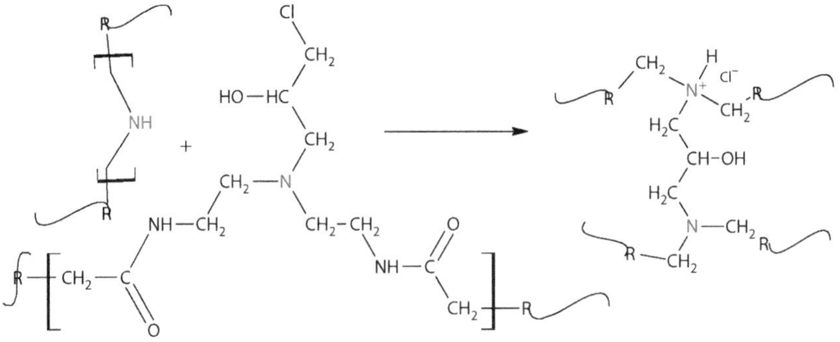

The process can be controlled by the ratio between secondary and tertiary amine [65,66] and the amount of cross-linker [42,90,91]. If reactive groups are still present, they are reacted with glycidyl-trimethyl-ammonium chloride [92]. The thermosetting PAE resin is converted to a non-thermosetting creping adhesive after a reaction with ammonia or amine [93,94].

The use of poly-secondary amines cross-linked with epichlorohydrin may pose environmental corrosion problems associated with the halogen moiety [46]. To reduce the residual by-products (1,3-Dichloro-2-propanol [DCP] and 3-chloropropane-1,2-diol [CPD]), the lower ratio Epi/amine is used and chlorinated compounds are consumed within an enzymatic posttreatment [95].

A halogen-free adhesive would prevent the corrosion of the Yankee drum surface [96]. In order to avoid that, the epichlorohydrin was replaced by 3-glycidoxypropyl trimethoxysilane [97]. The silyl-linked polyamidoamines are prepared with minimal levels of halide ion or with halide ion being completely absent.

As an alternative chemistry (chlorine is no longer involved), a di-aldehyde (glyoxal and/or glutar-aldehyde) [45,83] or ammonium zirconium carbonate (AZC) [82] is recommended as cross-linker for polyamine.

8.2.1.2 Reactive Self-Cross-Linkable Creping Adhesives

All the cross-linking reactions presented in the previous section were performed before the compound was used as creping adhesive. The lightly cross-linked polymers show higher molecular weights, and therefore their solutions are hard to handle. Sometimes it is better to develop the cross-linked structure directly on the Yankee dryer, which can be seen as a chemical reactor.

The changes in chemical structure of the adhesive are based on chemical or ionic interactions (such as between a cationic resin and an anionic polymer [69]). Polymers with at least two different reactive groups (such as chlorine pendant groups and secondary amines) are self-cross-linkable (homo-cross-linking process). Some other mono-functional adhesives may interact with a different cross-linker (co-cross-linking process) [73]. These reactions are designed to set the proper tackiness, adhesion, and film durability.

The PAE-type resins and ethylene–vinyl acetate–methylol acrylamide ternary copolymers [55] are self-cross-linkable creping adhesives. The copolymer latex is stabilized with a mixture (65:35) of anionic emulsifier (sodium lauryl sulfate containing 2–5 mol of ethylene oxide) and nonionic surfactant (a secondary alcohol ethoxylated with 12–20 mol of ethylene oxide), which makes the film more hydrophilic.

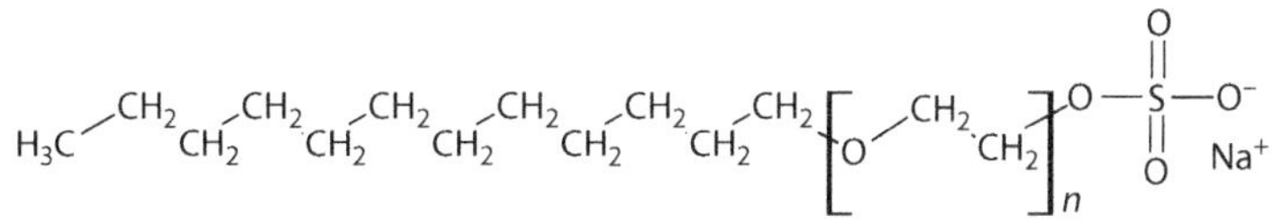

The properties of the PAE resin used as a creping adhesive [37], such as doctorability, rewettability, and the level of adhesion are controlled by the degree of cross-linking with epichlorohydrin. These resins have the same structure as the PAE wet-strength resins but have a much lower amount of cross-linking agent [80,96]. During the PAE resins synthesis, when the conversion of chlorine pendant groups is lower than 100%, the cross-linking process is completed on the drier surface. Self-cross-linkable PAE resins show a shorter shelf life.

Anionic (co) polymers (such as acrylamide/acrylic acid/methyl acrylate or styrene–methacrylic acid copolymer [48]) are also considered as creping adhesives [98] when the anionic groups are neutralized by tri-ethanol amine.

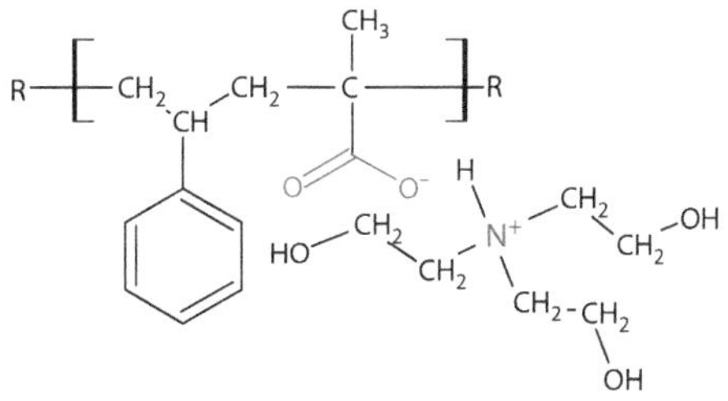

Tri-ethanol amine is a cationic group carrier, a plasticizer, and a potential cross-linker of the poly-carboxylic polymers.

8.2.1.3 Creping Adhesives with a Cross-Linker

The co-cross-linking process is much more common and easier to manipulate. In that case, at least two reactive polymers are involved in the creping process: a non-self-cross-linkable material and a cross-linking agent [44,71]. A typical reactive polymer is poly 2-ethyl-2-oxazoline, which is a polyamide with a weaker bond (the amide bond—in red). The amide bond can be hydrolyzed to poly-secondary amine or it can be involved in an interchange reaction with other amine (such as PEI), acid, or hydroxyl functionalities (see Table 8.1).

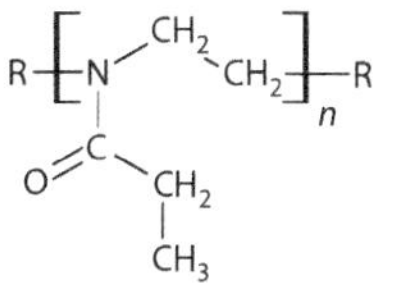

Polyamines, polyamidoamines, or poly-vinyl alcohol and/or vinyl amine copolymers are non-self-cross-linkable materials. An external cross-linker is used; for polyvinyl amine copolymers, the cross-linker is the ether of polyethylene glycol (PEG) diglycidyl [59], AZC, glyoxal [84], or 1,3,5-triglycidylisocyanurate [99]. The cross-linking density influences the mechanical properties and the T_g, which allow for the adjustment of the adhesion/release ratio of the fibrous substrate onto the dryer surface [82].

Chemical interactions developed on the Yankee dryer are even more interesting when a synergetic effect is present: the cross-linked polymer displays a better performance than each separate component and any of their physical blends. The existence of a synergetic effect is recorded when a response (R, such as adhesion) does not fit the simplest mixture rule [100]:

$$R = P_1\varphi_1 + P_2\varphi_2 \tag{8.2}$$

where φ_i is the concentration of the component i and $\varphi_1 + \varphi_2 = 1.0$. In order to fit the experimental data showing a synergetic effect, the mixture rule should be modified by adding a new term that is responsible to the synergetic interaction (I):

$$R = P_1\varphi_1 + P_2\varphi_2 + I\varphi_1\varphi_2 \tag{8.3}$$

In blends with PVA [74], the glyoxalated polyacrylamide (G-PAm) is a cross-linker. In that case, no synergism was found (Figure 8.1): there is no systematic deviation from the calculated adhesion values (the dotted line in Equation 8.2).

Two different PAE resins (with different Epi/s amine mole ratio PAE-I from 0.5 to 1.0 to about 1.8 to 1.0 and PAE-II less than 0.5–1.0) [80] are mixed and the synergetic effect (Figure 8.2) was found at 5%–15% PAE-I in a co-cross-linking reaction.

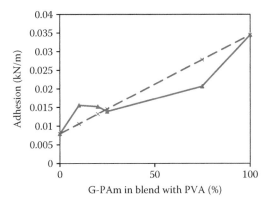

FIGURE 8.1   The adhesion for the PVA–G-PAm blends.

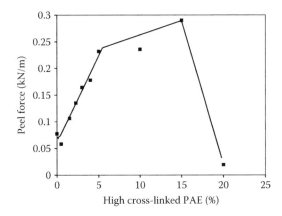

FIGURE 8.2   The effect of the highly cross-linked PAE in blends with lightly cross-linked resin.

There is a very good compatibility between PAE and PVA in solution and the degree of hydrolysis of PVA is important for the properties of the creping adhesive (Figure 8.3): a lower degree of hydrolysis and a lower molecular weight will reduce adhesion [101–104].

The PVA adhesion is higher than that of PAE [64,105], but their blends show a synergetic effect especially at higher PVA contents (Figure 8.4). The synthesis of the PAE resin (such as the reaction of polyamidoamine with epichlorohydrin) can be performed in the presence of PVA (Airvol 325 98% hydrolysis) or PEI [106].

Synergetic effects have encouraged more work to explore other cationic resins. If a polyamine (such as 1,6-hexamethylene-diamine) reacts with epichlorohydrin, that resin shows a higher adhesion than PVA [104]. However, the synergetic effect is noticed when a cationic resin is blended with PVA (Airvol 540) (Figure 8.5).

Polymeric blends as creping adhesives and the diversity of cationic resins and poly-hydroxyl compounds open the possibility to increase the number of components in an adhesive composition. Cationic starches are candidates for a ternary mixture along with PVA and PAE [61–63] for a soft, flexible, white film on the Yankee dryer.

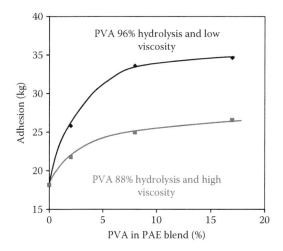

FIGURE 8.3　Adhesion forces for blends of PVA and PAE resins.

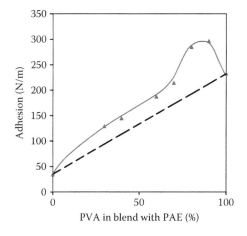

FIGURE 8.4　Peel adhesion as a function of the PVA–PAE composition.

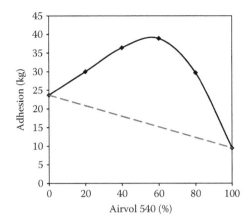

FIGURE 8.5   The synergetic effect on adhesion for blends of PVA and PAE resins.

If the PVA and PAE binary system is able to increase the dry integrity of the tissue (about 0.3%–0.5% PVA, based on dry fiber, is transferred to the paper during the creping process [107]), the ternary mixture (PAE, PVA, and cationic starch (S^+) in %) should have a complex influence on the web tensile (g/in.) [61,63]:

$$\text{Web tension} = 0.26[\text{PVA}] + 0.31[S^+] + 0.81[\text{PAE}] - 0.7[\text{PAE}][\text{PVA}] - 0.56[\text{PAE}][S^+] \quad (8.4)$$

This equation shows that each component contributes differently to the web tension but with a positive sign. However, there are strong interactions (with a negative sign!) between the PAE polymers and PVA and/or the cationic starch.

8.2.1.4   How to Control the Cross-Linking Reaction on the Yankee Dryer

There are several benefits to performing a chemical reaction on the Yankee dryer: the complex system with many variables allow for the development of specific formulas and synergetic effects to accommodate different conditions involved in the papermaking process. Moreover, cross-linkers (such as PAE and G-PAm-type resins) will also provide (temporary) wet strength to paper.

But the chemical reactions that occur on the dryer surface must be carefully kept under control. The cross-linking process helps by generating higher molecular weight compounds showing higher film durability. Unfortunately, the cross-linking process is hard to control and the film may turn into a brittle material with low water absorbability [11]. "Re-wettability" refers to the ability of the adhesive remaining on the Yankee dryer surface to be activated by the moisture contained in the fibrous structure of the paper [42]. The film rewettability prevents the excessive buildup of the creping aid on the drying surface.

The general techniques at controlling the cross-linking process on the Yankee dryer would be to change the reactivity of the participants and to adjust the ratio between components. These approaches were evaluated when a synergetic effect was studied (see the previous section). Several other methods were also investigated using a scavenger and/or diluting with a solvent to slow down the reaction.

The cross-linking process of self-cross-linkable adhesives can be controlled by changing the resin structure (lower cross-linking capabilities are preferred [108]) or it can be moderated with a secondary adhesive used as a controller or as a diluter [109]. Polyamidoamines are cross-linked with epichlorohydrin in the presence of vinyl amine–vinyl alcohol copolymers (low content of primary amine) [110].

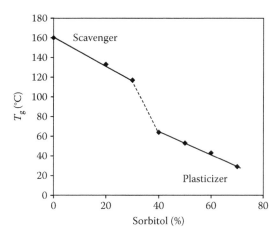

FIGURE 8.6 The effect of sorbitol concentration on T_g of PAAm.

A diluter may be a reasonable option to control the cross-linking of the PAE resins [106]. PVA and a solvent with low volatility (deep eutectic solvent with a melting point of less than 100°C [111]) are used to control the cross-linking process. PVA as a diluter for PAE resins can be added in one or two steps to the Yankee dryer (PVA first, PAE afterward through another nozzle) [68].

In order to control the cross-linking reaction performed with cationic G-PAm (cG-PAm) [112], a reactive small molecule, such a polyol, is used as a scavenger and a plasticizer. The effect of the concentration of the poly-hydroxylic compounds on the reduction in the T_g values shows two different regions (Figure 8.6). At low concentration, the polyol appears to behave as a scavenger, and its molecule is incorporated in the macromolecule. At higher concentrations of sorbitol (over 40%), the cross-linking process is slowed down and the plasticizer effect is obvious.

8.2.2 MODIFIERS

The creping adhesives are macromolecular compounds. Their properties change after blending with other polymers and by changing their chemical structure. The cross-linking process on the Yankee dryer surface may generate a harder and more hydrophobic film. Due to those limitations, small molecules are added in formulas in order to control chemical reactions (such as scavengers; see above), to change the film T_g (such as plasticizers), or to improve water absorption (such as humectants [113]).

For polymers with high T_g values, water may play the role of plasticizer [102]. The amount of water is limited by the drying process: residual water in the creping adhesive film is in equilibrium with the water content in paper. Humectants and hydrophilic solvents with high boiling point (such as ionic liquids) [111] may keep more water in the adhesive film.

Some small molecules, such as aliphatic polyols [108], their oligomers ($M_w < 600$), or polyalkanolamines [29] perform both as plasticizers and humectants: they change the coating T_g and tackiness [2,108] and are also partially transferred to paper.

Glycerol is the most common polyol [79], but ethylene glycol, diethylene glycol, propylene glycol, or PEG should also be used [29,65,66,108]. The same molecule of humectant can contain both hydroxyl and amide groups [113]. Water-soluble polyalkanolamides are prepared by the reaction of a polycarboxylic acid with an alkanolamine [114]. The mixture of mono- and di-alkanol amine reacts with adipic acid and a complex blend of polyamido-polyols is obtained.

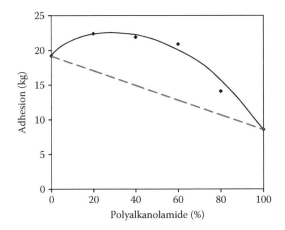

They are both polyols and polyamides, which makes them potential creping adhesive and ensures a very good compatibility with the PAE-type resin. In the case of blends of PAE resin and polyalkanolamide, adhesion shows a synergetic effect (Figure 8.7).

The adhesive layer should be chemically stable at over 100°C and in the presence of air. Regardless of the chemical composition, the chemical compounds with polar structure involve weaker chemical bonds. The thermoxidative process makes the film harder and less water absorbent. The addition of a phosphate salt to the creping adhesive formulation has proven very helpful

FIGURE 8.7   The synergetic effect of polyalkanolamide in blends with PAE resin.

in that respect. As a result, creping is more even, and the cleaning blades need to be changed less often [52,115,116].

Potassium polyphosphate is used in a concentration of 2% based on solids in a PVAc-PVA mixture [52]; 5% sodium hypophosphite is recommended for the PAE-type creping adhesive [115].

The chemicals incorporated in the creping adhesives help in film formation and improve adhesion to the dryer and paper. Moreover, the creping process involves interactions between the adhesive mixture and cellulose fibers and a double transfer of material: some fibers are transferred to the dryer and some chemicals from the adhesive film are transferred to paper. A release aid will control the fiber and the chemical transfer to paper, which improves various paper properties (such as softness).

8.2.3 RELEASE AIDS

Release aids help [37,66] by moderating the adhesive properties, primarily in terms of decreasing adhesion, and releasing the sheet from the dryer at the creping blade [54]. The release aid is important for both the paper quality and the dryer runnability.

The mechanism through which the release aid changes adhesion is still being studied. A change in surface energy and/or a phase separation when the new nonpolar surface interferes with the film adhesive are potential explanations for this mechanism. For instance, ethoxylated acetylenic diols [117] as release aids may change the dynamic surface tension of the creping adhesive; the uniformity of the coating is maintained for a longer period of operation before a doctor blade change is necessary.

On the other hand, the typical practice in the industry involves the use of an adhesive polymer in combination with release oil, i.e., mineral oil or terpene-type oil [44]. If the adhesive is very polar, water soluble, and cationic, the release aid displays reverse properties: nonpolar [66], insoluble in water, and neutral. An emulsifying surfactant package is often formulated to disperse the release aid in the water medium [2].

The preferred emulsifiers for mineral oils are nonionic (alkoxylated alcohol, alkoxylated alkylphenol) [30], cationic, or with both nonionic tail and cationic charges [118]. This type of emulsifier is able to make mineral oil emulsions stable both in $3\,N$ hydrochloric acid and in 5% potassium hydroxide solution.

The cationic emulsifier is a potential softener that adds more potential benefits to the release oil emulsions. The amino-functional poly-dimethyl-siloxane can also be included in that category of emulsifiers with double functions [119].

It is commonly believed that release oil (less soluble material) migrates from the hot Yankee dryer surface toward the air side, creating a gradient of the release oil concentration with the coating [2]. How evenly the release material is distributed on the coating surface is very important because coating uniformity is critical for consistent creping.

An oil release aid, which is not water soluble, will result in an opaque film and no change in the T_g of the adhesive is recorded [2]. Moreover, the oil transferred to paper can have a negative effect on the absorbency of the final product.

In an attempt to solve that problem, amphiphilic polymers have been used. The hydrophilic part is a block of polyethylene oxide (PEO) along with an alkyl tail or a hydrophobic block of propylene oxide. Creping adhesives of PVA or PAE resin (95%) with an amphiphilic release aid (such as alkyl-polypropylene glycol [PPG]) still provide hazy or cloudy films [57]. The haziness shows an incompatibility between components.

The nonpolar character of hydrocarbon oils is weakened by adding cationic charges. The most popular release agent is a cationic compound with a long hydrocarbon tail. The hydrocarbon being its major part, the cationic compound can be seen as a modified version of the hydrocarbon oil.

An interesting case is the lecithin (an amphoteric and amphiphilic compound), which inhibits the sticking of pulp fibers to the surface of the drier [120,121]. Lecithin is added at the wet end as aqueous emulsion in ammonia.

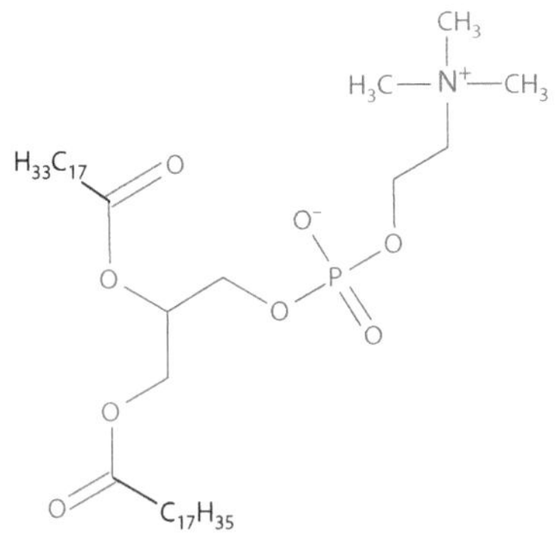

The cationic (or amphoteric) compounds with an amphiphilic structure are supposed to interact with anionic charges from cellulose fibers and change the interfiber bonding. Those types of interactions are also involved in the case of softeners.

8.3   DEBONDERS/SOFTENERS

Paper softness is a function of the stiffness and dry strength of the sheet. As a general rule, higher stiffness results in lower softness. The hydrogen bonds restrict the movement of adjacent cellulosic fibers, which results in a product that feels relatively stiff. The softening techniques disrupt the hydrogen bonds by creping the final paper web, adding a cationic softener, or exposing the dried paper web to electron beam radiation [122].

Paper softness depends on the pulp quality [123], the parameters of the papermaking process (the wet web consistency at the dryer, the pressure on rolls [124], the drying temperature, and the sheet speed), and the quality and amount of paper chemicals (dry-strength resin, creping adhesives, and softeners).

For better softness, tissue products are made from short-fiber pulp only [125] and/or multiple paper layers [126]. Multi-ply structure tries to solve the problem by using different furnishes and a different chemical treatment for each layer [127–130]. The inner layer contains northern softwood strengthened with a higher concentration of WSR (0.75%) and dry-strength resin (carboxymethyl cellulose [CMC] 0.2%). No softener was added to it. The external layer (Eucalyptus stock) contains less WSR (0.2%) and dry strength resin (DSR) (0.05%), but a softener blend is added in the amount of 0.25% at the wet end [128,131].

As far as the process parameters are concerned, there are limitations in the ranges in which the moisture content, temperature, and nip pressure can be adjusted. Drastic changes in the papermaking process (such as drying the wet web with high-pressure air, at high temperature [132]) are rare because those parameters always depend on the machine productivity and economics. These limitations are overcome by using chemical compounds: their chemical structure and molecular weight offer a virtually infinite number of options to improve the paper softness.

8.3.1   SOFTENERS RETENTION AND SOFTENING MECHANISM

The chemical structure and the molecular weight of the softener is important for both its effectiveness (and/or biodegradability) and the fact that it can cause excessive irritation upon contact with human skin [133]. The point of addition for a softener depends on its attributes and its chemical structure. In order to be effective, the softener must be located on paper and preferably on the paper surface. Softeners are added at the wet end [134,135], sprayed on the wet or dry web [135–138], or are included in the creping adhesive mixtures [84,139].

A cationic compound [140–145] or a reactive compound (with an aldehyde functionality on the oxidized cellulose [146]) can be retained at the wet end, but nonionic and nonreactive softeners should be added after the web formation. Modified fibers with more carboxylic groups will retain more cationic softener [147]. However, blends of cationic (di-hydrogenated Tallow di-methyl ammonium sulfate) and nonionic PEG 400 [128,131]) are added together at the wet end. The chemical softening composition is more effective when the components are first premixed together before being added to the furnish (the nonionic compound is retained more than 85%) [148].

The cationic softeners (debonder), added at the wet end, compete with other cationic compounds (wet strength or temporary wet-strength resins [TWSR]) for the anionic sites on the cellulose fibers. Anionic emulsions (such as high molecular weight, branched alcohols stabilized with anionic copolymers) are retained with cationic PAE resins [149].

The competition for anionic cellulose fibers is part of the softening mechanism. Cationic wet strength and/or TWSR increase the paper wet strength [150] but dry-strength resins enhance the number of interfiber hydrogen bonds. A debonder is an inhibitor of the rigid hydrogen bond formation and is believed to function by reducing the number of sites along the individual fibers susceptible to natural interfiber bonding [31]. Therefore, when the debonder is added at the wet end, the sheet density decreases [144,151] and the internal paper cohesion goes below the adhesion degree between the paper web and the drier surface [152].

The reduction in interfiber bonding results in lowering both dry and wet tensile strengths [27]. Surprisingly, for quaternized imidazoline [150], the debonder effect is relatively stronger on the paper dry strength than on its wet strength (Figure 8.8).

The addition of the softener at the wet end would help the creping process: the interfiber bonds are already weaker when the web reaches the Yankee dryer [153]. The softener (0.11% based on dry fiber) is added to the short-fiber stock before its blending with the long fibers. The entire stock is then treated with dry-strength resin and TWSR [154].

The strengthening resins make the paper web stiffer and more hydrophobic, and therefore the sheet shows a lower water absorption. Hydrophilic compounds (polyacrylic acid salt, cellulose derivatives, poly(vinyl pyrrolidone), polyethylene oxide, PVA, glycerol) are sprayed on paper to increase the water absorption [155]. If the lubrication effect is targeted, the softener should remain on the paper surface and must have a melting point at or below body temperature (about 37°C) [156].

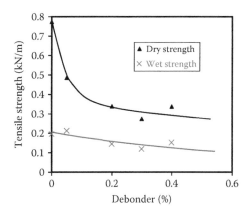

FIGURE 8.8 Debonder effect on the paper dry and wet strength.

8.3.2 PAPER SOFTNESS EVALUATION

The assessment of paper softness is a complex task. Softness can be enhanced by decreasing dry strength [157] and/or by increasing lubricity [149]. It is related to the paper's various degrees of stiffness (flexibility), lubricity [127], compressibility (high-bulk structure, see the double creping on both sides of the sheet [158]), and to the number of fibers protruding from the surface of a tissue per unit area [9].

It is therefore hard to assess the degree of softness of the paper, and the attempt to introduce a "softness evaluation gadget" [159] has not replaced, as yet the "experienced panelists." The "experienced panelist" is considered quite accurate, and thus it was possible to develop an empirical model for tissue softness. According to that model, tissue softness depends on the arithmetic mean of the surface roughness, the CD/MD tensile index, and the mean elastic modulus [160].

8.3.3 CHEMICAL STRUCTURE OF SOFTENERS/DEBONDERS

The chemical structure of the debonder–softener is similar to the release aids and humectants: a cationic charge, a hydrophobic tail [161], and—optionally—PEO segments (where $n \geq 10$ and p and $q \geq 1.0$). That type of debonder can be used as is or in blends with nonionic compounds [162].

The cationic charge of the debonder helps locate the new molecule on the fiber surface. The enhanced hydrophobic character of the fiber surface reduces its hydrogen bonding capabilities. The hydrophobic tail [163,164] has at least twelve carbon atoms [10,165–168] such as the cationic distearylamide of diethylene triamine (DETA) [144], the quaternary ammonium salt of fatty amines [169], the di-hydrogenated tallow alkyl dimethyl ammonium salt [136,157,170–174], and the di-oleyl imidazolinium ethyl quaternary sulfate [70,82,85,86,142,153,154,175–177].

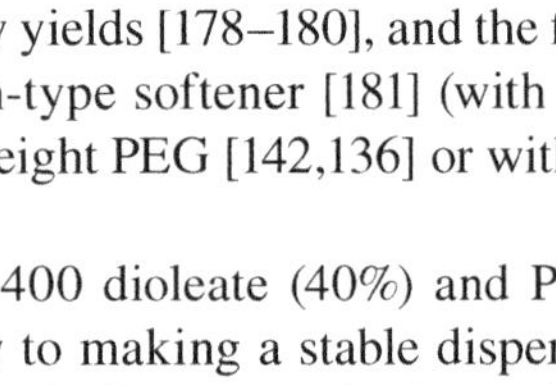

The synthesis of imidazoline derivatives (DETA condensation and fatty acids, followed by cyclization and quaternization) shows low yields [178–180], and the final product may contain intermediary compounds. The imidazolinium-type softener [181] (with a melting point of about 0°C–40°C [175,182]) is blended with 20% by weight PEG [142,136] or with nonionic surfactants [140] (such as silicone glycols [153]).

Amphiphilic molecules of PEG 400 dioleate (40%) and PEG 200 dilaurate (50%) are mixed with imidazoline (10%) with a view to making a stable dispersion with mixed micelles (cationic-nonionic) [177]. Due to these mixed micelles, the cationic-nonionic softener [170] is retained on the pulp at the wet end.

The poly-ether part can be incorporated in the imidazolinium molecule [183]. The new compound includes hydrophobic tails, a cationic charge, and a hydrophilic segment.

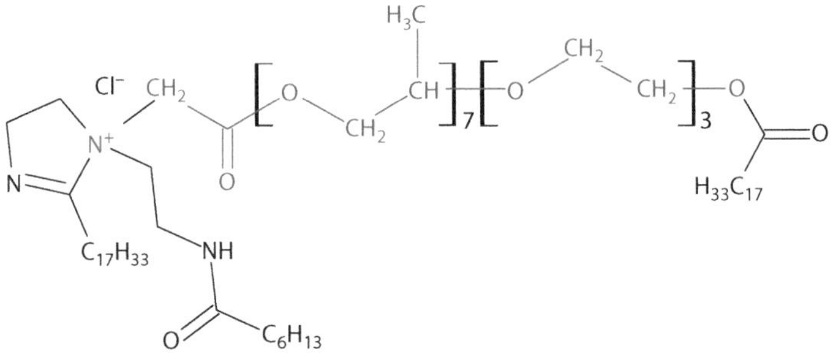

The different acyclic compound as softener is the salt of di-octyl quaternary ammonium [184], which was modified with poly-ether tails [185] (where $x+y<15$).

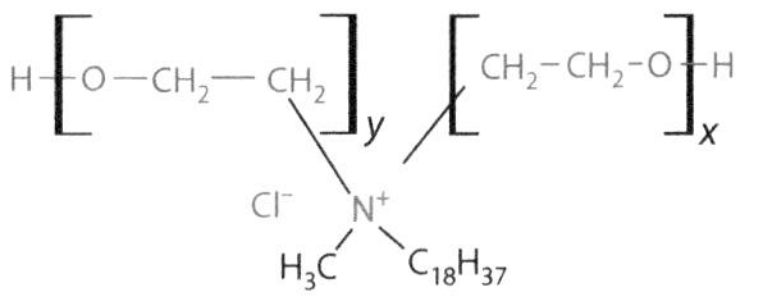

The architecture of the molecule can be changed by adding two hydrocarbon tails [186] and by having only one polyethoxylated group [172]. The polyethoxylated segment can also be located between the cationic charge and the hydrophobic tail [187–189]

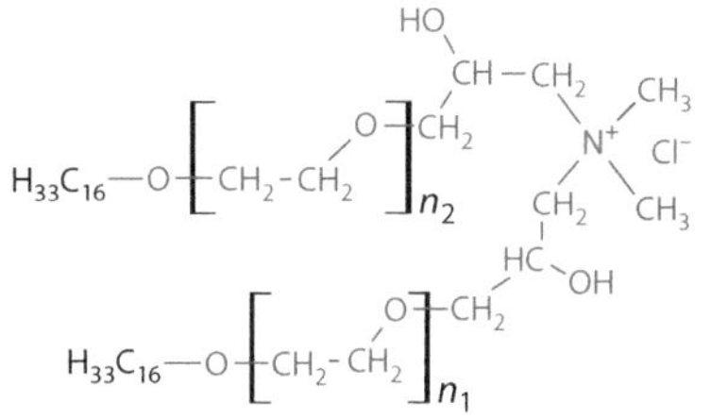

The hydrophilic cationic moiety can be placed in the same molecule with an anionic moiety (zwitterionic softener) [190].

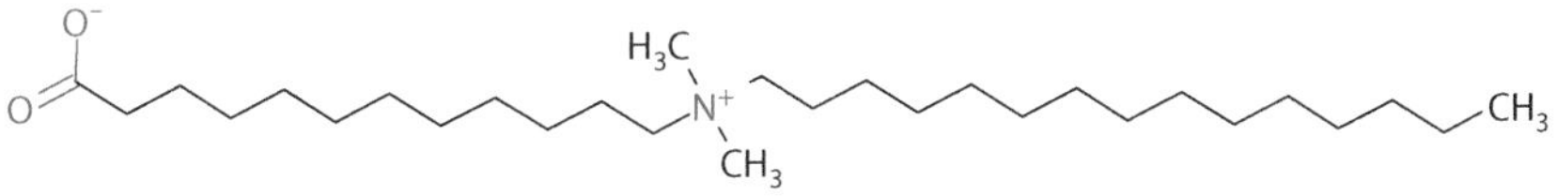

Cationic emulsifiers have two hydrocarbon tails bonded through ester groups [191–193], amino-amide [194–196], or di-amide group [84] and are also used as softeners.

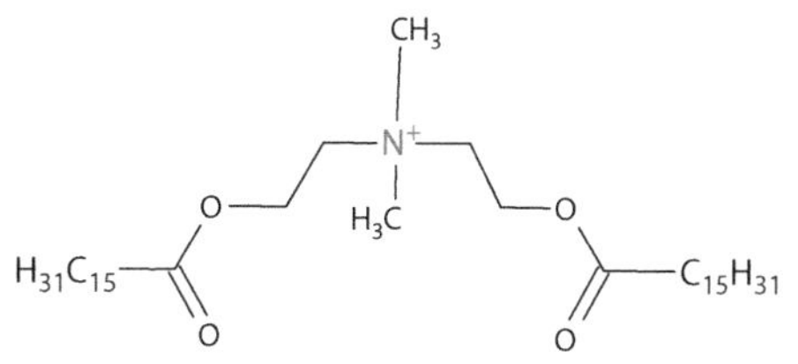

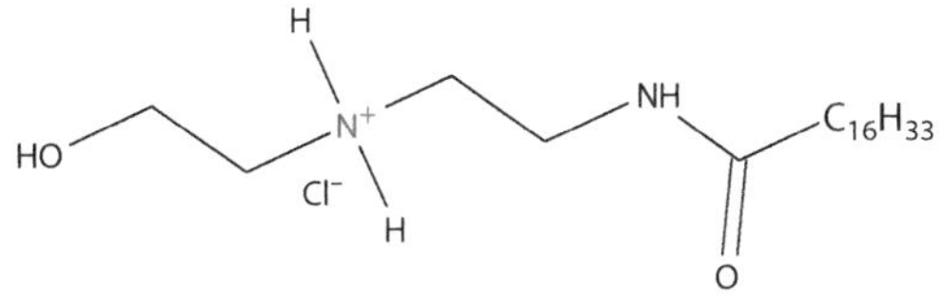

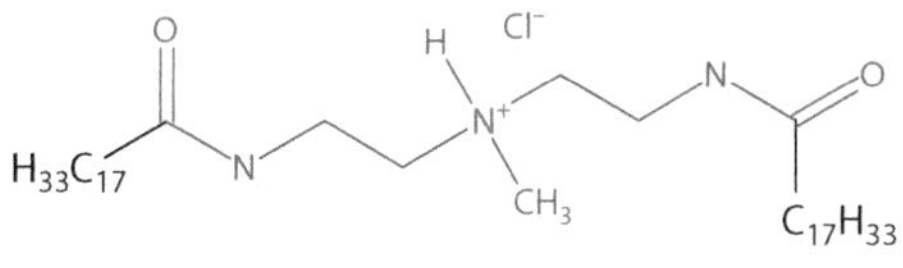

The amide functionality simulates the structure of wet-strength resins. Other functional groups can also be added [197]: an amphoteric structure after a Michael addition [198] or a reactive epoxy group [199].

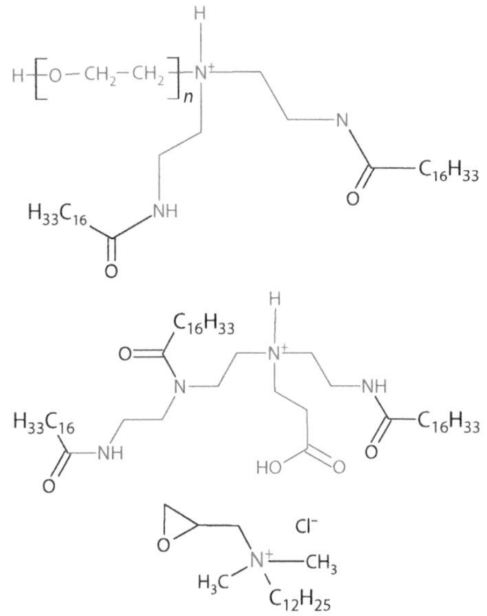

The ratio between the hydrophilic part, the cationic charges, and the hydrophobic tails is critical for the effectiveness of the softener. Quaternary ammonium salts with multiple cationic charges in their structure and hydrophobic tails [200–204] can reduce the paper internal bond factor by 30%. If the polycationic compound has no hydrophobic tails, the paper internal bond increases by 15%.

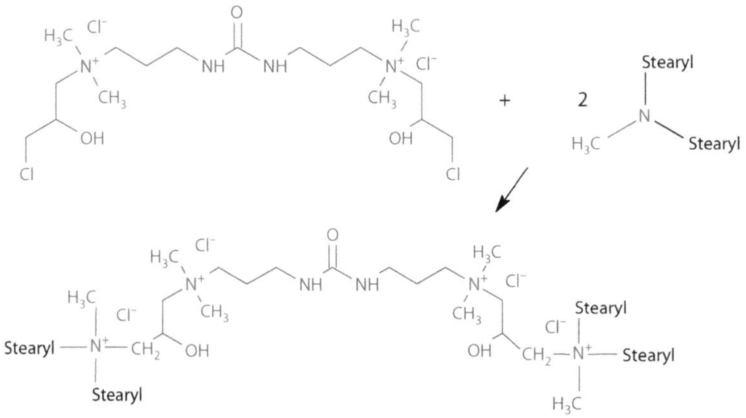

The amphiphilic part and the cationic charges can be located in different molecules, which perform as a blend. Di-isodecyl adipate is dispersed in water and the dispersion is stabilized with a cationic emulsifier [205]. The ratio of the average particle size of the dispersed softener to the average fiber diameter should be in the range of about 0.01 to about 15% [175]. The hydrophobic polysiloxane [206] is carried by cationic compounds [27,207,208] or is incorporated in the structure of the PAE resin [209].

Surprisingly, other chemicals used at the wet end have an impact on the paper softness: fillers and the hydrolyzed AKD [210,211] ("deactivated AKD" in hot water can increase the paper softness [212,213]). Fillers (such as kaolin) occupy the space between fibers, reduce hydrogen bonding, and improve paper softness [214,215]. The filler addition as a coating on the thin paper web, on the Yankee dryer [216], is difficult. Thus, filler (8%–20%) needs to be retained at the wet end.

There are limitations to the addition of softeners at the wet end. The increase in softness is accompanied by a significant decrease in paper strength. Moreover, the cationic surfactants are generally toxic to aquatic organisms and undesired in waste water [217]. The wet end chemistry must be modified to accommodate the presence of a softener and/or to perform the softener addition at a later stage in the papermaking process.

In order to change the wet end chemistry [218] in the presence of a softener, the paper wet strength can be improved by using an anionic wet-strength resin (acrylic latex and styrene-butadiene latex) and a cationic "deposition aid" [171]. Another option would be to replace the cationic compounds by nonionic softeners [218] or to separate the hydrophobic contribution and the cationic charge [219]: the coconut hydrophobic oil is emulsified with anionic (dodecyl benzene sulfonate) and nonionic (ethoxylated fatty alcohol) emulsifiers. The retention of the anionic particles is performed with poly-DADMAC or polyamine [217].

The dry strength of the weakened paper can be also improved by printing a bonding material (CMC, poly-N-methylol acrylamide or urea–formaldehyde [UF] resins [220]) on selected areas.

To avoid all the troubles at the wet end, softeners are added to the wet web (before pressure bonding [157]). They may be either mixed into the creping aids on the Yankee dryer [67] or spread on the dry paper (after Yankee dryer) [27,119,187,197] (especially for nonionic compounds [133,221]). The softeners added on the wet web have a simplified structure: the presence of cationic charges is no longer required. Surface softeners [206] are spread on the wet web (20%–80% consistency) [218,222–225]. Their retention is in excess of 90%.

Nonionic softeners have a long hydrophobic tail and a hydrophilic group, such as N-lauryl-N-methoxypropyl glucamide dispersed in water in the presence of ethoxylated fatty alcohols [226].

Blends that include softeners are complex mixtures with a water-dispersed organic phase (37%): 16% petroleum-based oil (saturated hydrocarbons C_{16}–C_{32}), 0.8% distearyl dimethyl ammonium chloride, 0.5% cetyl alcohol, and 82.7% glycerol [227].

The debonder–softener addition can be combined: the cationic compound blended with PEG-400 is added at the wet end and the polysiloxane emulsion is sprayed on the dry web (Eucalyptus layer) [129] or printed only on the small spots of the dried web [228]. The polysiloxane sequence can also be a side chain in a copolymer composition with other hydrophilic and/or hydrophobic comonomers [229].

The softener is dissolved or dispersed in the creping adhesive blend. The adhesive composition for the Yankee dryer includes 34% polyamide resin, 22% PVA, and 44% imidazoline softening agent [67]. For a heterogeneous, nonionic adhesive, the presence of sucrose (10% based on solids), in a blend with PVAc (80%) and PVA (10%) [53], is claimed to enhance the paper softness.

Multiple component mixtures (as creping adhesives or softeners) deal with the compatibility issues. Most of the components are not soluble, or partly soluble in water, they may lose their solubility due to some chemical reactions or due to the water evaporation on the Yankee dryer. A compatibilizer (ethoxylated sorbitan ester) is blended with a PEO (PEG-400) and a softener (sorbitan monostearatin) to make a stable dispersion [133,230]. The temperature and the concentration of the oil phase (sorbitan stearate and ethoxylated linear C_{12} alcohol or for cationic compounds and PEG 400) have a strong impact on the viscosity of the water dispersion [131,156].

Paper softness is improved after the sheet has left the Yankee dryer surface as dried paper. The dried paper surface is modified with softener and lubricants (such as linear fatty alcohols, ethoxylated fatty alcohols, dimethyl ditallow quaternary ammonium chloride [231]) or with compounds showing a low tendency to migrate inside the web (dimethyl-distearyl ammonium chloride, zinc stearate, and PEG) [135,139]. Migration is also prevented by a coupling agent, such as the ester of the polyhydroxy fatty acid [232].

Nonionic surfactants (alkyl glycoside ethers [233], alkyl polyethoxylate esters) or polyethoxylated compounds are associated with a polysiloxane softener [133] to improve water absorbency. That mixture is sprayed onto a heated transfer surface designed as a calender roll. The film formed on the heated surface is in contact with the creped paper and the molten paper chemicals are transferred to the paper surface [25].

Due to their effect on reducing the paper dry strength, softeners are added to the fibers in limited amounts. In an attempt to recover the paper dry strength, the use of a softener with a long hydrocarbon tail, a cationic charge, and a reactive moiety is suggested [234].

The reactive functionality can make a covalent bond with another fiber and thus, through ionic and covalent bonds the interfiber bonds are consolidated and the softness effect is also preserved.

REFERENCES

1. R.R. Borden, US 3,145,136 (1964).
2. V.A. Grigoriev, R.W. Cloud, G.S. Furman, and W. Su, *13th Fundamental Research Symposium*, Cambridge, U.K., September 2005, pp. 187–212.
3. G.S. Furman, *Tappi Proceedings Nonwovens Conference*, 1990, p. 271.
4. D.A.I. Goring, *Pulp Paper Mag. Can.*, **12**, 515–527 (1963).
5. J.F. Oliver, *Tappi*, **63(12)**, 91–95 (1980).
6. S.L. Edwards and R.J. Marinack, US 6,187,139 (2001).
7. C.T. Waldie and W.L. Hearn, US 3,018,214 (1962).
8. W.S. Pomplun and H.E. Grube, US 4,440,898 (1984).
9. J.E. Carstens, US 4,300,981 (1981).
10. M. Ryan and W. Brevard, US 2006/0249268 (2006).
11. S. Archer, G. Furman, V. Grigoriev, L. Bonday, and W. Su, *Tissue World*, February/March 2009, pp. 33–35.
12. R.F. Cook and D.S. Westbrook, US 5,048,589 (1991).
13. M.K. Ramasubramanian and D.L. Shmagin, *J. Manuf. Sci. Eng.*, **122**, 576–581 (2000).
14. G. Rose, *Paper Age*, June, 2004, pp. 22–26.
15. S. Freti and T. Eriksson, Creping blade and creping blade holder—key functions in tissue quality, *BTG Pulp & Paper Technology*, 1999.

16. M.F. Buccieri, Improvement of the final quality of tissue through a new chemical process, in *The 35th ABTCP Congress*, 2002.
17. J. Woodward, *Pulp & Paper*, February 2007, pp. 40–43.
18. D.L. Shaw, US 3,821,068 (1974).
19. C.J. Campbell, US 6,336,995 (2002).
20. S.L. Archer and G.S. Furman, US 7,048,826 (2006).
21. A.W. Winslow and J.C. Spicer, US 3,640,841 (1972).
22. F.D. Harper, T.P. Oriaran, and J.D. Litvay, US 6,372,087 (2002).
23. J.L. Salvucci and P.N. Yiannos, US 3,812,000 (1974).
24. T. Hassler, G. Sheridan, and H. Reuter, US 2007/0000630 (2007).
25. R.S. Ampulski and P.D. Trokhan, US 5,246,545 (1993).
26. P.V. Luu, T.P. Oriaran, D.W. White, A.O. Awofeso, G.L. Schroeder, and R.E. Fredricks, US 6,352,700 (2002).
27. P.D. Trokhan, D.V. Phan, W.W. Ostendorf, J.K. Monteith, B.S. Hersko, and R.S. Ampulski, US 5,538,595 (1996).
28. J.J. Latimer and T.E. Stevens, US 4,406,737 (1983).
29. A.J. Allen and G. Lock, US 5,833,806 (1998).
30. S.L. Archer, R.E. Dristas, and R.T. Gray, US 6,562,194 (2003).
31. S.R. Grossman, US 4,063,995 (1977).
32. N.W. Lazorisak, F.A. Christiansen, and J.M. Harriz, US 4,064,213 (1977).
33. J. Stitt, *Tissue World Magazine*, April/May 2005.
34. K. Kurosu and R. Pelton, *58th Appita Annual Conference and Exhibition*, 2004, Vol. 1, pp. 43–49.
35. D.A. Soerens, US 4,994,146 (1991).
36. B. Zhao and R. Pelton, *J. Adhesion Sci. Technol.*, **17(6)**, 815–830 (2003).
37. G.S. Furman, V.A. Grigoriev, W. Su, and C. Kaley, *Tissue World Americas Conference*, September 2004.
38. C.J. Roberts, US 3,903,342 (1975).
39. M.T. Goulet, S.L. Pomeroy, and M. Trimacco, US 7,189,307 (2007).
40. B.G. Klowak, W.L. Pauls, and F.J. Vermillon, US 4,125,659 (1978).
41. N.I. Salmeen and B.G. Klowak, US 4,610,743 (1986).
42. C.W. Neal, L.J. Forde-Kohler, and D.A. Salsman, US 6,187,138 (2001).
43. G.A. Bates, US 3,926,716 (1975).
44. N.S.Clungeon and J.J.Boettcher, US 7,404,875 (2008).
45. G.S. Furman, J.F. Kneller, K.M. Bailey, M.R. Finck, and W. Su, US 5,382,323 (1995).
46. P.V. Luu, C.M. Neculescu, and D.M. Mews, US 6,812,281 (2004).
47. W.B. Hill and J.B. Stitt, US 6,991,707 (2006).
48. D.T. Nguyen, US 2004/0162367 (2004).
49. B. Uner, M.K. Ramasubramanian, S. Zauscher, and J.F. Kadla, *J. Appl. Polym. Sci.*, **99**, 3528–3534 (2006).
50. K.H. Wassmer, U. Schroeder, and D. Horn, *Makromol. Chem.*, **192**, 553–565 (1991).
51. J.J. Latimer and T.E. Stevens, US 4,308,092 (1981).
52. P.P. Chen, T. Chiu, and J.R. Skerrett, US 4,883,564 (1989).
53. W.H. Pippin, GB 2,179,953 (1987).
54. H.E. Grube and T.D. Ries, US 4,304,625 (1981).
55. J.E. Goldstein and R.J. Pangrazi, US 6,908,524 (2005) and US 6,974,520 (2005).
56. D.A. Soerens, US 4,684,439 (1987).
57. N.S. Clungeon, B.J. Kokko, D.W. White, and J.J. Boettcher, US 2004/0211534 (2004).
58. D.A. Soerens, US 5,025,046 (1991).
59. V.A. Grigoriev, G.S. Furman, M. Wei, W. Su, and C.D. Kaley, US 2007/0000631 (2007).
60. F. Linhart, A. Esser, and D. Kannengiesser, US 7,604,713 (2009).
61. K.D. Vinson, H.T. Deason, and B.S. Hersko, US 5,944,954 (1999).
62. J.D. Lindsay and T.G. Shannon, US 6,911,114 (2005).
63. K.D. Vinson, H.T. Deason, and B.S. Hersko, US 6,207,734 (2001).
64. D.A. Soerens, US 4,501,640 (1985).
65. C.W. Neal, J.L. Forde-Kohler, and D.A. Salsman, US 6,048,938 (2000).
66. C.W. Neal, L.J. Forde-Kohler, and D.A. Salsman, US 5,942,085 (1999).
67. S.L. Drew, P.J. Allen, and S.J. McCullough, US 6,547,925 (2003).
68. K.D. Vinson, K.P. Mohammadi, and P.T. Weisman, US 5,865,950 (1999).

69. C.W. Neal, E. Aprahamian, and J.A. Cain, US 7,683,126 (2010).
70. J.J. Boettcher, N.S. Clungeon, B.J. Kokko, E.W. Post, P. Van Luu, G.L. Worry, and G.A. Wendt, US 2005/0006040 (2005).
71. D.H. Hollenberg, P. Van Luu, and S.R. Collins, US 5,246,544 (1993).
72. D.H. Hollenberg, P. Van Luu, and S.R. Collins, US 5,981,645 (1999).
73. P.V. Luu, S.R. Collins, and C.M. Neculescu, US 5,370,773 (1994).
74. G.S. Furman, C. Gruen, and P. Van Luu, US 5,179,150 (1993).
75. W.S. Pomplum and H.E. Grube, US 4,436,867 (1984).
76. J.F. Warchol and C.D. Walton-Bongers, US 5,633,309 (1997).
77. J.E. Warchol and C.D. Walton-Bongers, US 5,602,209 (1997).
78. J.F. Warchol and C.D. Walton, US 5,980,690 (1999).
79. L.H. Sanford and J.B. Sisson, US 3,301,746 (1967).
80. S.L. Archer, R.E. Dristas, and R.T. Gray, US 6,277,242 (2001).
81. V.A. Grigoriev and G.S. Furman, US 2009/0133846 (2009).
82. P. Van Luu, C.M. Neculescu, and D.M. Mews, US 6,663,942 (2003).
83. A.G. Sommese and G.S. Furman, US 5,374,334 (1994).
84. P. Van Luu, C.M. Neculescu, and D.M. Mews, US 5,961,782 (1999).
85. P.V. Luu, C.M. Neculescu, and D.M. Mews, US 6,689,250 (2004).
86. P.V. Luu, C.M. Neculescu, and D.M. Mews, US 6,699,359 (2004).
87. T.T. Nguyen, US 6,146,497 (2000).
88. J.W. Clark and C.C. DePugh, US 4,886,579 (1989).
89. J.J. Boettcher, N.S. Clungeon, H.L. Chou, C.E. Ringold, and D.C. Johnson, US 7,718,035 (2010).
90. A.J. Allen, US 5,786,429 (1998).
91. A.J. Allen, US 5,902,862 (1999).
92. W.W. Maslanka, US 6,352,613 (2002).
93. H.H. Espy and S.T. Putnam, US 3,966,694 (1976).
94. P. Knight and G.P. Sheridan, EP 0,856,083 (2000).
95. R.J. Riehle, R. Busink, M. Berri, and W. Stevels, US 7,303,652 (2007).
96. H.H. Espy and W.W. Maslanka, US 5,338,807 (1994).
97. A.J. Allen and A.C. Sau, US 5,990,333 (1999).
98. P. Knight and U. Welkener, US 5,234,547 (1993).
99. C.D. Walton and J.F. Warchol, US 2003/0114631 (2003).
100. L.E. Nielsen, *Predicting the Properties of Mixtures*, Marcel Dekker, New York, 1978, pp. 1–19.
101. D.A. Soerens, US 4,788,243 (1988).
102. G.S. Furman and W. Su, *Nord. Pulp Pap. Res. J.*, **8(1)**, 217–222 (1993).
103. B. Uner, Adhesion mechanism between polymer and metal interface, PhD Thesis, NCSU, Raleigh, NC, 2002.
104. A.J. Allen, US 6,808,597 (2004).
105. D.A. Soerens, US 4,528,316 (1985).
106. W.W. Maslanka, US 6,214,932 (2001).
107. M.J. Smith and S.J. McCullough, US 5,853,539 (1998).
108. A.J. Allen and G. Lock, US 5,660,687 (1997).
109. V.A. Grigoriev, G.S. Furman, S.L. Archer, W. Su, C.D. Kaley, and M. Wei, US 2007/0151684 (2007).
110. W.W. Maslanka, US 5,994,449 (1999).
111. B.J. Kokko, US 2008/0105394 (2008).
112. G.S. Furman, US 5,187,219 (1993).
113. T.P. Oriaran, B.T. Burrier, H.S. Ostrowski, E.W. Post, and J.H. Propp, US 6,207,012 (2001).
114. A.J. Allen, US 6,133,405 (2000).
115. A.J. Allen, US 6,280,571 (2001).
116. A.J. Allen, US 2002/096288 (2002).
117. F.M.C. Chen and F.G. Druecke, US 5,468,796 (1995).
118. K.L. Johnson, US 3,567,734 (1971).
119. R.S. Ampulski, US 5,389,204 (1995).
120. E.D. Martinez and T.F. Duncan, US 4,970,250 (1990).
121. E.D. Martinez and T.F. Duncan, US 5,034,097 (1991).
122. R.F. Ross and J.C. Lau, US 6,808,600 (2004).
123. T. Iwamida and Y. Sumi, *Cellul. Chem. Technol.*, **14(2)**, 253–268 (1980).
124. L.-S. Kuo and Y.-L. Cheng, *Tappi J.*, **83(12)**, 61–61 (2000).

125. K.D. Vinson, US 5,405,499 (1995).
126. S.L. Edwards, P.J. Allen, and O.P. Renier, US 5,494,554 (1996).
127. W.W. Ostendorf, S.R. Kelly, P.D. Trokhan, and D.V. Phan, US 5,397,435 (1995).
128. D.V. Phan and P.D. Trokhan, US 5,405,501 (1995).
129. R.S. Ampulski, J.K. Monteith, W.W. Ostendorf, D.V. Phan, and P.D. Trokhan, US 5,573,637 (1996).
130. D.V. Phan and P.D. Trokhan, US 5,223,096 (1993).
131. D.V. Phan and P.D. Trokhan, US 5,240,562 (1993).
132. R.I. Cole, D.E. Holcroft, E.A. Milligan, and S. Greenhalgh, US 3,432,936 (1969).
133. D.V. Phan and D. Trokhan, US 5,334,286 (1994).
134. D.M. Bhat, R.J. Marinack, H.S. Ostrowski, and S.D. Moesch, US 5,958,187 (1999).
135. L.E. Salaam and S. Prodoehl, US 2009/0188636 (2009).
136. E. Chan, B.M. Woods, and L.E. Salaam, US 2009/0188637 (2009).
137. G. Baumoeller, R. Kawa, A. Ansmann, S. Eichhorn, and A. Urban, US 7,622,021 (2009).
138. C.S. Benz, US 3,817,827 (1974).
139. K.W. Britt, US 3,305,392 (1967).
140. T.T.P. Oriaran, B.E. Burrier, H.S. Ostrowski, E.W. Post, and J.H. Propp, US 6,017,418 (2000).
141. J.H. Dwiggins, R. Ramesh, F.D. Harper, A.O. Awofeso, T.P. Oriaran, G.A. Schulz, and D.M. Bhat, US 6,334,931 (2002).
142. D.T. Nguyen, US 7,217,340 (2007).
143. P.A. Graef and F.R. Hunter, US 5,225,047 (1993).
144. Y. Nakamura, T. Tamai, and K. Asakura, US 7,344,621 (2008).
145. P.A. Graef and F.R. Hunter, US 5,399,240 (1995).
146. T.G. Shannon, G.B.D. Garnier, A.R. Negri, and M.T. Goulet, US 6,916,402 (2005).
147. T. Sun, US 6,361,651 (2002).
148. D.V. Phan and P.D. Trokhan, US 5,279,767 (1994).
149. B.J. Kokko and D.W. White, US 7,041,197 (2006).
150. B. Freimark and R.W. Schaftlein, US 3,755,220 (1973).
151. K. Asakura and A. Isogai, *Nord. Pulp Pap. Res. J.*, **18(2)**, 188–193 (2003).
152. B.G. Klowak, US 4,482,429 (1984).
153. G.A. Wendt, G.V. Anderson, K.S. Lehl, S.J. McCullough, and W.Z. Schroeder, US 5,730,839 (1998).
154. M.J. Smith, V.K. Rao, and G.L. Shanklin, US 5,785,813 (1998).
155. W.R. Watt, US 4,600,462 (1986).
156. S. Ferershtehkhou, L.N. Mackey, and D.V. Phan, US 5,527,560 (1996).
157. I.B. Strohbeen and J.H. Dinius, US 4,377,543 (1983).
158. C.E. Dunning, W.D. Lloyd, and J.G. Bicho, US 4,166,001 (1979).
159. J. Pearlman, US 3,060,719 (1962).
160. J. Liu and J. Hsieh, *Tappi J.*, **3(4)**, 3–8 (2004).
161. T.W. Osborn, US 4,351,699 (1982).
162. B.L. Baursen, US 4,303,471 (1981).
163. R. Pelton, *Appita J.*, **57(3)**, 181–190 (2004).
164. H. Diery, U. Cuntze, A. May, and E. Milewski, US 3,849,435 (1974).
165. L. Contor, Y. Lambremont, C. Courard, and P. Rivas, US 5,133,885 (1992).
166. J.H. Dwiggins, R. Ramesh, F.D. Harper, A.O. Awofeso, T.P. Oriaran, G.A. Schulz, and D.M. Bhat, US 6,328,849 (2001).
167. L.R.B. Hervey and D.K. George, US 3,395,708 (1968).
168. P.H. Schlosser and K.R. Gray, US 2,432,127 (1947).
169. P.H. Vossos, US 3,916,058 (1975).
170. D.V. Phan, US 5,217,576 (1993).
171. F.W. Meisel and K.C. Larson, US 3,844,880 (1974).
172. M.M. Waldman and A.E. Mariahazy, US 3,356,526 (1967).
173. L.R.B. Hervey and D.K. George, US 3,554,863 (1971).
174. L.R.B. Hervey and D.K. George, US 3,554,862 (1971).
175. T.P. Oriaran, B.E. Burrier, H.S. Ostrowski, E.W. Post, and J.H. Propp, US 6,190,499 (2001).
176. C. Poffenberger, *Pulp Paper*, **80(7)**, 40–43 (2006).
177. D.T. Nguyen, US 7,012,058 (2006).
178. Y. Wu and P.R. Herrington, *J. Am. Oil Chem. Soc.*, **74(1)**, 61–64 (1997).
179. D.F. Shumway and H.C. Robbins, US 4,855,440 (1989).
180. D. Bajpai and V.K. Tyagi, *J. Oleo Sci.*, **55(7)**, 319–329 (2006).

181. V.R. Gentile, R.R. Hepford, N.A. Jappe, C.J. Roberts, and G.E. Steward, US 3,879,257 (1975).
182. J.H. Dwiggins, R. Ramesh, F.D. Harper, A.O. Awfeso, T.P. Oriaran, G.A. Schulz, and D.M. Bhat, US 6,277,467 (2001).
183. J.E. Drach, R.D. Evans, J.J. Fanelli, and A.J. O'Lenick, US 4,720,383 (1988).
184. D.V. Phan and P.D. Trokhan, US 5,510,000 (1996).
185. J.A. Breese, US 4,432,833 (1984).
186. W.J. Zeman, C. Poffenberger, and Y.D.R. Deac, US 6,458,343 (2002).
187. T.W. Osborn, US 4,447,294 (1984).
188. J.G. Emanuelsson and S.L. Wahlen, US 4,144,122 (1979).
189. K.M.E. Hellsten, J.G. Emanuelsson, S.L. Wahlen, and A.I. Thebrin, US 4,476,323 (1984).
190. B.E. Chapman, US 3,617,439 (1971).
191. D.V. Phan and B.S. Hersko, US 5,262,007 (1993).
192. D.V. Phan and P.D. Trokhan, US 5,698,076 (1997).
193. D.V. Phan and P.D. Trokhan, US 5,415,737 (1995).
194. T.G. Shannon, N.S. Clungeon, and S.H. Hu, US 6,077,393 (2000).
195. G.S. Hutcheson, US 5,292,363 (1994).
196. G.S. Hutcheson, US 5,417,753 (1995).
197. T.P. Oriaran, A.O. Awofeso, T.N. Kershaw, P.V. Luu, C.M. Neculescu, and M.E. Huss, US 5,399,241 (1995).
198. D.T. Nguyen and S.S. Ashrawi, US 2005/0153866 (2005).
199. J.L. Keen and L.F. Elmquist, US 3,510,246 (1970).
200. J.D. Pera, US 4,851,532 (1989).
201. J.G. Fenyes and J.D. Pera, US 4,581,058 (1986).
202. J.G. Fenyes and J.D. Pera, US 4,778,813 (1988).
203. J.G. Fenyes and J.D. Pera, US 4,506,081 (1985).
204. J.G. Fenyes and J.D. Pera, US 4,970,211 (1986).
205. J.P. O'Brien, E.T. Reaville, and F.B. Erickson, US 3,296,065 (1967).
206. T.G. Shannon, J.D. Lorenz, D.A. Moline, T.M. Runge, and T.H. Schulz, US 6,964,725 (2005).
207. W.Z. Schroeder, D.A. Clarahan, M.T. Goulet, and T.G. Shannon, US 6,398,911 (2002).
208. T.G. Shannon, D.A. Clarahan, M.T. Goulet, and W.Z. Schroeder, US 6,596,126 (2003).
209. W.Z. Schroeder, D.A. Clarahan, M.T. Goulet, and T.G. Shannon, US 6,465,602 (2002).
210. J.C. Roberts, *Proceedings of SPIE-The International Society for Optical Engineering*, 1997, 3227 (Interactive Paper), pp. 20–39.
211. J.C. Roberts, in *Fundamentals of Papermaking Materials*, C.F. Baker (Editor), *Transaction of the Fundamental Research Symposium, 11th*, Cambridge, U.K., September. 1997, 1, pp. 209–263 (1997).
212. D.A. Jones and N.C. Clungeon, US 6,238,519 (2001).
213. D.A. Jones and N.C. Clungeon, US 6,458,243 (2002).
214. K.D. Vison, J.P. Erspamer, C.W. Neal, and J.P. Halter, US 5,611,890 (1997).
215. K.D. Vison, J.P. Erspamer, C.W. Neal, J.A. Ficke, and J.P. Halter, US 5,672,249 (1997).
216. W.F. Thiele and H.B. Richmond, US 2,216,143 (1940).
217. J. Liesen and K. Malmborg-Nystrom, US 7,604,715 (2009).
218. W.U. Spendel, US 4,940,513 (1990).
219. T.P. Oriaran, A.O. Awofeso, G.L. Schroeder, D.W. White, N.T. Luu, and B.J. Kokko, US 6,245,197 (2001).
220. H.E. Becker, A.L. McConnell, and R.W. Schutte, US 4,158,594 (1979).
221. D.D. Newkirk and D.L. Wilhoit, US 4,372,815 (1983).
222. J.H. Dwiggins, F.D. Harper, G.A. Schulz, B.J. Schuh, M.S. Heath, and T.P. Oriaran, US 6,365,000 (2002).
223. D.A. Soerens, L.K.H. Sauer, and G.A. Wendt, US 4,795,530 (1989).
224. J.F. Champaigne, US 3,556,931 (1971).
225. W.U. Spendel, US 4,959,125 (1990).
226. L.N. Mackey, S. Ferershtehkhou, and J.J. Scheibel, US 5,354,425 (1994).
227. P.D. Trokhan and D.V. Phan, US 5,575,891 (1996).
228. K.D. Vinson, P.T. Weisman, J.A. Ficke, and T.J. Klofta, US 5,814,188 (1998).
229. T.G. Shannon, D.A. Moline, J.D. Lorenz, and L.A. Flugge, US 6,893,537 (2005).
230. D.V. Phan, D. Trokhan, and D.C. Hippe, US 5,385,642 (1995).
231. B. Bret and J.F. Leboeuf, US 6,733,772 (2004).
232. J.A. Ficke and K.D. Vinson, US 6,179,961 (2001).
233. R.S. Ampulski and W.U. Spendel, US 5,059,282 (1991).
234. T.G. Shannon, WO 03/054298 (2003).

9 Chemicals for the Treatment of Paper Surface

The paper web produced without additional additives (water leaf) consists of hydrophilic fibers that make up a capillary system [1]. Due to both the capillarity and the hydrophilic nature of cellulose fibers, the paper sheet is very absorbent. The obvious nonuniformity of the paper can impact the product commercialization.

Paper shows a large surface/volume ratio and thus the surface quality becomes important. The surface treatment can improve the appearance and the printing capabilities of a preformed paper web [2]. Surface treatment is a procedure through which the paper surface becomes more even, with a better fiber-to-fiber interaction and with a higher hydrophobicity. That treatment promotes smoothness (due to a decrease of paper porosity) and improves surface strength (reduced linting [3]) and printability [4–6].

The surface treatment is actually an attempt to make an isotropic layer for a material (paper) that is anisotropic in bulk. However, the amount of paper chemicals added to the paper surface is quite low and, for most applications, the porosity changes only slightly [7].

The addition of chemicals to a dried paper web [8] involves a significant capital cost: size-press equipment, tanks for chemicals, and an extra drying of the sheet [9]. To reduce costs, some chemicals are added to the wet web [10] and the rest to the dry end [11]. However, the surface treatment shows several benefits and that is why the volume of surface sizing materials used in paper industries is increasing, while the amount of internal sizing agents shows a slow decrease [4].

The internal sizing agent is added at the wet end. The driving force for the retention process is the charge density of the cellulose fibers. After the saturation of cellulose fibers with cationic compound, some chemical compounds will remain in the white water. In order to reduce the consumption of water in the papermaking process, it is desirable to recycle water, which increases the concentration of un-retained chemicals in white water [12]. Side reactions (such as hydrolysis), ionic association, and precipitation may take place. Their by-products will reduce the machine performance and raise concerns about environmental issues. The risk of penalties charged on high levels of chemical oxygen demand in the mill effluent will impact the way the entire papermaking process is perceived. A cleaner wet end makes substantial savings possible on large, fast machines [13].

The paper surface treatment is beneficial because it is not influenced by pH or other factors that control the adsorption at the wet end. For instance, the anionic polyacrylamide (PAAm) is not retained at the wet end but can be added to the dry paper and in the presence of glyoxal can increase the paper stiffness [14].

The most common method for the application of chemicals to the surface of the paper web is by a size applicator, such as a size press, a calender water box [15], or a coater. The chemicals are supposed to remain on the paper surface; therefore, there is almost 100% retention with near-to-zero pollution [5]. In the size press, dry paper is passed through a flooded nip and a size-press solution contacts both sides of the paper. Excess liquid is squeezed out in the press and recycled. The coated paper is re-dried and cured.

Chemicals added for surface treatment of the dry sheet are the last chemicals used during the papermaking process and they usually influence only surface properties [16].

Surface size treatment can change paper performances. The same paper web can be converted to different paper qualities depending on the nature of the chemical compounds added in the size-press formulation. An appropriate surface treatment allows paper manufacturers to obtain a paper with good performances while reducing and even eliminating the chemicals added at the wet end (such as internal sizing agents) [17–19].

Surface treatment chemicals (or "surface modifiers" or "surface enhancers" [13]) do not need to have ionic charges but their molecular weight is important. Higher molecular weight seems to be a benefit for a surface treatment agent because the larger macromolecule has a lower mobility and will remain on the paper surface.

At the size press, the material pickup [20] is directly proportional to the chemical compound concentrations in water [13], the presence of a carrier (such as starch), the temperature and viscosity of the size-press solution, the moisture content of the sheet, the internal sizing agent concentration in the sheet, the type of surface applicator (film transfer or puddle size press), and the mechanical speed.

There is a limitation for adding chemicals at the dry end: the solution concentration is fairly low (<10%) and large amounts of water should be evaporated [21].

9.1 SURFACE SIZING AGENTS

The printing process requires holding out the ink on the paper surface and preventing characters from "feathering." Feathering is the result of ink absorption outside the boundaries of a character. The balance of repulsive and absorptive forces makes sizing optimization a difficult process. Although there is a concern regarding the improvement of the ink composition [22], the surface coating formulation should match the ink formulation: the surface sizing agents are the last added chemicals to paper but they are the first to start their service during the printing process. The ideal paper substrate quickly absorbs the ink-carrier liquid while retaining the dispersed dye (or pigment) on the surface [23,24].

Apparently, internal and surface sizing agents play the same role: increasing the hydrophobic character of the paper. Surface sizing also reduces paper porosity and improves surface strength. However, there are major differences in terms of sizing mechanism and molecular architecture.

Most of the internal sizing agents have small molecules ($M_w < 500$), which can migrate during the drying step and reorientate on the surface of the cellulose fiber. Therefore, they are evenly distributed in the bulk of sheet. The macromolecular compounds are less effective as sizing agents when added to the wet end prior to the sheet formation [25].

In order to be immobilized on the paper surface, the surface sizing agents must have a high molecular weight with a reduced mobility or must be reactive to the first layer of cellulose fibers (or starch) [3]. The tendency to excessively penetrate the paper sheet would be a major disadvantage limiting the efficiency of a surface size [15].

There is a permanent struggle to keep the surface sizing agents on the paper surface. Latex particles or expandable microspheres are added to a hydrophobic surface sizing solution to increase the amount of material on the paper surface [26]. Another option would be to apply multiple layers of sizing agents: first, oxidized starch (or wax) and, secondly, starch–clay mixture [27] or fluorine-containing polymers [28].

Molecular size is particularly important for a surface sizing agent and so are its chemical structure and macromolecular coil orientation in the fiber vicinity. Poly α-cyanoacrylates have a limited effect as sizing agents but the in situ polymerization of α-cyanoacrylate directly on cellulose fiber shows a much stronger sizing effect [29].

The presence of internal sizing agents improves the performance of surface sizing agents [5,30]. However, the application of starch as a carrier increases the hydrophilic character of the paper and the addition of surface size is designed to restore paper surface hydrophobicity. Therefore, the overall effect of internal and surface size is not a simple addition [31].

A very diverse portfolio of surface sizing agents is needed to fit the diversity of paper and printing process types [32]. In the case of paper for black ink-jet printer, a simply hydrophobic sizing agent is good enough. When it comes to color, we are dealing with a different type of ink and that dictates a different type of surface size. Sometimes the internally sized paper should be "de-sized" with hydrophilic compounds (polydimethyl siloxane and block copolymers Propylene Oxide-Ethylene Oxide [PO-EO]) in a size press, to improve the paper ink-jet printability [33]. The Hercules sizing test (HST) value decreases, for instance, from 68 s (before treatment) to 0.4 s after treatment.

Slurry used in surface sizing paper process contains water, starch, and sizing agents [34]. Sizing materials are typically in the form of aqueous solutions, emulsions, or suspensions [35]. The sizing agent is a nonreactive compound or can be a soluble polymer that becomes hydrophobic—through a chemical reaction—at the drying step.

For special applications, very different ingredients might be added to the sizing formula: modified zirconium salt [36], hyperbranched dendrimer with cationic moiety [37], polyvinyl alcohol (PVA) [38,39], polyurethane [40–42], cationic latexes [43], amphoteric copolymers [44], polyethylene glycol (PEG) ($M_w = 35,000$) [45], alum [46], casein and oxidized starch [47], hydroxyethylated starch and acylated gellan gum [48], and polyisocyanates [49].

The effect of the sizing agent is evaluated with different testing methods. The HST measures the time needed for ink to penetrate a sheet of paper, providing information about both internal size and surface size in only one measurement. Unfortunately, based on HST numbers alone we cannot separate the effect of the internal sizing agent and the effect of the surface sizing agent. The contact angle method [23] is a better tool, which closely simulates the "ink holdout" and provides surface size data.

Several other lab testing methods have been developed (Cobb, Bristow wheel, water drop, etc.). However, the ultimate test is, of course, the end user's requirements, i.e., the printing performance [50,51].

The interaction between sized fibers and water (or ink) is a dynamic process. The contact angle method [23] shows the variation of the contact angle vs. time, which most closely mimicked the print-quality response [30]. Figure 9.1 shows large experimental errors (see also Section 2.6) and the difference between ink and water as the fluid used for contact angle measurements (the slope is ≠ 1.0).

A smaller contact angle with ink than with water is always expected (the slope is smaller than 1.0) and the time is an aggravating factor: after 1 min the slope is even smaller.

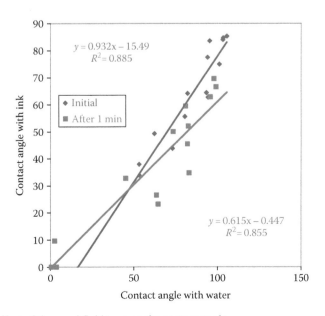

FIGURE 9.1   The effect of time and fluid type on the contact angle.

The printing performance of the coated paper depends on the amount of sizing agent on the paper surface. The following factors affect the pickup of the size-press formulation [13,52–54]: the size-press solution quality (the type of starch, the starch/sizing agent ratio, the surface tension, the solids content, the solution temperature, pH, and viscosity [1,7]), the sheet quality (moisture, alum content, filler content, internal sizing agent concentration, and sheet porosity), and the operating variables [55,56] (type of surface applicator—film transfer or puddle size press, speed, nip pressure, and type of roll). A better penetration of the sizing agent into the bulk of paper (if needed) is obtained by applying vacuum on the opposite side of the web [57].

9.1.1 Starches for Size-Press Solutions

Only in very few cases starch is not involved in paper surface sizing [58]. Starches (from different natural sources [59]) are biodegradable polymers [60] and they are largely used in the papermaking process: 5.2 million tons in 2000 (to be compared with about 0.58 million tons in 1960 [61]). About 70% from the starch used in the paper industry [62] is added in size-press solution [63–65].

The amount and the starch quality (molecular weight and charge density) are crucial for the performance of the size-press solution [31,66]. The starch grade used in surface sizing should always be carefully chosen to fit each particular production line (surface size agent—anionic/cationic, paper type, etc.) [67].

The compound used in surface treatment can interact with additives in size-press solution. Generally, cationic starch (or cationic amylose [68]) is a good quality surface size [69,70]. However, cationic starches [1,3] can react unfavorably with anionic macromolecular compounds. Anionic styrene–maleic anhydride copolymer (SMA) (amide ammonium salt), [71] 3% based on dry starch, shows an HST value of 95 s with oxidized corn starch and only 29 s with cationic corn starch. Cationic starches used for surface applications contain a lower charge density than cationic starches used at the wet end [72]. The charge density can be adjusted by cooking the anionic starch with cationic resins [70].

Starch, brought by the size-press solution, is also expected to add internal strength to the sheet through liquid penetration in the z-direction. Low molecular weight starch can penetrate into the fiber network more than higher viscosity starch remaining on the surface [8]. The starch molecular weight is an important factor for sizing and strength, and it can be controlled by starch cross-linking [73–76].

The thermal properties of starch are also important for paper surface treatment: in the drying step, the surface temperature reaches over 105°C. Modified starches (such as starch succinates) show a better thermal stability [77].

There are many procedure for cooking starch, [78,79] and a wide variety of starches are used in size-press solutions [8]. The final concentration, temperature, and pH are also very different from one mill to the other. As a general rule, for practical reasons, starch is cooked at a higher concentration (at high temperature or passed through a homogenizer [80]), and then it is diluted to a convenient concentration. Before the addition of any other chemical, a pH adjustment is made.

Corn starch (10% in water) was heated at 90°C for 30 min, and then the solution was diluted with water to 5%. The pH was adjusted with caustic soda to 8.0. The sizing agent solution (10 g of 10% concentration SMA half ester copolymer) was mixed with 200 g solution 5% starch. The size-press solution was held in a 55°C water bath until used in the size press [81].

Oxidized starch [82] (or ammonium persulfate converted starch [1,83–86]) is cooked at 20% solids and 95°C for 20–40 min, diluted to 3%, then NaOH solution is added to obtain a starch solution with a pH of 7.1–7.8 [87].

Cationic starch is recommended for cationic surface size [17]. Cationic starch (0.17% mole cationic groups) [88,89] slurry contains 20% by weight starch. The starch was cooked 30 min at 90°C–95°C and cooled down to 40°C. After dilution, the concentration was 3%.

One of the starches used in size press is enzymatically degraded [90]. After the starch degradation, the enzyme activity is stopped by adding acid (sulfuric or hydrochloric). The excess of acid is neutralized by caustic, which will generate relatively large amounts of electrolyte [91].

Regardless of the diversity of the starch-cooking procedure, several remarks can be made about starches as components of the size-press solution [8,92,93]:

1. Starches of different botanical origins (wheat, corn, waxy corn, tapioca, potato) have different amylose/amylopectin ratios and different amylose molecular weight [94,95]. Potato starch contains amylose with the highest molecular weight (M_w = 800,000), but potato starch with amylose M_w about 300,000 [96] gives the best sizing values [1]. The amylopectin molecular weight is up to 3 millions and shows a branched structure [97–99] by alpha 1,6 linkage [94]. Potato starch contains no fat or proteins [100], while cereal starches, due to their fat and/or protein content, will show a lower strength, and, therefore, a higher amount of sizing agent is needed for comparative results.

2. The number of hydrogen bonds between the macromolecules in a starch granule is so high that such granules are not soluble in cold water. However, due to the branched structure of amylopectin, a lower number of hydrogen bonds are developed in starch as compared with cellulose. This makes starch hot water soluble.

3. Uncooked granules of starch are not sticky. In order for starch to develop gluing properties, it has to be dissolved in water. Only the completely dissolved starch molecules exhibit their full properties such as charge density and a better interaction with cellulose fibers or paper chemicals.

4. Temperature and shear rate are of a high importance during starch cooking. Within a certain temperature range, the swollen granules begin to rupture and collapse to a viscous, colloidal dispersion of granule fragments, hydrated starch aggregates, and dissolved molecules. The swollen aggregates and the dissolved molecules have a much higher molecular coil volume, resulting in inter-aggregates and intermolecular interactions. The dispersion viscosity reaches a maximum. The viscosity of the solution is a function of the molecular weight factor of the macromolecules and—at high starch concentrations—of the intermolecular interaction factor.

5. Heating the starch slurry to temperatures below 100°C results in a dispersion still containing granule fragments and hydrated starch aggregates. A molecular solution cannot be obtained at such temperatures. In order to obtain a molecular solution, the starch dispersion should be heated at 110°C–150°C (cooking for a very short time, at higher pressure in jet cookers).

6. Solutions with higher starch concentration in water are obtained after starch hydrolysis with acids or enzymes (thinned starch). The viscosity of the starch solution cannot exceed about 150 cPs [101].

7. The ion concentration in water is of a high importance for starch solubilization. Starch dispersion in tap water shows a viscosity that is five times lower than in DI water. The viscosity of the ionic starch solution is lowered by an increase in the conductivity of the solution. (See the polyelectrolyte behavior in the presence of electrolytes.)

8. The final pH should be adjusted according to the nature of the paper chemicals used in size-press solution.

9. The starch solution is not stable: retrogradation takes place (starch molecules go from a dissolved and dissociated state to an associated form [62,94]). The processes of gelatinization and retrogradation are opposites [99]. The hydroxyl groups associate through interchain hydrogen bonds. The aggregates are stable even at high temperature conditions. Retrogradation does not occur at pH above 10 and is slow below pH 2. To prevent retrogradation, starches can be stabilized by oxidation [102,103], derivatization [104], and crosslinking [105,106]. A biocide is needed to protect against fungi.

The above rules are, more or less, related to starch viscosity because that is the most important liquid parameter determining its sorption rate. High-speed application requires starch products with narrow viscosity specifications. Starches with a mono-modal molecular weight distribution show a better runnability [92].

9.1.2 NONREACTIVE SURFACE SIZING AGENTS

Nonreactive chemical compounds should be stable in water and their structure must remain unchanged after deposition on paper. Water-soluble polymers as surface sizing agents should have a hydrophilic part (to help them get into the solution) and a hydrophobic part needed to provide hydrophobicity to the cellulose fiber. That double character makes them good surface-active compounds and their solutions will foam significantly.

A surface sizing agent performs better if its molecular weight is higher. As the molecular weight goes up, the solution viscosity goes up as well. That dependence is not linear, and for a given chemical composition, the high molecular compounds will have a very limited concentration for a reasonable viscosity value.

Starch has a large, hydrophilic molecule. In order to make it more hydrophobic, it needs to be blended with lignosulfonate and borax [107] or long alkyl tails must be attached to the starch molecule [106,108,109].

The need for higher concentrations of sizing material in water with low viscosity [110] has led to the conclusion that the hydrophobic compound should be dispersed in water. Dispersions have lower concentration of soluble material and therefore a lower tendency to generate foam [17].

The stabilizer of the dispersed phase is located at the water–organic compound interface: anionic, cationic, low molecular weight, or macromolecular compounds. In a blend with starch, the concentration of the stabilizer at the interface will change and the dispersed phase may coagulate. The hydrophobic particle stability depends on the compatibility between starch and stabilizer, pH, ionic strength, and temperature.

The dispersed organic compound form is characterized by molecular weight and molecular weight distribution, by hydrophilic–hydrophobic (lipophilic) balance (HLB) and by T_g (or minimum film formation temperature [MFFT] or the melting point in the case of a small molecule compound). T_g values are critical to ensure that appropriate processing and drying conditions are chosen for developing a latex film or to stick to the fiber as individual particles,

The hydrophobic character of the dispersed material and the ability of the system to "cover" as much of the paper surface as possible, convey the sizing effect to the hydrophobic dispersion. Due to the very small concentration of the sizing agent, even for a T_g below room temperature, the polymer does not generate a continuous film. It is a matter of number of particles: the larger the number of particles, the better the paper will be "covered." Smaller particles perform better in the printing process [110]. For a small and constant polymer amount, a smaller particle diameter means a larger number of particles.

However, there is a "geometric" area of the sheet and a specific surface area developed by fiber distribution. The pulp beating and refining are the two parameters able to change the specific fiber

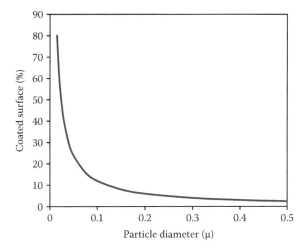

FIGURE 9.2   The paper surface area coated by polymer particles (1 kg/ton paper) as a function of the particle size.

area [111]. For a geometric area of 133 cm²/g (a sheet with a basis weight of 75 g/m²), the specific surface area is about 100 times higher (11,400–47,400 cm²/g) [112].

Figure 9.2 shows the "real" coated area by a dispersed particle system ($G = 1000$ g/T) as a function of particle size. In order to calculate the surface of that area (Equation 9.1)

$$\text{Area} = 2 \left[\frac{6G}{D} \right] (\text{cm}^2/\text{T}) \tag{9.1}$$

several assumptions were made: the dispersed particle density is 1.0 g/cm³, the specific surface area is taken at the lower limit suggested [112]: 11,000 cm²/g (see also the specific area for cellulose fibers: 20,000–450,000 cm²/g [113]), the T_g of the dispersed phase is below drying temperature (105°C), and—after drying—the area occupied by a flatted polymer particle is double the area of a circle with the initial particle diameter.

The polymer particles are typically 50–100 nm in diameter and are attached to the fiber by hydrogen bonds [13]. During the drying step, at temperatures higher than the polymer T_g, the polymer particles become soft and can be spread on the fiber surface. Therefore, less than 15% from the "real" paper surface can be coated. Any reduction in particle diameter will result in a tremendous increase of the coated area. Conversely, small particle sizes require a larger amount of stabilizer (usually wetting and foaming agents).

When the polymer dispersion is put in contact with paper a heterocoagulation process takes place: the particles deposit on the fibers. "Coated" paper does not refer to paper coated with a continuous film of hydrophobic material. "Coated" does not necessarily imply a continuous film of hydrophobic material; the surface sizing agent will only improve the fiber hydrophobicity. Even after drying at temperatures higher than the T_g values, when the hydrophobic compound is better spread, there is lot of room left between particles.

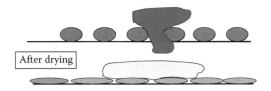

After drying

The surface size dispersion is added to the starch dispersion in the size-press solution. The stabilizer type, its desorption capabilities, and the presence of starch in the continuous phase have a strong effect on latex efficiency in paper sizing. Due to the dilution and the stabilizer desorption, the particles remain unprotected and a coagulation process takes place, which may change the sizing effectiveness due to the changing numbers of dispersed particles.

Starch, with its hydrophobic part (starch is also dispersed in water), can adsorb some of the stabilizer molecules. Thus, there is a competition between polymer particles and starch aggregates for the stabilizer molecules. The polymer particles get fewer stabilizer molecules and loose stability, i.e., homocoagulation occurs. That will result in larger aggregates and therefore in a lower number of particles.

In order to increase latex stability and prevent homocoagulation, more attention was paid to the immobilization of the stabilizer on the latex particles. Two options are available in order to increase latex stability in the presence of starch: (1) increasing the stabilizer molecular weight for slower desorption and (2) bonding the stabilizer to the particle surface through covalent bonds.

There are two ways to make dispersions: (a) by emulsification of a hydrophobic material with a reasonable low melting point, or (b) by emulsion polymerization.

9.1.2.1 Emulsions of Nonreactive Small Molecules as Sizing Materials

Dispersions of pentaerythritol, PEG [114], or glycerol esters of rosin and/or disproportionated rosin (with or without fortified rosin) are hydrophobic compounds used as surface sizing agents. The dispersions are obtained in the presence of cationic starch and sodium lignosulfonate [115]. The rosin derivatives can be blended with C_5 hydrocarbon resin, indene resin, terpene resins, and coumarone resins.

The emulsification is performed at high temperatures (145°C–160°C) in the presence of about 16% (based on organic phase) mixed stabilizer (cationic starch and sodium lignosulfonate (2:1)) and alkaline pH (7.0 < pH < 9.0) [12]. In order to reduce the mixing temperature, a solvent (methyl t-butyl ether) for rosin ester can be used. After emulsification, the solvent is distilled under reduced pressure.

Thermoplastic hydrocarbon resins (dropping point in the range 50°C–150°C) can be used without rosin as surface size [12,35,115]. The dropping point is the temperature at which the first drop of a melted sample under investigation flows through the 2.8 mm diameter bottom orifice of a standard cup on slow heating of 1°C/min.

Hydrocarbon resins are obtained by polymerization of deep cracking of petroleum distillates. There is an aliphatic stream (C_4–C_6), an aromatic stream (C_8–C_{10}), a cyclopentadiene, and a dicyclopentadiene stream. The dropping point of the hydrocarbon resins depends on the monomer structure and on their molecular weight. For the best HST value, the dropping point should be around 100°C (Figure 9.3), which is in the range of paper temperature. The thermoplastic resins having an optimum dropping point are equivalent as surface to the industrial standards: SMA-type solution and acrylic-type emulsions.

It seems that thermoplastic resins contribute to sizing by softening in the drying step and flowing over the surface of cellulose fibers to form a hydrophobic film [35]. Other water-dispersed thermoplastic resins that can be used as surface sizing agents include polyamide amide waxes (stearamide), anionic and cationic petroleum resin [116], polyester resin [117–121], paraffin wax [122], N,N′-alkyl substituted aspartamides [123], and polyvinyl acetate latex with dibutyl phthalate as plasticizer [124].

Amphoteric compounds are obtained by reacting stearic acid with triethylene-tetramine to form a di-amidoamine, which is reacted with maleic anhydride, fumaric acid, or acrylic acid [125]. A Michael addition is involved along with an amidation (the water is distilled) to form a compound with a molecular weight of about 2000. In the presence of ammonia, the free carboxylic groups are converted to their ammonia salts and wax becomes dispersible in water in the absence of any other stabilizer.

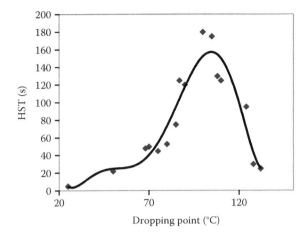

FIGURE 9.3   The effect of the hydrocarbon resin dropping point on the HST values.

9.1.2.2   Surface Size Obtained through Emulsion Polymerization

The emulsification of the low molecular weight compounds has several limitations: the melting point of the dispersed phase, the high emulsifier concentration for higher oil/water phase ratio, the particle size, and the particle size distribution are hard to adjust. Emulsification is a physical process and the stabilizer cannot be chemically bonded to the organic phase.

Emulsion copolymerization [126–128] is a better option for surface size synthesis. The power of the latex copolymers as surface sizing agents is well known [110,129,130]. That process can reach the required T_g and can control the particle size distribution and the stabilizer type and concentration.

At the starting point, the emulsion polymerization mixture includes droplets of monomers and a water-soluble initiator. Any organic monomer (styrene included) is at least slightly soluble in water. The free radical initiation starts in the water phase and the polymer particles are formed through "homogeneous nucleation" [131]. The number of nuclei (and therefore the final particle size) depends on the type of protection ensured.

The copolymer composition depends on the comonomer ratio and the type of process (batch wise, semi-continuous, or continuous). The emulsion copolymerization process can design the copolymer structure (in terms of T_g [132]), the particle composition (homogeneous, core-shell [133], "raspberry" structure, etc.), and the particle size distribution.

In the case of binary copolymerizations, the mole ratio between comonomers in the final copolymer determines the T_g of the copolymer. Figure 9.4 shows the T_g values for the styrene–butyl acrylate (St-BuA) copolymer calculated according to the Fox equation [134,135] (see Section 2.5.2). To get a T_g below room temperature ($T_g < 20°C$) [136], the copolymer composition must be richer than 40% in butyl acrylate.

The chemical composition distribution of the copolymer (CCD) depends on the reaction conversion, comonomer ratio, and the reactivity ratios of the comonomers. The comonomer reactivity ratios are included in the copolymerization equation [131,137] (see Section 4.3.1). The water-soluble comonomers are distributed between the organic phase and the water phase, and the resulting copolymer shows a very broad CCD. The copolymer richer in water-soluble comonomer is located on the particle surface.

The copolymerization equation is a differential equation: at every moment of the copolymerization conversion, the comonomer composition shifts from the initial composition, and a different copolymer composition is obtained as conversion goes up. Only for a very particular case when $r_1 = r_2$, the copolymer composition is identical to the comonomer composition.

The St-BuA copolymer is the most common surface sizing agent. The reactivity ratios in the binary copolymerization of styrene with butyl acrylate [138–140] are $r_{St} = 0.72$ and $r_{BuA} = 0.24$. For these reactivity ratios, the copolymer composition is shown in Figure 9.5 (at conversions of less than 5%) as a function of the feed composition. Only at 0.73 mole fraction of styrene (azeotropic composition) in the feed composition, a uniform copolymer composition (0.73 mole fraction of styrene in copolymer) can be obtained up to high conversions. Unfortunately, the St-BuA copolymer showing high performances as surface size must have a higher concentration of butyl acrylate (see Figure 9.4). All the copolymer compositions were calculated with the PROCOP computer program [131].

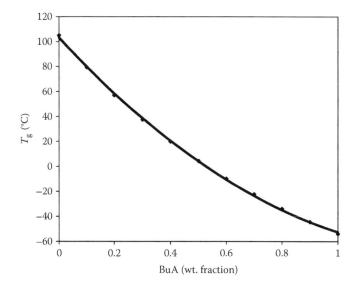

FIGURE 9.4   The glass transition temperature (T_g) for St-BuA copolymers.

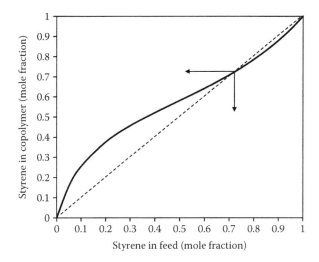

FIGURE 9.5 St-BuA copolymer composition as a function of comonomers feed composition.

St-BuA copolymers [135] with a composition of 0.5 mole fraction butyl acrylate were tested as surface size ($T_g = 20°C$ for a styrene/butyl acrylate 1:1 ratio [141]). This copolymer composition is far from the azeotropic composition and, in a batch process, both the instantaneous copolymer composition (in black) and the overall copolymer composition (in red) [142] start at about 0.42 mole fraction of butyl acrylate and shift to a copolymer richer in BuA as conversion goes up (Figure 9.6).

The spectacular change in instantaneous copolymer composition is recorded at high conversions where the processes are interesting from the commercial point of view. At high conversions, the copolymerization rate is changing [138], the unreacted monomer is a concern [143], and the particles have an outer layer rich in BuA. That may change the stabilizer adsorption, the latex stability, and the polymer performance as sizing agent.

The copolymer synthesized by using batch-wise emulsion copolymerization is often a blend of different copolymers, with different styrene concentrations and different T_g values. The copolymer fractions with T_g higher than the drying temperature will not melt and spread on the cellulose fiber and will lower the sizing agent efficiency.

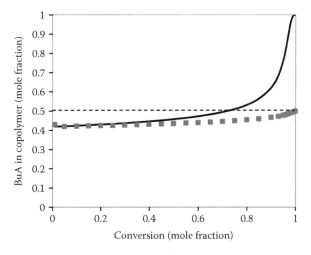

FIGURE 9.6 St-BuA copolymer composition vs. conversion.

That is why a "starved" semi-continuous process [144–146] was developed in order to get a homogeneous copolymer composition [147]. In a semi-continuous process, the overall copolymer composition at the end of the feeding was different from the feed composition and this difference was proportional to the monomer feed rate [147].

About 10%–15% of comonomers and 20% of the initiator solution is added in one shot at room temperature before heating at the reaction temperature [135,136], the rest of the comonomers and initiator solution could be added continuously [146]. At low feeding rates, the copolymerization proceeds at conversions of about 90%–95% and the copolymer composition is practically equal to that of the monomer feed.

The final conversion depends on the initial concentration of the initiator, the comonomer ratio [148], the chemical structure of the initiator, and time [143]. For a kinetic study, the unreacted monomer is not a concern, [149] but for a commercial reactor, a low residual monomer is a priority. Supplementary amounts of initiator are used to reduce the unreacted monomers.

The binary copolymers for surface size include one or more hydrophobic comonomers (styrene, methyl methacrylate [MMA], acrylonitrile, styrene derivatives, etc.) and at least one comonomer (such as butyl acrylate or butadiene [150]) for a lower T_g (20°C–30°C) [84].

The solids content could be increased from 10% to 50% but some floc formation may occur [151]. The copolymer particle size and the particle size distribution is related to the phase ratio, the amount of stabilizer, the polarity of the polymer formed in every stage, the temperature, pH, etc.

The particle size distribution is set by changing the monomer feed rate, the amount of stabilizer, and the timing of feed with respect to the overall conversion rate [144]. Particle size of 60–300 nm [130,152] is recommended. A larger particle size and/or a broader particle size distribution (bimodal) can also be achieved by seeded emulsion copolymerization [153].

The type of stabilizer and its concentration are crucial for the latex application as surface size [154]. There is an interrelationship between the particle size distribution, emulsion stability, and the stabilizer concentration at the interface. Hydrophobic materials are good for black ink-jet printing but for the color printing the hydrophobicity should be balanced by ionic charges [32]. The ionic charges are provided by an ionic comonomer and/or by the type stabilizer. Anionic charges make latex compatible with optical brighteners [32], and the presence of alum in the pre-sized paper [13,155] helps the sizing process.

9.1.2.2.1 Small Molecule Stabilizers

The low molecular weight emulsifiers (molecular weight <1000) are distributed between the particle surface and water. Stable latexes involve high concentration of stabilizer at the interface. The emulsifier concentration in water is high as well. That can lead to undesirable foaming [156] and lower sizing effect [87,157].

Due to their "pseudo-irreversible" adsorption, high molecular weight compounds may improve the adsorption of small molecules at the interface. The emulsion polymerization is performed in the presence of an anionic emulsifier (such as sodium alkyl sulfonate and alkyl diphenyl ether sulfonate) [58,135,150,158,159], and the latex stability is improved by adding a polymer (such as dextrin, starch, or PVA [160–162]).

The chemically bonded stabilizer is a broad concept, but it is an interesting attempt to increase the stabilizer concentration at the particle–water interface. As a result, the stabilizer concentration in the water phase is very low. That would open the opportunity to rethink the low molecular weight emulsifier application in surface size synthesis [163].

The bonded stabilizer can also be a potential sizing agent. Disproportionated rosin acid soaps (an internal sizing agent—see Section 7.2) are also emulsifiers used in emulsion polymerization [164,165]. Due to its reactivity as chain transfer agent (see the benzylic hydrogen atoms, in red, in the following figure), rosin is partially bonded to the polymer particles, which results in a better stability and lower foaming.

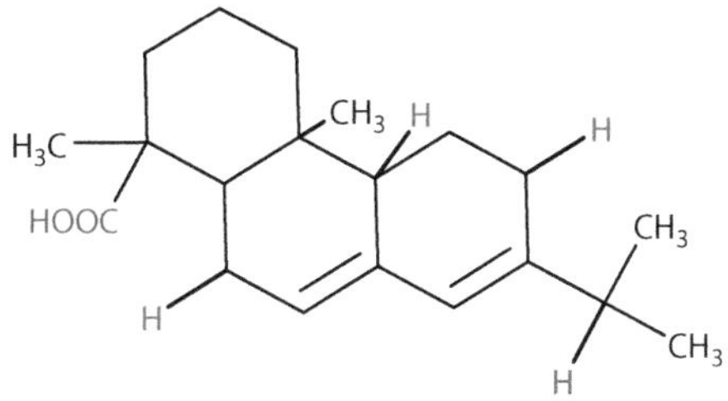

Within a free radical polymerization, new covalent bonds are developed both through the chain transfer reaction and during the propagation step. The potential stabilizers are small molecules or polymers and they must have reactive groups capable to react as chain transfer agents or as comonomers with carbon–carbon double bonds.

9.1.2.2.2 Emulsifier-Free Emulsion Polymerization

In order to eliminate any hydrophilic compound, the emulsifier used in emulsion copolymerization was taken out of the formula. The particle diameter obtained during the styrene polymerization in the absence of emulsifier follows the empiric equation [127]:

$$\log D = A\left[\log\left(\frac{[I][M]^{1.723}}{[P]}\right) + \frac{4929}{T}\right] - C \tag{9.2}$$

where
   I stands for the initial ionic strength
   P for the initiator concentration
   M for the monomer concentration
   T for the absolute temperature (A and C are constants)

The particle size increases with the ionic strength and the monomer/water ratio and decreases with the initiator concentration and the reaction temperature. The initiator brings the polar groups for the in situ synthesized stabilizer. The synthesis of that stabilizer is helped by polar comonomers copolymerized with the two major hydrophobic comonomers. Latex stability is increased by using ionic monomers (such as methacrylic acid [166], acrylic acid [167], or quaternized vinylimidazole [160]) or nonionic hydrophilic comonomers [87] (acrylamide, hydroxyethyl acrylate, N-acryloyl pyrrolidine [168], etc.).

As a general rule [58], a latex usable as surface size is a ternary copolymer: a very hydrophobic comonomer (styrene), an acrylic ester comonomer (for reduced T_g), and a water-soluble comonomer (acrylic acid, itaconic acid, methacrylic acid, dimethylaminomethyl methacrylate [127]). By copolymerization in the water phase (first step in emulsion copolymerization), a macromolecular stabilizer is formed [169]. The water-soluble comonomer (like acrylic acid) will be the major component in that copolymer [170]. The protection capabilities of the waterborne copolymers depend on the molar ratio of the comonomers. That copolymer remains in water or it is adsorbed on the hydrophobic polymer particle enhancing the latex stability. The higher initiator concentration makes the particle size smaller.

In the case of the St-BuA copolymerization in the presence of ionic comonomers (acrylic, methacrylic, or acid itaconic acids [169,171]), the amount of the acidic comonomer on the particle surface depends on the pH value. At low pH (pH=2.2), most of the acidic comonomer is undissociated and it is trapped in the particle, while at pH=6.0, most of the acidic comonomer is lost in water.

The experimental data on the evaluation of the carboxylic groups' concentration [149,172] provides information related to latex stability and the sizing properties of polymer particles.

Several other anionic water-soluble comonomers are suggested: fumaric acid, maleic acid, or 1-allyloxy-2-hydroxypropyl sulfonate [110,135,146,152,170]. The copolymer between a hydrophobic comonomer (major amount in the recipe) and the water-soluble comonomer will be located on the particle surface.

Higher ionic monomer concentration reduces the particle diameter [173]. During the emulsion copolymerization of styrene, butyl acrylate, and anionic sulfopropyl methacrylate (SPM) the following relationship was found between the polymer particle size and the concentration of the anionic comonomer [174]:

$$D \text{ (nm)} = K \frac{325.1}{[SPM]^{0.39}} \tag{9.3}$$

However, there is a lower limit for the particle size because there is an upper limit for water-soluble comonomer concentration: the copolymer with too high a concentration of water-soluble comonomer is hydrophilic and is a poor sizing agent. The particle size distribution can be improved by seeded emulsion polymerization [175].

The so-called "carboxylated latexes" have higher mechanical stability, freeze–thaw stability, and improved adhesion to polar substrates (like paper) [170]. Latexes with carboxylic groups on the particle surface are also useful as toner resins [176].

Cationic latexes show a good interaction with cellulose fibers (see Section 3.3.3.4). The cationic (or amphoteric [1]) latexes are also used in the size-press solutions. Cationic polymer additives enhance dot resolution and water fastness by associating with the anionic inks to form an immobilized colorant that retards bleeding and resists washing off with water [24].

Soap free emulsion polymerization can be used to obtain cationic latexes by polymerization of a hydrophobic (neutral) monomer in the presence of a cationic (water-soluble) comonomer and a water-soluble functional initiator [177–180]. Examples of hydrophobic monomers are St-BuA copolymer [177,181] or vinylidene chloride [182]. The following compounds are recommended as cationic comonomers: dimethyl aminoethyl methacrylate [177,179], vinyl-benzyl ammonium chloride [178,181,183], vinyl pyridines [182,184], *N,N*-dimethyl-*N*-butyl methacrylamidopropyl ammonium bromide [185], trimethyl aminopropyl maleimide ammonium iodate [186], trimethyl aminopropyl maleamic acid ammonium iodate [187], and *N*-vinyl imidazole [182].

In order to prevent any anionic–cationic interaction during the emulsion copolymerization process, the cationic water-soluble initiator is selected (such as 2,2′ Azobis(2-methylpropionamidine) dihydrochloride) [177]. The end group will remain on the particle surface and will increase the density of the cationic charge.

As in the case of anionic comonomers, higher concentrations of initiator and/or cationic comonomer will result in higher reaction rates and smaller particle sizes. Figure 9.7 shows the particle size as a function of the cationic comonomer (aminoethyl methacrylate hydrochloride) [180].

The cationic comonomer has a big impact on the particle size at very low concentrations. At higher concentrations, the curve reaches a plateau: at higher concentrations of cationic comonomer

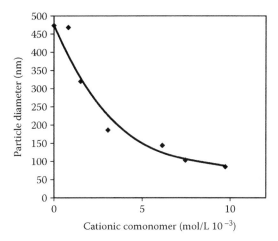

FIGURE 9.7   The dependence of the polymer particle size obtained in emulsion polymerization in the presence of cationic comonomer.

(over 5% based on hydrophobic comonomer), the particle size is about 100 nm (0.1 μm). The particle size distribution is very narrow: almost monodisperse. This is an ideal sizing agent: all particles have the same diameter, which will behave almost identically in the presence of starch and/or cellulose fiber.

Unfortunately, the latex concentration is about 10%. The addition of a cationic emulsifier alongside a nonionic emulsifier may increase the polymer concentration up to 50% [182] and the particle size is smaller [188].

The stabilizer synthesized during an emulsion copolymerization in the presence of a water-soluble comonomer shows a wide CCD: the fraction richer in hydrophilic comonomer remains in water and only a small fraction with an appropriate hydrophilic–hydrophobic balance is located on the particle surface [172]. The water-soluble polymer may damage the stability of the size-press solution [189] and its sizing effectiveness.

9.1.2.2.3   *Polymerizable Surfactants (Surfmers)*

In order to overcome that shortcoming, the water-soluble comonomer should be a polymerizable surfactant (surfmer) [190–192]. In order to prevent the formation of a rich fraction in surfmer through its homopolymerization in water, the reactivity ratio of the polymerizable surfactants should be $r_s = 0$ [193]. Thus, no homopolymer of surfmer is obtained.

Surfmers have a hydrophobic part, a cationic charge bonded to the hydrophilic part, and are called "polymerizable surfactants" because a polymerizable carbon–carbon double bond is present. Due to their surface tension properties, the surfmers are used in much lower concentration.

Surfmers can be ionic or nonionic compounds and their chemical structure includes a hydrophobic tail and an anion crotonic ester [194,195].

The ionic stabilizer can be pH sensitive, which might have an undesirable effect both on sizing capabilities and/or latex stability [87]. The latex stability and the hydrophilic–hydrophobic ratio of the final material can be balanced by using a nonionic stabilizer able to be involved in a copolymerization, or in a chain transfer reaction, as is the case of a long polyethylene oxide (PEO) branch (maleic ester) [196].

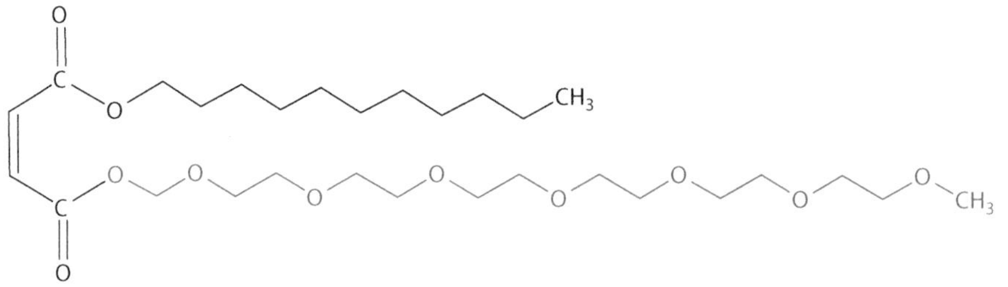

Neither the crotonic, nor the maleic esters homopolymerize, but both are ready to copolymerize with a hydrophobic monomer (such as styrene and/or butyl acrylate). Vinyl sugars (such as vinyl di-acetal of galactopyranose) can be copolymerized with any vinyl comonomer [197]; the final material is hydrolyzed to obtain a hydrophilic surface, which is supposed to be compatible with the starch solution of the size-press solution.

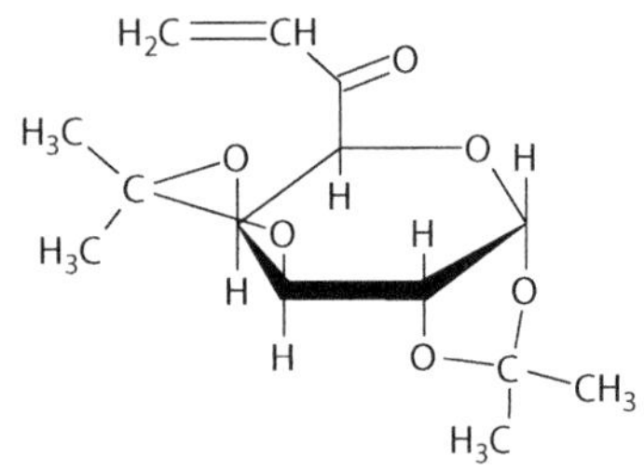

Polymeric surfactants [189,198] are another option for stable, high-concentration latexes. The amphiphilic macromolecule compound is a block copolymer consisting of a PEO segment (degree of polymerization about 31) and polybutadiene segment (degree of polymerization about 10) [154,199]. The butadiene units are involved in a chain transfer reaction at the allylic hydrogen.

9.1.2.2.4 Starch as Protective Colloid

Starch (in a modified form) [135,159,200,201] is suggested frequently as an appropriate stabilizer based on its capability of "pseudo-irreversible" adsorption. Moreover, starch is less sensitive to electrolytes, which is a very important benefit in paper industry [58].

The stabilizer for a colloidal system should be a macromolecular compound having a low molecular weight [152,158] to prevent an increase in the dispersion viscosity (dextrose: number average molecular weight 3300–5900) [150].

To reduce their molecular weight, starches are subjected to oxidative, thermal, acid or enzymatic hydrolysis. Cationic waxy maize starch (20% in water) was oxidized with ammonium persulfate (5% based on starch) for 130 min at 90°C [87], or with hydrogen peroxide [152,160].

Low molecular weight dextrin (<3000 $M_n < 6000$) is formed by an incomplete hydrolysis of starch with diluted acids or by the action of heat. The final product may have a bimodal distribution for their molecular weight (6500–13,000) [135]. Dextrin (33% based on solids) [136,150] is used to make emulsions with higher solids content (>45%).

Enzymatic starch degradation was performed at 85°C for 30 min with calcium acetate and a commercial α-Amylase solution [152,160]. The excess of acid used to control the starch degradation was neutralized by caustic, which will generate relatively large amounts of electrolyte [91]. That electrolyte may reduce latex stability.

The weight ratio of comonomers to the stabilizing agent (modified starch) is ranging from about 1:1 [87,202] to about 4:1 [152,203,204]. The amount of starch should be optimized because a high stabilizer concentration improves the latex compatibility with the size-press solution but it also increases the amount of hydrophilic on paper surface.

During the free radical emulsion polymerization, starch as stabilizer can be involved in grafting reactions (see Section 2.4.2); the newly formed grafted starch is a better stabilizer [156]. In order to obtain a grafted starch during emulsion copolymerization, a redox initiation system is used: hydrogen peroxide and iron (II) sulfate. The properties of a graft copolymer formed with gelatinized starch are substantially different from the properties of a graft copolymer of granular starch that is treated to gelatinize it after copolymerization [203].

The effectiveness of starch as a stabilizer can be improved by adding another nonionic emulsifier (such as nonyl-phenol ether of PEO) [202] or by using a water-soluble comonomer [146,205] to generate more stabilizer in situ. Examples of cationic comonomers are [201] quaternized vinylimidazole (**9.8**), quaternized dimethylaminoethyl acrylate, acrylamidoethyltrimethyl ammonium chloride, and methacrylamidopropyltrimethyl-ammonium chloride.

Cationic starch used as stabilizer for cationic surface size [204,206] should be combined with a cationic starch as a carrier in the size-press solution. This type of latex is recommended for ink-jet printer paper (see Section 9.1.3.1.4).

The protective colloid remains solidly anchored to the polymer core. It functions as a coupling agent between starch (used as a carrier in the size-press solution) and the hydrophobic polymer core (interpenetrating network) [204] and prevents tearing at the hydrophilic–hydrophobic interfaces.

A better control of the particle size is performed by seeded emulsion polymerization [150]: the seed particles (about 2%) have a particle size of 40 nm for a final particle size of about 250 nm.

An anionic latex used as surface size (Chromaset™ 600) [84] has the following characteristics: zeta potential (anionicity) between −40 and −55 mV, $5.0 < pH < 9.0$, particle size ranging from 80 to 150 nm, and polymer T_g ranging from −15°C to 50°C. The addition rates are approximately 3%–5% by weight of starch and 0.1%–0.25% polymer/dry paper, both calculated as active solids.

9.1.2.3 Surface Treatment for Oil-Resistant Paper

There are fluorochemicals used as surface size and grease-resistant materials [207]. The fluorochemicals are small molecules (such as fluoroalkyl phosphates) [208–210], medium molecular weight compounds [211], or copolymers. Anionic perfluoropolyethers (with medium molecular weight) are used in the size-press formulation along with nonionic starch [212]. In order to make a cationic compound, perfluoroaliphatic acrylates and methacrylates are copolymerized (in solution or emulsion) with cationic methacrylates [213–217]. The copolymer solution is used to size paper, which is not only hydrophobic but also oleophobic (greaseproof).

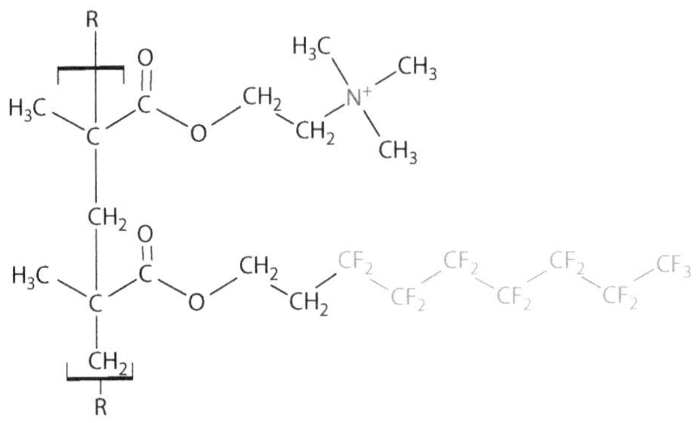

Unexpected water and oil resistance was noted in the case of paper treated with anionic blends (thinned starch, sodium lauryl sulfate, wax emulsion [101], branched polylactic acid [218]) or with cationic hexamethoxymethyl melamine modified with long chain alcohols (C_{16}–C_{24}) in the presence of hydrochloric acid [219]. The average molecular weight of this cationic sizing agent is about 1200, and it is also used as an active extender for fluorochemicals.

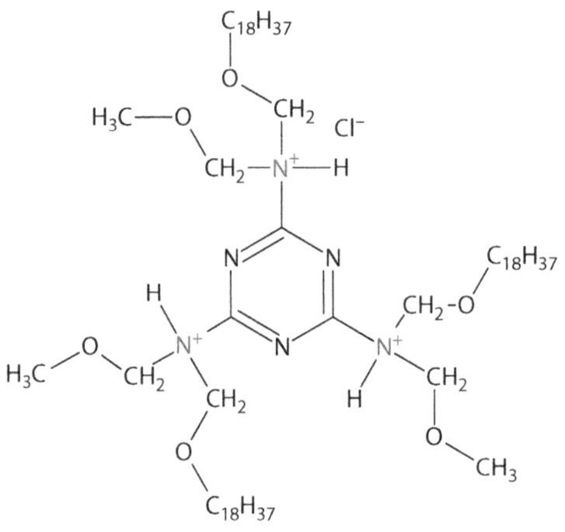

The surface sizing agents presented in this chapter do not show any chemical reactivity during the paper drying step. Their hydrophobic properties are set during their synthesis and remain unchanged throughout the papermaking process.

9.1.3 "REACTIVE" SURFACE SIZING AGENTS

The reactivity of a surface sizing compound means that the hydrophobicity comes not only from the sizing agent itself, but also from different types of interactions or chemical reactions with cellulose fibers, or from intramolecular changes in the chemical structure of the sizing agent during the drying step. The hydrophobic character is improved by developing new bonds and by reorienting the big molecule. For example, the chemical reaction is developed between a urea–formaldehyde (UF) thermosetting resin and acrylic latex [220] or between a polyamine and formaldehyde [221].

Cationic and reactive fluorochemicals show higher performances than the nonreactive fluorochemicals. Polyamines react with fluorinated epoxides at a mole ratio (amine hydrogen/epoxy) >1.0. The un-reacted amine groups are treated with epichlorohydrin to make a reactive azetidinium cycle within a fluorinated surface sizing agent [222].

The concept of "reactive" sizing compound may include those chemicals able to interact with ink: cationic or anionic polyelectrolytes (polyaminoamine epichlorohydrin polymer (PAE) or polyguanidine) [223–225].

9.1.3.1 Anionic Water-Soluble Polymers

In order to be dissolved in water, hydrophobic surface sizing agents must be amphiphilic compounds: some hydrophilic segments are also present in their molecule [226]. Low molecular weight propylene glycols esterified with isophtalic and trimellitic acids [227] are surface sizing agents when several end groups are unreacted carboxyl functionalities.

The hydrophilic units are acids or salts of those acids. Since acids and their salts often have different solubilities, the inclusion of these units in the surface size backbone may have an undesirable effect of increasing the pH sensitivity of the sizing agent [87]. Carboxylated water-soluble copolymers cannot be used at low pH because the carboxyl groups are protonated and the copolymer solubility is lost. However, that capability of a carboxylic group to change its hydrophilicity when converted from a salt to the undissociated acid functionality is part of the explanation of the success of polycarboxylic compounds as surface sizing agents.

9.1.3.1.1 Styrene–(Meth)Acrylic Acid Copolymers

Styrene copolymers with unsaturated acids are obtained [83] by solution polymerization in a mixture of water and iso-propanol. Sodium, potassium, or ammonium salts (or blends thereof) are used as surface size.

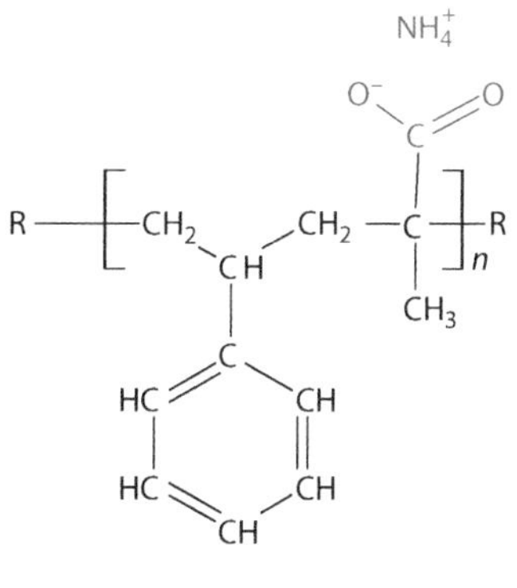

The copolymer of the styrene–methacrylic acid (1:1) has a balanced ratio between its hydrophobic (in black) and hydrophilic parts (in blue), which makes it soluble in water (as sodium, potassium, or ammonium salt) while being still hydrophobic.

The amount of styrene (the hydrophobic component) in the copolymer composition is crucial for good print performances (Figure 9.8). There is an optimum composition close to 1:1 mole ratio [85]: a higher number of carboxylic group makes the copolymer very hydrophilic, a higher styrene content makes the copolymer (in its salt form) insoluble in water.

The solution copolymerization of styrene and (meth)acrylic acid is performed in blended solvents (water and isopropanol). The copolymer composition depends on the comonomer ratio and the solvent. The alcohol–water mixture is preferred because it is easier to synthesize a copolymer with a homogeneous composition ($r_{ST}=1.11$ and $r_{MA}=1.13$ [83,85], therefore $r_{ST}=r_{MA}$). In bulk, the comonomer reactivity ratios are different ($r_{MA}>r_{ST}$ for $r_{ST}=0.124$ and $r_{MA}=0.60$ [228]).

The copolymer composition is critical for its performance as sizing agent (see Figure 9.8). While it is intended to produce a polymer having a substantially random composition, copolymers obtained at lower conversions have a different composition than that obtained at the end of the reaction [229]. In order to get a narrow CCD, a carefully chosen ratio between water and IPA (1:2–1:3) and an optimum molar ratio between styrene and methacrylic acid should be used.

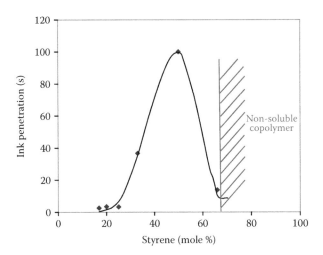

FIGURE 9.8 The effect of the composition of styrene–methacrylic acid copolymer on ink penetration.

Styrene–(meth)acrylic acid copolymers may incorporate a third and a fourth comonomer: acrylonitrile [230], MMA, hydroxyethyl acrylate, vinyl acetate, etc. [85,231]. As the copolymerization process becomes more complex, more attention should be paid to the homogeneity of the copolymer [232].

9.1.3.1.2 Styrene–Maleic Anhydride Copolymers

SMA is a very interesting molecule [233]: the anhydride is very reactive with alcohols, and amines, and it can be hydrolyzed with a water solution of sodium (potassium and/or ammonium) hydroxide.

SMA copolymers show the same sizing potential as the styrene–acrylic acid copolymers and have already been tested as internal size as well (Section 7.2). Moreover, maleic anhydride brings two carboxylic groups per mole and does not homopolymerize, which simplifies its synthesis.

SMA is obtained by copolymerization in bulk or in solution [234–238]. The ratio between styrene and maleic anhydride in a copolymer is designed by changing the styrene/maleic anhydride (or fumaric derivatives [239]) ratio and/or by using a proper copolymerization procedure [240–247]. Because maleic anhydride does not homopolymerize (the MA-MA sequence does not exist in the copolymer structure), there is a limitation of maleic content in the SMA copolymer: the maximum amount of anhydride incorporated in SMA is 50%. A large number of carboxylic groups can be incorporated by a ternary copolymerization of acrylic acid–styrene–maleic anhydride [247].

The SMA copolymers obtained from a styrene/maleic anhydride ratio higher than 1.0, at high conversions, will show a strong compositional drift [248]: copolymer fractions with high content of maleic will coexist with fractions with very high styrene content. The fractions with low maleic content (below 15% MA) may not be soluble in caustic or ammonia solution.

The SMA copolymers with an alternant structure are easy to obtain [240]. The 1:1 copolymer composition results from a 1:1 monomer ratio and the copolymers with a homogeneous composition are obtained up to high conversions. The copolymer molecular weight can be controlled by the concentration of some inorganic salts (such as lithium chloride) [249] or by the addition of di-vinyl benzene (0.3%–5%) [250].

When an SMA copolymer richer in styrene is targeted (SMA 2:1 and 3:1), the copolymerization conversion should be kept at a lower level (about 50% and recycling styrene), which increases the cost of the product.

The SMA copolymers are water soluble in alkaline pH. An aminic acid (amide–ammonium salt) is obtained with ammonium hydroxide (**9.13**). These SMA salts are anionic polyelectrolytes and their viscosity in water solution depends on the SMA concentration, the SMA molecular weight [18], and the presence of electrolytes.

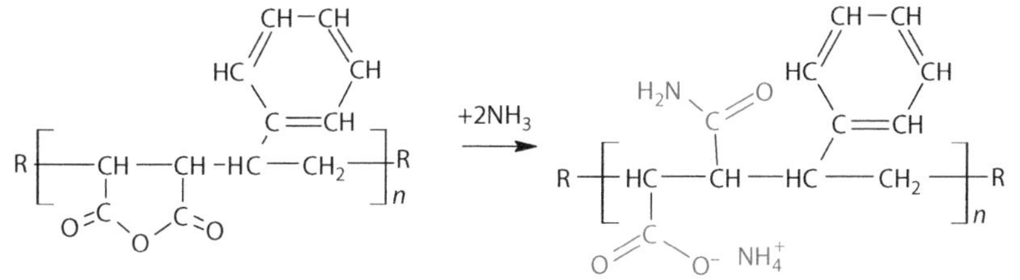

In acid (pH < 7.0), high-molecular-weight SMA is not water soluble, and its precipitation diminishes the effectiveness as sizing agent [71]. Lower molecular weight SMA ($M_w = 1700$ and maleic content over 33%) is soluble in water at pH = 4.7, to a solids contain of 16.5% [251].

SMA copolymers have been of special interest [71] because they offer a number of potential benefits to fine paper mills: increased sizing, decreased use of internal size, reduced porosity, increased coefficient of friction, increased optical brightener efficiency, improved ink-jet printability, and increased xerographic toner adhesion.

A commercially attractive SMA solution should have over 20% solids content. However, that concentration can be reached (for a reasonable viscosity level: $\eta < 1000$ cPs) only for very low molecular weights ($M_w < 10,000$). However, for better size, the SMA should have increased styrene content for higher molecular weight ($M_w > 50,000$ [252,253] or even $M_w > 150,000$ [81]). Higher molecular weight should be associated with higher hydrophobicity: 25% maleic SMA shows an HST value four times higher than 50% maleic anhydride SMA.

Obviously, optimum solution viscosity, solids content, production costs, and print performances should be reached. The trend is to increase the hydrophobic properties for a high molecular weight copolymer dissolved or dispersed in water at higher solids content. To avoid high viscosity in solution, the molecular weight of SMA (ammonium salt) is increased directly on paper with a cross-linker such as di-glycidyl ether of glycerol [254].

The imidized SMA can be obtained [255–257] by heating the aminic acid at temperatures over 105°C and right ammonia/MA ratio [258]. A controlled imidization reaction of the aminic acid of SMA (molecular weight about 80,000 and a maleic content of 26%) changes the solution in dispersion with a moderate viscosity at a concentration of 20%. The imidization process requires higher temperatures, high pressure, and a long reaction time. The imidized SMA is a more hydrophobic compound and shows better performances as surface sizing agent.

The imidized SMA is a dispersion of ternary copolymer (maleamic acid, styrene, maleimide). The dispersion stability can be improved by the addition of an emulsifier. The potassium salt of the SMA can also be used as dispersant (3% based on SMA) [259,260].

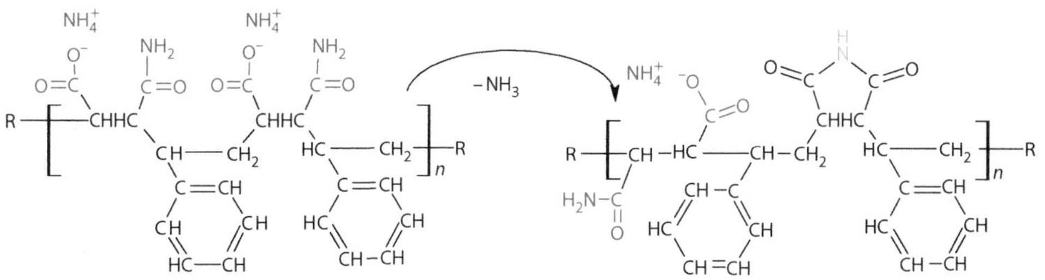

The degree of imidization is ranging from 50% to about 90% [257,260]. In order to control desizing effectiveness, the dispersion of imidized SMA must have a very low particle diameter. For a degree of imidization of 70% or lower, the imidized SMA particles in water are soft because they are swollen with water.

Hard particles are obtained at a higher degree of imidization (over 90%). Complete imidization may lead to the formation of an unstable dispersion with a very broad particle size distribution. The complete imidized SMA copolymer is very hydrophobic and shows higher T_g values (190°C–200°C). Small particles of paraffin oil (20–200 nm diameters) can be encapsulated by shells of imidized SMA [261].

9.1.3.1.3   Other Anionic Surface Sizing Agents

The copolymers of styrene with acrylic acid and maleic anhydride were presented in the previous chapter. There is a wide variety of other participants able to replace styrene or the acrylic acid in those copolymers and the final amphiphilic copolymer is neutralized with a base.

Styrene copolymers are modified by changing their synthesis (new comonomers) or by polymer analogous reaction performed on the preformed SMA. In the copolymerization reaction with maleic anhydride, styrene is replaced by diisobutylene [262–266], C_{20}–C_{24} olefins [267], dicyclopentadiene [268,269], or by (meth)acrylic esters [270–273] to make sizing agents.

The copolymer of ethylene–acrylic acid (EAA) (15%–20% acrylic acid) [25,274,275] is a surface sizing agent that is more effective in the presence of aluminum sulfate. EAA (ammonium/potassium salt) is used in size-press solution by itself [274] or in blends with the ammonium salt of the SMA [276,277].

The type of cation used to form the salt of water-soluble copolymers is also expected to influence its properties. The ammonium salt gives a better sizing response, but the sodium salt has a better film-forming ability (in blends with starch—see Section 9.3) [13]. However, ammonia is recommended to neutralize at least 25% of carboxylic groups [229]. Ammonia can be replaced by any primary amine. By changing the amine structure [278] a very different product can be obtained (lower T_g).

The SMA is also altered by adding a third comonomer but keeping the styrene/MA ratio 1:1. The third comonomer is either hydrophobic, like isobutyl vinyl ether [264,279], or hydrophilic as the acrylic acid [280]. Ternary copolymers are obtained by direct copolymerization [81,281] of styrene or diisobutylene [282,283] and half of the ester or amide of maleic acid. Half of the ester of maleic anhydride is easily obtained by direct reaction between maleic anhydride (1 mol) and 1 mol alcohol (or a blend of different alcohols [281]).

The polymer analogous reactions are developed on the reactive anhydride functionality of SMA. The anhydride group is reacted with a fatty amine [279] and half of the amide is obtained. The reaction with amine is performed to an amine/anhydride ratio lower than 1.0 and the extra carboxylic groups are converted to ammonium salt [263,264].

The reaction between the SMA copolymer with the corresponding alcohol results in a binary copolymer (styrene–maleic half ester, **9.15**).

In terms of HST, contact angle and contact angle decay, straight SMA amide ammonium salt performs better than the ester salt SMA [50]. The aminic acid can undergo imidization, while the half ester–ammonium salt only loses ammonia that leaves a carboxylic group.

The contact angle measurements (Figure 9.9) show different properties for different hydrophobic copolymers having about 50% mol maleic anhydride derivates (S-MA(HE): styrene–maleic half

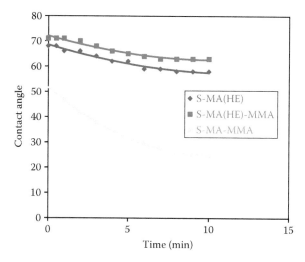

FIGURE 9.9 The variation of the contact angle with time for different copolymers: S-MA(HE), S-MA(HE)-MMA, and S-MA-MMA.

ester, ammonium salt, S-MA(HE)-MMA: styrene–maleic half ester–MMA ammonium salt, and S-MA-MMA copolymer amide—ammonium salt [81,284]).

The contact angle measurements performed for very short time intervals (up to 10 s, to simulate the printing process) [23] show that half of the SMA ester (ammonium salt) is a better surface size than regular SMA (amide ammonium salt). However, the curing parameters (90°C for 10 min) may be too low to enhance the hydrophobicity of the SMA by imidization. The imidization reaction has a significant reaction rate only at temperatures over 100°C.

The orientation of macromolecules on paper at the molecular level needs further examination. Hydrophilic–hydrophilic or hydrophobic–hydrophobic interactions might be involved, which leaves the hydrophobic segments toward the paper surface [23].

Quite unexpectedly, when anionic surface sizes (SMA, styrene–maleic half esters, styrene–(meth)acrylic acid copolymers) were mixed with anionic acrylamide copolymers, a synergetic effect was found [6,88,89,285]. The surface size and the PAAm must be premixed prior to application to paper. It is most desirable to adjust the pH of the aqueous solution of PAAm to a pH that is substantially the same as that of the surface size.

Some blends of anionic surface sizing agents contain a cross-linking agent (ammonium zirconium carbonate [AZC], hafnium salts, titanium salts, formaldehyde, glyoxalated polyacrylamide (G-PAm), melamine–formaldehyde polymers), which is added separately to the size-press solution [220,286]. The complexity of those blends requires precautions to prevent phase separation or coagulation [13]. In order to prevent the phase separation, PAAm is used as stabilizer [89].

9.1.3.1.4 Achieving Paper Printability with Concepts beyond Hydrophobicity

Hydrophobicity is only one important part of surface treatment. The printing process (multicolor print, ink-jet technology) requires ink absorption, fast drying, anchoring the ink dye on the sheet surface, high image resolution, good colorfastness, etc. Paper must have an ink receiving layer including a diversity of ingredients to match the diversity of ink components (two or more ink drops may overlap on the paper surface). From the chemistry point of view, a balance between hydrophobic and hydrophilic compounds [287] along with a specific ionic charge density is needed [288].

The hydrophobic character of a (petroleum) resin [289] is adjusted by using saponified, maleated petroleum resin or by adding hydrophilic compounds (such as polyvinyl pyrrolidone; the copolymer of vinyl acetate–vinyl pyrrolidone) [290] or amphiphilic block copolymers [291]. The hydrophobic character can also be controlled by cross-linking a hydrophilic polymer [292]. The complexity of the recording surface may be simplified by adding two coating layers with sophisticated distributions of pigments, binders, and sizing materials [293].

Cationic additives retard ink-jet dye movement [294]. The cationic charges are brought by cationic hydrophobic compounds, which are either solutions or latexes. The cationic version of maleic anhydride copolymers [295–297] or cationic ternary copolymers [298] are used in solution, while styrene–divinyl benzene–quaternary ammonium acrylate [299] or vinyl acetate–cationic methacrylate [300] are latexes. The hydrophobic character can be obtained after cross-linking the soluble polymers (blends of PVA and chitosan) with glyoxal [301].

A cationic compound is added to a sized paper substrate [302] or is used alongside an anionic hydrophobic molecule: rosin or styrene–acrylic acid copolymer [303]. Dual systems include anionic and cationic copolymer [304,305] and copolymers with an amphoteric character, such as copolymers of acrylamide (60%), methacrylate PEG ester (25%) and a betaine comonomer (15%) [306]. For betaine monomer synthesis [307] see Section 3.3.3.5.1.

The presence of a divalent soluble metal salt (chloride of calcium, magnesium, barium zinc) improves the printing process [308,309]. Water-soluble polyvalent metal salts (aluminum or zirconium) work better in the presence of a cationic small molecule [293] and/or a polyelectrolyte (such as poly-DADMAC) [293,310]. The chemical reactivity of hydrophilic, cationic polymers (such as PAE-type resins or cationically modified PVA [311]), improves the printing image when it is applied along with starch or PVA [312].

9.1.3.2 Dispersions of Nonreactive Sizing Agents Stabilized with Reactive Sizing Agents

Dispersions of nonreactive surface sizing agents are more effective as the particle size is smaller. The small particle size dispersions display a very high specific area. To protect that surface, a higher concentration of stabilizer is needed. Regular surfactants are very poor sizing agents. A higher amount of stabilizer can also generate a foaming issue.

Water-soluble amphiphilic compounds and a dispersed hydrophobic polymer are both potential surface sizing agents. Their mixture (latex blended with SMA half ester, ammonium salt) [313] shows synergetic effects. However, the polymer latex may include some hydrophilic surfactants, which can damage the sizing performance of the blend (see Section 9.1.2). On the other hand, the amphiphilic structure makes them potential stabilizers for hydrophobic dispersions in water. To enhance the performance of latex particles, the amphiphilic sizing agent is used as stabilizer in the polymer dispersion synthesis.

Types of "stabilizer" could be [151,198,314,315]: preexisting high molecular weight polymers or polymers created "in situ" during emulsion copolymerization. The water-soluble polymer used as a stabilizer with sizing capabilities, may enlarge the area of the hydrophobic material (in green, below) after the drying of the dispersed phase (in red, below).

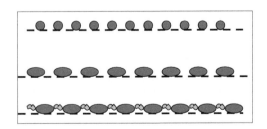

The dispersed hydrophobic material is generated during an emulsion polymerization performed in the presence of a stabilizer, which is a sizing agent as well. The stabilizer is one of the anionic or cationic compounds presented in Table 9.1.

TABLE 9.1

Emulsion Copolymerization of Hydrophobic Monomers (Styrene, Acrylic Esters, Acrylonitrile) in the Presence of Potential Sizing Agent as Stabilizer

Anionic or Nonionic Stabilizer	Cationic Stabilizer	Particle Size (nm)	References
SMA ammonium salt	—	25	[146]
SMA, 2-butoxyethanol half ester, ammonium salt	—		[317]
Styrene–acrylic acid copolymer	—	300	[318,319]
Maleimide and olefin copolymer	—	66	[155,158,262]
Polyurethane	—	59–105	[320]
Starch (30%)	—	80–110	[87,152,160]
Degraded starch	—	75	[156]
Dextrin	—	300–400	[136]
Dextrin and acrylic acid	—	180–250	[150]
—	Cationic SMA	98	[17,321]
—	Styrene-AcN-dimethyl-aminomethyl methacrylate copolymer	30	[322]
—	Cationic polyamidoamine	110–160	[91]

The dispersed phase is a hydrophobic copolymer with a T_g value lower than the drying temperature. The regular drying temperature of the sheet is 100°C–105°C; thus, the soft polymers (with low T_g) can readily develop their full potential. Polymers with high T_g (such as polystyrene) have a poor sizing effect after drying at 105°C. However, by increasing the drying temperature to 150°C, the hard polystyrene ($T_g = 100$°C) can reach the performance of the soft polymer [316].

Some systems may contain a third comonomer that helps the dispersion stability and the anchoring step on paper. The effect of the amount of stabilizer on the particle size and the stability of the emulsion is the critical parameter of this type of surface sizing agents.

Hydrophilic polymers (such as polyacrylic acid) may be stabilizers for the emulsion polymerization [323], but stabilizers as sizing agents are hydrophobic copolymers. Effective stabilizers and sizing agents are the copolymers of styrene–acrylic acid [318,319] and the copolymers of SMA [146,251,315,324]. To provide a hydrophobic character, the styrene/acrylic acid ratio ranges from 5:1 to 2.5:1.

The copolymer of maleic anhydride–diisobutylene ($m=n$ and a molecular weight of about 11,000) was reacted with Na 2-aminoethanesulfonate in the presence of water to make it a strongly anionic polyelectrolyte [155,158]. The reaction temperature is 100°C for the opening of the anhydride ring and 140°C for imidization.

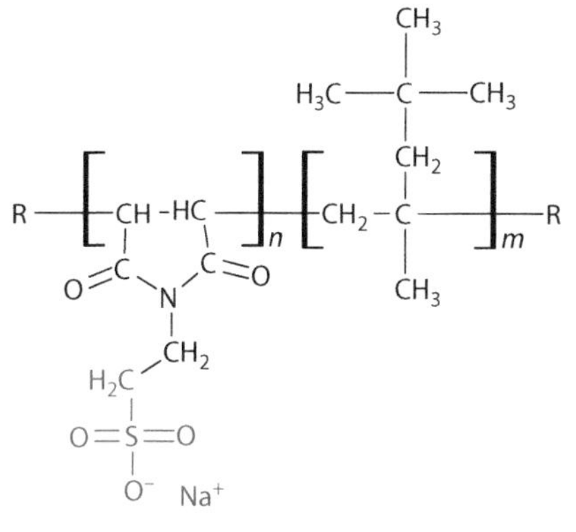

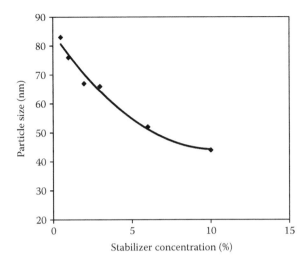

FIGURE 9.10   Butyl acrylate–MMA emulsion copolymerization in the presence of α-olefin–maleic anhydride copolymer as stabilizer.

Copolymers of alpha olefins (C_{10}–C_{18}) and maleic anhydride ($M_w = 16,000$) dissolved in ammonia are effective stabilizers even at lower concentration [190]. In the absence of a stabilizer, the particle size is 517 nm and only 0.5% of stabilizer reduces the particle diameter to 83 nm. However, with very small particle sizes the concentration of stabilizer exceeds 10% (Figure 9.10). If the data from Figure 9.10 are extrapolated to particles having a diameter of 30 nm, the calculated amount of stabilizer is 26%. That high concentration of stabilizer shows why a stabilizer, which is also a sizing agent, is needed.

The final latex concentration includes two sizing agents: the stabilizer and the hydrophobic latex particles obtained during the emulsion copolymerization. The solids concentration can easily reach over 30% [252], but the latex stability depends on the ratio between the hydrophilic and hydrophobic units of the stabilizer and on its molecular weight [325]. The smallest particles are obtained at a very high concentration of the stabilizer [146].

Many types of latex used as paper-sizing agents are prepared in the presence of modified starch (see Section 9.1.2.2). Starch is supposed to be grafted in the presence of free radical initiators (see Section 2.3.2); the hydrophobic branches improve its capabilities as sizing agent.

The anionic molecules or particles show less attachment to cellulose fibers, which are also slightly anionic. In order to obtain good sizing properties, a high dosage is necessary because a portion of the dispersion does not remain at the surface but spreads throughout the bulk of the paper sheet. Furthermore, such anionic emulsions are not compatible with cationic starches [17].

The attractive forces between polymer particles and paper sheet [326] may involve hydrophobic forces, forces caused by non-adsorbing or adsorbing molecules, and the electrostatic attractive force. Cationic particles strongly interact with cellulose fibers (see Section 3.2.3); these particles are obtained by emulsion polymerization in the presence of an amphiphilic cationic stabilizer.

The St-BuA copolymer stabilized with a cationic amphiphilic molecule particle is much more effective at lower dosages than the dispersion stabilized with anionic polymers [17]. In other words, the stabilizer is more than a compound able to protect polymer particle, it is an important component of the sizing system.

A cationic stabilizer is synthesized through a polymer analogous reaction or by copolymerization. A reactive macromolecule like SMA is modified with a nonsymmetrical diamine [321]: one primary amine and one tertiary amine groups in the same molecule (such as 1,1-dimethyl propylene diamine).

The primary amine reacts with anhydride to make an imid and, in a further step, the tertiary amine is quaternized [327]. The new cationic molecule has an important hydrophobic part due to the styrene units and the imid structure. That makes it a very attractive stabilizer for an emulsion copolymerization of styrene and acrylic esters. The stabilizer can be at a concentration of up to 20% [321] or even up to 50% [17] based on the organic phase. The emulsion shows a very good stability for 41% solids and particle sizes of about 95 nm.

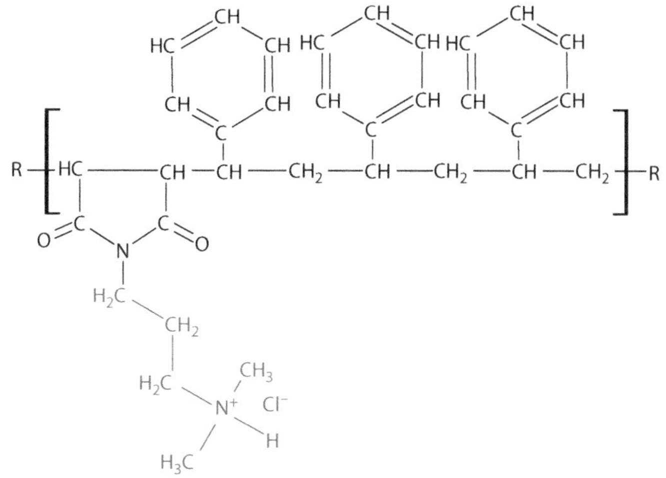

The chemical composition of SMA is broad (see Section 9.1.3.1.2) and a polymer analogous reaction makes the chemical composition of the cationic SMA even broader. That extended CCD may reduce its effectiveness both as a sizing agent and a stabilizer for the emulsion polymerization process. Cationic copolymers obtained through a ternary copolymerization have a narrower CCD.

The cationic stabilizer is prepared by solution copolymerization (blend of IPA and water as solvent) of a mixture of three comonomers: N,N-dimethylaminoethyl (meth)acrylate (20%), styrene (64%, and acrylonitrile (16%). The ternary copolymer has a statistical distribution of comonomer units. The organic solvent (IPA) is distilled at the end of copolymerization and recycled [91,328,329]. The quaternary ammonium salt is formed by reacting with acid [328] or with a hydrophobic epoxide ($R = C_5-C_7$) [329].

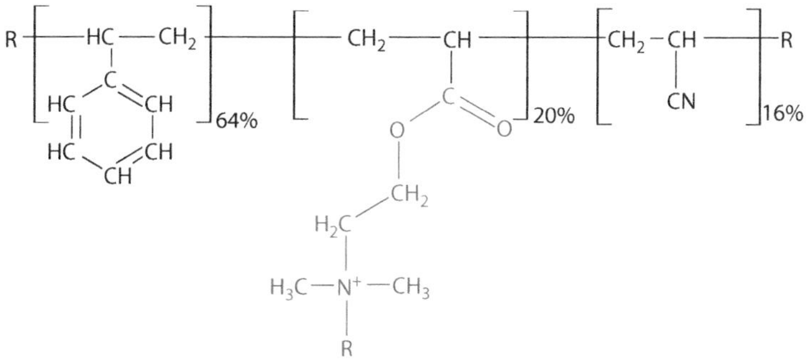

The copolymerization of acrylonitrile and butyl acrylate in emulsion is performed at total solids of 25% [322]. In order to provide the stability for the polymer particle, the cationic stabilizer is used in a high concentration (33% based on the total solids content). This type of stabilizer and its high concentration reduce the particle size down to 0.03–0.04 μm [328].

The hydrophobic segments of the stabilizer macromolecules are located on the particle surface and they form an extended hydrophobic "shell" (the area under the dashed line), which results in an extended hydrophobic character for that polymer particle. The hydrophilic units are oriented toward water and they help the interaction with cellulose fibers during the sizing step and with ink during the printing process.

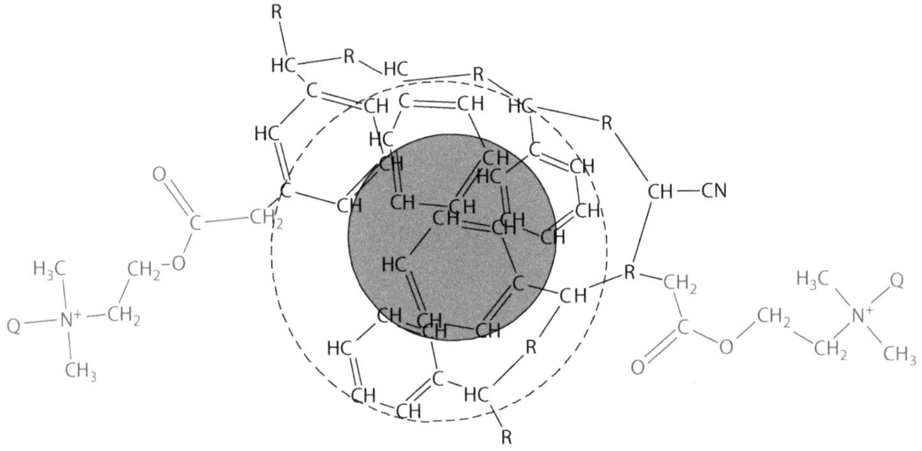

The cationic stabilizer can also be obtained by a polycondensation process. The hydrophobic part is polyester obtained by condensation of 12-hydroxystearic acid with palmitic acid. The excess of carboxylic groups are converted to acid chloride. In a separate step a polyamidoamine is synthesized from adipic acid and bis(3-aminopropyl)methylamine. The polyamidoamine has primary amine as end groups and several tertiary amines in its backbone. In the presence of an acid, the tertiary amine will change to ammonium salt (the hydrophilic part). This stabilizer has multiple cationic charges in the middle of the macromolecule and the hydrophobic end group. It is used in emulsion copolymerization of butyl acrylate and acrylonitrile. The dispersion with mixed sizing agents is added at the wet end as internal size and/or in the size-press solution as surface sizing agent [91].

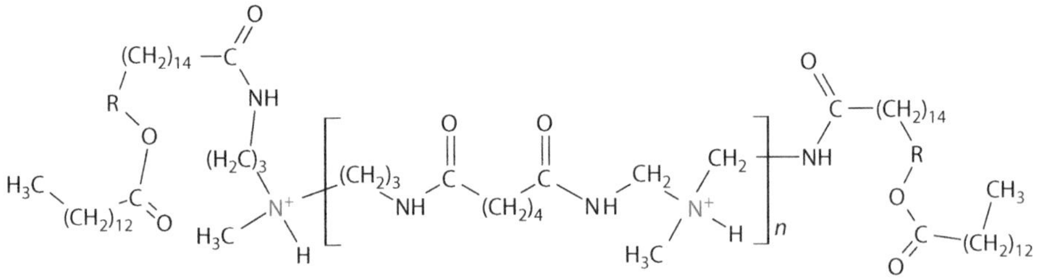

9.1.3.3   Internal Sizing Agents for Surface Treatment

The chemical compounds added at the wet end as internal sizing agents have proved their performances in making paper more hydrophobic. However, when internally added (rosin, alkenyl succinic anhydride [ASA] and/or alkenyl ketene dimer [AKD]) they are never totally retained on the cellulose fibers. The un-retained materials may hydrolyze and create problems during the papermaking process.

In order to increase retention up to 100%, the addition of reactive internal sizing agents (AKD, ASA, rosin, isocyanates, see Sections 7.2 through 7.4) in size-press solution was explored [5,330–336]. The size applied at the size press allows for a reduction in dosages of the total size [19].

As a general strategy, the internal sizing compounds can be added directly to the starch dispersion in the size-press solution. However, the surface sizing process and the environment in the size-press solution differ greatly from those at the wet end. That is why the majority of formulas include, along with internal sizing compounds, other surface sizing agents.

There is also another reason why rosin should be used at the dry end: rosin as an internal sizing agent can hardly be used at the wet end in the presence of alkaline fillers. In the size press, the cationic resins are not needed as retention aids for rosin [337] or for rosin blended with nonreactive diluters (such as paraffin wax with a low melting point [338] or hydrocarbon resins [115]). The anionic rosin is more effective in the presence of alum (see Section 7.2.4). Another option would be to use rosin and alum as surface sizing agents on the alkaline paper [339]. For other types of paper, alum may be present only in the preformed sheet [336].

Other carboxylated compounds are used in blends with rosin. Thus, synergetic effects were noted between rosin salt and the copolymers of maleic anhydride (styrene, olefins, and vinyl esters) [340], styrene–acrylic acid copolymers [341], or the copolymer of the ethylene–acrylic acid ammonium salt [342,343].

Rosin and fatty acids are used as "active solvents" in the free radical copolymerization of styrene with acrylic acid [341]. The final copolymer may include molecules of rosin through a chain transfer reaction.

Anionic carboxylated compounds (including rosin) may interact with the anionic cellulose fibers through an aluminum cation. AKD and ASA have reactive functionalities capable to form a covalent bond with the cellulosic hydroxyl group.

The 2-oxetanone ketene multimer [344] is used both as an internal and a surface sizing agent. This multimer (see Section 7.3.1) [345] is a liquid (melting point below $20^\circ C$), and its chemical structure includes hydrophobic tails (in black) and reactive groups (n is ranging from about 2 to about 5).

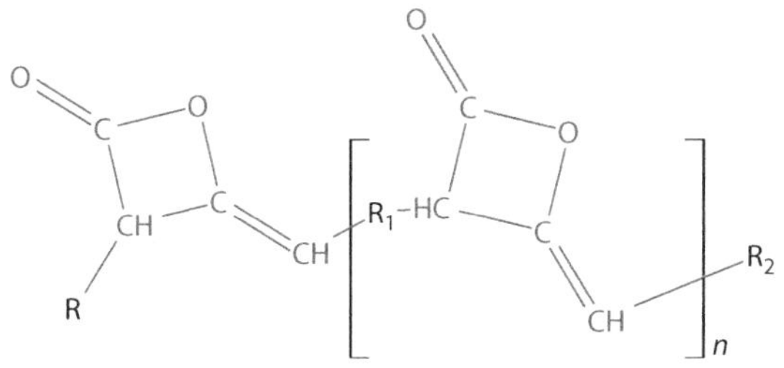

When $n = 0$, AKD has a much higher melting point. R and R_2 are hydrophobic in nature (hydrocarbons of at least 4 carbon atoms in length). R_1 is also a hydrocarbon having 4–8 carbon atoms (for instance, from azelaic acid, 1,9-Nonanedioic acid).

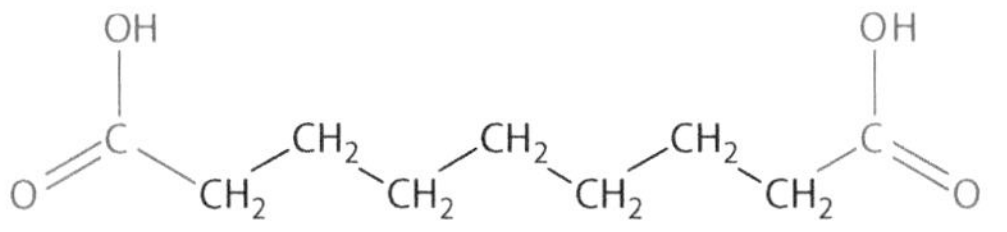

The AKD dispersions are stabilized with cationic macromolecular compounds, which are sizing promoters as well [332]. For instance, AKD emulsion is stabilized with cationic polyacrylamide (cPAAm) end caped with a long hydrocarbon tail [331]. Blends of AKD and fluorochemicals are used in the presence of a PAE-type resin [346].

The reactive AKD is used alongside other surface sizing agents. PVA (99% hydrolyzed) [347], SMA ammonium salt [10], and polymer latexes enhance AKD performances. The nonreactive surface sizing particles (latexes) and AKD are emulsified together [348,349] or simply blended in the size-press solution with a cationic polymer as promoter [350] in the presence of an oxidized starch [351].

Nonreactive hydrophobic waxes are blended with latex of the styrene–acrylate copolymer, and the mixture is passed through a homogenizer to obtain a mixed dispersion [319]. The addition of

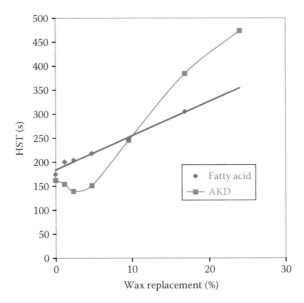

FIGURE 9.11 HST values for paper coated with blends (0.38% based on paper) of wax and fatty acid or AKD.

fatty acids or AKD improves the performance as surface sizing agents of the hydrocarbon wax dispersion [352] (cationic polyelectrolyte as stabilizer) (Figure 9.11).

The reactive compounds used in size-press solution [335,353] may have anhydride functionalities (such as ASA—see Section 7.4—or maleinized polybutadiene [330]).

or azetidinium cycle as in polyamine with hydrophobic tails (R—from fatty acids) [354]:

The need for high molecular weight compounds in the size-press solution was addressed by using reactive derivatives to modify preexisting polymers. A macromolecular compound with hydrophobic pendant moieties has been obtained. Surface sizing material is obtained from poliamidoamine reacted with AKD, ASA, rosin isocyanate, or rosin acid chloride and then with epichlorohydrin [355–357].

The reactive sizing agents are a source of many problems in terms of emulsification, shipping, and storage. In most of those formulations (for AKD and ASA dispersions), starch is used as a protective colloid. In order to solve the stability problem, starch was used as a carrier for hydrophobic tails. AKD is reacted (in the presence of sodium carbonate) with starch (dissolved or dispersed in water) [358] to form hydrophobic ester pendant groups (in black). The cationic starch reacts with cyclic dicarboxylic anhydrides [359] (such as tetrapropenyl succinic anhydride [360]). The new sizing agents are good emulsifying agents, they can make a water-repellent film and they are used in the size-press solution [360,361].

9.1.3.4 Surface Sizing Mechanism

Compounds with very different chemical structures provide multiple mechanisms through which hydrophobicity is delivered to the paper surface. There are hydrophobic macromolecular compounds without ionic charges. There are reactive or ionic small molecules with long hydrophobic tails, and there are also water-soluble polymers with high molecular weights. The number of possible surface sizing mechanisms is also related to the diversity of both physical processes and chemical reactions involved in a particular mechanism.

Hydrophobic macromolecular compounds such as latex particles are simply deposited on the cellulose fiber. The ionic particle stabilizer may help deposit formation (see cationic latexes) through ionic interactions. Hydrophobic small molecules are anchored on the cellulose fibers through ionic or covalent bonds. Water-soluble macromolecules may also be anchored through ionic or covalent bonds but their hydrophobicity changes through chemical reactions.

Due to its overwhelming concentration (as compared with the concentration of the surface sizing agent), starch is a key component of the sizing process: it controls liquid penetration into paper, competes in the adsorption step, may develop ionic interaction with cellulose fibers, and is a potential partner in chemical reactions.

The properties of the size-press solution that have the outmost effect on surface-size compatibility are the pH, the presence of polyvalent metal ions, and/or polyelectrolytes [13]. Anionic starches should be used at alkaline pH, cationic starches should be used in acidic pH. Polyelectrolytes (such as starches and sizing agents) are also sensitive to the presence of salt in solution.

In the size press the dried paper sheet is in contact with the sizing solution, which has a water content of about 85% or higher. At the moment when the size-press solution is in contact with the dry paper sheet, a "mini wet section" divided into the interphase and the bulk of paper is formed [362]. The components of the size-press solution are involved in several physical processes: diffusion, adsorption, molecule migration, macromolecular coil reconfiguration. Some ionic interactions may also be expected.

The goal of paper surface sizing is to enhance paper properties by keeping the hydrophobic compound on the paper surface. Liquid penetration in the paper z-direction due to the nip pressure and the capillary transport competes with polymer adsorption (water-soluble species and particles). Retention is 100%. However, the size distribution and the penetration depth of high molecular compounds depend on the solution viscosity and the ionic interaction.

The liquid penetration is controlled by the molecular weights of starch and the sizing agent [56,252]: higher molecular weights and higher solution viscosity increase the chance for paper chemicals to remain on the paper surface. Polyvalent metal salts (such as alum and AZC) are involved in increasing the size of the polyelectrolyte aggregates, can change the solution viscosity, prevent the penetration of the surface sizing agent through the paper sheet, and can be a partner in anchoring the sizing agent on the fiber surface.

The wet sheet passes through the drying section where high temperature and water evaporation accelerate the physical and chemical processes. At this point, it is possible to reach the melting point (or T_g) of the sizing agent and to develop ionic and/or covalent bonds through unimolecular or bimolecular reactions. Some of these interactions are similar to those developed during the internal sizing process.

Based on the HST measurements [18,252], only the ammonium salt of SMA can improve hydrophobicity (SMA sodium salt is a poor sizing agent). The sodium salt of SMA (1:1 and a molecular weight of 120,000) shows an HST value of about 4 s, while the ammonium salt of SMA (with the same composition and molecular weight) has an HST of 124 s.

Only an ammonia solution of SMA has that capability. In order to be effective, the ammonium salt of the SMA resin (or EAA or styrene–acrylic acid copolymer) must be heated at regular temperature and converted to acid for the paper machine dryer [276]. The free ammonia is flushed out and the equilibrium moves to the acid form. The undissociated carboxylic group makes the copolymer more hydrophobic.

The SMA sodium salt does not change its composition at the drying section and the sizing properties are not improved.

The SMA amide–ammonium salt also change into succinic imide units. The imidization of SMA takes place at higher temperatures [363]. However, for short lapses of time, it can occur at low conversions in the drying section.

Copolymers of SMA with half ester–ammonium salt cannot change their structure to imide, and thus the drying temperature does not have any impact on the HST numbers [71].

Imidization is a unimolecular reaction. Reactions of the bimolecular type would involve a surface sizing agent and cellulose or a cross-linker. Cellulose reacts with alkyl silicon halide [364–366], lactone, or anhydrides (such as AKD and ASA). The new hydrophobic tails are oriented away from the cellulose fiber [1].

The polyglycidyl ether [367,368], as a cross-linker for SMA, converts the carboxylic group to ester and the resulting molecules have a much higher molecular weight. For both uni- and bimolecular reactions, temperature and moisture are critical.

SMA, EAA, and styrene–acrylic acid copolymers are polyanionic electrolytes very likely to be involved in ionic interactions with polycations during the papermaking process. Whether or not the SMA performance as a surface sizing agent in the presence of alum can be improved is still under debate [13,18,252]. Moreover, many patents describe styrene copolymers with (meth)acrylic acid as being able to perform as sizing agents in the absence of alum [83,229,276]. For instance, the absence of alum is critical for the best performance of the maleated petroleum resin [289].

On the other hand, cationic SMA [296] is an effective surface sizing agent [297] in the absence of alum. However, polyfunctional cationic species seem to help the carboxylic groups to "bridge" to the anionic cellulose fibers [13,15,270,271]. Aluminum, zirconium compounds, and multifunctional reactive organic compounds (such as G-PAm) [286] or polyamines [369] are examples of polyfunctional cationic compounds.

The sizing performance of anionic polyurethanesis improved in the presence of alum [18,320]. Larger aggregates are recommended as surface sizing agents [370]. For white paper, it is recommended to use aluminum sulfate (0.1%–1.5% by weight of the fibrous material). Adding the aluminum salt at the wet end is beneficial [81], but it is also possible to impregnate the finished paper with a solution of alum prior to treatment with the surface size [25].

The two polyelectrolytes are put in contact in the size-press solution or direct on paper. It seems that at least a fraction of their hydrophobic properties are developed based on those interactions. The zirconium salt immobilizes the surface sizing compound, prevents its penetration onto the paper sheet [371,372], and orients the hydrophobic portion of the SMA macromolecule away from the surface of the sheet [373] in the wet state, before the sheet enters the drying section of the paper machine. Polyelectrolyte complexes [369] include both cationic and anionic residual charges, which may help the printing process (interactions with the ink components).

In order for HST to improve significantly, the amount of AZC [71] must be quite high (Figure 9.12): 0.15% SMA (amide ammonium salt) asks for about 3% AZC (based on dry starch).

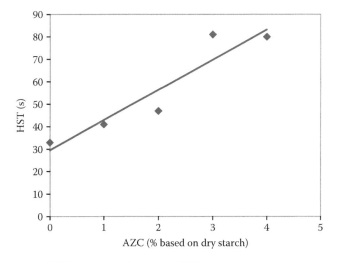

FIGURE 9.12 The effect of AZC concentration on the HST values.

AZC can interact with anionic polymers in alkaline pH [15]. The anionic oxidized starch is the preferred starch; AZC enhances the HST values especially when SMA is an ammonium salt.

AZC is a soluble alkaline salt of zirconium. Because zirconium Zr^{4+} salts are extensively hydrolyzed in water, ammonium cations and anionic hydroxy-bridge zirconium polymer containing carbonate [374] are present in the AZC solution. The AZC solution in water is stable at temperatures below 40°C and in the 7.5–11.0 pH range. The AZC solution can be added to a polymer solution having the same pH range. In order to improve the stability of the AZC solution at higher temperatures (up to 70°C), a hydroxy-carboxylic acid (tartaric, dihydroxy tartaric, mucic, and gluconic acid) can be used [375].

The chemical structure of the zirconium salt is still under discussion [371]. Its cationicity and the number of cationic charges able to bond the anionic groups of the surface sizing polymers are currently under study. The aqueous solution of Zr^{4+} in carbonate media involves [376] several complexes having the following stoichiometries: $[Zr(CO_3)_4]^{4-}$, $[Zr(CO_3)_3(OH)]^{3-}$, $[Zr(CO_3)_2(OH)_2]^{2-}$, $[Zr(CO_3)(OH)_3]^{-}$, and $[Zr(OH)_4]$. The general formula is

$$[Zr(CO_3)_x(OH)_{4-x}]^{x-}$$

where $0 \leq x \leq 4$. The relative concentrations of those species are related to the initial HCO_3^- ion concentration.

Zirconium salts can undergo a "condensation" process, and generate a complex structure. The increase of the extent of polymerization of Zr^{4+} complexes can be correlated with the formation of more hydrolyzed species. Condensation can be attributed to the formation of hydroxo bridges $Zr–(OH)–Zr$. Decreasing the CO_3^{2-} concentration, either by dilution or by addition of alkaline reagents, leads to the formation of more condensed species.

In the $[Zr(CO_3)_2(OH)_2]_n$ polymeric unit, carbonate ions replace H_2O in the coordination sphere of Zr^{4+}, and so, for $n=4$, the formula $[Zr_4(CO_3)_8(OH)_8]^{8-}$ may have the following structure [376,377]:

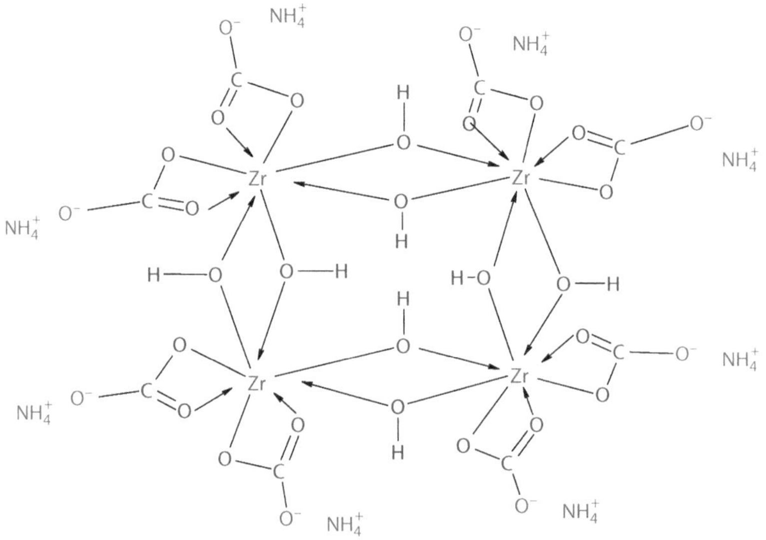

Ammonium cations are used for a closer structure of AZC. In this polymeric structure there are eight carbonate groups, which make AZC a polyfunctional compound. A linear structure is also proposed in which a different ratio between hydroxylic groups and carbonate groups is used [24,374].

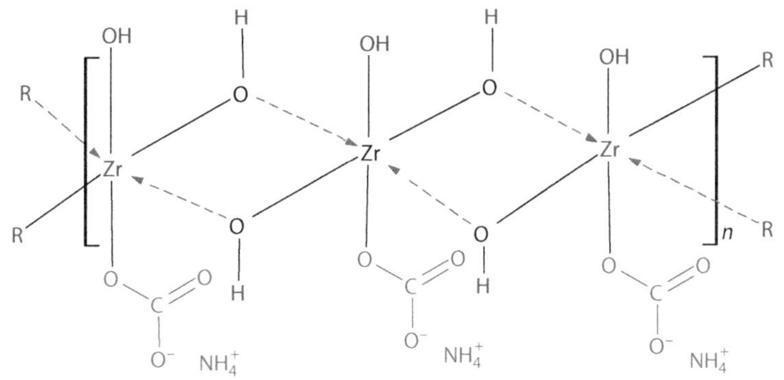

The carbonic acid is a very weak acid and it can be replaced by an organic acid (carboxyl group). If the carboxylic group is attached to a macromolecular structure (like surface sizing agents: the styrene–acrylic acid or the SMA copolymers), a cross-linking process takes place [374]. The polycarboxylic compound can be the water-dispersible latex of a hydrophobic material. The carboxylic groups are located on the particle surface and can interact with zirconium derivatives [372]. The carboxylic group can also be attached to the cellulose fiber and AZC makes it possible to anchor the macromolecular compound to the cellulose fiber.

Carbonate ions are in excess in order to confer stability to the polymeric compound. If the concentration of the carbonate ion changes, the molecular size, the molecular size distribution, and the overall solution properties will change as well.

The migration is controlled [371] by alginate salt (ammonium is preferred) and 0.01%–2% AZC. That is a very elegant way to control the amount and the "reactivity" of the zirconium complex when the migration control agent is blended with a polycarboxylic compound.

AZC has exhibited certain useful characteristics as an additive in the size-press solution [373]: it reduces dusting from internal ash during printing, reduces linting derived from surface fibers during printing, increases Gurley porosity, and improves performances of low-cost starch derivatives.

There are some downsides of using alum and AZC in conjunction with anionic macromolecular compounds as surface sizing agents. To mix the solutions of two different polymers is often a problem [378,379]: a phase separation can be developed in the size-press solution (insoluble aggregates—see Chapter 3). The AZC compatibility with the size-press solution can create the same difficulties: precipitation, phase separation, increase in viscosity, etc. An acceptable range for pH is between 5 and 10. It is recommended [71] to add the cationic compound (AZC, alum, cationic copolymers) to starch followed by mixing before adding the SMA-type resin.

The aluminum salt added at the wet end creates several problems as well: it is eluted from paper into the sizing solution and it can precipitate the anionic polyelectrolyte in the size-press solution [271].

9.1.4 Effect of the Defoamer

The sizing agent is dissolved or dispersed in a starch solution or dispersion. During the drying process, the sizing material coats the cellulose fiber surface and—at the macroscopic scale—the paper surface. That is mainly a "surface" phenomenon. Most surface sizing agents are also surface active compounds, which can generate foaming. That is why size-press solutions usually include a defoamer. Defoamers are often surface active materials, which cause the paper surface to be easily wetted [318] and reduce the size effectiveness.

The SMA solution with solids content of 10%–20% needs defoamer in a concentration of about 350 ppm to prevent foam formation during the synthesis process. In the size-press solution (starch plus surface sizing agent), for the same defoamer concentration, the defoamer/surface size ratio is

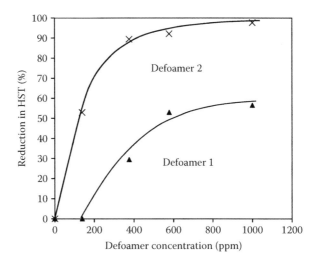

FIGURE 9.13 The defoamer effect on HST and HST decay. (From Latta, J.L., *Tappi Papermakers Conference*, 1994, p. 399. With permission.)

about 100 times higher than that in the SMA solution. Starch, the sizing agent, and the defoamer are 100% transferred to the paper sheet, and therefore the defoamer content on the paper surface is about 15% based on the sizing agent.

The effect of two different commercially available defoamers on the performance of the half ester ammonium salt of SMA is shown in Figure 9.13 [71]. Low concentrations of defoamer reduce with about 50% of the HST value, and—at higher concentrations—the size effectiveness decreases close to zero.

9.2 SURFACE STRENGTH AGENTS

The paper web is a capillary system made from hydrophilic fibers. The printing and coating processes result in a significant stress to paper surface. The cellulose fibers located in the bulk of sheet are involved in an interfiber bonding; fibers interact with one another in the three-dimensional space (x, y, and z). The average number of interfiber bonds calculated for one cellulose fiber is lower for the fibers located on the paper surface (a two-dimensional area). Linting, picking, blistering, delaminating, and feathering should be prevented by enhancing the paper surface strength with specially designed chemicals. The effect of those chemicals on the surface strength should be accurately evaluated [380].

Wet end additives (such as dry-strength resins) improve surface strength as well. For instance, lower molecular weight, cationic starches (charge density 0.7–1.1 meqv/g) [381,382], cationic (imidized) and anionic SMA [383] reduce linting and dust formation during the printing process. However, the chemicals added at the wet end are supposed to be evenly distributed in the bulk of paper; therefore, only a small portion of the compound is located closer to the paper surface. Because an important part of the resin does not contribute to the surface strength, the surface treatment seems to be a better choice.

Even during the surface treatment with the starch solution, the polymer does not remain solely on the paper surface but penetrates to a certain extent into the bulk of sheet. The penetration depth depends on the solution viscosity (the higher the viscosity, the lower the penetration), on the surface tension of the solution (the lower the surface tension of the solution, the deeper the penetration), on the ionic charge density, and on the paper porosity [52,362].

The treatment of the paper surface with chemical compounds should be performed in a way as to prevent their penetration into the bulk of sheet [384]. A special equipment with a vacuum blade unit was designed to keep starch on the paper surface. That device allows for the same pickup of starch but a lower linting than the size-press treatment.

Surface strength is required for paper designed to go into repetitive printing process in which the mechanical stress is applied in the presence of water or other aggressive liquids (such as ink). The presence of water [385] (or any other polar solvent) may change the paper surface strength: the hydrogen bonding is reduced, a strong fiber swelling results in a stress relaxation, which allows for fiber rising [386].

Various printing technologies require paper surface with high integrity and resistance to the abrasive effects of the printing processes. Higher printing speed, higher printing pressure, increases in ink tack (or ink viscosity) result in more linting [387]. Fiber picking, linting, and filler dusting all contribute to decreased runnability of paper.

Paper chemicals able to provide surface strength for the paper sheet combine the properties of dry strength resins and surface sizing agents. Any chemicals able to reduce the paper porosity, to improve the interfiber bonding (such as PAAm [388,389]) and to increase the fiber hydrophobicity, may have an impact on the surface strength of paper [34]. The sizing agent makes the paper more hydrophobic and thus more resistant to the water involved in the printing process [384].

Starch as paper surface treatment [20,390] improves the interfiber bonding through newly developed hydrogen bonds. During the drying step, other polymers can interact with cellulose fibers to form ionic or covalent bonds. The improvement of the surface strength through covalent bond formation should be considered in the case of PAAm and a carrier for glyoxal (glyoxal-monoureine [391], starch-glyoxal [392]), polyisocyanate [393], or acrylamide/N-methylol–acrylamide copolymer [394].

Ionic bonds are developed at the contact between chemicals and paper, such as between anionic cellulose fibers and any cationic polyelectrolyte. Polyvinyl amine and its copolymers [395], cationic starch [72], or cationic amylose [68]) are designed as surface strength agent for newsprint grade paper, which is involved in many rewetting processes during the printing. AZC added to the size-press solution [373] reduces dusting from internal ash during printing, reduces linting derived from surface fibers during printing, increases Gurley porosity, and improves performances of low-cost starch derivatives.

Potential surface strength additives may include ionic bonds before they reach the paper surface. Blends of anionic and cationic compounds [370,396,397] make polyelectrolyte complexes (dual system, see Section 6.5). Thus, cPAAm or polyethylene imine (PEI), is blended with hydrophobic copolymer (such as anionic SMA) [398,399] or anionic (oxidized) starch. The cationic groups and the anionic groups may be in the same molecule of amphoteric starch [400,401], which is able to provide a cross-linked structure by itself, and the film is not sensitive to water [402].

The ionic component can be located in different layers [403]: at the top of a polycationic primer coating, a second coating is added, which incorporates anionic polyelectrolytes [404]. In order to consolidate the presence of lignin (or lignosulfonate [405]) into the paper structure, a successive addition of a cationic PAE resin is used [406–411].

The right balance between the hydrophilic and hydrophobic parts on one hand and the formation of the hydrogen, covalent, and ionic bonds on the other hand, may be combined in sophisticated blends. Both strength and sizing functions are improved by premixing, prior to application on paper, hydrophilic PAAm [88,89,285,286] (or its copolymers) with a hydrophobic surface size agent (SMA type) in the presence of a cross-linking agent (zirconium salts, formaldehyde, G-PAm, melamine-formaldehyde polymers).

The paper surface treatment with strengthening resins can be performed by adding them to the size-press solution [412], spraying them on the wet web surface, [413] or applying them on the Yankee drier [390].

A paper product that is both strong and highly dispersible is obtained by adding a cationic or a nonionic strengthening agent [414] only to limited areas of the paper surface, which results in the formation of an interlocking serpentine pattern.

9.3 POROSITY BUILDERS

Paper is a porous material and fluids may pass through its structure [415]. Ink and air penetration are parts of the printing processes. They function under vacuum and improve capabilities to reduce the rate of the air flow passing through paper.

Paper porosity depends on the quality and the amount of the chemicals added on the paper surface. In other words, paper porosity can be controlled by paper chemicals and by cellulose fiber modification. For instance, the size of the paper pores is increased by cross-linking the cellulose fibers with formaldehyde, epichlorohydrin, or glyoxal [416].

The porosity builder does not develop a continuous film: it is only a reduction in the pore size, and, consequently, in the rate of air flow through paper. The film-forming glue [417] can reduce paper porosity by enlarging the area covered by that film (patches at the crossing point of the cellulose fibers).

Pore distribution depends on the thickness of the paper. The degree of penetration of the sizing material is thus a critical factor. Because starch and sizing agents have a high molecular weight structure, their penetration into the bulk of dried paper is limited. However, the solution penetration into a wet web is much easier. Therefore, the sizing agent is divided in two portions: one is sprayed on the wet web. The size penetration into the wet web depends on the web consistency and on the running speed between the addition point and the dryer. The second portion is added in the size press on the dry paper [11].

Paper surface sealing may combine different approaches: the pores are filled with inorganic filler and the macromolecular aggregates remain on the surface in small patches of film [67]. Some paper chemicals, such as starches, are film-forming materials. There have been numerous attempts at finding a synergetic effect between starch (or PVA) and other paper chemicals. In order to reduce paper porosity, blends of cationic starch and SMA ammonium salt are cooked at 120°C–150°C and used at the wet end [418].

Starch is the major component of the size-press solution. Because the properties of the starch solution in water pose many problems in terms of stability, changes in viscosity vs. time, profile, etc., PVA was chosen as a substitute for starch in order to study the interaction with polycarboxylic compounds [419]. Intermolecular interaction through hydrogen bonds was observed with the PVA solution in the presence of polyacrylic acid. That experimental data were confirmed for the EAA solution in the presence of starch: the solution viscosity shows a maximum at about 1:1 ratio [276].

There is a good chance to get such synergetic effects due to the intra- and intermolecular hydrogen bonds in the water solution of PVA [420] or to the bonds between PVA and polystyrene sulfonic acid or its sodium salt [421]. The presence of intermolecular interactions was measured by the viscosity of the solution. The intermolecular interaction between alginate and a polysaccharide (or polycarboxylic compound and a polyether [422]) is noticed by the increase in Brookfield viscosity of their solution [423].

Those findings were confirmed for polymers used in the paper industry. The hydrogen bonds formed between the carboxylic groups (EAA, carboxymethyl cellulose, or alginate [424]) and the hydroxyl functionality (PVA) improve film strength and reduce porosity [417,425]. There are also interactions between poly(methacrylic acid) and hydroxyethyl cellulose [426], starch [67] and SMA [18,427].

A porosity builder may have the carboxyl and hydroxyl groups located in the same macromolecule. Carboxylic groups are added to starch by esterification with tetra-propenyl succinic anhydride [360]. Carboxylated PVA is obtained [428] by copolymerization of vinyl acetate with carboxylic monomers, followed by hydrolysis with sodium hydroxide. PVA modified with itaconic acid shows a significant reduction in paper porosity.

If those interactions are strong enough to develop in water solution (solvent able to disrupt the hydrogen bonding) it is obvious that the same types of interactions are possible in the solid state (films or aggregates between starch—or cellulose—and polycarboxylic compounds).

The strength of those interactions depends not only on the molecular weight of the polymers (a higher molecular weight is better [18,252,373]), but also on the type of cation associated with the carboxylic polymer. The SMA sodium salt provides higher Gurley values than the ammonium salt [15,18,253]. AZC (or alum [276]) also improves porosity, through the immobilization of the surface sizing compound near the surface.

Chitosan is a biopolymer, a polysaccharide with amino groups. The chitosan solutions are prepared in acetic acid 1% with different molecular weights ($M_w = 573,000$ or $M_w = 1,080,000$) [429]. The cationic polymer interacts with anionic fibers and improves the paper porosity only at concentrations higher than 1% (Figure 9.14). The higher molecular weight chitosan is much more effective as a porosity builder.

The structure of chitosan (as a potential cationic polyelectrolyte) opens the opportunity to interact with the common anionic paper chemicals, such as oxidized starch. In that case, the addition of

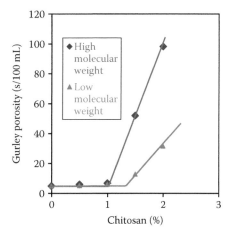

FIGURE 9.14   The effect of chitosan molecular weight on paper porosity.

only 0.4% chitosan results in an increase of about 400 times in the effectiveness of oxidized starch as porosity builder [429] (from 6.6 to 2480s). The interaction between a cationic and an anionic polyelectrolyte results in a "dual system" aggregate with a very large macromolecular size. Due to its size, the aggregate remains on the paper surface and reduces porosity.

9.4 POLYMERS IN PAPER COATINGS

A leak-proof barrier is required for paper that comes into contact with oils, greasy or watery food. Straight polymeric barriers (such as extruded polyethylene film) are costly to apply to paper, interfere with paper repulpability after their intended use [430], and are being increasingly restricted by regulations [213]. Coatings with aqueous dispersions of organic binders and inorganic fillers provide water resistance, glue, ink, and dye compatibility through a less expensive process.

During a surface coating process, the cellulose fibers act as a carrier and the coating layer imparts the major performance characteristics of the coated paper [7]: a continuous film covers the sheet and no cellulose fibers are left exposed at the surface. Within an ideal application, the coating material on the sheet does not penetrate the paper at all.

The composition of a paper surface coating includes two major components: about 75% inorganic filler and about 20% organic binder (calculated based on solids [431]). Dispersants, defoamers, preservatives [432], colorants [433], and lubricants [434–436] are used in smaller amounts. The coating process may also need chemicals for an improved runnability (such as release composition [437]).

9.4.1 NATURAL AND SYNTHETIC BINDERS

The binder is the key material able to make a continuous layer in which all other components are incorporated. It forms smooth, adherent coatings, which uniformly bind the pigment particles to the paper and simultaneously reduces the tendency of the coating to stick to the polished surface in the finishing process [438]. Regardless of the filler functionality in the printing process or as a vapor barrier, it is seen mainly as a polymeric binder substitute.

The chemical structure of the binder is designed to improve its mechanical properties, printability, repulpability, etc. The amount of coating material and the coating layer quality depends also on the quality of the paper surface (smoothness, absorptivity, and size) [439,440].

The reduction of the amount of binder [441] and the water content in the coating formula is targeted to reach 70% solids. If the inorganic pigments are about 75% based on solids, the binder polymeric material is about 20% from the entire coating formula. The binder, a waterborne polymer, and, based on the above-mentioned percentages, its concentration in water should be around 40%. As shown in Section 2.5.2, with high molecular weights, the viscosity of a water-soluble polymer at 40% concentration is very high (gellike). Solutions with 40% concentration of polymer show a reasonable viscosity only for polymers with low molecular weight.

Low molecular weight polymers may be successfully used as co-binders [442], but the binder effectiveness and the mechanical strength are provided by larger molecules. In other words, if a small molecule is used as a binder, the molecular weight should be reconstructed—for high strength—by cross-linking during the curing step. The cross-linking reaction performed at high solids, high viscosity, and large interface with inorganic material is a challenging protocol. In order to bring the polymer content over 40%, dispersions of polymers are a better choice.

The dispersed phase can be a hydrophobic polymer (such as in latexes) or a combination of dispersions and solutions of a polymer [443,444]. High molecular compounds (such as starch) are dissolved at low concentrations (cooked starch) and the resulting solution is used as a suspending environment (slurry) for raw starch [445] or for PVA particles [446].

Commonly used binders include natural macromolecular compounds (such as casein and its oxidized form [447], gelatin [438], modified soy protein [448–450], modified starch—with more carboxylic groups [105,401,402,451–456]), synthetic polymers (PVA [457,458], acrylamide copolymers [459,460], hydrolyzed methyl acrylate–vinyl acetate copolymer [461], wax, and polyethylene wax [440,462–464], ethylene–vinyl chloride–acrylamide terpolymer [465]), blends of natural polymers (starches or casein [466]) and synthetic polymers [467] (high amylopectin starch and SMA ammonium salt [468]), or blends of synthetic polymers [130] (SBR latex and PVA [469–472]).

Natural compounds are susceptible to insect infestation and are water sensitive. Synthetic polymers overcome the disadvantages of natural binders: the properties of casein are improved by adding an emulsion of synthetic polymer [438] and modified starch is made more hydrophobic with organosiloxane polymers [454], SMA blended with polyester amide [473] or encapsulated oil in an imidized SMA [261].

The final properties of the binder are obtained in the drying section by water evaporation [474] or/and by a chemical reaction (for instance, digested proteins are reconstituted with zinc cations [475]). The drying temperature should be correlated to the glass transition temperatures (T_g) or the film-forming temperature [476,477] of the binder. In the drying section, the adhesion between binder and filler is improved and the mechanical properties of the coating are set.

The dispersed binder in water must have a T_g lower than 55°C [444]. The major types of paper-coating binders are: homopolymer of butadiene [478] (T_g <0°C), styrene–butadiene copolymers [443] (–25°C< T_g <50°C), styrene–acrylic ester copolymers (–10°C< T_g <40°C), and polyvinyl acetate (T_g=30°C). Polymer particles with higher T_g (such as polystyrene) do not change their shape during the drying process and are considered as organic pigments [479].

Organic pigments [480–482] are obtained by styrene polymerization in emulsion in the presence of the copolymer of acrylamide–acrylic acid as stabilizer [481], by selective precipitation of starch esters [483], or from cellulose powder [484]. Due to their lower density, high T_g, and particle size of 100–600 nm [485,486], organic pigments are used to replace inorganic pigments [479,487].

A balance should be reached between T_g and the coating properties: the pick strength of the coating shows a maximum at T_g of about 10°C, the wet strength is higher at lower T_g (higher T_g latexes have larger void volumes [488]) and the coating gloss is higher at higher T_g [126,489,490].

The most common binder, the styrene–butadiene latex, shows some limitations: water sensitivity, poor adhesion to metals (such as steel), inability to air dry. The properties of styrene–butadiene latexes are improved by using minor amounts of polar comonomer (up to 10%) [491,492], monomethyl maleate, acrylamide [493], acrylonitrile, MMA, acrylic and methacrylic acids [126,494]. The polar comonomers are located on the particle surface.

As the hydrophobic part of the binder used in paper coatings, styrene [495] is replaced by ethylene [272,496] in copolymers with acrylic, methacrylic, or maleic acid [497] (or with itaconic acid [498,499]), or by acrylates and acrylonitrile [489,500]. Hydrophobic binders are not only copolymers of styrene, but also of ethylene–vinyl acetate copolymers [501], vinyl acetate–maleic half ester [502,503], butadiene–styrene–methacrylamide [497], vinylidene chloride–styrene–butadiene [499], or with ethylene–vinyl acetate–acrylic acid copolymers [504]. Polyester resins [505–509] are also considered as hydrophobic binders.

Acidic comonomer copolymers (such as EAA) are dispersed in water in the presence of caustic or ammonia (alkaline pH). Sodium hydroxide makes the coatings more sensitive to water, but the ammonia salt of EAA, in blends with SBR latexes, shows better coating properties [496].

Complex binders are obtained by simply blending different polymers (a rubbery copolymer of butadiene or acrylic copolymers [510] and a hard polymer such as PVC, PVAc, PMMA [510,511]) or by synthesis of one polymer in the presence of a preexisting macromolecular compound. Through seeded emulsion copolymerization [133,512–514], a soft copolymer styrene-butadiene-MMA-acrylic acid is synthesized in the presence of a hard copolymer (latex of styrene–acrylic acid copolymer) [515]. The resulting particles have a core-shell structure and the shell can also be a carboxylated styrene copolymer [445].

The dispersion of binder is characterized not only by the chemical structure of the particle, but also by the particle size, the particle size distribution, and the properties of the interface (nanoparticles can also be added to the formula [516]). Most of those parameters must reach an optimum for a designed application. The inorganic filler and the polymer latex form two different types of interface with the water phase. The water phase may contain electrolytes and polyelectrolytes. The coating viscosity, stability, and runnability depend on the composition of the water phase and the concentration of stabilizers at the interfaces.

The pigment and the binder dispersions should be colloidally stable and compatible with each other: they are mixed without formation of coagulum or excessive increase in viscosity. Electrolytes in the water phase may reduce the binder stability. That stability is improved by a nonionic, polar comonomer (such as 2-hydroxyethyl acrylate (12%) and styrene–butadiene copolymer [517]) located on the particle surface [494] or by a nonionic stabilizer [500].

The most common method to improve the filler stability is to add a dispersant. Examples of dispersants are anionic polyelectrolytes [518], such as bisulfite-terminated oligomers of the acrylonitrile-co-acrylic acid [519], sodium polyacrylate (for clay and calcium carbonate) [520], or sodium polyacrylates, along with sodium carboxymethyl cellulose (or starch [443]) for talc, in the presence of styrene–butadiene latex as binder [521].

The existence of a polymeric shell around the inorganic filler would simplify the nature of the interface: only polymer–water interface remains to be stabilized. The pigment particles are encapsulated in a polymer shell through emulsion polymerization [522] with imidized SMA [523] or by controlled adsorption of polymeric latex particles [524].

Coatings with a high content of solids are obtained with concentrated latexes generated by polymerization and water evaporation [525]. The high solids content for the coating formula and the presence of polymers, in the water phase or at the interface, result in a change in the mixture viscosity and runnability. There is a critical concentration when the particles (pigment and polymer binder) get in contact with each other.

As the average particle size of the monodisperse latexes decreases ($D < 1000 \text{\AA}$), the solids content gets limited to about 50% [526,527]. For a latex with an average particle size of $2100 \text{\AA}$, the viscosity reaches 1000 cPs at a concentration of about 60%, while a latex with particle size of $720 \text{\AA}$ reaches the same viscosity at about 50% solids content. Thus, the particle size distribution and the way the particles fill the volume are important parameters for the coating viscosity. The bimodal distribution of the particle size shows a better distribution in the latex volume, which reduces latex viscosity.

A better rheology for coatings was confirmed for the bimodal distribution of the particle size [525,528]. For latexes of styrene (63%)–butadiene (35%)–acrylic acid copolymer [525] a large reduction in viscosity is obtained at about 25% particles with smaller diameter (Figure 9.15). That also confirms the finding for polystyrene latexes [529].

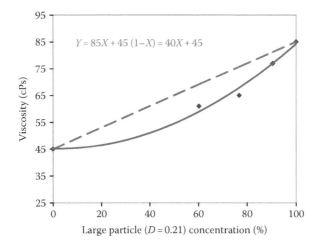

FIGURE 9.15   The viscosity of blended latexes vs. concentration of larger particles.

Colloidal interactions between pigment and binders [126,530] depend on the electric double layer of each component, on the molecular weights of water-soluble macromolecular compounds, on the presence of electrolytes, on the particle size distribution, etc. Polyvinyl acetate latexes interact with clay particles (hydrogen bonding between the hydroxyl groups of the PVA used as stabilizer and the silanol groups of clay particles), while the coating viscosity is higher [487]. A kaolin dispersion stabilized with anionic PAAm shows a higher viscosity in the presence of electrolytes (the repulsive forces are reduced), which may change the coating structure, the decrease in gloss, and the increase in fluid adsorption [531].

Associative thickeners (water-soluble or water-dispersible polymers) control the rheology of the dispersed aqueous systems. Their chemical composition [532] consists of a hydrophilic polymer (polysaccharides, PEO, PAAm, acrylamide–acrylic acid copolymer) or several hydrophobically modified water-soluble polymers [533]. To balance the effect of the thickeners, they are combined with a viscosity reducing agent (nonionic or ionic surfactants) [534].

9.4.2 BINDER MIGRATION

Paper coatings are double heterogeneous systems: (1) an organic material (latex—in blue) and (2) an inorganic material (filler and pigments—in yellow) dispersed in water. Some small molecules and/ or polymers are also dissolved in the continuous phase (water solution). The coating composition passes through a high shear rate [535] and then through a drying step.

Coating properties depend on the component type and relative concentration, on the preliminary treatment of the paper surface [536], and on the physical and chemical processes developed during the drying step. Because a circulation of the coating components takes place, it is hard to predict the evolution of the system during the drying step [537].

The particles in the dispersed system can be involved in the coagulation processes: heterocoagulation (latex-pigment) and/or homocoagulation (latex-latex). The heterocoagulation process may result in a good coating, but homocoagulation results in phase separation. Phase separation starts with the binder migration [530]: a differential movement of the binder with respect to pigment particles. The binder migration can occur toward the base paper during drainage and toward the coating surface during the drying step.

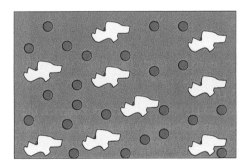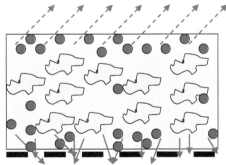

Binder migration results in a lack of uniformity, it may produce spots and affect both coating adhesion to paper and surface properties, such as printability. The extent of the binder migration depends on the paper properties (porosity, water absorbency, fiber dimensions), coating composition (pigment type, binder chemistry, solids level, viscosity, etc.), and the coating application process (coating application point, coating type, shear rate, time intervals between the applicator roll and the metering nip and between the metering nip and the dryer, etc.).

There are technological and compositional approaches aimed at preventing binder migration. The technique of multiple layers and a specific curing temperature profile (below and above T_g)

[538] are designed to freeze the homogeneous coating composition as it was obtained after formula preparation.

Double coating process, when a pre-coating process is involved, is able to minimize the binder migration toward the base paper, and, therefore, the second layer is more uniform and always improves the coating properties [521] because the prime coat provides a tie ply to the paper web for the top coat [510]. The binder molecules from the prime coat migrate into the paper structure and the top coat. The surface treatment of paper with SMA and PVA blend solution reduces wax penetration into paper [427].

Those two layers can incorporate the same binder or different binders. The prime coat may use a binder with a T_g of about 20°C, while the second includes a binder with a T_g of about 43°C [539]. Thus, a protein binder (zein or starch and gelatin blends [540]) is added to the top layer, while the colloidal clay and starch (clay has the ability to deposit within the pores on a normal paper surface) are added to the primer coat [541].

Less binder migration is expected for less stable colloidal systems. In order to control the binder migration, the stable colloidal systems should be partly coagulated (controlled destabilization with flocculants [487]), and thus the binder is immobilized, to some extent, in the coating structure. Lower T_g and the presence of a multivalent salt favor the latex particle immobilization. The colloidal system destabilization is faster at higher temperature, at higher solids content, and in the presence of an electrolyte. The presence of carboxylic groups in the chemical structure of water-soluble polymers (such as anionic cellulosic polymers) has an effect in preventing binder migration [542].

Mixed binder systems (styrene–butadiene latex and wax emulsion) provide a more flexible formulation [543]. Most of the coating formulas have a complex binder system: two or more water-soluble polymers (starches, proteins, synthetic polyelectrolytes as dispersants, anionic starch, and cationic synthetic polymers [544], etc.) and two or more latexes. Those components show different interactions with pigments and different behaviors during the migration process. Chemical bonds between different binders slow down the migration process. Latexes of the styrene–butadiene copolymer are obtained in the presence of starch when several grafting processes take place [492]. Grafted polyvinyl acetate onto cationic starch [545] is a substitute (over 50%) for protein binder.

9.4.3 HYDROPHOBIC AND CROSS-LINKED BINDERS

After the drying step, the coating is a continuous film in which filler particles are immobilized. The hydrophobic properties of the continuous film are obtained by simply blending with a hydrophobic compound (with wax emulsion [543], ethylene–vinyl acetate copolymer, and rosin esters [546]) or by chemical reactions: starch modified with ASA [547].

In order to turn the binder into a more hydrophobic and effective water vapor barrier, the latexes of styrene, MMA, and acrylic copolymers [146,548] are stabilized with a more hydrophobic stabilizer. The hydrophobic stabilizer is synthesized by the copolymerization of styrene–acrylic acid mixture in a high melting point "active" solvent (hydrophobic fatty acids and rosin blends [549]), which is incorporated in the copolymer through a chain transfer reaction.

The binder can be involved in a cross-linking process aimed at making coating stronger and stiffer. Cross-linkers are also called "insolubilizers" [550–554]. They are able to prevent the methacrylic acid copolymer from dissolving in alkaline pH (such as 0.5% ethylene glycol di-methacrylate) [130]. A cross-linked structure increases the hydrophobicity of the coating (polyamine with formaldehyde [221]) and also reduces the binder migration.

The chemical structure of the cross-linker should be at least difunctional. The cross-linking reaction may involve covalent bond formation (starch and SMA [555,556]) or ionic interactions (the amphoteric starches [400–402], amphoteric latexes [441], SMA inorganic salts [554], or blends of anionic starch and cationic PEI resins [399]). The cross-linker chemical structure is designed to react with the particular reactivity of the binder: carboxylated compounds are cross-linked with diphenylmethane-bis-4,4′-*N*,*N*′-ethyleneurea [557] or copolymers of 2-alkyl-2-oxazoline [558,559].

Other polyfunctional compounds are glyoxal [560–564], di-aldehyde starch [565–567], urea-glyoxal oligomers [568–573], acrylic polymers and glyoxal [561,574], amphoteric starch [575,576], diphenolic acid [577], polyethyleneimine–epichlorohydrin resins [578], or with salts of polyvalent metals (aluminum, antimony, titanium, stannum, zirconium, or chromium) [564,579–584]. Cross-linked starch (or PVA, casein, and soy protein) makes clay-based coatings less sensitive to water (higher wet-rub resistance) [580,585]. Binders from two reactive latexes [586] are able to cure even at room temperature.

In order to make a stronger binder, the ideal coating composition would have low viscosity during its shelf life and application and increased viscosity only during the drying step. Irrespective of the benefits of a cross-linked structure, cross-linking reactions should be controlled to avoid variations in viscosity of the coating formula before its application. Usual cross-linkers, such as formaldehyde, glyoxal, and multivalent salts seriously thicken the coating composition [456,520].

There are several ways to control the cross-linking process: changing the addition point of the cross-linker (glyoxal at the water box) [456]; applying two layers (one with the binder and one with a cross-linker) [587]; using a carrier for formaldehyde (UF resins [588]; modifying polyamine resins for starch and SBR latex [520,589], polyformals or methylol dimethyl hydantoin for proteins and polysaccharides [590–592]); involving cyclic bis-hemiacetal able to release glyoxal only at higher temperatures [550–553,593];

changing the pH for a latent reactive starch [594,595] (R is H or alkyl) or starch propionamide reacted with polyaldehydes [596].

9.4.4 COATING HYDROPHOBICITY AND ITS REPULPABILITY

The curing (cross-linking) process is designed to increase mechanical properties and to make the coating water repellant. The hydrophobic properties of the coatings make it difficult for the paper to be repulped. The repulping of coated paper (i.e., making the coating water-soluble or dispersible) is a step developed to reverse the drying and cross-linking processes. Moreover, the residual organic material resulting after repulping must not interfere with the next cycle of papermaking.

Paraffin wax is the most common water repellant. It may also incorporate dihydro-naphtalene resin [597], hydrogenated copolymers of ethylene and α-olefin [598], or Fischer–Tropsch wax (high melting point polyethylene wax) [599]. Despite the fact that the water resistance properties of waxed paper begin to decrease within about 2 months [600], wax makes paper repulping more difficult. That is why, in the year 2000, about 60% of waxed board was sent to a landfill [601].

Although wax replacement is currently under study by paper scientists [601,602], there are several other approaches in developing a repulpable coating: changing the binder melting point, adding breakable bonds in the binder chemical structure, and using a repulpable polymer matrix [603] to help with the dispersibility of other components (such a wax).

Repulpable coatings incorporate binders with breakable bonds or which are dispersible [604,605] in water. Thermoplastic polymers (crystalline and amorphous) or hot melt binders [606–609] are dispersed in water at higher temperature [610].

Binders made from compounds with more sensitive bonds to alkaline conditions are recommended for repulpable coatings. In order to break down the polymeric film, copolymers of vinyl acetate [611], hydrophobic styrene copolymers [612–614] with carboxylic groups, copolyesters [615], esters of fatty acids [604,616], or inorganic compounds (such as aluminum silicate) [617], are added to the coating formulas, which modify the repulping process [618].

Wax replacement involves materials with the same hydrophobic character but with breakable bonds: ester-type waxes [619], polyester resins [121,507–509], and acrylic latexes with anionic stabilizers (such as SMA and ammonium salts of fatty acids) [600].

In the cross-linked structure, ionic bonds are involved between the binder particles (carboxylated polymers [614,620–623]) and AZC [545,603] and/or zinc ammonium carbonate [624]. Those ionic bonds are easier to break at high pH (2% caustic at 60°C) [603,614,624,625].

A primer coat with different composition than the top coat may help both the repulpability and the biodegradation of coated paper [626,627]. The two components may also form a single layer: the repulpable cross-linked matrix (zinc or zirconium salt as cross-linkers) of carboxylated styrene–butadiene polymers is used to encapsulate a hydrophobic paraffin wax [603]. Fully hydrolyzed PVA (2%–3%) is used in a binder composition [493] to facilitate the re-dispersion in water of the major binder (styrene copolymer latexes) during the repulping step.

9.4.5 COATING SURFACE PROPERTIES

Surface gloss is a function of the coating roughness [628,629] and brightness, opacity and print density are functions of the coating void structure [474].

The coating gloss increases at lower levels of pigment concentration and at higher binder concentrations [480]. The coating gloss is also a function of the pigment type, the pigment particle size [602], and the particle size distribution [457,630]. Blends of pigments (clay, calcium carbonate, TiO_2, and polymeric pigments) can perform based on optimization studies [631]. A higher gloss is obtained with oriented "plate" pigments. The pigment is oriented during the drying cycle by surface tension effects. The higher the T_g values, the longer it takes for the latex–pigment coalescence to occur, thereby increasing the time for pigment particle orientation.

In order to improve the coating gloss, paper is dried at a temperature below the binder T_g, and then the coating is calendered at a temperature above the film-forming temperature [490]. The sintering of the latex particles that occurred during the thermal treatment, led to the formation of larger voids, which proved more effective in scattering the light [485,632]. In terms of coating properties [474,633], the temperature–time profile is crucial.

Low-gloss-coated paper is produced by using large particle size pigments and carboxylated latexes (or polystyrene particles [537]) able to swell during the preparation of the coating composition and shrink during the drying of the coated paper [634]. The carboxylated latex contains a styrene-butadiene-acrylic acid with up to 14% acrylic acid (based on solids) [635]. Ink gloss increases with an increasing polymer gel content and with decreasing latex particle sizes. Therefore, the coating partial flocculation may result in lower gloss [487].

The use of the hollow particles reduces the pigment concentration (such as TiO_2) without adding undesirable weight to the coating [636]. Polymer dispersions with hollow particles are useful as opacifying agents, because light scattering properties are related to the difference in refractive index between the shell and the internal void. They also improve the paper coating gloss.

The void particles are obtained by hydrolyzing the soft core (ethyl acrylate–methyl acrylate–allyl methacrylate cross-linked copolymer) and keeping the hard shell ($T_g > 80°C$), which is a copolymer of the styrene–MMA–methacrylic acid [637,638].

All these properties can hardly be achieved by just one binder; blends of binders with different chemical structures are involved in a coating: either latexes of styrene (or acrylic copolymers) with minor amounts of water-soluble polymers [639,640] or two blended latexes (butyl acrylate-styrene-acrylonitrile-acrylamide-acrylic acid copolymer latex [641] and 10% of the acrylic anionic latex [642]).

Two of the main factors that result in good printing behavior and pigment incorporation are the T_g of the copolymer and the presence of carboxylic groups [643]. Moreover, the degree of interaction between the binder and the ink solvent has a significant effect on the printing process [644]. Since the sheet is separated from the environment by a substantial film (such as nonpolar wax), aqueous dyes and aqueous inks are prevented from penetrating it [462,645].

Ink receptivity and absorption depends on the binder concentration (lower binder, higher ink receptivity) and on its chemical structure: polyvinyl acetate is more ink receptive than SBR [487].

It is well documented that cationic additives retard the ink-jet dye movement [294]. That is why the following dispersions are recommended for a better coating printability: cationic (or amphoteric) dispersions (such as quaternized vinyl imodazole [160] copolymerized in the presence of oxidized cationic starch as stabilizer [646]), anionic copolymers and PEIs [647], or the copolymers of styrene-2-vinylpyridine-methacrylic acid [648,649].

Coatings can be multilayered: while the first layer ensures a better interaction with the cellulosic support (or improves the coating dispersibility during pulping [605]), the second targets a better interaction with the exterior environment. The butadiene latex is the first layer; a thicker second layer is paraffin wax for better water resistance [650]. The ink-receptive material containing an acrylic copolymer is added on top of the sizing agent applied as first layer [26,651].

Greaseproof coatings are developed either by making a continuous film of polymer able to hold out the greasy materials or by incorporating a fluorinated compound in the coating composition. A high amount of polymer is needed to make a continuous film; the film effectiveness is a function of time.

Fluorochemicals are used in a smaller amount and show a more robust resistance to grease. Fluorochemicals can be either small moleculecular [652–654] or macromolecular compounds (fluoroalkyl-poly(oxyalkylene) [655]). Fluoroalkyl acrylate copolymers [656] can incorporate non-fluorinated comonomers, glycidyl methacrylate, or stearyl methacrylate as diluters [213,214,657].

The dispersion of cross-linking polymers is also recommended as greaseproof coating: SBR with MF resin, protein and clay [658], polyacrylic latex [600], or latex of acrylic copolymers cross-linkable with polyvalent metals (Al, Ca, Cu, Mg, Zr) [659] or mica, TiO_2, $CaCO_3$, silica [660].

However, it is difficult to repulp the greaseproof coatings made with those dispersions. Polymers with weak ester bonds in the backbone, such as the emulsion of low molecular weight polyester (stearic acid, phthalic anhydride, glycerol or polylactide) [661,662] are used in order to make the coating hydrolysable.

REFERENCES

1. J. Bung, *Ippita J.*, **16(3)**, 29–31 (2004).
2. P.H. Brouwer, T. Wielema, B. Ter Veer, and J. Beugeling, *PTS-Symposium*, 14/1–14/214 (2001).
3. R. Exner, *Paper Technol.*, **43(6)**, 45–51 (2002).
4. S. Iwasa, *Paper Technol.*, **43(3)**, 31–37 (2002).
5. B.W. Ranson, *Pulp Paper*, **78(7)**, 50–54 (2004).
6. F.R. Lehman, L.H. Silvernail, and E.K. Stilbert, US 2,945,775 (1960).
7. D.R. Dill and P.R. Graham, *Tappi*, **50(3)**, 133–136 (1967).
8. J. Lipponen, J. Gron, S.E. Bruun, and T. Laine, *J. Pulp Paper Sci.*, **30(3)**, 82–90 (2004).
9. L. Gardlund, J. Forsstrom, and L. Wagberg, *Nord. Pulp Paper Res. J.*, **20(1)**, 36–42 (2005).
10. A.L. Tammi and M. Laitila, US 2007/0107865.
11. J. Lipponen, J. Gron, S. Anttilainen, and J.S. Kinnunen, US 7,045,036 (2006).
12. R. Bates, G.J. Broekhuisen, E.R. Hensema, and M.J. Welch, US 5,972,094 (1999).
13. J. Anderson, in *Surface Application of Paper Chemicals*, J. Brander and I. Thorn (Eds.), Blackie, London, U.K., 1997, pp. 138–155.
14. R.B. Wasser, US 4,966,652 (1990).
15. V.E. Pandian, D. van Calcar, and B.W. Wolff, US 5,362,573 (1994).
16. J.C. Roberts, *Proceedings of SPIE—The International Society for Optical Engineering*, Vol. 3227 (Interactive Paper), pp. 20–39, 1997.
17. I. Betremieux, C. Dumousseaux, B. Feret, and J.J. Flat, US 6,830,657 (2004).
18. G.L. Batten, *Tappi Papermakers Conference*, 1992, 159–167.
19. H.A. Goldsberry, K.C. Dilts, C.R. Hunter, M.P. O'Toole, R.J. Proverb, L. Pawlowska, G.E. Baikow, K. Komarowska, D.L. Dauplaise, and M.J. Scanlon, US 2006/0049377.
20. C.T. Beals, Surface applications in *Dry Strength Additives*, W.F. Reynolds (Ed.), Tappi Press, Atlanta, GA, 1980, pp. 33–65.
21. B.T. Hofreiter, G.E. Hamerstrand, and C.L. Mehltretter, US 3,087,852 (1963).
22. T. Yui, A. Suzuki, N. Ichizawa, K. Yamashita, and K. Hashimoto, US 5,948,155 (1999).
23. P.D. Garrett and K.I. Lee, *Tappi J.*, **81(4)**, 198–203 (1998).
24. R.Y. Ryu, R.D. Gilbert, and S.A. Khan, *Tappi J.*, **82(11)**, 128–134 (1999).
25. E. Scharf, H. Naarmann, and F. Reichel, US 3,983,268 (1976).
26. R. Williams, P.M. Froass, and S.Kulkarni, US 7,279,071 (2007), US 7,666,272 (2010) and US 7,666,273 (2010).
27. A.J. Aycock and H.G. Booth, US 3,413,190 (1968).
28. E.R. Sandstrom, K.J. Shanton, T.P. Hartjes, and D.P. Swoboda, US 5,876,815 (1999).
29. I. Rosenthal and P.J. McLaughlin, US 3,017,290 (1962).
30. L.J. Barker, R.J. Proverb, W. Brevard, I.J. Vazquez, O.S. dePierne, and R.B. Wasser, *Tappi Papermakers Conference*, 1994, pp. 393–397.
31. B. Brungardt, *Pulp Paper Canada*, **98(12)**, 152–155 (1997).
32. J.M. Shreier, *Paper Age*, June 1997, p. 26.
33. S.L. Malhotra, D.F. Rutland, and A.Y. Jones, US 5,302,249 (1994).
34. I. Reinbold and H. Wilfinger, *Pulp Paper Mag. Can.*, April, 110–112 (1967).
35. R. Bates, G.J. Broekhuisen, E.R. Hensema, and M.J. Welch, US 6,165,320 (2000).
36. Y. Abe and T. Noda, US 5,372,884 (1994).
37. K.B. Lawrence and J.-S. Wang, US 6,347,867 (2002).
38. J.J. Rabasco, E.H. Klingenberg, G.P. Dado, R.K. Pinschmidt, and J.R. Boylan, US 6,096,826 (2000).
39. A. Kobayashi, M. Tokita, M. Nakagawa, and K. Yasuda, US 5,126,010 (1992).
40. T. Saitoh and M. Sasaki, US 6,140,412 (2000).
41. P. Loewrigkeit and K.A. Van Dyk, US 4,644,030 (1987).
42. D. Dieterich and O. Bayer, US 3,479,310 (1969).
43. J.P. Miknevich, US 6,824,840 (2004).
44. P. Confalone and R. Farwaha, US 6,734,244 (2004).
45. D.W. Park and M.J. Dougherty, US 6,416,626 (2002).
46. L.A. Bays, B.J. Kokko, M.A. Schmelzer, and G.L. Schroder, US 2009/0017235.

47. H. Oshima, M. Daisaku, K. Yutaka, K. Yukio, and N. Koichi, US 4,442,172 (1984).
48. L.E. Rooff, C. Bennett, and F.J. Miskiel, US 6,290,814 (2001).
49. T. Jung, S. Ryu, and J. Kim, US 6,855,383 (2005).
50. M. Sklarewitz, *Tappi Papermakers Conference*, 1994, pp. 407–412.
51. D. Antos, *Paper Age*, August 1997, pp. 11–12.
52. P.H. Brouwer, J. Beugeling, and B.C.A. Ter Veer, *PTS-Symposium*, 1999 pp. 1–15/28.
53. B.U. Cho and G. Garnier, *Tappi J.*, **83(12)**, 60 (2000).
54. T.W. Tompkins and J.A. Shepler, *Tappi Papermakers Conference*, 1991, pp. 191–201.
55. D.R. Dill and K.A. Pollart, Surface sizing, in *The Sizing of Paper*, 2nd edn., W.F. Reynolds (Ed.), Tappi Press, Atlanta, GA, 1989, pp. 63–78.
56. P. Wilson, Surface sizing, in *The Sizing of Paper*, J.M. Gess and J.M. Rodriquez (Eds.), Tappi Press Atlanta, GA, 2005, pp. 211–233.
57. J. Lipponen, P. Pakarinen, J. Pakarinen, K. Holopainen, J. Vestola, and A. Tuomi, US 7,540,941 (2009).
58. Y.G. Tsai, US 6,087,457 (2000).
59. A.E. McPherson and J. Jane, *Carbohydr. Polym.*, **40**, 57–70 (1999).
60. D.R. Lu, C.M. Xiao, and S.J. Xu, *eXPRESS Polym. Lett.*, **3(6)**, 366–375 (2009).
61. J.W. Swanson, *Tappi*, **44(1)**, 142–181 (1961).
62. P.H. Brouwer, M.A. Johnson, and R.H. Olsen, Starch and retention, in *Retention of Fines and Fillers during Papermaking*, J.M. Gess (Ed.), Tappi Press, Atlanta, GA, 1998, 199–242.
63. N.O. Bergh, Starches in *Surface Application of Paper Chemicals*, J. Brander and I. Thorn (Eds.), Blackie Academic & Professional, London, U.K., 1997, pp. 69–108.
64. J. Baas, P.H. Brouwer, and T.A. Wielema, *PTS-Symposium*, 38/1–38/30, 2002.
65. P.H. Brouwer, *PTS-Manuskript*, 2/1–2/11. 2003.
66. C.L. Brungardt and D.F. Varnell, *Tappi Papermaking Conference*, 1996, pp. 99–113.
67. T. Kimpimaki and S. Rennes, *Paperi ja Puu*, **82(8)**, 525–528 (2000).
68. G.H. Brown and E.D. Mazzarella, US 3,671,310 (1972).
69. B.T. Hofreiter, H.D. Heath, M.I. Schulte, and B.S. Phillips, *Starch/Staerke*, **33(1)**, 26–30 (1981).
70. P.D. Buikema, US 4,029,885 (1977).
71. J.L. Latta, *Tappi Papermakers Conference*,1994, p. 399.
72. D. Glittenberg and A. Becker, *Paperi ja Puu*, **79(4)**, 240–243 (1997).
73. K.R. Anderson, US 5,122,231 (1992).
74. H. Ketola, T. Andersson, A. Karppi, S. Laakso, and A. Likitalo, US 6,187,144 (2001).
75. K.R. Anderson and D.E. Garlie, US 6,524,440 (2003).
76. K.R. Anderson and D.E. Garlie, US 6, 451,170 (2002).
77. E. Rudnik, G. Matuschek, N. Milanov, and A. Kettrup, *Thermochim. Acta*, **427**, 163 (2005).
78. D.W. Boling, US 5,964,950 (1999).
79. D.W. Boling, US 6,619,076 (2003).
80. L.H. Rees, US 2,635,068 (1953).
81. K.I. Lee, US 5,525,661 (1996).
82. R.P.W. Kesselmans, B.C.A. ter Veer, P.H. Brouwer, and T.A. Wielema, US 6,777,548 (2004).
83. O.S. dePierne, D.L. Dauplaise, and R.J. Proverb, US 5,139,614 (1992).
84. D.F. Varnell, US 6,051,107 (2000).
85. O.S. dePierne, D.L. Dauplaise, and R.J. Proverb, US 5,122,568 (1992).
86. P.B. Davidson, US 2,549,177 (1951).
87. M.P. O'Toole and O.S. dePierne, US 6,114,417 (2000).
88. G. Guerro, D. Dauplaise, and R. Bazaj, US 5,824,190 (1998).
89. R. Bazaj, G. Guerro, and D. Dauplaise, US 6,034,181 (2000).
90. J.C. Rankin, A.J. Ernst, B.S. Phillips, B.T. Hofreiter, and W.M. Doane, *Tappi*, **58(1)**, 106–108 (1975).
91. M. Muller, J. Probst, J. Alberts, J. Konig, H. Baumgen, and F. Puchner, US 5,314,721 (1994).
92. J. Beugeling, P. Brouwer, and B. Ter Veer, *PTS-Manuskript*, 20/1–20/14 (2000).
93. P.H. Brouwer and F. Naumann, *PTS-Manuskript*, 15/1–15/36 (2002).
94. T.J. Schoch, *Wallerstein Lab. Commun.*, **32(109)**, 149–71 (1969).
95. L.A. Bello-Perez, S.L. Rodriguez-Ambriz, M.M. Sanchez-Rivera, and E.A. Acevedo, Starch macromolecular structure, in *Starches: Characterization, Properties and Applications*, A.C. Bertolini (Ed.), CRC Press, Boca Raton, FL, 2010, pp. 33–58.
96. Y. Takeda, T. Shitaozono, and S. Hizukuri, *Starch/Starke*, **40(2)**, 51–54 (1988).
97. H.F. Zobel, *Starch/Starke*, **40(2)**, 44–50 (1988).
98. I. Hanashiro, J. Abe, and S. Hizukuri, *Carbohydr. Res.*, **283**, 151–159 (1996).

99. H.F. Zobel and A.M. Stephen, *Food Sci. Technol.*, **67**, 19–66 (1995).
100. P.H. Brouwer, *PTS-Manuskript*, 12/1–12/37 (2003).
101. S. Shen, P.A. Patton, M. Hostetler-Schrock, and M.D. Harrison, US 2009/0238925.
102. R.L. Mellies, C.L. Mehltretter, and I.A. Wolff, *Ind. Eng. Chem.*, **50(9)**,1311–1314 (1958).
103. R.L. Mellies, C.L. Mehltretter, and F.R. Senti, *J. Chem. Eng. Data*, **5(2)**, 169–171 (1960).
104. G.F. Bramel, *Tappi Papermaking Conference*, 1985, pp. 79–81.
105. H. Ketola and P. Hagberg, US 6,670,470 (2003).
106. M.G. Fitton, L.A. van der Auwera, and J.L. Hemmes, EP 620 315 (1994).
107. W.E. Mowry, US 3,644,167 (1972).
108. B. Mesic, L. Jarnstrom, C. Hjarthag, and M. Lestelius, *Appita J.*, **57(4)**, 281–285 (2004).
109. T.G. Shannon, D.A. Clarahan, M.T. Goulet, and W.Z. Schroeder, US 6,596,126 (2003).
110. N.-O. Bergh, *Pulp Paper Europe*, **2(9)**, 13 (1997).
111. J.H. Young, Fiber preparation and approach flow, in *Pulp and Paper Chemistry and Chemical Technology*, Vol. 2, 3rd edn., J.P. Casey (Ed.), John Wiley & Sons, New York, 1980, p. 827.
112. C.E. Brandon, Properties of paper, in *Pulp and Paper Chemistry and Chemical Technology*, Vol. 3, 3rd edn., J.P. Casey (Ed.), John Wiley & Sons, New York, 1981, p. 1884.
113. W.E. Scott, *Principles of Wet End Chemistry*, Tappi Press, Atlanta, GA, 1996.
114. M.F. Jones, EP 0 296 729 (1988).
115. R. Bates, G.J. Broekhuisen, E.R. Hensema, and M.J. Welch, US 6,074,468 (2000).
116. T. Kitaoka, A. Isogai, and F. Onobe, *Nord. Pulp Paper Res. J.*, **16(2)**, 96–101 (2001).
117. M.E. Bateman and E.M. Holda, US 3,931,422 (1976).
118. R.B. Login and D.R. Dutton, US 4,263,094 (1981).
119. R.B. Login and D.R. Dutton, US 4,210,685 (1980).
120. R.B. Login and D.R. Dutton, US 4,251,402 (1981).
121. R.K. Salsman, US 4,977,191 (1990).
122. J.F. Jackson, R.C. Williams, P.M. Froass, and S. Kulkarni, US 2006/0254736.
123. H. Beck, US 3,900,335 (1975).
124. L.E. Herdle and G.C. Milligan, US 3,062,679 (1962).
125. W. von Bonin, P. Mummenhoff, U. Beck, and H. Baumgen, US 4,499,153 (1985).
126. D.I. Lee, Latex, in *Papermaking Science and Technology*, Vol. 11, 2000, pp. 196–217.
127. D.I. Lee, and Y. Chonde, Ionic and ionogenic polymer colloids, in *Microspheres, Microcapsules and Liposomes*, Vol. 4, Citrus Books, London, U.K., 2002, pp. 137–170.
128. J. Anderson, Synthetic latex binders for paper manufacture in *Surface Application of Paper Chemicals*, J. Brander and I. Thorn (Eds.), Blackie, London, U.K., 1997, pp. 48–68.
129. A. Victoria, D.A. Winey, M.P. De Grandpre, and O.C. Ziemann, EP 0623659 (1994).
130. M. Joanicot, J.P. Lavalee, and R. Reeb, EP 0509878 (1992).
131. C. Hagiopol, *Copolymerization: Towards a Systematic Approach*, Kluwer-Academic Press, New York, 1999.
132. P.R.R. Carel, R.P.A. Decamp, and J. Perronin, US 4,112,155 (1978).
133. X. Liu, X.-D. Fan, M.-F. Tang, and Y. Nie, *Int. J. Mol. Sci.*, **9**, 342–354 (2008).
134. T.G. Fox, *Bull. Am. Phys. Soc.*, **1**, 123 (1956).
135. K. Wendel, T. Schwezel, and G. Hirsch, US 5,358,998 (1994).
136. F. Krause and R. Kniewske, US 5,004,767 (1992).
137. G. Odian, *Principles of Polymerization*, Wiley-Interscience, New York, 2004.
138. V. Chrastova, P. Citovicky, and J. Bartus, *J. Macromol. Sci. Pure Appl. Chem.*, **A31(7)**, 835–846 (1994).
139. M. Fernandez-Garcia, M. Fernandez-Sanz, E.L. Madruga, R. Cuervo-Rodriguez, V. Hernandez-Gordo, and M.C. Fernandez-Monreal, *J. Polym. Sci. Part B Polym. Phys.*, **38(1)**, 60–67 (2000).
140. G. Chambard, B. Klumperman, and A.L. German, *Polymer*, **40**, 4459–4463 (1999).
141. S. Abele, C. Gauthier, C. Graillat, and A. Guyot, *Polymer*, **41**, 1147–1155 (2000).
142. A.L. Herzog Cardoso, J.M. Moita Neto, A. Cardoso, and F. Galembeck, *Colloid Polym. Sci.*, **275**, 244–253 (1997).
143. L.L. Anderson and W.M. Brouwer, *J. Coat. Technol.*, **68(855)**, 75–79 (1996).
144. K.J. Abbey, US 4,254,004 (1981).
145. F. Galimberti, A. Siani, M. Morbidelli, and G. Storti, *Chem. Eng. Comm.*, **163**, 69–95 (1998).
146. C. Hagiopol, US 6,734,232 (2004).
147. J. Snuparek and F. Krska, *J. Appl. Polym. Sci.*, **20**, 1753–1762 (1976).
148. A.M. Santos and F.M.B. Coutinho, *Polym. Bull.*, **29**, 309–315 (1992).
149. G.W. Ceska, *J. Appl. Polym. Sci.*, **18**, 427–437 (1974).

150. G. Rinck, K. Moller, S. Fullert, F. Krause, and H. Koch, US 5,147,907 (1992).
151. A. Guyot and A. Goux, *J. Appl. Polym. Sci.*, **65**, 2289–2296 (1997).
152. H.J. Degen, F. Reichel, U. Riebeling, and L. Hoehr, US 4,835,212 (1989).
153. R. Reeb and B. Chauvel, EP 0089877 (1983).
154. I. Noda, *Nature*, **350**, 143–144 (1991).
155. G. Sackmann, J. Konig, and H. Baumgen, US 4,931,510 (1990).
156. J. Konig, E. Wenz, G. Sackmann, T. Roick, B. Thiele, G. Kinkel, J. Kijistra, and B. Hauschel, US 6,426,381 (2002).
157. T. Inoue and Y. Nakano, US 6,130,288 (2000).
158. G. Sackmann, J. Konig, F. Puchner, and H. Baumgen, US 5,258,466 (1993).
159. A. Hamunen and K. Nurmi, WO 93/11300.
160. H.J. Degen, F. Reichel, U. Riebeling, and L. Hoehr, US 4,855,343 (1989).
161. R. Brueckmann, H. Schoepke, T. Wirth, and J. Hartmann, US 5,240,771 (1993).
162. R. Brueckmann, H. Schoepke, T. Wirth, and J. Hartmann, US 5,231,145 (1993).
163. C.M. Lambert, US 3,135,698 (1964).
164. M.J.J. Mayer, J. Meuldijk, and D. Thoenes, *J. Appl. Polym. Sci.*, **56**, 119–126 (1995).
165. M.J.J. Mayer, J. Meuldijk, and D. Thoenes, *J. Appl. Polym. Sci.*, **59**, 1047–1055 (1996).
166. T. Aslamazova and S. Bogdanova, *Colloids Surf. A Physicochem. Eng. Aspects*, **104**, 147–55 (1995).
167. L.W. Morgan, US 4,628,071 (1986).
168. H. Kawaguchi, F. Hoshino, and Y. Ohtsuka, *Makromol. Chem. Rapid Commun.*, **7**, 109–114 (1986).
169. E. Vaclavova, A. Hrivik, and V. Chrastova, *Makromol. Chem.*, **193**, 2243–2250 (1992).
170. S.-T. Wang and G.W. Poehlein, *J. Appl. Polym. Sci.*, **50**, 2173–2183 (1993).
171. A.M. Dos Santos, T.F. Mckenna, and J. Guillot, *J. Appl. Polym. Sci.*, **65**, 2343–2355 (1997).
172. B.R. Vijayendran, *J. Appl. Polym. Sci.*, **23**, 893–901 (1979).
173. J.L. Guillaume and C. Pichot, *Polym. Mater. Sci. Eng.*, **54**, 530–534 (1986).
174. A. Andre and F. Henry, *Colloid Polym. Sci.*, **276**, 1061–1067 (1998).
175. L.L. de Arbina, M.J. Barandiaran, L.M. Gugliotta, and J.M. Asua., *Polymer*, **38**,143–148 (1997).
176. G.L. Burroway and R.D. Mate, US 4,868,259 (1989).
177. Z. Xu, C. Yi, S. Cheng, and J. Zhang, *J. App. Polym. Sci.*, **66**, 1–9 (1997).
178. F. Sauzedde, F. Ganachaud, A. Elaissari, *J. Appl. Polym. Sci.*, **65**, 2331–2342 (1997).
179. K. Sakota and T. Okaya, *J. Appl. Polym. Sci.*, **20**, 1725–1733 (1976).
180. F. Ganachaud, F. Sauzedde, A. Elaissari, and C. Pichot, *J. Appl. Polym. Sci.*, **65**, 2315–2330 (1997).
181. H.S. Killam, US 4,152,201 (1979).
182. H. Fikentscher, B. Durkheim, K. Herrle, G. Winter, A. Mueller, J.G. Reich, and H. Voss, US 3,174,874 (1965).
183. Z. Liu, H. Xiao, and N. Wiseman, *J. Appl. Polym. Sci.*, **76**, 1129–1140 (2000).
184. Y. Ohtsuka, H. Kawaguchi, and S. Hayashi, *Polymer*, **22**, 658–662 (1981).
185. Z. Xu, C. Yi, G. Lu, J. Zhang, and S. Chang, *Polym. Int.*, **44**, 149–155 (1997).
186. W.-F. Lee and G.-Y. Hwong, *J. Appl. Polym. Sci.*, **59**, 599–608 (1996).
187. W.-F. Lee and Y.-M. Chen, *J. Appl. Polym. Sci.*, **80**, 1619–1626 (2001).
188. H.J. Yang and C.H. Yang, *J. Appl. Polym. Sci.*, **69**, 551–563 (1998).
189. R.A. Wessling and D.M. Pickelman, US 4,337,185 (1982).
190. S.M. Hurley and F.C. Hansen, US 6,020,061 (2000).
191. D. Joynes and D.C. Sherrington, *Polymer*, 38(6), 1427–1438 (1997).
192. E.T.W.M. Schipper, O. Smidt, T. Hamaide, P. Lacroix Desmazes, B. Muller, A. Guyot, M.J.W.A. van den Enden, F. Vidal, J.J.G.S. van Es, A.L. German, A. Montaya Goni, D.C. Sherrington, H.A.S. Schoonbrood, J.M. Asua, and M. Sjoberg, *Colloid Polym. Sci.*, **276**, 402–411 (1998).
193. J.M. Asua, and H.A.S. Schoombrood, *Acta Polym.*, **49**, 671–686 (1998).
194. S. Roy, P. Favresse, A. Laschewsky, J.C. de la Cal, and J.M. Asua, *Macromolecules*, **32**, 5967–5969 (1999).
195. H.A.S. Schoonbrood, M.J. Unzue, O.J. Beck, J.M. Asua, A.M. Goni, and D.C. Sherrington, *Macromolecules*, **30**, 6024–6033 (1997).
196. O. Sindt, C. Gauthier, T. Hamaide, and A. Guyot, *J. Appl. Polym. Sci.*, **77**, 2768–2776 (2000).
197. G. Wulff, H. Schmidt, and L. Zhu, *Macromol. Chem. Phys.*, **200**, 774–782 (1999).
198. D. Cochin, A. Laschewsky, and F. Nallet, *Macromolecules*, **30**, 2278–2287 (1997).
199. I. Noda, US 4,735,843 (1988).
200. H. Brabetz, H. Eck, R. Jira, and H. Hopf, US 4,560,724 (1985).
201. A. De Clercq, L. Hohr, and U. Riebeling, US 5,672,392 (1997).

202. C.E. Brockway, US 3,061,472 (1962).
203. C.E. Brockway, D.W. Christman, and R.R. Estes, US 3,061,471 (1962).
204. M. Niinikoski, K. Nurmi, and A.L. Tammi, US 6,753,377 (2004).
205. K. Sakai and Y. Hayashi, US 2010/0016480.
206. K. Matsunaga, K. Ohira, and Y. Nakagawa, US 4,376,177 (1983).
207. L. Yang, R. Pelton, F. McLellan, and M. Fairbank, *Tappi J.*, **82(9)**,128–135 (1999).
208. R.F. Heine, US 3,094,547 (1963).
209. R.F. Deutsch, J.W. Eberlin, and R.F. Kamrath, US 5,271,806 (1993).
210. N.O. Brace and A.K. Mackenzie, US 3,083,224 (1963).
211. M. Visca, S. Modena, S. Fontana, and G. Gavazzi, US 6,221,434 (2001).
212. P. Iengo and P. Gavezotti, US 7,534,323 (2009).
213. J.J. Fitzgerald, US 5,674,961 (1997).
214. S. Raynolds, US 4,147,851 (1979).
215. P.C. Hupfield, T. Masutani, S. Minami, and I. Yamamoto, US 2010/0018659.
216. A.L. Dessaint, US 4,366,299 (1982).
217. M.J. Lina, C. Collette, J.M. Corpart, and A. Dessaint, US 5,439,998 (1995).
218. M.D. Harrison, G.A.R. Nobes, P. Patton, S. Shen, and H. Westerhof, US 2010/0021751.
219. W.W. Carlin, S.H. Hui, and J.D. Mansell, US 5,059,675 (1991).
220. J.E. Katchko, C.W. Strobel, and T.H. Plaisance, US 4,797,176 (1989).
221. M.E. Cupery, US 2,526,638 (1950).
222. K.J. Bottorff, US 6,951,962 (2005).
223. A. Nigam, US 6,686,054 (2004).
224. J.J. Waldmann, US 5,659,011 (1997).
225. J.J. Waldmann, US 4,902,779 (1990).
226. R.W. Staxkman and S.M. Hurley, *Polym. Mater. Sci. Eng.*, **57**, 830–834 (1987).
227. M.E. Bateman and E.M. Holda, US 4,006,112 (1977).
228. J. Brandrup and E.H. Imergut (Eds.), *Polymer Handbook*, 4th edn., John Wiley & Sons, New York, 1999.
229. O.S. dePierne, D.L. Dauplaise, and R.J. Proverb, US 5,138,004 (1992).
230. J. Briskin and E.C. Chapin, US 2,772,252 (1956).
231. M. Haruta, T. Hamamoto, and S. Toganoh, US 4,481,244 (1984).
232. R.J. Slocombe and G.L. Wesp, US 2,851,448 (1958).
233. M. Ratzsch, *Prog. Polym. Sci.*, **13**, 277–337 (1988).
234. J. Szalay, I. Nagy, I. Banyai, G. Deak, G. Bazsa, and M. Zsuga, *Macromol. Rapid Commun.*, **20**, 315–318 (1999).
235. Q. Lin, M. Talukder, and C.U. Pittman, *J. Polym. Sci. Part A Polym. Chem.*, **33**, 2375–2383 (1995).
236. Z. Yao, B.-G. Li, W.-J. Wang, and Z.R. Pan, *J. Appl. Polym. Sci.*, **79**, 615–622 (1999).
237. S. Belkhiria, T. Meyer, and A. Renken, *Chem. Eng. Sci.*, **49(24B)**, 4981–4990 (1994).
238. D.J.T. Hill, J.H. O'Donnell, and P.W. O'Sullivan, *Macromolecules*, **18**, 9–17 (1985).
239. L.F. Vander Burgh and C.E. Brockway, *J. Polym. Sci.*, **A3**, 575–581 (1965).
240. M.H. Gray and H.F. Sparks, US 3,297,657 (1967).
241. A. Voss and E. Dickhauser, US 2,047,398 (1936).
242. H.L. Gerhart, US 2,230,240 (1941).
243. C.A. Vana, US 2,430,313 (1947).
244. H.L. Gerhart, US 2,297,351 (1942).
245. F.E. Condo and C.J. Krister, US 2,286,062 (1942).
246. G.R. Barrett, US 2,838,475 (1958).
247. G.R. Barrett, US 2,675,370 (1954).
248. M.S. Montaudo, *Macromolecules*, **34**, 2792–2797 (2001).
249. S. Evani and R.J. Raymond, US 4,200,720 (1980).
250. W.D. Niederhauser, US 2,725,367 (1955).
251. Hoechst, IE 50353 (1986).
252. G.L. Batten, *Tappi J.*, **78(1)**, 142–146 (1995).
253. G.L. Batten, *Tappi Papermakers Conference*, Atlanta, GA, 1994 pp. 413–418.
254. C.F. Bishop and O.P. Cohen, US 3,002,860 (1961).
255. H.J. Van Den Berg, M.H.G. Maassen, and L.W. Steenbakkers, US 6,407,197 (2002).
256. H.J. Van den Berg, S.M.P. Mutsers, and A.A.M. Stroeks, EP 0728767 (1996).
257. H.J. Van Den Berg, M.H.G. Maassen, and L.W. Steenbakkers, WO 99/45039 (1999).
258. H.J. Van den Berg and M.H. Maassen, WO 00/34362.

259. J.P. Friederichs and H.J.F. Van den Abbeele, EP 1,405,865 (2004).
260. H.J.F. Abbeele Van Den, J.P. Friederichs, and H.J.G. Luttikhedde, WO 2004/031249.
261. H.J.F. Abbeele Van Den, WO 2008/014903.
262. W. von Bonin, H.L. Honig, and W. Theuer, US 3,726,822 (1973).
263. G. Sackmann, G. Kolb, J. Probst, F. Muller, and H. Baumgen, US 4,481,319 (1984).
264. R. Topfl, US 4,147,585 (1979).
265. G. Sackmann, W. von Bonin, F. Muller, and G. Kolb, US 4,152,312 (1979).
266. W.E. Hanford, US 2,378,629 (1945).
267. A. DeClercq, W. Denzinger, N. Greif, H. Hartmann, K. Oppenlaender, and U. Riebeling, US 5,266,165 (1993).
268. F. Gude, H. Haferkorn, and F. Schulde, US 4,316,977 (1982).
269. K. Peterlein, F. Schulde, K.M. Diedrich, and V.W. Kulisch, US 4,200,559 (1980).
270. Y. Tominaga and Y. Shibahara, US 4,030,970 (1977).
271. Y. Tominaga and Y. Shibahara, US 4,115,331 (1977).
272. T.J. Drennen and L.E. Kelley, US 2,999,038 (1961).
273. Y. Shibahara and Y. Tominaga, US 3,925,328 (1975).
274. M.F. Finlayson, K.E. Springs, J.J. Gathers, J.L. Cooper, W.L. Vaughn, B.H. Schumann, and S.M. Oliver, US 6,482,886 (2002).
275. R.D. Jenkinson, US 3,677,989 (1972).
276. C.S. Maxwell, *Tappi*, **53(8)**, 1464–1466 (1970).
277. C. Hagiopol, J.W. Johnston, S.M. Tyler, and M.J. Bush, US 2008/0214738.
278. J.P. Friederichs, WO 03/016361.
279. R. Topfl, US 4,221,886 (1980).
280. W. von Bonin, F. Muller, and N. Schon, US 4,001,193 (1977).
281. D.E. Allberry and K.I. Lee, US 5,237,024 (1993).
282. G. Sackmann, U. Beck, J. Konig, H. Baumgen, and K.D. Albrecht, US 4,771,097 (1988).
283. G. Sackmann, U. Beck, and H. Baumgen, US 4,614,759 (1986).
284. K.I. Lee, US 5,290,849 (1994).
285. R. Bazaj, G. Guerro, and D. Dauplaise, US 6,494,990 (2002).
286. R. Bazaj, G. Guerro, and D. Dauplaise, US 6,281,291 (2001).
287. E. Robbart, US 4,551,385 (1985).
288. S.A. Swanson, N.M. Weinshenker, R.E. Wingard, and D.J. Dawson, US 4,381,185 (1983).
289. S. Miyamoto and Y. Watanabe, US 4,335,184 (1982).
290. M. Murakami, Y. Hiromori, and H. Naito, US 4,425,405 (1984).
291. J. Seppala and A. Salminen, US 2009/0255641.
292. K. Hayama and A. Yamashita, US 4,741,969 (1988).
293. E. Akutsu, T. Fujii, K. Murakami, and T. Aruga, US 4,740,420 (1988).
294. D.W. Donigian, P.C. Wernett, M.G. McFadden, and J.J. McKay, *Tappi J.*, **82(8)**, 175–182 (1999).
295. W.D. Schroer, J. Probst, I. Kolb, P. Mummenhoff, and H. Baumgen, US 4,784,727 (1988).
296. H. Schurmann, J.J. Weissen, and K. Schloter, US 4,839,415 (1989).
297. W. von Bonin, P. Mummenhoff, and H. Baumgen, US 4,381,367 (1983).
298. K. Inaoka, T. Nakata, and Y. Hashiguchi, US 2009/0272507.
299. M. Sugiyama, I. Nakanishi, A. Ogawa, and M. Maekawa, US 4,371,582 (1983).
300. K. Yasuda, A. Kobayashi, M. Tokita, M. Maekawa, M. Kitajima, and H. Maruyama, US 4,944,988 (1990).
301. H. Maruyama, I. Ono, S. Shiraga, and J. Yamauchi, US 5,132,146 (1992).
302. A. Nigam, US 6,197,880 (2001).
303. M. Fryberg, S. Schuttel, and H. Tomimasu, US 7,235,284 (2007).
304. M. Sakaki, R. Arai, T. Akiya, S. Toganoh, M. Higuma, N. Eto, H. Mouri, M. Tobita, M. Ishida, and S. Kono, US 4,877,680 (1989).
305. J.F. Oliver and R.E. Sandborn, US 5,270,103 (1993).
306. H. Dungworth, A. Naisby, J. Suhadolnik, and D.A. Yale, US 7,618,693 (2009).
307. P.W. Carter, P.G. Murray, L.E. Brammer, and A.J. Dunham, US 6,313,246 (2001).
308. D.F. Varnell, US 6,207,258 (2001).
309. M.F. Koenig, J.P. John, T.R. Arnson, and B.T. Liguzinski, US 2009/0274855.
310. M.J. Cousin, L.O. Hill, and R.G. Justice, US 4,554,181 (1985).
311. S. Kono, H. Mouri, M. Tobita, M. Ishida, N. Eto, T. Akiya, S. Toganoh, M. Higuma, M. Sakaki, and R. Arai, US 4,801,497 (1989).

312. A. Nigam, US 6,761,977 (2004).

313. C.F. Bishop, US 3,396,135 (1968).

314. A.M. Santos, and F.M.B. Coutinho, *Polym. Bull.*, **30**, 407–414 (1993).

315. I. Betremieux, B. Feret, and C. Verge, US 6,054,526 (2000).

316. B. Alince, *J. Appl. Polym. Sci.*, **23**, 549–560 (1979).

317. G.J. Anderson and J. Fantl, US 3,396,049 (1968).

318. L.R. Dragnder and J.P. Farewell, US 5,591,489 (1997).

319. L.R. Dragner, D.W. Harper, and N.H. Thompson, US 5,795,932 (1998).

320. G. Sackmann, W. Henning, J. Probst, J. Konig, and H. Baumgen, US 5,288,787 (1994).

321. J.A. Verdol and H.J. Gonzalez, US 3,444,151 (1969).

322. J. Probst, G. Kolb, P. Mummenhoff, and H. Baumgen, US 4,434,269 (1984).

323. T. Corner, *Colloid and Surfaces*, **3**, 119–129 (1981).

324. B. Feret and I. Betremieux, EP 0810323 (1997).

325. D.T. Nzudie, V.L. Dimonie, D.E. Sudol, and M.S. El-Aasser, *J. Appl. Polym. Sci.*, **70(13)**, 2729–2747 (1998).

326. T.G.M. van de Ven, *Colloid Surf.*, **A-138**, 207–216 (1998).

327. D.T. Nzudie and C. Collette, US 6,225,395 (2001).

328. J. Probst, U. Beck, and H. Baumgen, US 4,659,431 (1987).

329. J. Probst, B. Bomer, J. Konig, and R. Mottweiler, US 4,806,591 (1989).

330. I.R. Hiskens, D.C. Johnson, and M.J. Caudwell, US 4,222,820 (1980).

331. H. Oikawa, M. Ogawa, K. Iwai, and M. Narushima, US 5,013,775 (1991).

332. D.H. Dumas, US 4,243,481 (1981).

333. C.L. Brungardt, R.J. Riehle, and J.J. Zhang, US 5,846,663 (1998).

334. W. Henning, W. Meckel, U. Beck, H. Baumgen, and J. Konig, US 4,670,100 (1987).

335. L. Pawlowska, K.C. Dilts, C.R. Hunter, M.P. O'Toole, R.J. Proverb, T.T. Long, G. Garro, D.L. Dauplaise, K. Komarowska, H.A. Goldsberry, and G.E. Baikow, US 2006/0060814.

336. H.L. Jones, US 3,421,976 (1969).

337. E. Dowthwaite and I.R. Hiskens, US 3,906,142 (1975).

338. E. Dowthwaite and I.R. Hiskens, US 4,323,425 (1982).

339. H.R. Rafton, US 1,976,744 (1934).

340. P.R. Graham and A.F. Ottinger, US 3,468,823 (1969).

341. R. Shah and S.C. Adams, US 6,172,149 (2001).

342. Y.C. Huang, M.B. Lyne, and J.H. Stark, US 5,741,889 (1998).

343. Y.C. Huang, M.B. Lyne, and J.H. Stark, US 6,048,439 (2000).

344. D.H. Dumas, US 4,317,756 (1982).

345. C.L. Brungardt, R.J. Richle, and J.J. Zhang, US 6,048,392 (2000).

346. C.A. Schwartz, US 4,426,466 (1984).

347. A. Famili and G.D. Miller, US 5,484,509 (1996).

348. M.F. Cenisio, E.R. Hensema, A. Mears, D.F. Varnell, US 6,162,328 (2000).

349. R.V. Lauzon, US 6,414,055 (2002).

350. D.F. Varnell, US 2010/0018660.

351. A. Brockmeyer, R. Ettl, and R. Dyllick-Brenzinger, US 2010/0016478.

352. P.H. Aldrich, US 4,017,431 (1977).

353. P. Peutherer, M. Waring, and L. Collett, US 6,284,099 (2001).

354. S. Tamagawa and T. Fuchizawa, US 4,749,678 (1988).

355. P.H. Aldrich and D.H. Dumas, US 4,087,395 (1978).

356. P.H. Aldrich and D.H. Dumas, US 3,922,243 (1975).

357. P.H. Aldrich and D.H. Dumas, US 3,990,939 (1976).

358. C.G. Caldwell and O.B. Wurzburg, US 2,661,349 (1953).

359. W. Maliczyszyn and H.R. Hernandez, US 4,872,951 (1989).

360. P. Fuertes, A. Lambin, and J.L. Dreux, US 5,647,898 (1997).

361. L.A. Gaspar, M. Tessler, and A.R. Malcom, US 4,239,592 (1980).

362. M. Shirazi, N. Esmail, G. Garnier, and T.G.M. van de Ven, *J. Disp. Sci. Technol.*, **25(4)**, 457–468 (2004).

363. M.K. Hargreaves, J.G. Pritchard, and H.R. Dave, *Chem. Rev.*, **70(4)**, 439–469 (1970).

364. E. Robbart, US 3,856,558 (1974).

365. E. Robbart, US 4,339,479 (1982).

366. W.I. Patnode, US 2,306,222 (1942).

367. W.H. Griggs, A.H. Gray, and R.D. Zaffrann, US 3,103,462 (1963).

368. W.H. Griggs, A.H. Gray, and R.D. Zaffrann, US 3,132,944 (1964).
369. A. Nigam, US 6,171,444 (2001).
370. A. Rodriguez and C. Hagiopol, EP 1,649,105 (2006).
371. A.H. Drelich and G.J. Lukacs, US 3,930,074 (1975).
372. M.J. Shaw, US 4,400,440 (1983).
373. V.E. Pandian and D. van Calcar, *Tappi Papermakers Conference*, Atlanta, GA, 1994 pp. 419–426.
374. I. McAlpine, *Paper Technol. Ind.*, **26(4)**, 194–198 (1985).
375. D.T. Stewart and I. McAlpine, US 3,741,782 (1973).
376. A. Veyland, L. Dupont, J. Rimbsult, J.-C. Pierrard, and M. Aplincourt, *Helv. Chim. Acta*, **83**, 414–427 (2000).
377. E. Hollink and D.W. Stephan, *Comprehensive Coord. Chem.*, **II-4**, 105–173 (2004).
378. K.W. Suh and J.M. Corbett, *J. Appl. Polym. Sci.*, **12**, 2359–2370 (1968).
379. D. Braun and M.I. Meyer, *Macromol. Chem. Phys.*, **199**, 735–744 (1998).
380. B. Zhao and R. Pelton, *Tappi J.*, **3(7)**, 3–7 (2004).
381. T. Vihervaara, K. Luukkonen, and V. Niinivaara, US 6,398,912 (2002).
382. T. Vihervaara, K. Luukkonen, and V. Niinivaara, US 6,716,313 (2004).
383. E. Valton, J. Schmidhauser, and M. Sain, *Tappi J.*, **3(4)**, 25–30 (2004).
384. I. Fineman and M. Hoc, *Tappi J.*, **61(5)**, 43–46 (1978).
385. T.J. Senden and M.B. Lyne, *Nord. Pulp Paper Res. J.*, **15(5)**, 554–563 (2000).
386. J.S. Aspler and M.C. Beland, *J. Pulp Paper Sci.*, **20(1)**, J27–J32 (1994).
387. P.J. Mangin, *J. Pulp Paper Sci.*, **17(5)**, J156–J163 (1991).
388. T. Takano, M. Fukuda, T. Satake, and M. Taniguchi, US 6,013,359 (2000).
389. T. Takano, M. Fukuda, and T. Satake, US 5,849,154 (1998).
390. M.J. Smith and S.J. McCullough, US 5,853,539 (1998).
391. H. Tsutumi, Y. Ikemoto, M. Maehama, T. Oyanagi, and T. Matsubara, US 5,470,918 (1995).
392. C. Trouve, US 2009/0269603.
393. D.F. Hiscock, J.A. Jacomet, and D.H. Carter, US 4,617,223 (1986).
394. M. Suzuki, J. Furuhata, M. Fukuda, S. Hatano, T. Takano, H. Umeda, and H. Hashimoto, US 5,698,305 (1997).
395. R. Pelton and J. Hong, *Tappi J.*, **10(1)**, 21–25 (2002).
396. C.L. Burdick, US 6,359,040 (2002).
397. J. Kotz, *Nord. Pulp Paper Res. J.*, **8(1)**, 11–14 (1993).
398. T. Satake, T. Takano, M. Fukuda, and Y. Uehori, US 5,750,253 (1998).
399. K.G. Taylor, US 3,719,514 (1973).
400. L.H. Elizer, US 3,673,171 (1972).
401. R.M. Powers and R.W. Best, US 3,649,624 (1972).
402. H. Benninga, US 3,467,647 (1969).
403. S.R. Wagg, W.L. Wagg, and J.B. Wagg, US 1,032,973 (1912).
404. C.C. Olson, R.C. Kumar, and J.C. Chang, US 5,492,599 (1996).
405. P.C. Lucas, R.J. Messenger, E. Macklem-Nathan, and L.C. Nykwest, US 5,338,404 (1994).
406. B.A. Owens, D.I. Collias, and A.J. Wnuk, US 6,114,471 (2000).
407. B.A. Owens, D.I. Collias, and A.J. Wnuk, US 6,211,357 (2001).
408. B.A. Owens, D.I. Collias, and A.J. Wnuk, US 6,620,461 (2003).
409. B.A. Owens, D.I. Collias, and A.J. Wnuk, US 6,281,350 (2001).
410. B.A. Owens, D.I. Collias, and A.J. Wnuk, US 6,623,806 (2003).
411. B.A. Owens, D.I. Collias, and A.J. Wnuk, US 6,458,419 (2002).
412. M.A. Hubbe, Pulp and paper chemical additives, in *Encyclopedia of Forest Sciences*, J. Burley, J. Evans, and J. Youngquist, (Eds.), Elsevier, Amsterdam, the Netherlands, 2005.
413. E.M. Williston, A.S. Gregory, and C.C. Heritage, US 3,305,435 (1967).
414. M. Ryan and D. Dauplaise, WO 2004/001127 (2004).
415. A.J. Stamm, *Tappi*, **32(5)**, 193–203 (1949).
416. J.A. Harpham and H.W. Turner, US 3,069,311 (1962).
417. K. Clare, P.E. Wiston, H.D. Dial, and T.M. Ortega, US 5,079,348 (1992).
418. M. Nurminen and M. Niinikoski, US 6,939,441 (2005).
419. L. Daniliuc, C. De Kesel, and C. David, *Eur. Polym. J.*, **28(11)**, 1365–1371 (1992).
420. H. Maeda, T. Kawai, and S. Sekii, *J. Polym. Sci.*, **35**, 288–292 (1959).
421. C.O. M'Baseck, M. Metayer, D. Langevin, and S. Roudesli, *J. Appl. Polym. Sci.*, **62**, 161–165 (1996).
422. W. Shimokawa, US 3,759,861 (1973).

423. J.C. Racciato and R.I. Yin, US 4,257,768 (1981).
424. T. Caykara and S. Demirci, *J. Macromol. Sci. A Pure Appl. Chem.*, **43**, 1113–1121 (2006).
425. C.S. Maxwell, *Tappi*, **54(4)**, 568–570 (1971).
426. D. Lath and M. Sivova, *Makromol. Chem. Macromol. Symp.*, **58**, 181–187 (1992).
427. F.A. Bonzagni, US 3,251,709 (1966).
428. T. Moritani and K. Kajitani, *Polymer*, **38(12)**, 2933–2945 (1997).
429. P. Lertsutthiwong, M.M. Nazhad, S. Chandrkrachang, and W.F. Stevens, *Appita J.*, **57(4)**, 274–280 (2004).
430. J.S. Michelman and D.M. Capella, *Polymer, Laminates & Coatings Conference, Tappi Proceedings*, Atlanta, GA, 1991, pp. 197–201.
431. R.E. Popil and M.K. Joyce, *Tappi J.*, **8(4)**, 1–8 (2008).
432. W.J. Long, US 3,998,944 (1976).
433. L.G. Andersson, in *Paper Coating Additives*, Tappi Press, Atlanta, GA, 1995, pp. 1–10.
434. W. Floyd, D. Harper, and N. Thompson, US 6,103,308 (2000).
435. D.W. Brown, K. Breindel, R.W. Broadbent, and M.S. Wiggins, US 6,770,129 (2004).
436. H.C. Schwalbe, R.C. Hydell, and J.J. O'Conner, US 2,700,621 (1955).
437. C. Soremark and I. Olsson, US 4,857,126 (1989).
438. J.C. Rice, US 3,028,258 (1962).
439. T. Koskinen, K. Koskinen, P. Aho, and S. Sarela, US 7,387,703 (2008).
440. P. Muller and O. Herrmann, US 2,252,091 (1941).
441. R. Gagnon and D. Hiscock, US 7,625,441 (2009).
442. G.D. Miller and M.M. Baas, US 5,057,570 (1991).
443. T. Kimpimaki, K. Niinikoski, and K. Nurmi, US 7,214,728 (2007).
444. G. Luebke and E.P. Pauley, US 5,800,870 (1998).
445. A.N. Leadbetter and J.G. Galloway, US 4,812,496 (1989).
446. H. Maruyama, K. Terada, and T. Katayama, US 5,527,852 (1996).
447. H.K. Brooks, US 750,048 (1904).
448. C.E. Coco, P.M. Graham, and T.L. Krinski, US 4,474,694 (1984).
449. T.L. Krinski, T.H. Tran, and J.J. Gambaro, US 4,961,788 (1990).
450. P.B. Davidson, US 2,414,858 (1947).
451. T.A. Wielema and P.H. Brouwer, *Paper Technol.*, **44(11)**, 27–40 (2003).
452. D. Gilttenberg, R.J. Tippett, and P. Leonhardt, *Paper Technol.*, **45(6)**, 27–33 (2004).
453. J. Kronfeld, US 3,052,561 (1962).
454. J.K. Smith, US 4,495,226 (1985).
455. R.W. Cescato, US 3,654,263 (1972).
456. J.T. Loomer, US 2,632,714 (1953).
457. H.K. Lee, M.K. Joyce, and P.D. Fleming, *Final Program and Proceeding of International Conference on Digital Printing Technologies*, Salt Lake City, UT, 2004, pp. 934–939.
458. M.C. Drake and A.J. Miklasiewicz, US 5,352,503 (1994).
459. J.L. Azorlosa, US 2,661,309 (1953).
460. J.L. Azorlosa, US 2,616,818 (1952).
461. J.L. Azorlosa, US 2,661,308 (1953).
462. R.M. Hume, US 4,714,727 (1987).
463. J.B. Hyde, US 2,582,037 (1952).
464. A.H. Boenau, US 3,105,823 (1963).
465. P.R. Graham and A.F. Ottinger, US 3, 632,424 (1972).
466. T. Shimada, H. Hirata, and Y. Iimori, US 5,281,467 (1994).
467. N.O. Bergh, J.L. Hemmes, and L. Olander, US 5,399,193 (1995).
468. R.C. Mavis, US 4,243,564 (1981).
469. E. Echt, P.A. Morken, and G.M. Collins, US 7,371,796 (2008).
470. E. Echt, P.A. Morken, and G.M. Collins, US 7,608,660 (2009).
471. E. Echt, P.A. Morken, and G.M. Collins, US 7,608,661 (2009).
472. E. Echt, P.A. Morken, and G.M. Collins, US 7,608,662 (2009).
473. H.J.F. Abbeele Van Den and E.M.J. Jonsson, US 2004/0054037.
474. J. Watanabe and P. Lepoutre, *J. Appl. Polym. Sci.*, **27**, 4207–4219 (1982).
475. P.M. Graham and T.L. Krinski, US 4,421,564 (1983).
476. P. Lepoutre and B. Alince, *J. Appl. Polym. Sci.*, 26(3), 791–798 (1981).
477. R. Uhlemayr, US 2009/0256347.

478. R.A. Dratz, US 3,047,427 (1962).
479. P. Canard and A. Levy, US 4,010,207 (1977).
480. D.I. Lee, *Coating/Papermakers Conference*, Vol. 1, Atlanta, GA, 1998, pp. 311–335.
481. G.G. Spence, US 4,196,253 (1980).
482. I.T. Badejo and D.J. Rice, US 6,066,203 (2000).
483. H. Mikkonen, M. Miettinen, K. Kataja, S. Luukkanen, S. Peltonen, and P. Qvinus-Leino, US 2009/0255441.
484. G.A. Richter, US 1,880,045 (1932).
485. B. Alince and P. Lepoutre, *J. Appl. Polym. Sci.*, **26(3)**, 799–808 (1981).
486. B. Alince and P. Lepoutre, *Tappi*, **63(5)**, 49–53 (1980).
487. D.I. Lee, *NATO ASI Ser E Appl. Sci.*, **335**, 497–513 (1997).
488. C.S. Kan, L.H. Kim, D.I. Lee, and R.L. Van Gilder, *Coating Conference Proceedings*, Nashville, TN, May 19–22, 1996, pp. 49–60.
489. R.A. Wamsley, J.E. Barnett, K.R. Rose, and G.S. Jones, US 3,634,298 (1972).
490. J.H. Vreeland, US 3,873,345 (1975).
491. J.F. Vitkuske, US 3,399,080 (1968).
492. C.C. Nguyen, V.J. Martin, and E.P. Pauley, US 5,003,022 (1991).
493. D.E. Stark and C.D. Rowe, US 5,897,411 (1999).
494. S. Moulay and S. Medjadji, *J. Soc. Alger. Chim.*, **14(2)**, 269–279 (2004).
495. T. Ashida, F. Saito, and M. Akamatsu, US 4,173,669 (1979).
496. J.F. Hoover, *Tappi*, **54(2)**, 257–261 (1971).
497. D. Distler, M. Mueller, H.G. Bubam, and G. Addicks, US 4,298,513 (1981).
498. R. Reeb and B. Chauvel, US 4,448,923 (1984).
499. J.E. Carmichael, J.J. Scobbo, E.M. Vincent, C.T. Romer, and C.E. Miller, EP 0 290 115 (1988).
500. W.F. Hill, US 3,242,121 (1966).
501. D.L. Funck and V.C. Wolff, US 3,296,172 (1967).
502. W.F. Fallwell, US 3,567,503 (1971).
503. W.F. Fallwell, US 3,395,131 (1968).
504. H. Watanabe, S. Takahashi, A. Shingyochi, and N. Watanabe, US 3,337,482 (1967).
505. R.K. Salsman, US 5,958,601 (1999).
506. R.K. Salsman, US 5,858,551 (1999).
507. M.O. Ogilvie, P.M. Whatley, and M.W. Olvey, US 6,113,981 (2000).
508. R.K. Salsman, US 5,726,277 (1998).
509. R.K. Salsman, US 5,281,630 (1994).
510. R.E. Weber, US 3,132,042 (1964).
511. J.C. Rice, US 3,281,267 (1966).
512. D.I. Lee and T. Mundorf, US 4,478,974 (1984).
513. G.H. Mueller and K. Wendel, US 3,847,856 (1974).
514. G.H. Redlich and R.W. Novak, EP 0 203 724 (1986).
515. T. Kawamura, K. Tazaki, T. Sakakiyama, and T. Hasegawa, US 4,265,977 (1981).
516. G. Mason, T. Kimpimaki, and K. Sundberg, US 7,285,182 (2007).
517. R.J. Pueschner and F.A. Miller, US 3,404,116 (1968).
518. C. Auschra, E. Eckstein, M.O. Zink, and A. Mühlebach, US 6,849,679 (2005).
519. J.T. O'Brien and W.W. White, US 4,115,435 (1978).
520. S. Kawakami, T. Saka, H. Takagishi, S. Ura, M. Iwata, Y. Tokugawa, and N. Jinno, EP 0081994 (1983).
521. K. Nurmi, K. Santamaki, and T. Carne, US 6,545,079 (2003).
522. R.W. Martin, US 4,608,401 (1986).
523. E. Jonson, and H.J.F. Abbeele Van Den, WO 2007/014635.
524. W.D. Emmons, M. Vogel, E.C. Kostansek, J.C. Thibeault, and P.R. Sperry, US 6,080,802 (2000).
525. R.L. Van Gilder and D.I. Lee, US 4,474,860 (1984).
526. D.I. Lee, *Polpu, Chongi Gisul (J. Tappi K)*, **15(2)**, 17–23 (1983).
527. R. Van Gilder, D.I. Lee, R. Purfeerst, and J. Allswede, *Tappi J.*, **66(11)**, 49–53 (1983).
528. R.L. Van Gilder and D.I. Lee, US 4,567,099 (1986).
529. B. Alince and P. Lepoutre, *Polym. Mater. Sci. Eng.*, **55**, 26–30 (1986).
530. D.I. Lee, Fundamentals and strategies in paper coating, in *Binder Migration in Paper and Paperboard Coatings*, 1993, pp. 19–38.
531. P. Lepoutre and D. Lord, *J. Colloid Interface Sci.*, **134(1)**, 66–73 (1990).
532. T. Ghosh, M. Skinner, T. Luttikhedde, C. Hossenlopp, and L. Brauge, US 2009/0298996.

533. A.C. Sau, US 6,162,877 (2000).
534. K.N. Bakeev, J.G. Binder, T. Klootwijk, D. Kruythoff, B. Maagdenberg-Roels, T.T. Nguyen, J.K. Politis, and W.G. Salomons, US 7,531,591 (2009).
535. W. Windle and K.M. Beazley, *Tappi*, **50(1)**, 1–7 (1967).
536. T. Hirabayashi, S. Fujiwara, and T. Fukui, US 6,458,413 (2002).
537. G. Eksi and D.W. Bousfield, *Tappi J.*, **80(2)**, 125–135 (1997).
538. N. Yasuda, M. Masuda, and T. Morita, US 4,301,210 (1981).
539. A. Kennedy, US 3,583,881 (1971).
540. R.L. Billmers, V.L. Mackewicz, and R.M. Trksak, US 6,790,270 (2004).
541. M.O. Schur, US 2,304,287 (1942).
542. H. Hamada, T. Enomae, I. Shibata, A. Isokai, and F. Onabe, *PITA Coating Conference*, Edinburgh, Great Britain, 2003, pp. 41–45.
543. H. Gotoh, A. Igarashi, R. Kobayashi, and K. Akiho, US 4,117,199 (1978).
544. P.D. Buikema and T. Aitken, US 4,146,515 (1979).
545. C.C. Nguyen and D.E. Tupper, US 5,536,764 (1996).
546. H.C. Moyer, US 3,522,081 (1970).
547. V.L. Mackewicz, R.L. Billmers, and D.J. Hanchett, US 6,372,361 (2002).
548. G.F. Hutter and A.J. Conte, US 6,849,681 (2005).
549. R. Shah and S.C. Adams, US 6,329,068 (2001).
550. W.C. Floyd, US 4,537,634 (1985).
551. W.C. Floyd, US 4,547,580 (1985).
552. W.C. Floyd and B.F. North, US 4,625,029 (1986).
553. W.C. Floyd, US 4,656,296 (1987).
554. D.A. Newman and A.T. Schlotzhauer, US 2,696,783 (1954).
555. A.P. Hudson and J.E. Nevin, US 6,300,393 (2001).
556. W.G. Evans, M.A. Nisely, and P.D. Harper, US 4,361,669 (1982).
557. M. Nakanishi, A. Sugiyama, T. Sakakiyama, and A. Matsueda, EP 0062338 (1982).
558. D.L. Schmidt, R.F. Harris, and C. Coburn, US 5,470,908 (1995).
559. S.J. Sargeant and J.D. Rundus, US 6,127,037 (2000).
560. G.W. Buttrick, G.B. Kelly, and N.R. Eldred, *Tappi*, **48(1)**, 28–33 (1965).
561. W.C. Floyd and S.H. Hui, US 4,695,606 (1987).
562. B.R. Bobsein, J.T. Brown, Z. Fu, and J.D. Windisch, US 6,547,929 (2003).
563. B.R. Bobsein, J.T. Brown, Z. Fu, and J.D. Windisch, US 6,863,775 (2003).
564. E.D. Mazzarella and E. Dalton, US 3,320,080 (1967).
565. H. Suzumura and H. Miyahara, US 3,324,057 (1967).
566. M.F. Zienty, US 3,145,116 (1964).
567. J.H. Curtis, US 3,395,106 (1968).
568. S.A. Lipowski, US 4,471,087 (1984).
569. B.F. North, US 4,345,063 (1982).
570. W.C. Floyd and B.F. North, US 4,505,712 (1985).
571. W.C. Floyd and B.F. North, US 4,455,416 (1984).
572. J.A. Dodd and R.L. Lane, US 4,343,655 (1982).
573. N. Clungeon, G.J. Flynn, Z. He, and J.M. Rodriguez, US 5,435,841 (1995).
574. S.H. Hui, US 4,544,609 (1985).
575. J.L. Zimmerman, US 3,884,853 (1975).
576. J. Fernandez, D. Solarek, and J. Koval, US 5,093,159 (1992).
577. G.H. Perkins, US 3,125,455 (1964).
578. G.W. Strother and W.P. Coker, US 3,607,331 (1971).
579. R.K. Iler, US 2,359,858 (1944).
580. K.G. Taylor, US 3,137,588 (1964).
581. R.T. Hart, US 3,311,492 (1967).
582. R.T. Hart, US 3,425,896 (1969).
583. Y.E. Dikler, V.E. Guryanov, and B.D. Korolkov, US 4,146,669 (1979).
584. R.T. Hart, US 3,332,794 (1967).
585. K.G. Taylor, US 3,081,199 (1963).
586. C. Verge and I. Betremieux, US 6,107,391 (2000).
587. L.W. Hoel and D.L. Wolfe, US 2,860,073 (1958).
588. C.G. Landes, US 2,399,489 (1946).

589. D.D. Ritson and Y. Jen, US 2,918,438 (1959).
590. B.H. Kress, US 2,868,773 (1959).
591. G.S. McKnight and R.C. Brown, US 2,865,773 (1958).
592. B.H. Kress, US 2,968,581 (1961).
593. Y. Nobuo, US 3,847,948 (1974).
594. M.M. Tessler, US 4,153,585 (1979).
595. E.F. Paschall, F.D. Thayer, and WH. Minkema, US 3,132,113 (1964).
596. F.D. Thayer, US 3,127,393 (1964).
597. P. La Frone Magili, US 2,091,180 (1937).
598. L.E. Kidwell, US 4,049,893 (1977).
599. R.B. Porter, US 3,985,932 (1976).
600. S. Zhang, C.A. Rumble, W. Srisiri-Sisson, M.S. Moisa, and J.G. Hayden, US 6,713,548 (2004).
601. R.E. Popil and M. Schaepe, *Tappi J.*, **4(8)**, 25–32 (2005).
602. H.K. Lee, M.K. Joyce, P.D. Fleming, and J.E. Cawthorne, *Tappi J.*, **4(2)**, 11–16 (2005).
603. T.C. Ma, A.L. Berzins, C.J. Davis, and B.T. Watson, US 6,066,379 (2000).
604. D.K. Pattilloch, US 2,658,828 (1953).
605. A.K. Druckrey, J. MacKay Lazar, and M.H. Lang, US 7,235,308 (2007).
606. P.E. Sandvick and C.J. Verbrugge, US 5,491,190 (1996).
607. P.E. Sandvick and C.J. Verbrugge, US 5,587,202 (1996).
608. P.E. Sandvick and C.J. Verbrugge, US 5,599,596 (1997).
609. P.E. Sandvick and C.J. Verbrugge, US 5,700,516 (1997).
610. S.U. Ahmed, A. Emiru, L.J. Clapp, M.S. Kroll, and G.J. VanLith, US 6,103,809 (2000).
611. A. Takahira and Y. Yoshii, US 5,527,623 (1996).
612. H.J.F. van den Abbeele and T. Kimpimuki, US 7,244,510 (2007).
613. S. Berube, US 6,441,080 (2002).
614. J.A. Dooley, R.D. Vieth, and H. Burrell, US 3,287,149 (1966).
615. J.S. Michelman, US 6,255,375 (2001).
616. A. Hassan, A. Hassan, and G. Borsinger, US 6,811,824 (2004).
617. S. Berube, US 5,929,155 (1999).
618. R.E. Locke, G.N. Prentice, and C.M. Vitori, US 6,053,439 (2000).
619. L. Auer, US 2,406,336 (1946).
620. J.R. Quick, D.J. Wenzel, G.W. Bartholomew, M.S. Delozier, and M. Klass-Hoffman, US 5,763,100 (1998).
621. W.A. Wittosch, D.J. Romanowicz, B. Rose, D.J. Wenzel, G.W. Bartholomew, J.R. Quick, M.S. Delozier, and M. Klass-Hoffman, US 6,548,120 (2003).
622. W.A. Wittosch, D.J. Romanowicz, B. Rose, D.J. Wenzel, G.W. Bartholomew, J.R. Quick, M.S. Delozier, and M. Klass-Hoffman, US 5,989,724 (1999).
623. D.J. Wenzel, G.W. Bartholomew, J.R. Quick, M.S. Delozier, and M. Klass-Hoffman, US 5, 837,383 (1998).
624. T.C. Ma, A.L. Berzins, C.J. Davis, and B.T. Watson, US 5,635,279 (1997).
625. A.L. Berzins, T.C. Ma, and C.J. Davis, US 5,626,945 (1997).
626. S. Koutitonsky, US 5,562,980 (1996).
627. D.J. Wenzel, G.W. Bartholomew, J.R. Quick, M.S. Delozier, and M. Klass-Hoffman, US 5,654,039 (1997).
628. Z.R. Zhang, R.W. Wygant, A.V. Lyons, and F.A. Adamsky, *Proceeding of Tappi Coating Conference*, Toronto, Ontario, Canada, 1999, pp. 275–285.
629. Z.R. Zhang, R.W. Wygant, and A.V. Lyons, *Tappi J.*, **84(3)**, 48–71 (2001).
630. E.J. Sare, T.L. Adkins, and S.C. Raper, US 7,611,575 (2009).
631. R.W. Wygant, *Coating and Graphic Arts Conference and Exhibit*, Baltimore, MD, 2004, pp.301–311.
632. P. Lepoutre, N. Pauler, B. Alince, and M. Rigdahl, *J. Pulp Paper Sci.*, **15(5)**, J183–J185 (1989).
633. C.S. Kan, L.H. Kim, D.I. Lee, and R.L. Van Gilder, *Tappi J.*, **80(5)**, 191–201 (1997).
634. D.I. Lee and R.E. Hendershot, US 4,751,111 (1988).
635. F.T. Stollmaier, D.M. Elsaesser, and P.J. Salminen, US 5,837,762 (1998).
636. D.I. Lee, M.R. Mulders, D.J. Nicholson, and A.N. Leadbetter, WO 94/04603.
637. D.I. Lee, M.R. Molders, D.J. Nicholson, and A.N. Leadbetter, US 5,521,253 (1996).
638. D.I. Lee, M.R. Molders, D.J. Nicholson, and A.N. Leadbetter, US 5,157,084 (1992).
639. J.C. Song and S. Yang, US 7,608,338 (2009).
640. J.M. Kruse and D. Kimball, US 4,892,787 (1990).

641. B.R. Bobsein, J.C. Chiang, and S.L. Egolf, US 7,217,443 (2007).
642. J.C. Chiang and T.E. Stevens, US 4,421,902 (1983).
643. P.J. McLaughlin and B.B. Kine, US 2,790,735 (1957).
644. R.L. Van Gilder and R.D. Purferst, *Tappi J.*, **77(5)**, 230–239 (1994).
645. R.N.S. Sodhi, L. Sun, M. Sain, and R. Farnood, *J. Adhes.*, **84**, 277–292 (2008).
646. W. Auhorn, H.J. Degen, L. Hoehr, and U. Riebeling, US 4,908,240 (1990).
647. G. Veaute, US 4,226,749 (1980).
648. S.C. Sharma, C.M. Kausch, R.D. Mohan, and R.J. Weinert, US 5,693,732 (1997).
649. S.C. Sharma, C.M. Kausch, R.D. Mohan, and R.J. Weinert, US 5,872,200 (1999).
650. E.F. Knights, US 3,874,905 (1975).
651. S. Kulkarni, Y. Ling, R. Williams, V.P. Holbert, US 6,645,642 (2003).
652. H.A. Brown, US 2,934,450 (1960).
653. R.A. Falk and K.P. Clark, US 5,091,550 (1992).
654. H. Aoyama and Y. Amimoto, EP 0 280 115 (1988).
655. F. Fieuws, K. Allewaert, and D. Coppens, US 5,370,919 (1994).
656. Y. Amimoto, H. Aoyama, and A. Chida, US 4,728,707 (1988).
657. M. Usugaya and M. Matsuda, US 2009/0155600.
658. M.J. Shaw and R.J. Thiessen, US 4,272,569 (1981).
659. G.L. Brown and B.B. Kine, US 2,757,106 (1956).
660. D.D. Ritson, US 2,889,299 (1959).
661. C. Montclair, US 2,086,903 (1937).
662. J. Nangeroni, M.H. Hartmann, M.L. Iwen, C.M. Ryan, J.J. Kolstad, and K.T. McCarthy, US 6,183,814 (2001).

Index

A

Acrolein, 105–106, 109, 124, 165, 172
   acrolein copolymers, 125, 169, 171, 180–181, 208, 241, 246
   polyacrolein, 71, 127, 165, 208
Acrylamide copolymerization, 101, 114, 251
Acrylamide copolymers, 31, 64, 68, 101, 106–107, 125, 169, 173, 198, 223, 247, 254, 363, 387, 391
   amphoteric polyacrylamide, 255, 280, 374
   anionic polyacrylamide, 29, 31, 34, 80, 82, 118, 198–200, 244, 246–248, 258, 275, 277, 280, 287, 333, 351, 373, 391, 393, 397
   cationic polyacrylamide, 31, 33–34, 76, 80, 113–121, 127, 224, 249, 276, 285–286, 301, 379
     as dry strength resins, 241, 251–252, 254
     in paper surface treatment, 387
     in retention process, 56, 58–59, 67, 73, 76, 78, 82–84, 280
     as stabilizer for AKD, 379
     as wet strength resins, 130, 144, 168–172, 188, 196–199
Acrylamide, N-(2,2 dimethoxyethyl), 106, 110
Acrylamide, N-(3-dimethylaminopropyl), 78, 250, 288
Acrylamide, N-Methylol, 144, 171, 177, 252, 345, 387
Acrylamide, N,N-dimethyl, 106, 108, 119, 127
Acrylamide, N,N'-Methylene bis-, 20, 77, 101, 149, 151, 246, 315
Acrylamide polymerization, 28, 120
   inverse emulsion polymerization, 117–118, 122, 166
Acrylamide reactions, 20, 28, 106, 112–114, 124, 127, 144
Acrylic esters, 102
   copolymers, 118–119, 160, 366
     anionic, 32, 55, 246, 281, 333, 372, 391
     cationic, 76, 82, 106–108, 114, 116, 118, 168–169, 171, 173, 288, 301, 308, 312, 367, 374, 377
     grafted copolymers, 25, 200, 249, 352
     latexes, 76, 80, 176–177, 200, 331, 360–364, 367, 376–379, 392, 397
   homopolymers, 164–165, 352
   reactions, 144, 148, 151, 170, 174, 177, 312
Adsorption,
   equilibrium, 58, 204, 283
   mechanism, 67, 205
Alginate, 71, 385, 388
Alkenyl succinic anhydride (ASA), 7, 13, 35, 128, 268, 282, 297
   ASA hydrolysis, 302–304
   emulsification, 300–302
   in paper surface treatment, 378–381
   reactions, 175, 286, 298, 313
   retention at wet end, 61, 297, 300–305
   sizing mechanism, 304–307
   synthesis, 298–300

Alkyl ketene dimer (AKD), 7, 35, 218, 268, 275, 282, 310
   deactivated AKD, 345
   emulsification, 283–287
   liquid AKD, 283, 287, 295, 307
   in paper surface treatment, 378–381
   reactions, 175, 286, 296–297, 331, 381, 383
   retention at wet end, 13, 61, 287–289
   sizing mechanism, 289–297
   synthesis, 282–283
Aluminum compounds, 53, 55–57, 65–66, 196
   alum, 6–7, 37–38, 53, 62, 65–66, 298
   aluminum ion structures, 268–271
   as crosslinkers, 162, 395
   precipitation inhibitor, 270
   reactions, 74, 221, 274, 277–278, 287, 299, 310
   as retention aid, 52, 70–73, 79, 82
     for DSR, 247–248, 258–259
     for internal size, 267, 271, 273, 275–276, 280–282, 287, 301, 305, 309, 312–313
     for WSR, 106, 153, 163, 176, 178–179, 198, 208
   in sizing mechanism, 278–279, 289, 293–294, 304–306, 316
   as stabilizer for WSR, 191
   as stabilizers for AKD, 284–285, 287
   in surface treatment, 353–354, 362, 372, 374, 379, 382–383, 385, 389, 396
Amidine, 182, 364
Amidol bond, 112–113, 121–122, 124–125, 127, 129–130
Amphoteric compounds, 34, 269, 340, 353, 358
   dry strength resins, 255–256, 258–259
   in paper surface treatment, 364, 374, 387, 394, 397
   retention aids, 68, 78–83
   stabilizers for internal sizing dispersions, 275–276, 280, 289, 300
   starch, 18–19, 34, 79, 394–395
   wet strength resins, 106, 118, 126, 154, 163, 169, 199–200, 207
Amphoteric starch, 18, 34, 78–79, 255, 387, 394–395
Amylopectin, 57, 74, 258, 355, 391
Anionic rosin size, 274–276
Anionic starch, 71, 81–84, 245–258, 284, 354, 381, 394
Anionic trash, 13, 34, 51–53, 55, 57, 64–65, 68–69, 71, 74, 77, 206, 214, 270–271, 285, 288
Anionic trash catcher, 13, 74
Azetidinium functionality, 75
   crosslinking mechanism, 190, 210, 212–214, 219
   in dry strength resins, 252–253
   identification by NMR, 182
   in monomers, 164, 170
   on polyamidoamine, 177, 183–186, 188–192, 194, 196, 199–200
     on pendant group, 171, 222
   on polyethers, 163

B

Binder for coatings, 29, 222, 390–398
Binder migration, 393–394
Block copolymers, 163, 275, 331, 353, 366, 374
Bridging flocculation, 67, 70, 77
Butadiene copolymers, 61
    as binder in coatings, 391–392, 394, 396–397
    as dry strength resins, 254
    as internal sizing agent, 275, 281, 312
    in paper surface treatment, 362
    as wet strength resins, 160, 176–178, 345

C

Campbell effect, 241
Canadian standard freeness, 72–73
Casein, 6–7, 276, 280, 313, 353, 390–391, 395
Cationic cellulose, 19
    fibers, 63, 155
Cationic demand, 8, 53, 62–63, 65, 67, 288
Cationic rosin size, 61, 276, 288
Cationic starch, 34
    adhesive and binder, 331, 335–336, 394
    adsorption, 56–57, 60, 63
    dispersion stabilizer, 276, 280, 283, 285–288, 300,
        303–305, 309–311, 314, 358, 397
    improving recycled fibers, 224
    porosity builder, 388
    reactions, 78, 111–112, 200, 255, 381
    retention aid, 73–74, 78, 81–84, 165, 281, 284,
        286–287, 301
    size-press solution, 224, 354, 367, 376, 381
    strengthening resin, 64, 130, 155, 213, 244–245,
        248–249, 252, 257–258
    surface strength agent, 386–387
    synthesis, 18–19, 26, 119
Cellulose depolymerization (peeling), 11, 22
Cellulose pulp, 19, 51, 254–255, 268, 281, 294
    bleaching process, 8, 11–12, 14, 285
    Kraft pulp, 6, 8, 11, 37–39, 146, 205–206, 248
    reactions, 24
    refining, 11, 52, 61, 204, 356
    sulfite pulping, 6, 8–11
    thermo-mechanical pulp, 7–8
Cellulose structure, 19, 52, 54, 57, 242
Charge neutralization, 66–67, 218, 259
Charge titration, 55, 62
Chelating agents, 277, 280
Chitosan, 17–18, 22, 201, 288, 374, 389–390
Chloracetamide, N-(2,2-dimethoxyethyl)-N-methyl-2-,
    20, 109
3–Chloropropane-1,2-diol (CPD), 170, 193–196, 332
Coagulation, 70–71, 83, 274, 297, 358, 373
    hetero-coagulation, 51, 53, 61, 78, 176, 257, 277, 393
    homo-coagulation, 176, 358, 393
Consistency, 35, 58, 61, 63–64, 327, 340, 345, 388
Contact angle, 36–37, 267, 289, 295, 353, 372–373
Copolymerization, 28–29
    binary copolymerization, 80–81, 116, 119, 171, 246, 360
    in emulsion, 29, 160, 177–178, 246, 254, 359–365,
        367, 374, 376–378
        seeded emulsion copolymerization, 177, 362, 364,
            367, 391

grafting copolymerization, 26, 70, 249, 301
    reactivity ratios in copolymerization, 116–117, 173,
        246, 360, 369
    ternary copolymerization, 80, 119, 171, 370, 377
Creping process, 327–330, 333, 336, 339, 341
    creping adhesives, 325
        external crosslinkers for, 333
        nonreactive, 330–332
        reactive, 332–333
    synergetic effect, 330, 333–336, 338
Crosslinked starch, 395
Cyclo-polymerization, 165–166

D

Debonders, 138, 207, 327, 342, 345
Detrimental substances, 13
Dextran, 59, 110, 130, 154–155, 218, 243
Di-aldehyde cellulose, 155
Di-aldehyde starch, 23, 37, 39, 130, 152–154, 200,
    258, 395
1,3-Dichloro-2-propanol, 170, 184, 193–196, 332
Dispersions, 7, 11
    emulsions of synthetic polymers, 28–29, 33, 38,
        80, 139
    of paper chemicals, 51, 53, 55
    pulp at wet end, 12, 64–67, 78, 82
Dry strength resins, 28, 35, 53, 85, 241–260, 340–341,
    386–387
    amphoteric, 255–260
    anionic, 65, 245–248
    cationic, 144, 248–254
    dry strength mechanism, 241–245
    dual systems, 256–260
    latexes, 254–255
Dust reduction, 385–387

E

Emulsification, 311–312, 359, 381
    for AKD, 282–287, 289
    for ASA, 300, 302–303
    for rosin derivatives, 275, 358
Emulsion copolymerization, 29, 160, 177–178, 246, 254,
    359–365, 367, 374, 376–378, 391
Emulsion stability
    AKD emulsion, 283–285, 287, 289
    ASA emulsion, 300, 303
    rosin dispersion, 275
Environmental issues, 1, 8, 185, 194, 221, 251, 332, 351
Enzyme as catalyst, 20, 23, 67, 107, 195–196, 222, 355
Epichlorohydrin
    reaction with cellulose, 19, 241
    reaction with polyamine, 74–75, 80, 119, 121, 163, 165,
        168, 174, 180
    reaction with starch, 23, 25, 74
    in wet strength resins, 99, 138, 142–146, 149, 151,
        170–171, 175, 182–198
Esterification, 21–22, 24, 78–79, 161–162, 175, 268, 298,
    300–301, 313–314, 389
Etherification, 18–20, 24–25, 52, 78
Ethylene–acrylic acid copolymers, 276, 281, 372, 379
Experimental errors, 36–38, 214, 302, 353

F

Fatty acids, 13, 22, 52, 274, 281, 283, 286, 298, 301, 314, 342, 346, 379–380, 394, 396
   fatty acid anhydride, 7, 282, 284
   fatty acid chloride, 145, 151, 284
   polymeric fatty acids, 146
   as sizing agents, 281
Fillers, 1, 6, 13, 15, 17, 53, 282, 328, 345, 354, 379, 387
   in coatings, 388, 390–394
   in dry strength mechanism, 242, 245
   effect on internal size, 287, 290–291, 302
   effect on rosin size, 280
   in paper testing, 35
   polyelectrolyte adsorption on, 56, 58
   in repulpable paper, 224
   retention, 61, 63–68, 71–72, 74, 78, 82–84
First pass retention, 63–65, 302–303
Flocculation, 12, 35, 51, 63–65, 244
   with aluminum compounds, 70–71
   with amphoteric compounds, 82–83
   with anionic compounds, 71–72
   with cationic compounds, 72–76
   co-flocculation, 63, 66–67
   flocculation mechanism, 65–70
   homo-flocculation, 63
Formaldehyde, 104, 121, 137, 139
   as crosslinker, 25, 151–152, 178, 208, 210, 212, 241, 373, 387–388, 394–395
   formaldehyde based resins
      AKD stabilizer and promoter, 284–285, 289
      as dry strength resins, 249, 251, 258, 345
      reactive sizing agents, 368
      resin retention, 33, 66, 248
      as retention aids, 74, 78, 284
      retention enhancer, 68–69
      synthesis, 139–142, 144–146
      as wet strength resin, 7, 98, 201
   formaldehyde emission, 241
   reactions, 8, 10, 31, 79, 122–123, 127, 180, 217, 243, 251–254, 272–273, 280
Fortified rosin, 272–276, 278, 280, 358

G

Gemini surfactants, 284
Glass transition temperature, 32–33, 176, 208, 216, 360, 391
Glycosidic bond, 11, 20, 22
Glyoxal as crosslinker, 104, 111, 121–125, 151, 332–333, 395
Glyoxalated hydroxyethyl cellulose, 152
Glyoxalated polyacrylamide, 7, 25, 75, 117–131, 154, 198, 200, 223, 252, 260, 334, 373
Glyoxalated poly β-alanine, 119
Glyoxalated polyvinyl amine, 252
Glyoxalated starch, 112
Glyoxalation reaction, 112, 117–125, 127, 252, 301
Grafted copolymers, 25–26, 68, 70, 77, 80, 119, 121, 200, 243, 253, 329
   grafted starch, 22, 25–27, 107, 153, 173, 198, 200, 246, 249–251, 300–301, 367, 376, 394
Greaseproof, 367, 397

H

Hemiacetal, 5, 102–107, 109, 111–112, 121, 127–130, 155, 210, 223, 249, 395
Hemicellulose, 6–9, 11–15, 17, 33, 52, 56, 62, 64, 97, 209
   in creping process, 327–329
   hemicellulose reactivity, 11, 22
   hemicellulose structure, 243
   in strengthening mechanism, 217, 243–244, 248, 258
Hofmann degradation, 31–32, 171, 211, 252, 255
Hydrocarbon resins, 38, 275–276, 281, 286, 313, 358–359, 379
Hydrogen bonding, 1, 5, 9, 21, 26–27, 33, 35, 53, 60, 291, 306, 355, 393
   in adhesives for Yankee dryer, 329, 340–342, 345
   in paper dry strength mechanism, 13, 37, 241–244, 246, 248–249, 252, 254–255, 259, 387–388
   in paper structure, 7, 14–17
   in paper wet strength mechanism, 38, 97, 100, 104, 106, 113, 121, 127, 137, 200–204, 207–210, 214–220, 222
   in retention mechanism, 67–70, 357

I

Ionic liquids, 8, 337
Imidization, 175, 268, 312, 371–373, 375, 382–383
Ink-jet technology, 352–353, 362, 367, 371, 373–374
Internal size, 38, 56, 268, 271, 353, 370–371, 378
   AKD, 282–297
   ASA, 297–307
   reactive internal size, 282–311
   rosin, 271–282
Intrinsic viscosity, 33–35, 68, 120, 188
Itaconic acid, 147
   copolymers, 80, 116, 176–177, 246, 255, 363, 389, 391
   polyamide, 146

L

Latent aldehyde group, 107
Lignin, 6–9, 12–13, 15, 17, 52, 56, 64, 76, 209, 246, 258, 312
   associations with PEO, 70
   Kraft lignin, 11, 63, 69, 82
   lignin reactions, 8, 11–12, 79–80
   lignin sulfonic acid, 198
   lignosulfonate, 10, 179, 258, 283–285, 288, 310, 331, 356, 358, 387
   structure, 9–10

M

Macro-monomer, 77, 178
Maleated hydrocarbon resins, 281, 373, 383
Mannich reaction, 8, 76, 79, 181, 252–255, 280
Melamine
   di-allyl-melamine, 251
   melamine–formaldehyde resins, 7, 33, 66, 74, 141–142, 179, 248, 285, 373, 387
   reactions, 141, 153, 368
Michael addition, 20, 31, 75, 102, 109, 112–114, 144, 146, 148–152, 170, 174, 177, 211, 255, 309, 312, 314–315, 344, 358
Multilayer polyelectrolyte structure, 61, 82, 199, 257

N

Nonionic emulsifier, 284, 301, 309–310, 365, 367
Nonionic stabilizer, 61, 276, 284–285, 287, 301, 366, 392

O

Olefins, 281, 286, 298–299, 308, 312, 372, 376, 379
One-shot rosin emulsion, 279
2-Oxetanone multimer, 379
Oxidized starch, 19, 23, 57, 81, 153, 200, 224, 352–354, 379, 384, 387, 389–390

P

Paper impregnation, 100, 152, 217, 243, 271
Paper porosity, 351–352, 386–389
Paper strength, 7, 12, 15, 32, 37, 39, 61, 63, 85, 111, 129, 141, 160, 221
    crosslinkers effect on, 162–163
    decay, 111, 118, 130, 157, 202
    effect of internal size on, 38
    latexes effect on, 254
    softeners effect on, 345
    strengthening mechanism, 15–17, 209–219, 241, 244, 329
    water effect on, 97–98, 104, 137, 202–203
Paper structure, 2, 14–17, 35, 96–97, 136, 202, 206, 209, 217, 220, 387, 394
Paper surface strength, 386–387
Paper testing, 35–39, 120
Papyrus, 5
Paraffin wax, 6, 176, 242, 286, 313, 328, 358, 379, 396–397
Pectin, 20
Performance chemicals, 13
Permanent wet strength, 98–99, 128, 137, 152, 197, 219–220
Pitch, 12–13, 224
Polyacrylamide, 31–34, 68, 113, 119, 171, 181, 254
Polyacrylic acid, 31, 35, 55, 62, 72, 81–83, 118, 164, 200, 206, 224, 257–258, 331, 341, 375, 388
Polyalkylene polyamine, 113, 185, 274, 290, 313–314
Polyaluminium chloride (PAC), 277
Polyamide, 30, 68, 76, 113, 122, 146, 184, 190, 274, 330–331, 333, 358
Polyamidoamine, 7
    in creping adhesives, 329, 331–333, 335–336
    as retention aids, 68, 74–75, 77, 80
    in wet strength resin synthesis, 7, 99, 113, 138, 146–152, 182–184
Polycondensation reaction, 30–31, 139, 142, 147
Polyelectrolyte
    adsorption, 53, 57–58, 62
    complex, 53, 55–56, 61–63, 82–83, 199, 218, 258–259, 276, 383, 387
    precipitation, 35, 55, 62, 82, 258–259
Polyester resins, 281, 358, 391, 396
Polyethylene glycol, 82, 129, 146, 162, 173, 217, 273, 301, 333, 353
Polyethyleneimine (PEI), 52, 55, 60–61, 63, 181, 254, 285
    branched PEI, 57, 206
    PEI-epichlorohydrin resin, 216, 331, 395
    PEI retention, 60, 66, 73, 204–207, 218

in polyelectrolyte complex, 258, 387, 394, 397
    reactions, 145, 154, 171, 199, 221, 312, 333, 335
    as retention aid, 64–67, 77–78, 309, 313
        dual retention system, 83–84
    synthesis, 145
    as trash catcher, 74
    as wet strength resin, 146, 154, 214–215, 217
    as Yankee adhesive, 331
Polyionene, 58, 63, 75, 254, 329
Polyisocyanates, 25–26, 150, 155–161
    blocking agents for polyisocyanates, 159, 161
Polymer analogous reactions, 28, 31–32, 78, 164, 177, 210, 258
    on acrylamide (co)polymers, 80, 118, 251–252, 255
    on cellulose, 107
    on maleic anhydride copolymers, 308, 372, 376–377
    on polyvinyl alcohol, 250
    on secondary amine, 144, 150, 163, 307, 312
    on starch, 74, 112
    on N-vinyl formamide copolymers, 76, 171, 173–174
Polymerization process, 25, 28–29, 31
Polysaccharide, 8, 14, 70, 121
    chemistry of, 17–27, 145
    depolymerization, 11, 23–24
    as prepolymer for WSR, 152–155
Polyurea, 70, 145
Polyurethanes, 315, 353, 375, 383
Polyvinyl alcohol, 31, 77, 100, 103, 122, 152, 173, 175, 199, 243, 304, 329, 353
Polyvinyl amine, 113, 172, 174, 181, 197, 199–201, 215, 249, 252, 258, 280, 285, 329, 333, 387
Porosity builders, 388–390
Process chemicals, 1, 13
Pulping yield, 8–9
Pulp oxidation, 23, 204

Q

Quaternization, 77, 151, 166, 183, 191, 330, 342

R

Recycled fibers, 9, 220, 224, 241, 256
Recycled paper, 9, 121, 219, 222–224
Repulpability index, 223
Repulping yield, 221
Retention aid
    amphoteric, 78–79, 199
    anionic, 71
    cationic, 72–78
    dual retention systems, 78, 81–84, 281, 285, 288
    nonionic, 67–70
    retention enhancers, 64, 68–70, 83
Retention mechanism, 61, 65–66, 70, 77, 206–207, 245
Reverse sizing, 279
Ring opening polymerization, 26, 296
Rosin, 11
    reactions, 272–274
    rosin adsorption, 277, 279–280
    rosin esters, 273, 275, 277–278, 358, 394
    rosin sizing mechanism, 276–279
    in surface treatment, 358, 362, 374, 378–379, 381, 394

S

Selective adsorption, 53, 196
Semi-continuous process, 117–118, 360, 362
Size-press solution, 351–352
    defoamer effect, 386
    ionic strength, 356
    starches for, 354–356
Sizing enhancers, 284, 288
Sizing mechanism, 276–282, 289–295, 304–307, 311–312
    surface sizing mechanism, 352, 381–385
Sodium aluminate, 270, 273–274
Softeners, 25, 328, 330, 339–346
Solvent for cellulose, 7–8
Starch oxidation, 23, 107
Starch retrogradation, 27, 111, 249, 355
Starch xanthate, 154, 200, 207, 223, 246, 258, 260
Styrene-Maleic Anhydride copolymers (SMA), 29, 33, 198, 246
    as binder in coatings, 391–392, 394, 396
    in blends with cationic resins, 218, 223, 256, 258, 282, 386–388
    cationic SMA, 375, 377
    in creping adhesives, 331
    as internal size, 281–282, 312
    as porosity builder, 388–389
    reactions, 312, 371–372, 376
    sizing mechanism, 382–385
    as stabilizer for dispersions, 177, 286, 375
    as surface size, 354, 358, 370–373
        blends with synergetic effect, 373–374
    synthesis, 370
Sulfamic acid, 21, 274, 280
Surface sizing agents, 352
    nonreactive, 356–368
    reactive, 368–381
    surface enhancer, 352
Surface tension, 5, 13, 17, 167, 176, 241, 245, 267, 339, 354, 365, 386, 396

T

Temporary wet strength resins, 97–131
Thickeners, 393
Thinned starch, 22, 355, 368
Trans-amidation reaction, 30, 76, 118, 149, 171, 174
Tub-size, 5, 178

U

Unsaturated fatty acids, 22, 283
Urea
    cyclic urea, 111
    dimethylol, 140, 178, 241
    mono-methylol, 140
    sulfate, 13
    in trans-amidation, 30, 141, 145
Uronic acid, 23, 243

V

Vinyl acetate copolymers, 25, 34, 55, 78, 116, 171, 173–174, 176, 200, 247, 250, 331, 370, 374, 389, 391, 394, 396
Vinylidene chloride copolymers, 312, 364, 391
Virgin fibers, 1, 224

W

Washburn equation, 267
Waste water, 345
Water adsorption, 85, 204, 218–219, 267, 294
Wax emulsion, 176, 313, 368, 394
Weak chemical bonds, 100, 105, 111–112, 297, 306
Wet end chemistry, 6, 51–53, 55, 58, 63, 68, 222, 270, 289, 345
    cellulose fibers
        carboxylic groups on fibers, 52, 56, 58
        external surface, 15, 52, 204
        fiber size distribution, 51, 209
        fiber swelling in water, 9, 11–12, 37–38, 51
        internal surface, 14, 51–52, 204
        specific surface area, 52, 67
    electrophoretic mobility, 52, 61, 65–67, 207, 256, 270, 277
    ionic strength, 55–56, 62, 65–67, 71, 187, 207, 209
    paper chemicals adsorption,
        electrolyte effect, 52–54, 56, 58, 61, 64, 67, 70–71, 73, 76, 82
        fiber saturation point, 52, 73, 205, 207, 256
        hydrophobic interactions, 51, 53, 59, 62
        ionic interactions, 53–54, 60, 178
        pseudo-irreversible adsorption, 54–55, 60, 66–67, 206–207, 215, 293
    pH, 7, 13, 52, 54–57, 61, 65–67, 71–72
Wet strength decay, 98–99, 104–106, 111, 118–120, 128, 130, 137, 252
Wet strengthening mechanism, 201, 203, 209, 211, 214, 219
Wet strength resins, 7, 17, 29, 53, 137–224
    in alkaline pH, 142, 144, 146, 169, 181, 206, 214
    curing reaction of, 146, 210
    dual system, 7, 198–199, 218, 224
    formaldehyde based resins, 139–142
    polyisocyanates, 155–160
    polysaccharides, 152–155
    reactive methylol in, 140
    in repulpable paper, 219–221, 223
    as retention aids, 64, 85
    temporary, 7, 17, 53, 97–131, 223
Wood preservation, 1

Z

Zipper effect, 202–203, 209, 216–217, 219
Zirconium salts, 52, 70, 102, 153, 162, 201, 332, 353, 373–374, 383–384, 387, 395–396
Zwitterion, 78, 215, 255, 343

T - #0883 - 101024 - C432 - 254/178/19 - PB - 9781032099262 - Gloss Lamination